# THE ONCOGENE AND TUMOUR SUPPRESSOR GENE *FactsBook*

**Second Edition**

**Other books in the FactsBook Series:**

A. Neil Barclay, Albertus D. Beyers, Marian L. Birkeland, Marion H. Brown, Simon J. Davis, Chamorro Somoza and Alan F. Williams
***The Leucocyte Antigen FactsBook, 1st edn***

Robin Callard and Andy Gearing
***The Cytokine FactsBook***

Steve Watson and Steve Arkinstall
***The G-Protein Linked Receptor FactsBook***

Rod Pigott and Christine Power
***The Adhesion Molecule FactsBook***

Shirley Ayad, Ray Boot-Handford, Martin J. Humphries, Karl E. Kadler and C. Adrian Shuttleworth
***The Extracellular Matrix FactsBook***

Grahame Hardie and Steven Hanks
***The Protein Kinase FactsBook***
***The Protein Kinase FactsBook CD-Rom***

Edward C. Conley
***The Ion Channel FactsBook***
***I: Extracellular Ligand-Gated Channels***

Edward C. Conley
***The Ion Channel FactsBook***
***II: Intracellular Ligand-Gated Channels***

Kris Vaddi, Margaret Keller and Robert Newton
***The Chemokine FactsBook***

Marion E. Reid and Christine Lomas-Francis
***The Blood Group Antigen FactsBook***

A. Neil Barclay, Marion H. Brown, S.K. Alex Law, Andrew J. McKnight, Michael G. Tomlinson and P. Anton van der Merwe
***The Leucocyte Antigen FactsBook, 2nd edn***

Jeff Griffiths and Clare Sansom
***The Transporter FactsBook***

# THE ONCOGENE AND TUMOUR SUPPRESSOR GENE *FactsBook*

## Second Edition

**Robin Hesketh**
*Department of Biochemistry*
*University of Cambridge, Cambridge, UK*

**Academic Press**
**Harcourt Brace & Company, Publishers**
SAN DIEGO LONDON BOSTON NEW YORK
SYDNEY TOKYO TORONTO

This book is printed on acid-free paper



First edition published 1995



Academic Press
525 B Street, Suite 1900, San Diego, California 92101-4495, USA
http://www.apnet.com

Academic Press Limited
24–28 Oval Road, London NW1 7DX, UK
http://www.hbuk.co.uk/ap/

ISBN 0-12-344548-5

**Library of Congress Cataloging-in-Publication Data**

Hesketh, Robin.
The oncogene and tumor suppressor gene factsbook / by T. Robin Hesketh. – 2nd ed., [completely rev.]
p. cm. – (Factsbook series)
Includes index.
ISBN 0-12-344548-5 (alk. paper)
1. Oncogenes. 2. Antioncogenes. I. Title. II. Series.
RC268.42.H47 1997
616.99'4071–dc21 97-21738
CIP

A catalogue record for this book is available from the British Library

Typeset in Great Britain by Alden, Oxford, Didcot and Northampton
Printed in Great Britain by WBC, Bridgend, Mid Glamorgan

97 98 99 00 01 02 EB 9 8 7 6 5 4 3 2 1

# Contents

## Contents

## Section III THE ONCOGENES



## Section IV TUMOUR SUPPRESSOR GENES



## Section V CYCLIN-DEPENDENT KINASE INHIBITORS



## Section VI DNA TUMOUR VIRUSES

# Preface

Material published in The Oncogene Handbook and in the first edition of The Oncogene FactsBook inevitably forms a significant component of this book and I remain deeply grateful to all those colleagues who commented on sections of those books and who are acknowledged in The Oncogene Handbook. However, the second edition of The Oncogene Factsbook is not merely up to date for the beginning of 1997 but has also been radically revised and expanded to cover the many new developments in the field since publication of the first edition. For critical comment and encouragement in the writing of this second edition of The Oncogene FactsBook I am particularly indebted to Chi Dang (Johns Hopkins University School of Medicine), Paul Kemp (Department of Biochemistry, University of Cambridge), Adam Lacy-Hulbert (Department of Medicine, University College, London) and Jane Rogers (The Sanger Centre, Hinxton Hall, Cambridge). I remain, of course, solely responsible for any errors of fact or omission and I am happy to receive any constructive comments from readers. Please write either to The Editor, *The Oncogene and Tumour Suppressor Gene FactsBook*, Academic Press, 24–28 Oval Road, London, NW1 7DX, UK or to Robin Hesketh, Department of Biochemistry, Cambridge, CB2 1QW, UK (Fax: (0044) 1223 333345; E-mail: t.r.hesketh@-bioc.cam.ac.uk

# Abbreviations

| | |
|---|---|
| Ab-MuLV | Abelson murine leukaemia virus |
| ABML | Abelson virus-induced myeloid lymphosarcoma |
| ADPRT | ADP-ribosyltransferase inhibitor |
| ALL | Acute lymphocytic leukaemia |
| ASV | Avian sarcoma virus |
| BM | Basement membrane |
| CAR | Cell Adhesion Regulator |
| CBF | Core binding factor |
| CDI | Cyclin-dependent kinase inhibitor |
| CDK | Cyclin-dependent kinase |
| CHRPE | Congenital hypertrophy of the retinal pigment epithelium |
| CML | Chronic myeloid leukaemia |
| Con A | Concanavalin A |
| CRE | cAMP response element |
| CSF1 | Colony stimulating factor 1 |
| DMBA | Dimethylbenzanthracene |
| DSE | Dyad symmetry element (or SRE) |
| EBV | Epstein–Barr virus |
| EGF | Epidermal growth factor |
| ENU | Ethylnitrosourea |
| ERK | Extracellular signal-regulated kinase (also called MAP kinases) |
| ETF | Epidermal growth factor (EGF) receptor transcription factor |
| FAP | Familial adenomatosis polyposis |
| FGF | Fibroblast growth factor |
| FIRE | *Fos* intragenic regulatory element |
| FMTC | Familial medullary thyroid carcinoma |
| GADD | Growth arrest on DNA damage |
| GAP | GTPase activating protein |
| GM-CSF | Granulocyte-macrophage colony-stimulating factor |
| GPI | Glycosyl phosphatidylinositol |
| GRB | Growth factor receptor (protein) |
| GRF | Growth hormone-releasing factor |
| GVBD | Germinal vesicle breakdown |
| HBGF | Heparin-binding growth factor |
| HBV | Hepatitis B virus |
| HCP | Haematopoietic cell phosphatase (also called PTP1C) |
| HIV | Human immunodeficiency virus |
| HMBA | Hexamethylene bisacetamide |
| HNPCC | Hereditary non-polyposis colon cancer |
| HOB1, HOB2 | Homology box 1, 2 |
| HPV | Human papillomaviruses |
| HTLV | Human T cell leukaemia/lymphoma viruses |
| HZ2-FeSV | Hardy–Zuckerman 2 feline sarcoma virus |
| IFN | Interferon |
| IL | Interleukin |

| | |
|---|---|
| JNK | JUN N-terminal kinases (or SAPK) |
| LFA-1 | Lymphocyte function-associated antigen 1 |
| LOH | Loss of heterozygosity |
| LPS | Lipopolysaccharide |
| LTR | Long terminal repeat |
| MAP | Microtubule-associated protein |
| MAP kinases | Mitogen-activated protein kinases (also called ERK1, ERK2) |
| MAPKK | Mitogen-activated protein kinase kinase (or MEK) |
| MDR | Multidrug resistance |
| MEK | MAP kinase or ERK kinase |
| MHC | Major histocompatibility complex |
| MMP | Matrix metalloproteinase |
| MMTV | Mouse mammary tumour virus |
| MNNG | *N*-Methyl-*N′*-nitro-*N*-nitrosoguanidine |
| Mo-MuSV | Moloney sarcoma virus |
| MuSV | Murine sarcoma virus |
| NCAM | Neural cell adhesion molecule |
| NFAT-1 | Nuclear factor of activated T cells |
| NGF | Nerve growth factor |
| NLS | Nuclear localization signal |
| NMU | Nitrosomethylurea |
| NRE | Negative regulatory element |
| NSCLC | Non-small cell lung carcinoma |
| ODN | Oligodeoxynucleotide |
| OMGP | Oligodendrocyte-myelin glycoprotein |
| ORF | Open reading frame |
| PA | Plasminogen activator |
| PAF | Platelet activating factor |
| PAI | Plasminogen activator inhibitor |
| PDGF | Platelet-derived growth factor |
| PHA | Phytohaemagglutinin |
| PKC | Protein kinase C |
| PKR | RNA-activated protein kinase |
| PLC | Phospholipase C |
| PTC | Papillary thyroid carcinoma (also called TPC) |
| PtdIns | Phosphatidylinositol |
| PTP1B | Protein tyrosine phosphatase 1B |
| PTP1C | Protein tyrosine phosphatase 1C (also called SH-PTP1, SHP or HCP) |
| PTP1D | Protein tyrosine phosphatase 1D (also called SYP, SH-PTP2, PTP2C or SH-PTP3) |
| $RER^+$ | Replication error positive |
| REV | Reticuloendotheliosis virus |
| SAPK | Stress-activated protein kinase (or JNK) |
| SCLC | Small cell lung carcinoma |
| SM-FeSV | Susan McDonough feline sarcoma virus |
| SRE | Serum response element (or DSE) |
| SRF | Serum response factor |
| STAT | Signal transducer and activator of transcription |
| STK1 | Stem cell tyrosine kinase 1 |
| SV40 | Simian vacuolating virus 40 |

| | |
|---|---|
| $T_3$ | Triiodothyronine |
| TCR | T cell receptor |
| TGF | Transforming growth factor |
| TIMP | Tissue inhibitor of metalloproteinase |
| TNF | Tumour necrosis factor |
| TPA | 12-*O*-Tetradecanoylphorbol-13-acetate (also called PMA) |
| TRE | TPA response element |
| VEGF-VPFR | Vascular endothelial growth factor |
| VSMC | Vascular smooth muscle cell |

# Introduction

## ORGANIZATION

In the short period between my submitting the manuscript of the first edition of The Oncogene FactsBook and its publication there was a significant burst of activity in the identification and sequencing of novel cancer genes. These included *BRCA1*, the family of DNA mismatch repair genes and several cyclin-dependent kinase inhibitors, each of which received scant or no attention in the first edition. The substantial space required to accommodate these and other new genes, together with the continuing expansion of our knowledge of the entire oncogene field, has caused the second edition to be some 20% larger than its predecessor. If, on first acquaintance this seems a slightly depressing statistic, some encouragement may be gained from the fact that, when I came to write a broad overview of the molecular biology of cancer as an introduction to the second edition, I was struck by the much more integrated picture it is now possible to paint than seemed to be possible just three or four years ago. Thus Section I is an attempt to draw together several major areas, including control of the cell cycle and apoptosis, with what is known of the role of interacting oncogene and tumour suppressor gene products. Although this is not a completely comprehensive summary, it is intended to illustrate that these separable areas form a linked network regulating cell growth, cell death and cell location and to focus on the key proteins involved, thereby emphasizing functional interactions between oncoproteins, an aspect that is more difficult to convey in the individual sections that concentrate on the specific properties of individual oncogenes. This overview is extended to include molecular aspects of angiogenesis and metastasis and the subsequent chapter on the multistep nature of cancer reviews the patterns of genetic abnormalities that are associated with the major cancers. Section I concludes with a summary of developments in genetically based approaches to cancer therapy.

After Section I, the format of the book follows that of its first edition predecessor. Section II comprises summary tables of the major categories of oncogenes and tumour suppressor genes, defined chromosomal translocations and chromosomal loci. In Section III the individual entries for each oncogene include the following headings (with minor modifications for individual genes as appropriate): Identification, Related genes, Table of properties (Nucleotides, Chromosome, Protein molecular mass, Cellular location, Tissue distribution), Protein function (including sections on Cancer, Transgenic animals, In animals and *In vitro*), Gene structure, Transcriptional regulation, Protein structure, Amino acid sequence, Domain structure, Database accession numbers, References (with reviews indicated by bold type).

Sections IV and V comprise individual entries with the same layout as in Section III for the tumour suppressor genes and the cyclin-dependent kinase inhibitors, respectively. Section VI, DNA tumour viruses, details human papillomaviruses and Epstein–Barr virus.

## NOMENCLATURE

The recommendations of the International Standing Committee on human gene nomenclature have been followed. Thus human genes are written in italicized capitals, their gene products in non-italicized capitals [1]. For genes of other species the nomenclature recommended for murine genes has been followed. Thus genes are italicized with an initial capital letter followed by lower case letters: the corresponding proteins are written in non-italicized capitals [2]. In the individual gene entries (Section III) the official designation is given at the outset but for some genes the commonly accepted form is used thereafter to avoid confusion. These are human *ABL1* (*ABL*), the avian *Erb* genes (*ErbA* and *ErbB*), *EPHT* (*EPH*), *TP53*/TP53 (*P53*/p53), *RB1* (*RB1*/pRb) and *NME1*/*NME2* (*NM23*).

Viral oncogenes are referred to by trivial names of the form v-*onc* (e.g. v-*myc*), that is, the names do not imply target cell specificity or function [3]. The prefixes "p", "gp", "pp" or "P" followed by the molecular mass in kilodaltons indicate "protein", "glycoprotein", "phosphoprotein" or "polyprotein" respectively. An additional italicized superscript indicates the gene encoding the protein (e.g. p105$^{RB1}$). Hyphenated superscripts denote polyproteins derived from two genes (e.g. gp180$^{gag\text{-}src}$). Suffixes -a, -b, etc. denote inserts in the same virus that can code for different proteins *via* distinct RNAs (e.g. *Erba*, *Erbb*).

### *References*


1. Shows, T.B. et al. (1987) Guidelines for human gene nomenclature. Cytogenet. Cell Genet. 46, 11–28.
2. Lyon, M.F. (1984) Rules for nomenclature of genes, chromosome anomalies and inbred strains. Mouse News Lett. 72, 2–27.
3. Coffin, J.M. et al. (1981) Proposal for naming host cell-derived inserts in retrovirus genomes. J. Virol. 40, 953–957.

# Section I

# THE INTRODUCTORY CHAPTERS

# 1 Genetic Events in Cancer Development

Tumours result from subversion of the processes that control the normal growth, location and mortality of cells. This loss of normal control mechanisms arises from the acquisition of mutations in three broad categories of genes:

1 proto-oncogenes, the normal products of which are components of signalling pathways that regulate proliferation and which, in their mutated form, become dominant oncogenes;
2 tumour suppressor genes, which generally exhibit recessive behaviour, the loss of function of which in cancers leads to deregulated control of cell cycle progression, cellular adhesion, etc.; and
3 DNA repair enzymes, mutations in which promote genetic instability (Fig. 1).

Mutations in these genes are also presumed to determine the changes in cell surface protein expression, protein secretion and cell motility that contribute to the metastatic phenotype, because no mutations have yet been specifically associated with metastatic development.

The notion that cancer might be caused by genetic abnormality originated in the early nineteenth century when it was noted that predisposition to cancer seemed to run in families. By the turn of the century it had been observed using light microscopy that the chromosomes from cancer cells were frequently of abnormal length or shape when compared with those from normal cells. More recent discoveries have shown that there is a connection between susceptibility to cancers and an impaired ability of cells to repair damaged DNA and that the mutagenic potential of a substance is related to its carcinogenicity, all of which is consistent with the general concept that cellular genes (proto-oncogenes) in another form (oncogenes) cause neoplastic growth[1].

These observations suggest that cancers arise solely from deregulated proliferation and whilst it is certainly true that abnormally high proliferation can give rise to cancers, it is now recognized that modulation of the normal processes that lead to

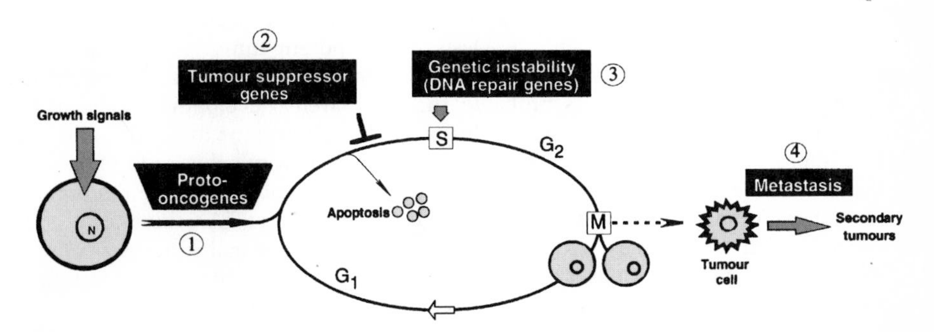


**Figure 1** *Genetic signals in cancer development. Three broad functional categories may be distinguished within which mutations may arise: (1) and (2) pathways driving cell proliferation and controlling cell cycle progression and apoptosis (proto-oncogenes and tumour suppressor genes (see Chapters 2, 3, and 5), (3) in the promotion of genetic instability through mutations in DNA repair genes (see Chapter 6), and (4) effects associated with metastasis (see Chapter 7).*

cell loss (programmed cell death or apoptosis – see Chapter 5) is also sufficient, although not necessary, for tumour development. Mathematical modelling of cellular behaviour predicts that tumours may indeed arise when exponential cell growth occurs, usually when stem cells fail to die or differentiate. At other times a new, higher equilibrium is reached in which an abnormally high number of cells are present within a population as a result of fully differentiated cells failing to undergo apoptosis. This may correlate with the formation of early pre-malignant lesions, detectable, for example, in cervical intraepithelial neoplasia, that can regress spontaneously. This model also predicts the occurrence of long lag phases of growth that persist until mutations arise that eventually cause exponential growth[2].

A further consequence of the activation of oncogenes or the ablation of tumour suppressor genes may be that the growth arrest process associated with ageing (senescence) may be over-ridden. Senescence usually correlates with decrease in the length of telomeres which are short, tandem repeats of the hexanucleotide 5′-TTAGGG-3′ at the ends of each chromosome. Telomeres are synthesized by telomerase, a ribonucleo-protein DNA polymerase that is active in many tumours but is not common in normal tissues[3]. In tumour cell lines telomerase activity reaches a maximum in S phase and a variety of inhibitors of cell cycle progression (e.g., transforming growth factor $\beta_1$: see Chapter 4) also inhibit telomerase activity[4].

## IDENTIFICATION OF ONCOGENES

Oncogenes were first directly identified in viruses capable of inducing tumours in animals and/or of transforming cells *in vitro*. Many such viruses have RNA genomes and this family of "retroviruses" replicate through a DNA intermediate in infected cells (e.g. avian leukosis virus (ALV) and mouse mammary tumour virus (MMTV)). The oncogenes carried by such viruses (e.g. v-*src*, v-*myc*) are strongly homologous in sequence to normal cellular genes (proto-oncogenes, e.g. *Src, Myc*) that are highly conserved in evolution. Retroviruses that do not themselves carry an oncogene may activate cellular proto-oncogenes by insertion of their proviral DNA at appropriate sites in the host genome, the effect being either to modify the proto-oncogene directly or to modify its expression by insertion of powerful regulatory elements present in the viral genome. Despite their potent tumorigenic capacity in appropriate animal hosts, no retrovirus has yet been shown to be directly oncogenic in humans. However, it seems probable that the latent development of cancers commonly associated with infection by human T-lymphotropic viruses (HTLV-I and HTLV-II) or human immunodeficiency virus (HIV) arises from subversion of normal cellular control mechanisms by the transcription factors encoded in their genomes. Thus, for example, HTLV-I or its *trans*-activating gene product TAX transforms cells *in vitro* and this correlates with activation of JUN N-terminal kinases (JNKs or stress-activated protein kinases, SAPKs), which are components of the mitogen-activated protein kinase (MAPK) signalling pathway (see Fig. 4)[5].

The double-stranded DNA viruses of the adenovirus, herpesvirus, poxvirus and papovavirus families also possess oncogenic potential. Some DNA tumour viruses are oncogenic in humans (hepatitis B virus, Epstein–Barr virus and some types of papillomavirus) but most are only tumorigenic in other species. However, like many proto-oncogenes that encode transcription factors, most DNA tumour virus oncoproteins can immortalize primary cells and co-operate with other oncoproteins (typically RAS) to transform cells. A major difference between DNA viruses and retroviruses is that the former may replicate autonomously without being integrated into the host chromosome. DNA viral genes, including those with oncogenic potential, have therefore evolved to encode proteins that are essential for the continuation of the life cycle of the virus. Hence the oncogenes of DNA viruses differ from those of retroviruses, possession of which confers no advantage on the virus, and their transforming genes have not yet been shown to have proto-oncogene homologues within the normal genome, save for the presence of *BCL2* sequences in the Epstein–Barr virus *BHRF1* gene and SV40-like sequences that have been detected in some human bone tumours[6] (see Section VI, **Epstein–Barr virus**).

There are presently nearly 200 known oncogenes that, under certain conditions, can contribute to the release of cells from the normal controls of proliferation, death, migration and adhesion to cause neoplastic transformation. It is probable that the majority of genes that possess oncogenic potential have now been identified and this figure of approximately 200 proto-oncogenes from about 60 000 functional human genes thereby sets an upper limit to the number of points at which the biochemical pathways controlling normal cell growth might be subverted by oncoproteins. Fortunately, the actual number of general mechanisms is probably much smaller, as oncoproteins fall into groups of similar activity (e.g. tyrosine kinases, guanine nucleotide binding proteins, etc., see Table 1), and the members of each group are

**Table 1** *Classes of major oncoproteins (see Table V, page 86 for complete list)*

*Class 1 Growth factors*
HSTF1/HST-1, INT2, PDGFB/SIS, WNT1, WNT2, WNT3
*Class 2 Tyrosine kinases*
Receptor-like tyrosine kinases
CSF1R/FMS, EGFR/ERBB, FMS, KIT, MET, HER2/NEU, RET, TRK
Non-receptor tyrosine kinases
ABL1, FPS/FES
Membrane-associated non-receptor tyrosine kinases
SRC and SRC-related kinases
*Class 3 Receptors lacking protein kinase activity*
MAS
*Class 4a Membrane-associated G proteins*
HRAS, KRAS, NRAS
*Class 5 Cytoplasmic protein serine kinases*
BCR, MOS, RAF/MIL
*Class 6 Protein serine, threonine (and tyrosine) kinase*
AKT1, AKT2
*Class 7 Cytoplasmic regulators*
BCL1, CRK
*Class 8 Cell cycle regulators*
INK4A, INK4B, Cyclin D1, CDC25A, CDC25B
*Class 9 Transcription factors*
E2F1, ERBA, ETS, FOS, JUN, MYB, MYC, REL, TAL1, SKI
*Class 10 Transcription elongation factor*
ELL
*Class 11 Intracellular membrane factor*
BCL2
*Class 12 Nucleoporins*
NUP98, NUP214
*Class 13 Adapter proteins*
SHC
*Class 14 RNA binding proteins*
EWS

presumed to act at corresponding points in signalling pathways. In the main, oncogene activation is the result of somatic events (i.e. what we do to ourselves) rather than hereditary genetic causes transmitted by mutation in the germline. It is, in other words, a consequence of evolution (mutation and selection) within the body of one animal.

The epidemiological evidence supporting this conclusion is now overwhelming and it is widely known, for example, that in the Western world smoking and dietary factors each contribute to approximately 30% of all cancer deaths. Although the molecular mechanisms are far from understood, it seems probable that the major effects are manifested by the promotion of damage to DNA. Thus, for example, the beneficial effects of an adequate consumption of fruit and vegetables appear to derive from their anti-oxidant content and its protective effect against oxidative damage to proteins and DNA [7]. Oxidative lesions in DNA are normally repaired with great efficiency by DNA repair enzymes (see Chapter 6) but nevertheless accumulate with time and constitute the principal cause of cancers, essentially diseases of old age. Oxidative damage to DNA is also increased by the actions of the immune system in destroying infectious agents (viruses, parasites and bacteria). Thus, although proteins encoded by the hepatitis B virus (HBV) genome appear to modulate proto-oncogene-controlled proliferation pathways, the chronic liver inflammation that follows infection is also associated with increased DNA damage arising from nitric oxide and superoxide generated by cells of the immune system. Similar effects may arise from the inflammatory response to a variety of parasites, infection with which can be associated with specific cancers. These include *Schistosoma japonicum* (bladder cancer, hepatocellular carcinoma and/or colorectal cancer) [8], *Schistosoma haematobium* (bladder cancer) [9], *Opisthorchis viverrini* and *Chlonorchis sinensis* (biliary tract carcinoma) [10].

It has recently emerged that genes homologous to the eukaryotic superfamily of protein kinases are also widely expressed in prokaryotes [11]. Remarkable evidence indicates that some pathogenic organisms may be able to transfer the products of such genes to their target cells. *Yersinia pseudotuberculosis* is a pathogenic bacterium the cytotoxicity of which involves translocation of three of its gene products, YopE, YopH and YpkA, through the plasma membrane of infected cells from extracellularly located *Yersinia* cells [12]. YopH is a tyrosine phosphatase and YpkA is a serine/threonine kinase with homology to the eukaryotic family that is essential for the virulence of the organism. Other pathogenic bacteria adhere to target cells and cause localized actin accumulation beneath their adhesion sites. These include enteropathogenic *E. coli* (EPEC), enterohaemorrhagic *E. coli* (EHEC, which causes haemorrhagic colitis), *Citrobacterfreundii* and *Hafnia alvei*. EPEC colonizes epithelial cells and the resultant underlying cytoskeletal structures contain α-actinin, myosin light chain, ezrin and talin. The recruitment of these components depends on the action of intimin, a bacterially encoded protein that appears to bind to Hp90, which acts as a receptor on the surface of the target cell. Hp90 is only active as an intimin receptor after tyrosine phosphorylation induced by the initial interaction of bacterium with host cell [13]. Thus the bacterial pathogen appears to create its own receptor. Tyrosine phosphorylated Hp90 subsequently functions to recruit the cytoskeletal components to the region beneath the bacterium but it also mediates an increase in inositol phosphates and in the concentration of free intracellular calcium. The significance of the interference with intracellular signalling pathways by extracellular pathogens is unknown but it is notable that the activities identified so far include serine/threonine and tyrosine kinases, phospholipase C and tyrosine phosphatase, each of which could critically modulate control of cell proliferation (see below). The strongest evidence implicating bacterial infection with any cancer is for *Helicobacter pylori* and stomach cancer (see Chapter 8) but it is not known whether this bacterium interacts with target cells in ways that resemble the pathogenic actions just described.

## ONCOGENE ACTIVATION IN HUMANS

The fact that human cancers do not generally appear to be caused by retrovirally activated oncogenes raises the question of how proto-oncogenes become activated in human cancers. In normal cells proto-oncogene activation may occur by mutation, DNA rearrangement or gene amplification (Fig. 2). Point mutations may arise from the action of chemicals or radiation. For example, the transfection experiments that originally revealed activated *RAS* genes in human tumours led to the finding that, for *RAS*, the transformation from normal proto-oncogene to oncogene arose from substitution of a single base, commonly resulting in the exchange of valine for glycine or glutamine for lysine at residues 12 or 61 respectively (see ***RAS***).

The mechanisms of amplification and chromosome translocation can give rise to elevated cellular concentrations of the normal gene product or to the expression of new proteins created by in-frame fusion of coding sequences from separate genes.

An increase in gene copy number by amplification of specific DNA sequences occurs frequently in tumour cells although it has not been shown to occur during normal mammalian cell development. Regions up to several megabases in length can be involved and these often include proto-oncogenes. The process expands the number of copies of a gene which can lead to excess production of oncogene message and protein. Amplification appears to be mainly associated with the *EGFR*, *MYC* and *RAS* families and the 11q13 locus (*BCL1, CCND1, EMS1, INT2, MEN1, SEA,*

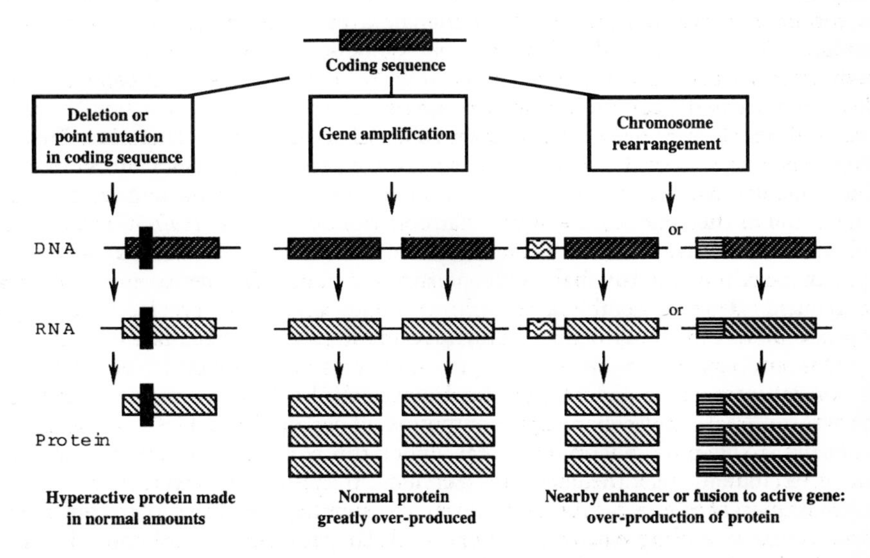


**Figure 2** *Schematic representation of mechanisms of oncogene activation.*

*HSTF1, CALM, LPC, MLL, PLZF, ETS1, RCK, CBL, FLI1*). Less frequently involved genes include *ETS1, GLI, MYB, MET* and *RAF1*. Amplification is thought to play a role in the later stages of cancer, generally appearing in cells that have metastasized. This is consistent with evidence that loss of the tumour suppressor gene *P53*, also generally a late event in tumour development, may be permissive for amplification. Thus, fibroblasts from patients with the Li–Fraumeni syndrome that lack functional p53 do not arrest in $G_1$ but show induction of gene amplification by *N*-(phosphonacetyl)-L-aspartate (PALA). PALA specifically inhibits uridine synthesis and resistance to PALA arises by amplification of the *CAD* gene (which encodes a single polypeptide encoding *c*arbamyl phosphate synthetase, *a*spartate transcarbamylase and *d*ihydroorotase). Expression of wild-type p53 in these cells restores $G_1$ arrest and resistance to *CAD* amplification[14]. p53-dependent growth arrest is exquisitely sensitive to the presence of only a few double-strand breaks[15]. Inactivation of p53 permits immortalized non-tumorigenic cells and primary fibroblasts to replicate in the presence of chromosome breaks and to exhibit high rates of gene amplification.

The chromosomal region including the cyclin D1 gene is amplified in ~20% of human breast carcinomas and also in bladder tumours. About 50% of breast tumours examined immunohistochemically have revealed excessive levels of cyclin D1. Surprisingly, however, enhanced cyclin D1 expression appears to correlate with extended periods of remission and increased overall survival rates, whereas the worst prognosis correlates with reduced levels of cyclin D1[16]. This suggests that cyclin D1 may be a useful prognostic indicator but also that other factors, most probably *RB1* status, are more critical regulators of cell behaviour (see Chapters 4 and 8).

Amplification and/or over-expression of members of the EGFR family also occurs frequently in breast cancer (see Chapter 8). Amplification of *MYC* and *MYCL* has frequently been observed in small cell carcinoma of the lung. Amplified genes may also have undergone mutation and, for *KRAS2* at least, there is evidence that amplification of the normal gene may occur in parallel with mutation of that gene.

The exchange of genetic material can occur between homologous or non-homologous chromosomes and can either be a balanced, reciprocal event or can involve loss of material from one or both junctions. Alternatively, inversion of segments within a chromosome may occur without net loss, or interstitial deletions may give rise to shortened chromosomes. The genes involved in over 60 chromosome translocations have now been defined (see Table III, page 78). The majority of these encode transcription factors, the activities of which are altered as a consequence either of the formation of novel, chimeric proteins or of the over-expression of the gene driven by anomalous regulatory sequences. Other classes of genes may be involved, however, and fusion proteins containing regions of cytokines, small G proteins, nucleophosmin, stress response proteins, docking proteins, a clathrin assembly protein, mitochondrial proteins, proteases, cell death proteins, RNA binding proteins and high-mobility group DNA binding proteins have been defined.

Burkitt's lymphoma and human chronic myeloid leukaemia are characterized by chromosome exchange between non-homologous chromatids. In some B cell leukaemias the proto-oncogene comes under the control of an immunoglobulin promoter and enhancer. In Burkitt's lymphoma the gene involved is *MYC* and this releases transcription of *MYC* from normal controls so that it may be expressed at inappropriate times as well as being over-expressed. The translocated gene may also suffer damage that increases mRNA stability and/or acquire mutations that modulate function (see ***MYC***). Translocation involving the cyclin D1 gene in which it too is brought under the control of an immunoglobulin promoter occurs in a variety of non-Hodgkin's lymphomas and B cell leukaemias (see **D-cyclins**). The consequent over-expression of cyclin D1 is a marker for some forms of lymphoma and its effect is to over-ride normal control of cell cycle progression (see Chapter 4).

***References***


[1] **Bishop, J.M. (1995) Genes Dev. 9, 1309–1315.**
[2] Tomlinson, I.P.M. and Bodmer, W.F. (1995) Proc. Natl Acad. Sci. USA 92, 11130–11134.
[3] Mehle, C. et al. (1996) Oncogene 13, 161–166.
[4] Zhu, X. et al. (1996) Proc. Natl Acad. Sci. USA 93, 6091–6095.
[5] Xu, X. et al. (1996) Oncogene 13, 135–142.
[6] Carbone, M. et al. (1996) Oncogene 13, 527–535.
[7] **Ames, B.N. et al. (1995) Proc. Natl Acad. Sci. USA 92, 5258–5265.**
[8] **Ishii, A. et al. (1994) Mutat. Res. – Fund. Mol. Mechs. Mutagenesis 305, 273–281.**
[9] Weintraub, M. et al. (1995) Int. J. Oncol. 7, 1269–1274.
[10] **Holzinger, F. et al. (1995) Digestive Surg. 12, 208–214.**
[11] **Zhang, C.-C. (1996) Mol. Microbiol. 20, 9–15.**
[12] Håkansson, S. et al. (1996) Mol. Microbiol. 20, 593–603.
[13] Rosenshine, I. et al. (1996) EMBO J. 15, 2613–2624.
[14] Yin, Y. et al. (1992) Cell 70, 937–948.
[15] Huang, L. et al. (1996) Proc. Natl Acad. Sci. USA 93, 4827–4832.
[16] Gillett, C. et al. (1996) Int. J. Cancer (Pred. Oncol.) 69, 92–99.

# 2 Proliferation Control Pathways

The regulation of normal cell proliferation occurs through the activation of biochemical pathways by growth factors (or mitogens) interacting with their receptors on the plasma membrane (Fig. 3). There are many different growth factors (e.g. epidermal growth factor (EGF), platelet-derived growth factor (PDGF), insulin, bombesin) and a single cell often possesses a variety of types of receptor. There are, however, only four known basic signalling mechanisms that can be activated by transmembrane receptors:

1 Activation of adenylate cyclase, a membrane-bound enzyme that generates cyclic AMP and is stimulated by many different receptor agonists (e.g. prostaglandin $E_2$). The activity of adenylate cyclase is regulated either positively or negatively by specific heterotrimeric G proteins. Mutations in G proteins occur in some human cancers (e.g. in GSP/$G\alpha_s$ giving rise to a sustained elevation of cyclic AMP in some thyroid carcinomas and ovarian tumours [1,2] and in GIP2/$G\alpha_{i2}$ in adrenal cortex and endocrine ovarian tumours [3]). Cyclic AMP-dependent protein kinases regulate many cellular processes.
2 Activation of guanylate cyclase to generate cyclic GMP (e.g. cyclic GMP concentrations are modulated in retinal cells in response to light). Cyclic GMP directly regulates membrane cation channels in photoreceptor cells and mediates other processes including the relaxation of smooth muscle.
3 Activation of phospholipase enzymes (principally PtdIns-PLC$\beta_1$) by agonists interacting with receptors having seven transmembrane segments that are coupled to specific heterotrimeric G proteins. This increases the free, intracellular concentration of $Ca^{2+}$ ($[Ca^{2+}]_i$) as a result of the action of inositol 1,4,5-trisphosphate

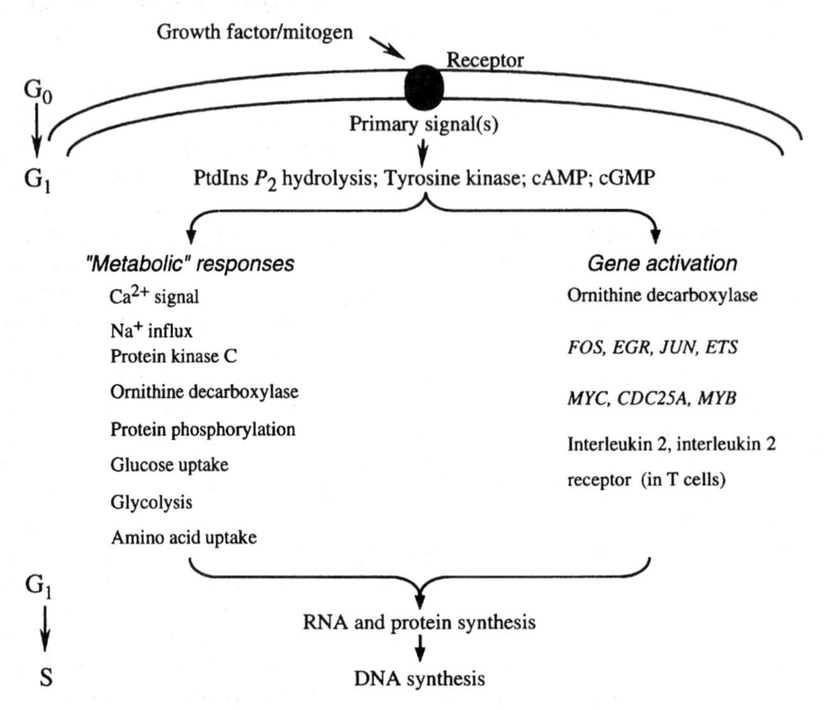


**Figure 3** *Biochemical events during proliferation in eukaryotic cells.*

released during the hydrolysis of phosphatidylinositol 4,5-bisphosphate (PtdIns(4,5)$P_2$). In addition to the oncoproteins indicated in Fig. 4, the over-expression of various seven transmembrane spanning receptors or of mutated G proteins (e.g. GIP2/G$\alpha_{i2}$)[3], G$\alpha_{12}$, G$\alpha_{13}$[4,5], GSP/G$\alpha_s$[1,2], G$\alpha_q$[6]) renders some cultured cells tumorigenic.

4 Activated protein kinases, the most important group of which are tyrosine kinases, often intrinsic components of the receptor molecule. These include the receptors for many growth factors (e.g. EGF, PDGF, nerve growth factor). For some receptors (e.g. the T cell receptor) an equivalent signalling pathway is activated, even though the receptor itself lacks intrinsic kinase activity, by association of a protein tyrosine kinase with the activated receptor. Many growth factors cause PtdIns(4,5)$P_2$ hydrolysis, equivalent to that promoted by the G protein-coupled response, as a consequence of the stimulation of phospholipase C$\gamma$ by activated tyrosine kinase receptors (e.g. PDGF, bombesin, anti-T cell receptor antibody). Activated receptor tyrosine kinases may also associate with phosphatidylinositol 3-kinase, which promotes phosphorylation of the D-3 position of inositol phospholipids and generates PtdIns-3 phosphate, PtdIns(3,4)$P_2$ and PtdIns(3,4,5)$P_3$. These lipids are present in normal cells including non-growing cells (neutrophils, brain) at very low levels (1–2% of non-3-phosphorylated PtdIns polyphosphates) and in fibroblasts and platelets their concentrations are increased in response to growth factors (e.g. EGF, PDGF, CSF1 or insulin). They are not hydrolysed by phospholipase C enzymes and have been implicated in cytoskeletal reorganization[7]. Receptors with intrinsic serine/threonine kinase activity are also important regulators of cell growth, most notably the transforming growth factor $\beta$ (TGF$\beta$) receptor family that form heterodimeric complexes following cytokine binding and generally transmit signals that inhibit growth (see ***TGFBR1*** and ***TGFBR2***).

One growth factor may recognize both receptors that activate adenylate cyclase and receptors that cause elevation of $[Ca^{2+}]_i$: such receptors are often, but not invariably, on different types of cell. Furthermore, the activation of tyrosine kinase(s) may be coupled to the activation of multiple intracellular signalling pathways (see Figs 3 and 4).

The interaction of growth factors with their receptors on the cell surface causes quiescent, somatic cells to leave $G_0$, traverse $G_1$ and enter S phase, whereupon cells are normally committed to at least one round of the cell cycle. Following the generation of one or more primary signals a sequence of "metabolic" events occurs that includes ionic changes, increased ornithine decarboxylase activity, protein phosphorylation and enhanced glycolytic flux. In parallel with and independent of these events the coordinated transcription of ~100 genes is activated within 6 hours[8,9]. A substantial proportion of these genes is activated within the first hour and includes ornithine decarboxylase, the over-expression of which may cause transformation and which is mutated in some cancers[10], and the proto-oncogene families of *JUN* (*JUN*, *JUNB* and *JUND*) and *FOS* (*FOS*, *FRA1*, *FRA2* and *FOSB*). Other immediate early response genes are *ETS1* and *ETS2*, serum response factor (SRF), the steroid hormone receptors *Nurk77*, *N10*, *NGFI-B*, *T1*, *TIS1* and the early growth response family (*EGR1*, *EGR2*, *EGR3* and *EGR4*), *WAF1*, *TIS11*, *TTP*, *Nup475*, fibronectin, fibronectin receptor $\beta$ subunit, $\beta$-actin, $\alpha$-tropomyosin, *NGFI-A*, *CEF-4* (or *9E3*, related to interleukin 8), *CEF-5 d-2*, *c25*, *rIRF-1*, p27, *Mtf*, the glucose transporter, *KC*, *N51*, *JE*, *TIS10*, *Cyr61*, *PC4*, *TIS7*, *Snk* (serum-inducible kinase) and *Pip92*[11]. The proto-oncogenes *MYC*, *CDC25A* and *MYB* are also transcriptionally activated, approximately 2 hours and 6 hours respectively after cell stimulation. This large number of genes, identified in cultured cells, probably does not represent a physiological response and it is notable that a number of the genes induced by stimulation with serum or growth factors are also activated by growth inhibitory agents, for example, TGF$\beta$.

In normal, untransformed cells the consistent pattern of a correlation between the early stages of proliferation and expression of the proto-oncogenes *ETS*, *FOS*, *JUN* and *MYC* clearly suggests that these proto-oncogenes function as essential mediators of the biochemical pathways that regulate proliferation and that their corresponding oncogenic forms may act *via* sustained perturbation of normal growth control mechanisms. It should also be noted that normal growth factors synthesized in an appropriate setting may act as "promoters" in the early stages of the development of cancers. Gastric-releasing peptide (GRP or mammalian bombesin) functions as an autocrine growth factor in small cell lung cancer: these tumour cells produce large amounts of the peptide which causes PtdIns(4,5)$P_2$ hydrolysis and an increase in $[Ca^{2+}]_i$ [12], characteristic responses of cells entering the cell cycle. In this situation the growth factor may act selectively to promote the proliferation of a clone of tumour cells.

## ONCOPROTEIN FUNCTION

In principle, any component in a pathway linking a growth factor to a critical biochemical step that regulates cell division has oncogenic potential. As Table 1 and Fig. 4 indicate, oncoproteins have now been identified that function at most, although not all, of the known steps in such signalling pathways. Table 1 should be considered together with the summary showing that many tumour suppressor genes exert controlling effects on cell cycle progression (Table 4). Aspects of the major classes are summarized here.

The locations of some of the major components in proliferation control pathways are represented in Fig. 4. Oncoproteins may arise as extracellular growth factors (SIS), as

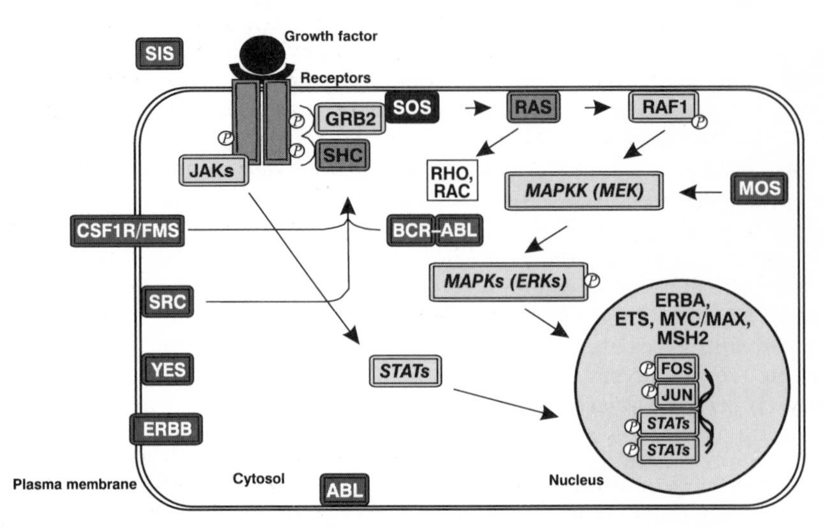


**Figure 4** *Cellular location of some proto-oncoproteins/oncoproteins.*

ligand-independent transmembrane proteins (CSF1R/FMS, ERBB, RET), as membrane-associated proteins (SRC, YES, RAS, RAF1), as cytosolic factors (ABL, MOS) or in the nucleus as transcription factors (ERBA, ETS, FOS, JUN, MYB, MYC) or DNA repair enzymes (MSH2). A major signal transduction pathway in which growth factors or oncoproteins cause the activation of mitogen-activated protein serine/threonine kinases (MAPKs or extracellular signal-regulated kinases (ERKs)) is also shown. In this pathway, growth factor-activated receptors or oncoproteins (SRC, BCR/ABL) interact with SH2 domain adaptor proteins (SHC, GRB2, SOS) to activate RAS to its GTP-bound form. Activated RAS then causes the initiation of a cascade of protein phosphorylation by serine/threonine kinases including RAF, MAPKK (or MEK MAP kinase or ERK kinase)) and MAPKs[13]. In germ cells MOS activates MAPKK by serine phosphorylation. Transcription factors that are targets for phosphorylation by MAPKs include EGR, ETS family proteins, FOS and JUN.

A variety of cytokines (including interleukins, interferons, erythropoietin, growth hormone, prolactin and granulocyte colony-stimulating factor) signal by activating members of the Janus family of protein kinases (JAK1, JAK2, JAK3 and TYK3) that phosphorylate on tyrosine residues various STAT proteins (signal transducers and activators of transcription) which then dimerize and are translocated to the nucleus to direct gene transcription[14]. STATs may form stable homodimers or heterodimers that are active as transcription factors. For example, STAT1$\alpha$ is a cytoplasmic monomer that undergoes tyrosine phosphorylation in response to interferon $\alpha$ (IFN-$\alpha$) or IFN-$\gamma$ which promotes stable homodimerization through SH2-phosphotyrosyl peptide interactions: the homodimers then bind to specific DNA sequences to direct transcription. Members of the interleukin 6 (IL-6) family of cytokines activate receptors that cause tyrosine phosphorylation of STAT1 and STAT3. STAT3 also undergoes serine phosphorylation and, in some types of cell, inhibition of serine phosphorylation prevents the formation of STAT3–STAT3 DNA binding complexes but not of STAT3–STAT1 or STAT1–STAT1 dimers. STATs may also form dimers with members of other transcription families, for example with JUN, which cooperates with STAT3$\beta$, a truncated form of STAT3, to activate the IL-6-responsive element of the rat $\alpha_2$-macroglobulin gene.

In addition to cytokines, growth factors can also activate the JAK/STAT signalling pathway. Thus, EGF causes the tyrosine phosphorylation of STAT1 which in turn mediates the activation of *FOS* transcription *via* interaction with the serum-inducible element of the *FOS* promoter. Intraperitoneal injection of EGF into mice results in the appearance in liver nuclei of multiple tyrosine phosphorylated proteins, including STAT5. Thus EGF is not only capable of activating the RAS signalling pathway but can also activate the JAK/STAT pathway. A number of cytokines (e.g. IL-3 and granulocyte-macrophage colony-stimulating factor) are also able to activate both pathways and the phosphotyrosyl moieties of some activated cytokine receptors can provide binding sites for the regulatory subunit of phosphatidylinositol 3-kinase, as do EGFR family members.

## Growth factor-related proteins

Class 1 (Table 1) includes the v-*sis* gene product referred to earlier that appears to function as an autocrine growth factor providing sustained activation of proliferation *via* a normal plasma membrane receptor. v-*sis* encodes the B chain subunit of PDGF (see ***PDGFB/Sis***) and in some cells a v-*sis* homodimer of PDGF is released that has a structure and activity similar to that of PDGF-BB. Hence, one possibility is that the cells proliferate indefinitely by an autocrine mechanism, although no increase in tyrosine phosphorylation corresponding to that caused by activation of the PDGF receptor has yet been detected.

## Receptor tyrosine kinases

Oncogenes that encode tyrosine kinase receptor-like proteins are generally considered to exert their effects through their sustained kinase activity. All tyrosine-specific protein kinases share sequence homology over ~300 amino acids although some (e.g. PDGFR, KIT, CSF1R/FMS) contain an insert region (Fig. 5).

The EGFR family (see ***EGFR***, ***HER2***, ***HER3*** and ***HER4***) form a complex set of transmembrane signalling proteins. HER2 can form heterodimers with EGFR, HER3 or HER4 and, although no specific ligand has been identified for HER2, it appears to be a critical component in the signal relay system activated by ligand binding to the other family members. Activated EGFR family proteins associate *via* their phosphotyrosyl residues with a wide variety of cytosolic proteins, including phospholipase C$\gamma$, phosphatidylinositol 3-kinase, phosphatidylinositol 4-kinase, GRB2/SOS, SHC, GAP, SRC and various phosphatases. Interaction with GRB2/SOS, either directly or *via* tyrosine-phosphorylated SHC proteins that associate with the receptor *via* their SH2 domains, couples receptor activation to the stimulation of RAS and the MAP kinase (ERK) and JUN kinase (JNK/SAPK) pathways.

The transforming activity of avian erythroblastosis virus (AEV) arises from v-*erbB* which encodes a truncated form of the EGFR that has lost the ligand binding domain and remains constitutively active as a protein tyrosine kinase, independent of EGF. v-ERBB has also lost a C-terminal region that includes an autophosphorylation

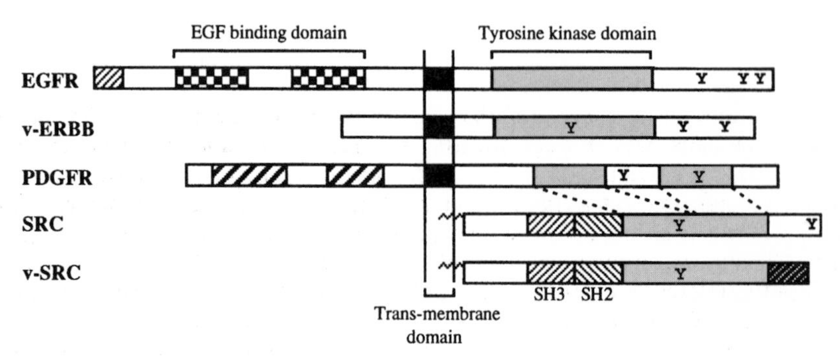


**Figure 5** *Epidermal growth factor receptor (EGFR), v-ERBB, platelet-derived growth factor receptor (PDGFR), SRC and v-SRC. The kinase domains are homologous but that of the PDGF receptor is divided by a kinase insert region. The product of the* v-erbB *gene lacks the extracellular epidermal growth factor binding domain and a tyrosine autophosphorylation site in the truncated C-terminus by comparison with the normal EGFR. Active SRC and v-SRC are N-terminally myristylated which promotes their association with the plasma membrane. The viral oncoprotein form of SRC differs from the normal protein by having a different C-terminus that does not contain a regulatory tyrosine residue: some point mutations also occur in v-SRC.*

site, presumed important for normal function. Major deletions in the human EGFR that remove ligand binding capacity and confer constitutive phosphorylation on the protein have been detected in brain and lung tumours and in a high proportion of breast and ovarian tumours examined. This form of the receptor thus resembles v-ERBB. It is constitutively associated with GRB2 although its expression does not increase RAS-GTP and only slightly activates the MAPK pathway.

Activated PDGF receptors resemble those of the EGFR family in that they bind and activate a similar, wide range of intracellular signalling molecules. PDGF and PDGF receptor genes are coexpressed in a variety of sarcomas and astrocytomas and cellular proliferation is promoted by autocrine PDGF stimulation (see ***PDGFB/Sis***). In general, however, abnormal expression of PDGF or its receptors is rare in human cancers and there are no reports of point mutations in these genes. In a subgroup of chronic myelogenous leukaemia the balanced t(5;12)(q33;p13) translocation results in the expression of a fusion transcript encoding the tyrosine kinase domain of the PDGFB receptor coupled to the putative helix-loop-helix region encoded by the ETS-related *TEL* gene [15]. Both the PDGF and EGF receptors undergo degradation mediated by ubiquitin acting on their C-termini and factors masking these sites (e.g. PLC$\gamma$) may increase the half-life of the proteins.

The RET receptor tyrosine kinase is the first oncoprotein detected in which dominantly acting point mutations initiate human hereditary neoplasia. Point mutations in RET, the receptor for glial-cell-line-derived neurotrophic factor (GDNF), occur in familial medullary thyroid carcinoma and in multiple endocrine neoplasia types 2A and 2B (see ***RET***). A number of mutations have been detected that result in constitutive kinase activation either by direct modulation of catalytic activity or by induction of ligand-independent homodimerization. Mutations giving rise to *RET* oncogenes are particularly prevalent in thyroid tumours of children exposed to fallout from the Chernobyl reactor accident.

## SRC

It has been known for some time that the tyrosine kinase activity of SRC increases markedly as cells pass through mitosis (see ***SRC***). This appears to be an indirect effect of CDC2 (CDK1)/cyclin B1 which phosphorylates the N-terminus of SRC causing a conformational change that renders the C-terminal regulatory tyrosine residue ($Tyr^{527}$) susceptible to phosphatase activity, thereby activating SRC kinase. Transgenic mouse studies indicate some functional redundancy between the closely related members of the SRC family, SRC, FYN and YES and the activity of the latter two also rises during mitosis. Potentially important substrates of SRC are SAM68, which may be involved in RNA processing, and the serine/threonine kinase RAF1 which is activated *via* RAS by many growth factors (see below) but which is also activated during mitosis and is stimulated by activated SRC [16]. As RAF1 is a MAP kinase kinase kinase (MAPKKK), this may reflect the fact that MAP kinase activity, necessary for both meiotic maturation and growth factor signalling, may also be required for mitotic progression. More recently it has been shown that active SRC is essential for the stimulation of DNA synthesis in response to PDGF [17]. Blockade of the $G_1$ to S phase transition by inhibition of SRC is reversible by over-expression of *Myc* but not by the early response genes *Fos* and/or *Jun*, suggesting that SRC kinases may control *Myc* transcription.

These findings imply that SRC may play a critical role in both progression through $G_1$ phase and in mitosis. This possibility, together with the powerful tumorigenic capacity of oncogenic SRC in animals, suggests that *SRC* might be a frequent target for anomalous expression in human cancers. However, no mutations have yet been detected in *SRC* although increased SRC kinase activity has been reported in some colon cancers, skin tumours and breast carcinomas.

## RAS and RAF

RAS proteins are molecular switches that are transiently activated in response to ligand-stimulated receptor tyrosine kinases by conversion from the GDP- to the GTP-bound form (see ***RAS***). As well as the receptors for EGF and PDGF, a large number of tyrosine kinase receptors also activate RAS including those for colony-stimulating factor-1 (CSFR1/FMS), neurotrophins (TRK), hepatocyte growth factor (MET), stem cell factor (KIT) and fibroblast growth factor. RAS can interact directly with RAF1, BRAF, RALGDS (RAL guanine nucleotide dissociation stimulator), RLF (RALGDS-like factor), the catalytic subunit of phosphatidylinositol-3-OH kinase and protein kinase Cζ, as well as with proteins of the GTPase-activating family (GAPs). RAS-GTP activates RAF1 by complex formation, promoting the translocation of RAF1 to the plasma membrane and its hyperphosphorylation. RAF1 kinase is a component of the cascade that transmits signals from activated growth factor receptor tyrosine kinases to MAPKs (see ***RAF1***). The phosphorylation of MAPK is catalysed by the protein kinase MEK which is in turn activated by RAF1. Other members of the RAF family (RAFA1 and RAFB1) also phosphorylate MEK1. However, RAS-independent events that may include interaction with other proteins of the small GTP-binding superfamily (e.g. RAP1A), proteins of the 14-3-3 family, lipids and proteins kinases also appear to be necessary for full RAF1 activation.

As Fig. 4 indicates, in addition to the linear pathway represented by RAS, RAF1, MEK and MAPKs, alternative signalling mechanisms may emanate from RAS or bypass it. Thus in some cell types at least full transformation by oncogenic RAS requires the activation by RAS of members of the RHO/RAC family that are involved in the organization of the cytoskeleton. In *C. elegans* a multivulval phenotype is induced by activated RAS. The recently identified suppressor of this phenotype, KSR-1, may interact directly with RAS to exert its effects. KSR-1 bears some sequence homology to the *C. elegans* RAF protein LIN-45 and may therefore be activated directly by RAS to control a pathway paralleling that involving MEK and MAPK that is also required for induction of the full effects of activated RAS. However, KSR-1 may itself be activated by tyrosine kinase receptors and activate RAF independently, bypassing RAS [18].

The *RAS* family (*HRAS1*, *KRAS2* and *NRAS*) are involved in a wide range of cancers. In all cancers the average incidence of *RAS* mutations is approximately 15% but in pancreatic carcinomas, for example, *KRAS2* mutations occur in 95% of tumours and in colorectal carcinoma the reported range is from 20% to 50%. In general, mutations in *RAS* reduce its GTPase activity, the oncogenic protein thus remaining in an active, GTP-bound state. One consequence of this would be the sustained activation of the RAF–MAP kinase pathway (Fig. 4).

The minisatellite region of *HRAS1* also appears to be mutated in some cancers. Minisatellites are tandem arrays of between 14 and 100 bp in length of a locus-specific consensus sequence. They are often polymorphic in the number of tandem repeats (hence referred to as variable number of tandem repeats (VNTRs) or variable tandem repetitions (VTRs). Minisatellites are dispersed throughout the genome and often occur just upsteam or downstream of genes (or within introns). Many VNTR loci display dozens of alleles. The *HRAS1* minisatellite (VTR) lies ~1 kbp downstream of the poly-A signal in the gene and comprises 30–100 copies of a 28 bp consensus repeat. There are four common and 25 rare alleles, the latter being more than twice as common in the genotypes of cancer patients than they are in normal individuals[19]. The VTR binds the constitutively expressed forms of NF-κB, suggesting that, by providing a tandem array of binding sites, this region may regulate transcription of both *HRAS1* and possibly other genes. Individuals with rare alleles of the *HRAS1* VTR who also carry an hereditary mutation in the *BRCA1* gene have an increased risk of ovarian cancer but not of breast cancer[20].

*RAF1* is amplified in some non-small cell lung cancers and it is over-expressed in many small cell lung cancers. Deletions and translocations in the region of the *RAF1* locus (3p25) correlate with some human malignancies, including small cell lung carcinoma and renal carcinoma.

## Nuclear oncoproteins

The proto-oncogenes that encode transcription factors (Table 2) are of particular interest in that they may regulate, either positively or negatively, genes that are directly involved in growth control. They fall into two broad categories: those that only interact with DNA as complexes with other proteins (FOS/JUN, MYC and REL) and those that, in monomeric form, possess a high affinity for specific DNA sequences (ETS, MYB and ERBA). The activity as transcription factors of proteins of both classes is regulated by phosphorylation. The modulation of the activity of relevant kinases (e.g. casein kinase II) and phosphatases by growth factors provides one mechanism by which transcriptional regulation mediated by these proto-oncogene products may be coupled to proliferation.

## ETS

The ETS family of transcription factors show strong homology in their DNA binding domains across a wide range of species. ETS protein binding sequences occur in many cellular and viral promoters and enhancers and members of the family may activate or repress transcription depending on the target gene and cell type (see ***ETS1* and *ETS2***).

**Table 2** *Transcription factor oncoproteins*

| Gene family | Transcription factors | DNA target sequence |
|---|---|---|
| *ETS* | ETS1, ETS2 | GC$^{C}/_{G}$GGAAGT |
| *FOS* | FOS, FOSB, FRA1, FRA2 | TGA$^{C}/_{G}$TCA (**AP1**) |
| | | TGACGTCA (**CRE**) |
| *JUN* | JUN, JUNB, JUND | TGA$^{C}/_{G}$TCA (**AP1**) |
| *MYB* | MYB | C$^{A}/_{C}$GTT$^{A}/_{G}$ |
| *MYC* | MYC/MAX | CACGTG |
| *REL* | REL/NF-κB | GGG$^{A}/_{G}$NT$^{T}/_{C}$$^{T}/_{C}$CC |
| *THRA/ErbA* | THRA/ERBA/$T_3$ | TCAGGTCATGACCTGA |

AP1: AP1 consensus site; CRE: cAMP response element.

The expression of ETS genes is complex and may be controlled by initiation of transcription, alternative splicing, post-translational modification and protein stability. It is differentially modulated by growth stimuli during differentiation, indicating that ETS proteins are tissue-specific regulators of gene expression. In particular ETS1 and ETS2 regulate lymphoid-specific genes *via* ETS binding domains. Specific DNA binding is abolished by phosphorylation of ETS1 or ETS2, most probably carried out by $Ca^{2+}$-dependent myosin light chain kinase. ETS1 contains a potential MAP kinase phosphorylation site and the ETS proteins ELK1 and ΔELK1 are both activators of and substrates for MAP kinases.

ETS1 and ETS2 activate transcription through the PEA3 motif (CACTTCCT) that is present in the oncogene responsive domain (ORD) of the polyoma enhancer and they cooperate with AP1 that binds to an adjacent PEA1 site (GTTAGTCA). ETS1 can also interact directly with the basic domain of JUN proteins, giving rise to ternary complexes at AP1-ETS binding sites. Trimeric complexes between ETS1, PEBP2α and lymphoid enhancer-binding factor (LEF1) occur at the T cell receptor α enhancer[21]. β-Catenin, which links cadherin cell-adhesion molecules to the cytoskeleton and associates with the product of the *APC* tumour suppressor gene, also interacts with LEF1 to regulate gene expression[22]. Other partners known to associate with ETS1 include GATA1, PU.1 and CBF and transcriptional activation of the HTLV-I LTR may occur through the synergistic action of ETS1 with SP1. The ETS family members ELK1 and SAP1 form ternary complexes with serum response factor (SRF) that modulate transcription from the serum response element (SRE) present in a number of genes including *FOS*, *JUNB*, vinculin, *EGR1* and *EGR2* (see ***FOS***). Thus ETS family proteins appear to function as components of transcription factor complexes that regulate the expression of cellular and viral genes.

The ETS genes *ETS1* and *FLI1/ERGB* map to a region of chromosome 11, disruption of which is implicated in a number of leukaemias (Table 3). The 21q22 region (*ETS2* and *ERG*) is involved in the translocation (8;21)(q22;q22) that occurs in acute myelogenous leukaemia and Down's syndrome[23]. The t(11;22) translocation substitutes the RNA binding region of EWS (22q12) for the ETS DNA binding domain present in the C-terminus of FLI1[24]. In Ewing's sarcoma up-regulation of *MYC* arises from *trans*-activation of the *MYC* promoter by EWS/FLI1[25] which also activates transcription of stromelysin-1[26]. A variant Ewing's sarcoma translocation (7;22) fuses *EWS* to *ETV1* (ETS translocation variant), the human homologue of the murine ETS gene *ER81*, resulting in a fusion protein having sequence-specific DNA binding[27]. In the

**Table 3** *Chromosomal translocations in cancers involving ETS family genes*

| ETS family gene | Locus | Translocation | Cancer |
|---|---|---|---|
| *ETS2* | 21q22.3 | (8;21)(q22;q22) | Acute myelogenous leukaemia |
| *ERG* | 21q22.3 | | |
| *FLI1* | 11q24 | (11;22)(q24;q12) | Ewing's sarcoma |
| *EWS* | 22q12 | | |
| *ERG* | 21q22.3 | (16;21)(p11;q22) | Chronic myelogenous leukaemia |
| *TLS/FUS* | 16p11 | | |
| *ERG* | 21q22.3 | (22;21)(q12;q22) | Ewing's sarcoma |
| *EWS* | 22q12 | | |
| *ETV1* | 7p22 | (7;22)(p22;q12) | Ewing's sarcoma |
| *EWS* | 22q12 | | |

t(16;21)(p11;q22) translocation found recurrently in myeloid leukaemia, *ERG* is fused to the *TLS/FUS* gene to produce a predicted protein product similar to the EWS/FLI1 chimeric protein responsible for Ewing's sarcoma[28]. *ELK* genes (*ELK1* (Xp11.2) and *ELK2* (14q32.3)) map close to the translocation breakpoint characteristic of synovial sarcoma [t(X;18)(p11.2;q11.2)] and the 14q32 breakpoints seen in ataxia telangiectasia and other T cell malignancies, respectively[29,30].

The general consequence of these translocations appears to be the generation of aberrant transcription factors, as exemplified by EWS/FLI1. EWS/FLI1 binds to the same *ETS2* consensus sequence as FLI1 but is a more potent *trans*-activator that, in contrast to FLI1, transforms NIH 3T3 cells[31]. EWS/FLI1 but not FLI1 forms a ternary complex with SRF on the *Fos* SRE[32]. In addition to these translocations, a non-conservative mutation has been detected in *ETS1* in a T cell acute lymphoblastic leukaemia (T-ALL), suggesting that the normal gene product may also possess suppressor activity[33].

ETS proteins may be involved in tumour-associated angiogenesis and invasiveness (see Chapter 7). The type I collagenase (*MMP-1*), type IV collagenase (*MMP-9*) and stromelysin promoters are activated both by AP1 and by the ETS-related E1AF protein that promotes the invasive phenotype of human cancer cells[34,35]. *MMP-1* can be activated synergistically by AP1 and PEA3 and stromelysin may also be activated by ETS1 or ETS2. *ETS1* expression can be induced by tumour necrosis factor $\alpha$ or basic fibroblast growth factor. Hence the action of cytokines or angiogenic factors involved in tumour progression may be mediated by the induction of ETS protein expression, either in tumour cells or in adjacent normal fibroblasts. However, most epithelial tumours do not express *ETS1* and its over-expression in human colon carcinoma cell lines reduces their tumorigenic capacity[36].

## FOS and JUN

*FOS* and *JUN* are members of a multigene family that includes the FOS-related (FOSB, FRA1, FRA2) and JUN-related (JUNB, JUND) nuclear transcription factors (see ***FOS***, ***FOSB***, ***FRA1*** and ***FRA2***, ***JUN***, ***JUNB*** and ***JUND***). These proteins have in common a hydrophobic leucine zipper domain that mediates dimerization and a basic DNA binding domain. FOS and JUN family proteins are the major components of the complex that binds to DNA at AP1 consensus sequences to mediate gene expression in response to a variety of extracellular stimuli. The FOS/JUN system, thus far unique in its resolved complexity, illustrates the way in which the transcriptional control exerted by various heterodimeric combinations may undergo subtle modulation during cell development or proliferation as the extent of transcription and translation of different members of the families changes.

AP1 activity is enhanced as a result of either increased translation of FOS and JUN family proteins and/or their post-translational modification by phosphorylation. The protein kinases that phosphorylate FOS and JUN, FRK and JNK, respectively have been identified and it is established that the target of JNK is the N-terminal activation domain of JUN, phosphorylation of which increases *trans*-activation by JUN. However, at least 10 isoforms of JNK have now been discovered and these have differing effects both on JUN family members and also on other transcription factor targets, e.g. ATF2 and ELK1. Thus, JNK binds to both JUN and JUNB but only phosphorylates JUN, whereas weak binding to the substrate is sufficient to permit some phosphorylation of JUND. Furthermore, for at least one transcription factor

(ELK1), JNK binding is not necessary for phosphorylation. Thus, the role of JNK in regulating transcription is complex and in addition to its direct effects on AP1 and other transcription factors, complex formation with JNK may serve to target its kinase activity to yet other proteins[37].

The cellular expression of several oncogenes (*Fos*, *Mos*, *Ras*, *Raf*, *Src* or polyoma middle T) induces AP1 activity and AP1-binding enhancer elements occur in the *cis* control regions of a number of genes that are strongly expressed in transformed cells, including collagenase, metallothionein IIA and stromelysin. FOS cannot dimerize with itself and dimerization between FOS and JUN or between members of the JUN family is necessary for DNA binding. Eighteen different dimeric combinations can thus be formed from within the FOS and JUN families. DNA binding occurs by interaction between the leucine zipper regions (see ***JUN***) of the proteins that lie next to basic, DNA binding domains. JUN proteins also bind with high affinity to the cAMP response element (CRE) and JUN, but not FOS, proteins dimerize with some members of the CREB family, the dimers formed binding preferentially to the CRE. The capacity of JUN to interact with cAMP signalling pathways by forming JUN/CREB dimers thus provides an additional group of transcription factors. The basic and leucine zipper regions are present in both the normal and oncogenic forms of FOS and JUN and mutations that prevent dimerization decrease the transforming potential of v-FOS and v-JUN. In fibroblasts that have been stimulated by serum, all the products of the *Fos* and *Jun* gene families coexist but the extent and temporal pattern of expression varies. These genes are also differentially expressed during development. This suggests that these proteins have distinct functions that are probably mediated by the wide range of affinities for AP1 sites of the different dimeric forms. Thus JUN dimers have a 10-fold greater affinity for AP1 than JUNB or JUND: in JUN/FOS dimers the affinity and the half-life of interaction with DNA depends on the specific FOS protein involved (FOSB > FRA1 > FOS). Furthermore, single base substitutions in the regions immediately flanking the AP1 sequence (or the CRE element) can cause a 10-fold change in binding affinity. These *in vitro* determinations of relative affinities are consistent with the finding that *Jun* with *Hras-1* is a more potent transforming combination than *JunB* and *Hras-1* and that *JunB* expressed at high levels inhibits the transforming potential of *Jun*.

The principal difference between the oncogenic form of JUN and the cellular form is the loss of a 27 amino acid N-terminal region ($\delta$) that is not well conserved in JUNB and JUND. The $\delta$ region appears to facilitate or stabilize the interaction of an inhibitor protein present in some cells with the A1 transcription activation domain of JUN. v-JUN and JUN bind equally well to the AP1 site but in HeLa cells that express the inhibitory protein, v-JUN is a much stronger transcriptional activator. Thus the deletion of $\delta$ in v-JUN releases the oncoprotein from regulation by the inhibitory protein. However, JUN from which $\delta$ has been deleted is still less effective at activating transcription than v-JUN, indicating the importance of the three additional mutations present in v-JUN. JUN/FOS heterodimers do not appear to bind the inhibitory protein.

In the FOS family, although the leucine zipper and basic domains are essential for DNA binding, the regulation of transcription from AP1 sites is also dependent on net negative charge in a C-terminal region, usually conferred by phosphorylation[38]. In both v-FOS and the truncated form of the normal protein FOS, FOSB2, the C-terminus is deleted: such proteins form dimers with JUN that bind to AP1 but do not activate transcription from AP1 promoters or suppress *Fos* transcription. Thus the effect of FOSB2 is to compete with normal FOS thereby decreasing the concentration of active FOS/JUN dimers.

The exquisite sensitivity of FOS activation to a wide variety of stimuli in practically every type of cell suggests that its activity as a transcription factor might be essential for cell proliferation. However, it is evident from transgenic studies that essentially normal development occurs in *Fos*$^{-/-}$ mice, although they develop osteopetrosis as a consequence of the absence of terminally differentiated osteoclasts. Consistent with this, transgenic mice over-expressing *Fos* develop osteosarcomas as a result of abnormal bone deposition by osteoblasts whilst other tissues remain largely unaffected. In view of these findings, it is unsurprising that there are few reports of involvement of *FOS* in human cancers, although in one study its over-expression was detected in 60% of osteosarcomas. Over-expression of *Jun* in transgenic mice does not give rise to an abnormal phenotype but does appear to have a cooperative effect with *Fos* to accelerate the rate at which osteosarcomas arise in *Jun/Fos* double transgenics. Deletion of *Jun* in mice is embryonically lethal. Anomalous expression of *JUN* in human cancers appears to be rare although there are reports of substantial over-expression in some lung cancers.

## MYB

The transcription factor MYB appears to be involved in controlling cell cycle progression and differentiation (see ***MYB***). Thus, immature proliferating haematopoietic cells express high levels of MYB and *Myb*$^{-/-}$ mice are non-viable due to the failure of hepatic erythropoiesis. The critical genes controlled by MYB remain unidentified, in part because a large number of promoters have been shown to bind MYB directly and also because an even greater number of genes possess potential MYB binding sites. However, in lymphocytes two bursts of MYB expression occur as cells proceed through $G_1$ and S phases and the expression of high levels of exogenous MYB can activate the transcription of *CDC2*. MYB also regulates genes required for DNA synthesis (PCNA and DNA polymerase $\alpha$), indicating its importance in cell cycle progression. Consistent with this role is the evidence that MYB expression inhibits differentiation, although the mechanism by which this is effected is unknown.

MYB contains transcriptional activation and negative regulatory domains as well as a domain that binds directly to the DNA sequence PyAAC$^{G}/_{T}$G to activate transcription. MYB DNA binding activity is negatively regulated *via* phosphorylation by casein kinase II. The cellular activity of this enzyme is increased by the action of some growth factors (e.g. insulin), suggesting that proliferation may involve the suppression of MYB-controlled genes, although PDGF, for example, does not activate casein kinase II. The casein kinase II phosphorylation site is in the N-terminal region of MYB but in all oncogenic forms of MYB N-terminal mutations cause the deletion of this site together with most of the negative regulatory domain. These changes result in DNA binding that is independent of a normal regulatory mechanism and in an enhanced level of transcription caused by loss of the inhibitory region. An alternative product of *Myb* (*Mbm2*) encodes a truncated protein that retains the DNA binding and nuclear localization regions but has lost the regulatory regions required for transcriptional activation. Expression of normal *Myb* blocks differentiation whereas the effect of *Mbm2* is to enhance differentiation. This is reminiscent of FOSB2 that possesses normal FOS binding activity but not *trans*-activation potential. Thus *Mbm2* interferes with *Myb* function during mouse erythroid leukaemia cell differentiation and FOSB inhibits *trans*-activation by v-FOS, FOS or FOSB, acting effectively as a *trans*-negative regulator.

In addition to regulating positively its own transcription, MYB can stimulate transcription of a large number of genes including *MYC*, and MYB also *trans*-activates the LTRs of human immunodeficiency virus type 1 (HIV-1) and human T cell leukaemia virus type I (HTLV-I). Other genes containing MYB consensus binding sequences in their promoters, many of which are involved in regulating cell cycle progression, include cyclins B and D1, *Src, Bcr, Ets, Jun, Kit, P53* and IL-2.

Amplification of *MYB* has been detected in a small proportion of leukaemias, colon carcinomas, melanomas and breast carcinomas. In the latter *MYB* expression may correlate inversely with that of *HER2* and constitute a good prognostic factor (see Chapter 8). Abnormally high levels of *MYB* transcripts have been detected in some ovarian and cervical carcinomas.

## MYC

The MYC proteins (MYC, MYCN and MYCL) each contain basic, helix-loop-helix and leucine zipper domains. Each forms heterodimers with the protein MAX (which also contains all three motifs) that bind specifically to DNA as *trans*-activating complexes (see ***MYC*** and ***MAX***). MAX is the only known helix-loop-helix protein that forms dimers with MYC. MAX expression is independent of MYC and MAX may thus regulate transcription of genes independently of MYC. However, other leucine zipper proteins (MAD/MXI1) form heterodimers with MAX that repress transcription by binding to the MYC/MAX consensus sequence, and so the action of MYC/MAX may be regulated indirectly by the abundance of these proteins.

The principal functions of MYC are the induction of proliferation and the inhibition of terminal differentiation in adipocytes and myeloid cells and, in the absence of appropriate growth factors, the induction of programmed cell death (apoptosis). Although MYC is essential for cell proliferation and it has been shown to regulate transcription of a wide variety of genes *in vitro*, the biochemical connections between proto-oncogene signalling pathways and the machinery of cell cycle control have been difficult to resolve. MYC has long been known to activate ornithine decarboxylase, which is necessary for cell proliferation, but other critical targets have remained elusive. Recently, however, it has emerged that MYC can *trans*-activate both cyclin D1 and, indirectly, cyclin A, both of which are components of cyclin-dependent kinases that control passage through the cell cycle. Additionally, MYC/MAX directly trans-activates *CDC25A* that encodes a phosphatase mediating the activation of cyclin-dependent kinases. CDC25A itself has oncogenic capacity and its expression pattern during the cell cycle closely resembles that of MYC (see Chapter 4). Consistent with these critical roles in driving DNA replication and cell division, *MYC* is repressed as a component of p53-mediated growth arrest.

Amplification and/or over-expression of *MYC* commonly occurs in a wide range of tumours. *MYCN*, which is primarily expressed in neuronal cells during embryogenesis, is amplified in neuroblastomas and also in retinoblastomas, astrocytomas, gliomas and small cell lung carcinomas. A high level of *MYCN* expression is associated with the advanced stages of cervical intraepithelial carcinoma. Amplification of *MYCL1* has been detected in small cell lung carcinoma.

The most common genetic rearrangements in B cell lymphomas involve *MYC* (8q24) or *BCL2* (18q21.3). The translocations in Burkitt's lymphoma, mentioned earlier, give rise to a *MYC* gene adjacent to one of three immunoglobulin (Ig) loci, most commonly IgH (14q32). In the less common variants one of the Ig light chain loci (Ig$\kappa$ (2p12) or Ig$\lambda$ (22q11)) translocates to a position 3′ of *MYC* on chromosome 8. In each type of translocation *MYC* expression is deregulated as a result of juxtaposition to Ig constant region gene segments, non-translocated *MYC* being transcriptionally silent. Translocations involving *BCL2* and IgH generally activate *BCL2*, promoting cell survival and resistance to apoptosis. Three-way translocations involving *BCL2, MYC* and IgH also occur in which the IgH allele not associated with the t(8;14) translocation drives the expression of *BCL2*. However, translocations in which *MYC* is activated and *BCL2* is inactivated also occur. Thus, although the deregulation of *BCL2* would be expected to prolong survival of affected pre-B cells and hence increase the probability of other oncogenes, including *MYC*, being activated subsequently, leukaemias can develop in the absence of functional BCL2. A translocation involving *MYC* and the T cell receptor $\alpha$ chain has been detected in a B cell lymphoma that presumably arose from inappropriate recombinase activity during B cell development[39].

## REL

The REL-related family, which includes REL, RELA, RELB, NF-$\kappa$B1 and NF-$\kappa$B2, function as homo- and heterodimeric transcription factors *via* response elements related in sequence to the NF-$\kappa$B binding consensus (see ***REL***). NF-$\kappa$B is a complex of the REL family proteins p50 (NF-$\kappa$B1) and p65 (RELA). REL proteins possess conserved DNA binding and dimerization domains, the former being strongly homologous to the DNA binding subunits of the NF-$\kappa$B transcription factor family. The activity of REL proteins is primarily regulated by their subcellular localization: inactive forms are retained in the cytoplasm, undergoing translocation to the nucleus in response to external stimuli. Cytoplasmic NF-$\kappa$B exists as a complex with an inhibitor protein I$\kappa$B (major form: I$\kappa$B-$\alpha$; minor form: I$\kappa$B-$\beta$). Phosphorylation of I$\kappa$B dissociates the complex and releases NF-$\kappa$B to translocate to the nucleus. A family of I$\kappa$B-like proteins, including MAD-3 and pp40, regulate the activity of REL proteins. v-REL appears to be unique in that it transforms cells equally effectively whether it is located in the cytoplasm or the nucleus. v-REL has, however, lost the strong transcription-activating element in the C-terminus of REL and appears by itself to be a weak activator of transcription. C-terminal loss may promote movement to the nucleus (C-terminal removal does allow the protein to move to the nucleus but v-REL to which the C-terminus of REL has been attached remains cytoplasmic and yet still transforms). v-REL forms complexes with REL and with three proteins to which REL binds, two of which (p115/p124) may be precursors of NF-$\kappa$B. Thus v-REL may deplete the cell of transcription factor precursors or bind directly to NF-$\kappa$B or REL DNA target sequences. In any of these mechanisms v-REL would be acting as a dominant negative oncogene.

Amplification of *REL* occurs in some lymphomas and rearrangement of this chromosomal region (2p14–15) is associated with some non-Hodgkin's lymphomas. Rearrangement of *NF-$\kappa$B2* (10q24) occurs occasionally in lymphoid neoplasms in which the the REL homology domain and $\kappa$B binding is retained by the chimeric protein. Deletion of the C-terminal coding region has been detected in cutaneous T cell leukaemia. Over-expression of NF-κB1 has been detected in non-small cell lung carcinoma. Abnormally high levels of the NF-κB2 precursor occurs in some primary breast tumours which may promote NF-κB (NF-κB1/RELA) heterodimer formation. Over-expressed members of the REL family may contribute to abnormal cell proliferation by sequestering members of the IκB family and thereby activating κB-binding transcription factors.

## THR/ERBA

The v-*erbB* and v-*erbA* genes are both carried by the avian erythroblastosis virus AEV-ES4. v-*erbA* is the oncogenic homologue of *ErbA-1* that encodes thyroid hormone receptor α (human homologue *THRA1*), and it increases the transforming potential of other oncogenes, including v-*erbB*. ERBA appears to be permanently bound to the THR response element (see ***THR/ErbA***). The physiological ligand for ERBA (triiodothyronine, $T_3$) stimulates transcription by over 50-fold. v-ERBA represses transcription to a level similar to that caused by ERBA but is insensitive to $T_3$. Thus v-ERBA causes sustained repression and thus apparently acts as a dominant negative oncoprotein. The N- and C-terminal truncations occurring in the conversion of *ErbA* to v-*erbA* affect only its hormone responsiveness, not its capacity to bind to DNA. An additional mutation removes a phosphorylation site from ERBA, the loss of which is essential for the effects of v-*erbA* on transformation.

Loss of heterozygosity at the *THRA1* locus has been detected in sporadic breast cancers and in a breast cancer cell line (BT474) *THRA1* undergoes fusion at exon 7 to the *BTR* gene[40]. The resultant truncated form of THRA1 resembles v-ERBA. It thus appears that involvement of thyroid hormone receptors in human cancers is rare but it may be noted that genetic rearrangements of the structurally related retinoic acid receptor α gene occur in acute promyelocytic leukaemia (see Table III, page 78) the fusion protein produced being a cell- and sequence-specific transcription factor[41].

The oncoprotein forms of many transcription factors have deleted C-terminal regions. This does not affect their DNA binding capacity but it invariably alters the effect that the bound protein can exert on transcription. Thus v-FOS and v-ERBA bind normally to promoter regions but v-FOS is unable to suppress *Fos* transcription and v-ERBA is insensitive to $T_3$. The effect of each is that of a dominant negative oncogene and resembles the dominant negative mutations that may occur in the tumour suppressor gene *P53* and cause the formation of complexes between mutant and wild-type p53, inhibiting the function of the latter.

***References***


[1] Michiels, F.-M. et al. (1994) Proc. Natl Acad. Sci. USA 91, 10488–10492.
[2] Muca, C. and Vallar, L. (1994) Oncogene 9, 3647–3653.
[3] Rudolph, U. et al. (1995) Nature Genet. 10, 143–149.
[4] Voyno-Yasenetskaya, T.A. et al. (1994) Oncogene 9, 2559–2565.
[5] Vara Prasad, M.V.V.S. et al. (1994) Oncogene 9, 2425–2429.
[6] De Vivo, M. et al. (1992) J. Biol. Chem. 267, 18263–18266.
[7] Tolias, K.F. et al. (1995) J. Biol. Chem. 270, 17656–17659.
[8] Almendral, J.M. et al. (1988) Mol. Cell. Biol. 8, 2140–2148.
[9] Zipfel, P.F. et al. (1989) Mol. Cell. Biol. 9, 1041–1048.
[10] Tamori, A. et al. (1995) Cancer Res. 55, 3500–3503.
[11] **Herschman, H.R. (1991) Annu. Rev. Biochem. 60, 281–319.**

[12] Minna, J.D. (1988) Lung Cancer 4, P6–P10.
[13] **Crews, C.M. et al. (1992) Science 258, 478–480.**
[14] **Ihle, J.N. (1996) Adv. Cancer Res. 68, 23–65.**
[15] Golub, T.R. et al. (1994) Cell 77, 307–316.
[16] **Taylor, S.J. and Shalloway, D. (1996) BioEssays 18, 9–11.**
[17] Barone, M.V. and Courtneidge, S.A. (1995) Nature 378, 509–512.
[18] **Downward, J. (1995) Cell 83, 831–834.**
[19] Trepicchio, W.L. and Krontiris, T.G. (1992) Nucleic Acids Res. 20, 2427–2434.
[20] Phelan, C.M. et al. (1996) Nature Genet. 12, 309–311.
[21] Giese, K. et al. (1995) Genes Dev. 9, 995–1008.
[22] Behrens, J. et al. (1996) Nature 382, 638–642.
[23] Papas, T.S. et al. (1990) Am. J. Med. Genet. Suppl. 7, 251–261.
[24] Delattre, O. et al. (1992) Nature 359, 162–165.
[25] Bailly, R.-A. et al. (1994) Mol. Cell. Biol. 14, 3230–3241.
[26] Braun, B.S. et al. (1995) Mol. Cell. Biol. 15, 4623–4630.
[27] Jeon, I.-S. et al. (1995) Oncogene 10, 1229–1234.
[28] Ichikawa, H. et al. (1994) Cancer Res. 54, 2865–2868.
[29] Kocova, M. et al. (1985) Cancer Genet. Cytogenet. 16, 21–30.
[30] Rowley, J.D. (1983) Nature 301, 290–291.
[31] May, W.A. et al. (1993) Mol. Cell. Biol. 13, 7393–7398.
[32] Magnaghi, L. et al. (1996) Nucleic Acids Res. 24, 1052–1058.
[33] Collyn-d'Hooghe, M. et al. (1993) Leukemia 7, 1777–1785.
[34] Higashino, F. et al. (1995) Oncogene 10, 1461–1463.
[35] Kaya, M. et al. (1996) Oncogene 12, 221–227.
[36] Suzuki, H. et al. (1993) Int. J.Oncol. 3, 565–573.
[37] Gupta, S. et al. (1996) EMBO J. 15, 2760–2770.
[38] **Hunter, T. and Karin, M. (1992) Cell 70, 375–387.**
[39] Park, J.K. et al. (1989) Genes Chromosom. Cancer 1, 5–22.
[40] Futreal, P.A. et al. (1994) Cancer Res. 54, 1791–1794.
[41] Jansen, J.H. et al. (1995) Proc. Natl Acad. Sci. USA 92, 7401–7405.

# 3 Tumour Suppressor Genes

The existence of what are now variously known as tumour suppressor genes, recessive oncogenes, anti-oncogenes or growth suppressor genes was originally inferred from the finding that when tumorigenic and non-tumorigenic cells were fused in culture the resulting hybrids were generally non-tumorigenic. When such hybrid cells do give rise to tumours in animals, this usually involves the loss of a specific chromosome derived from the non-tumorigenic cell. In many spontaneously arising tumours, individual chromosomes or specific regions of a chromosome are lost or deleted.

These observations suggest the existence of a substantial class of tumour suppressor genes, presumed to comprise approximately 50 genes on the basis of the number of different hereditary cancers, the normal function of which is to govern cell proliferation. When non-tumorigenic hybrids are formed from two different tumorigenic cells genetic complementation may be occurring. The two best understood tumour suppressor genes are the retinoblastoma (*RB1*) gene and *TP53* (generally called *P53*). Retinoblastoma provides the classical model for a recessive tumour suppressor gene in that both paternal and maternal copies of *RB1* must be inactivated for the tumour to develop. For *TP53* and some other tumour suppressor genes, mutation at one allele may be sufficient to give rise to the altered cell phenotype (Fig. 6).

The protein products of these genes were first isolated through their association with the oncoproteins of DNA tumour viruses. The general mechanism by which many DNA virus oncoproteins cause transformation is by forming complexes with host cell proteins. The adenovirus E1A proteins and the T antigens of simian vacuolating virus 40 (SV40) and polyoma form specific complexes with the *RB1* gene product (pRb) and these and other viral proteins share a sequence motif essential for this interaction. SV40 T antigen also forms complexes with p53 at a site distinct from that binding pRb and p53 also associates with adenovirus E1B. These observations suggest that the transforming potential of SV40 T antigen lies in its capacity to bind to and inactivate pRb and p53, both of which regulate the proliferation of normal cells. Human papillomavirus proteins E6 and E7 also bind to p53 and pRb, respectively. The polyoma antigens do not bind p53 but it is probable that the increasing number of proteins being detected in complexes with these and other DNA viral antigens will include tumour suppressor gene products that are presently unknown (see **DNA Tumour Viruses**). The brief summaries below indicate the important roles of pRb and p53 in the control of cell proliferation, considered in more detail in Chapter 4.

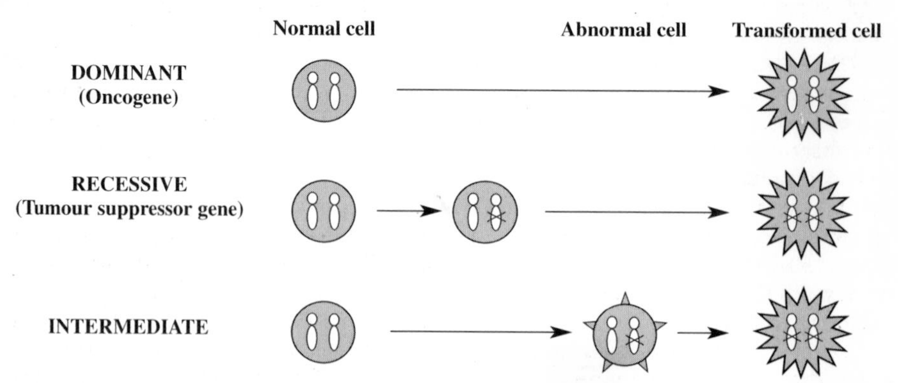


**Figure 6** *Types of mutation in tumorigenesis.*

## RB1

pRb functions as a signal transducer, connecting the cell cycle clock with transcriptional control mechanisms mediating progression through the first phase of the cell cycle (see ***RB1***). The broad transcriptional effects of pRb are mediated by its inhibition of transcription factors that are required for the expression of genes involved in DNA replication and by repressing transcriptional activation by RNA polymerases I and III that are responsible for ribosomal and transfer RNA synthesis. Thus pRb regulates the overall biosynthetic capacity of the cell and its loss not merely releases a brake at the $G_1$/S phase transition but drives proliferation and hence chromosome duplication and the accumulation of mutations.

pRb binds to many cellular proteins but the most critical appear to be the E2F family and UBF. pRb binding inhibits the *trans*-activation function of at least three (E2F1, E2F2, E2F3) of the five known members of the E2F family of transcription factors that are required for S phase. E2F1 is an oncoprotein in that its coexpression with that of other oncogenes can cause transformation and E2F1 alone can induce DNA synthesis in serum-starved cells. Remarkably, however, E2F1 also has the characteristics of a tumour suppressor gene. Mice in which E2F has been deleted show tissue-specific hyperplasia and develop a variety of tumours. The molecular basis for this complex behaviour is unclear, although it presumably reflects tissue-specific regulation of transcription by E2F1. E2F1 itself is not essential for normal development but its loss leads to subtle aberrations with ageing in some tissues, notably testicular atrophy. However, in the reproductive tract and the lung, loss of E2F1 promotes aggressive tumorigenesis with lymphomas also developing apparently as a result of loss of thymocyte apoptosis [1]. Major proliferation-associated genes regulated by E2F include *MYC, MYCN, MYB, CDC2*, dihydrofolate reductase, thymidine kinase and *EGFR*.

## P53

p53 is a transcription factor (see ***P53***) that regulates the normal cell growth cycle by activating transcription of genes that control progression through the cycle and of other genes that cause arrest in $G_1$ when the genome is damaged and, in some cell types, promote apoptosis [2]. Hypoxia, occurring in areas of tumours with poor blood supply, promotes p53-dependent apoptosis: cells with mutated p53 are resistant to killing by hypoxia [3]. Genes activated by p53 include *WAF1, MDM2, GADD45, BAX* and cyclin G, whilst p53 represses *MYC* and *BCL2*. In addition to these activities which are mediated by transcriptional control, p53 also possesses 3′ to 5′ DNA exonuclease capacity. This $Mg^{2+}$-dependent activity may be involved in replication-associated DNA repair, p53 acting as a proof-reading enzyme for DNA polymerases.

Mutations in the single copy *P53* gene are the most frequent genetic changes yet shown to be associated with human cancers and point mutations, deletions or insertions in *P53* occur in ~70% of all tumours. The point mutations occur almost entirely in the central region of the protein that has been revealed by X-ray diffraction to bind to DNA. The most frequently mutated residues lie in the regions that directly contact DNA but other mutations modulate function by affecting the overall three-dimensional structure of the protein [4]. The rare, autosomal dominant Li–Fraumeni syndrome arises from *P53* mutations inherited through the germline [5]:

**Table 4** *Major tumour suppressor genes (see Table IV, page 83)*

| Gene | Function | Recessive/ intermediate | Principal cancer associations |
|---|---|---|---|
| *RB1* | Cell cycle regulator<br>Protects from apoptosis | R | Retinoblastoma, lung, bladder, breast, pancreatic carcinomas |
| *P53* | Cell cycle regulator<br>Promotes growth arrest and apoptosis | I | Most – Sarcomas, breast carcinomas, leukaemias |
| *E2F1* | Transcription factor | ? | Leukaemias ? |
| *INK4A/ MTS1* | Cyclin-dependent kinase inhibitor | R | Most – Melanoma, acute lymphoblastic leukaemia, pancreatic carcinomas |
| *APC* | Binds $\alpha$- and $\beta$-catenin: may mediate adhesion, cell cycle progression | R | Colon carcinoma |
| *BRCA1* | Transcription factor? | R | Breast, ovarian |
| *CDH1/* E-cadherin | $Ca^{2+}$-dependent intercellular adhesion, signalling | R | Many: breast, ovarian |
| *NF1* | RAS-GTPase-activating protein | R | Neurofibromas, chronic myelogenous leukaemia |
| *WT1* | Transcription factor | I | Renal cell carcinoma |
| *VHL* | Modulates RNA polymerase II *via* elongin | R | Renal cell carcinoma |
| *MSH2, MLH1* | DNA mismatch repair | R | Hereditary non-polyposis colon cancer |
| *TGFBR2* | Cell growth inhibitor | I | Colon, lung |
| *DPC4/SMAD4* | Cell growth inhibitor | ? | Pancreas, colon, bladder and biliary |
| *PTC* | Transcription repressor | ? | Nevoid basal cell carcinoma syndrome |

50% of the carriers develop diverse cancers by 30 years of age, compared with 1% in the normal population. In general, however, *P53* mutations are somatic and occur with high frequency in all types of lung cancer, in over 60% of breast tumours and in ~40% of brain tumours (astrocytomas), frequently in combination with the activation of oncogenes (Table 4). The region deleted in chromosome 17p in most colorectal neoplasms includes *P53*. Nevertheless, as observed previously and despite the high frequency of *P53* mutations, absolute patterns of mutations do not characterize individual forms of cancer. Nevertheless, certain combinations of genetic defects do occur with relatively high frequency in association with some types of cancers (Table 4: see also Chapter 8).

The other major tumour suppressor genes in Table 4 comprise an extremely varied group both in terms of what is known of their functions and the range of tissues in which they are associated with tumorigenesis. The DNA mismatch repair genes, cyclin-dependent kinase inhibitors, signalling through the TGF$\beta$ type II receptor, *DPC4/SMAD4*, *PTC* and E-cadherin are discussed in the next three sections. The other tumour suppressor gene products in Table 4 are notable for their highly restricted tissue involvement. Thus familial adenomatosis polyposis which, when untreated, leads to colorectal cancer arises from the inheritance of one abnormal *APC* (adenomatous polyposis coli) allele, *BRCA1* is associated with breast and ovarian cancers, *NF1* (and *NF2*) with neuroectodermal tumours, *VHL* is the gene responsible for Von Hippel–Lindau disease that predisposes individuals to renal cell carcinomas and *WT1* is inactivated in Wilms' tumour of the kidney.

## APC

APC binds as a homodimer to the catenins that mediate signal transduction from E-cadherin cell surface adhesion proteins (see ***APC***). The significance of this is unclear but regulation of cell proliferation may be involved as ectopic expression of *APC* reduces CDK2 activity and causes concomitant $G_1$ arrest. The introduction of wild-type APC also reduces $\beta$-catenin expression and suppresses tumorigenicity in some human cells expressing only mutant forms of the protein, suggesting that APC resembles p53 in being able to exert dominant negative behaviour.

## BRCA1

Germline mutations in the *BRCA1* gene appear to be responsible for ~50% of families that have a dominant predisposition to breast cancer and between 80 and 90% of those in which multiple cases of both breast and ovarian cancer occur (see ***BRCA1, BRCA2***). *BRCA1* encodes a large protein (1863 amino acids) the function of which is unknown, although it contains a zinc finger region and it seems probable that it is a nuclear protein whose activity is regulated by phosphorylation.

There is a significant correlation between the location of mutations and the relative risk of breast or ovarian cancer within a family. Mutations in the N-terminal four-fifths of the protein carry a significant predisposition to ovarian cancer whereas C-terminal mutations are strongly associated with breast cancer. This suggests that mutant BRCA1 may function in a tissue-specific manner as a tumour suppressor protein or alternatively that mutant BRCA1 is capable of exerting dominant negative effects in a manner similar to that observed with p53 or APC.

Mice expressing *BRCA1* from which exon 11 has been deleted survive only until midgestation whereas deletion of exons 5 and 6 cause death before day 7.5 of embryogenesis. These findings suggest that BRCA1 positively regulates proliferation in early embryos. It is noteworthy that a form of *BRCA1* lacking exon 11 is produced in normal animals and alternative splicing of this gene may be an important mechanism in development, as established for the Wilms' tumour suppressor gene *WT1*.

## NF1

The *NF1* gene product (neurofibromin) contains a GTPase-activating protein (GAP)-related domain so that one consequence of *NF1* inactivation is to render cells exquisitely sensitive to agents that raise cellular levels of RAS-GTP (see ***NF1, NF2***). This occurs in tumour cells from patients with type 1 neurofibromatosis, even though the $p120^{GAP}$ protein is present, and presumably causes constitutive activation of RAS signalling pathways. This may be a consequence of the much higher affinity (300-fold) for RAS of NF1 by comparison with $p120^{GAP}$. However, it is evident from transgenic studies that $p120^{GAP}$ and neurofibromin cooperate during embryonic development. Thus the functional balance between NF1 and other GAP proteins is unclear and it is possible that NF1 may operate by alternative mechanisms, for example, by mediating differentiation in a RAS-regulated manner so that loss of NF1 promotes cell proliferation.

## VHL

The expression of a number of genes implicated in cancers is regulated in part by control of transcriptional elongation (e.g. *MYC, MYCN, MYCL1, FOS*). The transcription elongation factor elongin that activates transcription by RNA polymerase II is negatively regulated by VHL (see ***VHL***). Specifically, VHL binds to the elongin B and C subunits that are positive regulators of the catalytic elongin A subunit. VHL activity is probably normally regulated by casein kinase II phosphorylation and the major deletions or insertions in the gene that occur with high frequency in Von Hippel–Lindau disease are presumed to inactivate VHL, thereby minimizing RNA polymerase pausing and increasing the rate of transcription of genes involved in proliferation.

## WT1

Four WT1 zinc finger proteins arise as a consequence of alternative splicing of *WT1*, the most abundant form of the protein in both Wilms' tumour and normal kidney being that with both insertions (see ***WT1***). WT1 is a potent repressor of EGR1-mediated transcription and a number of genes have been identified that contain potential WT1 binding sequences. These include *BCL2*, insulin-like growth factor II (*IGF-II*), *MYC*, TGF$\beta$, *EGFR* and ornithine decarboxylase. Deletions or point mutations in *WT1* occur in Wilms' tumour and in Denys–Drash syndrome that predisposes to Wilms' tumour. These abolish DNA binding and hence repression of target genes. *IGF-II* is considered to be of particular importance in this context because it encodes an autocrine growth factor expressed at high levels in this pediatric tumour. *IGF-II* is expressed from the paternal allele in normal human fetal tissue but relaxation of genomic imprinting appears to contribute to the development of Wilms' tumour. The various isoforms of WT1 can form dimers and germline mutations that give rise to Denys–Drash syndrome do so by antagonizing transcriptional repression by wild-type WT1. Like p53, therefore, WT1 can exhibit dominant negative behaviour.

***References***


[1] **Weinberg, R.A. (1996) Cell 85, 457–459.**
[2] **Lane, D.P. (1992) Nature 358, 15–16.**
[3] Graeber, T.G. et al. (1995) Nature 379, 88–91.
[4] **Milner, J. (1995) Trends Biochem. Sci. 20, 49–51.**
[5] Srivastava, S. et al. (1990) Nature 348, 747–749.

# 4 Cell Cycle Control

Progression around the cell cycle is governed by a family of cyclin-dependent kinases (CDKs) and their regulatory subunits the cyclins [1]. As cells enter the cell cycle from $G_0$, D- and E-cyclins are synthesized sequentially and both are rate limiting for S phase entry (Fig. 7). A key role of CDKs is to inactivate by phosphorylation negative regulators of progression, notably pRb and its relatives p107 and p130, to permit exit from $G_1$ and entry to S phase. D-cyclins bind Rb directly and Rb is the critical substrate of CDK4 and CDK6 although cyclin E/CDK2 may also phosphorylate pRb. Phosphorylation of Rb relieves its inhibitory effect on the *trans*-activation function of both E2F family transcription factors and UBF (which regulates RNA polymerase I) that are required for S phase. p107 binds to and suppresses the *trans*-activation capacity of E2F4 and p107 associates with cyclin E/CDK2 and cyclin A/CDK2 and its phosphorylation by cyclin D kinases in late $G_1$ promotes the dissociation of p107/E2F4 [2].

## CYCLIN-DEPENDENT KINASE INHIBITORS (CDIs)

The activities of the CDK family are in turn regulated by CDK inhibitors (CDIs). These include WAF1, the INK4 family, KIP1 and KIP2 (Fig. 8). WAF1 mediates p53-dependent $G_1$ arrest by inhibiting cyclin D/CDK4, cyclin E/CDK2 and cyclin A/CDK2 through interaction with the cyclin component of the complexes (see ***WAF1***). WAF1 exists as a quaternary complex with a cyclin/CDK and proliferating cell nuclear antigen (PCNA), a subunit of DNA polymerase $\delta$ involved in DNA replication and repair. These complexes are active when they contain only one molecule of WAF1 but are inactivated by the presence of additional WAF1 proteins. Independent effects of WAF1 on PCNA are the inhibition of its capacity to activate DNA polymerase $\delta$ and the prevention of its interaction with GADD45 which is probably involved in stimulating DNA excision repair. Thus the induction of WAF1 by p53 causes both cell cycle arrest and inhibition of DNA synthesis. WAF1 also inhibits the JUN N-terminal kinases (JNKs, also called stress-activated protein kinases (SAPKs)), MAP kinases that are activated by a variety of oncoproteins and cytokines (Fig. 4), heat or UV irradiation.

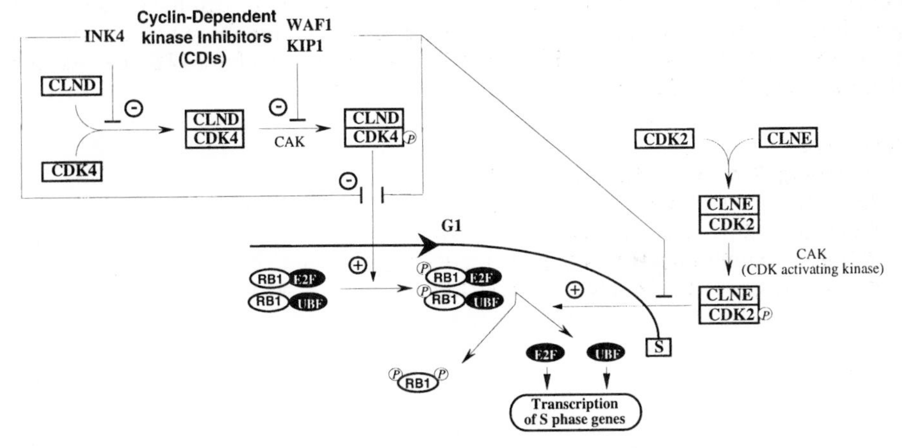


**Figure 7** *Cyclin-dependent kinases in the control of cell cycle progression.*

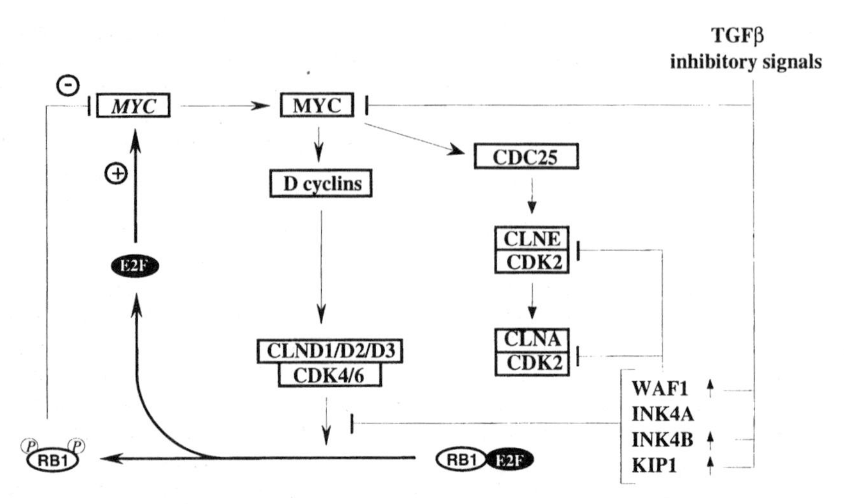


**Figure 8** *Basic regulation of the $G_1$–S transition by cyclin/CDKs and negative signals activated by TGFβ. One effect of TGFβ is to increase INK4B and hence the level of INK4B/CDK4 complexes: this decreases the concentration of cyclin D/CDK4, releasing KIP1 to inhibit cyclin E/CDK2.*

The cyclin kinase inhibitor WAF1 plays a central role in mediating the transition from a proliferative to a differentiated phenotype through its capacity to inhibit cyclin D1 activity and to prevent the $G_1$/S transition. Maintenance of the differentiated state requires the abrogation of signals that normally drive proliferation, as these in combination with a block to cell cycle progression often promote apoptosis. Thus, the expression of MYC (which is required for proliferation) will result in apoptosis in the absence of serum or other survival factors or in combination with p53 (which inhibits cell cycle progression). Ectopic expression of MYC can over-ride p53-induced $G_1$-arrest by inducing an inhibitor of WAF1. MYC expression can cause activation of both cyclin D1- and cyclin E-associated kinase activities without significant increase in the amounts of the complexes, together with hyperphosphorylation of Rb[3]. It seems probable that this is a result of the direct activation by MYC/MAX of CDC25 transcription[4]. Like MYC, CDC25A (or CDC25B) cooperates with oncogenic RAS or with loss of *RB1* to render primary fibroblasts tumorigenic and CDC25A also induces apoptosis in the absence of growth factors. Thus CDC25A is a critical target of MYC in regulating both cell cycle progression and apoptosis. MYC shares with E2F1 the distinction of being able to drive quiescent cells through $G_1$ into S phase and the inactivation of CDIs by MYC may also contribute to its capacity to drive proliferation and immortalization and to inhibit differentiation[5].

WAF1 has an additional function in that it is required for the coordination of the S and M phases of the cell cycle. Thus, in the absence of WAF1, cells with damaged DNA arrest not in $G_1$ but in a $G_2$-like state and can then pass through additional S phases without intervening normal mitoses. The deformed, polyploid cells that result then die by apoptosis[6]. The loss of WAF1, or of its regulator p53, can therefore permit the accumulation of gross nuclear abnormalities. The uncoupling of S and M phases may contribute to the acquisition of chromosomal abnormalities manifested in most tumour cells when apoptotic pathways have been circumvented.

INK4 proteins bind to CDK4 and CDK6 in competition with D cyclin to block CDK activity (see ***INK4A/MTS1/CDK41/CDKN2, INK4B/MTS2, INK4C, INK4D***). Ectopic expression of INK4A causes $G_1$ arrest of normal cells but is without effect in $Rb^{-/-}$ cells. This is consistent with the hypothesis that phosphorylation of pRb is the sole function of cyclin D/CDKs and with the finding that INK4A levels are elevated in tumour cells lacking functional Rb. In the latter the competition between INK4A and cyclin D1 for binding to CDKs enhances the more rapid turnover of free cyclin D1 (half-life ~10 min, cf. >3 h for INK4A), leading to an overall reduction in the cellular concentration of the cyclin [7]. However, INK4A can also form ternary complexes with pre-formed cyclin D1/CDKs which are then inactivated [8].

Three *INK4A*-related genes are known (*INK4B, INK4C* and *INK4D*) the products of which have broadly indistinguishable activity to that of INK4A, although INK4C interacts more strongly with CDK6 than with CDK4. INK4B synthesis is stimulated in epithelial cells by TGF$\beta$ and INK4B may mediate TGF$\beta$-induced growth arrest.

KIP1 is also an inhibitor of cyclin D-, E-, A- and B-dependent kinases that has significant homology with WAF1 and that cooperates with INK4B to cause TGF$\beta$-induced arrest (see ***KIP1***). Both KIP1 and WAF1 are expressed in cycling cells, although the former accumulates in quiescent cells and decreases following mitogenic stimulation whereas WAF1 exhibits the converse behaviour. The combined levels of WAF1 and KIP1 must therefore be exceded by those of the cyclin/CDK complexes to permit cell cycle progression.

In principle, as is true of components of growth factor signalling pathways (Fig. 4), any cell cycle regulator is a potential oncogene or tumour suppressor gene. In addition to loss of *RB1* and the high frequency of mutations in *P53*, mutation and deletion of *INK4A* is one of the most common genetic events in primary tumours. Mutations have also been detected in *WAF1* and *KIP1* but with only very low frequency. Anomalous expression of cyclin D1 occurs in a considerable variety of cancers, amplification of cyclin D2 has been detected in colorectal carcinoma and anomalously high expression of both cyclins D1 and D3 occurs in some primary breast carcinomas. Mutant forms of CDK4 have been detected and CDK4 and CDK6 are over-expressed or amplified in some primary tumours and derived cell lines. The cyclin-dependent kinase activator CDC25B is over-expressed in ~30% of primary human breast cancers examined and this correlates with a less favourable prognosis [9].

### *References*


[1] **Sherr, C.J. and Roberts, J.M. (1995) Genes Dev. 9, 1149–1163.**

[2] Xiao, Z.-H. et al. (1996) Proc. Natl Acad. Sci. USA 93, 4633–4637.

[3] Steiner, P. et al. (1995) EMBO J. 14, 4814–4826.

[4] Galaktionov, K. et al. (1996) Nature 382, 511–517.

[5] Hermeking, H. et al. (1995) Oncogene 11, 1409–1415.

[6] Waldman, T. et al. (1996) Nature 381, 713–716.

[7] Bates, S. et al. (1994). Oncogene, 9, 1633.

[8] Hirai, H. (1995) Mol. Cell. Biol. 15, 2672.

[9] Galaktionov, K. et al. (1995) Science 269, 1575–1577.

# 5 Apoptosis

Apoptosis or programmed cell death is a mechanism by which cells commit suicide. It is characterized by chromatin condensation, internucleosomal fragmentation of DNA by a $Mg^{2+}$-dependent endonuclease, blebbing of the cell membrane and vesicularization of the cell contents. Apoptosis plays an important role in normal development and as a defence against viral infection. Inhibition of apoptosis can lead to cancer [1].

The product of the *BCL2* gene is an important regulator of apoptosis, first identified from its involvement in the most common chromosome translocation in B cell follicular lymphoma (see ***BCL2***) as a result of which the *BCL2* oncogene is activated which enhances cell survival, rather than promotes proliferation. BCL2 is now recognized as a survival factor for many types of cell, notably for neurons. In T cells BCL2 confers resistance to apoptosis normally induced by glucocorticoids, radiation and other agents. Consistent with this behaviour, expression of *BCL2* is widespread during embryogenesis but is restricted to long-lived cells in the adult. A critical mediator of BCL2-regulated apoptosis is interleukin-1$\beta$-converting enzyme (ICE), a cysteine protease that processes IL-1$\beta$ during the inflammatory response. Mammalian cells express several cell death cysteine proteases that form a proteolytic cascade [2] capable of activation without *de novo* protein synthesis. What initiates the cascade and the identity of the targets essential for apoptosis are not understood. However, over-expression of ICE (or the nematode *Caenorhabditis elegans* homologue CED-3) in mammalian cells causes apoptosis that is inhibited by BCL2 and in general the activity of a family of ICE-related genes appears critical in driving apoptosis. A number of ICE family substrates have been identified in addition to IL-1$\beta$, including poly(ADP) ribose polymerase (PARP) and nuclear lamins (Fig. 9).

In addition to causing the over-expression of *BCL2*, there is direct evidence that genetic rearrangements in cancers can inhibit apoptosis. The translocation t(17;19)(q22;p13) that occurs in pre-B cell acute lymphocytic leukaemia fuses the *trans*-activation domain of E2A to the basic leucine zipper (bZIP) domain of HLF, a protein not normally expressed in lymphoid cells. Expression of a dominant negative suppressor of the E2A/HLF fusion protein promotes rapid apoptosis in human leukaemia cells carrying the t(17;19) translocation. E2A/HLF also reverses p53- and IL-3-dependent apoptosis in murine pro-B lymphocytes. The bZIP region of HLF is highly homologous to that of the *C. elegans* CES-2 cell death specification protein that specifically regulates apoptosis of neural progenitors in the developing worm. Thus E2A/HLF may dominantly interfere with an apoptotic pathway analogous to the CES/CED pathway that operates in *C. elegans* [3].

*BCL2* is a member of a multigene family, highly conserved in evolution, that includes a number of viral homologues. Other proteins of the family, e.g. $BCLX_S$, BAD, BAK, BAX and BIK, antagonize inhibition of apoptosis by binding to BCL2. Hence the balance of expression of various members of the BCL family determine the extent to which cell death is promoted or prevented. This model is consistent with the detection of high levels of BCL2 in a variety of solid tumours (including prostate carcinomas, colorectal cancer, squamous cell lung cancers, breast and nasopharyngeal cancers) and also with the observation that high levels of *BAX* expression relative to that of *BCL2* is a good prognostic indicator for bladder carcinomas [4]. One consequence of BCL2 expression is that cells expressing MYC in the absence of appropriate growth factors that would normally undergo apoptosis are prevented from doing so. Other transforming oncogenes, for example, *Hras*, can also protect cells from apoptosis, possibly by modulation of BCL2 function.

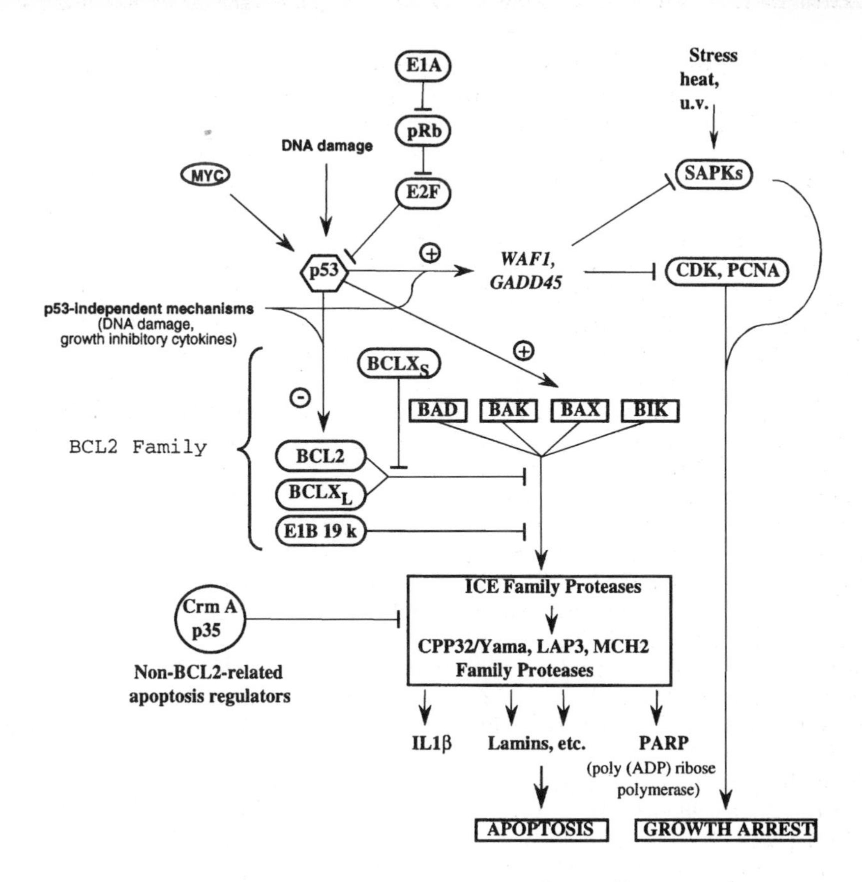


**Figure 9** *Pathways regulating apoptosis in mammalian cells.*

Growth arrest and apoptosis can be induced by a variety of cytokines including the TGF$\beta$ family (TGF$\beta_1$, TGF$\beta_2$, TGF$\beta_3$), members of which inhibit the proliferation of a wide variety of cell types that may also undergo concomitant cell death (see ***TGFBR1*** and ***TGFBR2***). Similar effects are also caused by a range of exogenous stimuli including genotoxic agents and some tumour cells appear more susceptible than their normal counterparts to apoptotic induction by such agents. TGF$\beta$-induced apoptosis is blocked in myeloblastic leukaemia cells by BCL2 expressed at a level that does not block but merely delays p53-induced apoptosis [5]. This may reflect the fact that both TGF$\beta$ and p53 suppress *BCL2* but only p53 has the capacity to activate *BAX*, thus deflecting the expression pattern towards apoptosis, as discussed above. In these cells the coexpression of MYB prevents the rapid growth arrest caused by TGF$\beta_1$ but accelerates apoptosis [6]. MYB is a transcription factor oncoprotein the expression of which is activated during proliferation and suppressed during differentiation and which can also promote apoptosis when coexpressed with p53 by a mechanism in which MYB indirectly increases transcription of *BAX* [7].

The mechanisms by which TGF$\beta$ causes arrest of cell growth and contributes to apoptosis are not understood. However, the elements that mediate signalling from the activated TGF$\beta$ receptor complex are beginning to be identified (see ***TGFBR1***, ***TGFBR2***). One downstream effector of activated TGFBR1 is TAK1, a mitogen-activated protein kinase kinase kinase (MAPKKK). Directly or indirectly, TAK1 promotes the phosphorylation of members of the MAD family that translocate to the nucleus upon phosphorylation and have transcription factor activity. The product of the tumour suppressor gene *DPC4/SMAD4* is a member of the MAD family and is non-functional in a high proportion of pancreatic carcinomas (see ***DPC4/SMAD4***). Other MAD-related human genes have been identified, some of which acquire somatic mutations in colorectal carcinomas. The first MAD protein was identified in *Drosophila* in which the TGF$\beta$ homologue *decapentaplegic* (dpp) acts as a morphogen involved in embryonic wing patterning. In *Drosophila* the *mad (mothers against dpp)* gene product mediates Dpp signalling and homozygous mutation of *mad* gives rise to defects resembling those of *dpp* mutant phenotypes. *PTC* is a recently discovered human tumour suppressor gene, inactivation of which is associated with nevoid basal cell carcinoma syndrome (NBCC). The *Drosophila* homologue of *PTC* specifically represses *dpp* and PTC therefore controls development by cell-type specific *trans*-repression of genes encoding members of the TGF$\beta$ and WNT families of signalling proteins.

Growth arrest and apoptosis are components of haematopoietic differentiation and the activation of a restricted set of genes, the myeloid differentiation primary response (MyD) genes, correlates with the induction of terminal differentiation of myeloid precursor cells. One member of this set is interferon regulatory factor 1 (IRF-1) that has been shown to be a growth suppressor[8] and another, *MyD32*, is closely related to *GADD45*, mentioned earlier as one of the group of genes induced in a wide variety of mammalian cells by DNA damaging agents or other causes of growth arrest that is directly upregulated by p53[9,10].

Entry into S phase and apoptosis are both preceded by increases in the activity of cyclin D1- and cyclin E-dependent kinases which is necessary and sufficient for induction of cyclin A by MYC. Protection from apoptosis by BCL2 also correlates with increased cyclin C, D1 and E transcription[11]. However, MYC can induce apoptosis in serum-starved cells in the presence of CDK inhibitors, indicating that apoptosis is independent of the activation of $G_1$ CDKs[12].

## TUMOUR SUPPRESSOR GENES AND APOPTOSIS

Expression of the tumour suppressor gene *P53*, which can arrest cell proliferation as summarized above, may also promote apoptosis. Thus p53 levels increase in response to a variety of DNA-damaging treatments and, in addition to *trans*-activating *BAX*, wild-type p53 can repress *BCL2* expression. Thus, for example, in $P53^{-/-}$ mice the levels of BCL2 are increased whilst BAX is reduced, whereas the ectopic expression of p53 has the converse effects. p53 also appears to be able to induce apoptosis by transcription- and translation-independent mechanism(s). The latter may depend on the capacity of the C-terminal region of p53 to recognize damaged DNA. This may modulate DNA repair mechanisms in an anomalous manner that promotes apoptosis. Wild-type p53 can induce cell cycle arrest in $G_1$ and apoptosis within the same cell but susceptibility to these processes is cell-type specific and apoptosis may occur without detectable arrest (e.g. in E1A-transformed kidney cells)[13] or cells may arrest in $G_1$ without undergoing apoptosis (e.g. primary rat embryo fibroblasts)[14]. This balance between arrest and apoptosis has been revealed in myeloblastic leukaemia cells in which ectopic expression of *BCL2* permits temperature-sensitive p53-induced $G_1$ arrest by delaying apoptosis[15].

p53 also positively regulates exit from $G_2$ and in myeloblastic leukaemia cells the rate of irradiation-induced apoptosis correlates with the rate of exit from $G_2$ [16]. This, together with studies of Burkitt's lymphoma cells, suggests that, in the absence of functional p53, the primary point of entry of cells into apoptosis following DNA damage is at $G_2$/M [17]. Regulation of the $G_2$/M checkpoint does not appear to involve transcriptional regulation by p53 and may also reflect interaction with damaged DNA. A further role for p53 may be in the spindle checkpoint during mitosis. $P53^{-/-}$ fibroblasts continue to progress through the cell cycle in the presence of inhibitors of spindle formation without completing an intervening mitosis [18].

The key role of the tumour suppressor gene product pRb in inhibiting cell cycle progression in its hypophosphorylated form is consistent with the observation that $Rb^{-/-}$ mice die prenatally as a consequence of inappropriate DNA synthesis and apoptosis in cells of the nervous system. The abrogation of pRb function by adenovirus E1A or by over-expression of E2F also correlates with stimulation of DNA synthesis and induction of apoptosis and in each of these cases apoptosis is largely p53-dependent. However, p53-independent apoptosis also occurs and is particularly important during normal development. In some cells during the early stages of DNA damage-induced, p53-independent apoptosis, for example in response to anti-cancer drugs such as araC, pRb becomes the target for a serine/threonine phosphatase, the action of which contributes to cell cycle arrest. As DNA fragmentation proceeds, the hypophosphorylated Rb undergoes rapid cleavage by an ICE-like protease [19]. Many cancers exhibit loss of function of both Rb and p53 and functionally equivalent consequences arise from infection by DNA tumour viruses including HPV. By whatever mechanism, the ablation of pRb and/or p53 activity may be expected to diminish substantially the capacity for apoptotic elimination of abnormal cells. This hypothesis is supported by the multistep development of tumours in pancreatic islet $\beta$ cells of transgenic mice induced by the expression of the SV40 large T antigen. Tumorigenesis is initiated by the inactivation of p53 and pRb by large T antigen but multiple stages are subsequently involved before solid islet cell carcinomas become detectable after about 12 weeks. During these stages there is a progressive increase in apoptosis that reaches a maximum in angiogenic progenitor islets. The subsequent transition of these cells to tumours correlates with increased expression of $BCLX_L$ and a dramatic decrease in the incidence of apoptosis [20].

***References***


1. **Vaux, D.L. and Strasser, A. (1996) Proc. Natl Acad. Sci. USA 93, 2239–2244.**
2. Shimizu, S. et al. (1996) Oncogene 12, 2251–2257.
3. Inaba, T. et al. (1996) Nature 382, 541–544.
4. Gazzaniga, P. et al. (1996) Int. J. Cancer (Pred. Oncol.), 69, 100–104.
5. Selvakumaran, M. et al. (1994) Mol. Cell. Biol. 14, 2352–2360.
6. Selvakumaran, M. et al. (1994) Oncogene 9, 1791–1798.
7. Sala, A. et al. (1996) Cancer Res. 56, 1991–1996.
8. Harada, H. et al. (1994) Oncogene 9, 3313–3320.
9. Zhan, Q. et al. (1994) Oncogene 9, 3743–3751.
10. Zhan, Q. et al. (1994) Mol. Cell. Biol. 14, 2361–2371.
11. Hoang, A.T. et al. (1994) Proc. Natl Acad. Sci. USA 91, 6875–6879.
12. Rudolph, B. et al. (1996) EMBO J. 15, 3065–3076.
13. Chiou, S.K. et al. (1994) Mol. Cell. Biol. 4, 2556–2563.
14. Martinez, J. et al. (1991) Genes Dev. 5, 151–159.

[15] Guillouf, C. et al. (1995) Blood 85, 2691–2698.
[16] Guillouf, C. et al. (1995) Oncogene 10, 2263–2270.
[17] Allday, M.J. et al. (1995) EMBO J. 14, 4994–5005.
[18] Cross, S.M. et al. (1995) Science 267, 1353–1356.
[19] An, B. and Dou, Q.-P. (1996) Cancer Res. 56, 438–442.
[20] Naik, P. et al. (1996) Genes Dev. 10, 2105–2116.

# 6 DNA Repair

Errors that arise during DNA replication and that are not corrected by the proofreading activity of DNA polymerase may be rectified by a process called mismatch repair. This involves the recognition of mismatched bases or single base insertions or deletions in the newly replicated strand, followed by correction of the error. In *E. coli* the genes critical for this process include *mutH*, *mutL* and *mutS*, mutation in any of which generates a mutator phenotype in which spontaneous mutagenesis is enhanced at many loci [1].

Six human DNA mismatch repair genes have been identified, *MSH2*, *MLH1*, *PMS1*, *PMS2* (see ***MSH2*, *MSH3*, *MSH6/GTBP*, *MLH1***). *MSH2* is homologous to *mutS* and *MLH1*, *PMS1* and *PMS2* are homolgous to *mutL*. Mutations in these genes were first detected in hereditary non-polyposis colon cancer (HNPCC) and mutations in each of the four human genes have now been identified in the germline of some HNPCC families. At least 40% of HNPCC kindreds are associated with germline mutations in *MSH2*. Mutations in mismatch repair genes lead to an overall increase of the mutation rate, predispose to cancer development and are associated with a phenotype of length instability of microsatellite loci.

Microsatellites (or simple repeated sequences, SRSs) are short sequences of DNA (up to 6 bp), repeated between 10 and 50 times, that are stably inherited, vary from individual to individual and have a relatively low inherent mutation rate. A mutator mechanism for cancer arises from ubiquitous somatic mutations (USMs) at SRSs caused by failures of the strand-specific mismatch repair system to recognize and/or repair replication errors due to slippage by strand misalignment. Microsatellite length heterogeneity is present in colon tumours, classified as replication error positive ($RER^+$), showing that somatic genomic instability (SGI) is a very early event in the development of these tumours. A significant proportion of sporadic colorectal cancers with microsatellite instability have somatic mutations in *MSH2* and microsatellite instability has also been detected in gynaecological sarcomas.

It is noteworthy that mutations in *P53*, *APC* and *KRAS* have been shown to correlate with absence of microsatellite instability [2]. This suggests that although loss of function of mismatch repair genes clearly promotes genetic instability, the acquisition of mutations in major growth regulatory genes may equally well enhance the incidence of chromosomal aberrations.

***References***


[1] **Kolodner, R. (1996) Genes Dev. 10, 1433–1442.**

[2] Heinen, C.D. et al. (1995) Cancer Res. 55, 4797–4799.

# 7 Angiogenesis and Metastasis

Metastasis is the spread through the body of malignant cells that have detached from a primary tumour to give rise to secondary tumours and it is the major cause of death from cancers. The transition from pre-invasive tumour to invasion occurs in parallel with tumour-induced neovascularization/angiogenesis which permits dissemination of tumour cells into the vascular system.

The molecular events that determine whether a cell becomes metastatic may broadly be divided into two categories: (1) intracellular signalling pathways that regulate proliferation, apoptosis and the synthesis of cytokines and (2) changes in cell surface proteins and secreted proteins involved in cell adhesion, cell regulation and proteolysis. The first category may be regarded as including events that determine whether an individual cell evolves into a tumour clone. The second group refers to cellular properties that may directly determine whether tumour cells metastasize and individual gene products that fall under this heading are considered below (see **Metastasis: correlations with gene expression**). Metastasis is a multistep process that requires many different tumour–host cell interactions. Crucial among these are the events associated with invasion of the basement membrane (BM) by the tumour cell which occurs at several points during the development of metastases. These may include the escape from the primary neoplasm of epithelial tumour cells, entry to and escape from the circulatory system during haematogenous dissemination and perineural and muscular invasion.

## ANGIOGENESIS

Angiogenesis, the sprouting of capillaries from pre-existing vessels, occurs during embryonic development but is almost absent in adult tissues [1]. However, transient, regulated angiogenesis occurs in adult tissues during the female reproductive cycle and during wound healing. Pathological angiogenesis is characterized by the persistent proliferation of endothelial cells and is a prominent feature of a number of diseases including rheumatoid arthritis, psoriasis and proliferative retinopathy. In addition, many tumours are able to attract blood vessels from neighbouring tissues and the induction of new blood vessel growth is necessary if solid tumours are to grow beyond a minimal size.

### Angiogenic promoters

A wide variety of factors can stimulate angiogenesis (Table 5), some acting directly on endothelial cells, others stimulating adjacent inflammatory cells. Those acting on endothelial cells may cause migration but not division (e.g. angiotropin, macrophage-derived factor, TNFα) or stimulate proliferation (e.g. EGF, aFGF, bFGF, VEGF, PD-ECGF, TGFα). A number of angiogenic factors including aFGF, bFGF, VEGF, PD-ECGF and midkine bind heparin and heparin itself can enhance or inhibit the actions of angiogenic factors, depending on the agent and receptor type involved as well as on the concentration of heparin or heparan sulfate [2]. The expression of midkine is strongly elevated in bladder cancers and this correlates with a poor prognosis for invasive tumours [3]. Two VEGF receptors have been identified, FLK1/KDR and FLT1, and inhibition of FLK1 signalling inhibits the growth of a variety of solid tumours [4]. Injection of a neutralizing antibody to VEGF strongly suppresses the

**Table 5** *Factors regulating angiogenesis*

| Angiogenic promoters | Angiogenic inhibitors |
|---|---|
| Angiogenin, angiotropin | Angiostatin, cartilage-derived inhibitor (CDI)[23] |
| Acidic FGF, basic FGF, FPS/FES, HST-1, IL-8, INT2 | Chondrocyte-derived inhibitor (ChDI) |
| EGF, TPA, TGF$\alpha$, TGF$\beta$, TNF$\alpha$, HGF, FLK1 | Glioma-derived angiogenesis inhibitory factor (GD-AIF) |
| Heparinase, LERK-1 (B61), midkine | Heparinases I and III, interferons $\alpha$, $\beta$ |
| Basic FGF binding protein, integrins $\alpha_V\beta_3$, $\alpha_V\beta_5$ | 2-Methyloxyoestradiol, platelet factor 4 |
| HIV TAT protein | Prolactin fragment, protamine |
| 12($R$)-Hydroxyeicosatrienoic acid | Proliferin-related protein (mouse placenta) |
| Macrophage-derived factor, fibrin, nicotinamide | Suramin, fumagillin and analogues, AGM 1470 |
| Platelet-derived endothelial cell growth factor (PD-ECGF) | SPARC |
| Placenta growth factor, pleiotropin | Tamoxifen |
| Proliferin (mouse placenta)[24] | Thrombospondin (TSP) |
| Prostaglandins $E_1$, $E_2$ | TIMP-1, TIMP-2 and other MMP inhibitors |
| Sodium orthovanadate | |
| Vascular endothelial growth factor (VEGF) | |

growth of solid tumours of the subcutaneously implanted HT-1080 human fibrosarcoma cell line[5]. In a variety of human cell lines hypoxia causes the activation of transcription of VEGF by a mechanism that involves accumulation of *SRC* mRNA and activation of SRC kinase[6]. The proliferative and permeabilization effects of VEGF on vascular endothelium are strongly potentiated by placenta growth factor but in thyroid tumours, although high tumorigenic capacity is associated with elevated VEGF expression, placenta growth factor expression is reduced in most thyroid tumour cell lines and tumours[7].

Some factors have multifunctional capacity: for example, bFGF stimulates both endothelial cell proliferation and migration and also the secretion of proteases[8]. In NIH 3T3 cells bFGF is non-transforming but when expressed at high levels or with an added signal sequence it renders the cells highly tumorigenic[9]. The oncogenes *Int-2* and *Hst-1* are also members of the FGF family and there is evidence that both can manifest angiogenic activity[10,11]. Thus members of this family may exert both growth promoting and transforming activities and act to promote angiogenesis. In gliomas a paracrine mechanism appears to operate in which vascular endothelial growth factors (VEGF and PD-ECGF) are expressed in the tumour cells and genes encoding their receptors are coordinately induced in adjacent endothelial cells[12]. Complexes of laminin–entactin also stimulate angiogenesis *in vitro* in an effect enhanced by bFGF, although high concentrations of the complex are inhibitory[13]. Expression of a high-affinity FGF binding protein can promote angiogenesis and the mRNA for this protein is expessed at high levels in squamous cell carcinomas but not in normal tissues[14].

Angiogenin, a 14 kDa protein with homology to pancreatic ribonuclease, displays weak RNAase activity, is secreted *in vitro* by a variety of human tumour cell lines and promotes endothelial cell adhesion, growth and invasiveness[15]. Anti-angiogenin antibody suppresses tumour growth in athymic mice[16]. Endocytosis and nuclear translocation of angiogenin in proliferating endothelial cells appears to be necessary for its angiogenic activity and these cells express an angiogenin-binding protein (AngBP/smooth muscle $\alpha$-actin). Exogenous actin, anti-actin antibody, heparin, and heparinase treatment all inhibit the internalization of angiogenin, suggesting the involvement of cell surface AngBP/actin and heparan sulfate proteoglycans in this process[17]. Angiogenin has also been shown to interact with actin to form a complex which, like actin itself, can accelerate plasmin generation by tissue plasminogen activator[18].

The expression in mice of a gain-of-function mutant allele of human *FPS/FES* that encodes an N-terminally myristylated, membrane-associated protein causes widespread hypervascularity progressing to multifocal haemangiomas[19]. This protein has increased tyrosine kinase activity and weak transforming potential relative to the wild-type, cytosolic form and is specifically expressed in endothelial cells. LERK-1 (B61), the ligand for the ECK receptor tyrosine kinase, has angiogenic activity *in vivo* and acts as a chemoattractant for endothelial cells: anti-B61 antibody attenuates angiogenesis induced by TNF$\alpha$ but not by bFGF[20]. 12(*R*)-Hydroxyeicosatrienoic acid increases *Fos, Jun* and *Myc* transcription in rabbit endothelial cells by a mechanism that, in part, appears to mediated by activation of NF$\kappa$B[21]. HIV TAT protein has growth factor activity for spindle cells derived from Kaposi's sarcoma-like lesions and TAT has angiogenic properties. It binds to heparin with high affinity and induces endothelial cell growth, migration and invasion *in vitro*[22].

## Angiogenic inhibitors

The activities of a variety of endogenous angiogenic inhibitors are also presumed to regulate tumour vascularization (Table 5). These include thrombospondin-1 (TSP) which contains a putative TGF$\beta$-binding domain and inhibits endothelial cell proliferation, migration, BM proteolysis and tumorigenesis, acting in part by increasing expression of PAI-1[25,26]. p53 can positively regulate the expression of TSP and in fibroblasts from Li–Fraumeni patients the loss of p53 function appears to mediate the switch to an angiogenic phenotype[27]. However, conflicting evidence as to the function of TSP suggests that it can indirectly promote angiogenesis by stimulating the proliferation of myofibroblasts[28]. Early loss of wild-type *P53* activity also occurs in the malignant progression of astrocytoma towards end-stage glioblastoma, a process characterized by dramatic neovascularization of the neoplasms. Glioblastoma cells that do not express *P53* show strong angiogenic activity, whereas on induction of wild-type, but not mutant, *P53* they secrete glioma-derived angiogenesis inhibitory factor (GD-AIF) which neutralizes the angiogenic factors produced by the parental cells and also the angiogenic activity of bFGF[29].

Angiostatin is a 38 kDa plasminogen fragment (kringles 1–3) that blocks neovascularization and growth of metastases[30]. Chondrocyte-derived inhibitor[31], platelet factor 4[32] and heparinases I and III, but not heparinase II, inhibit both angiogenesis *in vivo* and proliferation of capillary endothelial cells mediated by bFGF *in vitro*[33]. Interferon suppression may arise indirectly *via* effects on cytokine secretion[34]. 2-Methoxyoestradiol is the first steroid shown to possess high anti-angiogenic activity, inhibiting endothelial cell proliferation and migration *in vitro* and suppressing neovascularization and growth of solid tumours in mice[35].

Suramin, which competes for heparin binding and antagonizes bFGF-induced angiogenesis *in vitro*[36], has been used in humans to treat hormone-refractory prostate carcinoma[37] and analogues of suramin with improved therapeutic ratios have also been developed[38]. Other synthetic angiogenic inhibitors include the antibiotic fumagillin and analogues thereof[39], AGM-1470 which inhibits endothelial cell proliferation *in vitro* and *in vivo* and prevents tumour growth *in vivo*[40] and several heparin mimetics and heparin–steroid conjugates that have been shown to inhibit tumour growth in mice[41,42]. Tamoxifen causes arrest and regression of tumours of MCF7 human breast cancer cells implanted in nude mice as a consequence of inhibition of angiogenesis and neovascularization[43]. SPARC (also called osteonectin, BM-40 and 43K protein) is an acidic, cysteine-rich, extracellular matrix protein the expression of which is regulated by progesterone and dexamethasone and is indirectly controlled by TGF$\beta$, IL-1 and CSF1. SPARC is expressed at reduced levels in ovarian carcinoma cells and expression of exogenous SPARC in such cells reduces their tumorigenicity in nude mice[44].

## METASTASIS: CORRELATIONS WITH GENE EXPRESSION

The expression of a considerable number of genes in a variety of experimental systems has been shown to affect metastatic capacity (Table 6). However, it seems probable that many of these effects were indirect and attention has therefore focused on changes specific to tumour cells in the expression of cell surface adhesion molecules and of proteolytic enzymes that might be involved in tissue degradation. The major families of such proteins are (a) cadherins, (b) the immunoglobulin superfamily, (c) integrins, (d) CD44, (e) matrix metalloproteinases (MMPs), (f) tissue inhibitors of metalloproteinases (TIMPs), (g) the serine, cysteine and aspartic proteinases and heparanase and (h) the putative metastasis-suppressor genes *NME1* and *NME2*.

### Cadherins

Cadherins are a multigene family of transmembrane glycoproteins that mediate $Ca^{2+}$-dependent intercellular adhesion, cytoskeletal anchoring and signalling, and are thought to be essential for the control of morphogenetic processes, including myogenesis[46]. The family includes B-cadherin, E-cadherin (also known as uvomorulin (*UVO*), *Arc-1* or cell-CAM 120/80, see **E-cadherin/CDH1**), EP-cadherin, K-cadherin, M-cadherin, N-cadherin (ACAM), P-cadherin, R-cadherin, T-cadherin, U-cadherin, cadherins 4 to 11 and LCAM. Cadherin function is regulated by cytoplasmic proteins including a vinculin-like protein, $\alpha$-catenin (or CAP102), and $\beta$-catenin (a homologue of plakoglobulin). Catenins undergo tyrosine phosphorylation in cells transformed by v-*src* that correlates with metastatic potential and may

**Table 6** *Gene products modulating metastasis* in vivo *or* in vitro

| Metastasis enhancement | Metastasis suppression |
|---|---|
| BRN-2, CD44, IL-8, MMPs, TGF$\beta$, cathepsin D, JUNB, FMS, FPS, MOS, MYC, NEU, RAS, MTS1, hepatocyte growth factor (HGF), HSTF1 | FOS, JUN, IL-2, IL-6, IFN-$\gamma$, KAI1, ME491, MRP1, NCAM, NME1/NME2, TIMPs |
| Acidic fibroblast growth factor (aFGF), v-SRC | Plasminogen activator inhibitors (PAIs) |
| Autocrine motility factor (AMF), AT2.1, MTLn3 | |
| Cancer cell chemotactic factor, E1AF, TIAM1 | $\beta$m-Actin, gelsolin, suramin, tropomyosin 1 |
| Invasive stimulatory factor (ISF) | |
| Migration stimulatory factor, integrins | Adenovirus 5 E1A (*Neu*-transformed cells) |
| 5T4 antigen[45] | |
| SV40 T antigen, HPV E6, HPV E7 | E-cadherin, integrins |

reflect the modulation of cell–cell adhesion[47]. Furthermore, the tumour suppressor gene product APC binds to the catenins and catenin–APC complexes may play a role in regulating cell growth and/or metastasis.

Taken together, these observations suggest that E-cadherin and possibly other members of the family act as invasion suppressors (see **E-cadherin/CDH1**), although some cadherins, for example K-cadherin, are over-expressed in some human cancers[48].

## The immunoglobulin superfamily (IgSF): neural cell-adhesion molecule (NCAM)

NCAMs arise from a single gene by alternative splicing and they contain an extracellular domain of five C2-set IgSF domains and two fibronectin type II domains. They mediate homophilic or heterophilic $Ca^{2+}$-independent intercellular adhesion. The three subunits comprise A and B (membrane phosphoproteins) and C (PtdIns-anchored). NCAM expression modulates the adhesive phenotype of glioma cells, invasion by $NCAM^{+}$ cells involving degradation of normal tissue and its replacement by tumour cells. $NCAM^{-}$ cells are highly invasive *in vivo*, migrating along the vascular elements of the brain and not displaying a well-defined tumour mass. Expression of the B (transmembrane) subunit of NCAM appears to cause marked downregulation of expression of *MMP-1* and *MMP-9*[49]. CCAM1 exogenous expression decreases the tumorigenicity of the human prostatic cancer cell line PC-3 and anti-sense ODN against CCAM1 increases tumour formation in nude mice by the non-tumorigenic NbE cell line[50].

## Integrins

The integrins are a family of cell surface proteins that mediate cell–substratum and cell–cell adhesion[51,52]. Integrins are heterodimers of non-covalently linked $\alpha$ and $\beta$ subunits, each of which is a transmembrane protein. Eleven $\alpha$ and six $\beta$ subunits have been identified that give rise to at least 16 distinct integrins and, in addition, a single $\alpha$ or $\beta$ subunit can associate with more than one $\beta$ or $\alpha$ chain, respectively. Although no integrin has yet been shown to be a tumour suppressor gene, they mediate some of the processes involved in metastasis and changes in integrin expression accompany malignant transformation (Table 7). Thus, loss of $\alpha_1$, $\alpha_2$ or $\alpha_3$ expression and enhanced expression of $\alpha_v$ occur frequently in lung cancer and both loss and aberrant expression of various integrins has been detected in breast and colon cancers. However, thus far it has been difficult to discern correlations between altered patterns of integrin expression and tumorigenicity. Major associations of specific integrin expression with cancers are summarized in Table 7. The urokinase-type plasminogen activator receptor (uPAR) forms stable complexes with $\beta_1$ integrins that lead to uPAR binding to vitronectin ($\alpha_v\beta_3$ integrin) and inhibition of normal integrin adhesion[53]. The common association of over-expression of the uPAR with metastasis (see **Matrix metalloproteinases (MMPs)** below) may in part account for the loss of stable cellular adhesion during tumour expansion.

## CD44

The ubiquitous 90 kDa cell surface glycoprotein CD44 exerts a diverse range of actions, being implicated in cell–cell and cell–matrix adhesion, lymphocyte

**Table 7** *Integrin expression: correlations with metastasis*

| Decreased expression (or inhibitory effect on metastasis) | Increased expression | Associated cancer |
|---|---|---|
| $\alpha_1, \alpha_2, \alpha_3$ | $\alpha_v$ | Lung [54] |
| $\alpha_2, \alpha_3, \alpha_6$ | $\beta_4$ | Squamous cell carcinoma [55,56] |
| $\alpha_2/\beta_1$ | | Cervical intraepithelial carcinoma [57,58] |
| | $\alpha_1/\beta_1, \alpha_2/\beta_1, \alpha_3/\beta_1$ | E1A-transformed + TGF$\beta$ |
| | $\alpha_2/\beta_1$ | Rhadomyosarcoma [59–61] |
| | $\alpha_v/\beta_3$ | Melanoma [62–65] |
| | $\alpha_v/\beta_5$ | [66] |
| | $\alpha_v/\beta_6$ | Carcinoma cell lines [67] |
| | $\alpha_{RLC}$ | Lung [68] |
| $\alpha_4/\beta_1$ | | B16 melanoma [69] |
| $\alpha_5\beta_1$ (FN receptor) | | Cervical intraepithelial carcinoma [70–72] |
| $\alpha_6/\beta_4, \alpha_6/\beta_1$ | | Breast [73] |

activation, lymph node homing and metastasis. There are a number of isoforms generated by alternative splicing of 10 "variant" exons that are differentially expressed in a specific manner in normal tissues in response to a variety of cytokines. Alternatively spliced forms are also differentially expressed on various malignant cell types and the capacity of isoforms to adhere to each other may be responsible for the apparent involvement of CD44 in metastasis [74,75]. CD44 ligands include addressin, hyaluronate, osteopontin (OPN or Eta-1) and gp600 (serglycin) and the finding that anti-CD44 antibody can inhibit metastasis *in vivo* suggests that interactions between CD44 on tumour cells and its ligands may be necessary for tumour growth and metastasis [76].

Many primary carcinomas express high levels of CD44 and this enhanced expression appears to correlate with high metastatic capacity in bladder carcinomas, non-Hodgkin's lymphoma and in melanoma cell lines. The serum concentration of soluble CD44 is elevated >8-fold in patients with advanced gastric or colon cancers and this increase correlates with tumour burden and metastasis [77]. Disseminated large cell lymphomas show increased expression of both the "haematopoietic" isoform (CD44H) and of some variant forms [78] and gross over-production of alternatively spliced forms occurs in breast and colon cancers and is associated with poor prognosis [79]. Abnormal retention of intron 9 in mRNA has been detected in oesophageal, colon and breast carcinoma cell lines [80]. In certain ovarian cancer cell lines CD44 variants predominate but nevertheless a critical level of CD44H appears to determine binding to mesothelium or hyaluronate [81]. Reintroduction of CD44H into colon carcinoma cells reduces their growth rate and tumorigenicity in an effect dependent on hyaluronate binding [82]. Osteopontin also binds to CD44 to induce cellular chemotaxis but not homotypic aggregation, the inverse of the responses induced by hyaluronate [83].

*In vitro* transformation of fibroblasts by SV40, *Neu* or RSV results in increased expression of *CD44* and, in general, the introduction of *CD44* into *CD44*$^-$ cell lines promotes tumour growth and metastasis *in vivo* [84]. The *in vitro* invasiveness of human glioma cell lines is strongly inhibited by anti-CD44 monoclonal antibodies (MAbs) or by expression of a *CD44*-specific anti-sense oligonucleotide [85]. The tumour growth promoting property of CD44 may be dependent on its ability to mediate cell attachment to hyaluronate[86].

## Matrix metalloproteinases (MMPs)

At least 14 members of the zinc-dependent MMP family have been detected (Table 8)[87]. They are synthesized by connective tissue cells and act synergistically to digest the major macromolecules of connective tissue matrices. The synthesis and secretion of high levels of MMPs correlates with invasion and metastasis for a variety of human cancers and it may be noted that cartilage, a tissue relatively resistant to invasion, expresses proteinase inhibitors and an endothelial growth inhibitor. Cell lines with high metastatic capacity synthesize abnormally large amounts of MMPs. Normal cells may synthesize MMPs constitutively (e.g. MMP-9 by human synovial fibroblasts) or in response to a variety of developmental or stress-induced signals including those generated by IL-1$\beta$, PDGF, TNF$\alpha$, TNF$\beta$ or TGF$\beta$. Furthermore, several MMPs cleave TNF$\alpha$ to its active form[88]. MMPs are secreted as zymogens and activated by plasmin. Plasmin is itself generated from its plasma zymogen by plasminogen activators, the two major forms of which are tissue plasminogen activator (tPA) and urokinase-type plasminogen activator (uPA), both of which are serine proteinases (Fig. 10). No clear distinctions between the expression patterns of tPA in tumour and benign cells have been detected. However, high expression of uPA (and also of PAI-1) has been detected in a variety of malignant tissues, including breast tumours, brain tumours and melanomas. uPA expression is a powerful prognostic indicator for breast and possibly for other cancers[89] and blockade of the uPA receptor (uPAR) has been shown to inhibit *in vivo* tumour cell invasiveness[90–95]. The uPAR is over-expressed in the more malignant grades of astrocytoma, tumours that are particularly highly vasculated[96] and metastases of ovarian tumours may display elevated expression of uPA, PAI-1 uPAR and PAI-2 by comparison with primary tumours[97]. Nevertheless, invasion appears to depend on a balance between the activities of proteases and their inhibitors that is more subtle than these observations suggest. Thus Bowes' melanoma cells that synthesize large amounts of tPA and are poorly invasive in amniotic membranes and HT1080 fibrosarcoma cells that produce large amounts of uPA both show increased invasiveness in the presence of plasmin inhibitors or of anti-tPA or anti-uPA antibodies, respectively[98]. Osmond cells, however, that express low levels of uPA show highly invasive behaviour in amniotic membranes that is inhibited by plasmin inhibitors or anti-uPA antibody. Agents that induce angiogenesis *in vitro* (e.g. bFGF, TPA or sodium orthovanadate) stimulate the expression of both uPA and PAI-1 in endothelial cells, as does TGF$\beta$.

In addition to the plasmin-activation pathway, MMP-2 is specifically activated *in vitro* by MT1-MMP/MMP-14 (membrane-type matrix metalloproteinase), a transmembrane protein that enhances cellular invasion of reconstituted BM and is expressed in invasive lung carcinoma cells[99]. Thus the proteolytic activity of MT1-MMP is localized to the cell surface but may be responsible for activating an extracellular protease and thus amplifying the invasive process. In these respects the behaviour of MT1-MMP parallels that of uPA which, in binding to its glycosylphosphatidylinositol-anchored receptor, focuses the activation of plasminogen at the periphery of the tumour cell. Three other MT-MMPs have been isolated (MT2-MMP/MMP-15[100], MT3-MMP/MMP-16[101] and MT4-MMP/MMP-17), each of which is a transmembrane protein synthesized as an inactive proenzyme with a zinc

**Table 8** *Metalloproteinases and their substrates*

| Enzyme group | Other names | MMP classification | $M_r$ (kDa) | Substrates |
|---|---|---|---|---|
| Membrane MMP | | MT1-MMP/MMP-14 | 66 | Pro-gelatinase A |
| | | MT2-MMP/MMP-15 | | |
| | | MT3-MMP/MMP-16 | | |
| | | MT4-MMP/MMP-17 | 70 | |
| Collagenases | Interstitial collagenase | MMP-1 | 55 | Collagens I, II, III, VII, VIII and X, gelatin, PG core protein |
| | PMN collagenase | MMP-8 | 75 | As MMP-1 |
| | Collagenase-3 | MMP-13 | 54 | Fibrillar collagens |
| Gelatinases | Gelatinase A, 72 kDa gelatinase, type IV collagenase | MMP-2 | 72 | Collagens IV, V, VII, X and XI, gelatin, PG core protein, FN |
| | Gelatinase B, 92 kDa gelatinase, type IV collagenase | MMP-9 | 92 | Collagens IV and V, gelatin, PG core protein |
| | PUMP-1, matrilysin, uterine metalloproteinase | MMP-7 | 28 | Collagen IV, type I gelatin, proteoglycan, elastin, FN, LN-1 |
| Stromelysins | Stromelysin-1, transin (rat), procollagenase activator | MMP-3 | 54 | Collagens III, IV, V, IX and X, entactin, gelatin, FN, LN, pro-MMP-1, casein |
| | Stromelysin-2, transin-2 | MMP-10 | 54 | As MMP-3 |
| | Stromelysin-3 | MMP-11 | 45 | Serpins |
| | Metalloelastase | MMP-12 | 54 | Elastin, FN |

FN, fibronectin; LN, laminin.

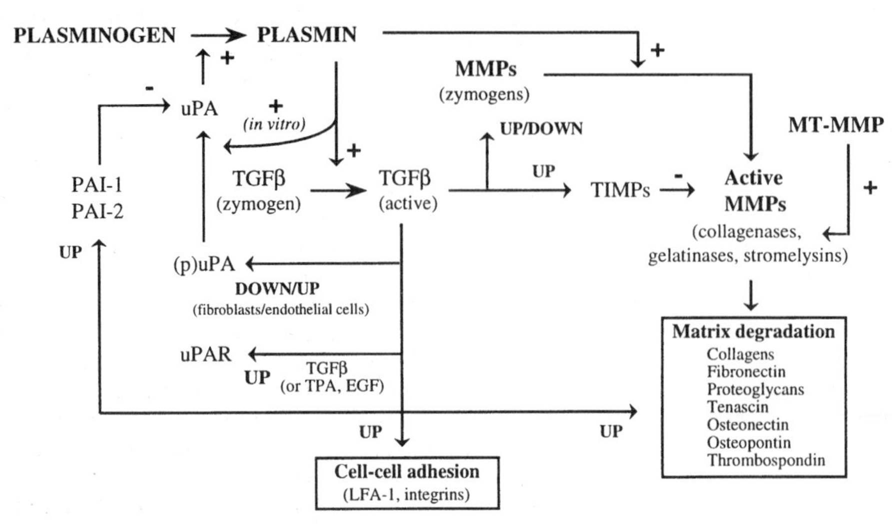


**Figure 10** *Regulation of the activity of cell adhesion proteins by plasmin and transforming growth factor β (TGFβ). MMPs: matrix metalloproteinases; TIMPs: tissue inhibitors of metalloproteinases; (p)uPA: (proenzyme precursor) of urokinase-type plasminogen activator; uPAR: uPA receptor; PAI: plasminogen activator inhibitor; LFA-1: lymphocyte function-associated antigen 1; + : activator; – : inhibitor; UP: upregulation; DOWN: downregulation. TGFα stimulates the secretion of PAI-1 and downregulates the expression of (p)uPA.*

binding site and a haemopexin domain. MT4-MMP/MMP-17 is expressed mainly in the brain, leukocytes, colon, ovary and testis and also appears to be expressed with high frequency in breast carcinoma [102].

Antibodies to MMPs can abolish the *in vitro* invasiveness of human cells and specific inhibition of MMPs (e.g. by batimastat, aranciamycin, estramustine, galardin, matlystatin, minocycline or SC 444463) has been shown to block tumour development and metastasis in both animals and humans. Thus batimastat, a peptide-like analogue of collagen, inhibits MMP-1, MMP-2 and MMP-9 with an $IC_{50}$ of ~4 nM [103] *via* binding to the catalytic $Zn^{2+}$ ion [104]. Peptide inhibitors based on the highly conserved PRCGXPDV sequence in the cleaved N-termini of pro-MMPs are effective inhibitors of invasion *in vitro* [105]. Antibodies to MMP-1 or the addition of a zinc chelator can inhibit angiogenesis in model systems [106] and the MMP inhibitor N-[2*R*-2-(hydroxamidocarbonylmethyl)-4-methylpentanoyl)]-L-tryptophan methylamide (GM6001) reduces angiogenesis in rat corneas [107]. Inhibition of angiogenesis, *MMP-2* expression and tumour growth has also been demonstrated using the carboxyamide aminoimidazole (CAI) compounds that modulate cellular calcium homeostasis [108,109].

## Tissue inhibitors of metalloproteinases (TIMPs)

The activity of MMPs is regulated by transcriptional control of MMP genes, the activation of secreted proenzymes and by the actions of TIMPs. TIMPs are present in normal bone and cartilage cells and are secreted by a variety of cells in culture [110].

TIMPs form 1:1 inactivating complexes with active MMPs. TIMP-1 and TIMP-2 inhibit all MMPs and bind to the proenzyme forms of gelatinase B (MMP-9) and gelatinase A (MMP-2), respectively, whereas TIMP-3 binds to the extracellular matrix (ECM). TIMPs are potent inhibitors of several steps in invasion and metastasis, including the proteolytic degradation of the extracellular matrix, invasion of surrounding connective tissue, extravasion and subsequent tumour growth[111]. Other inhibitors of metalloproteinase (including IMP-a and IMP-b) have been detected in human embryonic and neuroblastoma cell lines[112].

The importance of MMPs in metastasis is emphasized by the finding that TIMP-1 and TIMP-2 can block cell invasion *in vitro* and that TIMP-1 inhibits lung colonization *in vivo*[113]. In steroid-sensitive breast cancer cells oestradiol and progestin coordinately activate expression of 67LR and suppress the expression of *TIMP-2*[114]. The transfection of Swiss 3T3 fibroblasts with a vector conferring constitutive expression of TIMP anti-sense deoxyoligonucleotide renders the cells tumorigenic and metastatic in mice. Transfection of tumour cells with a vector expressing TIMP-2 decreases the activity of secreted MMPs and the growth rate of the cells *in vivo* as well as suppressing the invasive capacity of the cells for surrounding tissue[115,116]. TIMP-2 also inhibits the development of Kaposi's sarcoma-like lesions in mice[117].

These observations suggest that TIMPs may have tumour suppressor activity. However, TIMP-1 expression is elevated in some malignant non-Hodgkin's lymphomas, the amounts correlating with tumour aggressiveness[118] and increased serum levels of TIMP-1 and collagenase occur in metastatic prostatic cancer. Although lower levels of TIMP-2, which has a particularly high affinity for type IV collagenase, have been reported in patients with prostatic cancer[119], TIMP-2 is produced by many tumour cells and high levels of expression of *TIMP-3* have been detected in breast carcinomas[120]. The expression of TIMPs also increases transcription of several proteinases, and of the $Ca^{2+}$-binding proteins osteopontin (OPN), SPP and calcyclin[121]. Although the significance of these effects is unclear, OPN expression has been detected in TPA-treated epidermal cells, in metastasizing rat cells[122] and in *Ras*-transformed cell lines, the tumorigenicity of which is reduced by expression of anti-sense OPN DNA, as is that of the high OPN producing B77-Rat 1 malignant cell line[123–125]. Thus the role of TIMPs is clearly more complex than that of inhibiting invasion by tumour cells. It is possible that TIMPs may have additional functions as cytokines or growth factors. Thus, TIMP-1 appears to inhibit angiogenesis by blocking ECM degradation[126], although TIMP-1 also directly blocks endothelial cell migration *in vitro*[127] and TIMP-2 has been shown to inhibit bFGF-stimulated proliferation of human endothelial cells[128]. However, both TIMP-1 and TIMP-2 have also been reported to be potent growth factors for a wide variety of cell types[129] and TIMP-2 stimulates fibroblast proliferation *via* the activation of cAMP-dependent protein kinase[130].

Comparison of MMP-1, MMP-2, MMP-3, MMP-9 and TIMP-1 activities in a variety of tumours[131] indicates that astrocytomas, which are usually slow growing, show little of any of these activities, in common with normal tissue: however evolution into anaplastic astrocytomas correlates with strong expression of at least four MMPs, together with TIMP-1. Other CNS and epithelial tumours also have distinctive patterns of expression. Significantly higher mean MMP-2 : TIMP-2 ratios have been measured in a sample of patients with recurrent urothelial cancer than in those with non-recurrence, indicating the possible prognostic value of this measurement[132].

## Cysteine and aspartic proteinases and heparanase

The cysteine proteinases degrade matrix protein *in vitro* and there is evidence that one of these proteinases, cathepsin L, is over-expressed on metastatic melanoma cells compared with normal melanocytes and that carcinoma-derived cells can secrete mature forms of cathepsins L and B[133–135]. Cathepsin L is also highly expressed in *Ras*-transformed fibroblasts. The lysosomal aspartic proteinase cathepsin D is over-expressed in most primary breast cancers, undergoing abnormal secretion as a proenzyme. Its expression is induced by oestrogen acting *via* an oestrogen receptor/SP1-like consensus sequence and has been used as an indicator of poor prognosis for breast cancer[136]. Over-expression of the human gene increases metastasis in nude mice[137]. Heparanase, a heparan sulfate-specific endo-β-D-glucuronidase, is involved in metastasis of malignant melanoma cells and is inhibited by suramin[138].

## NME1 and NME2

*NME1* (non-metastatic cells 1, expressed) and *NME2* (formerly *NM23* (non-metastatic 23) also called PuF, NM23-H2 and NDPK-B) encode nucleoside diphosphate (NDP) kinases expressed on the cell surface. The concentration of NDP kinase A (encoded by human *NME1* and 88% identical to NDP kinase B (*NME2*)) increases in proliferating normal cells[139] and reduced *NME2* expression correlates with high metastatic potential in some tumours (e.g. hepatocellular carcinoma[140]) and cell lines. The transfection of human or murine *NME1* cDNA reduces primary tumour formation and metastasis *in vivo* and also eliminates the capacity of TGFβ to stimulate colony formation *in vitro*[141]. However, *NME2* expression is increased in some neoplastic tissues, although in ductal breast carcinomas this increase does not appear to correlate with tumour size, oestrogen receptor or progesterone receptor expression, lymph node metastases or other prognostic factors[142]. Enhanced expression of *NM23-1* and *NM23-2* has been detected in the early stages of ovarian cancer, correlating with enhanced expression of *HER2*[143]. In transient transfection assays, NM23-2 *trans*-activates *MYC*[144]. Amplification and over-expression of *NME2* has been detected in childhood neuroblastomas, but this is accompanied by a point mutation in the gene and mutations have also been detected in colorectal adenocarcinoma[145,146]. The mutation $Ser^{120} \rightarrow Gly$, that decreases enzyme stability, has been detected in ~25% of advanced but not limited stage tumours[147]. *DR-nm23* is ~70% similar to *NME1* and *NME2* and its over-expression inhibits granulocyte differentiation and induces apoptosis in 32Dc13 myeloid cells, suggesting that it may contribute to differentiation arrest in CML[148]. *WDNM1* and *WDNM2* are rat genes having human homologues that are also downregulated in metastatic cells[149].

### *References*


1. **Fidler, I.J. and Ellis, L.M. (1994) Cell 79, 185–188.**
2. Tessler, S. et al. (1994) J. Biol. Chem. 269, 12456–12461.
3. O'Brien et al. (1996) Cancer Res. 56, 2515–2518.
4. Millauer, B. et al. (1996) Cancer Res. 56, 1615–1620.
5. Asano, M. et al. (1995) Cancer Res. 55, 5296–5301.
6. Mukhopadhyay, D. et al. (1995) Nature 375, 577–581.
7. Viglietto, G. et al. (1995) Oncogene 11, 1569–1579.
8. Moscatelli, D. et al. (1986) Proc. Natl Acad. Sci. USA 83, 2091–2095.

9. Rogelj, S. et al. (1988) Nature 331, 173–175.
10. Brustle, O. et al. (1992) Oncogene 7, 1177–1183.
11. Costa, M. et al. (1994) Cancer Res. 54, 9–11.
12. Plate, K.H. et al. (1994) Brain Pathol. 4, 207–218.
13. Nicosia, R.F. et al. (1994) Dev. Biol. 164, 197–206.
14. Czubayko, F. et al. (1994) J. Biol. Chem. 269, 28243–28248.
15. Hu, G.-F. et al. (1994) Proc. Natl Acad. Sci. USA 91, 12096–12100.
16. Olson, K.A. et al. (1995) Proc. Natl Acad. Sci. USA 92, 442–446.
17. Moroianu, J. and Riordan, J.F. (1994) Proc. Natl Acad. Sci. USA 91, 1677–1681.
18. Hu, G.F. and Riordan, J.F. (1993) Biochem. Biophys. Res. Commun. 197, 682–687.
19. Greer, P. et al. (1994) Mol. Cell. Biol. 14, 6755–6763.
20. Pandey, A. et al. (1995) Science 268, 567–569.
21. Laniado-Schwartzman, M. et al. (1994) J. Biol. Chem. 269, 24321–24327.
22. Albini, A. et al. (1996) Oncogene 12, 289–297.
23. Moses, M.A. (1993) Clin. Exp. Rheumatol. 11 (Suppl. 8), S67–S69.
24. Jackson, D. et al. (1994) Science 266, 1581–1584.
25. Bagavandoss, P. et al. (1993) Biochem. Biophys. Res. Commun. 92, 325–332.
26. Sheibani, N. and Frazier, W.A. (1995) Proc. Natl Acad. Sci. USA 92, 6788–6792.
27. Dameron, K.M. et al. (1994) Science 265, 1582–1584.
28. Nicosia, R.F. and Tuszynski, G.P. (1994) J. Cell Biol. 124, 183–193.
29. Van Meir, E.G. et al. (1994) Nature Genet. 8, 171–176.
30. O'Reilly, M.S. et al. (1994) Cell 79, 315–328.
31. Moses, M.A. et al. (1992) J. Cell Biol. 119, 475–482.
32. Maione, T.E. et al. (1990) Science 247, 77–79.
33. Sasisekharan, R. et al. (1994) Proc. Natl Acad. Sci. USA 91, 1524–1528.
34. Sidky, Y.A. and Borden, E.C. (1987) Cancer Res. 47, 5155–5161.
35. Fotsis, T. et al. (1994) Nature 368, 237–239.
36. Pesenti, E. et al. (1992) Br. J. Cancer 66, 367–372.
37. Myers, C. et al. (1992) J. Clin. Oncol. 10, 881–889.
38. Harris, A.L. et al. (1994) Cancer 74, 1021–1025.
39. Yanase, T. et al. (1993) Cancer Res. 53, 2566–2570.
40. Antoine, N. et al. (1994) Cancer Res. 54, 2073–2076.
41. Zugmaier, G. et al. (1992) J. Natl. Cancer Inst. 84, 1716–1724.
42. Thorpe, P.E. et al. (1993) Cancer Res. 53, 3000–3007.
43. Haran, E.F. et al. (1994) Cancer Res. 54, 5515–5517.
44. Mok, S.C. et al. (1996) Oncogene 12, 1895–1901.
45. Myers, K.A. et al. (1994) J. Biol. Chem. 269, 9319–9324.
46. Takeichi, M. (1991) Science 251, 1451–1455.
47. Hamaguchi, M. et al. (1993) EMBO J. 12, 307–314.
48. Xiang, Y.-Y. et al. (1994) Cancer Res. 54, 3034–3041.
49. Rucklidge, G.J. et al. (1994) Trans. Biochem. Soc. 22, 63–68.
50. Hsieh, J.-T. et al. (1995) Cancer Res. 55, 190–197.
51. **Hynes, R.O. (1992) Cell 69, 11–25.**
52. **Pigott, R. and Power, C. (1993) The Adhesion Molecule FactsBook. Academic Press, London.**
53. Wei, Y. et al. (1996) Science 273, 1551–1555.
54. Arrick, B.A. et al. (1992) J. Cell Biol. 118, 715–726.
55. Morino, N. et al. (1995) J. Biol. Chem. 270, 269–273.
56. Damsky, C. et al. (1992) Matrix Suppl. 1, 184–191.

57 Waleh, N.S. et al. (1994) Cancer Res. 54, 838–843.
58 Shaw, L.M. et al. (1996) Cancer Res. 56, 959–963.
59 Hodivala, K.J. et al. (1994) Oncogene 9, 943–948.
60 Riikonen, T. et al. (1995) J. Biol. Chem. 270, 13548–13552.
61 Zutter, M.M. et al. (1995) Proc. Natl Acad. Sci. USA 92, 7411–7415.
62 Seftor, R.E.B. et al. (1992) Proc. Natl Acad. Sci. USA 89, 1557–1561.
63 Brooks, P.C. et al. (1994) Science 264, 569–571.
64 Brooks, P.C. et al. (1994) Cell 79, 1157–1164.
65 Lafrenie, R.M. et al. (1992) Cancer Res. 52, 2202–2208.
66 Friedlander, M. et al. (1995) Science 270, 1500–1502.
67 Agrez, M. et al. (1994) J. Cell Biol. 127, 547–556.
68 Hibi, K. et al. (1994) Oncogene 9, 611–619.
69 Qian, F. et al. (1994) Cell 77, 335–347.
70 Wang, X. et al. (1995) Cancer Res. 54, 4726–4728.
71 Albeda, S.M. et al. (1990) Cancer Res. 50, 6757–6764.
72 Zhang, Z. et al. (1995) Proc. Natl Acad. Sci. USA 92, 6161–6165.
73 Natali, P.G. et al. (1992) Br. J. Cancer 66, 318–322.
74 Droll, A. et al. (1995) J. Biol. Chem. 270, 11567–11573.
75 Gansauge, F. et al. (1995) Cancer Res. 55, 5499–5503.
76 Guo, Y. et al. (1994) Cancer Res. 54, 1561–1565.
77 Guo, Y.J. et al. (1994) Cancer Res. 54, 422–426.
78 Salles, G. et al. (1993) Blood 82, 3539–3547.
79 Finn, L. et al. (1994) Biochem. Biophys. Res. Commun. 200, 1015–1022.
80 Yoshida, K. et al. (1995) Cancer Res. 55, 4273–4277.
81 Cannistra, S.A. et al. (1995) Clin. Cancer Res. 1, 333–342.
82 Takahashi, K. et al. (1995) Oncogene 11, 2223–2232.
83 Weber, G.F. et al. (1996) Science 271, 509–512.
84 Zhu, D. and Bourguignon, L. (1996) Oncogene 12, 2309–2314.
85 Merzak, A. et al. (1994) Cancer Res. 54, 3988–3992.
86 Bartolazzi, A. et al. (1994) J. Exp. Med. 180, 53–66.
87 **Stetler-Stevenson, W.G. et al. (1993) FASEB J. 7, 1434–1441.**
88 Gearing, A.J.H. et al. (1994) Nature 370, 555–557.
89 Duffy, M.J. (1996) Clin. Cancer Res. 2, 613–618.
90 Jankun, J. et al. (1993) J. Cell. Biochem. 53, 135–144.
91 Montgomery, A.M. et al. (1993) Cancer Res. 53, 693–700.
92 Crowley C.W. et al. (1993) Proc. Natl Acad. Sci. USA 90, 5021–5025.
93 Landau, B.J. et al. (1994) Cancer Res. 54, 1105–1108.
94 Quattrone, A. et al. (1995) Cancer Res. 55, 90–95.
95 **Blasi, F. (1993) BioEssays 15, 105–111.**
96 Yamamoto, M. et al. (1994) Cancer Res. 54, 5016–5020.
97 Schmalfeldt, B. et al. (1995) Cancer Res. 55, 3958–3963.
98 Tsuboi, R. and Rifkin, D.B. (1990) Int. J. Cancer 46, 56–60.
99 Sato, H. et al. (1994) Nature 370, 61–65.
100 Will, H. and Hinzmann, B. (1995) Eur. J. Biochem. 231, 602–608.
101 Takino, T. et al. (1995) J. Biol. Chem. 270, 23013–23020.
102 Puente, X.S. et al. (1996) Cancer Res. 56, 944–949.
103 Wang, M. et al. (1994) Cancer Res. 54, 2492–2495.
104 Botos, I. et al. (1996) Proc. Natl Acad. Sci. USA 93, 2749–2754.
105 Fotouhi, N. et al. (1994) J. Biol. Chem. 269, 30227–30231.
106 Mignatti, P. et al. (1989) J. Cell Biol. 108, 671–682.

[107] Galardy, R.E. et al. (1995) Cancer Res. 54, 4715–4718.
[108] Kohn, E.C. and Liotta, L.A. (1995) Cancer Res. 55, 1856–1862.
[109] Kohn, E.C. et al. (1995) Proc. Natl Acad. Sci. USA 92, 1307–1311.
[110] **Hayakawa, T. (1994) Cell Struct. Funct. 19, 109–114.**
[111] Ogata, Y. et al. (1995) J. Biol. Chem. 270, 18506–18511.
[112] Kishnani, N.S. et al. (1994) Matrix Biol. 14, 479–488.
[113] Goldberg, G.I. and Eisen, A.Z. (1990), In Regulatory Mechanisms in Breast Cancer, Ed. M.E. Lipman and R.B. Dickson. Kluwer Academic, Dordrecht, pp. 421–440.
[114] Van der Brûle, F.H. et al. (1992) Int. J. Cancer 52, 653–657.
[115] Curry, V.A. et al. (1992) Biochem. J. 285, 143–147.
[116] Imren, S. et al. (1996) Cancer Res. 56, 2891–2895.
[117] Albini, A. et al. (1994) AIDS 8, 1237–1244.
[118] Kossakowska, A.E. et al. (1991) Blood 77, 2475–2481.
[119] Baker, T. et al. (1994) Br. J. Cancer 70, 506–512.
[120] Uría, J.A. et al. (1994) Cancer Res. 54, 2091–2094.
[121] Khokha, R. et al. (1991) J. Cancer Res. Clin. Oncol. 117, 333–338.
[122] Oates, A.J. et al. (1996) Oncogene 13, 97–104.
[123] Behrend, E.I. et al. (1994) Cancer Res. 54, 832–837.
[124] Gardner, H.A.R. et al. (1994) Oncogene 9, 2321–2326.
[125] Su, L. et al. (1995) Oncogene 10, 2163–2169.
[126] Moses, M.A. and Langer, R. (1991) J. Cell. Biochem. 47, 230–235.
[127] Johnson, M.D. et al. (1994) J. Cell. Physiol. 160, 194–202.
[128] Murphy, A.N. et al. (1993) J. Cell. Physiol. 57, 351–358.
[129] Hayakawa, T. et al. (1994) J. Cell Sci. 107, 2373–2379.
[130] Corcoran, M.L. and Stetler-Stevenson, W.G. (1995) J. Biol. Chem. 270, 13453–13459.
[131] Nakagawa, T. et al. (1994) J. Neurosurg. 81, 69–77.
[132] Gohji, K. et al. (1996) Cancer Res. 56, 3196–3198.
[133] Rozhin, J. et al. (1989) Biochem. Biophys. Res. Commun. 164, 556–561 [published erratum appears in Biochem. Biophys. Res. Commun. (1989) 165, 1444].
[134] Metzelaar, M.J. et al. (1991) J. Biol. Chem. 266, 3239–3245
[135] **Keppler, D. et al. (1994) Biochem. Soc. Trans. 22, 43–49.**
[136] Krishnan, V. et al. (1994) J. Biol. Chem. 269, 15912–15917.
[137] Liaudet, E. et al. (1994) Oncogene 9, 1145–1154.
[138] Nakajima, M. et al. (1991) J. Biol. Chem. 266, 9661–9666.
[139] Keim, D. et al. (1992) J. Clin. Invest. 89, 919–924.
[140] Iizuka, N. et al. (1995) Cancer Res. 55, 652–657.
[141] Leone, A. et al. (1993) Oncogene 8, 2325–2333.
[142] Sastre-Garau, X. et al. (1992) Int. J. Cancer 50, 533–538.
[143] Mandai, M. et al. (1994) Cancer Res. 54, 1825–1830.
[144] Berberich, S.J. and Postel, E.H. (1995) Oncogene 10, 2343–2347.
[145] Leone, A. et al. (1993) Oncogene 8, 855–865.
[146] Wang, L. et al. (1993) Cancer Res. 53, 717–720.
[147] Chang, C.L. et al. (1994) Oncogene 12, 659–667.
[148] Venturelli, D. et al. (1995) Proc. Natl Acad. Sci. USA 92, 7435–7439.
[149] Dear, T.N. et al. (1989) Cancer Res. 49, 5323–5328.

# 8 The Multistep Nature of Cancer

The epidemiology of cancer strongly suggests that to drive a cell through the various stages preceding the production of a tumour *in vivo* requires the accumulation of several genetic lesions. It is not known how many mutations are required for the genesis of a specific cancer but it is generally supposed that approximately 12 may be involved, and in all but a small proportion of cases (<1%) these mutations are acquired somatically (Table 9).

In experimental animals the chemical induction of tumours generally requires the sequential application of two types of agent, an "initiator" and a "promoter". The initiator is usually a mutagen that has irreversible effects and must be administered before the promoter. These model systems indicate that the initiating event in tumorigenesis is mutation and that subsequent tumour progression may be mediated by either genetic or epigenetic mechanisms. *In vitro* cellular studies generally support the notion that transformation is at least a two stage process, one oncogene being needed for immortalization and another for transformation. Thus, primary fibroblasts can often be transformed to a focal growth phenotype (anchorage-independent) by expression of a single oncogene (e.g. *Src*, *Ras*, *Raf*, *Mos*, *Trk*, *Fos*, *Sis*, or SV40 large T antigen). Full transformation, however, generally requires the expression of two complementing oncogenes, typically *Ras* and *Myc*. For example, mutant *RAS* from human tumours only transforms embryonic rat cells when supplemented with other oncogenes in transfected cells, e.g. the E1A adenovirus gene, large T antigen of polyoma or *Myc*. However, massive over-expression of a single gene can probably over-ride this distinction, thus v-*myc* alone can transform embryonic cells.

In addition to the anomalous initiation of growth promoting events, the loss of the growth inhibitory activity of TGF$\beta$ is thought to contribute to the development of many types of tumours. The present evidence indicates that this arises mainly as the result of functionally impairing mutations in one of the receptors for this cytokine, TGFBR2 (see ***TGFBR1*** and ***TGFBR2***). Thus mutations causing anomalous kinase activity occur in cell lines from head and neck tumours and the complete absence of TGF$\beta$ receptors has been detected in cell lines from human retinoblastomas, hepatomas and gastric cancers. Colon carcinoma cells with microsatellite instability (RER$^+$ cells) may also lack functional TGFBR2 as a result of mutations causing truncation of the protein. A dominant inhibitory point mutation has been

**Table 9** *Combinations of oncogenes and tumour suppressor genes associated with some human cancers*

| Affected loci | Tumour |
|---|---|
| *HRAS, MYC, MYCN, P53, RB1* | Cervical carcinoma |
| *APC, MCC, DCC, KRAS2, P53, TGFBR2* | Colorectal carcinoma |
| *MYB, MYC*, cyclin D1, cyclin D3, *EGFR, HER2, HRAS, P53, RB1, BRCA1, HSTF1, INT2, YES1* | Breast carcinoma |
| *MET, HER2, P53, RB1* | Prostate carcinoma |
| *VHL, EGFR, HER2, MET* | Renal cell carcinoma |
| *MYC, MYCN, MYCL, HRAS, RB1, P53, RAF1, JUN* | Lung carcinoma |
| Cyclin D1, *BCL2, BCL3, HER2, REL, NFKB2, NPM, ALK, KRAS2, NRAS, YES1* | Lymphoma |
| *HER2, MET, K-SAM, BCL2, APC*, P-cadherin, *P53* (*KRAS, RB1*) | Stomach carcinoma |
| *ABL, BCR, MYC, NRAS, RB1, P53, ERG, TLS/FUS, HOXA9, NUP98, AML1, EAP, EVI1, MN1, TEL, MDS1* | Chronic myelogenous leukaemia |

detected in a cutaneous T cell lymphoma that prevents transport of functional receptor to the plasma membrane. The latter mutation correlates with onset of the aggressive stage of tumour development and it seems probable that TGFBR2 inactivation is a critical late step in tumour progression, rather than a coincidental consequence of genetic instability. TGFBR2 may resemble p53 in that mutant forms can oligomerize with wild-type proteins to exert a dominant negative effect. The *TGFBR2* locus (3p22) resides in a region that shows frequent loss of heterozygosity during tumour progression, notably in lung cancer[1] and mutations in this gene may therefore accompany the development of a wide variety of cancers. Furthermore, the tumour suppressor gene *DCP4/SMAD4*, the loss of which contributes to the development of some pancreatic carcinomas, is a member of the MAD family that are components of the TGF$\beta$ signalling pathway. Thus, there appear to be at least two general mechanisms by the inhibitory effects of TGF$\beta$ on proliferation can be overcome during tumorigenesis.

## Cervical carcinoma

In humans, the multistep nature of neoplastic development may be readily distinguished in some forms of the disease. In cervical cancer the initiating event is frequently infection with specific types of human papillomavirus. Subsequently cervical dysplasia (cervical intraepithelial neoplasia, CIN) may arise. CIN types I, II and III represent progressively more severe forms in the development of malignant cervical carcinoma. However, the development of an abnormal cell clone (CIN types I and II) is usually followed by its disappearance and only very infrequently does it give rise to a still more abnormal clone of fully invasive cells. The sustained expression of papillomavirus proteins including those that abrogate pRb and p53 function is necessary to maintain the transformed state and promote genetic instability. Amplification and/or over-expression of *MYC* family genes occurs in most cancers and abnormal expression of both *MYC* and *MYCN* is frequently associated with cervical carcinoma. In particular, the expression of *MYCN* is strongly correlated with the more advanced stages of the disease (CIN types II and III). The fact that smoking significantly increases the probability of cervical tumour progression indicates the importance of environmental factors that may not operate through genetic mechanisms.

## Colorectal carcinoma

The majority of malignant colorectal carcinomas develop from benign adenomas in a process in which DNA hypomethylation occurs together with the acquisition of multiple mutations, notably in *KRAS2*, *APC*, *MCC*, *DCC* and *P53*[2], although abnormal expression of other genes (e.g. *HER2* over-expression) may also occur. Point mutations in codons 12, 13 or 61 of *KRAS2* occur in ~50% of colorectal carcinomas and are an early event in tumour development. G-T transversions in the first or second bases of codon 12 are the most common and particularly characterize adenomas. However, the incidence of *KRAS2* mutations increases with tumour progression and multiple mutations occur in Dukes' B and C stages. In particular, G-C transversions in the second base of codon 12 show a trend of association with metastasized tumours when prognosis is poor[3]. The functional significance of these mutational variants is unknown but it may be noted that G-C transversions arise as a result of the action of alkylating agents, whereas G-A and G-T transversions can result from errors in DNA replication and repair[4]. Although mutations in *P53* usually appear at a relatively advanced stage of colorectal carcinoma development, the sequence in which they and the *APC, MCC* and *DCC* changes arise varies between different tumours. This fact has led to the suggestion that the development of malignant colorectal carcinoma requires an overall accumulation rather than a fixed sequence of somatic mutations. It may also be noted that, although the majority of colorectal carcinomas show mutations in *KRAS2* (>50%), *P53* (>75%) and *DCC* (>70%), this disease is typical of the vast majority of cancers in that there is no combination of genetic abnormalities that correlates absolutely with its development. In hereditary non-polyposis colon cancer multiple somatic mutations have been detected in *APC* (up to six) and *P53* (up to four in the same patient) following loss of functional *MSH2*. This is consistent with a mechanism by which inactivation of MSH2 promotes the progressive accumulation of mutations that are critical for tumour development.

## Breast carcinoma

Members of the epidermal growth factor (EGF) family are widely expressed and are thought to play important roles in both mammary gland development and in tumorigenesis. Over-expression of at least two members of this family, amphiregulin and cripto-1, has been detected in breast tumours. Either can transform immortalized human and mouse mammary epithelial cells and cripto-1 in particular may be a useful marker for breast tumorigenesis[5]. Consistent with these findings, members of the EGF receptor family (EGFR, HER2, HER3 and HER4) also frequently show abnormal expression in breast and other human cancers.

The involvement of EGFR over-expression in breast cancer remains controversial, in part because of the wide range of its reported incidence (14–91%). Nevertheless, many reports indicate that high levels of EGFR expression occur in ~20% of primary breast tumours and that this correlates with a particularly poor prognosis in both lymph node-positive and lymph node-negative patients (see ***EGFR/ErbB***). *EGFR* expression correlates with that of *P53* but is inversely correlated with *MYB* amplification: thus *EGFR* and *MYB* expression provide indicators of a poor or good prognosis, respectively. Although elevated transcription of *EGFR* correlates with the absence of steroid hormone (oestrogen and progesterone) receptors, it is reduced by tamoxifen, presumably acting *via* targets other than the oestrogen receptor[6–8].

No direct ligand has been identified for HER2 but this receptor can be *trans*-activated by ligands for either EGFR, HER3 or HER4 (see ***HER2/ErbB-2/Neu***). HER3 and HER4 are activated by members of the neuregulin family at least one of which, heregulin, stimulates proliferation of human breast and ovarian cell lines in a HER2-dependent manner[9]. In breast cancer, bladder cancer and some other adenocarcinomas there is a correlation between the degree of over-expression of *HER2*, undetectable in normal breast tissue, and tumour stage. In one study *HER2* expression has been shown to be inversely correlated with the expression of *MYB*. For lymph node-negative patients *HER2* amplification correlates with a significantly decreased relapse-free interval and reduced survival times. As with *EGFR*, there is an inverse correlation between *HER2* over-expression and expression of the oestrogen and progesterone receptors in invasive cancer[5]. *P53* expression also correlates with over-expression of *HER2*. The introduction of *HER2* cDNA into breast cancer cells expressing low levels of HER2 results in downregulation of the oestrogen receptor and oestrogen-independent growth that is insensitive to tamoxifen[10].

Amplification of *MYC* is also frequently associated with breast carcinoma and the combination of enhanced *MYC* and *HER2* expression is indicative of a particularly poor prognosis. Loss of heterozygosity at the *MYCL1* locus has also been detected in this disease. Abnormally high expression of either cyclin D1, cyclin D3 or both occurs in ~70% of breast carcinomas and the coexpression of cyclin D1 with either EGFR or pRb is a strong indicator of poor prognosis. Amplification of *HSTF1* and *INT2* occurs in ~20% of breast carcinomas but this does not appear to be reflected in enhanced synthesis of their gene products. Over-expression of the *SRC* family gene *YES1* has been detected and mutations in *HRAS1* also occur with low frequency in this disease.

## Prostate carcinoma

Loss of *P53* occurs in ~25% of advanced primary prostate cancer and in ~50% of metastases. *RB1* loss is less common but occurs in a minority of advanced stage tumours. *MET* expression may be associated with the early stages of this disease and over-expression of *HER2* with its development. Major changes in the extracellular matrix, cell surface integrin expression and the activity of matrix metalloproteinases also occur[11].

## Renal cell carcinoma

In renal cell carcinoma there is a close correlation between over-expression of both *EGFR* and *HER2* without amplification and de-differentiation and the development of metastatic disease, indicating the possible importance of cooperative effects between these two tyrosine receptor kinases in the progression of this cancer[12]. Expression of MET, the receptor for HGF/scatter factor, is low in normal kidney but is markedly increased in a high proportion of renal carcinomas[13] and over-expression of *YES1* has also been reported in some cases. The major tumour suppressor gene associated with this disease is *VHL*, which frequently undergoes functional inactivation. Chromosomal aberrations in the region of *RAF1* occur and enhanced expression of *YES1* has also been detected.

## Lung carcinoma

Amplification of *MYC, MYCN* or *MYCL1* is commonly associated with lung cancers and in particular with small cell lung carcinoma (SCLC). *HER2* over-expression or amplification and high levels of HER2 protein have also been correlated with poor prognosis in adenocarcinoma of the lung. Mutations in each member of the RAS gene family (*HRAS, KRAS2, NRAS*) can occur in lung cancers although there are variations between forms of the disease with, for example, *KRAS2* mutations being relatively common in squamous cell lung carcinomas in which *HRAS* mutations are rare. Abnormally high expression of *FES/FPS, JUN* or *YES1* has been detected in various types of lung cancers and chromosomal aberrations in the vicinity of *RAF1* occur in SCLC. Increased expression of some members of the WNT family of growth factors is associated with a number of cancers and over-expression of *WNT5A* has been detected in some lung cancers and also in primary breast cancers, prostate carcinomas and melanomas.

## Lymphomas

Chromosomal translocations in which the breakpoints lie adjacent to the cyclin D1, *BCL2* and *BCL3* genes are frequently associated with a variety of lymphoid neoplasms. These include the t(11;14) translocation present in some B cell non-Hodgkin's lymphomas, B cell chronic lymphocytic leukaemias and in multiple myeloma, as a result of which cyclin D1 over-expression is driven by the Ig heavy chain promoter. Such over-expression constitutes a marker for mantle cell lymphoma, a non-Hodgkin's B cell lymphoma. However, in some leukaemias over-expression of cyclin D1 occurs in the absence of translocations. Rearrangements involving the *REL* family genes *REL* and *NFKB2* also occur in various lymphomas which are presumed to result in anomalous regulation of genes *via* κB sequences. A translocation involving the nucleophosmin (*NPM*) and anaplastic lymphoma kinase (*ALK*) genes has also been detected in a non-Hodgkin's lymphoma that gives rise to a fusion product with *in vitro* transforming activity. Amplified or over-expressed *HER2* has been detected in a malignant lymphoma, as have high levels of *YES1* expression. Simultaneous mutations in *KRAS2* and *NRAS* may occur in multiple myeloma.

## Stomach carcinoma

There is considerable evidence that *Helicobacter pylori* infection is a risk factor for gastric cancer[14]. *H. pylori* is a gram-negative spiral-shaped bacterium that colonizes the stomach in humans, and in the UK half the population over the age of 50 are infected. *H. pylori* seropositivity may be elevated in the early stages of tumour development and there is evidence that infection is associated with suppressed immune responsiveness[15]. Thus by causing chronic inflammation and metaplasia and also modulating immune responsiveness, *H. pylori* infection may contribute to carcinogenesis. Genetic abnormalities associated with this disease include over-expression of HER2 and MET (as well as K-SAM see ***HSTF1/Hst-1, HST2/Hst-2***), of *BCL2* in some poorly differentiated type cancers and of *APC* in some well-differentiated types. Abnormal expression of P-cadherin has also been detected in gastric carcinogenesis[16] and p53 mutations commonly appear as late events[17]. Mutations in *KRAS* and *RB1* have been detected but at low frequency[18].

## Chronic myeloid leukaemia

The chromosome translocation t(9;22)(q34;q11) involving *ABL* occurs in over 95% of cases of chronic myeloid leukaemia (CML) and in 25% of adult acute lymphocytic leukaemias (see ***BCR***). The translocation involves the loss of N-terminal *ABL* sequences (corresponding to the retroviral transduction event when *Abl* is juxtaposed to viral *gag*). The hybrid *BCR/ABL* gene encodes the BCR/ABL fusion protein which has elevated tyrosine kinase activity and transforming capacity. The *ABL* gene is also amplified tenfold in CML but the product of the reciprocal translocation, the *ABL-BCR* hybrid gene, has not been detected. BCR/ABL can bind to $p120^{CBL}$, to the SH2/SH3 protein CRK, phosphatidylinositol 3′-kinase and GRB2/SOS1. GRB2/SOS1 promotes the activation of RAS and the JUN N-terminal kinase (JNK) pathway that appears to drive BCR/ABL transformation. This suggests that oncogenic activation of RAS could contribute to CML and indeed abnormal expression of *RAS* and *MYC* as well as mutations in the tumour suppressor genes

*RB1* and *P53* have been detected in the blast crisis of CML. Furthermore, a variety of other chromosomal rearrangements may also occur in CML, including those between *ERG* and *TLS/FUS*, *HOXA9* and *NUP98*, *AML1* and *EAP* or *EVI1*, *MN1* and *TEL* and inversion of *MDS1*, each of which generates putative aberrant transcription factor activity (see Table III, page 78). Thus diverse molecular defects may contribute to the pathogenesis of this disease.

The most notable exception to the pattern of variable expression of cancer genes in histologically identical tumours is provided by the retinoblastoma (*RB1*) gene, both alleles of which are defective in all retinoblastomas. However, inactivation of this tumour suppressor gene together with other genetic abnormalities occurs in a wide variety of other neoplasms (Table IV, page 83), e.g. Wilms' kidney tumour, breast cancer and small cell lung carcinoma (SCLC), and individuals with retinoblastoma are particularly susceptible to the development of sarcomas. The frequency of pRb inactivation in tumours other than retinoblastoma varies widely. It is mutated with high frequency in SCLCs and most osteosarcomas and soft tissue sarcomas. In non-SCLCs (e.g. adenocarcinomas, squamous cell carcinomas or large cell carcinomas), however, *RB1* mutations are infrequent although some tumours of this type have been shown to be defective in both *RB1* and *P53*. The *RB* gene is inactivated in ~20% of primary breast carcinomas. In ~65% of cases there is allelic loss of the *P53* gene on chromosome 17. The retained *P53* allele frequently contains point mutations although some tumours retain one normal allele. Between 20% and 50% of breast carcinomas contain non-random loss of heterozygosity for specific locations on chromosomes 1q, 3p or 11p at which specific genes have not been identified and virtually all chromosomes show some susceptibility to allelic imbalance either as an increase in allele copy number (~25%) or as loss of heterozygosity. *P53* mutations are common in some non-SCLCs (squamous cell carcinomas) but not in others (adenocarcinomas), whereas chromosomal deletions affecting *HRAS*, *RAF1* and *INT2* are relatively common (occurring in 18–50% of cases examined) in both these types of non-SCLC.

The foregoing summary indicates that patterns of expression of members of the EGFR family and other key genes involved in regulation of cell proliferation are emerging that are already providing useful diagnostic markers. Such patterns clearly suggest targets for gene therapy and *HER2* in particular has been the subject of a number of such approaches (see Chapter 9).

### *References*


[1] Hibi, K. et al. (1994) Oncogene 9, 611–619.
[2] **Fearon, E.R. and Vogelstein, B. (1990) Cell 61, 759–767.**
[3] Span, M. (1996) Int. J. Cancer (Pred. Oncol.) 69, 241–245.
[4] **Bos, J.L. (1989) Cancer Res. 49, 4682.**
[5] Panico, L. et al. (1996) Int. J. Cancer 65, 51–56.
[6] Le Roy, X. et al. (1991) Oncogene 6, 431–437.
[7] Diez Gibert, O. et al. (1995) Eur. J. Clin. Chem. Clin. Biochem. 33, 563–568.
[8] Seshadri, R. et al. (1996) Int. J. Cancer (Pred. Oncol.) 69, 23–27.
[9] Lewis, G.D. et al. (1996) Cancer Res. 56, 1457–1465.
[10] Pietras, R.J. et al. (1995) Oncogene 10, 2435–2446.
[11] Bostwick, D.G. et al. (1994) J. Cell. Biochem. 56, 283–289.
[12] Stumm, G. et al. (1996) Int. J. Cancer (Pred. Oncol.) 69, 17–22.
[13] Natali, P.G. et al. (1996) Int. J. Cancer (Pred. Oncol.) 69, 212–217.

[14] IARC Monographs on the Evaluation of Carcinogenic Risks to Humans. Infection with *Helicobacter pylori*, Vol. 61, pp. 177–240. IARC, Lyon, 1994.
[15] Klaamas, K. et al. (1996) Int. J. Cancer 67, 1–5.
[16] Shimoyama, Y. and Hirohashi, S. (1991) Cancer Res. 51, 2185–2192.
[17] Tamura, G. et al. (1991) Cancer Res. 51, 3056–3058.
[18] **Tahara, E. (1993) J. Cancer Res. Clin. Oncol. 119, 265–272.**

# 9 Gene Therapy for Cancer

Current statistics indicate that one person in three of the population of the UK will develop some form of cancer during their lifetime. Over 250 000 new cases are recorded each year and over 160 000 cancer deaths occur (Table 10). A high proportion of these are accounted for by a few types of cancer: 25% of all cancer deaths are caused by lung cancer, 12% by colorectal cancers and 7% by stomach cancer. Breast cancer causes 9% of all cancer deaths and prostate cancer 5%, the second most frequent cause of male cancer deaths after lung cancer (33%). In the USA prostate cancer has the highest incidence of cancers in men (excluding non-melanoma skin cancer) and accounts for 21% of all newly diagnosed cancers. The high mortality rate from lung cancer, which kills over 100 people per day in Britain, has shown a small overall decline in recent years, although in the older age group of women (over 55 years of age) rates are still rising. The UK has the highest mortality rate for breast cancer and it is estimated that 1 in 12 of all women in the country will contract the disease. Although the death rate in younger age groups may be marginally declining, the overall trend appears to be of increasing mortality. The disease is highly prevalent throughout the Western developed world but the international variation in incidence exceeds five-fold, indicating the critical importance of environmental factors in its aetiology. In the UK mortality from colorectal cancers has remained relatively constant over the last 20 years. The worldwide incidence of this cancer also shows great variation (up to 20-fold).

The international incidence and mortality rates of cancers indicate the urgent need for effective therapeutic measures. Traditional measures for treating cancers have involved appropriate combinations of surgery, radiotherapy, hormone therapy and chemotherapy. Quite apart from the obvious disadvantages of some of these approaches, their overall success in decreasing deaths from cancers has been extremely limited. However, for some cancers major strides in treatment have been

**Table 10** *Incidence and deaths from cancers in the UK 1988 (Cancer Research Campaign statistics)*

| | New cases | | Deaths | |
|---|---|---|---|---|
| | Men | Women | Men | Women |
| All malignant neoplasms | 135 956 | 135 877 | 83 765 | 76 824 |
| Bladder | 9 185 | 3 716 | 3 665 | 1 693 |
| Brain | 1 988 | 1 575 | 1 660 | 1 280 |
| Breast | 205 | 29 870 | 81 | 15 300 |
| Cervix | – | 4 943 | – | 2 170 |
| Colon | 8 697 | 10 639 | 5 855 | 7 114 |
| Hodgkin's lymphoma | 798 | 632 | 281 | 202 |
| Kidney | 2 653 | 1 677 | 1 607 | 1 060 |
| Leukemia | 3 173 | 2 659 | 2 174 | 1 910 |
| Liver | 913 | 594 | 938 | 669 |
| Lung | 30 166 | 13 627 | 27 968 | 12 255 |
| Melanoma | 1 716 | 2 722 | 532 | 660 |
| Non-Hodgkin's lymphoma | 3 635 | 3 346 | 2 035 | 1 886 |
| Ovary | – | 5 832 | – | 4 275 |
| Pancreas | 3 470 | 3 578 | 3 282 | 3 513 |
| Prostate | 13 974 | – | 8 234 | – |
| Stomach | 7 840 | 4 975 | 6 322 | 4 290 |
| Testis | 1 418 | – | 152 | – |

made over the last 20 years. Thus acute lymphocytic leukaemia is the most common of the childhood cancers, accounting for ~26% of cases, but the 5-year survival rate for this disease now exceeds 70%, principally as a result of the development of effective combination chemotherapy. For testicular cancer, the commonest cancer in 20–34-year-old males in the UK, the cure rate exceeds 90% as a consequence of the development of assays for early detection of this disease together with advances in combination chemotherapy. In general, however, currently available chemotherapeutic measures are not very effective against common solid cancers, particularly after metastasis has occurred, and it remains the case that the most effective general anti-cancer strategy would be the abolition of smoking because ~30% of all cancer deaths, together with a significant proportion of deaths from heart disease, are attributable to tobacco smoking.

As genetic diseases, cancers would appear to be suitable targets for gene-directed therapy and much work has been directed towards the suppression of oncogene activity or the replacement of defective tumour suppressor genes in specific cancers. Additional approaches have attempted either to enhance the endogenous response of the immune system to tumour cells or to target tumour cells directly with cytotoxic agents. The major categories of therapy methods may be summarized as follows:

1 Nucleotide sequence-targeted strategies, i.e. anti-sense methods using oligodeoxy-nucleotides, oligoribonucleotides or ribozymes
2 Tumour suppressor gene therapy
3 Cytokine gene therapy and tumour vaccination
4 Virally directed enzyme prodrug therapy (VDEPT)
5 Antibody-directed therapy.

Despite the ingenuity of many of these approaches, success has been severely limited, in part by the problems of targeting the therapeutic vehicle efficiently to tumour cells. Furthermore, additional mutations may develop that neutralize the treatment, as commonly occurs with traditional therapies, or mutations that contributed to the early stages of tumour development may not be essential for subsequent progression and thus cease to be relevant therapeutic targets.

## NUCLEOTIDE SEQUENCE-TARGETED STRATEGIES

The important role in cancer development of the expression of anomalous genes in the form of oncogenes or of novel genes arising from chromosomal aberrations suggests the potential usefulness of therapeutic methods by which the activity of such genes could be suppressed. This is the essential aim of oligonucleotide-based or "anti-sense" strategies that utilize oligodeoxynucleotides or oligoribonucleotides, generally of 18–24 bases in length, that are complementary in sequence to a region of the target gene or mRNA. Anti-sense molecules therefore function either by steric inhibition or by activation of RNAaseH or, for ribozymes (see below), through their endogenous catalytic RNAase activity.

### Oligodeoxynucleotides

Anti-sense oligodeoxynucleotides (ODNs) expressed from transduced retroviral vectors have been used to modulate the tumorigenicity or metastatic potential of a number of oncogenes including *ABL, AML1/MTG8, MYB, MYC, MYCN, RAF1, RAS, SCL, SRC,* HPV E7, *RB1* and *P53* [1]. ODNs directed against *HER2* have also been shown selectively to downregulate expression of the receptor by a mechanism that involves triplex formation, i.e. binding of a third strand of DNA or RNA [2]. The expression of anti-sense ODNs directed against human type I regulatory subunit ($RI_{\alpha}$) of the cAMP-dependent protein kinase arrests the growth of human and rodent cancer cells *in vitro* and anti-sense $RII_{\alpha}$ blocks cAMP-inducible growth inhibition and differentiation. Cell growth is also inhibited by retroviral vector expression of $RII_{\beta}$ [3].

A major problem with this approach is the achievement of reproducible and effective delivery of the specific ODN to the interior of the target cell. A recent development utilizes receptor-mediated endocytosis of an adenovirus-polylysine-ODN complex that appears to be rapidly (<4 h) translocated to the nucleus within which the ODN remains stable for prolonged periods [4].

## Ribozymes

Ribozymes are a class of small ribonucleic acids that can exert *cis* or *trans* catalytic activity against RNA [5]. The sequence specificity of their action makes them attractive anti-sense molecules for cancer therapy and ribozymes have been used as anti-sense agents against a variety of genes, including *RAS, BCR/ABL, FOS,* drug resistance genes and human immunodeficiency virus type 1 genes. Thus ribozymes have been designed that specifically mediate cleavage of *BCR/ABL* transcripts *in vitro* and eliminate $P210^{BCR/ABL}$ protein kinase activity in the chronic myelogenous leukaemia blast crisis cell line K562 [6]. Ribozyme-suppression of Moloney murine leukaemia virus and HIV production has been demonstrated in NIH 3T3 cells and human T cells, respectively [7]. Reversion of the neoplastic phenotype has been achieved using a ribozyme expressed from a recombinant adenovirus cassette and designed to cleave mutant *HRAS* [8]. The evidence that PDGFB and its receptor may be over-expressed in some cancers has prompted the design of a ribozyme that cleaves *PDGFB* mRNA and reduces growth of a malignant mesothelioma-derived cell line [9]. Hammerhead ribozymes (that require a GUC motif to cleave their target) have also been designed that target with high specificity chimeric transcripts arising from chromosomal translocations. These include *PML/RARA* [10] and *AML1/MTG8* that occurs in acute myeloid leukaemia (AML) with the t(8;21) translocation, the latter inhibiting the growth of AML cells *in vitro* [11].

As observed above, conventional chemotherapy is generally of limited use, particularly against disseminated neoplasms. The most fully characterized mechanism involves expression of the multiple drug transporter phosphoglycoprotein (P-GP) encoded by *MDR1* which appears to prevent intracellular accumulation of a wide range of drugs. One approach to this problem is to modulate the activity of *MDR1* in tumour cells and hammerhead ribozymes have been synthesized that show specific activity against this gene. For example, Kiehntopf et al. (1994) [12] obtained liposome-mediated uptake of ribozymes into >98% of cultured human pleural mesothelioma cells to produce *MDR1* mRNA cleavage, reduced expression of P-GP protein and restoration of the chemo-sensitive phenotype. However, in a human ovarian cancer cell line, an anti-*FOS* ribozyme has been shown to be more effective than one targeted against *MDR1* in the reversal of drug resistance [13]. *MDR1* is an AP1-responsive gene and the expression of *JUN, RAS,* topoisomerase I and mutant p53 is also decreased by the anti-*FOS* ribozyme. This implies a role for *FOS* in regulating drug resistance, previously suggested by evidence of its involvement in the acquisition of resistance to cisplatin, azidothymidine and 5-fluorouracil. Ribozymes and also anti-sense phosphorothioate oligonucleotides directed against *MDR1* have been designed that reduce the chemoresistance of cultured blast cells from patients with acute myeloid leukaemia [14].

In addition to focusing on oncogenes and drug resistance, anti-sense methodologies have begun to be directed towards steps in metastasis. In some cells *in vivo* tumorigenicity is reduced by the expression from retroviral vectors of the *JE*/MCP-1 gene [15], *NM23* [16] or $\alpha_5\beta_1$ integrin [17]. *In vitro* invasiveness is decreased when the highly metastatic cell line B16-F10 is transfected with a plasmid expressing $\beta$m-actin [17]. Anti-CD44 ribozymes reduce expression of the CD44 cell surface adhesion protein which is believed to contribute to invasion, particularly of glioblastomas, *via* its interaction with extracellular matrix proteins [19]. Integrin $\alpha_6\beta_1$, a major surface receptor for the basement membrane, has also been successfully targeted by a ribozyme, the expression of which in the human HT1080 fibrosarcoma cell line renders the cells non-metastatic in nude mice [20]. The expression of anti-sense RNA directed against the cell adhesion molecule E-cadherin or against tissue inhibitors of metalloproteinases (TIMPs) can promote metastasis [21,22].

## TUMOUR SUPPRESSOR GENE THERAPY

It is well established that insertion of a normal copy of a tumour suppressor gene can suppress the tumorigenicity in immunodeficient mice of cells defective for that gene. This has been achieved inserting wild-type genes to replace defective copies of retinoblastoma (*RB1*), *P53*, *WT1* or neurofibromatosis type 1, principally using adenovirus vectors, although other types of construct have proved effective *in vivo* [23]. Thus reversion to a normal phenotype of human retinoblastoma and osteosarcoma cells and also breast, bladder, lung and prostate carcinoma cells has been achieved by transfection of a normal copy of the *RB1* gene. There is evidence that loss of *RB1* may promote the synthesis of an angiogenic factor by affected cells [24], which is consistent with the high angiogenic capacity of $\beta$ cells in which pRb function is ablated by the action of SV40 T antigen (see Chapter 5). Furthermore, the expression *via* a replication-defective adenovirus of a constitutively active (non-phosphorylatable) form of Rb inhibits the proliferation of vascular smooth muscle cells *in vivo* and may offer a method of treating vascular proliferative disorders [25].

Recombinant adenoviral vectors expressing full length or N-terminally truncated retinoblastoma protein are effective in reducing the tumorigenicity of human non-small cell lung carcinoma and bladder carcinoma cells in nude mice [26]. Helper-free retroviruses have been constructed that contain the immediate early region of the human cytomegalovirus enhancer-promoter fused to the Moloney murine leukaemia virus long terminal repeat [27]. These vectors provide high levels of controlled expression of incorporated genes and have been used to produce high levels of pRb, p53 and INK4A in human glioblastoma cells, in each case causing cell cycle arrest.

Wild-type p53 expressed from recombinant adenoviral vectors promotes apoptosis and suppression of cell growth and direct injection of such vectors has been effective in causing tumour regression in animal model systems of human non-small cell lung cancer (NSCLC), SCLC, head and neck cancer and cervical cancer [28,29]. Clinical trials of an adenovirus vector expressing wild-type p53 in patients with NSCLC are currently in progress [30].

Vectors expressing *INK4A* inhibit growth of some human cancer cell lines and reduce tumorigenicity in nude mice, consistent with the role of INK4A as a tumour suppressor gene[31].

In addition to tumour suppressor genes the exogenous expression of other genes has been correlated with anti-tumour activity. Thus the adenovirus E1A gene inhibits *in vitro* growth of human ovarian cancer cells that over-express *HER2* and causes a significant reduction of tumorigenicity in mice[32]. An adenoviral vector that over-expresses $BCLX_S$ induces apoptosis and reduction in size of solid MCF-7 tumours in nude mice[33].

## CYTOKINE GENE THERAPY AND TUMOUR VACCINATION

Considerable efforts have been made to direct and amplify the cellular immune reactions activated as a response of the host to the development of cancers. In particular, these have focused on tumour infiltrating lymphocytes (TILs) which are lymphoid cells that infiltrate and accumulate within developing tumours. TILs may be grown *in vitro* in the presence of interleukin 2 (IL-2) and expanded populations of TILs that can be infused into patients have been obtained in this way from samples of a wide variety of human tumours, including colon, breast, bladder and renal carcinomas, melanomas, lymphomas and neuroblastomas. Genetic modification of TILs can be accomplished by transduction of a retroviral vector and infusion of such cells can give rise to a high local concentration of synthesized, diffusible cytokine encoded by the vector, delivered to the site of the tumour without requirement for high systemic doses[34]. This approach has been employed, for example, to introduce an active tumour necrosis factor (TNF) gene into cells from patients with melanoma[35]. The administration of TNF-TILs together with IL-2 has been effective in causing tumour regression in patients with melanoma and this technique is clearly suitable for the introduction of other cytokines and agents of potential therapeutic value into TILs.

The positive effects achieved using adoptive transfer of TILs together with injection of IL-2 in metastatic melanoma have promoted attempts at more widespread application of this technique. TILs isolated from resected NSCLC tumours expanded *in vitro* and then infused together with IL-2 have produced reversion and prolonged survival in a number of patients in one study[36]. TIL and IL-2 combination therapy has also shown signs of effectiveness against metastatic renal cell carcinoma (RCC) although careful analysis of responding patients and non-responders suggests that the natural immune status of individuals (e.g. the proportion of circulating natural killer cells) may critically determine the outcome of such treatment[37].

An alternative approach to using TILs for the delivery of cytokines involves the generation of retroviral producer cells, normally fibroblasts, that express a cytokine gene from a retroviral vector. This exploits the inability of retroviral vectors to transfer genes into non-dividing cells. When injected into the circulation retroviral producer cells target liver metastases in mice and the retroviral vectors produced *in situ* transduce neighbouring proliferating cells. Vectors encoding IL-2 or IL-4 direct tumoricidal inflammatory responses to established metastases and inhibit metastasis and tumour burden[38]. This represents a novel method for targeting gene transfer to multifocal tumour deposits and hence a potential advance over methods in which delivery systems are injected directly into single tumour deposits. In mice bearing metastatic melanomas the combined action of IL-2 and IL-6 gene-transfected tumour cells exceeds that of either gene alone in terms of the potency of evoked specific anti-tumour immune responses[39]. In animal model systems other genes (e.g. human ICAM-1) have also been shown to activate long-term, tumour-specific immunity[40].

These cytokine-based approaches rely on the well established observation that the expression of genes that stimulate the immune system (IL-2, IL-4, IL-7, TNF, IFN-$\gamma$, GM-CSF) can promote tumour rejection. However, there is evidence that the mere expression of an integrated *lacZ* gene can reduce tumorigenicity[41]. Furthermore, in this study the rejection of highly tumorigenic mastocytoma cells that expressed *lacZ* gave rise to prolonged immunity against unmodified cells. The mechanism of this remarkable observation is unknown but it emphasizes the complexity of the immune system in responding to tumour cells.

## VIRALLY DIRECTED ENZYME PRODRUG THERAPY (VDEPT)

The incorporation of specific promoters into retroviral vectors has led to the technique of virally directed enzyme prodrug therapy (VDEPT, also called gene-directed enzyme prodrug therapy (GDEPT)). This approach has principally relied upon the enzyme deoxypyrimidine kinase encoded by the genome of the herpesviruses. This enzyme was originally described as a thymidine kinase but is in reality a broad specificity nucleoside kinase that readily phosphorylates thymidine analogues, including ganciclovir, which then inhibit DNA replication. Ganciclovir and related analogues are poorly phosphorylated in normal cells so that selective growth inhibition may be achieved. The prodrug 6-methoxypurine arabinonucleoside (araM) resembles ganciclovir in that it is a good substrate for deoxypyrimidine kinase and it is converted to a cytotoxic metabolite (araATP) within tumour cells.

Deoxypyrimidine kinase has been expressed from a recombinant adenovirus vector in rat gliomas by direct injection into the tumour. Subsequent injection of ganciclovir caused complete tumour regression in >60% of the animals as assessed by magnetic resonance imaging[42]. Although long-term tumour recurrence was observed, repetition of this treatment promoted tumour rejection. This procedure has been extended to the treatment of human brain tumors using a retroviral vector to express deoxypyrimidine kinase[43]. Deoxypyrimidine kinase has also been specifically expressed in hepatoma cells through the activation of the $\alpha$-fetoprotein promoter[44]. Effective treatment of glial tumours in animals has also been achieved using adenoviral vector-mediated gene transfer of cytochrome P450 2B1 which converts the inactive prodrug cyclophosphamide to a cytotoxic form[45].

A development of VDEPT strategy involves tumour-specific expression of a chimeric gene comprised of the transcriptional regulatory sequence (TRS) of a tumour antigen and a non-mammalian gene. The TRS normally regulates expression of a tumour-associated antigen (human carcinoembryonic antigen gene (CEA)) and the gene it regulates in the artificial construct is *E. coli* cytosine deaminase (*CD*). Expression of *CD* occurs only in cells naturally transcribing CEA and results in conversion of the non-toxic prodrug 5-fluorocytosine (5-FCyt) to the toxic anabolite, 5-fluorouracil (5-FUra). Delivery of the chimeric gene *via* a replication-defective retroviral vector promotes regression of human liver xenografts and it is noteworthy that regression appears complete when only a small proportion (~4%) of tumour cells express the *CD* gene[46].

The variation in CD44 isoforms expressed as a result of alternative splicing in many metastatic tumours provides a further potential target for CD-mediated therapy. The *E. coli* gene has been fused to CD44 exons that undergo alternative splicing and shown to be activated as a result of this process, giving rise to the synthesis of functional enzyme *in vitro*[47].

## ANTIBODY-DIRECTED THERAPY

A variety of monoclonal antibodies have been shown to exert anti-tumour activities. Thus, murine anti-HER2 antibodies inhibit the growth of HER2-overexpressing cells *in vitro* and *in vivo*[48]. Chimeric mouse/human anti-CD19 antibodies reduce the size of human B lymphomas tumours in *scid/scid* mice[49] and in the same model system chimeric c30.6 antibodies that detect an antigen mainly expressed on colorectal adenocarcinoma cells greatly reduce tumour burden[50]. Monoclonal antibodies have also been frequently employed for tumour cell targeting of cytotoxic agents, including doxorubicin, methotrexate, ricin A chain toxin, *Pseudomonas* exotoxin and radioisotopes[51]. One method utilizes immunoliposomes to enhance drug delivery by incorporating into the lipid vehicle a monoclonal antibody directed against a specific tumour-associated antigen. Anti-HER2 immunoliposomes containing doxorubicin administered *in vivo* in *scid* mice bearing human breast tumour (BT-474) xenografts are specifically cytotoxic[52].

Bacterial and plant toxins expressed as part of a chimeric molecule that includes a binding site for a specific cell surface antigen offer an alternative cytotoxic approach. Antibodies specific for receptors have been linked to ricin A chain and to *Pseudomonas aeruginosa* exotoxin A and this method has been used to target immunotoxin gene products and also cytotoxic T cells to HER2-expressing cells with cytotoxic effects both *in vitro* and in nude mice[53]. This type of application is potentially powerful because bacterial toxins may be employed that kill target cells by direct lysis of the plasma membrane, e.g. the $\delta$-endotoxins of *Bacillus thuringiensis*, rather than requiring uptake into the cell of the cytotoxic agent.

An alternative strategy to VDEPT for targeting tumours with prodrug activating enzymes is to conjugate the enzyme to antibodies directed against tumour-associated proteins. After administration and binding to the target tumour-associated antigen, unbound conjugate is either allowed to clear naturally from the plasma or removed by use of a "clearance" antibody, before administration of the prodrug. This method of antibody-directed enzyme prodrug therapy (ADEPT) reduces non-specific toxicity and a number of systems have been developed that are effective against human cancer cell lines *in vitro*[54]. Tumour regression in mice has been achieved using some ADEPT systems in, for example, human colon carcinoma xenografts using $\beta$-glucuronidase active against a doxorubicin derivative[55] and $\beta$-lactamase with activity against a cephalosporin derivative[56].

A major limitation of the usefulness of many ADEPT systems is the immunogenicity of the bacterial enzymes used and attempts have been made to reduce this problem by synthesizing humanized catalytic antibodies ("abzymes") to function as the enzymatic component. These include an antibody-directed abzyme prodrug therapy (ADAPT) system using a carbamate prodrug that has cytotoxic efficacy against a human colon carcinoma cell line[57].

*References*


[1] **Scanlon, K.J. et al. (1995) FASEB J. 9, 1288–1296.**
[2] Benz, C.C. et al. (1995) Gene 159, 3–7.
[3] Sadano, H. et al. (1990) FEBS Lett. 271, 23–27.
[4] Ebbinghaus, S.W. et al. (1996) Gene Therapy 3, 287–297.
[5] **Long, D.M. and Uhlenbeck, O.C. (1993) FASEB J. 7, 25–30.**
[6] Kronenwett, R. et al. (1996) J. Mol. Biol. 259, 632–644.
[7] Sun, L.-Q. et al. (1994) Proc. Natl Acad. Sci. USA 91, 9755–9759.
[8] Feng, M. et al. (1995) Cancer Res. 55, 2024–2028.
[9] Dorai, T. et al. (1994) Mol. Pharmacol. 46, 437–444.
[10] Pace, U. et al. (1994) Cancer Res. 54, 6365–6369.
[11] Kozu, T. et al. (1996) J. Can. Res. Clin. Oncol. 122, 254–256.
[12] Kiehntopf, M. et al. (1994) EMBO J. 13, 4645–4652.
[13] Scanlon, K.J. et al. (1994) Proc. Natl Acad. Sci. USA 91, 11123–11127.
[14] Bertram, J. et al. (1995) Anti-Cancer Drugs 6, 124–134.
[15] **Rosenberg, S.A. (1992) J. Clin. Oncol. 10, 180–199.**
[16] Rollins, B.J. and Sunday, M.E. (1991) Mol. Cell. Biol. 11, 3125–3131.
[17] Leone, A. et al. (1991) Cell 65, 25–35.
[18] Giancotti, F.G. and Ruoslahti, E. (1990) Cell 60, 849–859.
[19] Ge, L. et al. (1995) J. Neuro-Oncol. 26, 251–257.
[20] Yamamoto, H. et al. (1996) Int. J. Cancer 65, 519–524.
[21] Vleminckx, K.L. et al. (1994) Cancer Res. 54, 873–877.
[22] DeClerck, Y.A. et al. (1992) Cancer Res. 52, 701–708.
[23] Roth, J. et al. (1996) Proc. Natl Acad. Sci. USA 93, 4781–4786.
[24] Dawson, D.W. et al. (1995) Proc. Am. Assoc. Cancer Res. 36, 88.
[25] Chang, M.W. et al. (1996) Science 267, 518–522.
[26] Xu, H.J. et al. (1996) Cancer Res. 56, 2245–2249.
[27] Naviaux, R.K. et al. (1996) J. Virol. 70, 5701–5705.
[28] Hamada, K. et al. (1996) Cancer Res. 56, 3047–3054.
[29] Clayman, G.L. et al. (1995) Cancer Res. 55, 1–6.
[30] Roth, J.A. (1996) Gene Therapy 7, 1013–1030.
[31] Spillare, E.A. et al. (1996) Mol. Carcinogenesis 6, 53–60.
[32] Zhang, Y. et al. (1995) Oncogene 10, 1947–1954.
[33] Ealovega, M.W. et al. (1996) Cancer Res. 56, 1965–1969.
[34] **Rosenberg, S.A. (1996) Annu. Rev. Med. 47, 481–491.**
[35] Huber, B.E. et al. (1991) Proc. Natl Acad. Sci. USA 88, 8039–8043.
[36] Melioli, G. et al. (1996) J. Immunotherapy, 19, 224–230.
[37] Belldegrun, A. et al. (1996) J. Immunotherapy, 19, 149–161.
[38] Hurford, R.K. et al. (1995) Nature Genet. 10, 430–435.
[39] Cao, X. et al. (1996) Gene Therapy 3, 421–426.
[40] Wei, K. et al. (1996) Gene Therapy 3, 531–541.
[41] Abina, M.A. et al. (1996) Gene Therapy 3, 212–216.
[42] Maron, A. et al. (1996) Gene Therapy 3, 315–322.
[43] Izquierdo, M. et al. (1996) Gene Therapy 3, 491–495.
[44] **Miller, A.D. (1992) Nature 357, 455–460.**
[45] Manome, Y. et al. (1996) Gene Therapy 3, 513–520.
[46] Huber, B.E. et al. (1995) Advanced Drug Delivery Rev. 17, 279–292.
[47] Asman, D.C. et al. (1995) J. Neuro-Oncol. 26, 243–250.
[48] Park, J.W. et al. (1995) Proc. Natl Acad. Sci. USA 92, 1327–1331.

[49] Pietersz, G.A. et al. (1995) Cancer Immunol. Immunother. 41, 53–60.
[50] Mount, P.F. et al. (1994) Cancer Res. 54, 6160–6166.
[51] **Pietersz, G.A. and Krauer, K. (1994) J. Drug Targeting 2, 183–215.**
[52] Hung, M.-C. et al. (1995) Gene 159, 65–71.
[53] Wels, W. et al. (1995) Gene 159, 73–80.
[54] **Melton, R.G. and Sherwood, R.F. (1996) J. Natl Cancer Inst. 88, 153–165.**
[55] Meyer, D.L. et al. (1993) Cancer Res. 53, 3956–3963.
[56] Bosslet, K. et al. (1994) Cancer Res. 54, 2151–2159.
[57] Wentworth, P. et al. (1996) Proc. Natl Acad. Sci. USA 93, 799–803.

# Section II

# TABLES

# Tables

**Table I** *Oncogenes transduced by retroviruses*

| Gene/locus | Activating virus | Associated tumours |
|---|---|---|
| ***Abl*** | Ab-MuLV/HZ2-FeSV | T lymphoid/sarcoma |
| *Akt* | AKT8 | Thymoma |
| ***Cbl*** | Cas NS-1 | B lymphomas |
| *Cyl-1* | MuLV | Lymphomas |
| ***Crk*** | ASV CT10 | Sarcoma |
| ***ErbA/ErbB*** | AEV-ES4 | Erythroid |
| ***ErbB*** | ALV/RPL25/RPL28 | Erythroid |
| ***Ets*** | AEV-E26 | Erythroid |
| ***Fgr*** | GR-FeSV | Sarcoma |
| ***Fms*** | SM-FeSV and HZ5-FeSV | Sarcoma |
| ***Fos*** | FBJ and FBR MuSV | Sarcoma |
| ***Fps/Fes*** | FSV | Sarcoma |
| ***Jun*** | ASV 17 | T lymphomas |
| ***Kit*** | HZ4-FeSV | Sarcoma |
| *Maf* | AS42 | Sarcoma |
| ***Mos*** | Mo-MuSV | B lymphoid/sarcoma |
| *Mpl* | MyLV | Erythroid |
| ***Myc*** | ALV, MuLV, REV, FeLV | T and B cell lymphomas |
| ***Myb*** | MuLV, ALV | B lymphoid, myeloid |
| ***Raf/Mil/****Mht* | MuSV | Carcinoma/lymphoma |
| ***Hras*** | ALV | Nephroblastic |
| ***Kras*** | F-MuLV | Erythroid |
| *Qin* | ASV 31 | Sarcoma |
| ***Rel*** | REV | B lymphomas |
| ***Ros*** | ASV UR2 | Sarcoma |
| *Ryk* | RPL30 | Sarcoma |
| ***Sea*** | AEV-S13 | Erythroid/sarcoma |
| ***Sis*** | SSV | Glioblastoma |
| ***Ski*** | SKV | Carcinoma |
| ***Src*** | RSV | Sarcoma |
| ***Yes*** | Esh and Y73 | Sarcoma |

Ab-MuLV: Abelson murine leukaemia virus; AEV: avian erythroblastosis virus; AKT8: leukaemia virus isolated from lymphomatous AKR mice; ALV: avian leukosis virus; ASV: avian sarcoma virus; FeLV: feline leukaemia virus; F-MuLV: Friend murine leukaemia virus; FSV: Fujinami sarcoma virus; GaLV: gibbon ape leukaemia virus; G-MuLV: Gross murine leukaemia virus; GR-FeSV: Gardner–Rasheed feline sarcoma virus; HZ4-FeSV: Hardy–Zuckerman 4 feline sarcoma virus; Mo-MuLV: Moloney murine leukaemia virus; MMTV: mouse mammary tumour virus; MuSV: murine sarcoma virus; MyLV: myeloproliferative leukaemia virus; REV: reticuloendotheliosis virus; RPL30: acute avian retrovirus; RSV: Rous sarcoma virus; SFFV: spleen focus forming virus; SKV: Sloan Kettering virus; SM-FeSV: Susan McDonough feline sarcoma virus; SSV: simian sarcoma virus. Genes shown in bold type are described in individual entries in Section III. For summaries/references for genes shown in plain type see R. Hesketh, The Oncogene Handbook, Academic Press, 1994.

**Table II** *Oncogenes activated by retroviral insertion*

**A. Mouse mammary tumour virus**

| Gene/locus | Chromosomal location | Activating virus/System | Associated tumours | Refs |
|---|---|---|---|---|
| ***Hst-1***/kFGF | 7 | BR6 | Mammary | 1 |
| ***Wnt-1*** (*Int-1*) and *Int-2* | 15 | BALB/cfC3H; BR6; C3H; GR; GRf; C3Hf; *Mus cervicolor* | Mammary | 2 |
| *Int-3* | 17 | BR6; Czech II | Mammary | 3 |
| ***Wnt-3*** (*Int-4*) | 11 | BALB/cfC3H; GR | Mammary | 4 |
| *Int-5* | 9 | BALB/c | Mammary | 5 |
| *Tpl-2/Cot* | | GR | Mammary | 6 |

**B. Murine leukaemia viruses**

| Gene/locus | Chromosomal location | Activating virus/System | Associated tumours | Refs |
|---|---|---|---|---|
| *Ahi-1* | 10 | Ab-MuLV | Pre-B cell | 7 |
| ***Akt*** | 12 | MuLV | Leukemia | 8 |
| *Bla-1* | ? | Eμ-*myc* transgenics | B cell | 9 |
| *Bmi-1/Bup* | 2 | Eμ-*myc* transgenics | B cell | 10 |
| *Pal-1* | 5 | Eμ-*myc* transgenics | B cell | 11 |
| *Gfi-1* | 5 | Eμ-*Lmyc/pim-1* transgenics | T cell | 12 |
| *CSF-1* | 3 | BALB/c eco | Monocytic | 13 |
| *Dsi-1* | 4 | Mo-MuLV (rat) | T cell | 14 |
| *Evi-2* | 11 | MuLV (BXH-2) | Myeloid | 15 |
| *Evi-5* | 5 | AKXD T cell | | 16 |
| *Hoxa7, Hoxa9* | 6 | MuLV (BXH-2) | Myeloid | 17 |
| *Meis1, Evi-9* | 11 | MuLV (BXH-2) | Myeloid | 18 |
| ***Nf-1*** | 11 | MuLV (BXH-2) | Myeloid | 19 |
| *Fim-1* | 13 | F-MuLV | Myeloid | 20 |
| *Fim-2*/***Fms*** | 18 | F-MuLV | Myeloid | 21 |
| *Fim-3* (or *Evi-1* or *CB-1*) | 3 | F-MuLV<br>Cas-Br-E MuLV | Myeloid<br>Non-T/B cell leukaemias | 22 |
| *Fis-1* (or *Cyl-1*) | 7 | F-MuLV | Myeloid, lymphoid | 23 |
| *Fli-1* | 9 | F-MuLV<br>Cas-Br-E MuLV | Erythroid<br>Non-T/B cell leukaemias | 24 |
| *Fli-2* | | F-MuLV | Erythroid | 25 |
| *Gin-1* | 19 | Gross A | T cell | 26 |
| ***Lck*** | 4 | Mo-MuLV | T cell | 27 |
| *Mlvi-1* (or *Pvt*, *Mis-1* or *RMO-int-1*) | 15 | AKR; AKXD; Mo-MuLV (rat) | T cell | 28 |
| *Mlvi-2, -3, -4* | 15 | Mo-MuLV (rat) | T cell | 29 |
| ***Myb*** | 10 | Ab-MuLV;<br>Cas-Br-M | Myeloid<br>NFS-60 cell line | 30 |
| ***Myc*** | 15 | AKR; AKXD; Gross A; MCF247; MCF69L1; Mo-MuLV (rat); Soule; Eμ-*pim*-1 transgenics | T cell | 31 |
| ***Nmyc-1*** | 12 | MCF247; Mo-MuLV Eμ-*pim*-1 transgenics | T cell | 32 |
| ***Pim-1*** | 17 | AKR; AKXD; MCF247; MCF1233; MCF69L1<br>AKXD<br>ΔMo-MuLV + SV | T cell<br><br>Non-T cell<br>B-lymphoblastic | 33 |

**Table II** *Continued*

**B. Murine leukaemia viruses – continued**

| Gene/locus | Chromosomal location | Activating virus/System | Associated tumours | Refs |
|---|---|---|---|---|
| *Pim-2/Tic-1* | 17 | AKXD; Mo-MuLV; Transplanted | T cell | 34 |
| ***Hras*** | 7 | Mo-MuLV | T cell | 35 |
| ***Kras*** | 6 | F-MuLV | Myeloid | 36 |
| *Sic-1* | 9 | Cas-Br-E MuLV | Non-B, non-T cell | 37 |
| *Spi-1* | 2 | SFFV | Erythroid | 38 |
| *P53* | 11 | F-MuLV; SFFV<br>Ab-MuLV | Erythroid<br>Lymphoid | 39 |
| *Til-1* | 17 | Mo-MuLV | T cell | 40 |
| *Tpl-1/**Ets-1*** | 9 | Mo-MuLV | T cell | 41 |
| *Vin-1* | 6 | RadLV | T cell | 42 |

**C. Avian retroviruses**

| Gene/locus | Activating virus/System | Associated tumours | Refs |
|---|---|---|---|
| *Bic* | UR2AV + RAV-2 | Lymphomas | 43 |
| ***Erbb*** | RAV-1 | Erythroblastosis | 44 |
| ***Myb*** | RAV-1; EU-8; UR2AV + RAV-2 | Lymphomas | 45 |
| ***Myc*** | ALV; REV (CSV); RPV<br>RPV<br>REV | B-lymphomas<br>Adenocarcinoma<br>T-lymphoma | 46 |
| *Nov* | MAV1 | Myeloblastosis | 47 |
| ***Hras*** | MAV | Nephroblastoma | 48 |
| *Blym* | ALV | B-lymphoma | 49 |
| ***Rel*** | ALV | B-lymphoma | 50 |

**D. Other systems**

| Gene/locus | Activating virus/System | Associated tumours | Refs |
|---|---|---|---|
| Erythropoietin receptor | SFFV | Erythroid | 51 |
| *Fit-1* | FeLV | T cell | 52 |
| *Flvi-1* | FeLV | T cell | 53 |
| *His-1, His-2* | CasBrMo-MuLV (IL-3-dependent) | Myeloid | 54 |
| *Hox-2.4* | IAP | WEHI-3B | 55 |
| *IL2* | GaLV | T cell line | 56 |
| *IL3* | IAP | WEHI-3B | 57 |
| *IL2R* | IAP | Lymphoma cell line | 58 |
| ***Mos*** | IAP | Plasmacytoma | 59 |
| ***Myc*** | IAP | Plasmacytoma | 60 |
| ***Myc*** | F-MuLV | T cell | 61 |
| ***Myc*** | Retrotransposon | Canine | 62 |
| ***Pim-1*** | F-MuLV | T cell | 63 |
| *Tsc2* | IAP | Hereditary renal carcinoma | 64 |
| ***Tiam-1*** | | T cell | 65 |

Abbreviations are given in the legend to Table I. RadLV: BL.VL3 radiation leukaemia virus; IAP: intracisternal A particle; MAV: myeloblastosis-associated virus. References for genes shown in bold type are given in the individual entries in Section III. Summaries/references for genes shown in plain type are given below.

Based on Peters, G. (1990) Oncogenes at viral integration sites. Cell Growth Differ. 1, 503–510.

*References*

[1] Brustle, O. et al. (1992) Oncogene 7, 1177–1183.
[2] Roelink, H. et al. (1992) Oncogene 7, 487–492.
[3] Robbins, J. et al. (1992) J. Virol. 66, 2594–2599.
[4] Roelink, H. et al. (1992) Oncogene 7, 487–492.
[5] Morris, V.L. et al. (1991) Oncogene Res. 6, 53–63.
[6] Erny, K.M. et al. (1996) Oncogene 13, 2015–2020.
[7] Poirer, Y. et al. (1988) J. Virol. 62, 3985–3992.
[8] Altomare, D.A. et al. (1995) Oncogene 11, 1055–1060.
[9] van Lohuizen, M. et al. (1991) Cell 65, 737–752.
[10] Tetsu, O. et al. (1996) Biochim. Biophys. Acta, 1305, 109–112; Alkema, M.J. et al. (1995) Nature 374, 724–727; Haupt, Y. et al. (1991) Cell 65, 753–763.
[11] Baxter, E.W. et al. (1996) J. Virol. 70, 2095–2100.
[12] Gilks, C.B. et al. (1995) Endocrinology 136, 1805–1808; Grimes, H.L. et al. (1996) Mol. Cell. Biol. 16, 6263–6272.
[13] Baumbach, W.R. et al. (1988) J. Virol. 62, 3151–3155.
[14] Vijaya, S. et al. (1987) J. Virol. 61, 1164–1170.
[15] Buchberg, A.M. et al. (1990) Cell. Biol. 10, 4658–4666.
[16] Liao, X. et al. (1995) J. Virol. 69, 7132–7137.
[17] Nakamura, T. et al. (1996) Nature Genet. 12, 149–153.
[18] Moskow, J.J. et al. (1995) Mol. Cell. Biol. 15, 5434–5443.
[19] Largaespada, D.A. et al. (1995) J. Virol. 69, 5095–5102.
[20] Sola, B. et al. (1988) J. Virol. 62, 3973–3978.
[21] Sola, B. et al. (1988) J. Virol. 62, 3973–3978.
[22] Bartholomew, C. and Clark, A.M. (1994) Oncogene 9, 939–942; Matsugi, T. et al. (1995) Oncogene 11, 191–198.
[23] Lammie, G.A. et al. (1992) Oncogene 7, 2381–2387.
[24] Howard, J.C. et al. (1993) Oncogene 8, 2721–2729;
[25] Lu, S.J. et al. (1994) Proc. Natl Acad. Sci. USA 91, 8398–8402.
[26] Villemur, R. et al. (1987) Mol. Cell. Biol. 7, 512–522.
[27] Shin, S. and Steffen, D.L. (1993) Oncogene 8, 141–149.
[28] Mengle-Gaw, L. and Rabbitts, T.H. (1987) EMBO J. 6, 1959–1965; Tsichlis, P.N. et al. (1989) Proc. Natl Acad. Sci. USA 86, 5487–5491; Villeneuve, L. et al. (1986) Mol. Cell. Biol. 6, 1834–1837.
[29] Tsichlis, P.N. et al. (1990) J. Virol. 64, 2236–2244; Kozak, C.A. et al. (1985) Mol. Cell. Biol. 5, 894–897.
[30] Baluda, M.A. and Reddy, E.P. (1994) Oncogene 9, 2761–2774.
[31] Verbeek, S. et al. (1991) Mol. Cell. Biol. 11, 1176–1179.
[32] van Lohuizen, M. et al. (1989) EMBO J. 8, 133–136; Moroy, T. et al. (1991) Oncogene 6, 1941–1948.
[33] van der Lugt, N.M.T. et al. (1995) EMBO J. 14, 2536–2544.
[34] van der Lugt, N.M.T. et al. (1995) EMBO J. 14, 2536–2544.
[35] Ihle, J.N. et al. (1989) J. Virol. 63, 2959–2966.
[36] Trusko, S.P. et al. (1989) Nucleic Acids Res. 17, 9259–9265.
[37] Bergeron, D. et al. (1991) J. Virol. 65, 7–15.
[38] Goebl, M.G. et al. (1990) Cell 61, 1165–1166; Pahl, H.L. et al. (1993) J. Biol. Chem. 268, 5014–5020; Ray, D. et al. (1992) Mol. Cell. Biol. 12, 4297–4304.
[39] Munroe, D.G. et al. (1990) Mol. Cell. Biol. 10, 3307–3313.
[40] Stewart, M. et al. (1996) J. Gen. Virol. 77, 443–446.

[41] Bellacosa, A. et al. (1994) J. Virol. 68, 2320–2330.
[42] Hanna, Z. et al. (1993) Oncogene 8, 1661–1666; Tremblay, P.J. et al. (1992) J. Virol. 66, 1344–1353 and 5176.
[43] Clurman, B.E. and Hayward, W.S. (1989) Mol. Cell. Biol. 9, 2657–2664.
[44] Vennstrom, B. et al. (1994) Oncogene 9, 1307–1320.
[45] Pizer, E. and Humphries, E.H. (1989) J. Virol. 63, 1630–1640.
[46] Chen, C. et al. (1989) J. Virol. 63, 5092–5100.
[47] Joliot, V. et al. (1992) Mol. Cell. Biol. 12, 10–21.
[48] Westaway, D. et al. (1986) EMBO J. 5, 301–309.
[49] Diamond, A. et al. (1984) Science 225, 516–519.
[50] Kabrun, N. et al. (1990) Mol. Cell. Biol. 10, 4788–4794.
[51] Lacombe, C. et al. (1991) J. Biol. Chem. 266, 6952–6956.
[52] Tsujimoto, H. et al. (1993) Virology 196, 845–848.
[53] Levesque, K.S. et al. (1991) Oncogene 6, 1377–1379.
[54] Askew, D.S. et al. (1994) Mol. Cell. Biol. 14, 1743–1751.
[55] Blatt, C. et al. (1988) EMBO J. 7, 4283–4290; Kongsuwan, K. et al. (1989) Nucleic Acids Res. 17, 1881–1892.
[56] Chen, S.J. et al. (1985) Proc. Natl Acad. Sci. USA 82, 7284–7288.
[57] Algate, P.A. and McCubrey, J.A. (1993) Oncogene 8, 1221–1232.
[58] Kono, T. et al. (1990) Proc. Natl Acad. Sci. USA 87, 1806–1810.
[59] Horowitz, M. et al. (1984) EMBO J. 3, 2937–2941.
[60] Connelly, M.A. et al. (1994) Proc. Natl Acad. Sci. USA 91, 1337–1341.
[61] Dreyfus, F. et al. (1990) Leukemia 4, 590–594.
[62] Katzir, N. et al. (1985) Proc. Natl Acad. Sci. USA 82, 1054–1058.
[63] van der Lugt, N.M.T. et al. (1995) EMBO J. 14, 2536–2544.
[64] Kumar, A. et al. (1995) Hum. Mol. Genet. 4, 2295–2298; Xiao, G.-H. and Yeung, R.S. (1995) Oncogene 11, 81–87.
[65] Habets, G.G.M. et al. (1994) Cell 77, 537–549.

**Table III** *Oncogenes at chromosomal translocations*

| Gene (chromosome) | Translocation | Leukaemia |
|---|---|---|
| ***ABL*** (9q34.1)<br>***BCR*** (22q11) | (9;22)(q34;q11) | CML |
| ***ABL*** (9q34.1) [1]<br>*TEL* (12p13) | (9;12)(q34;p13) | Common ALL |
| *AML1* (21q22) [2,3]<br>*ETO/MTG8* (8q22) | (8;21)(q22;q22) | AML |
| *AML1* (21q22) [4]<br>*EAP* (3q26)<br>*EVI1* (3q26) [5,6]<br>*MDS1* (3q26) [7] | (3;21)(q26;q22)<br><br>inv(3)(q22q26) | CML, myelodysplastic syndrome (MDS) |
| *AML1* (21q22) [8]<br>*TEL* (12p13) | (12;21)(p13;q22) | ALL |
| *ATF1* (12q13) [9]<br>*EWS* (22q12) | (12;22)(q13;q12) | Malignant melanoma of soft parts (MMSP) |
| ***BCL1*** (11q13.3)<br>IgH (14q32) | (11;14)(q13;q32) | B cell lymphomas, B-CLL, multiple myeloma |
| ***BCL2*** (18q21.3)<br>IgH (14q32) | (14;18)(q32;q21) | Non-Hodgkin's lymphoma |
| ***BCL3*** (19q13.1)<br>IgH (14q32) | (14;19)(q32;q13) | B-CLL |
| *BTG1* (12q22) [10]<br>Deletion | (8;12)(q24;q22) | B-CLL |
| *CBFB* (16q22) [11]<br>*MYH11* (16p13.12–p13.13) | inv(16)(p13q22) | AML |
| *CBP* (16p11) [12]<br>*MOZ* (8p11) | (8;16)(p11;p13) | AML |
| ***CCND3*** (6p21)<br>*PTH* (11p15) | inv(11)(p15q13) | Parathyroid adenoma |
| ***CCND3*** (6p21)<br>IgH (14q32) | (11;14)(q13;q32) | Non-Hodgkin's lymphoma |
| *CHN* (9q22–31) [13]<br>*EWS* (22q12) | (9;22)(q22–31;q11–12) | Myxoid chondrosarcoma |
| *CHOP* (12q13) [14,15]<br>*TLS/FUS* (16p11) | (12;16)(q13;p11) | Myxoid liposarcoma |
| *CHOP* (12q13) [16]<br>*EWS* (22q12) | (12;22)(q13;q12) | Myxoid liposarcoma |
| *D10S170* (10q21)<br>*ELE1* (10) | inv10(q11.2;q21) | Papillary thyroid carcinoma |
| ***ERG*** (21q22.3)<br>*EWS* (22q12) | (22;21)(q12;q22) | Ewing's sarcoma |
| ***FLI1*** (11q24)<br>*EWS* (22q12) | (11;22)(q24;q12) | Ewing's sarcoma |
| ***ERG*** (21q22.3) [17]<br>*TLS/FUS* (16p11) | (16;21)(p11;q22) | CML |
| *EWS* (22q12)<br>*ETV1* (7p22) [18] | (7;22)(p22;q12) | Ewing's sarcoma |

**Table III** *Continued*

| Gene (chromosome) | Translocation | Leukaemia |
|---|---|---|
| *HLF* (17q22)[19] | (17;19)(q22;p13) | Pre-B ALL |
| *E2A* (19p13.3) | | |
| *HOX11* (10q24)[20] | (10;14)(q24;q11) | T-ALL |
| *TCRD* (14q11.2) | | |
| *TCRB* (7q35) | (7;10)(q35;q24) | |
| *IL2* (4q26)[21] | (4;16)(q26;p13) | T cell lymphoma |
| *BCM* (16p13) | | |
| *IL3* (5q31)[22] | (5;14)(q31;q32) | Acute pre-B cell |
| IgH (14q32) | | |
| ***INK4A*** (9p21–p22)[23] | (9;14)(p21–p22;q11) | ALL |
| *TCRA* (14q11) | | |
| *LAZ3/BCL5/BCL6* (3q27)[24] | (3;14)(q27;q32) | Non-Hodgkin's lymphoma |
| IgH (14q32) | (3;22)(q27;q11) | |
| *LAZ3* (3q27)[25] | (3;4)(q27;p11) | Non-Hodgkin's lymphoma |
| *TTF* (4p11) | | |
| ***LCK*** (1p34) | (1;7)(p34;q35) | T-ALL |
| *TCRB* (7q35) | | |
| *LMO1 /RBTN1* (11p15)[26] | (11;14)(p15;q11) | T-ALL |
| *LMO2/RBTN2* (11p13) | (11;14)(p13;q11) | |
| *TCRA* (14q11) | | |
| *TCRB* (7q35) | (7;11)(p35;p13) | |
| *TCRD* (14q11.2) | | |
| *LPC* (11q23)[27] | (11;14)(q23;q32) | Large cell lymphoma |
| IgH (14q32) | | |
| *LYL1* (19p13.2)[28] | (7;19)(q35;p13) | T cell leukaemia line |
| *TCRB* (7q35) | | |
| *LYT10* (*NFKB2*: 10q24)[29] | (10;14)(q24;q32) | B lymphoma |
| IgH (14q32) | | T lymphoma |
| | | Multiple myeloma |
| *MLL /ALL1/HRX* (11q23)[30] | (4;11)(q21;q23) | ALL |
| *MLL*[31] | (1;11) | Gastric carcinoma cell |
| *AF4* (4q21)[32] | | |
| *AF5α* (5q12)[33] | | AML |
| *AF6* (6q27)[34] | (6;11)(q27;q23) | AML |
| *AF9* (9q22) | (9;11)(q22;q23) | AML |
| *AF10* (10p13) | | |
| *CALM* (11q14)[35] | (10;11)(p13;q14) | U937 cell line |
| *AF17* (17q21)[36] | (11;17)(q23;q21) | AML |
| *AF1P* (1p32)[37] | (1;11)(p32;q23) | AML |
| *ENL* (19p13.3)[38] | (11;19)(q23;p13) | AML |
| *MLL /ALL1/HRX* (11q23) | | |
| *ELL/MEN* (19p13.1)[39] | (11;19)(q23;p13.1) | AML |
| ***MYC*** (8q24) | (8;14)(q24;q32) | Burkitt's lymphoma |
| IgH (14q32) | | |
| Igλ (22q11) | (8;22)(q24;q11) | |
| Igκ (2p12) | (2;8)(p12;q24) | |
| *TCRA* (14q11)[40] | (8;14)(q24;q11) | B cell lymphoma |
| ***MYC*** (8q24)[41] | (8;14)(q24;q32) | ALL |
| ***BCL2*** (18q21.3) | (14;18)(q32;q21) | |
| IgH (14q32) | | |

**Table III** *Continued*

| Gene (chromosome) | Translocation | Leukaemia |
|---|---|---|
| *NPM* (5q25)[42]<br>*ALK* (2p23) | (2;5)(q25;p23) | Non-Hodgkin's lymphoma |
| *NPM* (5q25)[43]<br>*MLF1* (3q34) | (3;5)(q25.1;q34) | AML<br>Myelodysplastic syndrome |
| *NUP98* (11p15)[44,45]<br>*HOXA9* (7p15) | (7;11)(p15;p15) | AML, CML |
| *NUP214/CAN* (6p23)[46]<br>*DEK* (9q34)<br>*SET* (9q34)[47] | (6;9)(p23;q34) | AML |
| *PAX3* (2q35)[48]<br>*FKHR/ALV* (13q14) | (2;13)(q35;q14) | Alveolar rhabdomyosarcoma |
| *PAX7* (1p36)[49]<br>*FKHR/ALV* (13q14) | (1;13)(p36;q14) | Alveolar rhabdomyosarcoma |
| *PBX1/PRL* (1q23)[50]<br>*E2A* (19p13.3) | (1;19)(q23;p13) | Pre-B ALL |
| *PDGFRB* (5q33)[51]<br>*TEL* (12p13) | (5;12)(q33;p13) | CML |
| *PLZF* (11q23)[52]<br>*RARA* (17q21) | (11;17)(q23;q21) | APL |
| *PML* (15q21)[53]<br>*RARA* (17q21.1) | (15;17)(q11–22) | APL |
| *RCK* (11q23.3)[54]<br>*AFX1* (Xq13) | (11;14)(q23;q32)<br>(X;11)(q13;q23) | ALL<br>ALL |
| ***REL*** (2p13)[55]<br>*NRG* (2p11.2–14) | inv(2(p13;p11.2–14) | NHL |
| ***RET*** (10q11.2)<br>Protein kinase A RIα (17q23) | (10;17)(q11.2;q23) | Papillary thyroid carcinoma |
| *SYT* (18q11.2)[56]<br>*SSX1* (Xp11.2)<br>*SSX2* (Xp11.2)[57] | (X;18)(p11.2;q11.2) | Synovial sarcoma |
| ***TAL1*** (1p32)<br>*TCRA* (14q11) | (1;14)(p32;q11) | T-ALL |
| ***TAL1*** (1p32)[58]<br>*TCTA* (3p21) | (1;3)(p34;p21) | T-ALL |
| *TAL2* (9q34)[59,60]<br>*TCRB* (7q35) | (7;9)(q35;q34) | T-ALL |
| *TAN1* (9q34.3)[61]<br>Deletion | (7;9)(q34;q34) | T-ALL |
| *TCL1* (14q32.1)[62]<br>TCR-Cα (14q11) | inv14 & (14;14)(q11;q32) | T-CLL |
| *TCL1* (14q32.1)[63]<br>*TCRα/δ* (14q11)<br>*TCRβ* (7q35) | inv14(q11;q32.1)<br>(7;14)(q35;q32.1) | T-PLL<br>T cell leukaemia<br>T cell lymphoma |
| *TCRA* (14q11)[64]<br>IgH (14q32) | inv14(q11;q32) | T/B cell lymphoma |
| *TPR* (16q22)[65]<br>*HPR* (6p21) | (6;16)(p21;q22) | Prostate carcinoma |

**Table III** *Continued*

| Gene (chromosome) | Translocation | Leukaemia |
|---|---|---|
| ***TRK*** (1q23–1q24)[66]<br>*TPM3* (1q31) | inv1(q23;q31) | Colon carcinoma |
| ***WT1*** (11p13)[67]<br>*EWS* (22q12) | (11;22)(p13;q12) | Desmoplastic small round cell tumour (DSRCT) |
| *c6.1A/c6.1B* (Xq28)[68–70]<br>*TCRA* (14q11.2) | (X;14)(q28;q11) | T-PLL |

ALL: acute lymphoblastic leukaemia; AML: acute myeloid leukaemia; APL: acute promyelocytic leukaemia (a subtype of AML); ANLL: acute non-lymphocytic leukaemia; B-CLL: chronic B cell lymphocytic leukaemia; CML: chronic myelogenous leukaemia; T-ALL: acute T cell leukaemia; T-PLL: T cell prolymphocytic leukaemia.

Genes shown in bold type are described in individual sections in Section III.

### *References*


[1] Golub, T.R. et al. (1996) Mol. Cell. Biol. 16, 4107–4116.
[2] Chang, K.-S. et al. (1993) Oncogene 8, 983–988.
[3] Erickson, P.F. et al. (1994) Cancer Res. 54, 1782–1786.
[4] Nucifora, G. et al. (1993) Proc. Natl Acad. Sci. USA 90, 7784–7788.
[5] Tanaka, T. et al. (1995) Mol. Cell. Biol. 15, 2383–2392.
[6] Ogawa, S. et al. (1996) Oncogene 13, 183–191.
[7] Zent, C.S. et al. (1996) Proc. Natl Acad. Sci. USA 93, 1044–1048.
[8] Hiebert, S.W. et al. (1996) Mol. Cell. Biol. 16, 1349–1355.
[9] Fujimura, Y. et al. (1996) Oncogene 12, 159–167.
[10] Roualt, J.-P. et al. (1992) EMBO J. 11, 1663–1670.
[11] Liu, P. et al. (1993) Science 261, 1041–1044.
[12] Borrow, J. et al. (1996) Nature Genet. 14, 33–41.
[13] Clark, J. et al. (1996) Oncogene 12, 299–235.
[14] Crozat, A. et al. (1993) Nature 363, 640–644.
[15] Rabbitts, T.H. et al. (1993) Nature Genet. 4, 175–180.
[16] Panagopoulos, I. et al. (1996) Oncogene 12, 489–494.
[17] Ichikawa, H. et al. (1994) Cancer Res. 54, 2865–2868.
[18] Jeon, I.-S. et al. (1995) Oncogene 10, 1229–1234.
[19] Inaba, T. et al. (1996) Nature 382, 541–544.
[20] Zhang, N. et al. (1996) Oncogene 13, 1755–1763.
[21] Laabi, Y. et al. (1992) EMBO J. 11, 3897–3904.
[22] Meeker, T.C. et al. (1990) Blood 76, 285–289.
[23] Duro, D. et al. (1996) Cancer Res. 56, 848–854.
[24] Seyfert, V.L. et al. (1996) Oncogene 12, 2331–2342.
[25] Dallery, E. et al. (1995) Oncogene 10, 2171–2178.
[26] Fisch, P. et al. (1992) Oncogene 7, 2389–2397.
[27] Meerabux, J. et al. (1996) Cancer Res. 56, 448–451.
[28] Visvader, J. et al. (1991) Oncogene 6, 187–194.
[29] Fracchiolla, N.S. et al. (1993) Oncogene 8, 2839–2845.
[30] Prasad, R. et al. (1993) Cancer Res. 53, 5624–5628.
[31] Baffa, R. et al. (1995) Proc. Natl Acad. Sci. USA 92, 4922–4926.
[32] Corral, J. et al. (1993) Proc. Natl Acad. Sci. USA 90, 8538–8542.

[33] Taki, T. et al. (1996) Oncogene 13, 2121–2130.
[34] Tkachuk, D.C. et al. (1992) Cell 71, 691–700.
[35] Dreyling, M.H. et al. (1996) Proc. Natl Acad. Sci. USA 93, 4804–4809.
[36] Prasad, R. et al. (1993) Proc. Natl Acad. Sci. USA 91, 8107–8111.
[37] Bernard, O.A. et al. (1994) Oncogene 9,1039–1045.
[38] Nakamura, T. et al. (1993) Proc. Natl Acad. Sci. USA 90, 4631–4635.
[39] Shilatifard, A. et al. (1996) Science 271, 1873–1876.
[40] Park, J. K. et al. (1989) Genes Chromosom. Cancer 1, 15–22.
[41] Kiem, H. P. et al. (1990) Oncogene 5: 1815–1819.
[42] Morris, S.W. et al. (1994) Science 263, 1281–1284.
[43] Yoneda-Kato, N. et al. (1996) Oncogene 12, 265–275.
[44] Borrow, J. et al. (1996) Nature Genet. 12, 159–167.
[45] Nakamura, T. et al. (1996) Nature Genet. 12, 154–158.
[46] Fornerod, M. et al. (1996) Oncogene 13, 1801–1808.
[47] Adachi, Y. et al. (1994) J. Biol. Chem. 269, 2258–2262.
[48] Fredericks, W.J. et al. (1995) Mol. Cell. Biol. 15, 1522–1535.
[49] Davis, R.J. et al. (1994) Cancer Res. 54, 2869–2872.
[50] Dedera, D.A. et al. (1993) Cell 74, 833–843.
[51] Golub, T.R. et al. (1994) Cell 77, 307–316.
[52] Chen, Z. et al. (1993) EMBO J. 12, 1161–1167.
[53] Perez, A. et al. (1993) EMBO J. 12, 3171–3182.
[54] Akao, Y. et al. (1992) Cancer Res. 52, 6083–6087.
[55] Lu, D. et al. (1991) Oncogene 6, 1235–1241.
[56] Clark, J. et al. (1994) Nature Genet. 7, 502–508.
[57] de Leeuw (1995) Hum. Mol. Genet. 4, 1097–1099.
[58] Aplan, P.D. et al. (1995) Cancer Res. 55, 1917–1921.
[59] Xia, Y. et al. (1991) Proc. Natl Acad. Sci. USA 88, 11416–11420.
[60] Xia, Y. et al. (1994) Oncogene 9, 1437–1446.
[61] Ellisen, L.W. et al. (1991) Cell 66, 649–661.
[62] Mengle-Gaw et al. (1987) EMBO J. 6, 2273–2280.
[63] Fu, T. et al. (1994) Cancer Res. 54, 6297–6301.
[64] Denny, C.T. et al. (1986) Nature 320, 549–551.
[65] Veronese, M.L. et al. (1996) Cancer Res. 56, 728–732.
[66] Radice, P. et al. (1991) Oncogene 6, 2145–2148.
[67] Karnieli, E. et al. (1996) J. Biol. Chem. 271, 19304–19309.
[68] Fisch, P. et al. (1993) Oncogene 8, 3271–3276.
[69] Stern, M.H. et al. (1993) Oncogene 8, 2475–2483.
[70] Soulier, J. et al. (1994) Oncogene 9, 3565–3570.

**Table IV** *Tumour suppressor genes detected in human tumours*

| Gene | Chromosomal locus | Neoplasm |
|---|---|---|
| 3pK [1,2] | 3p21.3 | Lung carcinoma |
| ***APC*** | 5q21 | Colorectal cancer |
| *BAM22* [3] | 22q12 | Meningioma |
| *BCNS* [4] | 9q31 | Medulloblastoma |
| ***BRCA1*** | 17q21 | Familial breast cancer |
| ***BRCA2*** | 13q12–13 | Breast and ovarian carcinomas |
| *BRUSH1* [5] | 13q12–q13 | Breast carcinoma |
| *BWS* [6] | 11p15.5 | Wilms' tumour |
| *CMAR/CAR* [7] | 16q | Breast, prostate cancers |
| *CYLD1* [8] | 16q12–13 | Cylindromatosis |
| *DAN* [9] | 1p36.11–p36.13 | Neuroblastoma |
| ***DCC*** | 18q21 | Colon carcinoma |
| ***DPC4/SMAD4*** | 18q21.1 | Pancreatic carcinoma |
| **E-cadherin** | 16q22.1 | Breast, endometrial, ovarian and other carcinomas |
| ***E2F1*** | 20q11 | Lymphomas (murine) |
| *EXT1* [10] | 8q24.1 | Chondrosarcomas |
| *EXT2* [11] | 11p11–13 | Chondrosarcomas |
| *FHIT* [12]*/HRCA1* [13] | 3p14.2 | Hereditary renal carcinoma |
| *FWT1* | 17q12–q21 | Familial Wilms' tumour |
| *HUGL* [14] | 17p11.2–12 | ? |
| ***INK4A/MTS1/CDK41/ CDKN2*** | 9p21 | Many cancers |
| ***INK4B/MTS2*** | 9p21 | Non-small cell lung cancer and others |
| ***INK4C*** | 1p32 | Breast carcinoma |
| *IRF-1* [15] | 5q31.1 | Leukemia |
| *KAI1* [16] | 11p11.2 | Prostate |
| *LC1* [17] | 3p14–p21 | Lung carcinoma |
| *M6P/IGF2R* [18] | 6q26–27 | Breast cancer |
| ***MCC*** | 5q21 | Colon cancer, lung cancer? |
| *MEN1* [19] | 11q13 | Parathyroid, pancreatic and pituitary tumours |
| ***MLH1*** | 3p21.3–p23 | Hereditary non-polyposis colon cancer (HNPCC) |
| *MLM* [20] | 9p21 | Melanoma |
| ***MSH2*** | 2p22–p21 | Hereditary non-polyposis colon cancer (HNPCC) |
| *NB1* [21] | 1p36.1 | Neuroblastoma, pituitary and adrenal cortex tumours |
| ***NF1*** | 17q11.2 | Neurofibromatosis type 1 |
| ***NF2*** | 22q12 | Neurofibromatosis type 2 |
| *NME1, NME2 (NM23)* [22] | 17q21.3 | Neuroblastoma, colon carcinoma |
| *NRC1* [23] | 3p14–p12 | Non-papillary renal carcinoma |
| *OVCA1, OVCA2* [24,25] | 17p13.3 | Ovarian carcinoma |
| ***P53*** | 17p13.1 | Sarcomas, gliomas, carcinomas |
| ***PHB*** | 17q21 | Breast carcinoma |
| *PLANH1 (PAI-1)* [26] | 7q21–q31 | Ovarian carcinoma |
| *PMS1* (see ***MSH2***) | 2q31–33 | Hereditary non-polyposis colon cancer (HNPCC) |
| *PMS2* (see ***MSH2***) | 7p22 | Hereditary non-polyposis colon cancer (HNPCC) |
| *PRLTS* [27] (PDGFR β-like tumour suppressor) | 8p21.3–p22 | Hepatocellular carcinoma (HCC), colorectal cancer, non-small cell lung cancer |
| ***PTC*** | 9q22.3 | Nevoid basal cell carcinoma syndrome (NBCCS) |

**Table IV** *Continued*

| Gene | Chromosomal locus | Neoplasm |
|---|---|---|
| ***RB1*** | 13q14 | Retinoblastoma, sarcomas, carcinomas |
| ***TGFBR1*** | | Prostate and colon carcinoma, Kaposi's sarcoma |
| ***TGFBR2*** | 3p21 | Colon carcinoma, head and neck squamous cell carcinoma lines |
| *TSC1*[21] | 9q34 | Tuberous sclerosis |
| *TSC2*[28] | 16p13.3 | Tuberous sclerosis |
| ***VHL*** | 3p25–p26 | von Hippel–Lindau disease |
| ***WAF1*** | 6p21 | Prostate cancer |
| ***WT1*** | 11p12 | Wilms' tumour |
| α-Inhibin (*INHA*)[29] | | Mouse |
| *Mel-18*[30] | | Mouse |
| *PKR*[31] | | |
| [32] | 1p33–p35 | Breast carcinoma |
| [33] | 3p | Ovarian |
| [34] | 3p14.2 | Nasopharyngeal, gastric carcinoma |
| [35] | 5q35–qter | Hepatocellular carcinoma (HCC) without cirrhosis |
| [36,37] | 8p22, 8p21, 10q24, 16q | Prostate cancer |
| [38] | 12p | Non-small cell lung cancer |
| [39] | 13q14 | B cell chronic lymphocytic leukaemia |
| [40] | 6q22–23, 6q26–27, 15q11–21, 15q26–qter | Parathyroid adenomas |
| 322[38/41] | | Mouse |

Genes shown in bold type are described in individual sections in Section III. The remaining putative tumour suppressor genes have not been cloned.

## *References*


[1] Hibi, K. et al. (1994) Oncogene 9, 611–619.
[2] Sithandam, G. et al. (1996) Mol. Cell. Biol. 16, 868–876.
[3] Peyrard, M. et al. (1994) Hum. Mol. Genet. 3, 1393–1399.
[4] Bare, J.W. et al. (1993) Am. J. Hum. Genet. 53 (Suppl.), Abstract 274.
[5] Schott, D.R. et al. (1994) Cancer Res. 54, 1393–1396.
[6] Koufos, A. et al. (1993) Am. J. Hum. Genet. 44, 711–719.
[7] Durbin, H. et al. (1994) Genomics 19, 181–182.
[8] Biggs, P.J. et al. (1996) Oncogene 12, 1375–1377.
[9] Enomoto, H. et al. (1994) Oncogene 9, 2785–2791.
[10] Ahn, J. et al. (1995) Nature Genet. 11, 137–143.
[11] Stickens, D. et al. (1996) Nature Genet. 14, 25–32.
[12] Ohta, M. et al. (1996) Cell 84, 587–597.
[13] Boldog, F.L. et al. (1993) Proc. Natl Acad. Sci. USA 90, 8509–8513.
[14] Strand, D. et al. (1995) Oncogene 11, 291–301.
[15] Harada, H. et al. (1994) Oncogene 9, 3313–3320.
[16] Dong, J.-T. et al. (1995) Science 268, 884–886.
[17] Whang-Peng, J. et al. (1993) Genes Chromosom. Cancer 3, 168–188.
[18] Hankins, G.R. et al. (1996) Oncogene 12, 2003–2009.
[19] Larsson, C. et al. (1993) Nature 332, 85–87.
[20] Walker, G.J. et al. (1994) Oncogene 9, 819–824.
[21] Biegel, J.A. et al. (1993) Am. J. Hum. Genet. 52, 176–182.

[22] See Chapter 7.
[23] Sanchez, Y. et al. (1994) Proc. Natl Acad. Sci. USA 91, 3383–3387.
[24] Phillips, N.J. et al. (1996) Cancer Res. 56, 606–611.
[25] Schultz, D.C. et al. (1996) Cancer Res. 56, 1997–2002.
[26] Kerr, J. et al. (1996) Oncogene 13, 1815–1818.
[27] Fujiwara, Y. et al. (1995) Oncogene 10, 891–895.
[28] Wienecke, R. (1996) Oncogene 13, 913–923.
[29] Matzuk, M.M. et al. (1992) Nature 360, 313–319.
[30] Kanno, M. et al. (1995) EMBO J. 14, 5672–5678.
[31] **Proud, C.G. (1996) Trends Biochem. Sci. 20, 241–246.**
[32] Huynh, H.T. et al. (1995) Cancer Res. 55, 2225–2231.
[33] Rimessi, P. et al. (1994) Oncogene 9, 3467–3474.
[34] Kastury et al. (1996) Cancer Res. 56, 978–983.
[35] Ding, S.F. et al. (1993) Cancer Detect. Prev., 17, 405–409.
[36] Kagan, J. et al. (1995) Oncogene 11, 2121–2126.
[37] Gray, I.C. et al. (1995) Cancer Res. 55, 4800–4803.
[38] Takeuchi, S. et al. (1996) Cancer Res. 56, 738–740.
[39] Chapman, R.M. et al. (1994) Oncogene 9, 1289–1293.
[40] Tahara, H. et al. (1996) Cancer Res. 56, 599–605.
[41] Lin, X. et al. (1995) Mol. Cell. Biol. 15, 2754–2762.

**Table V** *Functions of oncoproteins*

*Class 1 Growth factors*
AIGF[1], **HSTF1/HST-1**, **INT2**, NOV[2], **PDGFB/SIS**, **WNT1**, **WNT2**, **WNT3**

*Class 2 Tyrosine kinases*
Receptor-like tyrosine kinases
**EPH** (CEK, HEK, MEK, NUK, SEK), **ECK**, **EEK**, **ELK**, **ERK**, **EGFR/ERBB**, **FMS**, **KIT**, TYK1/LTK[3], **MET**, **HER2/NEU**, **RET**, **ROS**, EYK/RYK[4], **SEA**, TIE[5], **TRK**, UFO[6]
Non-receptor tyrosine kinases
**ABL1**, (**ARG**), BRK[7], CSK/CYL[8], **FPS/FES** (FER/TYK3), TKF[9]
Membrane-associated non-receptor tyrosine kinases
**SRC**, SRC-related kinases: BLK[10], **FGR**, **FYN**, **HCK**, **LCK** (TKL[11]), **LYN/SYN**, **YES**

*Class 3 Receptors lacking protein kinase activity*
**MAS**, MPL[12]

*Class 4a Membrane-associated G proteins*
**HRAS**, **KRAS**, **NRAS**, TC21/RRAS2[13], GSP[14], GIP2/$G\alpha_{i2}$[15], $G\alpha_{12}$, $G\alpha_{13}$[16], $G\alpha_z$[17]

*Class 4b Guanine nucleotide exchange proteins*
SDC25[18], OST[19]

*Class 4c RHO/RAC binding proteins*
**BCR**, **DBL**, ECT2[20], NET1[21], **TIAM1**, TIM[22], **VAV**

*Class 5 Cytoplasmic protein serine kinases*
**BCR**, CLK (NEK)[23], EST/COT[24], **MOS**, **PIM1**, **RAF/MIL**, TPL-2[25], Protein kinase $C\varepsilon$[26]

*Class 6 Protein serine, threonine (and tyrosine) kinase*
**AKT1**, **AKT2**, STY[27]

*Class 7 Cytoplasmic regulators*
**BCL1**, **CRK**, NCK[28], ODC1[29], PEM[30]

*Class 8 Cell cycle regulators*
**INK4A**, **INK4B**, **INK4C**, **CCND1** (**Cyclin D1**), CDC25A, CDC25B[31]

*Class 9 Transcription factors*
**BCL3**, **CBL**, **E2F1**, **ERBA**, **ETS**, (**ELK**), **FOS** (**FOSB**, **ΔFOSB**, **FRA1**, **FRA2**), GLI[32], HOXA9[33], HOX-2.4[34], HOX-7.1[35], HOX11[36], IRF-2[37], **JUN** (**JUNB**, **JUND**), **MYB** (**MBM2**), **MYC**, **MYCL**, **MYCN**, QIN/FKH2[38], **REL**, **TAL1**, **SKI**, TRE[39], MZF1[40]

*Class 10 Transcription elongation factor*
ELL[41]

*Class 11 Intracellular membrane factor*
**BCL2**

*Class 12 Nucleoporins*
NUP98[42], NUP214[43]

*Class 13 Adapter proteins*
SHC[44]

*Class 14 RNA binding proteins*
EWS[45]

*Class 15 Function unknown*
CPH[46], DAN[47], DLK[48], LBC[49], LCO/LCA[50], MAF (MAFB, MAFK, MAFF, MAFG, p18, NRL)[51], MEL[52], MELF[53], N8[54], SCC[55], TLM[56], UNPH[57]

This table lists gene products that have been shown to be tumorigenic or that are specifically expressed in at least one type of tumour cell. Those in bold type are described in individual sections in Section III. Entries in brackets are discussed in the preceding bold type entry. Superscripts refer to the reference list below for recently discovered genes.

For a list of genes considered to be tumour suppressor genes see Table IV. For oncoproteins identified as a consequence of chromosomal translocations see Table III.

For details of other genes see R. Hesketh, The Oncogene Handbook, Academic Press, London, 1994.

*References*

[1] Kouhara, H. et al. (1994) Oncogene 9, 455–462.
[2] Martinerie, C. et al. (1996) Oncogene 12, 1479–1492; Scholz, G. et al. (1996) Mol. Cell. Biol. 16, 481–486.
[3] Snijers, A.J. et al. (1993) Oncogene 8, 27–35.
[4] Zong, C. et al. (1996) EMBO J. 15, 4515–4525.
[5] Partanen, J. et al. (1992) Mol. Cell. Biol. 12, 1698–1707.
[6] Fridell, Y.-W.C. et al. (1996) Mol. Cell. Biol. 16, 135–145; Goruppi, S. et al. (1996) Oncogene 12, 471–480.
[7] Mitchell, P.J. et al. (1994) Oncogene 9, 2383–2390.
[8] Brauninger, A. et al. (1993) Oncogene 8, 1365–1369; Hata, A. et al. (1994) Mol. Cell. Biol. 14, 7306–7313.
[9] Holtrich, U. et al. (1991) Proc. Natl Acad. Sci. USA 88, 10411–10415.
[10] Dymecki, S.M. et al. (1995) Oncogene 10, 477–486.
[11] Chow, L.M.L. et al. (1992) Mol. Cell. Biol. 12, 1226–1233.
[12] Alexander, W.S. et al. (1995) EMBO J. 14, 5569–5578; Drachman, J.G. et al. (1995) J. Biol. Chem. 270, 4979–4982.
[13] López-Barahona, M. et al. (1996) Oncogene 12, 463–470.
[14] Muca, C. and Vallar, L. (1994) Oncogene 9, 3647–3653.
[15] Ikezu, T. et al. (1994) J. Biol. Chem. 269, 31955–31961; Rudolph, U. et al. (1995) Genetics 10, 143–149.
[16] Voyno-Yasenetskaya, T.A. et al. (1994) Oncogene 9, 2559–2565; Zhang, Y. et al. (1996) Oncogene 12, 2377–2383; Vara Prasad, M.V.V.S. et al. (1994) Oncogene 9, 2425–2429.
[17] Wong, Y.H. et al. (1995) Oncogene 10, 1927–1933.
[18] Chevallier-Multon, M.-C. et al. (1993) J. Biol. Chem. 268, 11113–11118.
[19] Horii, Y. et al. (1994) EMBO J. 13, 4776–4786.
[20] Miki, T. et al. (1993) Nature 362, 462–465.
[21] Chan, A.M.-L. et al. (1996) Oncogene 12, 1259–1266.
[22] Chan, A.M.-L. et al. (1994) Oncogene 9, 1057–1063.
[23] Ben-David, Y. et al. (1991) EMBO J. 10, 317–325.
[24] Chan, A.M.-L. et al. (1993) Oncogene 8, 1329–1333; Miyoshi, J. et al. (1991) Mol. Cell. Biol. 11, 4088–4096.
[25] Salmerón, A. et al. (1996) EMBO J. 15, 817–826.
[26] Perletti, G.P. et al. (1996) Oncogene, 12, 847–854.
[27] Howell, B.W. et al. (1991) Mol. Cell. Biol. 11, 568–572.
[28] Rivero-Lezcano, O.M. et al. (1995) Mol. Cell. Biol. 15, 5725–5731.
[29] Tamori, A. et al. (1995) Cancer Res. 55, 3500–3503.
[30] Wilkinson, M.F. et al. (1991) Dev. Biol. 141, 451–455.
[31] Galaktionov, K. et al. (1996) Nature 382, 511–517.
[32] Ruppert, J.M. et al. (1991) Mol. Cell. Biol. 11, 1724–1728.
[33] Nakamura, T. et al. (1996) Nature Genet. 12, 149–153.
[34] Aberdam, D. et al. (1991) Mol. Cell. Biol. 11, 554–557.
[35] Song, K. et al. (1992) Nature 360, 477–481.
[36] Salvati, P.D. et al. (1995) Oncogene 11, 1333–1338; Tang, S. and Breitman, M.L. (1995) Nucleic Acids Res. 23, 1928–1935.
[37] Vaughan, P.S. et al. (1995) Nature 377, 362–365.
[38] Chang, H.W. et al. (1995) Proc. Natl Acad. Sci. USA 92, 447–451; Kastury, K. et al. (1994) Proc. Natl Acad. Sci. USA 91, 3616–3618.

[39] Nakamura, T. et al. (1992) Oncogene 7, 733–741.
[40] Hromas, R. et al. (1995) Cancer Res. 55, 3610–3614.
[41] Shilatifard, A. et al. (1996) Science 271, 1873–1876.
[42] Borrow, J. et al. (1996) Nature Genet. 12, 159–167; Nakamura, T. et al. (1996) Nature Genet. 12, 154–158.
[43] Kraemer, D. et al. (1994) Proc. Natl Acad. Sci. USA 91, 1519–1523.
[44] Pelicci, G. et al. (1995) Oncogene 11, 899–907.
[45] Ouchida, M. et al. (1995) Oncogene 11, 1049–1054.
[46] Avila, M.A. et al. (1995) Oncogene 10, 963–971.
[47] Enomoto, H. et al. (1994) Oncogene 9, 2785–2791.
[48] Laborda, J. et al. (1993) J. Biol. Chem. 268, 3817–3820.
[49] Zheng, Y. et al. (1995) J. Biol. Chem. 270, 10120–10124.
[50] Tokino, T. et al. (1988) Cytogenet. Cell Genet., 48, 63–64.
[51] Kataoka, K. et al. (1996) Oncogene 12, 53–62; Ho, I.-C. et al. (1996) Cell 85, 973–983; Johnsen, O. et al. (1996) Nucleic Acids Res. 24, 4289–4297.
[52] Nimmo, E.R. et al. (1991) Oncogene 6, 1347–1351.
[53] Misawa, Y. and Shibuya, M. (1992) Oncogene 7, 919–926.
[54] Chen, S.-L. et al. (1996) Oncogene 12, 741–751.
[55] Moen, C.J.A. et al. (1996) Proc. Natl Acad. Sci. USA 93, 1082–1086.
[56] Lane, M.A. and Tobin, M.B. (1990) Nucleic Acids Res. 18, 3410.
[57] Gray, D.A. et al. (1995) Oncogene 10, 2179–2183.

**Table VI** *Chromosome locations of human proto-oncogenes and tumour suppressor genes*

| | | | |
|---|---|---|---|
| 1 | *ECK* | 3q27 | *LAZ3* |
| 1 | *EEK* | 3q34 | *MLF1* |
| 1 | *ERK* | 4 | *FLK1/KDR* |
| 1 | *TPR* | 4p11 | *TTF* |
| 1p13 | *NRAS* | 4q11–q21 | *KIT* |
| 1p32 | *AF-1p* | 4q21 | *MLLT2/AF4* |
| 1p32 | *BLYM* | 4q26 | *IL2* |
| 1p32 | *MYCL1* | 5/17/18 (elements) | *TRE* |
| 1p32 | *TAL1* | 5q | *GAP* |
| 1p32–p31 | *JUN* | 5q11–13 | *MSH3* |
| 1p33 | *SIL* | 5q21 | *FER* |
| 1p33–p34 | *TIE* | 5q21 | *MCC* |
| 1p33–p35 | *MDGI* | 5q21–q22 | *APC* |
| 1p34 | *MPL* | 5q25 | *NPM* |
| 1p35–p32 | *LCK* | 5q31 | *IL3* |
| 1p36 | *E2F2* | 5q31.1 | *IRF1* |
| 1p36 | *NB1* | 5q31.3–q33.2 | *FGFA* |
| 1p36 | *PAX7* | 5q33 | *PDGFRB* |
| 1p36.2–p36.1 | *FGR* | 5q33.3–q34 | *CSF1R* |
| 1p36.11–p36.13 | *DAN* | 5q35 | *FLT4* |
| 1q22–q24 | *SKI* | 6p21 | *CCND3* |
| 1q23 | *PBX1* | 6p21 | *HPR* |
| 1q23–q24 | *TRK* | 6p21 | *PIM1* |
| 1q23–q31 | *TRKC* | 6p21 | *TRKE* |
| 1q24–q25 | *ABL2 (ARG)* | 6p21 | *WAF1* |
| 1q32 | *ELK4/SAP1* | 6p23–p22.3 | *FIM1* |
| 2p13–p12 | *REL* | 6q21 | *FYN* |
| 2p22–p21 | *MSH2* | 6q21–q22 | *ROS1* |
| 2p23–p12 | *MEIS1* | 6q22 | *E2F3* |
| 2p16 | *GTBP/MSH6* | 6q22–q23 | *MYB* |
| 2p23 | *ALK* | 6q23 | *DEK* |
| 2p24.1 | *MYCN* | 6q24–q27 | *MAS* |
| 2p25 | *ODC1* | 6q26–27 | *M6P/IGF2R* |
| 2q14–q21 | *HIS1* | 6q27 | *AF6* |
| 2q14–q21 | *LCO* | 7 | *TIM* |
| 2q31–33 | *PMS1* | 7p13–p12 | *EGFR* |
| 2q33.3–34 | *HER4* | 7p15 | *MYCLK1* |
| 2q33–qter | *INHA* | 7p15 | *HOXA9* |
| 2q35 | *PAX3* | 7p22 | *ETV1* |
| 3p14.2 | *FHIT/HRCA1* | 7p22 | *PMS2* |
| 3p14–p12 | *NRC1* | 7q31 | *MET* |
| 3p14–p21 | *LC1* | 7q31 | *WNT2* |
| 3p14.2–p21.1 | *WNT5A* | 7q32–q36 | *EPHT* |
| 3p21.3 | 3pK | 7q33–q36 | *RAFB1* |
| 3p21.3 | Integrin $\alpha_{RLC}$ | 8p11 | *MOZ* |
| 3p21.23 | *UNPH* | 8p12 | *FLT2* |
| 3p21.3–p23 | *MLH1* | 8p21.3–p22 | *PRLTS* |
| 3p21 | *TGFBR2* | 8q11–q12 | *MOS* |
| 3p21 | *TCTA* | 8q13–qter | *LYN* |
| 3p24.1–p22 | *THRB* | 8q22 | *MYBL1* |
| 3p25 | *RAF1* | 8q22 | *ETO/MTG8* |
| 3p25–p26 | *VHL* | 8q24.1 | *EXT1* |
| 3q22 | *RYK* | 8q24 | *MYC* |
| 3q26 | *EVI1* | 8q24 | *PVT1* |
| 3q26 | *EAP* | 8q24.1 | *NOV* |
| 3q26 | *MDS1* | 9p21 | *MLM* |

**Table VI** *Continued*

| | | | |
|---|---|---|---|
| 9p21 | *INK4A/MTS1/CDK4I, INK4B/MTS2* | 12q13–q14.3 | *GLI1* |
| | | 12q22 | *BTG1* |
| 9p34.1 | *ABL1* | 12q22–q23 | *ELK3/NET/ERP/SAP2* |
| 9q22 | *MLLT3/AF9* | 13q12 | *FLT1* |
| 9q22.1 | *TRKB* | 13q12 | *FLT3/FLK2* |
| 9q22–31 | *CHN* | 13q12–13 | *BRCA2* |
| 9q31 | *BCNS* | 13q12–q13 | *BRUSH1* |
| 9q34 | *CAN/NUP214* | 13q14 | *FKHR* |
| 9q34 | *SET* | 13q14.2 | *RB1* |
| 9q34 | *TAL2* | 14q13 | *FKH2/QIN* |
| 9q34 | *TAN1* | 14q24.3 | *FOS* |
| 9q34 | *TSC1* | 14q32 | *AKT1* |
| 10q11.2 | *ELE1* | 14q32.1 | *TCL1* |
| 10p11.2 | *EST* | 14q32.3 | *ELK2* |
| 10q11.2 | *RET* | 15 | *CSK* |
| 10p13 | *AF10* | 15q13–q21 | *LTK/TYK1* |
| 10p13 | *BMI1* | 15q21 | *PML* |
| 10p15 | *NET1* | 15q25–q26 | *FPS/FES* |
| 10q21 | *H4* | 16p11 | *CBP* |
| 10q24 | *HOX11* | 16p11 | *TLS/FUS* |
| 10q24 | *NFKB2/LYT10* | 16p13 | *BCM* |
| 10q24–q25 | *MXI1* | 16p13.12–p13.13 | *MYH11* |
| 11p11.2 | *KAI1* | 16p13.3 | *TSC2* |
| 11p11–13 | *EXT2* | 16p22–q23 | *MAF* |
| 11p12–p11.2 | *SPI1* | 16q | *CMAR/CAR* |
| 11p13 | *LMO2/RBTNL1* | 16q12–13 | *CYLD1* |
| 11p13 | *WT1* | 16q22 | *CBFB* |
| 11p15 | *LMO1/RBTN1* | 16q22 | *TPR* |
| 11p15 | *NUP98* | 16q22.1 | E-cadherin/*UVO* |
| 11p15 | *PTH* | 17p11.2–12 | *HUGL* |
| 11p15.5 | *BWS* | 17p13.1 | *TP53* |
| 11p15.5 | *KIP2* | 17p13.3 | *OVCA1, OVCA2* |
| 11p15.5 | *HRAS* | 17q | *BTR* |
| 11q13 | *BCL1* | 17q11.2 | *EVI2B* |
| 11q13 | *CCND1* | 17q11.2 | *NF1* |
| 11q13 | *EMS1* | 17q11.2–q12 | *THRA1* |
| 11q13 | *INT2* | 17q12–q21 | *BRCA1* |
| 11q13 | *MEN1* | 17q12–q21 | *FWT1* |
| 11q13 | *SEA* | 17q21 | *AF17* |
| 11q13.3 | *HSTF1* | 17q21 | *PHB* |
| 11q14 | *CALM* | 17q21 | *RARA* |
| 11q23 | *LPC* | 17q21–q22 | *HER2* |
| 11q23 | *MLL* | 17q21–q22 | *WNT3* |
| 11q23 | *PLZF* | 17q21.3 | *NME1, NME2* |
| 11q23–q24 | *FLI1/ERGB* | 17q22 | *HLF* |
| 11q23.3 | *ETS1* | 17q23 | *R1α* |
| 11q23 | *RCK* | 17q25 | *TIMP-2* |
| 11q23.3–qter | *CBL* | 18q11.2 | *SYT* |
| 12p12.1 | *KRAS2* | 18q21 | *BCL2* |
| 12p13 | *CCND2* | 18q21 | *DCC* |
| 12p13 | *KIP1* | 18q21.1 | *DPC4/SMAD4* |
| 12p13 | *TEL* | 18q21.3 | *YES1* |
| 12p13 | *HST2* | 19p13 | *E2A* |
| 12q13 | *ATF1* | 19p13.1 | *ELL/MEN* |
| 12q13 | *CDK4* | 19p13.2 | *LYL1* |
| 12q13 | *CHOP* | 19p13.2 | *JUNB* |
| 12q13 | *HER3* | 19p13.2 | *JUND* |
| 12q13 | *WNT1* | 19p13.2 | *VAV* |

**Table VI** *Continued*

| | | | |
|---|---|---|---|
| 19p13.3 | *MLLT1/ENL/LGT19* | 22q12 | *EWS* |
| 19CEN–p13.2 | *MEL* | 22q12 | *NF2* |
| 19q13.1 | *BCL3* | 22q12.3–q13.1 | *PDGFB* |
| 19q13.1 | *UFO* | 22q13.3 | *NMI* |
| 20q11 | *E2F1* | 22q13–qter | *TIMP-3* |
| 20q11 | *HCK* | Xp11.2–p11.1 | *ELK1* |
| 20q13.3 | *SRC* | Xp11.2 | *RAFA1* |
| 21q22 | *AML1* | Xp11.2 | *SSX1, SSX2* |
| 21q22.3 | *ERG* | Xp11.3–p11.23 | *TIMP1* |
| 21q22.3 | *ETS2* | Xq13 | *MYBL2* |
| 22q11 | *MN1* | Xq13 | *AFX1* |
| 22q11.2 | *BCR* | Xq27 | *DBL* |
| 22q12 | *BAM22* | Xq28 | *c6.1A/c6.1B* (*MTCP1*) |

# Section III

# THE ONCOGENES

## Identification

v-*abl* is the transforming gene of the replication defective Abelson murine leukaemia virus (Ab-MuLV), originally isolated from a mouse infected with Moloney murine leukaemia virus (Mo-MuLV) after chemical thymectomy. v-*abl* is also carried by the ABL/MYC murine retrovirus derived from Ab-MuLV and by the Hardy–Zuckerman 2 feline sarcoma virus (HZ2-FeSV). *ABL* (*ABL1*) is the human homologue of *Abl*[1,2].

## Related genes

ABL contains regions homologous to the kinase and homology regions 2 and 3 (SH2, SH3) of SRC (see ***SRC***). Other related genes: *ABL2/ARG* (Abelson-related gene), *EPH, NCP94, TKR11* and *TKR16, Drosophila melanogaster Dash* and *Caenorhabditis elegans Abl.*

| | *ABL* | v-*abl* |
|---|---|---|
| **Nucleotides (kb)** | 225 | 5.7(Ab-MuLV) |
| **Chromosome** | 9q34.1 | |
| **Mass (kDa): predicted** | 123 | |
| **expressed** | 150 (with alternative N-termini) | P160, P120, P90, P100 |
| **Cellular location** | p150: mainly nucleus, some cytoplasm/plasma membrane ARG: cytoplasmic[3] | Nucleus |

**Tissue distribution**
*ABL* expression is widespread but is particularly strong in the spleen, testis and thymus[4]. In germ cells there is a novel transcript, restricted to post-meiotic spermatogenic cells[5]. Neither *ABL* nor *BCR* are genomically imprinted in normal leukocytes nor is there preferential involvement of alleles in the formation of *BCR/ABL* fusion genes[6].

## Protein function

ABL is a non-receptor tyrosine kinase with weak enzymatic activity. Both ABL and transforming ABL proteins inhibit entry into S phase by a mechanism that requires nuclear localization and is both p53 and pRb dependent. Interaction of ABL proteins with pRb can promote E2F1-driven transcription, for example, of *Myc*[7]. Nuclear ABL, but not the cytoplasmic forms, has sequence-specific DNA binding activity for the EP element present in the enhancers of several viruses (hepatitis B virus, polyomavirus) and in the *Myc* promoter[8] and v-ABL has been shown to activate *Myc* transcription[9]. ABL is activated in NIH 3T3 cells or human U-937 cells in the stress response to DNA-damaging agents including *cis*-platinum, mitomycin C, ionizing radiation[10] and 1-$\beta$-D-arabinofuranosylcytosine (araC), the latter activating the stress-activated protein kinase (SAPK/JNK) in an ABL-dependent manner [11].

ABL tyrosine kinase activity and transformation potential may be activated by:

1 The formation of the fusion BCR/ABL protein in which BCR sequences replace the ABL first exon (see ***BCR***).
2 Deletion of the non-catalytic SH3 (negative regulatory) region [12]. The retention of an N-terminal myristylation site is essential for transforming activity. ABL differs from members of the SRC family of protein tyrosine kinases in that it does not contain a C-terminal regulatory tyrosine ($Tyr^{527}$ in SRC).
3 Hyper-expression of ABL (>500-fold more than the normal endogenous concentration). This may reflect the presence of cellular factors that normally inhibit the kinase [13].
4 N-terminal deletions or an in-frame deletion within the last exon [14].

The kinase activity of v-ABL is regulated *via* interaction with the retinoblastoma protein and it can efficiently phosphorylate RNA polymerase II and enhance transcription in collaboration with the VP16 transcription factor [15]. $P160^{v\text{-}abl}$ also phosphorylates $p93^{FES}$ and $P120^{v\text{-}ABL}$ associates with the cyclic AMP response element (CRE)-binding protein in myeloid cells to potentiate the activity of CREB [16]. Transformation by Ab-MuLV causes phosphorylation of enolase, vinculin and p42 and serine phosphorylation of ribosomal protein S6. The tyrosine kinase activity of v-ABL causes erythroid bursts in fetal liver cells in the absence of erythropoietin, renders the proliferation of mast cells and other types of cell independent of IL-3 and in fibroblasts reduces the number of EGFRs and stimulates secretion of an EGF-like growth factor. All myristylated, tyrosine kinase-active forms of ABL associate with PtdIns 3-kinase [17]. Activated ABL proteins may thus modulate the normal mechanisms that regulate growth factor-stimulated proliferation (for example, by activating PtdIns metabolism or the transcription of early genes including *Fos*), an inference strengthened by the finding that ABL is phosphorylated by $p34^{CDC2}$ [18].

**ABL binding proteins**

The ABL SH3 domain binds specifically to the sequence PPXΘ(aromatic)XPPPΨ(aliphatic)P [19]. 3BP-1, which has homology with GAP, binds to the SH3 domain of both ABL and SRC (see ***SRC***). ABL (residues 551–606) and ARG bind to the CRK $SH3_1$ domain and phosphorylate CRK on $Tyr^{221}$ [20,21].

ABI-2 (Abl interactor protein) is an SH3-containing protein that binds to ABL both *in vitro* and *in vivo* and is a substrate for the tyrosine kinase [22]. ABI-2 binds to both the SH3 and C-terminal regions of ABL: ABI-2 defective for SH3 domain binding but retaining C-terminal binding capacity activates the transformation by ABL, suggesting that ABI-2 may function as a tumour suppressor.

**Cancer**

Over 90% of chronic myeloid leukaemia (CML) cases involve the balanced translocation t(9;22)(q34;q11) of a fragment of the long arm of chromosome 9

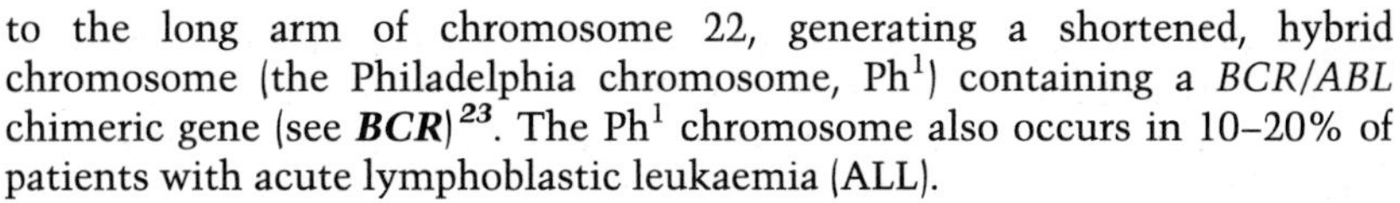


to the long arm of chromosome 22, generating a shortened, hybrid chromosome (the Philadelphia chromosome, $Ph^1$) containing a *BCR/ABL* chimeric gene (see ***BCR***)[23]. The $Ph^1$ chromosome also occurs in 10–20% of patients with acute lymphoblastic leukaemia (ALL).

Activation of ABL by fusion to the N-terminal sequence of TEL, an ETS-related helix-loop-helix protein, has been detected in a patient with common ALL[24].

**Transgenic animals**

Haematopoietic stem cells expressing v-*abl* initiate leukemogenesis in mice. Animals with a homozygous deletion of *Abl* ($Abl^{m1}$) have increased perinatal mortality and decreased levels of B and T cell precursors. Transgenic expression of either type I or type IV *Abl* rescues the lethality of $Abl^{-/-}$ *mice*[25]. Plasmacytoma development in *Abl* transgenic mice occurs after a translocation in *Myc* that causes over-expression of MYC protein and is accelerated by crossing with transgenic *Myc* mice[26].

**In animals**

Mo-MuLV induces lymphocytic leukaemias in mice and the acute transforming virus HZ2-FeSV causes feline sarcomas.

***In vitro***

*Abl* does not transform primary fibroblasts but does transform haematopoietic cells and lymphoid cells and is thus unique among murine retroviruses. v-ABL transforms NIH 3T3 fibroblasts and ABL may be activated to transform fibroblasts by a single point mutation ($Phe^{420} \rightarrow Val$). $Phe^{420}$ is adjacent to the predicted major site of tyrosine phosphorylation in ABL ($Tyr^{412}$) and is perfectly conserved among tyrosine kinases with N-terminal SH3 domains[27]. Point mutations in the ATP binding site ($Tyr^{272} \rightarrow Phe$ or $Tyr^{276} \rightarrow Phe$) also elevate tyrosine kinase activity and relieve growth dependence on IL-3[28].

Over-expression of *Myc* acts synergistically with the expression of *Abl* oncogenes in transformation, whereas dominant negative mutations in MYC that leave the dimerization motif intact (see ***MYC***) reduce transformation by v-*abl* or by $P185^{BCR/ABL}$[29].

The over-expression of ABL reversibly inhibits the growth of fibroblasts[30]. This requires tyrosine kinase activity, nuclear localization and an intact SH2 domain. The latter, rather than the catalytic domain, is largely responsible for determining the spectrum of proteins phosphorylated *in vivo*[31]. Dominant negative mutants of ABL disrupt cell cycle control and enhance transformation by tyrosine kinases, G proteins or oncoprotein transcription factors.

In pre-B lymphocytes v-ABL negatively regulates the activity of NFκB and specifically inhibits the NFκB/REL-dependent κ intron enhancer, which is implicated in promoting both transcription and rearrangement of the κ locus[32]. Inactivation of v-ABL promotes high-frequency rearrangement of both κ and λ light chain genes[33]. Thus v-ABL suppresses two of the steps that are essential for B cell differentiation. v-ABL also constitutively activates JAK1 and JAK3 Janus kinases and causes tyrosine phosphorylation of the STAT (signal transducers and activators of transcription) proteins that are normally activated by IL-4 or IL-7 in the absence of these ligands [34].

## Gene structure

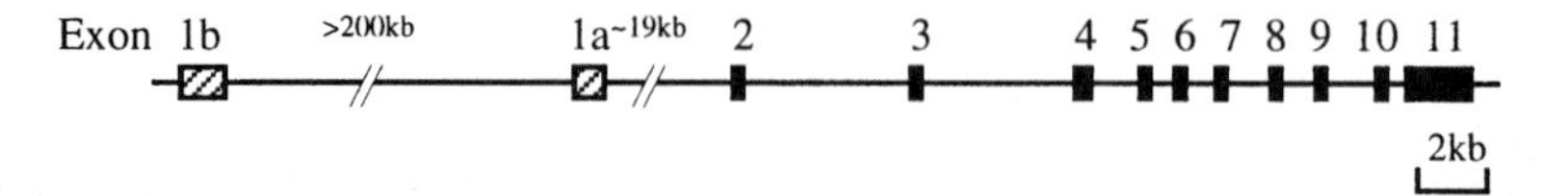


## Transcriptional regulation

Two specific promoters utilize exon 1a or 1b: mRNAs 1a and 1b are equivalent to mouse type I and type IV mRNAs, respectively [35]. There are four murine *Abl* mRNAs (types I, II, III and IV) that have alternative 5′ sequences encoded by differential splicing of exons 1a and 1b to a common 3′ sequence. Human 1a and 1b *ABL* promoter sequences show high homology [36]. The 1a promoter contains seven SP1 sites and a TTAA sequence that may function as a TATA box. The 1b promoter contains at least 12 protein binding elements, including seven SP1 motifs and four CCAAT boxes but no TATA box.

## Protein structure

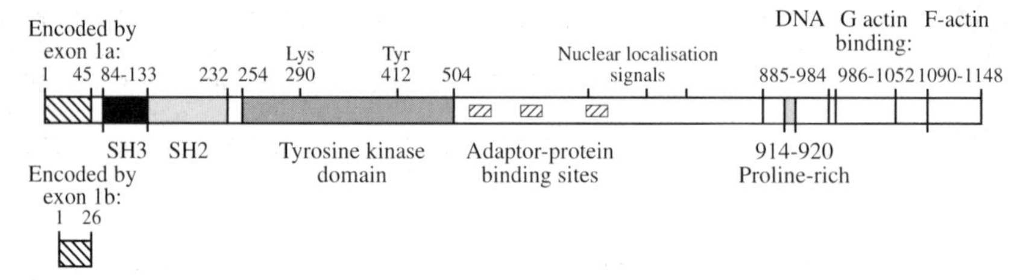


Lys$^{271}$ (Lys$^{290}$ in type 1b ABL) is essential for kinase activity. The highly conserved SH2 motif FLVRES is critical for binding to tyrosine phosphorylated proteins and transforming potential [37]. The affinity of ABL for F-actin is enhanced by the BCR sequences present in BCR/ABL (see ***BCR***). All ABL proteins (p150$^{ABL}$, p160$^{v\text{-}abl}$, P210$^{BCR/ABL}$) are phosphorylated at two C-terminal sites by protein kinase C [38], although this region is not required either for ABL kinase activity or for transformation by v-ABL. v-ABL proteins lack the GRB2 binding site present in BCR/ABL (see ***BCR***) but associate with SHC proteins [39].

The structure of the SH2 domain of ABL in solution has been shown by multidimensional NMR spectroscopy to comprise a compact sphere with a large three-stranded antiparallel $\beta$ sheet, a second smaller $\beta$ sheet and a C-terminal $\alpha$ helix enclosing the hydrophobic core [40]. The putative phosphotyrosyl binding site is formed by conserved residues in the large $\beta$ sheet and in a short amphipathic helix.

Murine *Abl*, from which v-*abl* was acquired, encodes a protein with 85% homology to human ABL. In v-ABL upstream fusion of *gag* provides a myristylated N-terminus that directs membrane localization and activates

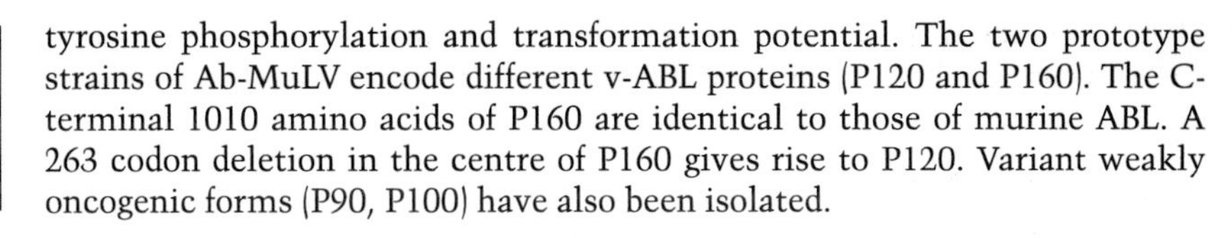

tyrosine phosphorylation and transformation potential. The two prototype strains of Ab-MuLV encode different v-ABL proteins (P120 and P160). The C-terminal 1010 amino acids of P160 are identical to those of murine ABL. A 263 codon deletion in the centre of P160 gives rise to P120. Variant weakly oncogenic forms (P90, P100) have also been isolated.

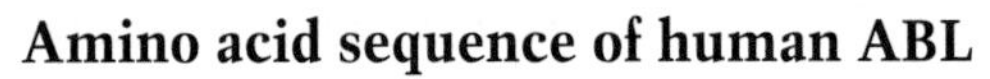

## Amino acid sequence of human ABL

```
   1 MLEICLKLVG CKSKKGLSSS SSCYLEEALQ RPVASDFEPQ GLSEAARWNS
  51 KENLLAGPSE NDPNLFVALY DFVASGDNTL SITKGEKLRV LGYNHNGEWC
 101 EAQTKNGQGW VPSNYITPVN SLEKHSWYHG PVSRNAAEYL LSSGINGSFL
 151 VRESESSPGQ RSISLRYEGR VYHYRINTAS DGKLYVSSES RFNTLAELVH
 201 HHSTVADGLI TTLHYPAPKR NKPTVYGVSP NYDKWEMERT DITMKHKLGG
 251 GQYGEVYEGV WKKYSLTVAV KTLKEDTMEV EEFLKEAAVM KEIKHPNLVQ
 301 LLGVCTREPP FYIITEFMTY GNLLDYLREC NRQEVNAVVL LYMATQISSA
 351 MEYLEKKNFI HRDLAARNCL VGENHLVKVA DFGLSRLMTG DTYTAHAGAK
 401 FPIKWTAPES LAYNKFSIKS DVWAFGVLLW EIATYGMSPY PGIDLSQVYE
 451 LLEKDYRMER PEGCPEKVYE LMRACWQWNP SDRPSFAEIH QAFETMFQES
 501 SISDEVEKEL GKQGVRGAVS TLLQAPELPT KTRTSRRAAE HRDTTDVPEM
 551 PHSKGQGESD PLDHEPAVSP LLPRKERGPP EGGLNEDERL LPKDKKTNLF
 601 SALIKKKKKT APTPPKRSSS FREMDGQPER RGAGEEEGRD ISNGALAFTP
 651 LDTADPAKSP KPSNGAGVPN GALRESGGSG FRSPHLWKKS STLTSSRLAT
 701 GEEEGGGSSS KRFLRSCSAS CVPHGAKDTE WRSVTLPRDL QSTGRQFDSS
 751 TFGGHKSEKP ALPRKRAGEN RSDQVTRGTV TPPPRLVKKN EEAADEVFKD
 801 IMESSPGSSP PNLTPKPLRR QVTVAPASGL PHKEEAEKGS ALGTPAAAEP
 851 VTPTSKAGSG APGGTSKGPA EESRVRRHKH SSESPGRDKG KLSRLKPAPP
 901 PPPAASAGKA GGKPSQSPSQ EAAGEAVLGA KTKATSLVDA VNSDAAKPSQ
 951 PGEGLKKPVL PATPKPQSAK PSGTPISPAP VPSTLPSASS ALAGDQPSST
1001 AFIPLISTRV SLRKTRQPPE RIASGAITKG VVLDSTEALC LAISRNSEQM
1051 ASHSAVLEAG KNLYTFCVSY VDSIQQMRNK FAFREAINKL ENNLRELQIC
1101 PATAGSGPAA TQDFSKLLSS VKEISDIVQR (1130)
```

**Domain structure**

| | |
|---|---|
| 1–26 | Type 1a N-terminus (underlined): corresponding N-terminus of type 1b ABL: MGQQPGKVLGDQRRPSLPALHFIKGAGKKESSRHGGPHCNVFVEH |
| 26–27 | Breakpoint for translocation to form BCR/ABL oncogene |
| 149–154 | FLVRES motif (underlined) |
| 211–291 | pRB binding domain |
| 248–256 and 271 | ATP binding |
| 363 | Active site |
| 393 | Autophosphorylation site |
| 61–121 | SH3 domain |
| 127–213 | SH2 domain (italics) |
| 551–606 | CRK binding [41] |
| 935–1067 | RNA polymerase II binding [42] |
| 544–637 | ABI-2 binds to the SH3 domain of ABL and to a proline-rich region (544–637) C-terminal of the NLS [22] |
| 601–615, 707–720, 759–772 | Nuclear localization signals (underlined) [43] |

In the *ABL-TEL* translocation the first 26 amino acids of ABL (underlined) are replaced by the N-terminus (1–336) of TEL (breakpoint after the first nucleotide of *ABL* codon 27)[24,44].

**Database accession numbers**

| | *PIR* | *SWISSPROT* | *EMBL/GENBANK* | *REFERENCES* |
|---|---|---|---|---|
| Human *ABL1* | A25582 | P00519 | M14752 | 35 |
| | | | X16416 | 45,46 |
| Human *TEL/ABL* | Z35761 | | | 24 |

***References***


1 **Rosenberg, N. and Witte, O. (1988) Adv. Virus Res. 35, 39–81.**
2 **Ramakrishnan, L. and Rosenberg, N. (1989) Biochim. Biophys. Acta 989, 209–224.**
3 Wang, B. and Kruh, G.D. (1996) Oncogene 13, 193–197.
4 Van Etten, R.A. et al. (1989) Cell 58, 669–678.
5 Ponzetto, C. and Wolgemuth, D.J. (1985) Mol. Cell. Biol. 5, 1791–1794.
6 Melo, J.V. et al. (1994) Nature Genet. 8, 318–319.
7 Birchenall-Roberts, M.C. et al. (1996) Oncogene 13, 1499–1509.
8 Dikstein, R. et al. (1996) Proc. Natl Acad. Sci. USA 93, 2387–2391.
9 Wong, K.-K. et al. (1995) Mol. Cell. Biol. 15, 6535–6544.
10 Kharbanda, S. et al. (1995) Nature 376, 785–788.
11 Kharbanda, S. et al. (1995) J. Biol. Chem. 270, 30278–30281.
12 Shore, S.K. et al. (1990) Proc. Natl Acad. Sci. USA 87, 6502–6506.
13 Pendergast, A.M. et al. (1991) Proc. Natl Acad. Sci. USA 88, 5927–5931.
14 Goga, A. et al. (1993) Mol. Cell. Biol. 13, 4967–4975.
15 Duyster, J. et al. (1995) Proc. Natl Acad. Sci. USA 92, 1555–1559.
16 Birchenall-Roberts, M.C. et al. (1995) Mol. Cell. Biol. 15, 6088–6099.
17 Varticovski, L. et al. (1991) Mol. Cell. Biol. 11, 1107–1113.
18 Kipreos, E.T. and Wang, J.Y.J. (1992) Science 256, 382–385.
19 Sparks, A.B. et al. (1996) Proc. Natl Acad. Sci. USA 93, 1540–1544.
20 Feller, S.M. et al. (1994) EMBO J. 13, 2341–2351.
21 Wang, B. et al. (1996) Oncogene 13, 1379–1385.
22 Dai, Z. and Pendergast, A.M. (1995) Genes Dev. 9, 2569–2582.
23 Nowell, P.C. and Hungerford, D.A. (1960) Science 132, 1497–1499.
24 Papadopoulos, P. et al. (1995) Cancer Res. 55, 34–38.
25 Hardin, J.D. et al. (1996) Oncogene 12, 2669–2677.
26 Rosenbaum, H. et al. (1992) EMBO J. 9, 897–905.
27 Jackson, P.K. et al. (1993) Oncogene 8, 1943–1956.
28 Allen, P.B. and Wiedemann, L.M. (1996) J. Biol. Chem. 271, 19595–19591.
29 Sawyers, C.L. et al. (1992) Cell 70, 901–910.
30 Sawyers, C.L. et al. (1994) Cell 77, 121–131.
31 Mayer, B.J. and Baltimore, D. (1994) Mol. Cell. Biol. 14, 2883–2894.
32 Klug, C.A. et al. (1994) Genes Dev. 8, 678–687.
33 Chen, Y.-Y. et al. (1994) Genes Dev. 8, 688–697.
34 Danial, N.N. et al. (1995) Science 269, 1875–1877.
35 Fainstein, E. et al. (1989) Oncogene 4, 1477–1481.
36 Zhu, Q.S. et al. (1990) Oncogene 5, 885–891.
37 Zhu, G. et al. (1993) J. Biol. Chem. 268, 1775–1779.
38 Pendergast, A.M. et al. (1987) Mol. Cell. Biol. 7, 4280–4289.

[39] Raffel, G.D. et al. (1995) J. Biol. Chem. 271, 4640–4645.
[40] Overduin, M. et al. (1992) Cell 70, 697–704.
[41] Feller, S.M. et al. (1994) EMBO J. 13, 2341–2351.
[42] Baskaran, R. et al. (1996) Mol. Cell. Biol. 16, 3361–3369.
[43] Wen, S. et al. (1996) EMBO J. 15, 1583–1595.
[44] Golub, T.R. et al. (1996) Mol. Cell. Biol. 16, 4107–4116.
[45] Groffen, J. et al. (1983) Nature 304, 167–169.
[46] Shtivelman, E. et al. (1986) Cell 47, 277–284.

# AKT1 (RAC$\alpha$), AKT2 (RAC$\beta$)

## Identification

*Akt* is the oncogene of the transforming retrovirus AKT8 isolated from a mouse thymoma that arose spontaneously in an AKR/J mouse [1]. Human *AKT1* and *AKT2* were cloned by screening a genomic DNA library with a v-*akt* probe under conditions of reduced stringency [1].

## Related genes

AKT1 and AKT2 are members of the RAC (related to the A and C kinases) subfamily of serine/threonine protein kinases. The homology of *AKT2* to *AKT1* and the non-viral portion of v-*akt* is 77.6% and 77.5% at the nucleotide level and 90.6% and 90.4% at the amino acid level, respectively. AKT1 is 98% homologous to the mouse AKT protein [2]. *Drosophila melanogaster Dakt1* is 64.6% identical to *AKT* at the nucleotide level and 76.5% similar at the amino acid level [3].

| | ***AKT1*** | ***AKT2*** |
|---|---|---|
| **Nucleotides** | Not fully mapped | Not fully mapped |
| **Chromosome** | 14q32 (mouse chromosome 12) | 19q13.1–13.2 |
| **Amino acids** | 479 | 481 |
| **Mass (kDa): predicted** | 55.7 | 55.8 |
| **expressed** | 59 | 56 |

### Cellular location

AKT is mainly cytosolic in distribution whereas v-AKT, which is myristylated, is distributed between the cytosol, nucleus and plasma membranes [4].

### Tissue distribution

Mouse *Akt-1* expression is detectable in the liver, spleen and testis [2]. Mouse $p56^{Akt2}$ shows variable tissue expression, being maximal in skeletal muscle [5]. v-*Akt* is highly expressed in the thymus [6].

## Protein function

AKT1 and AKT2 are serine/threonine protein kinases [7], the activity of which is primarily regulated by phosphorylation [8]. AKT1 and AKT2 are activated by PDGF *via* a pathway dependent on the PDGFR$\beta$ PtdIns 3-kinase binding tyrosines $Tyr^{740}$ and $Tyr^{751}$ and blocked by the PtdIns 3-kinase-specific inhibitor wortmannin and the dominant inhibitory N17Ras [9]. AKT is also activated by EGF and insulin. The only known function of AKT is to inhibit glycogen synthase kinase-3 [10].

### Cancer

Amplification of *AKT1* has been detected in a primary gastric adenocarcinoma [11].

Amplification of *AKT2* has been detected in 12.1% of samples of ovarian carcinomas and in 2.8% of breast carcinomas. Over-expression of *AKT2* can also occur in ovarian carcinomas negative for *AKT2* amplification. Alterations in *AKT2* may therefore play a specific role in ovarian oncogenesis and appear to be associated with a poor prognosis for ovarian cancer patients [12].

**In animals**

Anti-sense RNA to *AKT2* inhibits invasiveness and tumorigenicity of AKT2 in nude mice [13].

## Gene structure of mouse *Akt*

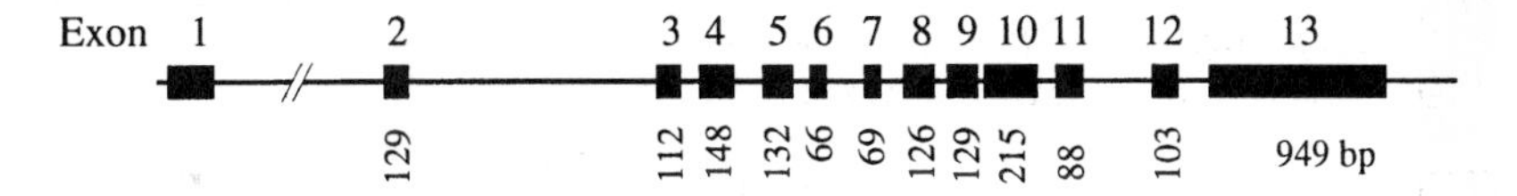


## Transcriptional regulation

Exon 1 contains a 5′ untranslated GC-rich region.

## Protein structure

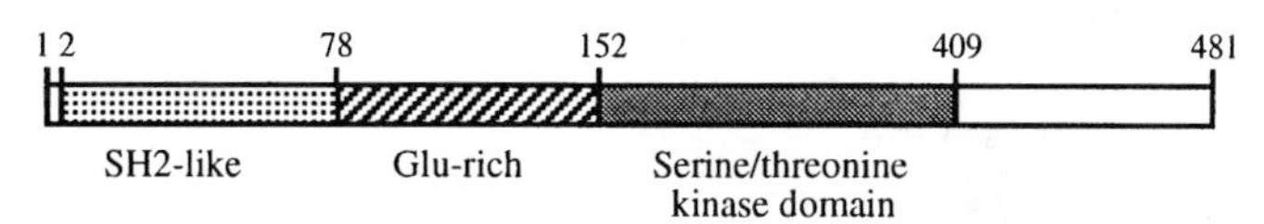


AKT contains an SH2 domain in its regulatory region, a glutamate-rich region, part of which is predicted to form an amphipathic helix, and a kinase domain having high homology with members of the protein kinase C (PKC) family.

## Amino acid sequence of human AKT1 and AKT2 (and v-AKT)

```
AKT2   1 MNEVSVIKEG WLHKRGEYIK TWRPRYFLLK SDGSFIGYKE RPEAPDQTLP
AKT1   1 .SD.AIV... .......... .......... N..T...... ..QDV..REA
v-AKT  1 ..D.AIV... .......... .......... N..T...... ..QDV..RES

      51 PLNNFSVAEC QLMKTERPRP NTFVIRCLQW TTVIERTFHV DSPDEREEWM
      51 ........Q. .......... ...I...... .......... ET.E.....T
      51 ........Q. .......... ...I...... .......... ET.E.....A

     101 RAIQMVANSL KQRAPGEDPM DYKCGSPSDS STTEEMEVAV SKARAKVTMN
     101 T...T..DG. .KQ--E.EE. .FRS.....N .GA.....SL A.PKHR....
     101 T...T..DG. .RQ--E.ET. .FRS.....N .GA.....SL A.PKHR....

     151 DFDYLKLLGK GTFGKVILVR EKATGRYYAM KILRKEVIIA KDEVAHTVTE
     149 E.E....... .......... .......... ...K....V. .......L..
     149 E.E....... .......... .......... ...K....V. .......L..
```

```
201 SRVLQNTRHP FLTALKYAFQ THDRLCFVME YANGGELFFH LSRERVFTEE
199 N....S.... .......S.. .......... .......... ........S.D
199 N....S.... .......S.. .......... .......... ........S.D

251 RARFYGAEIV SALEYLHSRD VVYRDIKLEN LMLDKDGHIK ITDFGLCKEG
249 .......... ...D....EKN....L... .......... ..........
249 .......... ...D....EKN....L... .......... ..........

301 ISDGATMKTF CGTPEYLAPE VLEDNDYGRA VDWWGLGVVM YEMMCGRLPF
299 .K........ .......... .......... .......... ..........
299 .K........ .......... .......... .......... ..........

351 YNQDHERLFE LILMEEIRFP RTLSPEAKSL LAGLLKKDPK QRLGGGPSDA
349 ......K... .......... ...G...... .S........ ......SE..
349 ......K... .......... ...G...... .S........ ......SE..

401 KEVMEHRFFL SINWQDVVQK KLLPPFKPQV TSEVDTRYFD DEFTAQSITI
399 ..I.Q....A G.V..H.YE. ..S....... ...T...... E.....M...
399 ..I.Q....A N.V....YE. ..S....... ...T...... E.....M...
451 TPPDRYDSLG LLELDQRTHF PQFSYSASIR E (481)
449 ....QD..ME CVDSER.P.. ........ST A (479)
449 ....QD..ME CVDSER.P.. ........GT A (479)
```

Dots indicate identity; dashes gaps introduced for alignment.

**Domain structure**

| | |
|---|---|
| 2–78 | SH2 domain (underlined) |
| 5–108 | Pleckstrin homology: part of the region responsible for homodimerization [14] |
| 152–409 | Kinase domain |
| 158–166, 181 | ATP binding |
| 275 | Active site |
| Conflict: | |
| 478–481 | SIRE → FREEKDLLMSLFVSLILFSDFSSLKSHSFSSNFCILLSFSSLKK [15] |

**Structure of v-AKT**

The entire coding region of *Akt* is identical to that of v-*akt* apart from five G to A transitions that do not alter the reading frame. There are three single-base differences in the 3′UTR. v-*akt* encodes a tripartite *gag* (p12, p15, Δp30)-X-*Akt* product arising from a recombination event between the virus at nucleotide 785 from the *gag* ATG codon and the 5′ untranslated region of *Akt* to 60 bp 5′ from the *Akt* ATG codon. Three nucleotides absent from both *gag* and *Akt* were inserted at the junction between the two genes. The outcome of these events was to place, in frame, a 63 bp fragment between *gag* and *Akt*[2].

An altered form of the protein is expressed following insertion of MMTV into the first exon of the gene[16]. Additional N-terminus: MKTLRPGRLGSAYREETL-SIIPGLPLSLGATDT preceding the normal first Met residue. The underlined 10 amino acids correspond to 269–278 of v-AKT.

**Database accession numbers**

| | *PIR* | *SWISSPROT* | *EMBL/GENBANK* | *REFERENCES* |
|---|---|---|---|---|
| Human *AKT2* | A46288 | P31751, | HSAKT2A M95936 | 15 |
| | | KRCB_HUMAN | M77198 | 17 |
| Mouse *Akt* | | | X65687 | 2 |
| Mouse *Akt2* | | | U22445 | 5 |
| Mouse v-*akt* (AKT8) | B40831 | P31748, | Aktakta M80675, | 6 |
| | | KAKT_MLVAT | M61767 | |

***References***


1 Staal, S.P. (1987) Proc. Natl Acad. Sci. USA 84, 5034–5037.
2 Bellacosa, A. et al. (1993) Oncogene 8, 745–754.
3 Franke, T.F. et al. (1994) Oncogene 9, 141–148.
4 Ahmed, N.N. et al. (1993) Oncogene 8, 1957–1963.
5 Altomare, D.A. et al. (1995) Oncogene 11, 1055–1060.
6 Bellacosa, A. et al. (1991) Science 254, 274–277.
7 **Bos, J.L. (1995) Trends Biochem. Sci. 20, 441–442.**
8 Kohn et al. (1996) J. Biol. Chem. 271, 21920–21926.
9 Franke, T.F. et al. (1995) Cell 81, 727–736.
10 Cross, D.A.E. et al. (1995) Nature 378, 785–789.
11 Staal, S.P. et al. (1988) Genomics, 2, 96–98.
12 Bellacosa, A. et al. (1995) Int. J. Cancer 64, 280–285.
13 Cheng, J.Q. et al. (1996) Proc. Natl Acad. Sci. USA 93, 3636–3641.
14 Datta, K. et al. (1995) Mol. Cell. Biol. 15, 2304–2310.
15 Jones, P.F. et al. (1991) Cell Regul. 2, 1001–1009.
16 Wada, H. et al. (1995) Cancer Res. 55, 4780–4783.
17 Cheng, J.Q. et al. (1992) Proc. Natl Acad. Sci. USA 89, 9267–9271.

# BCL2

## Identification

*BCL2* (B cell leukemia/lymphoma-2) was detected with a probe specific for chromosome 18 in DNA isolated from a pre-B cell acute lymphocytic leukaemia cell line [1].

## Related genes

Human BAX is 21% identical (43% similar) to BCL2, and human and murine BAX are 89% identical [2]. Two mRNAs are transcribed from human *BCLX* encoding $BCLX_L$ (41% identical to BCL2α) and $BCLX_S$ (identical to $BCLX_L$ save for a 63 amino acid deletion). BAD, BRAG-1 (brain-related apoptosis gene) and BFL1 (BCL2-related gene expressed in fetal liver) contain regions of strong homology to the BH1 and BH2 domains of BCL2. BAK and BIK share homology with the BH3 domain of BCL2 [3,4]. BCLW shares homology and functional similarity with BCL2 [5]. BCL2 has significant homology to avian NR-13 and weak homology to MCL1, murine A1, E1B 19kDa protein, LMW5-HL in African swine fever virus, Epstein–Barr virus protein BHRF1 (see Section VI, **Epstein–Barr virus**), *Xenopus xRI* and *xRII* and *Caenorhabditis elegans ced-9*.

### BCL2 family members

| Protein | Effect on apoptosis | Interacting proteins | References |
|---|---|---|---|
| BCL2 | Inhibits | BAX, BAK | 6,7 |
| $BCLX_L$ | Inhibits | BAX, BAK | 8 |
| $BCLX_S$ | Accelerates | BAX, BAK | 9–11 |
| BCLW | Inhibits | | 5 |
| BAD | Accelerates | BCL2, $BCLX_L$ | 12 |
| BAX | Accelerates | BCL2, $BCLX_L$, E1B 19 kDa | 2 |
| BAK | Accelerates/inhibits | BCL2, BCLX, E1B 19 kDa | 4,13,14 |
| BIK | Accelerates | BCL2, $BCLX_L$, EBV BHRF1 or E1B 19 kDa | 3,15,16 |
| BFL1 | Inhibits | BCL2 homologue | 17 |
| BRAG-1 | Inhibits | BCL2 homologue | 18 |
| MCL1 | Inhibits | | 19,20 |
| A1 | | BCL2 homologue | 21 |
| NR-13 | | BCL2 homologue | 22 |
| ASFV HMW%-HL | | BCL2 homologue | 23 |
| EBV BHRF1 | Inhibits | BCL2 homologue | 24 |
| E1B 19 kDa | Inhibits | BAX, BAK | 13 |
| CED-9 | Inhibits | BCL2 homologue | 25 |

| | *BCL2* |
|---|---|
| **Nucleotides (kb)** | >370 |
| **Chromosome** | 18q21 |
| **Mass (kDa): predicted** | 26 (BCL2α)/22 (BCL2β) |
| **expressed** | 26/30 |
| **Cellular location** | BCL2α: intracellular membranes (especially ER and nuclear membranes) BCL2β: cytoplasm |

**Tissue distribution**

*BCL2* is generally expressed in tissues characterized by apoptotic cell turnover and restricted to long-lived progenitor cells and post-mitotic cells that have an extended lifespan[26]. It is restricted within germinal centres to regions implicated in the selection and maintenance of plasma cells and memory B cells, and showing stage-specific expression during B cell development[27]. It is expressed in surviving T cells in the thymic medulla and in proliferating precursors, but not in post-mitotic maturation stages, of all haematopoietic lineages. Also expressed in glandular epithelium under hormonal or growth factor control, in complex differentiating epithelium characterized by long-lived stem cells and in some neurones[28].

## Protein function

BCL2 is the prototypic member of a family of cell death regulators and its expression inhibits apoptosis induced by a variety of stimuli in numerous types of cell. BCL2 may be functionally equivalent to the product of the Epstein–Barr virus early gene BHRF1 (see Section VI, Epstein–Barr virus)[7,29–32]. The *Caenorhabditis elegans ced-9* gene is the homologue of BCL2: it represses *ced-3* and *ced-4*, the former encoding a protein homologous to the protease IL-1$\beta$-converting enzyme (ICE). In mammalian cells, BCL2 and $BCLX_L$ function upstream of the ICE/CED-3-related cysteine proteases Yama and ICE-LAP3 (see Chapter 5)[33]. The cell-type-specific effects of targeting BCL2 to different intracellular membranes suggest that pathways mediating apoptosis may be preferentially enriched at different sites[34].

BCL2 can homodimerize but it can also form dimers with $BCLX_L$, $BCLX_S$, BAX and MCL1[35,36]. BAX is a conserved BCL2 homologue that is expressed in a predicted membrane form ($\alpha$) and in two cytosolic forms, $\beta$ and $\gamma$[2]. The BH1 and BH2 domains of BLC2 are necessary for dimerization with BAX and repression of apoptosis. BAX also heterodimerizes with $BCLX_L$, MCL1 and A1 and over-expression of BAX can accelerate apoptosis. BAX may be a common partner in competing dimerizations that modulate the death repressor activity of BCL2 and regulate susceptibility to apoptosis[37].

Both p53 and $TGF\beta_1$ downregulate *Bcl-2* expression in M1 myeloid leukaemia cells and p53, but not $TGF\beta_1$, activates *Bax* expression[38,39]. This may explain why p53-induced apoptosis occurs more rapidly in these cells than that induced by $TGF\beta_1$. The over-expression of BCL2 inhibits the p53-dependent expression of the cyclin-dependent kinase inhibitor WAF1[40].

The BCL2-related protein $BCLX_L$ is as effective as BCL2 in inhibiting cell death following growth factor withdrawal[8]. In MCF7 breast carcinoma cells $BCLX_L$ and BCL2 inhibit TNF and Fas-induced apoptosis and activation of phospholipase $A_2$[41] but in B lymphoid cell lines BCL2 provides little protection against Fas/APO-1-transduced apoptosis[42]. The truncated form of $BCLX_L$, $BCLX_S$, inhibits the ability of BCL2 or $BCLX_L$ to enhance the survival of growth factor-deprived cells.

BAG-1 is a BCL2 binding protein with no homology to the BCL2 family[43]. Coexpression of BAG-1 and BCL2 protects Jurkat cells and fibroblasts from apoptosis. BAD selectively dimerizes with $BCLX_L$ and BCL2 and reverses the death repressor activity of $BCLX_L$ but not that of BCL2[12]. Dimerization of BAD with BCL-$X_L$ displaces BAX and restores apoptosis.

BCL2 (and also E1B 19 kDa protein) interact with three different proteins, NIP1, NIP2 and NIP3 that may be involved in suppression of cell death[44]. The p53 binding protein 53BP2 also interacts with BCL2 to impede cell cycle progression at $G_2$/M[45].

MCL1 has functional homology with BCL2[18] whereas BIK promotes cell death in a manner similar to BAX and BAK and its activity can be suppresed by coexpression of BCL2, $BCLX_L$, EBV BHRF1 or E1B 19 kDa.

BCL2 or $BCLX_L$ prevent p53- or $\gamma$-irradiation-induced apoptosis and have been shown to function in an anti-oxidant pathway to inhibit lipid peroxidation[46,47]. However, BCL2 expression in superoxide dismutase-deficient *E. coli* indicates that it functions as a pro-oxidant, increasing transcription of a catalase-peroxidase, resistance to $H_2O_2$ and mutation rate[48]. In human B cells BCL2 expression inhibits the depletion of the endoplasmic reticulum calcium store in response to $H_2O_2$[49] and the inhibition of adriamycin-induced apoptosis in prostate carcinoma cells by BCL2 correlates with a reduction in the nuclear concentration of $Ca^{2+}$[50]. In a murine B cell line over-expressing BCL2, superoxide dismutase activity is increased by 73%. In addition to suppressing apoptosis, BCL2 or $BCLX_L$ have the effect of increasing the levels of radiation-induced mutations[51]. BCL2 and $BCLX_L$ have also been reported to inhibit cell death induced by chemical hypoxia by maintaining mitochondrial membrane potential after exposure to cyanide or other inhibitors of respiration[52].

Glucocorticoid-induced apoptosis of pre-B lymphocytes is blocked by the expression of high levels of BCL2 by a mechanism that requires the concurrent repression of *MYC* (see ***MYC***)[53]. In fibroblasts in which the over-expression of MYC is induced, the apoptotic cell death that would normally occur is prevented by the coexpression of BCL2[54,55]. Thus the proto-oncogenes *MYC* and *BCL2* can cooperate to prevent apoptosis and cause continuous cell proliferation in the absence of mitogens, although the cells do not appear to be transformed.

The induced expression of oncogenic RAS protein in haematopoietic cells up-regulates BCL2 and $BCLX_L$ but not BAX, which mechanism may underlie the IL-3/GM-CSF-induced inhibition of apoptosis *via* activation of the RAS pathway[56]. However, BCL2 can associate with RAS in cells in which protein kinase C has been downregulated and this interaction leads to the phosphorylation of BCL2 which prevents protection from apoptosis[57]. BCL2 also binds to the RAS-related protein R-RAS p23[58]. The cytotoxic drug taxol induces phosphorylation and inactivation of BCL2, both of which require the activity of RAF1. However, although RAF1 can be immunoprecipitated with BCL2, it may not be the kinase responsible for its phosphorylation[59,60]. These observations suggest that components of the RAS-RAF growth signalling pathway may also be involved in regulating apoptosis.

**Cancer**

The chromosomal translocation t(14;18)(q32;q21) is a specific abnormality of human lymphoid neoplasms that occurs in >85% of follicular small cleaved B cell lymphomas and ~20% of diffuse lymphomas[61]. The major breakpoint region (MBR) on chromosome 18, involved in 60% of these cases, is within the 3′ untranslated part of *BCL2* exon 3[62]. This region contains a target sequence, also present in the *Dxp* genes of the immunoglobulin (Ig) diversity ($D_H$) family, to which BCLF proteins bind to mediate chromosomal translocation[63]. The minor cluster region (MCR) [40%] is ~20 kb downstream of *BCL2*. Additional breakpoints occur at the 5′ and 3′ ends of *BCL2*[64]. These translocations create a *BCL2*/Ig fusion gene. However, although the neoplastic germinal centres in most follicular lymphomas express high levels of BCL2 protein, whereas normal germinal centres do not, BCL2 is also present in normal T and B cells and in hairy cell leukaemias and Ki-1 lymphomas that do *not* involve the 14;18 translocation[65,66].

In ~10% of chronic lymphocytic leukaemia (CLL) *BCL2* is translocated to the Ig$\kappa$ or Ig$\lambda$ light chain gene[67]. The chromosome 18 breakpoints involved are within the 5′ end of the *BCL2* gene, the variant cluster region (VCR). The simultaneous presence of MBR (or MCR) translocations and of minor rearrangements involving deletions in VCR occurs in some follicular lymphomas[68].

*BCL2* is expressed in normal and malignant plasma cells from myeloma patients[69,70] and in a number of human lymphoid and myeloid cell lines and tissues[71,72]. High expression of *BCL2* occurs in some human neuroblastoma and small cell lung carcinoma cell lines and lower levels of the protein occur in a variety of other neural crest-derived tumours and tumour cell lines, including some neuroepitheliomas, Ewing's sarcomas, neurofibromas and melanomas[73,74].

Expression of BCL2 has also been detected in colorectal adenomas and carcinomas[75] and an inverse correlation between BCL2 and p53 expression suggests that inhibition of apoptosis may be involved in colorectal cancers[76]. For breast cancer patients treated with adjuvant chemotherapy or tamoxifen, those with BCL2 protein positive tumours had significantly better survival rates[77].

BAX protein is identifiable by immunostaining in normal breast epithelium but reduced levels of BAX have been detected in 34% of metastatic breast tumours and this correlates with more rapid tumour progression and shorter overall patient survival time[78].

**Transgenic animals**

In transgenic mice that over-express *Bcl-2* the lifetime of immunoglobulin-secreting cells and memory B cells is extended and the proportion of $CD4^-8^+$ thymocytes is increased[79,80]. The response to immunization is enhanced, consistent with reduced death of activated T cells, and the overproduction of BCL2 substantially alters the $V_H$ gene repertoire in B cells[81,82]. The introduction of *Bcl-2* into *scid* mice (in which the failure to make productive rearrangements of Ig and TCR genes causes early abortion of lymphopoiesis) permits the accumulation of almost normal numbers of $Ig^-$ B lymphoid cells[83]. *TCR/Bcl-2/scid* mice develop normal numbers of $CD4^+8^+$ thymocytes, indicating that expression of the TCR is necessary for T cells to respond to BCL2. Thus BCL2 sustains immune responsiveness.

Mice transgenic for a *Bcl-2*/Ig fusion gene show an expansion of the lymphoid compartment and mature B cells from these mice show a survival advantage *in vitro*[84]. Mice doubly transgenic for *Bcl-2* and *Myc* under the control of the IgH enhancer show hyperproliferation of pre-B and B cells and develop tumours which appear at earlier times than in E$\mu$-*Myc* mice and display the phenotype of primitive haematopoietic cells having both B lymphoid and macrophage differentiation potential [85]. Thus, by extending cell survival, BCL2 may increase the chance of secondary genetic changes responsible for tumorigenicity. Expression of *Bcl-2* specifically in neurones protects against developmental and induced cell death [86].

*Nmyc* or *Lmyc* can also cooperate with *Bcl-2* to promote the development of B and T cell lymphomas. B cells from E$\mu$-*Nmyc*/*Bcl-2*-Ig double transgenic mice are resistant to the induction of apoptosis by some agents (e.g. $H_2O_2$) by a mechanism that involves suppression of any increase in the concentration of free intracellular $Ca^{2+}$ [87].

Targeted disruption of *Bcl-2αβ* gives rise to smaller mice with ~50% dying by 6 weeks [88]. There is a sequential loss of $CD8^+$ and then $CD4^+$ T cells, with B cells less affected. Abnormalities in non-lymphoid organs include smaller auricles, development of grey hair and polycystic kidney disease-like change of renal tubules. Thus BCL2 may be involved in morphogenesis but its deletion does not appear to affect the nervous system, the intestines or the skin, although in mice these normally express high levels of BCL2.

### In animals

The expression of *Bcl-2* in mice suppresses replication of alphavirus and protects against fatal encephalitis [89].

### *In vitro*

*BCL2* expressed from a retroviral vector does not morphologically transform NIH 3T3 cells, render FDCP-1 cells tumorigenic or immortalize normal bone marrow cells, but does immortalize pre-B cells from E$\mu$-*Myc* transgenic mice, some of which become tumorigenic. In addition, whilst not abolishing the growth factor requirements of some established haematopoietic cell lines, it promotes their extended survival in $G_0$ in the absence of growth factors [6].

A replication-defective adenoviral vector that over-expresses $BCLX_S$ induces apoptosis in MCF-7 cells (which over-express $BCLX_L$) and, after intra-tumoural injection, also causes apoptosis and tumour size reduction of solid MCF-7 tumours in nude mice [10].

Expression of HIV-1 TAT protein in haematopoietic cells causes down-regulation of BCL2 and increased expression of BAX, resulting in induction of apoptosis [90].

Anucleate cytoplasts have also been shown to undergo apoptosis when deprived of survival factors and are transiently protected from cell death by BCL2 [91]. This indicates that programmed cell death is effected by a cytoplasmic regulator.

## Gene structure

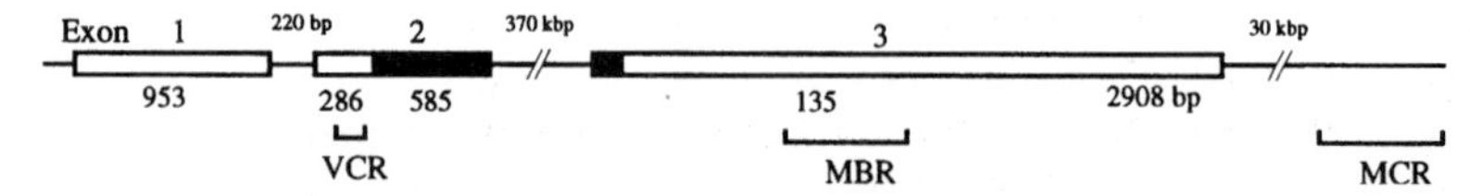


Alternative promoters utilize either exons 2 and 3 or exons 1, 2 and 3 [92].

## Transcriptional regulation

The major transcriptional promoter, P1, lies 1386–1423 bp upstream of the translation start site. It is TATA-less and GC-rich with multiple start sites and includes an SP1 site. P2 is a minor promoter at 1.3 kb downstream from P1 in some tissues. A 195 bp segment of a negative regulatory element (NRE) upstream of P2 controls p53-mediated downregulation[93–95]. Three $\pi$1 sites (−1781 to −1789, −1660 to −1668, −1032 to −1040) negatively regulate BCL2 transcription in the pre-B cell line Nalm-6 but not in the mature B cell line DHL-9[96]. Two potential triplex target sequences, Pur1 and Pur2 occur at −1464 and −1568, just 5′ of P1[97]. The regions −1552 to −1526 (DRE) and −1611 to −1552 (URE) constitute the major positive regulatory region of BCL2 in B cells and the DRE includes a cAMP response element[98]. There is a perfect MYB consensus sequence adjacent to the P2 promoter that can act as a strong positive regulatory element[99].

There is an 11 amino acid upstream ORF at −119 to −84 bp in the human gene which may be responsible for inhibiting *BCL2* expression at the translational level[100].

The immediate upstream region from the MBR (see **Cancer**) is a specific recognition site for single-strand DNA binding proteins on both sense and anti-sense strands, maximally active in late S and $G_2$/M phases. The downstream flank is a helicase binding site, maximally active in $G_1$ and early S phases. The Ku antigen is one component of the helicase complex[101].

Aberrant hypermethylation of a site within *BCL2* has been detected in non-small cell lung cancer[102].

## Protein structure

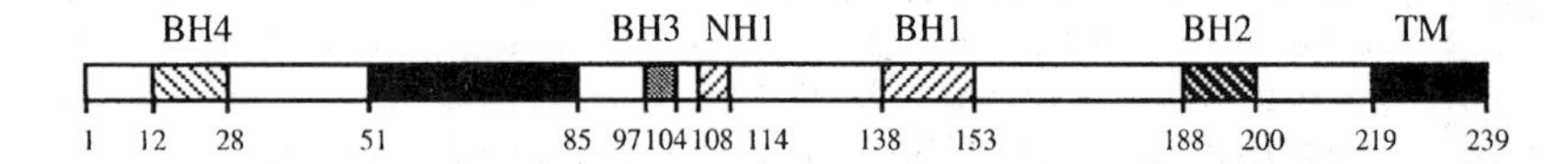


## Amino acid sequence of human BCL2$\alpha$

```
  1 MAHAGRTGYD NREIVMKYIH YKLSQRGYEW DAGDVGAAPP GAAPAPGIFS
 51 SQPGHTPHPA ASRDPVARTS PLQTPAAPGA AAGPALSPVP PVVHLTLRQA
101 GDDFSRRYRR DFAEMSSQLH LTPFTARGRF ATVVEELFRD GVNWGRIVAF
151 FEFGGVMCVE SVNREMSPLV DNIALWMTEY LNRHLHTWIQ DNGGWDAFVE
201 LYGPSMRPLF DFSWLSLKTL LSLALVGACI TLGAYLGHK (239)
```

**Domain structure**

| | |
|---|---|
| 34–85 | Domain II (italics): 64% conserved between mouse and human BCL2 but poorly conserved in the chicken protein |
| 6–31 | N-terminal domain essential for apoptosis |
| 51–85 | Deletion retains capacity to suppress apoptosis but also promotes cell proliferation[103] |
| 90–203 | C-terminal domain essential for apoptosis[104] |
| 97–105 | BCL2 homology 3 domain (BH3: underlined) |
| 136–155 | BCL2 homology 1 domain (BH1: underlined) |
| 187–202 | BCL2 homology 2 domain (BH2: underlined) |

| | |
|---|---|
| 145 | Mutation of Gly$^{145}$ in the BH1 domain or Trp$^{188}$ in the BH2 domain completely blocks BCL2 activity in IL-3 deprivation, $\gamma$-irradiation and glucocorticoid-induced apoptosis and prevents dimerization with BAX [105]. |
| 219–232 | Transmembrane domain (underlined italics) |
| 217–239 | Membrane targeting signal [106] |

Regions 42–50 and 108–114 (AAPAPGIFS and YRRDFAE) [44] show homology with E1B 19 kDa protein [107]. 108–114 (the NH1 domain) carries survival-promoting activity in BCL2: the homologous region in E1B (YKWEFEE) is functionally exchangeable between the two proteins.

The structure of the BCL2-related protein $BCLX_L$ comprises two central, primarily hydrophobic, α-helices surrounded by amphipathic helices. The three BCL2 homology domains (BH1, BH2 and BH3) form an elongated, hydrophobic cleft [108].

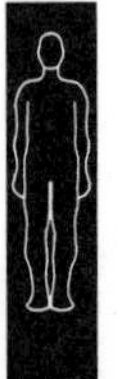

## Amino acid sequence of human BCL2β

```
  1 MAHAGRTGYD NREIVMKYIH YKLSQRGYEW DAGDVGAAPP GAAPAPGIFS
 51 SQPGHTPHPA ASRDPVARTS PLQTPAAPGA AAGPALSPVP PVVHLTLRQA
101 GDDFSRRYRR DFAEMSSQLH LTPFTARGRF ATVVEELFRD GVNWGRIVAF
151 FEFGGVMCVE SVNREMSPLV DNIALWMTEY LNRHLHTWIQ DNGGWVGASG
201 DVSLG (205)
```

### Domain structure

| | |
|---|---|
| 196–205 | Divergence between BCL2β and BCL2α (alternatively spliced versions of the same gene). |

### Database accession numbers

| | PIR | SWISSPROT | EMBL/GENBANK | REFERENCES |
|---|---|---|---|---|
| Human *BCL2α* | A24428, | P10415 | M13994 | 109 |
| | A29409 | | M14745 | 110 |
| Human *BCL2β* | B29409 | P10416 | M13995 | 95 |
| Human $BCLX_L$ | | | L20121 | 8 |
| Human $BCLX_S$ | | | L20122 | 8 |
| Human *BAXα* | | | L22473 | 2 |
| Human *BAXβ* | | | L22474 | 2 |
| Human *BAXγ* | | | L22475 | 2 |

### *References*


1 Tsujimoto, Y. et al. (1984) Science 224, 1403–1406.
2 Oltvai, Z.N. et al. (1993) Cell 74, 609–619.
3 Boyd, J.M. et al. (1995) Oncogene 11, 1921–1928.
4 Chittenden et al. (1995) EMBO J. 14, 5589–5596.
5 Gibson, L. et al. (1996) Oncogene 13, 665–675.
6 Vaux, D.L. et al. (1988) Nature 335, 440–442.
7 Hockenbery, D. et al. (1990) Nature 348, 334–336.
8 Boise, L.H. et al. (1993) Cell 74, 597–608.
9 Clarke, M.F. et al. (1995) Proc. Natl Acad. Sci. USA 92, 11024–11028.
10 Ealovega, M.W. et al. (1996) Cancer Res. 56, 1965–1969.
11 Minn, A.J. et al. (1996) J. Biol. Chem. 271, 6306–6312.
12 Yang, E. et al. (1995) Cell 80, 285–291.

[13] Farrow, S.N. et al. (1995) Nature 374, 731–733.
[14] Kiefer, M.C. et al. (1995) Nature 374, 736–739.
[15] Chittenden, T. et al. (1995) Oncogene 11, 1921–1928.
[16] Han, J. et al. (1996) Mol. Cell. Biol. 16, 5857–5864.
[17] Choi, S.S. et al. (1995) Oncogene 11, 1693–1698.
[18] Das, R. et al. (1996) Oncogene 12, 947–951.
[19] Reynolds, J.E. et al. (1994) Cancer Res. 54, 6348–6352.
[20] Kozopas, K. M. et al. (1993) Proc. Natl Acad. Sci. USA 90, 3516–3520.
[21] Lin, E. Y. et al. (1993) J. Immunol. 151, 1979–1988.
[22] Gillet, G. et al. (1995) EMBO J. 14, 1372–1381.
[23] Neilan, J.G. et al. (1993) J. Virol. 67, 4391–4394.
[24] Henderson, S. et al. (1993) Proc. Natl Acad. Sci. USA 90, 8479–8483.
[25] Hengartner, M.O. and Horvitz, H.R. (1994) Cell 76, 665–676.
[26] Villuendas, R. et al. (1991) Am. J. Pathol. 139, 989–993.
[27] Merino, R. et al. (1994) EMBO J. 13, 683–691.
[28] Hockenbery, D. et al. (1991) Proc. Natl Acad. Sci. USA 88, 6961–6965.
[29] **Korsmeyer, S.J. (1992) Blood 80, 879–886.**
[30] Borzillo, G.V. et al. (1992) Oncogene 7, 869–876.
[31] Garcia, I. et al. (1992) Science 258, 302–304.
[32] Allsopp, T.E. et al. (1993) Cell 73, 295–307.
[33] Chinnaiyan, A.M. et al. (1996) J. Biol. Chem. 271, 4573–4576.
[34] Zhu, W. et al. (1996) EMBO J. 15, 4130–4141.
[35] Sato, T. et al. (1994) Proc. Natl Acad. Sci. USA 91, 9238–9242.
[36] **Farrow, S.N. and Brown, R. (1996) Curr. Opin. Genet. Dev. 6, 45–49.**
[37] Sedlak, T.W. et al. (1995) Proc. Natl Acad. Sci. USA 92, 7834–7838.
[38] Selvakumaran, M. et al. (1994) Oncogene 9, 1791–1798.
[39] Miyashita, T. and Reed, J.C. (1995) Cell 80, 293–299.
[40] Upadhyay, S. et al. (1995) Cancer Res. 55, 4520–4524.
[41] Jäättela, M. et al. (1995) Oncogene 10, 2297–2305.
[42] Strasser, A. et al. (1995) EMBO J. 14, 6136–6147.
[43] Takayama, S. et al. (1995) Cell 80, 279–284.
[44] Boyd, J.M. et al. (1994) Cell 79, 341–351.
[45] Naumovski, L. and Cleary, M.L. (1996) Mol. Cell. Biol. 16, 3884–3892.
[46] Hockenbery, D. et al. (1993) Cell 75, 241–251.
[47] Schott, A.F. et al. (1995) Oncogene 11, 1389–1394.
[48] Steinman, H.M. (1995) J. Biol. Chem. 270, 3487–3490.
[49] Distelhorst, C.W. et al. (1996) Oncogene 12, 2051–2055.
[50] Marin, M.C. et al. (1996) Oncogene 12, 2259–2266.
[51] Cherbonnel-Lasserre, C. et al. (1996) Oncogene 13, 1489–1497.
[52] Shimizu, S. et al. (1996) Oncogene 13, 21–29.
[53] Alnemri, E.S. et al. (1992) Cancer Res. 52, 491–495.
[54] Bissonnette, R.P. et al. (1992) Nature 359, 552–554.
[55] Fanidi, A. et al. (1992) Nature 359, 554–556.
[56] Kinoshita, T. et al. (1995) Oncogene 10, 2207–2212.
[57] Chen, C.Y. and Faller, D.V. (1996) J. Biol. Chem. 271, 2376–2379.
[58] Fernandez-Sarabia, M.J. and Bischoff, J.R. (1993) Nature 366, 274–275.
[59] Wang, H.-G.(1994) Oncogene 9, 2751–2756.
[60] Blagosklonny, M.V. et al. (1996) Cancer Res. 56, 1851–1854.
[61] Tanaka, S. et al. (1992) Blood 79, 229–237.
[62] Limpens, J. et al. (1991) Oncogene 6, 2271–2276.

[63] Aoki, K. et al. (1994) Oncogene 9, 1109–1115.
[64] Weiss et al. (1987) New Engl. J. Med. 317, 1185–1189.
[65] Pezzella, F. et al. (1990) Am. J. Path. 137, 225–232.
[66] Pezzella, F. et al. (1992) Br. J. Cancer 65, 87–89.
[67] Tashiro, S. et al. (1992) Oncogene 7, 573–577.
[68] Seite, P. et al. (1993) Oncogene 8, 3073–3080.
[69] Hamilton, M.S. et al. (1991) Leukemia 5, 768–771.
[70] Pettersson, M. et al. (1992) Blood 79, 495–502.
[71] Delia, D. et al. (1992) Blood 79, 1291–1298.
[72] Haury, M. et al. (1993) Oncogene 8, 1257–1262.
[73] Reed, J.C. et al. (1991) Cancer Res. 51, 6529–6538.
[74] Ikegaki, N. et al. (1994) Cancer Res. 54, 6–8.
[75] Hague, A. et al. (1994) Oncogene 9, 3367–3370.
[76] Sinicrope, F.A. et al. (1995) Cancer Res. 55, 237–241.
[77] Gasparini, G. et al. (1995) Clin. Cancer Res. 1, 189–198.
[78] Krajewski, S. et al. (1995) Cancer Res. 55, 4471–4478.
[79] Nunez, G. et al. (1990) Nature 353, 71–73.
[80] Sentman, C.L. et al. (1991) Cell 67, 879–888.
[81] Yeh, T.M. et al. (1991) Int. Immunol. 3, 1329–1333.
[82] Katsumata, M. et al. (1992) Proc. Natl Acad. Sci. USA 89, 11376–11380.
[83] Strasser, A. et al. (1994) Nature 368, 457–460.
[84] McDonnell, T.J. et al. (1989) Cell 57, 79–88.
[85] Strasser, A. et al. (1996) EMBO J. 15, 3823–3834.
[86] Farlie, P.G. et al. (1995) Proc. Natl Acad. Sci. USA 92, 4397–4401.
[87] Zörnig, M. et al. (1995) Oncogene 11, 2165–2174.
[88] Nakayama, K. et al. (1994) Proc. Natl Acad. Sci. USA 91, 3700–3704.
[89] Levine, B. et al. (1996) Proc. Natl Acad. Sci. USA 93, 4810–4815.
[90] Sastry, K.J. et al. (1996) Oncogene 13, 487–493.
[91] Jacobson, M.D. et al. (1994) EMBO J. 13, 1899–1910.
[92] Seto, M. et al. (1988) EMBO J. 7, 123–131.
[93] Silverman et al. (1990) Proc. Natl Acad. Sci. USA 87, 9913–9917.
[94] Young, R.L. and Korsmeyer, S.J. (1993) Mol. Cell. Biol. 13, 3686–3697.
[95] Miyashita et al. (1994) Cancer Res. 54, 3131–3135.
[96] Chen, H.-M. and Boxer, L.M. (1995) Mol. Cell. Biol. 15, 3840–3847.
[97] Olivas, W.M. and Maher, L.J. (1996) Nucleic Acids Res. 24, 1758–1764.
[98] Wilson, B.E. et al. (1996) Mol. Cell. Biol. 16, 5546–5556.
[99] Taylor, D. et al. (1996) Genes Dev. 10, 2732–2744.
[100] Harigai, M. et al. (1996) Oncogene 12, 1369–1374.
[101] DiCroce, P.A. and Krontiris, T.G. (1995) Proc. Natl Acad. Sci. USA 92, 10137–10141.
[102] Nagatake, M. et al. (1996) Cancer Res. 56, 1886–1891.
[103] Uhlmann, E.J. et al. (1996) Cancer Res. 56, 2506–2509.
[104] Hunter, J.J. et al. (1996) Mol. Cell. Biol. 16, 877–883.
[105] Yin, X.-M. et al. (1994) Nature 369, 321–323.
[106] Nguyen, M. et al. (1993) J. Biol. Chem. 268, 25265–25268.
[107] Subramanian, T. et al. (1995) Oncogene 11, 2403–2409.
[108] Muchmore, S.W. et al. (1996) Nature 381, 335–341.
[109] Tsujimoto, Y. and Croce, C. (1986) Proc. Natl Acad. Sci. USA 83, 5214–5218.
[110] Cleary, M.L. et al. (1986) Cell 47, 19–28.

## Identification

*BCL3* (B cell leukemia/lymphoma-3) was detected by molecular cloning of the breakpoint junction of the 14;19 translocation [1].

## Related genes

BCL3 is a member of the IκB family. The seven ankyrin repeat regions in BCL3 are homologous to those in human MAD3, the β subunit of the heteromeric DNA binding protein GABP, TAN1, LYT10, *notch* (*D. melanogaster*), *Xnotch* (*Xenopus laevis*), *lin*-12 and *glp*-1 (*Caenorhabditis elegans*), SW14/SW16 (*Saccharomyces cerevisiae*), *cdc*10 (*S. pombe*).

| | ***BCL3*** |
|---|---|
| **Nucleotides (kb)** | ~10–11 |
| **Chromosome** | 19q13.1 |
| **Mass (kDa): predicted** | 46.8 |
| **expressed** | 28/30 |
| **Cellular location** | Nucleus |

**Tissue distribution**
*BCL3* expression increases seven-fold in normal human T cells between 15 min and 8 h after stimulation by PHA [1] and is abundant in B cells just prior to the Ig switch.

## Protein function

BCL3 is a probable transcription factor. Phosphorylated BCL3 functions as a form of IκB specific for the p50 subunit of NF-κB, rather than the NF-κB heterodimer (see ***REL***), inhibiting its translocation to the nucleus and binding to DNA [2–5], and increasing κB-dependent *trans*-activation in intact cells by acting as an anti-repressor of inhibitory p50/NF-κB homodimers [6] Conflicting data indicate that BCL3 does not redistribute into the cytoplasm, even in the presence of excess p50, but alters the subnuclear distribution of p50 [7]. BCL3 does not inhibit the DNA binding activity of REL protein or its ability to *trans*-activate genes linked to a κB motif [8]. However, BCL3 also associates tightly with homodimers of p50B, a protein closely related to p50 [9]. Formation of the BCL3/p50 ternary complex permits BCL3 directly to *trans*-activate transcription *via* κB sites.

BCL3 promotes *RB1* transcription and functions as a regulator of *RB1* expression during muscle cell differentiation [10].

**Cancer**
The (14;19)(q32;q13.1) translocation occurs in some cases of B cell chronic lymphocytic leukaemia (B-CLL): *BCL3* is adjacent to the breakpoints

involved and recombination occurs between the *BCL3* and IgH loci[11]. Rearrangement of a gene designated *BCL3* and *BCL5* (17q22) has been reported in prolymphocytic leukaemia but this gene is unrelated to the chromosome 19 B-CLL associated gene.

**Transgenic animals**

Expression of *Bcl-3* in mice increases the DNA binding activity of endogenous p50 (NF-κB1) homodimers by >10-fold without affecting p52 (NF-κB2) dimers[12].

## Gene structure

*BCL3* has not been fully mapped but consists of at least seven exons.

## Amino acid sequence of human BCL3

```
  1 MDEGPVDLRT RPKAAGLPGA ALPLRKRPLR APSPEPAAPR GAAGLVVPLD
 51 PLRGGCDLPA VPGPPHGLAR PEALYYPGAL LPLYPTRAMG SPFPLVNLPT
101 PLYPMMCPME HPLSADIAMA TRADEDGDTP LHIAVVQGNL PAVHRLVNLF
151 QQGGRELDIY NNLRQTPLHL AVITTLPSVV RLLVTAGASP MALDRHGQTA
201 AHLACEHRSP TCLRALLDSA APGTLDLEAR NYDGLTALHV AVNTECQETV
251 QLLLERGADI DAVDIKSGRS PLIHAVENNS LSMVQLLLQH GANVNAQMYS
301 GSSALHSASG RGLLPLVRTL VRSGADSSLK NCHNDTPLMV ARSRRVIDIL
351 RGKATRPAST SQPDPSPDRS ANTSPESSSR LSSNGLLSAS PSSSPSQSPP
401 RDPPGFPMAP PNFFLPSPSP PAFLPFAGVL RGPGRPVPPS PAPGGS (466)
```

**Domain structure**

| | |
|---|---|
| 1–112 | Proline-rich domain |
| 120–156, 157–189, 190–226, 227–260, 261–293, 294–326, 327–359 | Seven repeated ankyrin motifs (underlined) |
| 357–446 | Serine/proline-rich domain |

**Database accession numbers**

| | PIR | SWISSPROT | EMBL/GENBANK | REFERENCES |
|---|---|---|---|---|
| Human *BCL3* | A34794 | P20749 | M31731, M31732 | 1 |


***References***

1. Ohno, H. et al. (1990) Cell 60, 991–997.
2. Nolan, G.P. et al. (1993) Mol. Cell. Biol. 13, 3557–3566.
3. Hatada, E.N. et al. (1992) Proc. Natl Acad. Sci. USA 89, 2489–2493.
4. Naumann, M. et al. (1993) EMBO J. 12, 213–222.
5. Franzoso, G. et al. (1993) EMBO J. 12, 3893–3901.
6. Franzoso, G. et al. (1992) Nature 359, 339–342.
7. Zhang, Q. et al. (1994) Mol. Cell. Biol. 14, 3915–3926.
8. Kerr, L.D. et al. (1992) Genes Dev. 6, 2352–2363.
9. Bours, V. et al. (1993) Cell 72, 729–739.
10. Shiio, Y. et al. (1996) Oncogene 12, 1837–1845.
11. Ohno, H. et al. (1993) Leukemia 7, 2057–2063.
12. Caamaño, J.H. et al. (1996) Mol. Cell. Biol. 16, 1342–1348.

## Identification

*BCR* (breakpoint cluster region) is the first defined member of a small gene family localized on human chromosome 22, detected by screening DNA with a probe specific for the Philadelphia translocation breakpoint [1].

## Related genes

There are three *BCR*-related loci: *BCR2*, *BCR3* and *BCR4* [2]. Human *ABR* closely resembles BCR, having the same expression pattern, DBL homology region (that stimulates GTP binding to CDC42Hs > RHOA > RAC1 and RAC2) and $GAP^{rho}$ activity (both act as GAPs for RAC1, RAC2 and CDC42Hs but are inactive towards RHOA, RAP1A or HRAS). However, ABR lacks homology with the *BCR* serine/threonine kinase domain [3,4].

| | ***BCR*** |
|---|---|
| **Nucleotides (kb)** | 130 |
| **Chromosome** | 22q11.2 |
| **Mass (kDa): predicted** | 143 |
| **expressed** | 160<br>$P210^{BCR\text{-}ABL}$/$P185^{BCR\text{-}ABL}$ |
| **Cellular location** | Cytoplasm |

**Tissue distribution**

*BCR* is widely expressed in many types of human haematopoietic and non-haematopoietic cells and cell lines [5].

## Protein function

BCR has serine/threonine protein kinase activity [6], contains a central domain with homology to guanine nucleotide exchange proteins and has a C-terminal domain (absent in $P210^{BCR\text{-}ABL}$ and $P185^{BCR\text{-}ABL}$) that possesses *in vitro* GTPase-activating protein activity [7]. BCR has autophosphorylating activity *in vitro*. BCR/ABL, directly or indirectly, can phosphorylate the following proteins: BAP-1, casein, histones [8], CRKL, JAK1, JAK2, PTP1C, PTP1D, STAT5 [9], $p120^{CBL}$, $p95^{VAV}$, $p46^{SHC}$, $p93^{FES}$ [10], $p67^{SYP}$, paxillin, vinculin, $p125^{FAK}$, talin and tensin [11]. BAP-1 (14-3-3$\tau$), a member of the ubiquitously expressed 14-3-3 family, members of which are essential for cell proliferation, associates with BCR and with BCR/ABL, being phosphorylated by both on serine/threonine or tyrosine residues, respectively [12]. The 14-3-3$\beta$ isoform binds to the B box of BCR [13]. RAF does not bind directly to BCR but BCR, 14-3-3$\beta$ and RAF can be isolated as a ternary complex. In Ph-positive leukaemia and in haematopoietic cell lines transformed by BCR/ABL, tyrosine phosphorylated $p120^{CBL}$ occurs attached *via* CRK and CRKL to BCR/ABL [14]. These complexes also include the p85 subunit of phosphatidylinositol-3′

kinase[15]. Both p190$^{BCR/ABL}$ and P210$^{BCR/ABL}$ and also ABL may be present in these complexes.

BCR/ABL exists *in vivo* associated with the GRB2/SOS complex that links tyrosine kinases to the RAS signalling system[16,17]. Mutation of a major autophosphorylation site (Tyr$^{793}$) within the catalytic domain, the GRB2 binding site (Tyr$^{177}$) or the SH2 domain (Arg$^{552}$) blocks growth stimulation of fibroblasts by BCR/ABL without significantly affecting kinase activity, indicating that GRB2 is involved in normal development and mitogenesis, and that autophosphorylation within the BCR element of BCR/ABL modifies target specificity of the ABL kinase[16,17]. Loss of the the CRKL binding site reduces the fibroblast transforming capacity of BCR/ABL but inhibition of both CRKL and GRB2 binding has a synergistic effect on the reduction of transformation[18]. However, these mutations do not diminish the anti-apoptotic and transforming properties of BCR/ABL in haematopoietic cells. In myeloid cells expressing P210$^{BCR\text{-}ABL}$ the SHC protein that interacts with GRB2 is constitutively phosphorylated[19] and activation of SHC provides an alternative signal from BCR/ABL to RAS in haematopoietic cells[20]. GRAP (GRB2-related adapter protein) also associates with activated BCR/ABL in K562 cells[21]. Thus, in haematopoietic cells, BCR/ABL appears to activate multiple signalling pathways to RAS[20]. Nevertheless, a BCR/ABL deletion mutant (Δ176–427) activates RAS and blocks apoptosis but is only weakly tumorigenic, indicating that RAS activation is not sufficient for BCR/ABL-mediated transformation[22]. In fibroblasts and haematopoietic cells a consequence of activation of the RAS pathway by BCR/ABL is that the JUN N-terminal kinase pathway (JNK) is activated and this appears to be involved in promoting BCR/ABL transformation[23]. In the transformation of haematopoietic cells, BCR/ABL proteins inhibit apoptosis by inducing the expression of *BCL2*[24].

### Cancer

The formation of the Philadelphia chromosome (Ph$^1$) in acute lymphoblastic leukaemia (ALL) and chronic myeloid leukaemia (CML) generates a chimeric *BCR/ABL* gene (see ***ABL***). The proliferation *in vitro* of blast cells from CML patients is selectively suppressed by anti-sense oligodeoxynucleotides directed against the *BCR/ABL* junction[25].

### Transgenic animals

Mice transgenic for a P185$^{bcr\text{-}abl}$ DNA construct develop aggressive leukaemias with early onset and rapid progression, consistent with there being a critical role for the *BCR/ABL* gene product of the Ph$^1$ chromosome in human leukaemia[26].

### *In vitro*

In general, BCR/ABL proteins are non-transforming but P210$^{BCR\text{-}ABL}$ or P185$^{BCR\text{-}ABL}$ (or v-ABL) induces both lymphoid and myeloid colonies in bone marrow cells, although only the lymphoid colonies are tumorigenic[27].

Both BCR/ABL and v-ABL appear capable of cooperating with cyclin D1 to transform fibroblasts and haematopoietic cells *in vitro*, *via* a signal dependent on the ABL SH2 domain[28].

## Gene structure and the formation of the Philadelphia chromosome

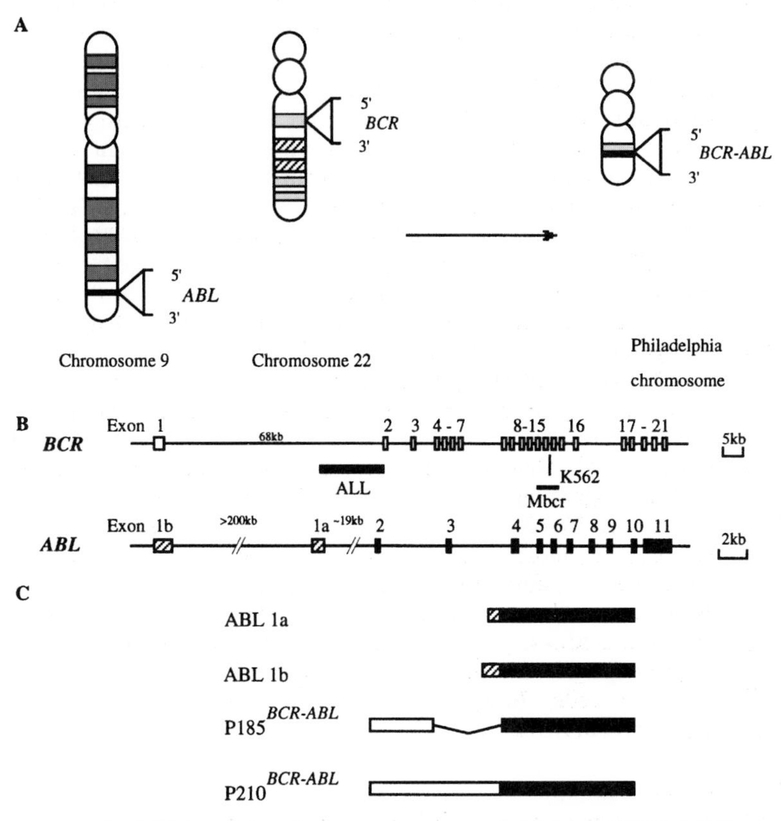


**A** The standard 9:22 reciprocal translocation occurs in about 92% of CML cases with a $Ph^1$ chromosome, resulting in a *BCR-ABL* chimeric gene on that chromosome. This translocation places *ABL* within the *BCR* gene on chromosome 22. The reciprocal recombination product is *5′-ABL/3′-BCR* on the $9q^+$ derivative which is also transcriptionally active[29]. Between 5 and 8% of CML cases carry variant translocations which are either (i) $Ph^1$ negative in which chromosome 22 is microscopically normal or (ii) complex variants in which up to three other chromosomes participate in translocation with chromosomes 9 and 22. The involvement of breakpoints on chromosomal regions other than 9q34 or 22q11 is non-random and all chromosomes except Y can participate. Chromosomal bands at 3p21, 6p21, 7p22, 11q13, 12p13, 17q21 and 19q13 are most commonly involved, an example being t(9;22;11)(q34;q11;q13) in which *BCR* 3′ of the breakpoint moves to 11q13, the breakpoint occurring 5′ of exon 4[30,31].

**B** $Ph^1$ chromosome breakpoints occur between exons 1a and 2 and also between exons 1a and 1b of *ABL*. The fused *BCR/ABL* gene contains the *BCR* exons 5′ of the breakpoint and all *ABL* exons except the first alternative exon. Thus in some CML patients the *ABL* gene located on the $Ph^1$ chromosome retains an intact exon 1a.

Mbcr (major breakpoint cluster region) indicates the region of the *BCR* gene in which breakpoints in CML are located[32]. Twenty-six codons from the 5′ end of *ABL* are replaced by 927 codons from the *BCR* gene and fused to 1104 amino acids from ABL, generating $P210^{BCR\text{-}ABL}$. Two minor breakpoint cluster regions (mbcr2 and mbcr3) are located within the 3′ half of the first *BCR* intron. K562 indicates the breakpoint in the K562 CML-derived cell line. The breakpoint in the *ABL* sequence is in close proximity to that occurring in the generation of v-*abl*: thus the mechanism of activation of *ABL* in CML and of *Abl* by Ab-MuLV is similar, although the N-terminal deletion in v-*abl* is much larger and internal or frame shift mutations do not occur in the *BCR/ABL* gene.

ALL indicates the 3′ region of intron 1 in which Mbcr negative, $Ph^1$-positive ALL breakpoints occur giving rise to a *BCR/ABL* mRNA that is translated into $P185^{BCR\text{-}ABL}$ (426 *BCR* encoded amino acids fused to 1104 from *ABL*)[33]. In ALL the breakpoints may also occur within Mbcr[34].

**C** Structures of types 1a and 1b ABL and the chimeric proteins $P185^{BCR\text{-}ABL}$ and $P210^{BCR\text{-}ABL}$, indicating the coding regions of the genes from which they are derived. Both forms of BCR/ABL retain the SH3 domain and lack a myristylation signal but tyrosine kinase activity is conferred by the additional N-terminal domain, specifically by sequences encoded by the first *BCR* exon[35]. $P210^{BCR\text{-}ABL}$ has a tyrosine kinase activity similar to that of v-ABL, potentially important substrates being RAS-GAP and its associated proteins p192 and p62[36]. $P185^{BCR\text{-}ABL}$ has a 5-fold greater tyrosine kinase activity, is a more potent transforming agent and is more often associated with acute than chronic leukaemia. ABL DNA binding activity is lost in the $P210^{BCR\text{-}ABL}$ protein.

## Transcriptional regulation

The *BCR* promoter occupies ~1 kb 5′ to exon 1 and contains six SP1 consensus sequences, two CCAAT boxes and no TATA-like boxes[37]. A zinc-finger protein engineered to bind specifically to a 9 bp region of *BCR/ABL* blocks transcription of the gene in transformed cells and causes them to revert to growth factor-dependent growth[38].

## Protein structure

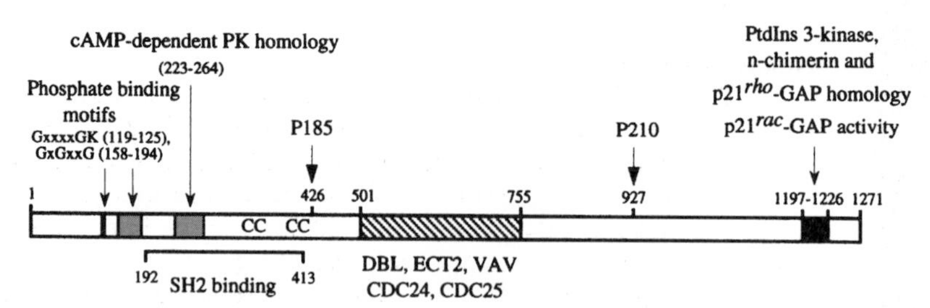


Cross-hatched box: homology with DBL, VAV, ECT2[39], *Saccharomyces cerevisiae* CDC24 (regulates cytokinesis in yeast) and CDC25[40]. Black box: GAP activity for $p21^{rac}$, homology with a $p21^{rho}$ GAP, PtdIns 3-kinase 85 kDa subunit and *n*-chimerin (which also has GAP activity for p21$^{rac}$) [7]. The C-termini of the portions incorporated into P185$^{BCR\text{-}ABL}$ and P210$^{BCR\text{-}ABL}$ are indicated (P185, P210). The N-terminal 63 amino acids mediate binding to actin [41].

Serine/threonine-rich regions of BCR (192–242 and 298–413) bind to the SH2 domain of ABL in a high-affinity, phosphotyrosine-independent interaction [42,43]. This interaction is essential for BCR/ABL-mediated transformation: for the chimeric BCR/ABL proteins the interaction may be inter- or intramolecular. The effect of the interaction may arise *via* interference with the binding of a cellular inhibitor to ABL regulatory domains.

## Amino acid sequence of human BCR

```
   1 MVDPVGFAEA WKAQFPDSEP PRMELRSVGD IEQELERCKA SIRRLEQEVN
  51 QERFRMIYLQ TLLAKEKKSY DRQRWGFRRA AQAPDGASEP RASASRPQPA
 101 PADGADPPPA EEPEARPDGE GSPGKARPGT ARRPGAAASG ERDDRGPPAS
 151 VAALRSNFER IRKGHGQPGA DAEKPFYVNV EFHHERGLVK VNDKEVSDRI
 201 SSLGSQAMQM ERKKSQHGAG SSVGDASRPP YRGRSSESSC GVDGDYEDAE
 251 LNPRFLKDNL IDANGGSRPP WPPLEYQPYQ SIYVGGMMEG EGKGPLLRSQ
 301 STSEQEKRLT WPRRSYSPRS FEDCGGGYTP DCSSNENLTS SEEDFSSGQS
 351 SRVSPSPTTY RMFRDKSRSP SQNSQQSFDS SSPPTPQCHK RHRHCPVVVS
                                   **
 401 EATIVGVRKT GQIWPNDGEG AFHGDADGSF GTPPGYGCAA DRAEEQRRHQ
 451 DGLPYIDDSP SSSPHLSSKG RGSRDALVSG ALESTKASEL DLEKGLEMRK
 501 WVLSGILASE ETYLSHLEAL LLPMKPLKAA ATTSQPVLTS QQIETIFFKV
 551 PELYEIHKEF YDGLFPRVQQ WSHQQRVGDL FQKLASQLGV YRAFVDNYGV
 601 AMEMAEKCCQ ANAQFAEISE NLRARSNKDA KDPTTKNSLE TLLYKPVDRV
 651 TRSTLVLHDL LKHTPASHPD HPLLQDALRI SQNFLSSINE EITPRRQSMT
 701 VKKGEHRQLL KDSFMVELVE GARKLRHVFL FTELLLCTKL KKQSGGKTQQ
 751 YDCKWYIPLT DLSFQMVDEL EAVPNIPLVP DEELDALKIK ISQIKSDIQR
 801 EKRANKGSKA TERLKKKLSE QESLLLLMSP SMAFRVHSRN GKSYTFLISS
 851 DYERAEWREN IREQQKKCFR SFSLTSVELQ MLTNSCVKLQ TVHSIPLTIN
                                     ++
 901 KEDDESPGLY GFLNVIVHSA TGFKQSSNLY CTLEVDSFGY FVNKAKTRVY
 951 RDTAEPNWNE EFEIELEGSQ TLRILCYEKC YNKTKIPKED GESTDRLMGK
1001 GQVQLDPQAL QDRDWQRTVI AMNGIEVKLS VKFNSREFSL KRMPSRKQTG
1051 VFGVKIAVVT KRERSKVPYI VRQCVEEIER RGMEEVGIYR VSGVATDIQA
1101 LKAAFDVNNK DVSVMMSEMD VNAIAGTLKL YFRELPEPLF TDEFYPNFAE
1151 GIALSDPVAK ESCMLNLLLS LPEANLLTFL FLLDHLKRVA EKEAVNKMSL
1201 HNLATVFGPT LLRPSEKESK LPANPSQPIT MTDSWSLEVM SQVQVLLYFL
1251 QLEAIPAPDS KRQSILFSTE V (1271)
```

### Domain structure

177 GRB2 SH2 domain binding site (Tyr[177]) in BCR/ABL. Tyr[177] is encoded by exon 1 of *BCR*: mutation blocks BCR/ABL-induced transformation, indicating that GRB2 is involved in oncogenesis as well as in normal development and mitogenesis [16]. Mutations in the highly conserved FLVRES SH2 motif in ABL also block transformation of fibroblasts by P185$^{bcr\text{-}abl}$ [44]. The overexpression of *Myc*, which synergizes with BCR/ABL to cause transformation, restores transformation by the FLVRES mutant but not by Tyr[177] (GRB2-binding) mutants, suggesting that BCR/ABL activates at least two independent transformation pathways.

| | |
|---|---|
| 197–239 | A box (Ser/Thr-rich region of kinase domain: underlined) |
| 299–385 | B box (Ser/Thr-rich region of kinase domain: underlined) |
| 426–427 | Breakpoint for translocation to form P185$^{BCR\text{-}ABL}$ oncogene (**) |
| 927–928 | Breakpoint for translocation to form P210$^{BCR\text{-}ABL}$ oncogene (++) |
| 501–755 | Homology with DBL, ECT2, VAV, TIM, CDC24 and CDC25 (underlined) |
| 1197–1226 | p21$^{rac}$-GAP activity region (italics) |

**Database accession numbers**

| | *PIR* | *SWISSPROT* | *EMBL/GENBANK* | *REFERENCES* |
|---|---|---|---|---|
| Human *BCR* | A26172 | P11274 | X02596, X52829 | 37 |
| | A26664 | | Y00661 | 45 |
| | A28765 | | M15025 | 46 |
| | | | M24603, M64437 | 47 |
| | | | | 48 |

*References*


1 **Konopka, J.B. and Witte, O.N. (1985) Biochim. Biophys. Acta 823, 1–17.**
2 Croce, C.M. et al. (1987) Proc. Natl Acad. Sci. USA 84, 7174–7178.
3 Heisterkamp, N. et al. (1993) J. Biol. Chem. 268, 16903–16906.
4 Chuang, T.-H. et al. (1995) Proc. Natl Acad. Sci. USA 92, 10282–10286.
5 Collins, S. et al. (1987) Mol. Cell. Biol. 7, 2870–2876.
6 Maru, Y. and Witte, O.N. (1991) Cell 67, 459–468.
7 Diekmann, D. et al. (1991) Nature 351, 400–402.
8 Liu, J. et al. (1993) Oncogene 8, 101–109.
9 Shuai, K. et al. (1996) Oncogene 13, 247–254.
10 Ernst, T.J. et al. (1994) J. Biol. Chem. 269, 5764–5769.
11 Salgia, R. et al. (1995) Oncogene 11, 1149–1155.
12 Reuther, G.W. et al. (1994) Science 266, 129–133.
13 Braselmann, S. and McCormick, F. (1995) EMBO J. 14, 4839–4848.
14 de Jong, R. et al. (1995) J. Biol. Chem. 270, 21468–21471.
15 Sattler, M. et al. (1996) Oncogene 12, 839–846.
16 Pendergast, A.M. et al. (1993) Cell 75, 175–185.
17 Puil, L. et al. (1994) EMBO J. 13, 764–773.
18 Senechal, K. et al. (1996) J. Biol. Chem. 271, 23255–23261.
19 Matsuguchi, T. et al. (1994) J. Biol. Chem. 269, 5016–5021.
20 Goga, A. et al. (1995) Oncogene 11, 791–799.
21 Feng, G.-S. et al. (1996) J. Biol. Chem. 271, 12129–12132.
22 Cortez, D. et al. (1995) Mol. Cell. Biol. 15, 5531–5541.
23 Raitano, A.B. et al. (1995) Proc. Natl Acad. Sci. USA 92, 11746–11750.
24 Sánchez-García, I. and Grütz, G. (1995) Proc. Natl Acad. Sci. USA 92, 5287–5291.
25 Szczylik, C. et al. (1991) Science 253, 562–565.
26 Heisterkamp, N. et al. (1990) Nature 344, 251–253.
27 Kelliher, M.A. et al. (1993) Oncogene 8, 1249–1256.
28 Afar, D.E.H. et al. (1995) Proc. Natl Acad. Sci. USA 92, 9540–9544.
29 Melo, J.V. et al. (1992) Blood 81, 158–165.
30 Koduru, P.R.K. et al. (1993) Oncogene 8, 3239–3247.
31 Morris, C. et al. (1996) Oncogene 12, 677–685.
32 Sowerby, S.J. et al. (1993) Oncogene 8, 1679–1683.

[33] Clark, S.S. et al. (1988) Science 239, 775–777.
[34] Hermans, A. et al. (1987) Cell 51, 33–40.
[35] Muller, A.J. et al. (1991) Mol. Cell. Biol. 11, 1785–1792.
[36] Carlesso, N. et al. (1994) Oncogene 9, 149–156.
[37] Shah, N.P. et al. (1991) Mol. Cell. Biol. 11, 1854–1860.
[38] Choo, Y. et al. (1994) Nature 372, 642–645.
[39] Miki, T. et al. (1993) Nature 362, 462–465.
[40] Cen, H. et al. (1992) EMBO J. 11, 4007–4015.
[41] McWhirter, J.R. and Wang, J.Y.J. (1993) EMBO J. 12, 1533–1546.
[42] Pendergast, A.M. et al. (1991) Cell 66, 161–171.
[43] Muller, A.J. et al. (1992) Mol. Cell. Biol. 12, 5087–5093.
[44] Afar, D.E.H. et al. (1994) Science 264, 424–426.
[45] Hariharan, I.K. and Adams, J.M. (1987) EMBO J. 6, 115–119.
[46] Heisterkamp, N. et al. (1985) Nature 315, 758–761.
[47] Lifshitz, B. et al. (1988) Oncogene 2, 113–117.
[48] Mes-Masson, A.M. et al. (1986) Proc. Natl Acad. Sci. USA 83, 9768–9772. [Correction: Proc. Natl Acad. Sci. USA 84, 2507.]

## Identification

v-*cbl* is the oncogene of the acutely transforming Cas NS-1 retrovirus (Casitas B-lineage lymphoma). *CBL* was detected by screening human haematopoietic cell lines with a v-*cbl* probe [1,2].

## Related genes

*CBL* has no homology with other known oncogenes but is 53% homologous to CBL-B [3] and 55% identical over 390 residues to *Caenorhabditis elegans sli-1*. It has sequence homology to the DNA binding and transcriptional activation domains of the yeast regulatory protein GCN4.

| | ***CBL*** | **v-*cbl*** |
|---|---|---|
| **Nucleotides** | 2718 | 7400 (Cas NS-1 genome) |
| **Chromosome** | 11q23.3–qter (*CBL2*) | |
| **Mass (kDa): predicted** | 100 | |
| **expressed** | p135 | $p100^{gag\text{-}cbl}$ |
| **Cellular location** | Cytoplasm | Cytoplasm/nucleus |

**Tissue distribution**

*Cbl* is expressed in cells of the B, T, erythroid, myeloid and mast cell lineages and is most readily detectable in the thymus and testis [4]. In proliferating fibroblasts and spleen cells *CBL* expression is constant throughout the cell cycle but its expression decreases in differentiating cells [5].

## Protein function

CBL participates in signal transduction from lymphokines and antigens in haematopoietic cells, including granulocyte-macrophage colony-stimulating factor (GM-CSF), erythropoietin (Epo) and EGF.

CBL associates with $p47^{nck}$ *via* the SH3 domains of NCK [6]. In activated Jurkat cells and B cells CBL undergoes tyrosine phosphorylation that correlates with the kinetics of dissociation of CBL from GRB2, tyrosine phosphorylated CBL then binding to all three forms of CRK [7]. CRKL is constitutively associated with the guanine nucleotide exchange protein C3G [8]. Subsequent association with PtdIns 3′-kinase and FYN has been detected in Jurkat cells [9] and also with SYK tyrosine kinase and the SHC adaptor protein in B cells [10]. In macrophages CSF-1 stimulation promotes association with SHC and pp80 and CBL simultaneously undergoes ubiquitination and translocation to the plasma membrane [11].

Rapid tyrosine phosphorylation of CBL occurs in Fcγ receptor-stimulated macrophages, in v-SRC transformed cells and in response to EGF in EGFR over-expressing cells when CBL forms a complex with the EGFR [12,13]. In fibroblasts and epithelial cells CBL appears to associate indirectly with the EGFR *via* the GRB2/ASH adaptor protein and it does not associate with other members of the EGFR family [14]. In human haematopoietic cells CBL is tyrosine phosphorylated by stimulation with GM-CSF or Epo and constitutively binds to the SH3 domain of GRB2/ASH [15].

v-CBL was generated by truncation of 60% of the C-terminus of CBL. This permits the *gag-cbl* fusion protein to enter the nucleus. v-CBL binds to DNA, whereas full-length CBL does not, and thus probably functions as a transcriptional activator [16].

**Cancer**

In Ph-positive leukaemia, tyrosine phosphorylated CBL occurs attached *via* CRKL to BCR/ABL [17] and in the human chronic myelogenous leukaemia cell line K562 CRKL and CRK-II both associate with CBL in a tyrosine phosphorylation-dependent manner [15].

**In animals**

Cas NS-1 virus is an acutely transforming murine retrovirus that induces pre-B (sIg$^-$, Lyb-2$^+$, Ly-5 (B220$^+$) and pro-B (Mac-1$^+$) cell lymphomas and transforms fibroblasts, the *gag-onc* fusion protein appearing to be the responsible agent [1]. v-*cbl* and v-*abl* induce histologically and phenotypically similar tumours although for v-*cbl* there is longer latency and resistance in adult mice.

***In vitro***

Fibroblasts transformed by Abelson MuLV have high levels of tyrosine-phosphorylated CBL with which ABL associates [19]. In HL60 cells CBL undergoes tyrosine phosphorylation in response to activation of the surface antigen CD38 [20].

## Protein structure

CBL has four potential *N*-linked glycosylation sites and many Ser/Thr stretches. v-CBL is a truncated form of mouse CBL containing 355 N-terminal amino acids fused behind 32 non-cellular residues [2]. The N-terminus of v-CBL is identical to the first 357 residues of human CBL with four deletions. v-CBL has lost a C-terminal leucine zipper and a 208 residue proline-rich region that occur in CBL and are similar to the activation domains of some transcription factors.

A truncated form of CBL occurs in the cutaneous T cell lymphoma line HUT78 in which 259 C-terminal amino acids are removed: the 72 kDa protein expressed lacks the leucine zipper region and part of the proline-rich domain [21].

A second oncogenic form of CBL detected in the 70Z/3 pre-B lymphoma has an internal deletion of 17 (EQYELYCEMGSTFQLCK: 366–382 in human CBL) amino acids [19].

## Amino acid sequence of human CBL

```
  1 MAGNVKKSSG AGGGTGSGGS GSGGLIGLMK DAFQPHHHHH HHLSPHPPGT
 51 VDKKMVEKCW KLMDKVVRLC QNPKLALKNS PPYILDLLPD TYQHLRTILS
101 RYEGKMETLG ENEYFRVFME NLMKKTKQTI SLFKEGKERM YEENSQPRRN
151 LTKLSLIFSH MLAELKGIFP SGLFQGDTFR ITKADAAEFW RKAFGEKTIV
201 PWKSFRQALH EVHPISSGLE AMALKSTIDL TCNDYISVFE FDIFTRLFQP
251 WSSLLRNWNS LAVTHPGYMA FLTYDEVKAR LQKFIHKPGS YIFRLSCTRL
301 GQWAIGYVTA DGNILQTIPH NKPLFQALID GFREGFYLFP DGRNQNPDLT
351 GLCEPTPQDH IKVTQEQYEL YCEMGSTFQL CKICAENDKD VKIEPCGHLM
401 CTSCLTSWQE SEGQGCPFCR CEIKGTEPIV VDPFDPRGSG SLLRQGAEGA
451 PSPNYDDDDD ERADDTLFMM KELAGAKVER PPSPFSMAPQ ASLPPVPPRL
501 DLLPQRVCVP SSASALGTAS KAASGSLHKD KPLPVPPTLR DLPPPPPPDR
551 PYSVGAESRP QRRPLPCTPG DCPSRDKLPP VPSSRLGDSW LPRPIPKVPV
601 SAPSSSDPWT GRELTNRHSL PFSLPSQMEP RPDVPRLGST FSLDTSMSMN
651 SSPLVGPECD HPKIKPSSSA NAIYSLAARP LPVPKLPPGE QCEGEEDTEY
701 MTPSSRPLRP LDTSQSSRAC DCDQQIDSCT YEAMYNIQSQ APSITESSTF
751 GEGNLAAAHA NTGPEESENE DDGYDVPKPP VPAVLARRTL SDISNASSSF
801 GWLSLDGDPT TNVTEGSQVP ERPPKPFPRR INSERKAGSC QQGSGPAASA
851 ATASPQLSSE IENLMSQGYS YQDIQKALVI AQNNIEMAKN ILREFVSISS
901 PAHVAT (906)
```

### Domain structure

| | |
|---|---|
| 124–127 | Nuclear localization signal (underlined) |
| 381–419 | $C_3HC_4$-type zinc finger |
| 357–476 and 689–834 | Acidic domains |
| 477–688 | Proline-rich region |
| 481–528 | GRB2 binding domain [22] |
| 857–892 | Leucine zipper (underlined italics) |
| 700, 774 | Major tyrosine phosphorylated sites in *Abl*-transformed cells [23] |

### Database accession numbers

| | *PIR* | *SWISSPROT* | *EMBL/GENBANK* | *REFERENCES* |
|---|---|---|---|---|
| Human *CBL* | A43817 | P22681 | X57110 | 2 |

### *References*


[1] Langdon, W.Y. et al. (1989) Proc. Natl Acad. Sci. USA 86, 1168–1172.
[2] Blake, T.J. et al. (1991) Oncogene 6, 653–657.
[3] Keane, M.M. et al. (1995) Oncogene 10, 2367–2377.
[4] Langdon, W.Y. et al. (1989) J. Virol. 63, 5420–5424.
[5] Mushinski, J.F. et al. (1994) Oncogene 9, 2489–2497.
[6] Rivero-Lezcano, O.M. et al. (1994) J. Biol. Chem. 269, 17363–17366.
[7] Buday, L. et al. (1996) J. Biol. Chem. 271, 6159–6163.
[8] Reedquist, K.A. (1996) J. Biol. Chem. 271, 8435–8442.
[9] Meisner, H. et al. (1995) Mol. Cell. Biol. 15, 3571–3578.
[10] Panchamoorthy, G. et al. (1996) J. Biol. Chem. 271, 3187–3194.
[11] Wang, Y. et al. (1996) J. Biol. Chem. 271, 17–20.
[12] Tanaka, S. et al. (1995) J. Biol. Chem. 270, 14347–14351.
[13] Bowtell, D.D.L. and Langdon, W.Y. (1995) Oncogene 11, 1561–1567.
[14] Levkowitz, G. et al. (1996) Oncogene 12, 1117–1125.
[15] Odai, H. et al. (1995) J. Biol. Chem. 270, 10800–10805.

[16] Blake, T.J. et al. (1993) EMBO J. 12, 2017–2026.
[17] de Jong, R. et al. (1995) Biol. Chem. 270, 21468–21471.
[18] Ribon, V. et al. (1996) Mol. Cell. Biol. 16, 45–52.
[19] Andoniou, C.E. et al. (1994) EMBO J. 13, 4515–4523.
[20] Kontani, K. et al. (1996) J. Biol. Chem. 271, 1534–1537.
[21] Blake, T.J. and Langdon, W.Y. (1992) Oncogene 7, 757–762.
[22] Donovan, J.A. et al. (1996) J. Biol. Chem. 271, 26369–26374.
[23] Andoniou, C.E. et al. (1996) Oncogene 12, 1981–1989.

# CDK4

## Identification

*CDK4* was originally cloned from HeLa cells on the basis of its homology to oligonucleotide probes representing highly conserved regions of other mammalian kinases [1].

## Related genes

CDK4 is a member of the cyclin-dependent kinase family of which there are 12 human proteins having high homology with $p34^{CDC2}$. CDK4 is 70% identical to CDK6.

| | *CDK4* |
|---|---|
| **Nucleotides (kb)** | 5 |
| **Chromosome** | 12q13 |
| **Mass (kDa): predicted** | 33.7 |
| **expressed** | 34 |
| **Cellular location** | Nuclear |
| **Tissue distribution** | Ubiquitous |

## Protein function

Cyclin-dependent kinases (CDKs) regulate progression through the cell cycle. They are activated by specific association with cyclins and those involved in $G_1$ progression and/or exit include the D-type cyclins (see **D-cyclins**) which associate with CDK2, CDK4, CDK5 and CDK6, of which CDK4 is the most abundant. The activation of CDK4 in complexes with D-type cyclins requires the phosphorylation of $Thr^{172}$, equivalent to $Thr^{160}$ in CDK2 that is located within a loop occluding the substrate binding site [2,3]. The action of the CDK-activating kinase (CAK) modifies the conformation to permit substrate access. The activation of CDK4/cyclin D in the cell cycle precedes that of other CDKs (D/CDK4 → D/CDK6 → E/CDK2 → A/CDK2 → A/CDC2 → B/CDC2). A critical substrate of D/CDK4 is pRb, phosphorylation of which is permissive for the $G_1$ to S transition.

CDK4 is induced by mitogens but dominant negative forms of CDK4 or CDK6 do not block in $G_1$ (whereas dominant negative CDK2 does). CDK4 may be a TGF$\beta$ target: TGF$\beta$ blocks serum induction of CDK4 in human keratinocytes and in mink lung epithelial (Mv1Lu) cells constitutive expression of CDK4 overcomes TGF$\beta$ growth inhibition. Disruption of pRb function decreases cyclin D abundance. In the absence of D cyclins, CDK4 (or CDK6) form an inactive binary complex with the inhibitory protein INK4A.

In keratinocytes and epithelial cells TGF$\beta_1$ regulates the activity of cyclin-CDK/CDC complexes, and thus possibly pRb phosphorylation, by suppressing the synthesis of CDK4 and CDK2 and by inhibiting cyclin A synthesis [4,5]. Constitutive synthesis of CDK4, but not CDC2, overcomes TGF$\beta_1$ growth inhibition.

**Cancer**

The mutation $Arg^{24} \rightarrow Cys$ has been detected as a somatic mutation in a melanoma giving rise to a tumour-specific antigen recognized by autologous cytotoxic T cells and also as a germline mutation in two unrelated melanoma families that do not carry germline mutations in *INK4A*[6]. This mutation does not affect cyclin D1 binding and formation of a functional kinase but disrupts the INK4A binding site[7]. This mutant form of *CDK4* is therefore a dominant oncogene, the product of which is resistant to inhibition by INK4A.

CDK4 and CDK6 are over-expressed in some tumour cell lines and CDK4 is amplified in some tumours (50% glioblastomas).

## Gene structure

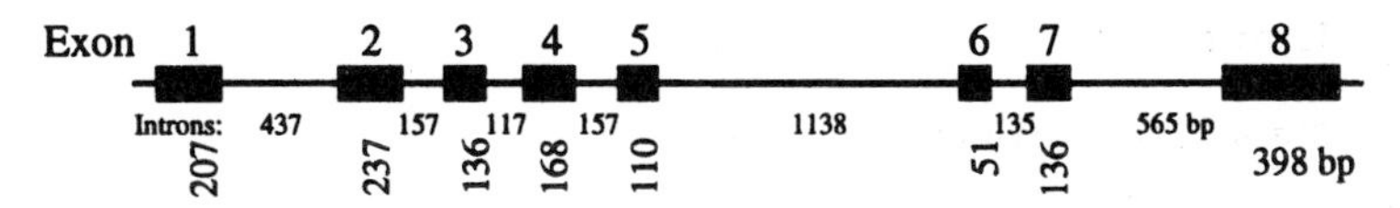


## Amino acid sequence of human CDK4

```
  1 MATSRYEPVA EIGVGAYGTV YKARDPHSGH FVALKSVRVP NGGGGGGGLP
 51 ISTVREVALL RRLEAFEHPN VVRLMDVCAT SRTDREIKVT LVFEHVDQDL
101 RTYLDKAPPP GLPAETIKDL MRQFLRGLDF LHANCIVHRD LKPENILVTS
151 GGTVKLADFG LARIYSYQMA LTPVVVTLWY RAPEVLLQST YATPVDMWSV
201 GCIFAEMFRR KPLFCGNSEA DQLGKIFDLI GLPPEDDWPR DVSLPRGAFP
251 PRGPRPVQSV VPEMEESGAQ LLLEMLTFNP HKRISAFRAL QHSYLHKDEG
301 NPE (303)
```

**Domain structure**

| | |
|---|---|
| 6–295 | Kinase domain |
| 172 | Phosphorylation of $Thr^{172}$ necessary for kinase activity |
| 12, 20 and 35 | ATP binding site |
| 140 | Active site |

**Database accession numbers**

| | *PIR* | *SWISSPROT* | *EMBL/GENBANK* | *REFERENCES* |
|---|---|---|---|---|
| Human *CDK4* | C26368 | P11802, | HSPSKC, M14505 | 8 |
| | | CDK4_HUMAN | S67448, U37022 | 6 |

***References***


1 **Hanks, S.K. (1987) Proc. Natl Acad. Sci. USA 84, 388–392.**
2 De Bondt, H.L. et al. (1993) Nature 363, 595–602.
3 Russo, A.A. et al. (1996) Nature 382, 325–331.
4 Slingerland, J.M. et al. (1994) Mol. Cell. Biol. 14, 3683–3694.
5 Ewen, M.E. et al. (1993) Cell 74, 1009–1020.
6 Zuo, L. et al. (1996) Nature Genet. 12, 97–99.
7 Wölfel, T. et al. (1995) Science 269, 1281–1285.
8 Khatib Z.A. et al. (1993) Cancer Res. 53, 5535–5541.

## Identification

v-*crk* (CT10 regulator of kinase) is the oncogene of avian sarcoma viruses CT10 and ASV-1 [1,2]. *CRK-I* and *CRK-II* were isolated from embryonic lung cells using the polymerase chain reaction and by screening a human placental cDNA library with a *CRK-I* probe, respectively [3].

## Related genes

CRK proteins contain SH2 and SH3 domains homologous with those present in SRC family proteins, FPS/FES and ABL, RAS-GAP, the PtdIns-specific phospholipase C$\gamma$ isozymes, $\alpha$-spectrin, the yeast actin binding protein ABP1p, myosin-I, p85$\alpha$ and p85$\beta$ (see ***SRC***). *CRKL* is 60% homologous to *CRK-II* and maps to chromosome 22q11.

| | *CRK* | v-*crk* |
|---|---|---|
| **Nucleotides (bp)** | Not fully mapped | 2407 |
| **Chromosome** | 17p13 | |
| **Mass (kDa): predicted** | 39.8 (CRK-II) | |
| **expressed** | 28 (CRK-I)<br>40/42 (CRK-II)<br>36 (CRKL) | $P47^{gag\text{-}crk}$ |
| **Cellular location** | Cytoplasm | Cytoplasm |

**Tissue distribution**
CRK-I (p28) is expressed in human embryonic lung cells. CRK-II (p42) is present in the osteosarcoma cell line 143B, A431 cells and the T cell line H9. CRK-II (p40) is present in a wide variety of cell lines [3].

## Protein function

CRK is an adapter protein composed of an SH2 domain and two SH3 domains. ABL and ARG bind to the CRK $SH3_1$ domain and phosphorylate CRK on $Tyr^{221}$ [4,5]. This tyrosine-phosphorylated residue also binds intramolecularly to the SH2 domain of CRK [6]. v-CRK also binds to ABL, although $Tyr^{221}$ is truncated from v-CRK. Other proteins shown to associate with CRK are p185 and p170 [7] and Eps15 and Eps15R [8].

The $SH3_2$ domain negatively regulates CRK-mediated tyrosine phosphorylation of p130 in rat 3Y1 cells [9]. p130 (CAS: Crk-associated substrate) itself is an SH3-containing protein with multiple putative SH2-binding motifs [10]. In mouse fibroblasts, $p130^{Cas}$ is associated with $p125^{FAK}$ and this interaction may be important in integrin-mediated signal transduction [11]. $pp125^{FAK}$-dependent tyrosine phosphorylation of paxillin creates a high-affinity binding site for CRK [12].

P47$^{gag\text{-}crk}$ does not itself possess kinase activity but promotes tyrosine phosphorylation in CT10-infected chick embryo fibroblasts, notably of proteins in the 135–155 kDa range including paxillin [13,14]. p47$^{gag\text{-}crk}$ immunoprecipitates with a wide range of tyrosine phosphorylated proteins, including the EGF and PDGF receptors and v-SRC, but association with the latter is prevented if autophosphorylation of Tyr$^{416}$ is blocked [15,16]. Mutational studies indicate that the SH2 domain of P47$^{gag\text{-}crk}$ binds specifically to tyrosine phosphorylated regions of peptides [17] and shows preferential affinity for binding to phosphotyrosine in the general motif pTyr-hydrophilic-hydrophilic-Ile/Pro [18]. v-CRK binding to paxillin can inhibit CSK binding and one effect of this may be to prevent CSK suppression of the kinase activity of SRC and SRC-related proteins [19].

**Cancer**
The region to which *CRK* maps (17p13) demonstrates frequent deletion or loss of heterozygosity in a wide range of human tumours [20].

**In animals**
CT10 and ASV-1 cause rapid tumour formation in chickens [2].

***In vitro***
ASV CT10 transforms chick embryo fibroblasts which are then tumorigenic when injected into chickens [1]. Cell lines expressing CRK-I are tumorigenic in nude mice. CRK-II does not transform cells *in vitro* and is non-tumorigenic [3]. Microinjection of CRK or stable expression of v-CRK causes differentiation of the rat pheochromocytoma cell line PC-12 that is abolished by mutations in either the SH2 or SH3 domains [21]. EGF and NGF both cause the tyrosyl phosphorylation of v-CRK which associates with the receptors for these ligands.

In T cells, CRK interacts with a tyrosine-phosphorylated p116 [22]. In PC-12 cells, CRK immunoprecipitates with the guanine nucleotide-releasing proteins (GNRPs) mSOS and C3G, and CRK binds to SHC *via* its SH2 domain after NGF stimulation [23]. This suggests that v-CRK may modulate growth factor-induced PC-12 cell differentiation and it has been shown to cause sustained activation of the RAS/MAPK pathway [24]. IGF-I stimulates tyrosine phosphorylation of endogenous CRK-II in human embryonic kidney carcinoma and in 3T3 cells [25].

## Protein structure

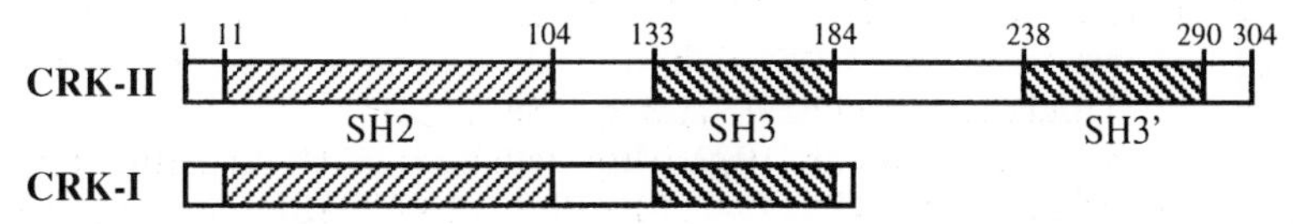


A member of the group of proteins of unrelated function that share homology with non-catalytic regions of SRC, P47$^{gag\text{-}crk}$ contains SRC homology regions 2 and 3 (SH2, SH3), transposed with respect to SRC. These regions also occur in phosphatidylinositol-specific phospholipase C$\gamma$ [26], where SH2 is duplicated, and P47$^{gag\text{-}crk}$ has strong homology with a 180 amino acid region of this protein. Human CRK-I and CRK-II each contain one SH2 domain together with one and two SH3 domains, respectively[3]. The B and C boxes of SH2, separated by a hinge region, coordinately form the functional phosphotyrosine binding domain[27]. The binding affinity of v-CRK for tyrosine phosphorylated proteins (e.g. EGFR, PDGFR, v-SRC) greatly exceeds that of CRK, even though both have identical SH2 domains. This is caused by the presence of a short region N-terminal to the SH2 domain in v-CRK[28].

The P47$^{gag\text{-}crk}$ transforming protein comprises 208 *gag*-encoded amino acids fused to 232 novel residues presumed to be from a cellular proto-oncogene. The cellular sequence lacks start and stop codons so is presumed to be truncated at both ends. The last three amino acids and the stop codon are provided by viral sequences.

## Amino acid sequence of human CRK-II

```
  1 MAGNFDSEER SSWYWGRLSR QEAVALLQGQ RHGVFLVRDS STSPGDYVLS
 51 VSENSRVSHY IINSSGPRPP VPPSPAQPPP GVSPSRLRIG DQEFDSLPAL
101 LEFYKIHYWD TTTLIEPVSR SRQGSGVILR QEEAEYVRAL FDFNGNDEED
151 LPFKKGDILR IRDKPEEQWW NAEDSEGKRG MIPVPYVEKY RPASASVSAL
201 IGGNQEGSHP QPLGPPEPGP YAQPSVNTPL PNLQNGPIYA RVIQKRVPNA
251 YDKTALALEV GELVKVTKIN VSGQWEGGCN GKRGHFPFTH VRLLDQQNPD
301 EDFS (304)
```

**Domain structure**

| | |
|---|---|
| 11–104 | CRK-II SH2 domain (underlined) |
| 133–184 and 238–290 | CRK-II SH3 domains (underlined): bind specifically to the sequence Ψ(aliphatic)PΨLPΨK |

The termination codon for CRK-I (204 amino acids, identical to CRK-II with the substitution of Arg$^{204}$ for Asn$^{204}$) is in the same position as that for v-CRK. *CRK-I* lacks a 170 bp sequence present in *CRK-II*: alternative splicing is presumed to remove this region, leaving a single SH3 domain in CRK-I. v-CRK is homologous to the N-terminal 203 amino acids of CRK-II with 47 scattered point mutations.

**Database accession numbers**

| | *PIR* | *SWISSPROT* | *EMBL/GENBANK* | *REFERENCE* |
|---|---|---|---|---|
| Human *CRK-I, CRK-II* | B45022, A45022 | | D10656 | 3 |

***References***


1 Mayer, B.J. et al. (1988) Nature 332, 272–275.
2 Tsuchie, H. et al. (1989) Oncogene 4, 1281–1284.
3 Matsuda, M. et al. (1992) Mol. Cell. Biol. 12, 3482–3489.
4 Feller, S.M. et al. (1994) EMBO J. 13, 2341–2351.
5 Wang, B. et al. (1996) Oncogene 13, 1379–1385.
6 Rosen, M.K. et al. (1995) Nature 374, 477–479.
7 Feller, S.M. et al. (1995) Oncogene 10, 1465–1473.
8 Schumacher, C. et al. (1995) J. Biol. Chem. 270, 15341–15347.
9 Ogawa, S. et al. (1994) Oncogene 9, 1669–1678.

[10] Sakai, R. et al. (1994) EMBO J. 13, 3748–3756.
[11] Polte, T.R. and Hanks, S.K. (1995) Proc. Natl Acad. Sci. USA 92, 10678–10682.
[12] Schaller, M.D. and Parsons, J.T. (1995) Mol. Cell. Biol. 15, 2635–2645.
[13] Birge, R.B. et al. (1993) Mol. Cell. Biol. 13, 4648–4656.
[14] Mayer, B.J. et al. (1989) Cold Spring Harbor Symp. Quant. Biol. 53, 907–914.
[15] Matsuda, M. et al. (1990) Science 248, 1537–1539.
[16] Mayer, B.J. and Hanafusa, H. (1990) Proc. Natl Acad. Sci. USA 87, 2638–2642.
[17] Matsuda, M. et al. (1991) Mol. Cell. Biol. 11, 1607–1613.
[18] Songyang, Z. et al. (1993) Cell 72, 767–778.
[19] Sabe, H. et al. (1995) J. Biol. Chem. 270, 31219–31224.
[20] Fioretos, T. et al. (1993) Oncogene 8, 2853–2855.
[21] Hempstead, B.L. et al. (1994) Mol. Cell. Biol. 14, 1964–1971.
[22] Sawasdikosol, S. et al. (1995) J. Biol. Chem. 270, 2893–2896.
[23] Matsuda, M. et al. (1994) Mol. Cell. Biol. 14, 5495–5500.
[24] Teng, K.K. et al. (1995) J. Biol. Chem. 270, 20677–20685.
[25] Beitner-Johnson, D. and LeRoith, D. (1995) J. Biol. Chem. 270, 5187–5190.
[26] Stahl, M.L. et al. (1988) Nature 332, 269–272.
[27] Matsuda, M. et al. (1993) J. Biol. Chem. 268, 4441–4446.
[28] Fajardo, J.E. et al. (1993) Mol. Cell. Biol. 13, 7295–7302.

## Identification

v-*fms* is the transforming oncogene of the Susan McDonough (SM-FeSV) and Hardy–Zuckerman 5 (HZ5-FeSV) strains of acutely transforming feline sarcoma virus [1,2]. *CSF1R/Fms* was detected by screening human placental cDNA with a v-*fms* probe [3].

## Related genes

*FMS* is closely related to *KIT*, the two receptors for PDGF, human *FRT* (*f*ms-*r*elated *t*yrosine kinase gene) and the product of the *FLT3/FLK2* gene (human stem cell tyrosine kinase 1 (STK1)). Each has a distinctive pattern of cysteine spacing in the extracellular domain that includes sequences characteristic of the immunoglobulin (Ig) gene superfamily and in each the kinase domain is interrupted by a hydrophilic spacer of between 64 and 104 amino acids.

FLT1 (FMS-like tyrosine kinase, also FLK3) is ~60% homologous to FMS but has a predicted seven Ig-like extracellular domains, compared to five in members of the CSF1R/FMS family. *FLK1* (human *KDR*, murine *Nyk*, rat *TKr-III*) and *FLT4* encode the other members of the class of receptor tyrosine kinases with seven Ig-like loops.

| | *FMS* | v-*fms* |
|---|---|---|
| **Nucleotides (kb)** | >30 | 8.2 (SM-FeSV genome) |
| **Chromosome** | 5q33.3–34 | |
| **Mass (kDa): predicted** | 108 | |
| **expressed** | gp150 | gp140$^{v\text{-}fms}$ (SM-FeSV) |

**Cellular location**

Transmembrane protein. Immature forms of the protein occur in the endoplasmic reticulum and Golgi apparatus. The ectodomain may be released from cells [4].

**Tissue distribution**

*CSF1R* is normally only expressed on monocytes, macrophages and their precursors [5] and at lower levels in normal and malignant B lymphocytes [6] but is also detectable in the human and murine uteroplacental unit [7] and in cell lines derived from human malignant placental trophoblasts. *FMS* is not expressed in normal vascular smooth muscle cells but is activated in intimal smooth muscle cells isolated from an experimental rabbit model of arteriosclerosis and is expressed at high levels in normal human medial smooth muscle cells in response to heparin binding epidermal growth factor-like growth factor or to PDGF-BB with either EGF or FGF [8,9].

## Protein function

*CSF1R* (*FMS*) encodes the receptor for colony-stimulating factor 1 (CSF1 or M-CSF) that cooperates with IL-2 or IL-3 to stimulate bone marrow cells during haematopoiesis[10]. The enhanced expression of *Fms* in pregnant mice indicates a possible role in embryogenesis. *Fms* may also be involved in the induction of macrophage differentiation[11,12]. Expression of v-*fms* may cause haematopoietic disorders[13].

FMS has intrinsic tyrosine kinase activity and RAF is a substrate *in vitro*. In macrophages CSF1 activates the $Na^+/K^+$-ATPase, the $Na^+/H^+$ exchanger and *Fos* and *Myc* transcription. CSF1 also activates the *Src* family kinases SRC, FYN and YES and causes these proteins to associate with the CSF1R[14].

NIH 3T3 fibroblasts are mitogenically stimulated by CSF1 after transfection with the human *FMS* gene, but CSF1 causes barely detectable tyrosine phosphorylation of PLC$\gamma$ and no early (5 min) change in $[Ca^{2+}]_i$[15]. It is possible that CSF1 activates other isoforms of PLC but it seems probable that, despite its structural similarity to the PDGF receptor, activation *via* FMS does not involve PtdIns(4,5)$P_2$ hydrolysis. CSF1 causes weak tyrosine phosphorylation of GAP and strong tyrosine phosphorylation of the GAP-associated proteins p62 and p190. Despite the contrast in effects on GAP, CSF1 and PDGF promote equivalent activation of RAS-GTP and stimulate mitogenesis to a similar extent in 3T3 cells[16]. The activated CSF1R associates with the SH2-containing adapter protein GRB2 and causes the tyrosine phosphorylation of SHC, consistent with the evidence that CSF1 activates a pathway involving RAS[17]. In the murine myeloid progenitor cell line FDC-P1 CSF1R binds to SOS1 and $p150^{SHIP}$. $p150^{SHIP}$ has inositol polyphosphate-5-phosphatase activity and after phosphorylation it associates with CSF1R *via* SHC[18,19]. Expression of $p150^{SHIP}$ in FDC-P1 cells inhibits M-CSF-activated growth.

Receptor-mediated endocytosis of FMS and v-FMS follows ligand binding but endocytosis of v-FMS does not require CSF1. The rapid internalization of the FMS receptor does not depend on the tyrosine kinase activity of the receptor nor on the presence of the kinase insert region of the cytoplasmic domain although degradation requires both these regions[20]. The kinase insert region is also unnecessary for the growth stimulating activity of the FMS kinase[21,22].

v-*fms*-mediated transformation appears to be caused by the sustained expression of CSF1-independent tyrosine kinase activity[23]. In SM-FeSV-transformed mink lung epithelial cells PtdIns(4,5)$P_2$ hydrolysis and PtdIns kinase activities are increased[24].

**Cancer**

Activating mutations of *FMS* have been detected in patients with primary acute myeloid leukaemia[25].

**In animals**

Inoculation of SM-FeSV-transformed NIH 3T3 fibroblasts or NRK cells into syngeneic animals or nude mice induces fatal tumours[1].

Two receptors for vascular endothelial growth factor (VEGF-VPF) have been identified, FLK1/KDR and FLT1, and inhibition of FLK1 signalling inhibits the growth of a variety of solid tumours[26].

### *In vitro*

SM-FeSV transforms NIH 3T3, MDCK, NRK, feline embryo fibroblasts and mink lung epithelial cells [27]. This requires tyrosine kinase activity, normal glycosylation and surface membrane expression. The presence of the complete ligand binding domain of CSF1 in v-FMS causes transformed cells to bind CSF1 specifically and CSF1 causes a 2–3-fold increase of phosphotyrosine in gp140$^{v\text{-}fms}$, thus enhancing the constitutive kinase activity of the transforming protein [15,23]. The expression of a constitutively activated tyrosine phosphatase suppresses transformation by v-*fms* [28].

## Gene structure

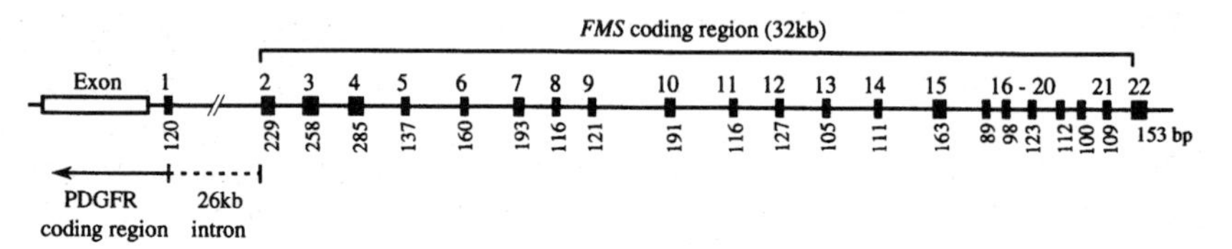


The overall gene structure and encoded amino acid sequence is similar to the PDGF receptor and the two genes may have arisen by duplication. Exon 2 encodes part of the 5′ untranslated sequence of *FMS* mRNA, the initiation codon and the signal peptide [29,30].

In the mouse, integration of Friend MuLV at the 5′ end of *Fms* directly upstream of the region corresponding to exon 2 of the human gene induces transcription as a component of the development of myeloid leukaemia. Proviral insertion is head-to-head with respect to the direction of *Fms* transcription, indicating that the promoter may lie close to exon 2 [31].

## Protein structure

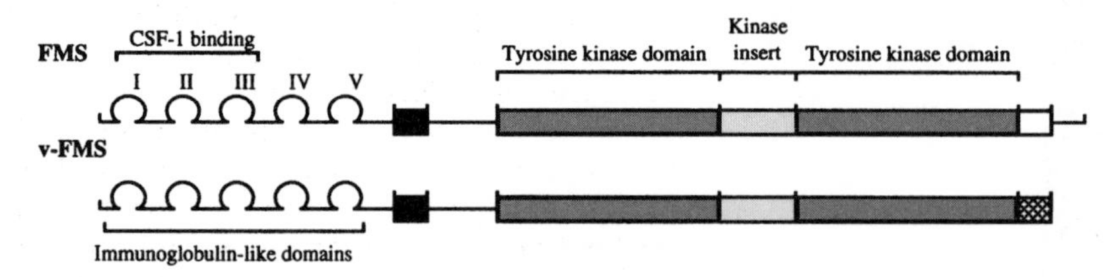


Human and feline FMS proteins share 80.5% identity. v-FMS and FMS differ only by scattered amino acid substitutions except that the 50 C-terminal amino acids of feline FMS are replaced by 14 residues in SM-FeSV v-FMS (cross-hatched box), encoded by a 3′ untranslated region of feline *Fms*, with the elimination of the C-terminal tyrosine of FMS [32]. The mutations that appear critical for the activation of v-FMS are the substitution of two serine residues in the extracellular domain of feline FMS, together with the modification of the C-terminus [33,34].

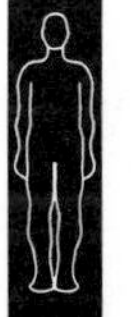

## Amino acid sequence of human CSF1R/FMS

```
  1 MGPGVLLLLL VATAWHGQGI PVIEPSVPEL VVKPGATVTL RCVGNGSVEW
 51 DGPASPHWTL YSDGSSSILS TNNATFQNTG TYRCTEPGDP LGGSAAIHLY
101 VKDPARPWNV LAQEVVVFED QDALLPCLLT DPVLEAGVSL VRVRGRPLMR
151 HTNYSFSPWH GFTIHRAKFI QSQDYQCSAL MGGRKVMSIS IRLKVQKVIP
201 GPPALTLVPA ELVRIRGEAA QIVCSASSVD VNFDVFLQHN NTKLAIPQQS
251 DFHNNRYQKV LTLNLDQVDF QHAGNYSCVA SNVQGKHSTS MFFRVVESAY
301 LNLSSEQNLI QEVTVGEGLN LKVMVEAYPG LQGFNWTYLG PFSDHQPEPK
351 LANATTKDTY RHTFTLSLPR LKPSEAGRYS FLARNPGGWR ALTFELTLRY
401 PPEVSVIWTF INGSGTLLCA ASGYPQPNVT WLQCSGHTDR CDEAQVLQVW
451 DDPYPEVLSQ EPFHKVTVQS LLTVETLEHN QTYECRAHNS VGSGSWAFIP
501 ISAGAHTHPP DEFLFTPVVV ACMSIMALLL LLLLLLLYKY KQKPKYQVRW
551 KIIESYEGNS YTFIDPTQLP YNEKWEFPRN NLQFGKTLGA GAFGKVVEAT
601 AFGLGKEDAV LKVAVKMLKS TAHADEKEAL MSELKIMSHL GQHENIVNLL
651 GACTHGGPVL VITEYCCYGD LLNFLRRKAE AMLGPSLSPG QDPEGGVDYK
701 NIHLEKKYVR RDSGFSSQGV DTYVEMRPVS TSSNDSFSEQ DLDKEDGRPL
751 ELRDLLHFSS QVAQGMAFLA SKNCIHRDVA ARNVLLTNGH VAKIGDFGLA
801 RDIMNDSNYI VKGNARLPVK WMAPESIFDC VYTVQSDVWS YGILLWEIFS
851 LGLNPYPGIL VNSKFYKLVK DGYQMAQPAF APKNIYSIMQ ACWALEPTHR
901 PTFQQICSFL QEQAQEDRRE RDYTNLPSSS RSGGSGSSSS ELEEESSSEH
951 LTCCEQGDIA QPLLQPNNYQ FC (972)
```

**Domain structure**

| | |
|---|---|
| 1–23 | Signal sequence |
| 24–512 | Extracellular domain |
| 513–537 | Trans-membrane region (underlined) |
| 538–972 | Cytoplasmic domain |
| 588–596 and 616 | ATP binding domain |
| 683–749 | Kinase insert region (underlined) |
| 699, 708, 809 | Autophosphorylation sites |
| 778 | Active site |
| 45, 73, 153, 240, 302, 335, 353, 412, 428 and 480 | Potential glycosylation sites |
| 723 | Binding site (when phosphorylated) for PtdIns 3-kinase *via* the two SH2 domains of its p85a subunit [35] |
| 809 | Substitution of Tyr$^{809}$ blocks ligand-dependent mitogenesis but does not inhibit the tyrosine kinase activity of the receptor nor its ability to associate with PtdIns 3-kinase and to induce *Fos* and *Jun* transcription [36] although it does decrease the binding and enzymatic activation of SRC, FYN and YES [13]. Activated Phe$^{809}$ mutant receptors do not, however, induce *Myc* expression but the coexpression of an exogenous *Myc* gene in cells expressing only the mutant form of the receptor restores the ability of the cells to proliferate in response to CSF1, implying the existence of an effector interacting with the domain including Tyr$^{809}$ that is required for *Myc* induction and CSF1-induced mitogenesis |
| 969 | Replacement of Tyr$^{969}$ with Phe enhances transforming potential [23] |

**Database accession numbers**

| | PIR | SWISSPROT | EMBL/GENBANK | REFERENCES |
|---|---|---|---|---|
| Human CSF1R | A24533 | P07333 | X03663 | 3 |
| Human FRT | | P16057 | D00133 | 37 |
| Human FLT1 | S09982 | P17948 | X51602 | 38 |
| Human FLT4 | | P35916, | HSFLT4 X68203 | 39 |
| | | VGR3_HUMAN | HSFLT4X X69878 | 40 |

*References*


1 McDonough, S.K. et al. (1971) Cancer Res. 31, 953–956.
2 Besmer, P. et al. (1986) J. Virol. 60, 194–203.
3 Coussens, L. et al. (1986) Nature 320, 277–280.
4 Downing, J.R. et al. (1991) Mol. Cell. Biol. 9, 2890–2896.
5 **Sherr, C.J. (1988) Biochim. Biophys. Acta, 948, 225–243.**
6 Baker, A.H. et al. (1993) Oncogene 8, 371–378.
7 Pampfer, S. et al. (1992) Biol. Reprod., 46, 48–57.
8 Inaba, T. et al. (1996) Mol. Cell. Biol. 16, 2264–2273.
9 Inaba, T. et al. (1996) J. Biol. Chem. 271, 24413–24417.
10 Stanley, E.R. et al. (1986) Cell 45, 667–674.
11 Rohrschneider, L.R. and Metcalf, D. (1989) Mol. Cell. Biol. 9, 5081–5092.
12 Borzillo, G.V. et al. (1990) Mol. Cell. Biol. 10, 2703–2714.
13 Heard, J.M. et al. (1987) Cell 51, 663–673.
14 Courtneidge, S. et al. (1993) EMBO J. 12, 943–950.
15 Downing, J.R. et al. (1989) EMBO J. 8, 3345–3350.
16 Heidaran, M.A. et al. (1992) Oncogene 7, 147–152.
17 van der Geer, P. and Hunter, T. (1993) EMBO J. 12, 5161–5172.
18 Lioubin, M.N. et al. (1994) Mol. Cell. Biol. 14, 5682–5691.
19 Lioubin, M.N. et al. (1996) Genes Dev. 10, 1084–1095.
20 Carlberg, K. et al. (1991) EMBO J. 10, 877–883.
21 Taylor, G.R. et al. (1989) EMBO J. 8, 2029–2037.
22 Shurtleff, S.A. et al. (1990) EMBO J. 9, 2415–2421.
23 Roussel, M.F. et al. (1987) Nature 325, 549–552.
24 Kaplan, D.R. et al. (1987) Cell 50, 1021–1029.
25 Ridge, S.A. et al. (1990) Proc. Natl Acad. Sci. USA 87, 1377–1380.
26 Millauer, B. et al. (1996) Cancer Res. 56, 1615–1620.
27 Donner, L. et al. (1982) J. Virol. 41, 489–500.
28 Zander, N.F. et al. (1993) Oncogene 8, 1175–1182.
29 Roberts, W.M. et al. (1988) Cell 55, 655–661.
30 Hampe, A. et al. (1989) Oncogene Res. 4, 9–17.
31 Gisselbrecht, S. et al. (1987) Nature 329, 259–261.
32 Woolford, J. et al. (1988) Cell 55, 965–977.
33 Roussel, M.F. et al. (1988) Cell 55, 979–988.
34 **Rohrschneider, L.R. and Woolford, J. (1991) Semin. Virol. 2, 385–395.**
35 Reedijk, M. et al. (1992) EMBO J. 11, 1365–1372.
36 Roussel, M.F. et al. (1991) Nature 353, 361–363.
37 Matsushime, H. et al. (1987) Jpn J. Cancer Res. 78, 655–661.
38 Shibuya, M. et al. (1990) Oncogene 5, 519–524.
39 Galland, F. et al. (1993) Oncogene 8, 1233–1240.
40 Aprelikova, O. et al. (1992) Cancer Res. 52, 746–748.

# D-cyclins *Cyclin D1 (CCND1), Cyclin D2 (CCND2), Cyclin D3 (CCND3)*

## Identification

*CCND1/PRAD1/BCL1* (B cell leukemia/lymphoma-1) was detected with a probe specific for chromosome 11 in DNA isolated from a chronic lymphocytic leukaemia (CLL) cell line [1].

*CCND2* and *CCND3* were cloned from cDNA libraries by cross-hybridization with *CCND1* [2,3].

## Related genes

Human cyclins have been grouped into five types, A to E, on the basis of sequence similarity. All contain a "cyclin box" which is evolutionarily conserved between cyclins of different types and between species from yeast to humans. The D-type cyclins are 39% identical to cyclin A, 36% identical to cyclin E, 29% identical to cyclin B and 21% identical to cyclin C. The murine homologue of cyclin D1 is *Cyl*-1 [4]. Herpesvirus saimiri encodes a D-type cyclin. *CCND2* is the human homologue of *Vin-1*, a site of MuLV integration in a mouse T cell leukaemia in which *Vin-1* is over-expressed [5].

| | *CCND1* | *CCND2* | *CCND3* |
|---|---|---|---|
| **Nucleotides (kb)** | ~15 | Not fully mapped | Not fully mapped |
| **Chromosome** | 11q13 | 12p13 | 6p21 |
| **Mass (kDa): predicted** | 33.4 | 33.4 | 32.5 |
| **expressed** | 34 | 36 | 34 |
| **Cellular location** | Nucleus | Predominantly nuclear in $G_1$: Nuclear and cytoplasmic from $G_1$/S onwards [6] | Nucleus |

**Tissue distribution**

Three D-cyclins are differentially and combinatorially expressed in various cell types in response to growth factor stimulation. Most cells express D3 and either D1 or D2. *CCND1* is expressed in proliferating macrophages (withdrawal of CSF1 from macrophages immediately stops D1 synthesis), but not in cells of other lymphoid or myeloid lineages. Expression is high in the mouse retina. Transcription varies during the cell cycle, being maximal in $G_1$ [4] and is essential for the progression through $G_1$ of at least some t(11;14) B cell tumours [7].

## Protein function

$G_1$-cyclins (cyclins D1–3, C and E) are rate-limiting controllers of $G_1$ phase progression in mammalian cells. $G_1$-cyclins were first detected in the budding yeast *Saccharomyces cerevisiae* as a class of proteins the accumulation of which was rate-limiting for progression from $G_1$ to S phase. D-cyclins are induced earlier than cyclin E and can activate CDK4 and in some cells also

CDK2, CDK5 and CDK6. Deregulation of cyclin D synthesis may render cell cycle progression less growth factor-dependent and thus contribute to oncogenesis. Binary cyclin/CDK complexes can fully phosphorylate their substrates *in vitro* but when isolated from proliferating mammalian cells they are associated with WAF1, a potent cyclin/CDK inhibitor, and proliferating cell nuclear antigen (PCNA)[2]. The stoichiometry of WAF1 binding determines whether it acts as an inhibitor of cyclin/CDK. KIP1 also associates with cyclin/CDK complexes and can inhibit CDK activity when present at high levels.

D-cyclins are crucial for pRb regulation, pRb being hypophosphorylated in $G_1$ and phosphorylated just before S, remaining so until late M. Hypophosphorylated pRb arrests cells in $G_1$: phosphorylation relieves arrest. pRb is phosphorylated *in vitro* by cyclin D1/CDK4 (which binds to pRb *via* N-terminal LXCXE), cyclin E/CDK2 and cyclin A/CDK2 and phosphorylation by any of these complexes disrupts E2F/RB1 complexes[8]. pRb and E2F are the only known *in vitro* substrates for cyclin D/CDK4 and D1/CDK4 phosphorylates most of the pRb sites phosphorylated in late $G_1$. The induction by p53 of WAF1 renders pRb hypophosphorylated and induces cyclin D1 synthesis[9].

Cyclin D1 expression (*via* an inducible promoter) in $G_1$ causes earlier pRb phosphorylation and progression through $G_1$ is accelerated. Anti-D1 antibody in early–mid $G_1$ causes arrest before S phase. However, cells do not arrest if functional pRb missing (either by mutation or *via* the action of DNA tumour virus oncoproteins). D1 is therefore dispensable for $G_1$ control in pRb-deficient cells, regardless of whether active cyclin D1/CDK4 is present[10]. This indicates that the key and perhaps only role of cyclin D1 is to inactivate pRb to permit cell entry into S phase.

D1 (and active D1/CDK4) are downregulated in $Rb^-$ cells, hypophosphorylated pRb stimulating D1 synthesis. Exogenously expressed pRb induces D1, consistent with D1 synthesis and pRb phosphorylation by D1/CDK4 being a negative feedback loop to shut off D1 expression in late $G_1$. In human fibroblasts, inactivation of pRb by adenovirus E1A is not sufficient to repress D1 gene expression but the action of E1B is also required[11]. E1B is a target of p53 and induction of p53 function increases D1 mRNA and protein synthesis. In mouse embryonic fibroblasts the constitutive expression of *Myc* is necessary for suppression of cyclin D1 synthesis in addition to absence of functional pRb[12].

Cyclin D1 blocks MyoD activation of muscle-specific genes and myoblasts differentiating in the absence of growth factors have decreased levels of cyclin D1 but increased expression of cyclin D3. D1 is induced by bFGF or TGF$\beta$, each of which inhibits D3[13]. These observations are consistent with other reports that growth factors induce D1 but suggest that D3 may have roles other than as a positive regulator of cell cycle progression.

$G_1$-cyclins stimulate the E2F-dependent adenovirus E2 promoter and D1 has been reported to repress SP1-mediated transcription and to increase the frequency of amplification of the *CAD* gene by >1000-fold[14]. Cyclin D1 expression is inhibited by TGF$\beta$ in epithelial cells.

**Cancer**

D1 is over-expressed in some gastric, breast (in one line by 3′ UTR truncation) and oesophageal cancers[15,16]. Enhanced expression of cyclin

D1 protein has been detected in ~50% of primary breast carcinomas examined immunohistochemically, with elevated expression of cyclin D1, cyclin D3 or both occurring in 69% [17]. However, increased levels of cyclin D1 appear to correlate with increased duration of remission and overall survival rates, whilst reduced levels of cyclin D1 are associated with the worst prognosis [18]. This suggests the importance of other factors and it is notable that the coexpression of cyclin D1 with EGFR or pRb correlates with a poor prognosis relative to that of patients expressing cyclin D1 alone. In some human breast cancer cell lines there is evidence that cyclin D1 and other cyclin genes are either over-expressed or encode transcripts with enhanced stability relative to that observed in normal cells [19,20]. In tumour lines where D1 is amplified and over-expressed it is necessary for $G_1$ progression. This evidence for a role in proliferation is consistent with the finding that ectopically expressed D2 or D3 can block differentiation of 32D myeloid cells and D1 blocks myocyte differentiation.

The t(11;14)(q13;q32) translocation, in which *CCND1* on chromosome 11 translocates to become juxtaposed to $J_H$ on chromosome 14, occurs in several B cell non-Hodgkin's lymphomas (diffuse, small and large cell lymphomas), B cell chronic lymphocytic leukaemia (CLL) and multiple myeloma. Most breakpoints map ~110 kb 5′ (centromeric) to the *CCND1* gene at the major translocation cluster (MTC), although in some cases of intermediately differentiated lymphoma breakpoints lie outside this region [21]. Over-expression of cyclin D1/BCL1 arising from t(11;14) is a marker for mantle cell lymphoma, a non-Hodgkin's B cell lymphoma comprising centrocytic lymphoma and intermediately differentiated lymphocytic lymphoma [22]. Over-expression of cyclin D1 occurs in some hairy cell leukaemias without (11;14) translocations [23]. *CCND1* is also over-expressed in B cell lines carrying the t(11;14)(q13;q32) translocation [24].

*CCND1* is part of an 11q13 region including *INT2* and *HSTF1* that is amplified in 15–20% of human breast and squamous cell carcinomas [25], and *EMS1*, frequently amplified with *CCND1*, in bladder tumours [26]. In breast tumours, however, *CCND1* expression is not invariably associated with that of *INT2* and *HSTF1* [27]. *CCND1* over-expression has been detected in ovarian cancers [28], in a malignant centrocytic lymphoma [29] and in a rare form of benign parathyroid tumour [30].

In addition to breast carcinomas, abnormal accumulation of cyclin D1 occurs in a significant proportion of other primary tumours including sarcomas, uterine and colorectal carcinomas and malignant melanomas [31]. Differential expression of cyclins D1 and D2 has been detected in B lymphoid cell lines [32] and amplification of cyclins D2 and E has been detected in colorectal carcinoma cell lines [33]. However, in a variety of tumour cell lines and primary tumours, rather than cyclins D1, D2 or D3, their catalytic partners CDK4 and CDK6 are over-expressed [34,35]. Aberrations in cyclin D1 expression can cooperate with loss of function of inhibitors of cyclin D-dependent kinase (see ***INK4A/MTS1/CDK41/CDKN2***).

The chromosomal inversion (inv(11)9p15;q13)) causes over-expression in some parathyroid adenomas because D1 comes under control of the parathyroid hormone promoter. The tumours are benign and non-invasive, so the effect probably is on proliferation.

A cyclin D-related gene, cyclin X, is expressed in Ewing's sarcoma and Wilms' tumour [36]. Herpesvirus saimiri infection gives rise to expression of a virally encoded cyclin D gene [37].

Cyclin D2 rearrangement has not been detected in human cancers but amplification has been detected in a colorectal carcinoma.

The cyclin D3 6p21 region is rearranged in several lymphoproliferative disorders. Elevated expression of cyclin D3 or cyclin D1 has been detected in 69% of a sample of primary breast carcinomas [17].

**Transgenic animals**

D1 over-expression (MMTV LTR) causes mammary hyperplasia and carcinomas in a pattern that correlates with age, level of expression and number of pregnancies [38]. The IgH E$\mu$ enhancer-directed coexpression of cyclin D1 and *Nmyc* or *Lmyc* in double transgenics reveals a strong cooperative effect causing the rapid development of clonal pre-B and B cell lymphomas [39]. Mice lacking cyclin D1 have reduced body size and viability with symptoms of neurological impairment. There is a large reduction in the number of retinal cells during embryonic development and the normal proliferation of mammary epithelium during pregnancy does not occur [40].

***In vitro***

Cyclin D1 over-expression transforms established fibroblasts but not primary cells. D1 cooperates with *Ras* to transform primary baby rat kidney cells or rat embryo fibroblasts (which become tumorigenic in nude mice) [41] and with *Myc* to cause B lymphomas in transgenic mice. The transforming action of CCND1 may be mediated by its overcoming the negative regulation normally exerted by pRb on D-type cyclins [42]. Cyclin D1 shortens $G_1$ and permits cells to complete the cycle in the absence of growth factors [43].

In human MCF-7 breast cancer cells cyclin D1 transcription is induced by 17$\beta$-oestradiol, promoting the rapid phosphorylation of pRb [44] and in HCE7 oesophageal cancer cells anti-sense cDNA to cyclin D1 inhibits growth and reverses the transformed phenotype [45].

Levels of cyclin D1 (and cyclin D3) together with those of CDK4, CDK5 and CDK6, as well as the *in vitro* kinase activity of the complexes, are unaffected by cell anchorage attachment. However, cyclin E/CDK2 complexes are inactive in unattached, normal fibroblasts or rat kidney cells but are active in transformed cells regardless of attachment. The inactive forms have decreased phosphorylation of $Thr^{160}$, the stimulatory phosphorylation site, that may result from increased levels of WAF1 and KIP1 in normal cells in suspension [46,47]. In NIH 3T3 cells, however, cyclin D1 expression and Rb hyperphosphorylation are fully anchorage-dependent [47]. Thus either cyclin E/CDK2 or cyclin D/CDK may be the target of signalling pathways that link the interaction of cell surface integrins with extracellular matrix proteins, depending on cell type.

In a number of cell lines the human T cell leukaemia virus type 1 (HTLV-I) TAX oncoprotein strongly induces expression of cyclin D2 although it appears to suppress transcription of cyclin D3 [48]. In HeLa cells and HT1080 fibrosarcoma cells the expression of cyclin D3 and MYC appear to be linked and both proteins are highly expressed in cells that show tumour necrosis factor-induced apoptosis [49].

## Gene structure

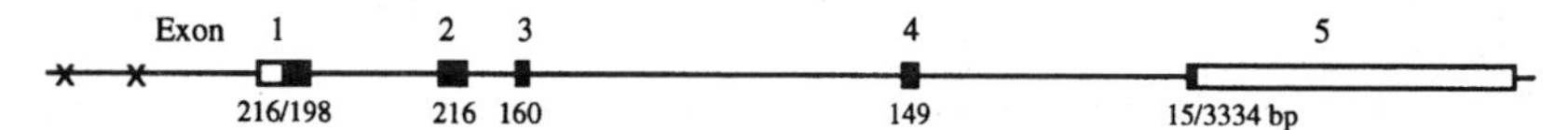


The *CCND1* promoter lacks a TATA box but is GC-rich and includes several SP1 consensus sites (+368 to −373, −108 to −119, −499 to −504 and −1738 to −1743) and an E2F site (−141 to −146). The two crosses indicate breakpoints in a parathyroid adenoma in which 717 bp of *CCND1* are deleted in a reciprocal rearrangement inv(11)(p15q13) with the *PTH* gene [50].

The exon–intron junctions of *CCND1* are similar to those of *CCND2* and *CCND3* [51]. Alternative splicing gives rise to a variant form (cyclin D1 b) that lacks exon 5. Both forms are normally expressed [52].

## Transcriptional regulation

The *CCND1* promoter includes an AP1 site at −935, an SRF site at −928 to −921, an oestrogen-responsive element between −944 and −136 and potential recognition sequences for E2F, MYC and OTF [44].

Elements positively regulating cyclin D2 transcription occur at −114 to −306 and −345 to −444. An element capable of reducing transcription occurs at −892 to −1624 [53].

Elements positively regulating cyclin D3 transcription occur at −167 to −366 [53].

## Amino acid sequence of human CCND1

```
  1 MEHQLLCCEV ETIRRAYPDA NLLNDRVLRA MLKAEETCAP SVSYFKCVQK
 51 EVLPSMRKIV ATWMLEVCEE QKCEEEVFPL AMNYLDRFLS LEPVKKSRLQ
101 LLGATCMFVA SKMKETIPLT AEKLCIYTDN SIRPEELLQM ELLLVNKLKW
151 NLAAMTPHDF IEHFLSKMPE AEENKQIIRK HAQTFVALCA TDVKFISNPP
201 SMVAAGSVVA AVQGLNLRSP NNFLSYYRLT RFLSRVIKCD PDCLRACQEQ
251 IEALLESSLR QAQQNMDPKA AEEEEEEEEE VDLACTPTDV RDVDI (295)
```

### Domain structure

| | |
|---|---|
| 5–9 | D-cyclin pentapeptide motif LLCCE shared with pRb-binding oncoproteins (consensus: LXCXE) |
| 56–165 | Region of homology with A-type cyclins (underlined) |
| 97, 197, 234 | Serines phosphorylated by protein kinase A [54] |
| 55–161 | Cyclin box |
| 272–281 | PEST destruction box |
| 242–295 | Exon 5 encoded region, replaced in the cyclin D1 b variant by VSEGDVPGSLAGAYRGRHLVPRKCRGWCQGPQG [52] |

## Sequence comparison of CCND1 with four A-type cyclins

```
[1](155) MRKIVATWMLEVCEEQKCEEEVFPLAMNYLDRFLSLEPVKKSRLQLLGATCMFVA
[2](209) --A-LVD-LV--G--Y-LQN-TLH--V--I-----SMS-LRGK---V-TAA-LL-
[3](195) --T-LVD-LV--G--Y-LHT-TLY-----------CMS-LRGK---V-TAAILL-
[4](194) --C-LVD-LV--S--D-LHR-TLF-GV--I-----KIS-LRGK---V--AS--L-
[5](234) --S-LID-LV--S--Y-LDT-TLY-SVF-------QMA-VR-K---V-TAA-YI-
```

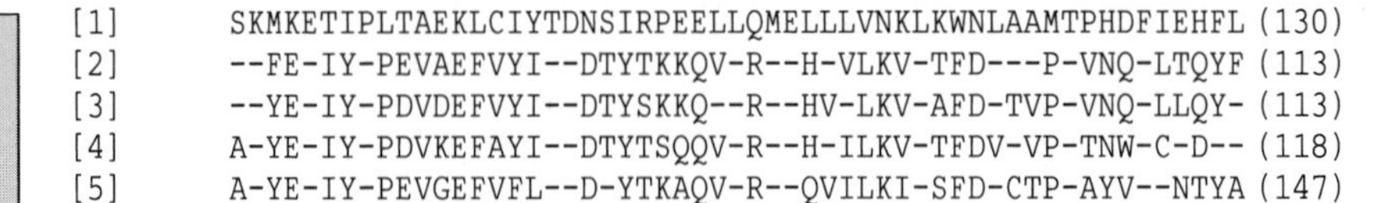

```
[1]   SKMKETIPLTAEKLCIYTDNSIRPEELLQMELLLVNKLKWNLAAMTPHDFIEHFL (130)
[2]   --FE-IY-PEVAEFVYI--DTYTKKQV-R--H-VLKV-TFD---P-VNQ-LTQYF (113)
[3]   --YE-IY-PDVDEFVYI--DTYSKKQ--R--HV-LKV-AFD-TVP-VNQ-LLQY- (113)
[4]   A-YE-IY-PDVKEFAYI--DTYTSQQV-R--H-ILKV-TFDV-VP-TNW-C-D-- (118)
[5]   A-YE-IY-PEVGEFVFL--D-YTKAQV-R--QVILKI-SFD-CTP-AYV--NTYA (147)
```

[1] Human CCND1, [2] Human cyclin A, [3] *Xenopus laevis* cyclin A, [4] *Spisula solidissima* cyclin A, [5] *Drosophila melanogaster* cyclin A. Dashes indicate identical residues. Numbers in round brackets refer to the number of amino acids preceding and following each sequence shown.

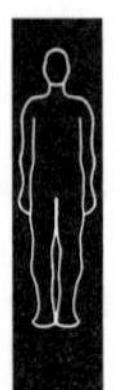

## Amino acid sequence of human cyclin D2

```
  1 MELLCHEVDP VRRAVRDRNL LRDDRVLQNL LTIEERYLPQ CSYFKCVQKD
 51 IQPYMRRMVA TWMLEVCEEQ KCEEEVFPLA MNYLDRFLAG VPTPKSHLQL
101 LGAVCMFLAS KLKETSPLTA EKLCIYTDNS IKPQELLEWE LVVLGKLKWN
151 LAAVTPHDFI EHILRKLPQQ REKLSLIRKH AQTFIALCAT DFKFAMYPPS
201 MIATGSVGAA ICGLQQDEEV SSLTCDALTE LLAKITNTDV DCLKACQEQI
251 EAVLLNSLQQ YRQDQRDGSK SEDELDQAST PTDVRDIDL (289)
```

## Amino acid sequence of human cyclin D3

```
  1 MELLCCEGTR HAPRAGPDPR LLGDQRVLQS LLRLEERYVP RASYFQCVQR
 51 EIKPHMRKML AYWMLEVCEE QRCEEEVFPL AMNYLDRYLS CVPTRKAQLQ
101 LLGAVCMLLA SKLRETTPLT IEKLCIYTDH AVSPRQLRDW EVLVLGKLKW
151 DLAAVIAHDF LAFILHRLSL PRDRQALVKK HAQTFLALCA TDYTFAMYPP
201 SMIATGSIGA AVQGLGACSM SGDELTELLA GITGTEVDCL RACQEQIEAA
251 LRESLREAAQ TSSSPAPKAP RGSSSQGPSQ TSTPTDVTAI HL (292)
```

Conflict: $A^{259} \rightarrow S$ [2].

**Database accession numbers**

| | *PIR* | *SWISSPROT* | *EMBL/GENBANK* | *REFERENCES* |
|---|---|---|---|---|
| Human *CCND1* | A40034, A41523, | CG1D_HUMAN P24385 | M64349, M74092 | 55 |
| | B40268, S14794 | | Hsprad1cy X59789, Hsbcl1 M73554, X59485, Z23022 | 30 |
| Human *CCND1b* | | | HSCCND1I4 X88930 | 52 |
| Human *CCND2* | A42822, S26580 | P30279, CGD2_HUMAN | HSCCND2A M90813, X68452 | 2,32 |
| Human *CCND3* | A44022, B42822 | P30281, CGD3_HUMAN | HSCYCD3A M92287 | 2,3 |
| Mouse *Bcl-1* | | CG1D_MOUSE P25322 | Mdcyl1 M64403 | 4 |
| *Xenopus Xic1* | | | U24434 | 56 |

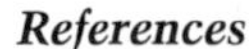

***References***


1 Tsujimoto, Y. et al. (1984) Science 224, 1403–1406.
2 Xiong Y. et al. (1992) Genomics 13, 575–584.
3 Motokura, T. et al. (1992) J. Biol. Chem. 267, 20412–20415.

[4] Matsushime, H. et al. (1991) Cell 65, 701–713.
[5] Hanna, Z. et al. (1993) Oncogene 8, 1661–1666.
[6] Lukas, J. et al. (1995) Oncogene 10, 2125–2134.
[7] Lukas, J. et al. (1994) Oncogene 9, 2159–2167.
[8] Suzuki-Takahashi, I. et al. (1995) Oncogene 10, 1691–1698.
[9] Chen, X. et al. (1995) Cancer Res. 55, 4257–4263.
[10] Lukas, J. et al. (1995) Mol. Cell. Biol. 15, 2600–2611.
[11] Spitkovsky, D. et al. (1995) Oncogene 10, 2421–2425.
[12] Marhin, W.W. et al. (1996) Oncogene 12, 43–52.
[13] Rao, S.S. and Kohtz, D.S. (1995) J. Biol. Chem. 270, 4093–4100.
[14] Zhou, P. et al. (1996) Cancer Res. 56, 36–39.
[15] Jiang, W. et al. (1992) Cancer Res. 52, 2980–2983.
[16] Lammie, G. et al. (1991) Oncogene 6, 439–444.
[17] McIntosh, G.G. et al. (1995) Oncogene 11, 885–891.
[18] Gillett, C. et al. (1996) Int. J. Cancer (Pred. Oncol.) 69, 92–99.
[19] Keyomarsi, K. and Pardee, A.B. (1993) Proc. Natl Acad. Sci. USA 90, 1112–1116.
[20] Buckley, M.F. et al. (1993) Oncogene 8, 2127–2133.
[21] de Boer, C.J. et al. (1993) Cancer Res. 53, 4148–4152.
[22] Bigoni, R. et al. (1996) Oncogene 13, 797–802.
[23] de Boer, C.J. et al. (1995) Oncogene 10, 1833–1840.
[24] Seto, M. et al. (1992) Oncogene 7, 1401–1406.
[25] Theillet, C. et al. (1990) Oncogene 5, 147–149.
[26] Bringuier, P.P. et al. (1996) Oncogene 12, 1747–1753.
[27] Faust, J.B. and Meeker, T.C. (1992) Cancer Res. 52, 2460–2463.
[28] Foulkes, W. D. et al. (1993) Br. J. Cancer 67, 268–273.
[29] Rosenberg, C.L. et al. (1993) Oncogene 8, 519–521.
[30] Motokura, T. et al. (1991) Nature 350, 512–515.
[31] Bartkova, J. et al. (1995) Oncogene 10, 775–778.
[32] Palmero, I. et al. (1993) Oncogene 8, 1049–1054.
[33] Leach, F.S. et al. (1993) Cancer Res. 53, 1986–1989.
[34] Tam, S.W. et al. (1994) Oncogene 9, 2663–2674.
[35] Ichimura, K. et al. (1996) Oncogene 13, 1065–1072.
[36] Williams, R.T. et al. (1993) J. Biol. Chem. 268, 8871–8880.
[37] Jung, J.U. et al. (1994) Mol. Cell. Biol. 14, 7235–7244.
[38] Wang, T.C. et al. (1994) Nature 369, 669–671.
[39] Lovec, H. et al. (1994) EMBO J. 13, 3487–3495.
[40] Sicinski, P. et al. (1995) Cell 82, 621–630.
[41] Lovec, H. et al. (1994) Oncogene 9, 323–326.
[42] Hinds, P.W. et al. (1994) Proc. Natl Acad. Sci. USA 91, 709–713.
[43] Musgrove, E.A. (1994) Proc. Natl Acad. Sci. USA 91, 8022–8026.
[44] Altucci, L. et al. (1996) Oncogene 12, 2315–2324.
[45] Zhou, P. et al. (1995) Oncogene 11, 571–580.
[46] Fang, F. et al. (1996) Science 271, 499–502.
[47] Zhu, X. et al. (1996) J. Cell Biol. 133, 391–403.
[48] Akagi, T. et al. (1996) Oncogene 12, 1645–1652.
[49] Jänicke, R.U. et al. (1996) Mol. Cell. Biol. 16, 5245–5253.
[50] Motokura, T. and Arnold, A. (1993) Genes, Chromosom. Cancer 7, 89–95.
[51] Inaba, T. et al. (1992) Genomics 13, 565–574.
[52] Betticher, D.C. et al. (1995) Oncogene 11, 1005–1011.

[53] Brooks, A.R. et al. (1996) J. Biol. Chem. 271, 9090–9099.
[54] Sewing, A. and Müller, R. (1994) Oncogene 9, 2733–2736.
[55] Withers, D.A. et al. (1991) Mol. Cell. Biol. 11, 4846–4853.
[56] Su, J.-Y. et al. (1995) Oncogene 10, 1691–1698.

# DBL

## Identification

*DBL* was identified by NIH 3T3 fibroblast transfection with DNA from a human diffuse B cell lymphoma. *MCF2* and *DBL* represent two activated versions of the same proto-oncogene [1,2].

## Related genes

The N-terminal region of DBL has homology with vimentin and the central region has homology with human ABR, BCR, ECT2, TIM, VAV, CDC24, CDC25 and RAS-GRF nucleotide exchange factor. DBS (Dbl's big sister) transforms fibroblasts and has high sequence homology to DBL in the CDC24 and guanine nucleotide exchange factor homology domain [3].

| | *DBL* |
|---|---|
| **Nucleotides (kb)** | 45 |
| **Chromosome** | Xq27 |
| **Mass (kDa): predicted** | 108 |
| **expressed** | p115 (DBL)<br>p66 (oncogenic) |

**Cellular location**
Cytoplasmic: associates with cytoskeletal actin [4].

**Tissue distribution**
Highly tissue specific: brain, adrenal glands, gonads.

## Protein function

Unknown. Microinjection of DBL causes germinal vesicle breakdown and activation of H1 histone kinase in oocytes [5]. *In vitro* it acts as a guanine nucleotide exchange factor for CDC42Hs, the human homologue of the *S. cerevisiae* protein CDC42Sc [6] and activates RHO family proteins.

Transformation by DBL (and also by VAV) involves activation of mitogen-activated protein kinases (MAPKs) independently of RAS activation [7].

**Cancer**
*DBL* is expressed in Ewing's sarcoma (tumours of neuroectodermic origin) [8] in which DBL appears to upregulate the expression of the DNA repair enzyme poly(ADP-ribose) polymerase [9].

**Transgenic animals**
*DBL* does not induce neoplasia in transgenic mice but animals that express the DBL protein in their lenses develop cataracts. *DBL* therefore appears capable of interfering with the ability of lens epithelial cells to differentiate into lens fibre cells [10].

***In vitro***

Transfection of NIH 3T3 cells with DNA from a human nodular poorly differentiated lymphoma (NPDL) generates *NPDL-DBL* that is homologous to oncogenic *DBL* and encodes a 76 kDa protein[11]. Transforming potential is enhanced 50–70-fold on substitution of 5′ sequences to form oncogenic *DBL*[12].

## Protein structure

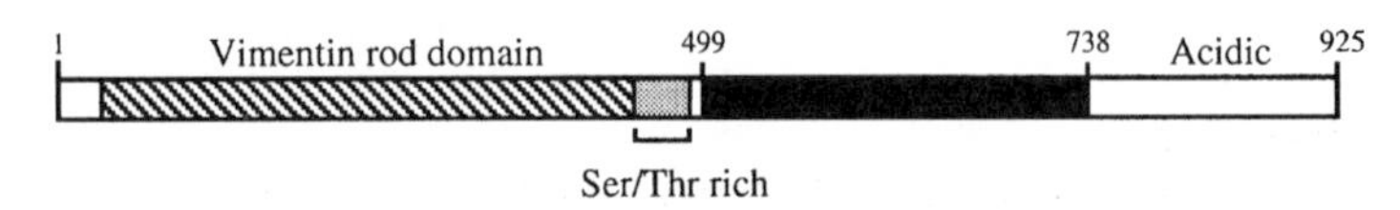


*DBL* and *MCF2* represent two activated versions of the same proto-oncogene. Activation of the *DBL* oncogene occurs by 5′ fusion of 50 amino acids encoded by an unrelated human gene to 428 amino acids encoded by the 3′ portion of *DBL*, with the loss of the first 497 amino acids of DBL. There is only one (conservative) substitution in the C-terminal 428 amino acids of oncogenic DBL. The loss of the N-terminal 497 amino acids of DBL, rather than the properties of the N-terminus acquired by oncogenic DBL, is responsible for the transforming activity[13]. In MCF2 397 amino acids are removed.

## Amino acid sequence of human DBL

```
  1 MAEANPRRGK MRFRRNAASF PGNLHLVLVL RPTSFLQRTF TDIGFWFSQE
 51 DFMPKLPVVM LSSVSDLLTY IDDKQLTPEL GGTLQYCHSE WIIFRNAIEN
101 FALTVKEMAQ MLQSFGTELA ETELPDDIPS IEEILAIRAE RYHLLKNDIT
151 AVTKEGKILL TNLEVPDTEG AVSSRLECHR QISGDWQTIN KLLTQVHDME
201 TAFDGFWEKH QLKMEQYLQL WKFEQDFQQL VTEVEFLLNQ QAELADVTGT
251 IAQVKQKIKK LENLDENSQE LLSKAQFVIL HGHKLAANHH YALDLICQRC
301 NELRYLSDIL VNEIKAKRIQ LSRTFKMHKL LQQARQCCDE GECLLANQEI
351 DKFQSKEDAQ KALQDIENFL EMALPFINYE PETLQYEFDV ILSPELK VQM
                                                       Δ
401 KTIQLKLENI RSIFENQQAG FRNLADKHVR PIQFVVPTPE NLVTSGTPFF
451 SSKQGKKTWR QNQSNLKIEV VPDCQEKRSS GPSSSLDNGN SLDVLKN HVL
                                                       ΔΔ
501 NELIQTERVY VRELYTVLLG YRAEMDNPEM FDLMPPLLRN KKDILFGNMA
551 EIYEFHNDIF LSSLENCAHA PERVGPCFLE RKDDFQMYAK YCQNKPRSET
601 IWRKYSECAF FQECQRKLKH RLRLDSYLLK PVQRITKYQL LLKELLKYSK
651 DCEGSALLKK ALDAMLDLLK SVNDSMHQIA INGYIGNLNE LGKMIMQGGF
701 SVWIGHKKGA TKMKDLARFK PMQRHLFLYE KAIVFCKRRV ESGEGSDRYP
751 SYSFKHCWKM DEVGITEYVK GDNRKFEIWY GEKEEVYIVQ ASNVDVKMTW
801 LKEIRNILLK QQELLTVKKR KQQDQLTERD KFQISLQQND EKQQGAFIST
851 EETELEHTST VVEVCEAIAS VQAEANTVWT EASQSAEISE EPAEWSSNYF
901 YPTYDENEEE NRPLMRPVSE MALLY (925)
```

**Domain structure**

398–925 MCF2 transforming protein (Δ)

498–925 DBL transforming protein (ΔΔ). The DBL breakpoint lies within an *MCF2* exon

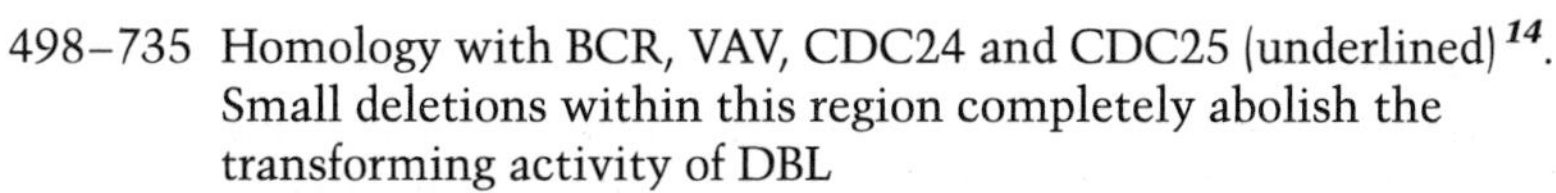

498–735 Homology with BCR, VAV, CDC24 and CDC25 (underlined)[14]. Small deletions within this region completely abolish the transforming activity of DBL
343 Putative palmitoylation site
742 Phosphorylation site

**Database accession numbers**

| | *PIR* | *SWISSPROT* | *EMBL/GENBANK* | *REFERENCES* |
|---|---|---|---|---|
| Human *DBL* | A30040 | P10911 | X12556 | [15] |
| | A28051 | P14919 | J03639 | [16] |
| | | | X13230 | [12] |

***References***


[1] Fasano, O. et al. (1984) Mol. Cell. Biol. 4, 1695–1705.
[2] Eva, A. and Aaronson, S.A. (1985) Nature 316, 273–275.
[3] Whitehead, I. et al. 1995) Oncogene 10, 705–711.
[4] Graziani, G. et al. (1989) Oncogene 4, 823–829.
[5] Graziani, G. et al. (1992) Oncogene 7, 229–235.
[6] Hart, M.J. et al. (1991) Nature 354, 311–314.
[7] Khosravi-Far, R. et al. (1994) Mol. Cell. Biol. 14, 6848–6857.
[8] Vecchio, G. et al. (1989) Oncogene 4, 897–900.
[9] Velasco, J.A. et al. (1995) Oncogene 10, 2253–2258.
[10] Eva, A. et al. (1991) New Biol. 3, 158–168.
[11] Eva, A. et al. (1987) Oncogene 1, 355–360.
[12] Noguchi, T. et al. (1988) Oncogene 3, 709–715.
[13] Ron, D. et al. (1989) Oncogene 4, 1067–1072.
[14] Ron, D. et al. (1991) New Biol. 3, 372–379.
[15] Ron, D. et al. (1988) EMBO J. 7, 2465–2473.
[16] Eva, A. et al. (1988) Proc. Natl Acad. Sci. USA 85, 2061–2065.

## Identification

v-*erbB* is the oncogene of avian *er*ythro*b*lastosis virus (AEV), strain AEV-H, detected by hybridizing viral mRNA with cDNA made against the unique sequences of AEV[1]. *EGFR* (*ErbB-1*) encodes the receptor for epidermal growth factor[2–4]. The P1, A3, B1, B8 and C2 retroviruses carrying *erbB* have been isolated from a chicken following infection with RAV-1[5].

## Related genes

The *EGFR* family includes *HER2/ErbB-2, HER3/Neu/ErbB-3, HER4/ErbB-4, Drosophila*: *DER*; *Xiphophorus maculatus: Xmrk*.

| | ***EGFR*** | **v-*erbB*** |
|---|---|---|
| **Nucleotides (kb)** | 110 | 7.8 (AEV-H) |
| **Chromosome** | 7p13–p12 | |
| **Mass (kDa): predicted** | 134 | 62 |
| **expressed** | gp170 | 68/74 |
| **Cellular location** | Plasma membrane | Plasma membrane |

**Tissue distribution**
Human *EGFR* is widely distributed, except in haematopoietic tissues. Most cells express between $2 \times 10^4$ and $2 \times 10^5$ receptors.

## Protein function

There are at least 15 EGFR family ligands[6,7], including epidermal growth factor (EGF), transforming growth factor α (TGFα), amphiregulin, cripto-1, betacellulin, heparin-binding EGF-like growth factor α (HB-EGF) and differentially spliced neuregulins, also called heregulins or neu differentiation factors (NRGs or NDFs: they include gp30, heregulin (HRG), NEU differentiation factor, glial growth factor, sensory and motor neurone-derived factor and acetylcholine receptor-inducing factor). Of this family, those that share an EGF motif bind to the EGFR: these include EGF, heparin-binding EGF-like factor (HB-EGF), TGFα, amphiregulin, betacellulin[8] and vaccinia virus growth factor. In epidermal cells the EGFR is activated by UV irradiation[9].

The EGFR can heterodimerize with each of the other members of the EGFR family (HER2, HER3 and HER4) and individual ligands for this family give rise to distinct patterns of *trans*-phosphorylation[10]. Thus, for example, in T47D breast tumour cells EGF, TGFα and HB-EGF stimulate differentially the tyrosine phosphorylation of heterodimerized receptors in the order EGFR > HER2 > HER3 > HER4. For NDF the order is reversed with phosphorylation of the EGFR being undetectable, whilst betacellulin promotes approximately equivalent levels of phosphorylation of all four receptor types[11].

The activated EGFR phosphorylates the RAS-GTPase-activating protein (GAP), a cytosolic tyrosine kinase[12], CBL, EPS8, EPS15, pp81, p91, lipocortins, pp42, protein kinase C$\delta$[13] and calmodulin and also stimulates substrate-selective protein tyrosine phosphatase activity directed towards the EGFR itself and HER2[14]. In addition to the GRB2/SOS complex, the activated EGFR physically associates with CBL, SHC proteins, EPS8, phosphatidylinositol 3-kinase and phospholipase C$\gamma$[15–17]. The EGFR may activate atypical PKC$\lambda$ (aPKC$\lambda$, which is not activated by diacylglycerol or phorbol ester) *via* PtdIns 3-kinase[18]. Protein kinase C also downregulates the kinase activity of the EGFR by an indirect mechanism that involves MAP kinase[19].

EGF activates a second pathway which involves the rapid activation of latent cytoplasmic transcription factors called signal transducers and activators of transcription (STATs). EGF or amphiregulin causes the tyrosine phosphorylation of STAT1, STAT3 and STAT5, which mediates the activation of *FOS via* interaction with the serum-inducible element (SIE)[20]. However, although JAKs may undergo concomitant phosphorylation, they do not appear to be essential for EGF activation of STATs. Furthermore, the kinase activity of EGFR is essential for activation of this pathway but tyrosine phosphorylation sites within the receptor are not, and the SIE may play only a minor role in STAT activation of *Fos* transcription[21,22]. Intraperitoneal injection of EGF into mice results in the appearance in liver nuclei of multiple tyrosine phosphorylated proteins, including STAT5[23].

Phosphorylated tyrosine residues on the activated EGFR associate with the SH2 domain of growth factor receptor-bound protein 2 (GRB2) which itself is bound to SOS1. This causes the activation of RAS (i.e. increase in GTP-RAS; see ***RAS***), leading to the activation of RAF1 and MAP kinase. This pathway can be downregulated by MAP kinase phosphorylation of SOS which promotes the dissociation of GRB2/SOS from SHC[24].

Activation of the MAP kinase pathway by EGF can activate the unliganded oestrogen receptor by direct phosphorylation[25]. This mechanism may therefore give rise to steroid-independent activation, either in the absence of oestrogen or of mutant receptors that arise during tumour development and are unable to bind their normal ligand.

EGFR activation increases the intracellular free concentration of $Ca^{2+}$ and causes transient membrane hyperpolarization[26]. In keratinocytes activation of the EGFR specifically stimulates transcription of *K6* and *K16* genes[27]. In GH4 neuroendocrine cells EGF activates the rat prolactin promoter by a pathway that is distinct from and antagonistic to the RAS/MAPK pathway[28].

In the heart, EGFR activates adenylyl cyclase *via* interaction with $G\alpha_s$, which is phosphorylated on tyrosine residues by the EGFR, leading to stimulation of the GTPase activity and GTP binding. The interaction involves a 13 amino acid, juxtamembrane region ($Arg^{646}$–$Arg^{658}$: RRREIVRKRTLRR) of EGFR[29].

Ligand/EGFR complexes but not ligand complexes with other EGFR family members, undergo receptor-mediated endocytosis[30]. The activated EGFR also undergoes covalent linkage to ubiquitin which may promote degradation, but it may reflect a role for ubiquitin as a chaperone to control the structure of the receptor[31].

ZPR1, a zinc finger protein, binds to EGFR in unstimulated cells but is released and migrates to the nucleus on EGF activation[32].

### Cancer

*EGFR* is amplified by up to 50-fold in some primary tumours, including ~20% of bladder tumours[33], and derived cell lines[34]. Amplification also occurs in ~20% of primary breast tumours where it is strongly associated with early recurrence and death in lymph node-positive patients[35–38]. Cells derived from human colon carcinomas show a correlation between high over-expression of EGFR mRNA and metastatic capacity[39]. Over-expression of EGFR occurs in pituitary adenomas and correlates with tumour aggressiveness[40]. Increased expression of EGFR (and of HER2, HER3 and HER4) has been detected in papillary thyroid carcinomas[41].

Three truncated forms of EGFR have been detected in some malignancies. Type I (EGFRvIII) lacks the majority of the extracellular domain (amino acids 6–273) and does not bind EGF but is constitutively associated with GRB2 although its expression does not increase RAS-GTP and only slightly activates the MAPK pathway[42]. Type II contains an in-frame deletion of 83 amino acids (520–603) in extracellular domain IV that does not prevent EGF and TGFα binding[43]. Type III has an in-frame deletion of 267 amino acids (29–296) in extracellular domains II and III[44]. Each of these mutations has been detected in glioblastomas. Type I mutations have been detected in brain and lung tumours and in over 70% of breast and ovarian tumours examined[45] and type III has also been detected in 16% of a sample of non-small cell lung carcinomas[46].

### Transgenic animals

*ErbB/EGFR*$^{-/-}$ mice survive for up to 18 days but have impaired epithelial development, particularly in the skin, lung and gastrointestinal tract[47]. Expression of a dominant negative mutant of EGFR in the epidermis that suppresses autophosphorylation of endogenous EGFR causes marked alterations in hair follicle development and skin structure[48].

### In animals

v-*erbB* is the transduced gene in avian retroviruses (e.g. avian leukosis virus) that causes rapid induction of erythroleukaemia and fibrosarcoma[49].

### *In vitro*

EGFR is a weak transforming agent in NIH 3T3 cells[50] but EGF acts synergistically with insulin to stimulate DNA synthesis and cell proliferation and can induce ligand-independent proliferation of myeloid precursor cells[51].

v-*erbB* transforms chick embryo fibroblasts and stimulates proliferation of erythrocyte progenitor cells although it does not completely block their differentiation.

## Gene structure

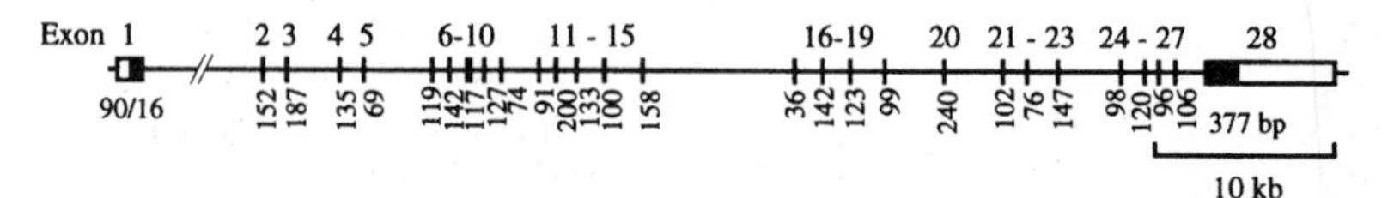

The chicken gene structure (shown above) is highly conserved among vertebrates [52]. The *EGFR* promoter is rich in GC regions and contains overlapping SP1 and $T_3$ receptor/RXRα binding sites but lacks TATA or CAAT boxes [53]. A sequence-specific transcription factor isolated from human malignant breast tissues binds to a core element (−22 to +9: GAAT-GAAGTTGTGAAGCTGAGATTCCCCTCCA) to activate transcription of *EGFR*: this factor is also mitogenic for NIH 3T3 cells [54]. *EGFR* transcription *in vivo* is positively activated by p53, EGF, cAMP or TPA and is inhibited by $T_3R$ or RAR [55,56].

## Protein structure

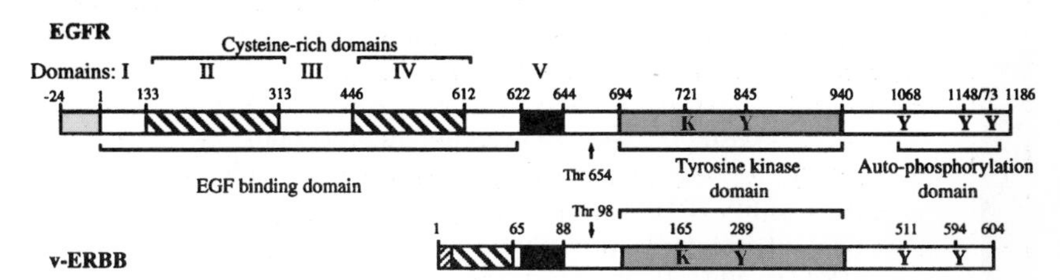


Similar cysteine-rich domains occur in other EGFR family members and in the receptors for insulin, insulin-like growth factor (IGF-I), nerve growth factor (NGF) and low-density lipoprotein (LDL). The 24 amino acid signal sequence (dotted box) is proteolytically cleaved from the EGFR. The *gag* encoded (six amino acids) of v-ERBB (lightly hatched box) also encodes a (putative, cleaved) signal sequence. v-*erbB* is derived from chicken *EGFR*. Chicken and human EGFR share 97% and 65% identity in the tyrosine kinase and C-terminal regions, respectively. In v-ERBB the receptor binding domain and 30 C-terminal residues are deleted and there are eight point mutations, relative to chicken EGFR.

## Amino acid sequence of human EGFR

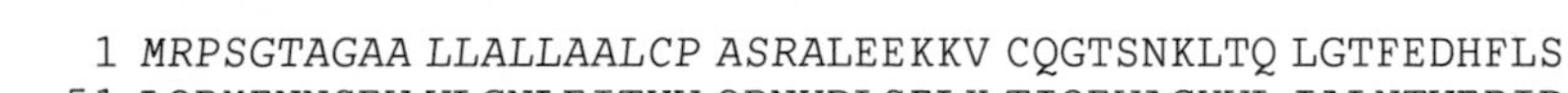

```
  1 MRPSGTAGAA LLALLAALCP ASRALEEKKV CQGTSNKLTQ LGTFEDHFLS
 51 LQRMFNNCEV VLGNLEITYV QRNYDLSFLK TIQEVAGYVL IALNTVERIP
101 LENLQIIRGN MYYENSYALA VLSNYDANKT GLKELPMRNL QEILHGAVRF
151 SNNPALCNVE SIQWRDIVSS DFLSNMSMDF QNHLGSCQKC DPSCPNGSCW
201 GAGEENCQKL TKIICAQQCS GRCRGKSPSD CCHNQCAAGC TGPRESDCLV
251 CRKFRDEATC KDTCPPLMLY NPTTYQMDVN PEGKYSFGAT CVKKCPRNYV
301 VTDHGSCVRA CGADSYEMEE DGVRKCKKCE GPCRKVCNGI GIGEFKDSLS
351 INATNIKHFK NCTSISGDLH ILPVAFRGDS FTHTPPLDPQ ELDILKTVKE
401 ITGFLLIQAW PENRTDLHAF ENLEIIRGRT KQHGQFSLAV VSLNITSLGL
451 RSLKEISDGD VIISGNKNLC YANTINWKKL FGTSGQKTKI ISNRGENSCK
501 ATGQVCHALC SPEGCWGPEP RDCVSCRNVS RGRECVDKCK LLEGEPREFV
551 ENSECIQCHP ECLPQAMNIT CTGRGPDNCI QCAHYIDGPH CVKTCPAGVM
601 GENNTLVWKY ADAGHVCHLC HPNCTYGCTG PGLEGCPTNG PKIPSIATGM
651 VGALLLLLVV ALGIGLFMRR RHIVRKRTLR RLLQERELVE PLTPSGEAPN
701 QALLRILKET EFKKIKVLGS GAFGTVYKGL WIPEGEKVKI PVAIKELREA
751 TSPKANKEIL DEAYVMASVD NPHVCRLLGI CLTSTVQLIT QLMPFGCLLD
801 YVREHKDNIG SQYLLNWCVQ IAKGMNYLED RRLVHRDLAA RNVLVKTPQH
851 VKITDFGLAK LLGAEEKEYH AEGGKVPIKW MALESILHRI YTHQSDVWSY
```

```
 901 GVTVWELMTF GSKPYDGIPA SEISSILEKG ERLPQPPICT IDVYMIMVKC
 951 WMIDADSRPK FRELIIEFSK MARDPQRYLV IQGDERMHLP SPTDSNFYRA
1001 LMDEEDMDDV VDADEYLIPQ QGFFSSPSTS RTPLLSSLSA TSNNSTVACI
1051 DRNGLQSCPI KEDSFLQRYS SDPTGALTED SIDDTFLPVP EYINQSVPKR
1101 PAGSVQNPVY HNQPLNPAPS RDPHYQDPHS TAVGNPEYLN TVQPTCVNST
1151 FDSPAHWAQK GSHQISLDNP DYQQDFFPKE AKPNGIFKGS TAENAEYLRV
1201 APQSSEFIGA (1210)
```

**Domain structure**

| | |
|---|---|
| 1–24 | Signal sequence (italics) |
| 25–645 | Extracellular domain |
| 627 | The mutation $Glu^{627} \rightarrow Val$, corresponding to the activating mutation in NEU, causes constitutive phosphorylation of the receptor and of SHC, association of GRB2 with the receptor and activation of MAPK [57] |
| 646–668 | Transmembrane (underlined) |
| 678 | Major protein kinase C phosphorylation site ($Thr^{678}$) |
| 669–1210 | Cytoplasmic domain |
| 75–300 and 390–600 | Approximate repeats |
| 718–726 and 745 | ATP binding |
| 678 | Threonine phosphorylation by protein kinase C |
| 693 | Threonine phosphorylation by MAP kinase and by EGFR $Thr^{669}$ (ERT) kinase [58] |
| 978 | Putative SH-PTP2/Syp binding site ($Tyr^{978}$) [59] |
| 1016 | Principal binding site for the SH2 domain of PLC$\gamma$ ($Tyr^{1016}$) |
| 1016, 1172 | High-affinity binding sites for PTP1B |
| 1110 | ABL SH2 domain binding site ($Tyr^{1110}$) [60] |
| 1172 | Principal binding site for the SH2 domain of GAP ($Tyr^{1172}$) |
| 1016 and 1172 | Binding site for protein tyrosine phosphatase 1B [61] |
| 1173 and 992 | $Tyr^{1173}$ and $Tyr^{992}$: major and minor SHC binding sites |
| 1068 and 1086 | $Tyr^{1068}$ and $Tyr^{1086}$: major and minor GRB2 binding sites [62] |
| 1026, 1070 and 1071 | Serine phosphorylation sites [63] |
| 1092, 1110, 1172, 1197 | Tyrosine phosphorylation sites (underlined) |
| 128, 175, 196, 352, 361, 413, 444, 528, 568, 603, 623 | Putative carbohydrate attachment sites |
| 908–958 | ZPR1 binding [32, 53] |

**Database accession numbers**

| | *PIR* | *SWISSPROT* | *EMBL/GENBANK* | *REFERENCES* |
|---|---|---|---|---|
| Human *EGFR* | A00641 | P00533 | X00588, X06370 | 4 |
| | A00642 | | K01885, K02047 | 2 |
| | A23062 | | M38425 | 3,64,65,66 |

***References***


1 Yamamoto, T. et al. (1983) Cell 35, 71–78.
2 Lin, C.R. et al. (1984) Science 224, 843–848.

[3] Simmen, F.A. et al. (1984) Biochem. Biophys. Res. Commun. 124, 125–132.
[4] Ullrich, A. et al. (1984) Nature 309, 418–425.
[5] Vennstrom, B. et al. (1994) Oncogene 9, 1307–1320.
[6] **Carpenter, G. (1987) Annu. Rev. Biochem. 56, 881–914.**
[7] **Laurence, D.J.R. and Gusterson, B.A. (1990) Tumor Biol. 11, 229–261.**
[8] Riese, D.J. et al. (1996) Oncogene 12, 345–353.
[9] Miller, C.C. et al. (1994) J. Biol. Chem. 269, 3529–3533.
[10] Riese, D.J. et al. (1995) Mol. Cell. Biol. 15, 5770–5776.
[11] Beerli, R.R. and Hynes, N.E. (1995) J. Biol. Chem. 271, 6071–6076.
[12] Filhol, O. et al. (1994) J. Biol. Chem. 268, 26978–26982.
[13] Denning, M.F. et al. (1996) J. Biol. Chem. 271, 5513–5518.
[14] Hernandez-Sotomayor, S.M.T. et al. (1993) Proc. Natl Acad. Sci. USA 90, 7691–7695.
[15] Zhu, G. et al. (1992) Proc. Natl Acad. Sci. USA 89, 9559–9563.
[16] Segatto, O. et al. (1993) Oncogene 8, 2105–2112.
[17] Fazioli, F. et al. (1993) EMBO J. 12, 3799–3808.
[18] Akimoto, K. et al. (1996) EMBO J. 15, 788–798.
[19] Morrison, P. et al. (1996) J. Biol. Chem. 271, 12891–12896.
[20] Sadowski, H.B. et al. (1993) Science 261, 1739–1744.
[21] Leaman, D.W. et al. (1996) Mol. Cell. Biol. 16, 369–375.
[22] David, M. et al. (1996) J. Biol. Chem. 271, 9185–9188.
[23] Ruff-Jamison, S. et al. (1995) Proc. Natl Acad. Sci. USA 92, 4215–4218.
[24] Porfiri, E. and McCormick, F. (1996) J. Biol. Chem. 271, 5871–5877.
[25] Bunone, G. et al. (1996) EMBO J. 15, 2174–2183.
[26] Peppelenbosch, M.P. et al. (1992) Cell 69, 295–303.
[27] Jiang, C.-K. et al. (1993) Proc. Natl Acad. Sci. USA 90, 6786–6790.
[28] Pickett, C.A. and Gutierrez-Hartmann, A. (1995) Mol. Cell. Biol. 15, 6777–6784.
[29] Poppleton, H. et al. (1996) J. Biol. Chem. 271, 6947–6951.
[30] Baulida, J. et al. (1996) J. Biol. Chem. 271, 5251–5257.
[31] Galcheva-Gargova, Z. et al. (1995) Oncogene 11, 2649–2655.
[32] Galcheva-Gargova, Z. et al. (1996) Science 272, 1797–1802.
[33] Proctor, A.J. et al. (1991) Oncogene 6, 789–795.
[34] **Gullick, W.J. (1991) Br. Med. Bull. 47, 87–98.**
[35] Horak, E. et al. (1991) Oncogene 6, 2277–2284.
[36] Borg, A. et al. (1991) Oncogene 6, 137–143.
[37] Paterson, M.C. et al. (1991) Cancer Res. 51, 556–567.
[38] Benz, C.C. et al. (1995) Gene, 159, 3–7.
[39] Radinsky, R. et al. (1995) Clin. Cancer Res. 1, 19–31.
[40] LeRiche, V.K. et al. (1996) J. Clin. Endocrinol. Metabol. 81, 656–662.
[41] Haugen, D.R.F. et al. (1996) Cancer Res. 56, 1184–1188.
[42] Moscatello, D.K. et al. (1996) Oncogene 13, 85–96.
[43] Humphrey, P.A. et al. (1991) Biophys. Res. Commun. 178, 1413–1420.
[44] Humphrey, P.A. et al. (1990) Proc. Natl. Adad. Sci. USA 87, 4207–4211.
[45] Moscatello, D.K. et al. (1995) Cancer Res. 55, 5536–5539.
[46] Garcia de Palazzo, I.E. et al. (1993) Cancer Res. 53, 3217–3220.
[47] Miettinen, P. et al. (1995) Nature 376, 337–341.
[48] Murillas, R. et al. (1995) EMBO J. 14, 5216–5223.
[49] Shu, H.-K.G. et al. (1991) J. Virol. 65, 6177–6180.
[50] Riedel, H. et al. (1988) Proc. Natl Acad. Sci. USA 85, 1477–1481.

[51] Segatto, O. et al. (1991) Mol. Cell. Biol. 11, 3191–3202.
[52] Callaghan, T. et al. (1993) Oncogene 8, 2939–2948.
[53] Xu, J. et al. (1993) J. Biol. Chem. 268, 16065–16073.
[54] Sarkar, F.H. et al. (1994) J. Biol. Chem. 269, 12285–12289.
[55] Deb, S.P. et al. (1994) Oncogene 9, 1341–1349.
[56] Ludes-Meyers, J.H. et al. (1996) Mol. Cell. Biol. 16, 6009–6019.
[57] Miloso, M. et al. (1995) J. Biol. Chem. 270, 19557–19562.
[58] Williams, R. et al. (1993) J. Biol. Chem. 268, 18213–18217.
[59] Case, R.D. et al. (1994) J. Biol. Chem. 269, 10467–10474.
[60] Zhu, G. et al. (1994) Oncogene 9, 1379–1385.
[61] Milarski, K.L. et al. (1993) J. Biol. Chem. 268, 23634–23639.
[62] Batzer, A.G. (1994) Mol. Cell. Biol. 14, 5192–5201.
[63] Kuppuswamy, D. et al. (1993) J. Biol. Chem. 268, 19134–19142.
[64] Haley, J. et al. (1987) Oncogene Res. 1, 375–396.
[65] Mroczkowski, B. et al. (1984) Nature 309, 270–273.
[66] Margolis, B.L. et al. (1989) J. Biol. Chem. 264, 10667–10671.

# ELK1

## Identification

*ELK1* was detected by screening a cDNA library with *ETS2* cDNA [1].

## Related genes

The *ETS* subfamily of *ELK1, ELK2, ELK3/NET/ERP/SAP2, ELK4/SAP1* are highly homologous (see ***ETS1, ETS2*** and ***FOS***) [2].

| | ***ELK1*** |
|---|---|
| **Chromosome** | Xp11.2 |
| **Mass (kDa): predicted** | 45 |
| **expressed** | 60 |
| **Cellular location** | Nucleus |

**Tissue distribution**
Restricted to lung and testis [1].

## Protein function

ELK proteins are transcription factors with DNA sequence specificity (CAGGA) similar to ETS1 and ETS2 [3]. ELK1 differs from other ETS proteins in having its ETS DNA binding domain located at the N-terminus of the protein. ELK1 (also called $p62^{TCF}$), is a component of the ternary complex that activates *FOS* transcription *via* the serum response element (SRE; see ***FOS***) [4].

ΔELK1 lacks 11 C-terminal amino acids of the ETS domain and all of the negative regulatory DNA binding domain of ELK1 [5]. The deleted region includes the SRF interaction domain (ESI) and ΔELK1 therefore has the potential to compete with ELK1 in binding to the SRE and thus inhibit transcriptional activation of *FOS* by SRF and ELK1. ELK1 and ΔELK1 are both activators of and substrates for MAP kinases.

**Cancer**
*ELK1* lies close to the t(X;18)(p11.2;q11.2) breakpoint characteristic of synovial sarcoma [6], *ELK2* is in the vicinity of the 14q32 breakpoints seen in ataxia telangiectasia and other T cell malignancies [7,8] and deletions in the region of *ELK3* (12q) occur in male germ cell tumours [9].

## Protein structure

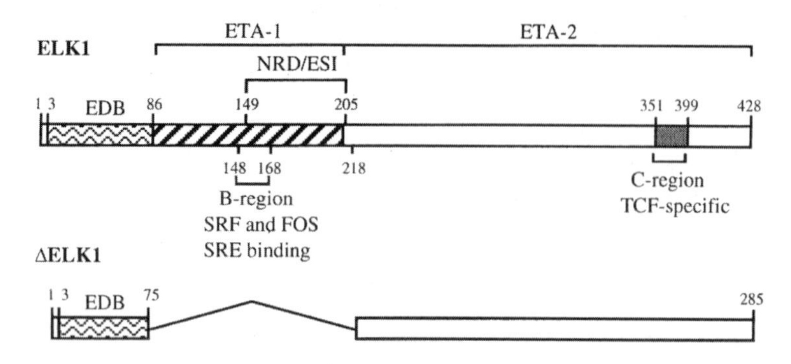


EDB: ETS domain (DNA binding); ETA1 and ETA2: ELK1 transcription activation domains; NRD: negative regulatory DNA binding domain; ESI: SRF interaction domain [10]. The efficient formation of a ternary complex at the *FOS* SRE requires the ELK1 (or SAP1) ETS domain and the conserved sequence of Box B [11]. The B-region mediates direct contact between ELK1 and SRF [12].

## Amino acid sequence of human ELK1

```
  1 MDPSVTLWQF LLQLLREQGN GHIISWTSRD GGEFKLVDAE EVARLWGLRK
 51 NKTNMNYDKL SRALRYYYDK NIIRKVSGQK FVYKFVSYPE VAGCSTEDCP
101 PQPEVSVTST MPNVAPAAIH AAPGDTVSGK PGTPKGAGMA GPGGLARSSR
151 NEYMRSGLYS TFTIQSLQPQ PPPHPRPAVV LPNAAPAGAA APPSGSRSTS
201 PSPLEACLEA EEAGLPLQVI LTPPEAPNLK SEELNVEPGL GRALPPEVKV
251 EGPKEELEVA GERGFVPETT KAEPEVPPQE GVPARLPAVV MDTAGQAGGH
301 AASSPEISQP QKGRKPRDLE LPLSPSLLGG PGPERTPGSG SGSGLQAPGP
351 ALTPSLLPTH TLTPVLLTPS SLPPSIHFWS TLSPIAPRSP AKLSFQFPSS
401 GSAQVHIPSI SVDGLSTPVV LSPGPQKP (428)
```

### Domain structure

| | |
|---|---|
| 3–86 | DNA binding ETS domain (EDB: underlined). The truncated protein (residues 1–89) binds autonomously to SRE, unlike the full-length protein, indicating the presence of a negative regulatory domain within ELK1 |
| 38 and 69 | Critical asparagine determinants of binding specificity [13] |
| 148–168 | B-region. |
| 133, 324, 389 | Optimal sites for MAP kinase phosphorylation. Other suboptimal sites are 336, 353, 363, 368, 383, 417 and 422 |

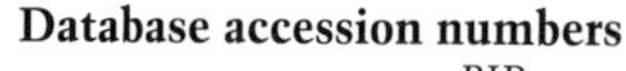

### Database accession numbers

| | *PIR* | *SWISSPROT* | *EMBL/GENBANK* | *REFERENCES* |
|---|---|---|---|---|
| Human *ELK1* | A41354 | P19419 | M25269 | 1,14 |
| Human *ELK3/NET/ERP/SAP2* | | | Z32815, Z36715 | 2 |
| Human *ELK4/SAP1* | | | M85165 | 2 |

***References***


[1] Rao, V.N. et al. (1989) Science 244, 66–70.
[2] Giovane, A. et al. (1995) Genomics 29, 769–772.
[3] Rao, V.N. and Reddy, E.S.P. (1992) Oncogene 7, 65–70.
[4] Hipskind, R.A. et al. (1991) Nature 354, 531–534.
[5] Bhattacharya, G. et al. (1993) Oncogene 8, 3459–3464.
[6] Janz, M. et al. (1994) Human Genet. 94, 442–444.
[7] Kocova, M. et al. (1985) Cancer Genet. Cytogen. 16, 21–30.
[8] Rowley, J.D. (1983) Nature 301, 290–291.
[9] Murty, V.V.V.S. et al. (1996) Genomics, 35, 562–570.
[10] Janknecht, R. et al. (1994) Oncogene 9, 1273–1278.
[11] Treisman, R. et al. (1992) EMBO J. 11, 4631–4640.
[12] Shore, P. and Sharrocks, A.D. (1994) Mol. Cell. Biol. 14, 3283–3291.
[13] Shore, P. et al. (1996) Mol. Cell. Biol. 16, 3338–3349.
[14] Janknecht, R. and Nordheim, A. (1992) Nucleic Acids Res. 20, 3317–3324.

## Identification

*EPH* (*EPHT*) is the prototype of a subfamily of receptor tyrosine kinases, initially isolated from a human genomic library with a v-*fps* probe, that is over-expressed in an erythropoietin producing human hepatocellular (*EPH*) carcinoma cell line [1].

## Related genes

The receptor tyrosine kinase family has been grouped into at least 14 subfamilies [2] of which the *EPH* family is the largest subgroup, being composed of ~15 members, not including homologous members from different species. Major related genes are: *ELK* (eph-like kinase); *EEK* (eph and elk-related kinase); *ECK* (epithelial cell kinase); *ERK* (elk-related kinase); *EPHT2, DRT* (developmentally regulated EPH-related tyrosine kinase gene); *HTK* (hepatoma transmembrane kinase). *ERK* is distinct from the *Erk-1, -2* and *-3* family of serine/threonine kinases [3]. EEK is 57% identical in sequence to EPH. ELK and ECK share >40% and >60% similarity in the extracellular and tyrosine kinase domains, respectively.

Other *EPH*-related genes are: murine *Myk-1* and *Myk-2* (mammary-derived tyrosine kinase *1* and *2*), *Nuk* (neural kinase) receptor kinase, murine *Sek*, murine *Mek4* and *Mep* (murine eph-family protein), and chicken *Cek* (chicken embryo kinases 4–10), chicken *Tyro-5*, rat *Ehk-1, Ehk-2, Xenopus Xek* and human *HEK* (*h*uman *e*mbryo *k*inase), *HEK2, HEK4, HEK5, HEK7, HEK8, HEK11* [4].

| | ***EPH*** |
|---|---|
| **Chromosome** | 7q32–q36 |
| **Mass (kDa): predicted** | 109 |
| **expressed** | 130 |
| **Cellular location** | Plasma membrane |

**Tissue distribution**

*EPH* is expressed in kidney, testis, liver, lung. *Mek4*, *Sek* and *Cek* family genes are expressed during embryonic development.

## Protein function

EPH is the prototype class IV transmembrane tyrosine kinase receptor. Chimeric EGFR/ELK proteins undergo autophosphorylation of the ELK cytoplasmic domain and stimulate the tyrosine phosphorylation of a set of cellular proteins in response to EGF, indicating that ELK and related proteins function as ligand-dependent receptor tyrosine kinases [5].

Two major subclasses of EPH receptors may be defined on the basis of ligand specificity: ECK/SEK2, EHK1/REK7, EHL2, EHK3/MDK1, HEK/MEK4 and SEK1/CEK8, ligands for which are attached to the cell membrane by glycosyl phosphatidylinositol (GPI) linkage (LERK1/B61, EHK1-L, ELF1, LERK4 and

AL-1/RAGS), and ELK, NUK/CEK5, HEK2/SEK4 and HTK that are receptors for transmembrane proteins (ELK-L/LERK2, ELK-L3 and HTK-L/ELF2)[6].

B61 (also called LERK1 (ligands for eph-related kinases), the product of an early response gene induced by TNFα), is the ligand for ECK, inducing its autophosphorylation[7]. LERK1 and LERK2 and the related LERK3 and LERK4 bind to human ELK and to the related HEK receptors[8–11]. ECK binds to the SH2 domain of the p85 subunit of PtdIns 3-kinase, the activity of which is increased by LERK1 binding. ECK also associates with Src-like adapter protein[12]. LERK1 has angiogenic activity *in vivo* and acts as a chemoattractant for endothelial cells: anti-LERK1 antibody attenuates angiogenesis induced by TNFα but not by bFGF[13]. Activated ELK receptors become tyrosine phosphorylated and associate with GRB2 and GRB10[14].

### Cancer

*EPH* is over-expressed in several carcinomas[15]. *HEK* is expressed in cell lines derived from human lymphoid tumours[16].

### *In vitro*

Over-expressed *EPH* renders NIH 3T3 cells tumorigenic[17].

## Protein structure

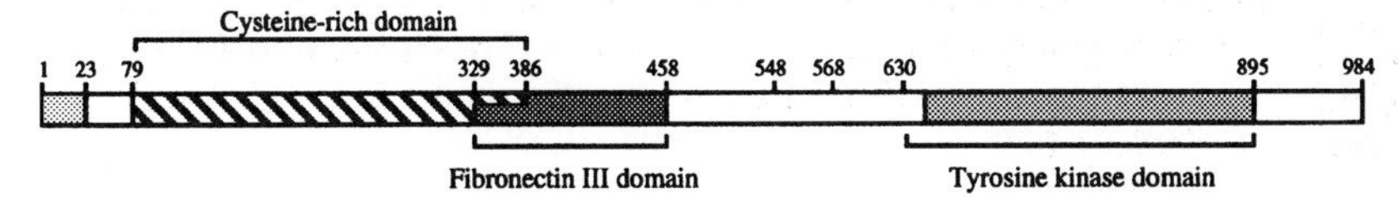


The extracellular domain contains a putative immunoglobulin-like loop at the N-terminus, a single cysteine-rich region and two fibronectin type III repeats[18].

## Amino acid sequence of human EPH

```
  1 MERRWPLGLG LVLLLCAPLP PGARAKEVTL MDTSKAQGEL GWLLDPPKDG
 51 WSEQQQILNG TPLYMYQDCP MQGRRDTDHW LRSNWIYRGE EASRVHVELQ
101 FTVRDCKSFP GGAGPLGCKE TFNLLYMESD QDVGIQLRRP LFQKVTTVAA
151 DQSFTIRDLA SGSVKLNVER CSLGRLTRRG LYLAFHNPGA CVALVSVRVF
201 YQRCPETLNG LAQFPDTLPG PAGLVEVAGT CLPHARASPR PSGAPRMHCS
251 PDGEWLVPVG RCHCEPGYEE GGSGEACVAC PSGSYRMDMD TPHCLTCPQQ
301 STAESEGATI CTCESGHYRA PGEGPQVACT GPPSAPRNLS FSASGTQLSL
351 RWEPPADTGG RQDVRYSVRC SQCQGTAQDG GPCQPCGVGV HFSPGARGLT
401 TPAVHVNGLE PYANYTFNVE AQNGVSGLGS SGHASTSVSI SMGHAESLSG
451 LSLRLVKKEP RQLELTWAGS RPRSPGANLT YELHVLNQDE ERYQMVLEPR
501 VLLTELQPDT TYIVRVRMLT PLGPGPFSPD HEFRTSPPVS RGLTGGEIVA
551 VIFGLLLGAA LLLGILVFRS RRAQRQRQQR HVTAPPMWIE RTSCAEALCG
601 TSRHTRTLHR EPWTLPGGWS NFPSRELDPA WLMVDTVIGE GEFGEVYRGT
651 LRLPSQDCKT VAIKTLKDTS PGGQWWNFLR EATIMGQFSH PHILHLEGVV
700 TKRKPIMIIT EFMENGALDA FLREREDQLV PGQLVAMLQG IASGMNYLSN
751 HNYVHRDLAA RNILVNQNLC CKVSDFGLTR LLDDFDGTYE TQGGKIPIRW
801 TAPEAIAHRI FTTASDVWSF GIVMWEVLSF GDKPYGEMSN QEVMKSIEDG
851 YRLPPPVDCP APLYELMKNC WAYDRARRPH FQKLQAHLEQ LLANPHSLRT
901 IANFDPRVTL RLPSLSGSDG IPYRTVSEWL ESIRMKRYIL HFHSAGLDTM
951 ECVLELTAED LTQMGITLPG HQKRILCSIQ GFKD (984)
```

## Domain structure

| | |
|---|---|
| 1–23 | Signal sequence (italics) |
| 24–547 | Extracellular domain |
| 79–386 | Cysteine-rich domain (cysteines underlined) |
| 548–568 | Transmembrane region (underlined) |
| 569–984 | Intracellular domain |
| 630–895 | Tyrosine kinase domain (underlined) |
| 638–646 and 664 | ATP binding region |
| 329–442 and 443–538 | Fibronectin type III domains |
| 757 | Active site |
| 789 | Potential autophosphorylation site |
| 338, 414 and 478 | Potential carbohydrate attachment sites |

## Database accession numbers

| | *PIR* | *SWISSPROT* | *EMBL/GENBANK* | *REFERENCES* |
|---|---|---|---|---|
| Human *EPH* | A34076 | P21709 | M18391 | 1 |
| Human *ECK* | | | M36395 | 19 |
| Human *EEK* | | | X59291 | 20 |
| Human *ERK* | | | X59292 | 20 |
| Human *HEK* | | | M83941 | 16 |

## *References*


[1] Hirai, H. et al. (1987) Science 238, 1717–1720.
[2] **van der Greer, P. et al. (1994) Annu. Rev. Cell Biol. 10, 251–337.**
[3] **Boulton, T.G. et al. (1991) Cell 65, 663–675.**
[4] Fox, G.M. et al. (1995) Oncogene 10, 897–905.
[5] Lhotak, V. and Pawson, T. (1993) Mol. Cell. Biol. 13, 7071–7079.
[6] Gale, N.W. et al. (1996) Oncogene 13, 1343–1352.
[7] Bartley et al. (1994) Nature 368, 558–560.
[8] Beckmann, M.P. et al. (1994) EMBO J. 13, 3757–3762.
[9] Fletcher, F.A. et al. (1994) Oncogene 9, 3241–3247.
[10] Kozlosky, C.J. et al. 1995) Oncogene 10, 299–306.
[11] Tessier-Lavigne, M. (1995) Cell 82, 345–348.
[12] Pandey, A. et al. (1995) J. Biol. Chem. 270, 19201–19204.
[13] Pandey, A. et al. (1995) Science 268, 567–569.
[14] Stein, E. et al. (1996) J. Biol. Chem. 271, 23588–23593.
[15] Maru, Y. et al. (1988) Mol. Cell. Biol. 8, 3770–3776.
[16] Wicks, I.P. et al. (1992) Proc. Natl Acad. Sci. USA 89, 1611–1615.
[17] Maru, Y. et al. (1990) Oncogene 5, 445–447
[18] Pasquale, E.B. et al. (1991) Cell Regul. 2, 523–534.
[19] Lindberg, R.A. and Hunter, T. (1990) Mol. Cell. Biol. 10, 6316–6324.
[20] Chan, J. and Watt, V.M. (1991) Oncogene 6, 1057–1061.

## Identification

*ERG* (ets-related gene) was detected by screening a cDNA library with *ETS2* cDNA[1,2].

## Related genes

*ERG* is a member of the *ETS* family (see ***ETS1*** and ***ETS2***).

| | ***ERG*** |
|---|---|
| **Chromosome** | 21q22.3 |
| **Mass (kDa): predicted** | 41 (ERG1)<br>52 (ERG2)<br>54.6 (ERG3) |
| **expressed** | 41 (ERG1)/57 (ERG2)/59 (ERG3) |
| **Cellular location** | Nucleus |

**Tissue distribution**
High levels of expression in human tumour-derived cell lines[2,3]. *ERG3* is expressed in a variety of cells[3].

## Protein function

The *ERG* family encode transcription factors that differ from ETS1 and ETS2 in their sequence specificity, binding weakly to the polyomavirus enhancer PEA3, Mo-MuSV LTR or PU box sequences, but strongly to E74 target sequences, although all of these regions contain a core GGAA motif[4]. In contrast to ETS1, ERG2 is phosphorylated by protein kinase C but not in response to $Ca^{2+}$ ionophore[5].

**Cancer**
The chromosomal translocation t(22;21)(q12;q22) that occurs in Ewing's sarcoma creates an ERG/EWS fusion protein[6].

## Protein structure

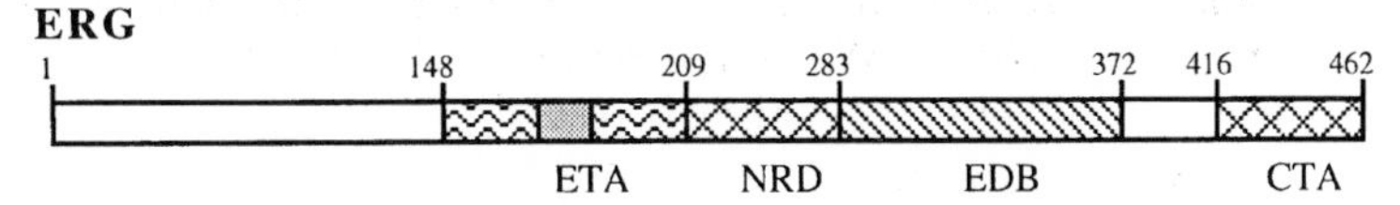


ERG proteins contain a C-terminal DNA binding domain (EDB), two autonomous transcriptional activation domains (ETA, N-terminal and CTA, C-terminal) and a negative regulatory transcriptional activation domain (NRD) N-terminal to the DNA binding domain[7].

Three ERG2-related isoforms ($p38^{ERG}$, $p49^{ERG}$ and $p55^{ERG}$) generated by alternative splicing of two exons (A81 and A72) have been detected in human COLO 320 cells [8]. p38$^{ERG}$ is colinear with ERG2 but has a truncation of the 5′ coding region that deletes 53 amino acids. The cDNA encoding p55$^{ERG}$ contains an additional 72 bp (exon A72) that adds 24 amino acids to the middle of the ORF that are not present in ERG1 or EGR2. The 5′ coding sequence of p55$^{ERG}$ also diverges from that of *ERG2*, giving rise to an alternative predicted N-terminus. p49$^{ERG}$ has an in-frame deletion of 81 bp (exon A81) and also of exon A72 relative to p55$^{ERG}$ [8].

*ERG3* encodes a 24 amino acid insert with respect to ERG2 [3].

## Amino acid sequence of human ERG

```
  1 MIQTVPDPAA HIKEALSVVS EDQSLFECAY GTPHLAKTEM TASSSSDYGQ
 51 TSKMSPRVPQ QDWLSQPPAR VTIKMECNPS QVNGSRNSPD ECSVAKGGKM
101 VGSPDTVGMN YGSYMEEKHM PPPNMTTNER RVIVPADPTL WSTDHVRQWL
151 EWAVKEYGLP DVNILLFQNI DGKELCKMTK DDFQRLTPSY NADILLSHLH
201 YLRETPLPHL TSDDVDKALQ NSPRLMHARN TDLPYEPPRR SAWTGHGHPT
251 PQSKAAQPSP STVPKTEDQR PQLDPYQILG PTSSRLANPG SGQIQLWQFL
301 LELLSDSSNS SCITWEGTNG EFKMTDPDEV ARRWGERKSK PNMNYDKLSR
351 ALRYYYDKNI MTKVHGKRYA YKFDFHGIAQ ALQPHPPESS LYKYPSDLPY
401 MGSYHAHPQK MNFVAPHPPA LPVTSSSFFA APNPYWNSPT GGIYPNTRLP
451 TSHMPSHLGT YY (462)
```

### Domain structure

| | |
|---|---|
| 1–462 | ERG2 (generated by alternative splicing of *ERG1* [1,2] |
| 100–462 | ERG1 |
| 290–374 | DNA binding ETS domain (underlined) |

| | |
|---|---|
| N-terminus of ERG2: | MIQTVPDPAAHIKEALS... |
| N-terminus of p55$^{ERG}$/p49$^{ERG}$: | MASTIKEALS... |
| Sequence encoded by exon A81: | TPLPHLTSDDVDKALQNSPRLMHARNT |
| Sequence encoded by exon A72: | GGAAFIFPNTSVYPEATQRITTRP |

### Database accession numbers

| | *PIR* | *SWISSPROT* | *EMBL/GENBANK* | *REFERENCES* |
|---|---|---|---|---|
| Human *ERG2* | A29515 | P11308 | M17254 | [2] |

### *References*


1. Reddy, E.S.P. et al. (1987) Proc. Natl Acad. Sci. USA 84, 6131–6135.
2. Rao, V.N. et al. (1987) Science 237, 635–653.
3. Prasad, D.D.K. et al. (1994) Oncogene 9, 669–673.
4. Reddy, E.S.P. and Rao, V.N. (1991) Oncogene 6, 2285–2289.
5. Murakami, K. et al. (1993) Oncogene 8, 1559–1566.
6. Zucman, J. et al. (1993) EMBO J. 12, 4481–4487.
7. Siddique, H.R. et al. (1993) Oncogene 8, 1751–1755.
8. Duterque-Coquillaud, M. et al. (1993) Oncogene 8, 1865–1873.

## Identification

v-*ets* (*E twenty-six specific*) and v-*myb* are the oncogenes of the acutely transforming avian erythroblastosis virus E26. v-*ets* was detected as a fusion gene with v-*myb* by screening cloned E26 provirus with a v-*myb* probe [1–5].

*ETS1* and *ETS2* were detected by screening DNA with a v-*ets* probe [6].

## Related genes

Related genes are *ERG, ELK1, ELK2, ELK3/NET/ERP/SAP2, ELK4/SAP1, ELF1, MEF, NERF, ERF, Erp, FLI1, TEL, GABPa, PU.1/Spi-1, Spi-1b, Pea-3, Tpl-1, E1AF, E74A* and *E74B* (*Xenopus*, sea urchin and *Drosophila*), *Ets-1* and *Ets-2* (*Xenopus*), D-*elg*, *Elf-1*, *pointed* and *yan* (*Drosophila*), *lin-1* (*C. elegans*), ERM, ER81 and Er71. There is ~65% identity between human ETS1 and ELK1, ETS2, ERG and v-ETS within the 80 amino acid C-terminal DNA binding domains. There is limited homology with the T cell receptor α enhancer binding protein (TCR1α).

| | ***ETS1*** | ***ETS2*** | **v-*ets*** |
|---|---|---|---|
| **Nucleotides (kb)** | 60 | 20 | 5.7 (E26) |
| **Chromosome** | 11q23.3 | 21q22.3 | |
| **Mass (kDa): predicted** | 50/55 | 53/54 | 75 |
| **expressed** | 52/48/42/39 (alternative splicing) | 58/62 | $P135^{gag\text{-}myb\text{-}ets}$ |
| **Cellular location** | Nucleus | Nucleus | Nucleus |

**Tissue distribution**

*ETS1*: Expression is high in the thymus, in endothelial cells differentiating or migrating during the development of blood vessels and in migrating neural crest cells [7]. Expression is high in quiescent T cells and declines on T cell activation [8]. It is abundantly expressed in fibroblasts and endothelial cells in invasive tumours.

*ETS2*: Expression occurs in most tissues. It is low in T cells but increases to a maximum ~2 h after T cell activation [8].

## Protein function

*ETS* family genes encode transcription factors that bind to the consensus sequence $^{C}/_{A}GGA^{A}/_{T}$ (ETS binding site (EBS)) [9,10]. The family sequence homology covers ~85 amino acids (the ETS domain), necessary and sufficient for specific binding to EBS *in vitro*, and located in the C-terminus of family members except for ELK1, SAP1 and ELF1 where it is N-terminal. The solution NMR structure of the DNA binding domain of ETS1 with a 17-mer DNA indicates that the C-terminal two thirds of the helix-loop-helix recognition region interacts with the major groove but $Trp^{28}$ intercalates with the minor groove. This distinctive interaction causes the minor groove to be widened over one half turn of DNA [11].

ETS-domain proteins interact with *TCRA* [12], *TCRB*, *MB-1*, *ENDOA*, keratin 18 and class II MHC genes, E74 target sequences, GATA-1, the human mitochondrial ATP synthase $\beta$ subunit gene, the polyomavirus enhancer and Mo-MuSV LTR [13]. ETS1 activates *MET* transcription [14] and ETS1 and ETS2 activate the stromelysin 1 and collagenase 1 genes [15] and over-expression of *ETS1* has been detected in fibroblasts and endothelial cells at sites of tumour invasion [16]. Putative ETS binding sites are present in several T cell-specific genes including *IL-2*, *IL-3*, *GM-CSF*, *CD2*, *CD3*, *CD34* [17,18], *TCRG* [19] and in human immunodeficiency virus type 2 (HIV-2), and the Moloney murine leukaemia virus enhancer [20].

ETS1 interacts directly with the basic domain of JUN proteins, giving rise to ternary complexes at AP1-ETS binding sites and trimeric complexes between ETS1, PEBP2$\alpha$ and lymphoid enhancer-binding factor (LEF1) occur at the T cell receptor $\alpha$ enhancer [21]. The ETS family members ELK1 and SAP1 form ternary complexes with serum response factor (SRF) that modulate transcription from the *FOS* serum response element (SRE: see ***FOS***) and ETS proteins form ternary complexes with PAX5 to activate the B cell-specific *MB-1* promoter. NET and ELK1 associate with PAX5 *in vitro* [22]. Transcriptional activation of the HTLV-I LTR requires the synergistic action of ETS1 and SP1 and a variety of other partners are known to associate with ETS1, including GATA1, PU.1 and CBF. In general, ETS proteins appear to function as components of transcription factor complexes to regulate the expression of viral and cellular genes [23,24].

ETS1 and ETS2 are phosphorylated in activated T cells, suggesting that they have a role in cell proliferation [25]. *Ets-2* expression is required for germinal vesicle breakdown in *Xenopus* oocytes [26]. Calmodulin-dependent protein kinase II phosphorylation of ETS1 (consensus site RXXS/T) diminishes *trans*-activation activity [27].

**Cancer**

At least four alternatively spliced forms of *ETS1* occur in haematological malignancies and a point mutation ($Tyr^{162} \rightarrow Cys$) has been detected in a case of acute lymphocytic leukaemia [28]. The *ETS1* and *ETS2/ERG* loci are involved in translocations associated with some leukaemias and with Down's syndrome that may result in activation of these genes [29]. *ETS1* is expressed in tumours of the peripheral nervous system (neuroblastomas and neuroepitheliomas) and in Ewing's sarcoma [30].

**Transgenic animals**

Deletion of *Ets-1* dramatically and differentially affects T and B cell development and function. T cell number is reduced and the cells are highly susceptible to apoptosis: B cells arise in normal numbers but a high proportion are IgM plasma cells [38,39].

**In animals**

E26 causes erythroblastosis and low level myeloblastosis in chickens. The potent leukaemogenicity of E26 derives from the fusion of the v-*ets* and v-*myb* genes [31,32]. v-*ets* cooperates with v-*erbA* to cause avian erythroleukaemia [33].

***In vitro***

E26 does not transform chick embryo fibroblasts (CEFs)[34] but does transform quail fibroblasts and stimulates the proliferation of CEFs[35], neuroretina cells[36] and NIH 3T3 fibroblasts[37]. v-*ets* can cause weak transformation of erythroid cells[31].

## Gene structure

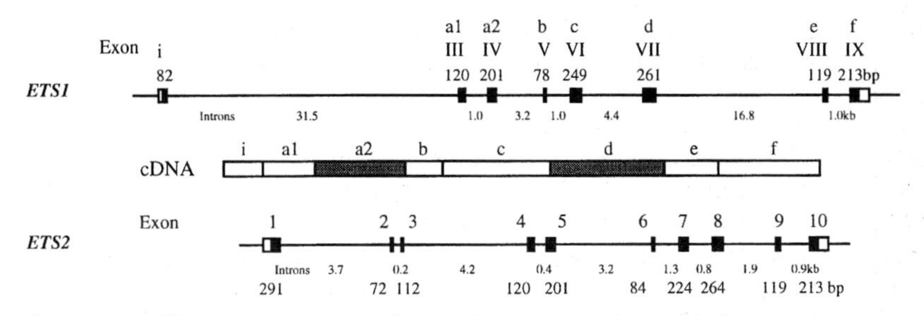


The exons of human *ETS1* are identical in size to those of chicken *Ets-1*, the progenitor of v-*ets*[40]. Alternative splicing generates human *ETS1* mRNAs lacking one or both of exons $a_2$/IV and d/VII (shown shaded in cDNA: nomenclature of refs 28 and 40). Differential polyadenylation site utilization creates a 2.7 kb *ETS1* mRNA. The absence of exon d/VII causes smaller ETS1 proteins (p39/p42) to be synthesized. The principal proteins have been denoted as ETS1a (~50 kDa: 441 amino acids) and ETS1b (~40 kDa: 354 amino acids)[41]. Scrambled splicing also occurs in which exon 5 or exon 4 splices with exon 2, which normally splices 3′ with exon 3[42]. Regions encoding identical or similar amino acids to ETS2 have retained their genomic organization: the most divergent regions have different organization.

The *ETS1* and *ETS2* promoters lack a TATA or CAAT box and have a high GC content. The *ETS1* promoter has six SP1 sites, AP1 and AP2 consensus sequences and binding site motifs for PEA3 and OCT as well as a palindromic region resembling the *FOS* SRE and an ETS1 protein binding site[43,44]. Two negative regulatory elements (NRE1 and NRE2) are present 230 nt and 350 nt 5′ of the promoter respectively[45]. The human *ETS2* promoter contains a region (−159 to +141) that includes one SP1 site and a GC-rich region and is essential for transcription[46].

## Protein structure of chicken p68$^{ets\text{-}1}$

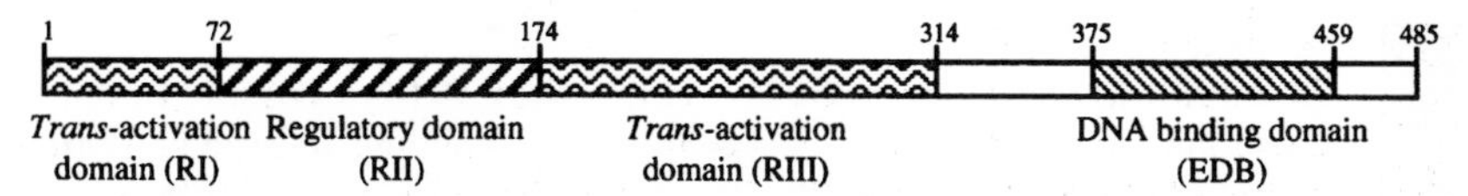


Human ETS1 and chicken p68$^{ets\text{-}1}$ are highly homologous but ETS1 lacks the RI *trans*-activation domain. RII negatively regulates the activity of RI and positively regulates that of RIII.

Human ETS1 and ETS2 are 98% and 95% homologous to v-ETS, respectively. In v-ETS the 13 C-terminal residues of chicken p68$^{ets}$ have been replaced by 16 C-terminal amino acids and there are two conservative point mutations[47].

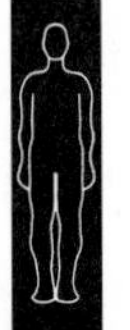

## Amino acid sequence of human ETS1

```
  1 MKAAVDLKPT LTIIKTEKVD LELFPSPDME CADVPLLTPS SKEMMSQALK
 51 ATFSGFTKEQ QRLGIPKDPR QWTETHVRDW VMWAVNEFSL KGVDFQKFCM
101 NGAALCALGK DCFLELAPDF VGDILWEHLE ILQKEDVKPY QVNGVNPAYP
151 ESRYTSDYFI SYGIEHAQCV PPSEFSEPSF ITESYQTLHP ISSEELLSLK
201 YENDYPSVIL RDPLQTDTLQ NDYFAIKQEV VTPDNMCMGR TSRGKLGGQD
251 SFESIESYDS CDRLTQSWSS QSSFNSLQRV PSYDSFDSED YPAALPNHKP
301 KGTFKDYVRD RADLNKDKPV IPAAALAGYT GSGPIQLWQF LLELLTDKSC
351 QSFISWTGDG WEFKLSDPDE VARRWGKRKN KPKMNYEKLS RGLRYYYDKN
401 IIHKTAGKRY VYRFVCDLQS LLGYTPEELH AMLDVKPDAD E (441)
```

**Domain structure**

| | |
|---|---|
| 28–130 | Homology with chicken p68*ets-1* RII domain |
| 38 | Thr38 is phosphorylated *in vitro* by ERK2 [48] |
| 97–130 | Helix-loop-helix domain (italics) |
| 130–270 | Homology with chicken p68*ets-1* RIII *trans*-activation domain |
| 280–331 | Negatively regulates DNA binding: undergoes DNA-induced unfolding in an α helical region [49] |
| 331–415 | DNA binding ETS domain (underlined) |
| 244–330 | Deleted by alternative splicing in ETS1B |
| 251–282 | PEST sequence which may target protein cleavage to this region |
| 338 | Trp338 is in the centre of a hydrophobic α helical region essential for the optimal conformation of the ETS binding domain [50] |
| 377–383 | Nuclear localization signal |
| 337–342 | Conserved region I (CRI) conserved in all ETS family proteins |
| 371–379 | Conserved region II (CRII) conserved in all ETS family proteins |
| 384–394 | Conserved region III (CRIII) conserved in all ETS family proteins |

The specific selectivity of ELF1 and E74 for GGAA core-containing sites is conferred on ETS1 by mutation of Lys396 within the conserved region III of ETS1 [51].

An 89 residue region adjacent to the DNA binding domains is conserved with 55% identity in ETS1 and ETS2 but does not occur in other members of the ETS family. This region inhibits DNA binding and activates transcription. The alternatively spliced form of human *ETS1*, lacking exon VII, does not have this inhibitory region and binds to DNA much more efficiently [52]. Deletion of a region C-terminal to the ETS domain, corresponding to one of the alterations in the E26-transduced form, also increases DNA binding by ETS1 [53].

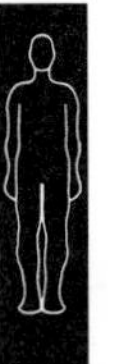

## Amino acid sequence of human ETS2

```
  1 MNDFGIKNMD QVAPVANSYR GTLKRQPAFD TFDGSLFAVF PSLNEEQTLQ
 51 EVPTGLDSIS HDSANCELPL LTPCSKAVMS QALKATFSGF KKEQRRLGIP
101 KNPWLWSEQQ VCQWLLWATN EFSLVNVNLQ RFGMNGQMLC NLGKERFLEL
151 APDFVGDILW EHLEQMIKEN QEKTEDQYEE NSHLTSVPHW INSNTLGFGT
201 EQAPYGMQTQ NYPKGGLLDS MCPASTPSVL SSEQEFQMFP KSRLSSVSVT
251 YCSVSQDFPG SNLNLLTNNS GTPKDHDSPE NGADSFESSD SLLQSWNSQS
301 SLLDVQRVPS FESFEDDCSQ SLCLNKPTMS FKDYIQERSD PVEQGKPVIP
351 AAVLAGFTGS GPIQLWQFLL ELLSDKSCQS FISWTGDGWE FKLADPDEVA
401 RRWGKRKNKP KMNYEKLSRG LRYYYDKNII HKTSGKRYVY RFVCDLQNLL
451 GFTPEELHAI LGVQPDTED (469)
```

## Domain structure

359–443 DNA binding ETS domain (underlined)

## Database accession numbers

| | PIR | SWISSPROT | EMBL/GENBANK | REFERENCES |
|---|---|---|---|---|
| Human *ETS1* | A32066 | P14921 | X14798 | 41 |
| | S10086 | | J04101 | 54 |
| Human *ETS2* | B32066 | P15036 | J04102, X55181 | 54,55 |

## *References*


[1] Ivanov, X. et al. (1962) Bulgaria Acad. Sci. Bull. Inst. Pathol. Comp. Anim. 9, 5–36.
[2] Ivanov, X. et al. (1964) Izv. Inst. Pat. Zhivotnite Sofia 10, 5–38.
[3] Nedyalkov, St. et al. (1975) Acta Vet. (Brno) 44, 75–78.
[4] Nunn, M.F. et al. (1983) Nature 306, 391–395.
[5] Leprince, D. et al. (1983) Nature 306, 395–397.
[6] Watson, D.K. et al. (1986) Proc. Natl Acad. Sci. USA 83, 1792–1796.
[7] Desbiens, X. et al. (1991) Development 111, 699–713.
[8] Bhat, N.K. et al. (1990) Proc. Natl Acad. Sci. USA 87, 3723–3727.
[9] **Macleod, K. et al. (1993) Trends Biochem. Sci., 17, 251–256.**
[10] Karim, F.D. et al. (1990) Genes Dev. 4, 1451–1453.
[11] Werner, M.H. et al. (1995) Cell 83, 761–771.
[12] Ho, I.-C. et al. (1990) Science 250, 814–818.
[13] Seth, A. et al. (1994) Oncogene 9, 469–477.
[14] Gambarotta, G. et al. (1996) Oncogene 13, 1911–1917.
[15] Woods, D.B. et al. (1992) Nucleic Acids Res. 20, 699–704
[16] Wernert, N. et al. (1994) Cancer Res. 54, 5683–5688.
[17] Melotti, P. and Calabretta, B. (1994) J. Biol. Chem. 269, 25303–25309.
[18] Thomas, R.S. et al. (1995) Oncogene 11, 2135–2143.
[19] Thompson, C.B. et al. (1992) Mol. Cell. Biol. 12, 1043–1053
[20] Gunther, C.V. and Graves, B.J. (1994) Mol. Cell. Biol. 14, 7569–7580.
[21] Giese, K. et al. (1995) Genes Dev. 9, 995–1008.
[22] Fitzsimmons, D. et al. (1996) Genes Dev. 10, 2198–2211.
[23] Gitlin, S.D. et al. (1991) J. Virol. 65, 5513–5523.
[24] Gegonne, A. et al. (1993) EMBO J. 12, 1169–1178.
[25] Fleischman, L.F. et al. (1993) Oncogene 8, 771–780.
[26] Chen, J.H. et al. (1990) Oncogene Res. 5, 277–285.
[27] Hodge, D.R. et al. (1996) Oncogene 12, 11–18.
[28] Collyn-d'Hooghe, M. et al. (1993) Leukemia 7, 1777–1785.
[29] Papas, T.S. et al. (1990) Am. J. Med. Genet. Suppl., 7, 251–261.
[30] Sacchi, N. et al. (1991) Oncogene 6, 2149–2154.
[31] Metz, T. and Graf, T. (1991) Genes Dev. 5, 369–380.
[32] Metz, T. and Graf, T. (1991) Cell 66, 95–105.
[33] Metz, T. and Graf, T. (1992) Oncogene 7, 597–605.
[34] Bister, K. et al. (1982) Proc. Natl Acad. Sci. USA 79, 3677–3681.
[35] Jurdic, P. et al. (1987) J. Virol. 61, 3058–3065.
[36] Amouyel, P. et al. (1989) J. Virol. 63, 3382–3388.
[37] Seth, A. and Papas, T.S. (1990) Oncogene 5, 1761–1767.
[38] Bories, J.-C. et al. (1995) Nature 377, 635–638.
[39] Muthusamy, N. et al. (1995) Nature 377, 639–642.

[40] Jorcyk, C.L. et al. (1991) Oncogene 6, 523–532.
[41] Reddy, E.S.P. and Rao, V.N. (1988) Oncogene Res. 3, 239–246.
[42] Cocquerelle, C. et al. (1992) EMBO J. 11, 1095–1098.
[43] Oka, T. et al. (1991) Oncogene 6, 2077–2083.
[44] Chen, J.H. and Wright, C.D. (1993) Oncogene 8, 3375–3383.
[45] Chen, J.H. et al. (1993) Oncogene 8, 133–139.
[46] Mavrothalassitis, G.J. et al. (1990) Oncogene 5, 1337–1342.
[47] Lautenberger, J.A. and Papas, T.S. (1993) J. Virol. 67, 610–612.
[48] Rabault, B. et al. (1996) Oncogene 13, 877–881.
[49] Petersen, J.M. et al. (1995) Science 269, 1866–1869.
[50] Mavrothalassitis, G.J. et al. (1994) Oncogene 9, 425–435.
[51] Bosselut, R. et al. (1993) Nucleic Acids Res. 21, 5184–5191.
[52] Wasylyk, C. et al. (1992) Genes Dev. 6, 965–974.
[53] Jonsen, M.D. et al. (1996) Mol. Cell. Biol. 16, 2065–2073.
[54] Watson, D.K. et al. (1988) Proc. Natl Acad. Sci. USA 85, 7862–7866.
[55] Watson, D.K. et al. (1990) Oncogene 5, 1521–1527.

## Identification

v-*fgr* is the oncogene of the acutely transforming Gardner–Rasheed feline sarcoma virus (GR-FeSV)[1]. *FGR* was detected by low stringency hybridization screening of human placental DNA with a v-*yes* probe[2].

## Related genes

*FGR* is a member of the *SRC* tyrosine kinase family (*Blk, FGR, FYN, HCK, LCK/Tkl, LYN, SRC, YES*). GR-FeSV has transduced portions of two distinct cellular genes, γ-actin and *Fgr*.

| | ***FGR*** | **v-*fgr*** |
|---|---|---|
| **Nucleotides (kb)** | ~15 | 4.6 (GR-FeSV) |
| **Chromosome** | 1p36.2–p36.1 | |
| **Mass (kDa): predicted** | 59.5 | 72 |
| **expressed** | 59 | $P70^{gag\text{-}actin\text{-}fgr}$ |

### Cellular location

Cytoplasm: associates with plasma membrane: present in the secondary granules of neutrophils.

### Tissue distribution

*FGR* expression is high in mature peripheral blood monocytes and granulocytes, alveolar and splenic macrophages, human natural killer cells and differentiating myelomonocytic HL60 and U937 cells[3–5]. It is expressed only in the later stages of differentiation of myeloid leukaemia cells, in contrast to *Src*[6].

## Protein function

FGR is a non-receptor tyrosine kinase. In normal bone marrow-derived monocytic cells *Fgr* mRNA expression is transiently increased by 20-fold by CSF1, GM-CSF or LPS[7] and agents that stimulate the differentiation of HL60 cells also activate transcription and translation[8]. The γ-actin domain inhibits both the kinase activity and oncogenicity of GR-FeSV[9]. FGR activity is downregulated by phosphorylation on $Tyr^{511}$ by CSK which prevents autophosphorylation at $Tyr^{400}$ (homologous to SRC $Tyr^{416}$)[10].

### Cancer

Fifty-fold amplification of *FGR* occurs in B cells transformed by Epstein–Barr virus[11].

### Transgenic animals

*Fgr*-deficient mice show normal haematopoietic development. Doubly homozygous $Hck^{-/-}/Fgr^{-/-}$ animals are immunodeficient, having increased susceptibility to infection with *Listeria monocytogenes*[12].

### In animals

GR-FeSV is highly infectious to cats, inducing differentiated fibrosarcomas and rhabdosarcomas in young animals [1].

### *In vitro*

*FGR* does not transform epithelial cells but transforms most mammalian fibroblasts including human [1].

## Gene structure

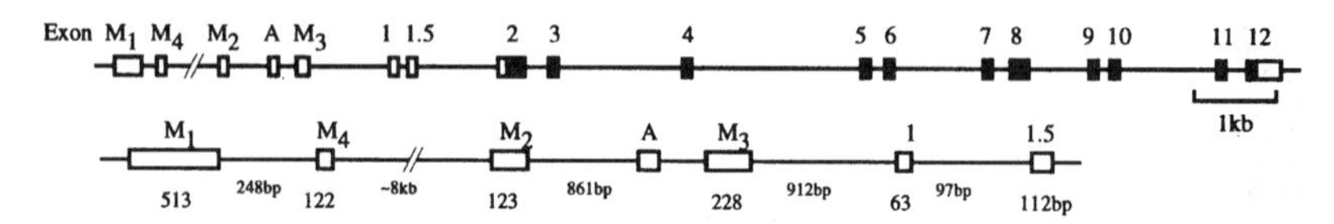


The expanded diagram represents the *FGR* promoter region. Exons 3–12 closely resemble those of avian *Src* and murine *Lck* [2]. Upstream exons differ and the 5′ untranslated region of *FGR* has an extra intron. In human myelomonocytes differential promoter utilization and alternative splicing gives rise to at least six distinct mRNAs that differ only in their 5′ untranslated regions [13].

The two predominant RNA species are *FGRA* (EBV-infected B cells) and *FGR4* (myelomonocytic cells). *FGRA* contains exon A linked to exon 1 and downstream exons and in *FGR4* exon $M_4$ is spliced to exon 1 and downstream exons [14]. Thus in EBV infection a cryptic exon A promoter is activated to regulate the expression of *FGR*, whereas the exon $M_4$ promoter is used exclusively in myelomonocytic cells. $M_1$ and $M_2$ are rarely utilized in mononuclear cells (monocytes, neutrophils and lymphocytes).

## Transcriptional regulation

There is a cluster of transcriptional start sites upstream of exon 1 that is rich in GC regions but lacks a TATA box.

## Protein structure

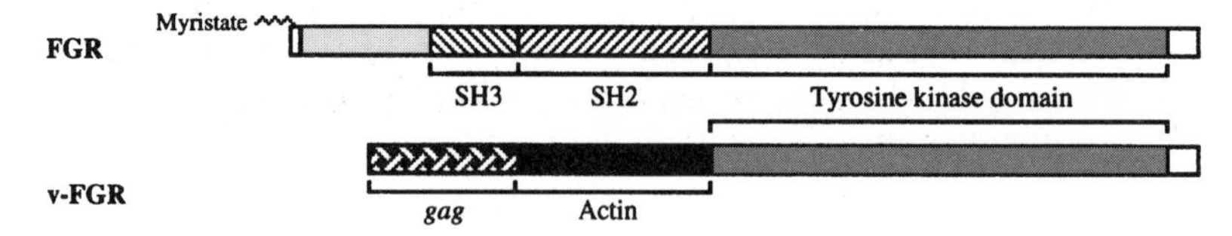


P70$^{gag\text{-}actin\text{-}fgr}$ (663 amino acids) is myristylated in the *gag* N-terminal region (118 amino acids) and includes 151 amino acids of the N-terminus of feline non-muscle actin, 389 of feline FGR and 5 *env* encoded C-terminal amino acids.

Deletion of neither the SH2 nor the SH3 domain activates transforming capacity, in contrast to the effects of the corresponding mutations in either FYN or SRC [15].

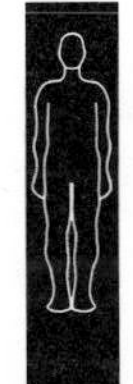

## Amino acid sequence of human FGR

```
  1 MGCVFCKKLE PVATAKEDAG LEGDFRSYGA ADHYGPDPTK ARPASSFAHI
 51 PNYSNFSSQA INPGFLDSGT IRGVSGIGVT LFIALYDYEA RTEDDLTFTK
101 GEKFHILNNT EGDWWEARSL SSGKTGCIPS NYVAPVDSIQ AEEWYFGKIG
151 RKDAERQLLS PGNPQGAFLI RESETTKGAY SLSIRDWDQT RGDHVKHYKI
201 RKLDMGGYYI TTRVQFNSVQ ELVQHYMEVN DGLCNLLIAP CTIMKPQTLG
251 LAKDAWEISR SSITLERRLG TGCFGDVWLG TWNGSTKVAV KTLKPGTMSP
301 KAFLEEAQVM KLLRHDKLVQ LYAVVSEEPI YIVTEFMCHG SLLDFLKNPE
351 GQDLRLPQLV DMAAQVAEGM AYMERMNYIH RDLRAANILV GERLACKIAD
401 FGLARLIKDD EYNPCQGSKF PIKWTAPEAA LFGRFTIKSD VWSFGILLTE
451 LITKGRIPYP GMNKREVLEQ VEQGYHMPCP PGCPASLYEA MEQTWRLDPE
501 ERPTFEYLQS FLEDYFTSAE PQYQPGDQT (529)
```

### Domain structure

| | |
|---|---|
| 269–277 and 291 | ATP binding region |
| 382 | Active site |
| 412 | Autophosphorylation site |
| 82–134 | SH3 domain (or A-box: underlined) |

### Database accession numbers

| | *PIR* | *SWISSPROT* | *EMBL/GENBANK* | *REFERENCES* |
|---|---|---|---|---|
| Human *FGR* (*SRC2*) | A27676, A28353 | P09769 | M12719–M12724 | 2,16,17 |

### *References*


[1] Rasheed, S. et al. (1982) Virology 117, 238–244.
[2] Nishizawa, M. et al. (1986) Mol. Cell. Biol. 6, 511–517.
[3] Ley, T.J. et al. (1989) Mol. Cell. Biol. 9, 92–99.
[4] Katagiri, K et al. (1991) J. Immunol. 146, 701–707.
[5] Biondi, A. et al. (1991) Eur. J. Immunol. 21, 843–846.
[6] Willman, C.L. et al. (1991) Blood 77, 726–734.
[7] Yi, T.L. and Willman, C.L. (1989) Oncogene 4, 1081–1087.
[8] Notario, V. et al. (1989) J. Cell Biol. 109, 3129–3136.
[9] Sugita, K. et al. (1989) J. Virol. 63, 1715–1720.
[10] Ruzzene, M. et al. (1994) J. Biol. Chem. 269, 15885–15891.
[11] Cheah, M.S.C. et al. (1986) Nature 319, 238–240.
[12] Lowell, C.A. et al. (1994) Genes Dev. 8, 387–398.
[13] Link, D.C. et al. (1992) Oncogene 7, 877–884.
[14] Gutkind, J.S. et al. (1991) Mol. Cell. Biol. 11, 1500–1507.
[15] Rivero-Lezcano, O.M. et al. (1995) Oncogene 11, 2675–2679.
[16] Katamine, S. et al. (1988) Mol. Cell. Biol. 8, 259–266.
[17] Inoue, K. et al. (1987) Oncogene 1, 301–304.

## Identification

*FLI1* (*Fli-1* (Friend leukemia integration-1)) was cloned by PCR using partially degenerate *ETS*-domain oligonucleotide primers [1].

## Related genes

*FLI1* is a member of the *ETS* family (see ***ETS1*** and ***ETS2***). It is 80% homologous to *ERG2*.

| | ***FLI1/ERGB2*** |
|---|---|
| **Chromosome** | 11q23–24 |
| **Mass (kDa): predicted** | 51 |
| **expressed** | 55 |
| **Cellular location** | Nucleus |

**Tissue distribution**

Human *FLI1* is expressed in a subset of erythroleukaemic cell lines. The 5′ untranslated regions of the human and mouse genes are 95% homologous, indicating that expression is post-transcriptionally regulated.

## Protein function

FLI1 has similar or identical DNA binding properties to ERG2, binding to the ETS2 binding site (GACCGGAAGTG) but not to that of PU.1 [2–4]. FLI1 can directly regulate the HIV-1 core enhancer, either positively or negatively depending on the presence of cell type specific accessory factors [5]. ETS family proteins including FLI1 form ternary complexes with PAX5 to activate the B cell-specific *mb-1* promoter [6].

**Cancer**

The chromosomal translocation t(11;22)(q24;q12) that occurs in Ewing's sarcoma creates an EWS/FLI1 fusion protein [7]. FLI1 is expressed in a subset of erythroleukaemic cell lines.

**In animals**

Mouse *Fli-1* is rearranged or activated by proviral insertion in 75% of Friend murine leukaemia virus-induced erythroleukaemias [8].

***In vitro***

EWS/FLI1 has the characteristics of an aberrant transcription factor. It binds to the same ETS2 consensus sequence as FLI1 but is a more potent *trans*-activator that, in contrast to FLI1, transforms NIH 3T3 cells [9].

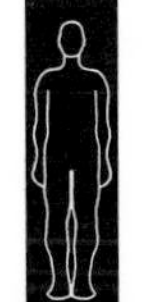

## Protein structure

**FLI1**

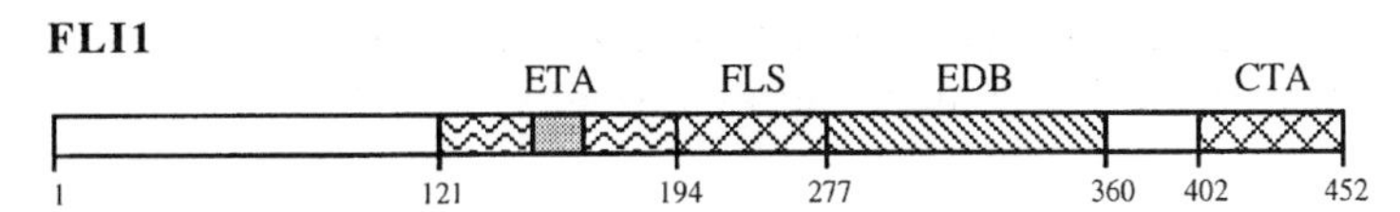


ERG proteins contain a C-terminal DNA binding domain (EDB) and two autonomous transcriptional activation domains (ETA, N-terminal and CTA, C-terminal) that are 98% homologous to the corresponding region of ERG2. FLI1 also contains an FLS domain (FLI1-specific domain, non-homologous to ERG).

## Amino acid sequence of human FLI1

```
  1 MDGTIKEALS VVSDDQSLFD SAYGAAAHLP KADMTASGSP DYGQPHKINP
 51 LPPQQEWINQ PVRVNVKREY DHMNGSRESP VDCSVSKCSK LVGGGESNPM
101 NYNSYMDEKN GPPPPNMTTN ERRVIVPADP TLWTQEHVRQ WLEWAIKEYS
151 LMEIDTSFFQ NMDGKELCKM NKEDFLRATT LYNTEVLLSH LSYLRESSLL
201 AYNTTSHTDQ SSRLSVKEDP SYDSVRRGAW GNNMNSGLNK SPPLGGAQTI
251 SKNTEQRPQP DPYQILGPTS SRLANPGSGQ IQLWQFLLEL LSDSANASCI
301 TWEGTNGEFK MTDPDEVARR WGERKSKPNM NYDKLSRALR YYYDKNIMTK
351 VHGKRYAYKF DFHGIAQALQ PHPTESSMYK YPSDISYMPS QHAHQQKVNF
401 VPPHPSSMPV TSSSFFGAAS QYWTSPTGGI YPNPNVPRHP NTHVPSHLGS
451 YY (452)
```

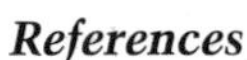

### Domain structure

121–194 5′ ETS homology domain (underlined)
277–360 DNA binding ETS domain (underlined)

### Database accession numbers

| | PIR | SWISSPROT | EMBL/GENBANK | REFERENCE |
|---|---|---|---|---|
| Human *ERGB/FLI1* | | | M98833 | 1 |

### *References*

[1] Watson, D.K. et al. (1992) Cell Growth Differ. 3, 705–713.
[2] Klemsz, M.J. et al. (1993) J. Biol. Chem. 268, 5769–5773.
[3] Rao, V.N. et al. (1993) Oncogene 8, 2167–2173.
[4] Zhang, L. et al. (1993) Oncogene 8, 1621–1630.
[5] Hodge, D.R. et al. (1996) Oncogene 12, 11–18.
[6] Fitzsimmons, D. et al. (1996) Genes Dev. 10, 2198–2211.
[7] Ouchida, M. et al. (1995) Oncogene 11, 1049–1054.
[8] Mélet, F. et al. (1996) Mol. Cell. Biol. 16, 2708–2718.
[9] May, W.A. et al. (1993) Mol. Cell. Biol. 13, 7393–7398.

## Identification

v-*fos* is the oncogene of the FBJ (Finkel, Biskis and Jinkins)[1] and FBR (Finkel, Biskis and Reilly) murine osteosarcoma viruses (MuSV) and of the avian transforming virus NK24[2]. Cellular *FOS* genes were detected by screening DNA with a v-*fos* probe[3,4].

## Related genes

The *FOS* family are members of the helix-loop-helix/leucine zipper superfamily (see ***JUN***). Closely related genes are *FRA1* and *FRA2* ("*FOS*-related antigens"), *FOSB*, Δ*FOSB*, r-*fos* (homologous to the third exon of *Fos*) and *Drosophila* dFRA and dJRA. The herpesvirus Marek disease virus (MDV) encodes a gene closely related to the *FOS/JUN* family that is expressed in MDV-transformed lymphoblastoid cells.

| | ***FOS*** | **v-*fos* FBJ-MuSV** | **v-*fos* FBR-MuSV** |
|---|---|---|---|
| **Nucleotides (bp)** | ~3500 | 4026 | 3791 |
| **Chromosome** | 14q24.3 | | |
| **Mass (kDa): predicted** | 41.6 | 49.6 | 60 |
| **expressed** | 55/62 | $55^{v\text{-}fos}$ | **P75**$^{gag\text{-}fos\text{-}fox}$ |
| **Cellular location** | Nucleus | Nucleus | Nucleus |

**Tissue distribution**

*FOS* expression is very low in most cell types but there is constitutive high expression in amnion, yolk sac, mid-gestation fetal liver, postnatal bone marrow, normal human skin[5] and in one human pre-B leukaemic cell line but not others of similar origin[6–13]. *FOS* is highly expressed in normal human skin, being localized in the nuclei of the upper spinous and granular layer cells[14].

*FOS* is an immediate early gene, the expression of which is induced rapidly (<5 min) and transiently by growth factors and mitogens. Protein expression is maximal after 90 min[10,15–19]. Other than during the $G_0$ to $G_1$ transition, *FOS* is barely detectable throughout the cell cycle but it may be induced at any stage other than during mitosis[20]. However, proliferation, cycling and re-entry of quiesent fibroblasts into the cell cycle following serum stimulation occur normally in $Fos^{-/-}$ cells and the lack of *Fos* expression is not compensated by activation of other *Fos* family or *Jun* family genes[21]. Transient transfection of HPV E6 or E7 proteins activates *Fos* transcription[22]. In many cell types inhibition of protein synthesis (e.g. by cycloheximide, actinomysin) causes superinduction of *Fos* transcription.

*FOS* is transiently expressed in IL-2- and IL-6-dependent mouse myeloma cell lines following withdrawal of growth factor and the onset of apoptosis[23] and in vitamin $K_3$-induced apoptosis of nasopharyngeal carcinoma cells[24].

FOS appears to induce apoptosis by a p53-dependent pathway that is blocked by over-expression of BCL2 and does not require protein synthesis[25].

FOS, FOSB (and JUN, JUNB and JUND) are degraded by $Ca^{2+}$-sensitive calpains. FOS is also degraded by the ubiquitin-proteasome pathway that requires the C-terminal PEST sequence of FOS. The degradation of FOS is promoted by JUN and further accelerated in the presence of casein kinase II, MAP kinase and CDC2 kinase[26].

## Protein function

FOS, FOSB, FRA1 and FRA2 form heterodimers with JUN proteins that function as positive or negative transcription factors by binding to specific DNA sequences (AP1 sites: consensus sequence $TGA^{C}/_{G}TCA$) *via* basic domains adjacent to the dimerizing helices[27,28]. FOS interacts cooperatively with JUN to inhibit its own transcription. However, for human metallothionein IIA, collagenase, collagen $\alpha_1$(III), transin (or stromelysin), proenkephalin and SV40 genes, FOS/JUN dimers form a potent transcription-activating complex and members of the FOS/JUN family (FRA1, JUNB and JUND) also activate human involucrin promoter transcription in keratinocytes[29]. FOS also activates the annexin II and IV, FOS-induced growth factor (FIGF), tyrosine hydroxylase, ornithine decarboxylase (*ODC1*), *T1/ST2* and *Fit-1* genes. The stability of dimers increases in the order FOS/FOS < JUN/JUN < FOS/JUN. Different combinations of FOS and JUN induce DNA bends of differing magnitude and orientation[30] and contradictory evidence that the AP1 target site is essentially straight in the region of the FOS/JUN complex[31] appears to have arisen from the use of probes of differing geometry.

FOS interacts with other transcription factors, including NF-κB to activate the HIV-1 LTR, and Fos-interacting protein (FIP), the transcriptional activity of which is enhanced in mast cells by aggregation of the high-affinity receptor for IgE[32]. FOS (and FOSB), but not FOSB/SF or FRA1, stably associates with the TATA box binding protein (TBP) and the multiprotein complex TFIID[33]. FOS (and also JUN and NF-κB p50 and p65) bind to the interferon-inducible p202 protein to inhibit *trans*-activation[34]. FOS/JUN dimerization also promotes association with the basal transcription factors TFIIE-34 and TFIIF to contribute to transcription initiation[35].

The transcriptional activity of FOS is augmented by RAS-stimulated phosphorylation at $Thr^{232}$ (the homologue of JUN $Ser^{73}$), mediated by a RAS and mitogen-responsive proline-directed protein kinase distinct from JNK or ERKs[36].

For FBR v-FOS there is evidence that N-terminal myristylation in the *gag* region is essential for transcriptional activation of collagen III[37]. Thus, although both collagen III and stromelysin promoters contain TRE sites, suggesting that v-FOS activates in an analogous manner to FOS, activation by v-FOS requires its N-terminal lipid modification and, for collagen III, occurs *via* a negative regulatory site that binds a key regulator of adipocyte differentiation.

The involvement of members of the *FOS/JUN* family in development is suggested by the observation that *Fos*, *Fosb* and *Fra-1* and *Jun*, *Junb* and *Jund* are differentially expressed during development and in adult tissues, consistent with their having distinct functions. In the newborn mouse a large transient (day 1) burst of *Fos* transcription occurs in all major tissues[38].

## Cancer

*FOS* over-expression has been detected in 60% of one sample of human osteosarcomas[39] and in a cell line derived from a pre-B cell acute lymphocytic leukaemia[40].

## Transgenic animals

Transgenic mice having a null mutation in *Fos* have low viability at birth (~40%). However, the homozygous mutants that survive have normal growth rates until severe osteopetrosis develops at approximately 11 days[41]. Thus *Fos* is not essential for the growth of most cell types although null mutants show delayed or absent gametogenesis and in most animals there is a reduction of ~75% in the levels of circulating T and B lymphocytes.

$Fos^{-/-}$ mice carrying a v-*Hras* transgene develop benign tumours: however, these do not undergo malignant conversion, in contrast to papillomas in wild-type animals. Thus FOS/AP1 activity may be essential for the development of malignant skin tumours[42].

Transgenic mice over-expressing *Fos* develop osteosarcomas: inactivation of the gene blocks osteoclast differentiation[43,44]. Thus FOS is a primary factor in directing cell differentiation along the osteoclast/macrophage lineages and thus regulates bone re-modelling[45]. Constitutive expression of *Fos* in transgenic mice induces transcription of collagenase type I (matrix metalloproteinase 1) in thymus, spleen and bone[46]. Neither collagenase type IV (MMP-2), stromelysin-1 (MMP-3) nor stromelysin-2 (MMP-10) transcription is affected. The expression of FBJ v-*fos* in the epidermis of transgenic mice results in the failure of regulated keratinocyte differentiation and the subsequent development of benign tumours[47].

## In animals

The expression of *Fos* by transfection into the highly metastatic murine 3LL cell line induces the expression of the major histocompatibility antigen H-2K$^b$ and reduces the metastatic properties of the cells *in vivo*[48]. FBJ- and FBR-MuSV induce sarcomas in animals[3,49]. NK24 causes avian nephroblastoma[2].

## *In vitro*

Over-expression of *FOS* does not transform human fibroblasts[50] but does transform rat fibroblasts if (1) an LTR is present to enhance transcription and (2) a 50–100 bp AU-rich element (ARE) 500 bp downstream of the chain terminator and 150 bp upstream of the poly-A addition signal in the 3′ UTR is removed. MOS can cooperate with FOS to transform NIH 3T3 cells[51].

FBJ- and FBR-MuSV both transform cells *in vitro* but FBR-MuSV is the more effective. FBR-MuSV immortalizes murine cells in culture: FBJ-MuSV does not. FBR-MuSV v-*fos* transforms human epidermal keratinocyte cells *in vitro*[52]. Internal deletions in the C-terminal half of the FBR-MuSV oncoprotein are responsible for its transforming potential (removing the *trans*-repression domain, preventing binding to the promoter and hence constitutively activating the *Fos* gene).

## Gene structure

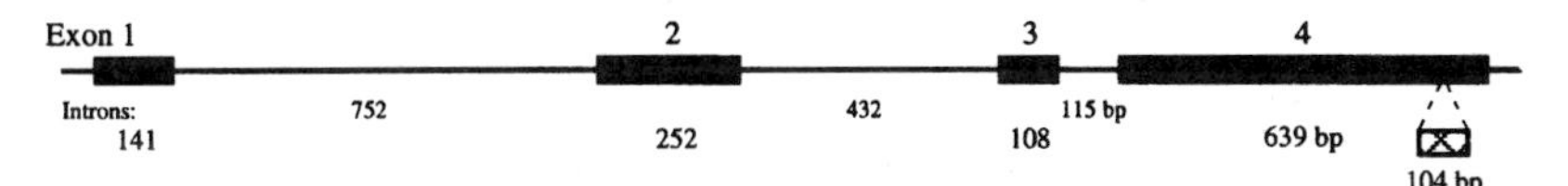


The deletion from murine exon 4 occurring in p55$^{v\text{-}fos}$ (FBJ-MuSV) is represented by the crossed box.

## Transcriptional regulation of human *FOS*

The *FOS* promoter includes a TATA box (−26 to −31 relative to the transcription initiation site), a cAMP response element (CRE: −56 to −63 – also activatable by cGMP)[53], a retinoblastoma control element (−73 to −102), a region *trans*-activated by the hepatitis B virus (HBV) pX (−120 to −220), which also modulates transcription through interactions with the SRE and AP1 sites[54,55], two regions of homology with the HSP70 promoter (−235 to −244 and −252 to −260), an AP1 site (−291 to −297), a 20 bp dyad symmetry element (DSE: −299 to −318) and a v-*sis*-conditioned medium inducible element (−335 to −347) that binds cytokine- and growth factor-regulated transcription factors of the STAT family. A vitamin D response element (VDRE: −178 to −144) can bind the vitamin D receptor, the retinoid X receptor α and a CCAAT binding transcription factor/nuclear factor 1 (CTF/NF-1) family member[56]. The ubiquitous DNA-binding protein YY1 binds to at least three sites in the *FOS* promoter, inducing bends of ~80°, and facilitating SRF binding to the SRE site[57].

The AP1 site is the target for FOS/JUN heterodimers that inhibit *FOS* transcription. The retinoblastoma gene product (pRb) acts *via* the retinoblastoma control element to repress *Fos* transcription following stimulation by serum and to lower the concentration of *Fos* mRNA in cells growing in the presence of serum. Deletion of the putative leucine zipper region of pRb abolishes transcriptional suppression. The DSE (or serum response element (SRE)) is required for activation by serum, PDGF, TPA or EGF or by pressure in cardiac muscle.

Serum response factor (p67$^{SRF}$) binds constitutively as a dimer to the SRE *via* the element $CC(A/T)_6GG$, termed the CArG box. SRF forms a ternary complex

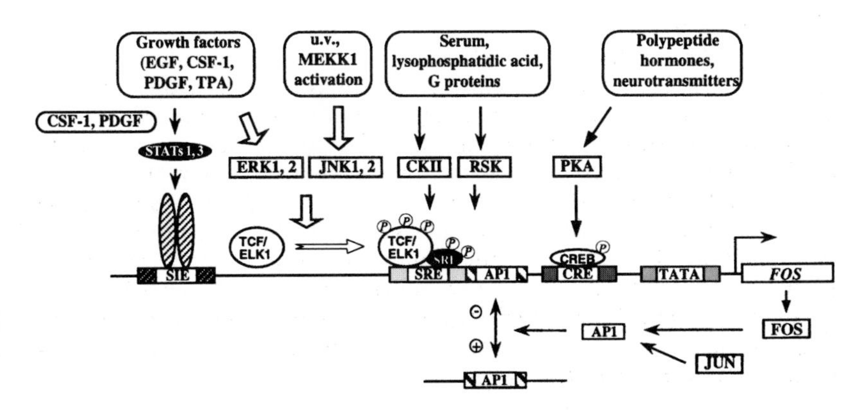

at the *Fos* SRE with members of a family of ETS domain accessory proteins, the ternary complex factors (TCFs), which bind to a conserved ETS motif [58,59]. The TCF family includes SAP1, ELK1/$p62^{TCF}$, and ERP1/NET/SAP2: all contain two conserved N-terminal ETS domains (required for DNA binding and ternary complex formation with SRF), a B region of 20 amino acids (necessary and sufficient for binding to SRF), and a C-terminal domain (essential for *trans*-activation) containing potential MAP kinase consensus sites [60]. These accessory proteins bind to DNA *via* an ETS motif ($^{C}/_{A}{}^{C}/_{A}GGA^{A}/_{T}$) that may vary in both its orientation and separation from the SRE [61]. The coactivator CBP binds directly to SAP1 and is necessary for *trans*-activation by phosphorylated SAP1 [62]. ELK1 interacts directly with the 5′ region of the DSE, and ternary complex formation requires both DSE-bound $p67^{SRF}$ and sequences both within and outside the DSE [63,64]. Only ELK1 and SAP1a efficiently bind the *Fos* SRE *in vivo*: ternary complex formation by SAP2 is weak and is substantially unaffected by serum stimulation or v-*ras* coexpression [65]. A quaternary SRE complex including an SRF dimer and two ELK1 molecules has also been detected [66].

ELK1 is phosphorylated *in vitro* by MAP kinase (ERK1 and ERK2) and this increases its transcriptional activation capacity [67] and promotes ternary complex formation [68]. *Fos* transcription is also induced by stimuli such as UV irradiation or activation of MEKK1: these scarcely affect ERK1 or ERK2 activity but strongly activate two other MAPKs, JNK1 and JNK2, that phosphorylate ELK1 on the same sites as ERK1/2 [69]. The activation of MAPK pathway by v-*ras*, v-*src* or v-*raf* is consistent with the finding that expression of these oncoproteins activates the SRE in the absence of growth factor stimulation.

$p67^{SRF}$ is also transcriptionally activated *via* its C-terminal half by HTLV-I TAX1 [70]. $p67^{SRF}$ is phosphorylated by casein kinase II but this does not appear to be significant in the growth factor activation of *Fos* expression [71]. The chimeric EWS/FLI protein that arises in Ewing's sarcoma forms a ternary complex with $p67^{SRF}$ on the *Fos* SRE, although the ETS family protein FLI1 does not [72].

The SRE also confers inducibility on a heterologous promoter (the $\beta$ globin promoter) but only in response to serum, not PDGF or TPA. This suggests that additional cooperative signals are required for transcriptional activation in response to PDGF or TPA. Sequences closely similar to that of SRE occur in other genes (e.g. actin and *Krox20)*.

SRE activation occurs independently of TCF binding *via* a heterotrimeric G protein-mediated pathway stimulated by lysophosphatidic acid or aluminium fluoride [73]. Activated forms of RhoA, Rac1 and CDC42Hs activate transcription *via* SRF and synergize with signals activating TCF [74]. RhoA mediates signalling by serum, lysophosphatidic acid or aluminium fluoride but not by Rac1 or CDC42Hs.

The SRE contains multiple overlapping enhanson elements [75] with which members of both the helix-loop-helix and the CCAAT/enhancer binding protein (C/EBP) transcription factor families [76] and the zinc finger protein SRE-ZBP interact [77]. An additional SRE binding protein (SRE BP) is required for maximal serum induction of *Fos* [78]. The binding site for SRE BP coincides with that for rNF-IL-6 but the proteins are distinct [76]. The SRE also includes binding sites for DBF/MAPF1 [79,80], Phox1 [81] and E12 [76]. The

FAP site, immediately 3′ to the SRE, is the principal target for regulation of initiation by calcium, which also requires an intact SRE [82]. In mesangial cells endothelin 1 (ET-1) stimulates *Fos* expression *via* $Ca^{2+}$ influx which requires the presence of the FAP site in the *Fos* promoter [83]. Calcium also promotes the elongation of FOS transcripts to generate the mature mRNA form.

*Fos* expression is greatly diminished during muscle cell differentiation and, consistent with this observation, the muscle-specific transcription factor MyoD binds to a region overlapping the SRE to function as a negative regulator of *Fos* transcription [84].

The v-*sis*-conditioned medium inducible element (SIE) binds cytokine- and growth factor-regulated transcription factors of the STAT family. The SIE, DSE, DR and CRE can mediate *trans*-activation by the HTLV-I TAX protein [85,86].

The *Fos* gene contains an intragenic regulatory element (FIRE) at the end of exon 1 that can cause premature termination of transcription. Intragenic regulation of transcription also occurs in the *Myc* and *Myb* genes [87].

Human and mouse *FOS* genes contain a functional oestrogen response element [88].

## RNA stability

The AU-rich element (ARE) in many rapidly induced mRNAs is thought to mediate selective degradation. The *Fos* ARE comprises two domains: (I) within the 5′ 49 bp (three AUUUA motifs) that can function as an RNA destabilizer by itself, and (II) a 20 bp U-rich sequence in the 3′ part of the ARE that enhances the destabilizing capacity of domain I. A second region that regulates mRNA stability is present within the coding sequence. This comprises the 5′ 0.38 kb region which contains multiple destabilizing elements that can function independently to promote both de-adenylation and degradation of mRNA. Within this region two major elements have been defined (CD1 and CD2) that require ribosome assembly and translation to direct mRNA decay. This region also includes a 56 bp purine-rich segment with which at least two protein factors associate to promote rapid degradation [89].

## Protein structure

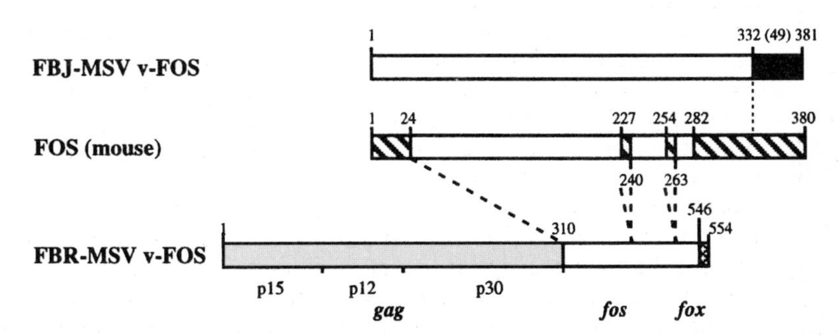


The cross-hatched boxes in FOS represent regions deleted in p75$^{gag\text{-}fos}$. The C-terminal 49 amino acids of p55$^{v\text{-}fos}$ (black box) differ from those of FOS.

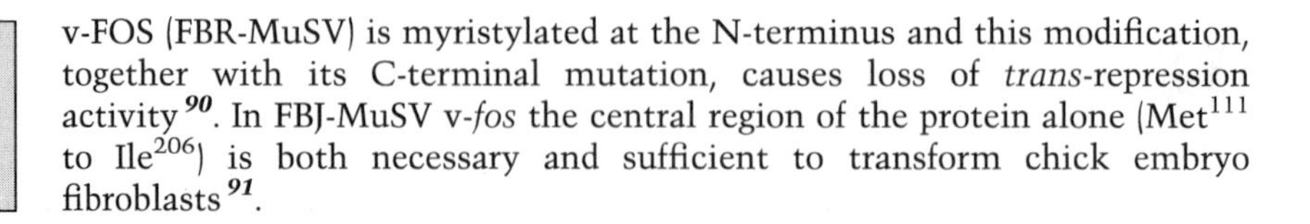

v-FOS (FBR-MuSV) is myristylated at the N-terminus and this modification, together with its C-terminal mutation, causes loss of *trans*-repression activity[90]. In FBJ-MuSV v-*fos* the central region of the protein alone ($Met^{111}$ to $Ile^{206}$) is both necessary and sufficient to transform chick embryo fibroblasts[91].

## Transcriptional autoregulation by FOS/JUN dimers

The regulatory activity resides in the C-terminal region of the FOS proteins (mutated in the viral proteins of FBJ-MuSV and FBR-MuSV). One of the major post-translational modifications undergone by FOS is serine phosphoesterification in the C-terminus and the negative charge thus conferred on the molecule is crucial for suppression of transcription[92]. Transformation-defective FOS proteins that lack either the leucine zipper region or the N-terminal 110 amino acids inhibit transformation by either v-*fos* or *Ras*[93].

The basic motif KCR (amino acids 153–155 in human FOS) is conserved in FOS, JUN and ATF/CREB family members and redox-regulation of the cysteine residue mediates DNA binding. Nevertheless, the KCR regions of FOS and JUN are not identical in DNA binding and non-basic amino acids (e.g. ISI) in this motif can function efficiently in binding[94]. Reduction of this cysteine by the ubiquitous nuclear redox factor REF-1 stimulates the DNA binding activity of FOS/JUN and JUN/JUN dimers, MYB, NF-κB and ATF/CREB proteins. Replacement of $Cys^{154}$ by serine enhances the transforming activity of the FOS protein[95].

## Sequence homology between FOS, FRA1, FRA2 and FOSB

| | | | | | |
|---|---|---|---|---|---|
| FOS (chicken) | 14-22 | 60-79 | 123-208 | 308-334 | 344-367 |
| FRA-1 (rat) | 11-13 | 39-58 | 93-179 | 217-242 | 251-275 |
| FRA-2 (chicken) | 12-20 | 46-65 | 111-197 | 256-282 | 300-323 |
| FOSB (mouse) | 11-19 | 54-73 | 142-227 | 285-309 | 315-338 |

The five homologous regions of the proteins are indicated by shaded regions. The basic and leucine zipper regions occur between FOS residues 123 and 208. The aligned numbers below indicate the amino acids at the extremes of the corresponding regions in the other proteins[96].

## Amino acid sequence of human FOS

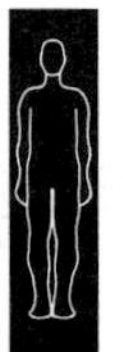

```
  1 MMFSGFNADY EASSSRCSSA SPAGDSLSYY HSPADSFSSM GSPVNAQDFC
 51 TDLAVSSANF IPTVTAISTS PDLQWLVQPA LVSSVAPSQT RAPHPFGVPA
101 PSAGAYSRAG VVKTMTGGRA QSIGRRGKVE QLSPEEEEKR RIRRERNKMA
151 AAKCRNRRRE LTDTLQAETD QLEDEKSALQ TEIANLLKEK EKLEFILAAH
201 RPACKIPDDL GFPEEMSVAS LDLTGGLPEV ATPESEEAFT LPLLNDPEPK
251 PSVEPVKSIS SMELKTEPFD DFLFPASSRP SGSETARSVP DMDLSGSFYA
301 ADWEPLHSGS LGMGPMATEL EPLCTPVVTC TPSCTAYTSS FVFTYPEADS
351 FPSCAAAHRK GSSSNEPSSD SLSSPTLLAL (380)
```

**Domain structure**

143–160 Basic motif (underlined)
165–193 Leucine zipper (underlined)
139–161 Basic nuclear targeting signal. However, this region is not essential for FOS nuclear localization [97]. Other FOS sequences that resemble known nuclear targeting signals are ineffective in directing pyruvate kinase to the nucleus: FOS may therefore contain a novel nuclear targeting sequence.
362 S6 kinase (RSK) phosphorylation site ($Ser^{362}$)
362/374 MOS/MAP kinase phosphorylation sites ($Ser^{362/374}$)

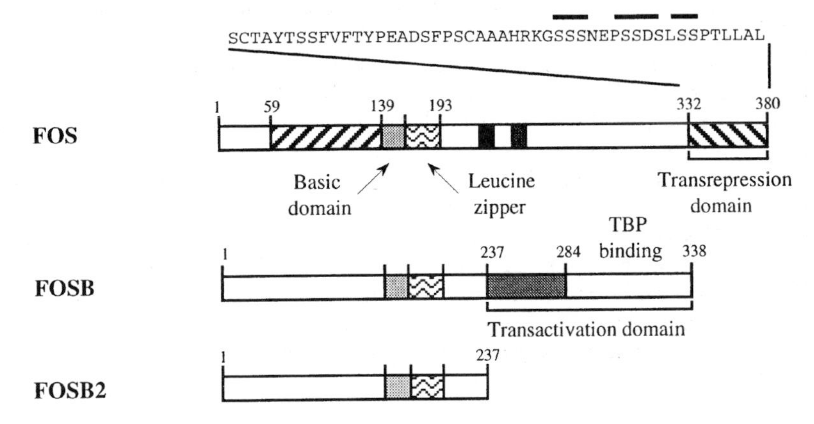


The two hatched boxes represent regions encompassing phosphorylation sites. The bars mark the three groups of C-terminal serine residues in FOS ($Ser^{362-364}$, $Ser^{368}$, $Ser^{369}$, $Ser^{371}$ and $Ser^{373-374}$). The N-terminal domain (59–139) is phosphorylated *in vitro* by protein kinase C, $p34^{CDC2}$ and a nuclear kinase: the C-terminal region by cAMP-dependent protein kinase, $p34^{CDC2}$ [98], S6 kinase (RSK) and MAP kinase [99]. C-terminal phosphorylation promotes transformation activity [100].

## Relationship between FOS and JUN

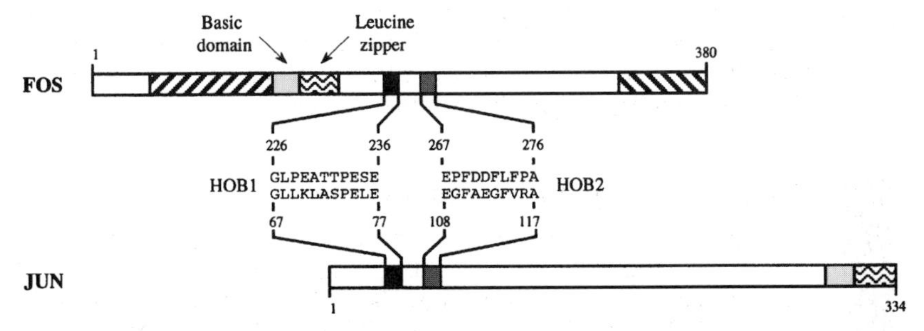


FOS contains five activation modules (FAM1 (1–47), FAM2 (48–81), FAM3 (210–244), FAM4 (245–308) and FAM5 (309–380)). The two homology box regions (HOB1 and HOB2) of FOS lie within a *trans*-activating domain and are conserved in the A1 activation domain of JUN [101]. The FOS sequence shown is for rat, which is identical to that of human FOS except for the substitution of alanine and threonine by valine and alanine at positions 230 and 231. HOB1 and HOB2 regions also occur in C/EBP and they act co-operatively to stimulate transcription of reporter gene constructs. HOB1 contains a phosphorylation site for MAP kinase. A second HOB motif [102]. The N-terminus contains an inhibitor domain (ID1, 120–134, which includes an inhibitor motif (IM1, 124–128) conserved between FOS and FOSB and is required for inhibitor function) which silences HOB1 activity. $Thr^{232}$ of HOB1 (analogous to JUN HOB1 $Ser^{73}$) is phosphorylated by a MAP kinase, increasing *trans*-activation capacity [103].

The C-terminal 55 amino acids of FOSB interact with TATA binding protein (TBM domain). The C-terminal activation domain of FOS (210–380) binds CBP that stimulates transcription of cAMP-responsive genes by binding to the CREB transcription factor. CBP binds in a phosphorylation-independent manner to stimulate FOS activity. This interaction competes with E1A binding to FOS and the repression exerted by E1A may occur *via* displacement of CRB from FOS [104].

## Database accession numbers

| | *PIR* | *SWISSPROT* | *EMBL/GENBANK* | *REFERENCES* |
|---|---|---|---|---|
| FBJ-MuSV v-*fos* | A01344 | P01102 | V01184 | 3 |
| | | | JO2084 | |
| FBR-MuSV v-*fos* | | | KO2712 | 3 |
| Human *FOS* | A01342 | P01100 | V01512 | 4 |
| | | | K00650, M16287 | 105,106 |

## *References*


1 Finkel, M.P. et al. (1966) Science 151, 698–701.
2 Nishizawa, M. et al. (1987) J. Virol. 61, 3733–3740.
3 van Beveren, C. et al. (1983) Cell 32, 1241–1255.
4 van Straaten, F. et al. (1983) Proc. Natl Acad. Sci. USA 80, 3183–3187.
5 Basset-Seguin, N. et al. (1994) Oncogene 9, 765–771.
6 Muller, R. et al. (1983) EMBO J. 2, 679–684.
7 Gonda, T.J. and Metcalf, D. (1984) Nature 310, 249–251.
8 Mason, I. et al. (1985) Differentiation 30, 76–81.
9 Mitchell, R.L. et al. (1985) Cell 40, 209–217.
10 Muller, R. et al. (1985) Nature 314, 546–548.
11 Conscience, J.-F. et al. (1986) EMBO J. 5, 317–323.
12 Kreipe, H. et al. (1986) Differentiation, 33, 56–60.
13 Panterne, B. et al. (1992) Oncogene 7, 2341–2344.
14 Welter, J.F. and Eckert, R.L. (1995) Oncogene 11, 2681–2687.
15 Muller, R. et al. (1984) Nature 312, 716–720.
16 Bravo, R. et al. (1987) Cell 48, 251–260.
17 Tannenbaum, C.S. et al. (1988) J. Immunol. 140, 3640–3645.
18 Sariban, E. et al. (1985) Nature 316, 64–66.
19 Kovary, K. and Bravo, R. (1991) Mol. Cell. Biol. 11, 4466–4472.
20 Bravo, R. et al. (1986) EMBO J. 5, 695–700.
21 Brüsselbach, S. et al. (1995) Oncogene 10, 79–86.
22 Morosov, A. et al. (1994) J. Biol. Chem. 269, 18434–18440.
23 Colotta, F. et al. (1992) J. Biol. Chem. 267, 18278–18283.
24 Wu, F.Y.-H. et al. (1993) Oncogene 8, 2237–2244.

[25] Preston, G.A. et al. (1996) Mol. Cell. Biol. 16, 211–218.
[26] Tsurumi, C. et al. (1995) Mol. Cell. Biol. 15, 5682–5687.
[27] **Distel, R.J. and Spiegelman, B. (1990) Adv. Cancer Res. 55, 37–55.**
[28] **Ransone, L.J. and Verma, I.M. (1990) Annu. Rev. Cell Biol. 6, 539–557.**
[29] Welter, J.F. et al. (1995) J. Biol. Chem. 270, 12614–12622.
[30] Kerppola, T.K. (1996) Proc. Natl Acad. Sci. USA 93, 10117–10122.
[31] Sitlani, A. and Crothers, D.M. (1996) Proc. Natl Acad. Sci. USA 93, 3248–3252.
[32] Lewin, I. et al. (1996) J. Biol. Chem. 271, 1514–1519.
[33] Metz, R. et al. (1994) Mol. Cell. Biol. 14, 6021–6029.
[34] Min, W. et al. (1996) Mol. Cell. Biol. 16, 359–368.
[35] Martin, M.L. et al. (1996) Mol. Cell. Biol. 16, 2110–2118.
[36] Deng, T. and Karin, M. (1994) Nature 371, 171–175.
[37] Jotte, R.M. et al. (1994) J. Biol. Chem. 269, 16383–16396.
[38] Kasik, J.W. et al. (1987) Mol. Cell. Biol. 7, 3349–3352.
[39] Wu, J.X. et al. (1990) Oncogene 5, 989–1000.
[40] Tsai, L.-H. et al. (1991) Oncogene 6, 81–88.
[41] Okada, S. et al. (1994) Mol. Cell. Biol. 14, 382–390.
[42] Saez, E. et al. (1995) Cell 82, 721–732.
[43] Ruther, U. et al. (1989) Oncogene 4, 861–865.
[44] Grigoriadis, A.E. et al. (1994) Science 266, 443–448.
[45] **Jacenko, O. (1995) BioEssays 17, 277–281.**
[46] Gack, S. et al. (1994) J. Biol. Chem. 269, 10363–10369.
[47] Greenhalgh, D.A. et al. (1993) Oncogene 8, 2145–2157.
[48] Kushtai, G. et al. (1990) Int. J. Cancer 45, 1131–1136.
[49] Van Beveren, C. et al. (1984) Virology 135, 229–243.
[50] Alt, M. and Grassmann, R. (1993) Oncogene 8, 1421–1427.
[51] Okazaki, K. and Sagata, N. (1995) EMBO J. 14, 5048–5059.
[52] Lee, M.-S. et al. (1993) Oncogene 8, 387–393.
[53] Gudi, T. et al. (1996) J. Biol. Chem. 271, 4597–4600.
[54] Natoli, G. et al. (1993) Mol. Cell. Biol. 14, 989–998.
[55] **Treisman, R. (1995) EMBO J. 14, 4905–4913.**
[56] Candeliere, G.A. et al. (1996) Mol. Cell. Biol. 16, 584–592.
[57] Natesan, S. and Gilman, M.Z. (1995) Mol. Cell. Biol. 15, 5975–5982.
[58] Hipskind, R.A. et al. (1991) Nature 354, 531–534.
[59] Dalton, S. and Treisman, R. (1992) Cell 68, 597–612.
[60] Hill, C.S. et al. (1994) EMBO J. 13, 5421–5432.
[61] Treisman, R. et al. (1992) EMBO J. 11, 4631–4640.
[62] Janknecht, R. and Nordheim, A. (1996) Oncogene 12, 1961–1969.
[63] Shaw, P.E. (1992) EMBO J. 11, 3011–3019.
[64] Sharrocks, A.D. et al. (1993) Mol. Cell. Biol. 13, 123–132.
[65] Price, M.A. et al. (1995) EMBO J. 14, 2589–2601.
[66] Gille, H. et al. (1996) Mol. Cell. Biol. 16, 1094–1102.
[67] Marais, R. et al. (1993) Cell 73, 381–393.
[68] Gille, H. et al. (1992) Nature 358, 414–417.
[69] Cavigelli, M. et al. (1995) EMBO J. 14, 5957–5964.
[70] Fujii, M. et al. (1992) Genes Dev. 6, 2066–2076.
[71] Manak, J.R. and Prywes, R. (1993) Oncogene 8, 703–711.
[72] Magnaghi, L. et al. (1996) Nucleic Acids Res. 24, 1052–1058.
[73] Hill, C.S. and Treisman, R. (1995) EMBO J. 14, 5037–5047.

[74] Hill, C.S. et al. (1995) Cell 81, 1159–1170.
[75] **Treisman, R. (1992) Trends Biol. Sci. 17, 423–426.**
[76] Metz, R. and Ziff, E. (1991) Oncogene 6, 2165–2178.
[77] Attar, R.M. and Gilman, M.Z. (1992) Mol. Cell. Biol. 12, 2432–2443.
[78] Boulden, A.M. and Sealy, L.J. (1992) Mol. Cell. Biol. 12, 4769–4783.
[79] Ryan, W.A. et al. (1989) EMBO J. 8, 1785–1792.
[80] Walsh, K. (1989) Mol. Cell. Biol. 9, 2191–2201.
[81] Grueneberg, D. et al. (1992) Science 257, 1089–1095.
[82] Lee, G. and Gilman, M. (1994) Mol. Cell. Biol. 14, 4579–4587.
[83] Wang, Y. and Simonson, M.S. (1996) Mol. Cell. Biol. 16, 5915–5923.
[84] Trouche, D. et al. (1993) Nature 363, 79–82.
[85] Alexandre, C. and Verrier, B. (1991) Oncogene 6, 543–551.
[86] Wagner, B.J. et al. (1990) EMBO J. 9, 4477–4484.
[87] Lamb, N.J.C. et al. (1990) Cell 61, 485–496.
[88] Hyder, S.M. et al. (1992) J. Biol. Chem. 267, 18047–18054.
[89] Schiavi, S.C. et al. (1994) J. Biol. Chem. 269, 3441–3448.
[90] Kamata, N. and Holt, J.T. (1992) Mol. Cell. Biol. 12, 876–882.
[91] Yoshida, T. et al. (1989) Oncogene Res. 5, 79–89.
[92] **Hunter, T. and Karin, M. (1992) Cell 70, 375–387.**
[93] Wick, M. et al. (1992) Oncogene 7, 859–867.
[94] Ng, L. et al. (1994) Nucleic Acids Res. 21, 5831–5837.
[95] Okuno, H. et al. (1993) Oncogene 8, 695–701.
[96] Nishina, H. et al. (1990) Proc. Natl Acad. Sci. USA 87, 3619–3623.
[97] Tratner, I. and Verma, I.M. (1991) Oncogene 6, 2049–2053.
[98] Abate, C. et al. (1991) Oncogene 6, 2179–2185.
[99] Chen, R.-H. et al. (1993) Proc. Natl Acad. Sci. USA 90, 10952–10956.
[100] Chen, R.-H. et al. (1996) Oncogene 12, 1493–1502.
[101] Sutherland, J.A. et al. (1992) Genes Dev. 6, 1810–1819.
[102] Brown, H.J. et al. (1995) EMBO J. 14, 124–131.
[103] Bannister, A.J. et al. (1994) Nucleic Acids Res. 22, 5173–5176.
[104] Bannister, A.J. and Kouzarides, T. (1995) EMBO J. 14, 4758–4762.
[105] Treisman, R. (1985) Cell 42, 889–902.
[106] Verma, I.M. et al. (1986) Cold Spring Harb. Symp. Quant. Biol. 51, 949–958.

## Identification

*FOSB* was detected by screening cDNA libraries with *FOS* DNA as a probe [1].

## Related genes

*FOSB* is a member of the helix-loop-helix/leucine zipper superfamily (see ***JUN***). Closely related genes are *FOS*, *ΔFOSB* (*FOSB/SF* or *FOSB2*), *FRA1* and *FRA2* and r-*fos*.

| | ***FOSB*** |
|---|---|
| **Nucleotides (kb)** | ~4.6 |
| **Mass (kDa): predicted** | 36 |
| **expressed** | 52<br>37 (FOSB2) |
| **Cellular location** | Nucleus |

**Tissue distribution**
*FOSB* is an immediate early gene. In stimulated fibroblasts the induction of its transcription is slightly delayed with respect to that of *FOS* [1]. In serum-stimulated Rat-1A cells the expression of ΔFOSB parallels that of FOSB [2]. In developing mice, FOSB is widely expressed with particularly high levels in bone and cartilage [3]. FOSB is present in the nuclei of all layers of normal human epidermis [4].

## Protein function

FOSB forms heterodimers with JUN proteins that function as positive or negative transcription factors (see ***FOS***). FOSB activates transcription of the fibroblast *T1* early response gene [5]. *Fosb* is a more potent *in vitro* transforming agent than *Fos*, having a transforming capacity equivalent to that of v-*fos* [6].

**Transgenic animals**
*Fosb*$^{-/-}$ mice develop normally although expression of the AP1-regulated stromelysin 1 and collagenase genes is reduced [3]. However, they have a specific behavioural defect as a result of which animals are unable to nurture their young [7].

## Gene structure

The overall structure of the *Fosb* gene is very similar to that of *Fos* [8,9]. The four exons of *Fosb* encode 42, 107, 36 and 153 amino acids, respectively and alternative splicing of exon 4 removes 140 bp to generate a truncated form, ΔFOSB, of 237 amino acids [8].

The murine *Fosb* promoter contains SRE and AP1 binding sites located in positions that are identical relative to those found in the *Fos* promoter and the activity of the *Fosb* promoter is downregulated by FOS or FOSB [8].

## Protein structure

FOS and FOSB are 70% homologous and 42% identical in sequence (see ***FOS***). FOSB has a truncated C-terminus relative to FOS but retains the leucine zipper and basic DNA binding regions of FOS. The truncated form of FOS, ΔFOSB (FOSB2), also retains the capacity to form dimers with JUN that bind to AP1 sites but no longer activate transcription from AP1-containing promoters or repress the *Fos* promoter[10]. The expression of FOSB2 suppresses transformation by v-*fos*, *Fos* or *Fosb*, presumably by interfering with *trans*-activation events required for transformation[11]. However, over-expression of *Fosb2* does not prevent normal cells from entering the cell cycle in response to serum, indicating that *Fosb2* is not a negative regulator of cell growth[12].

## Amino acid sequence of mouse FOSB

```
  1 MFQAFPGDYD SGSRCSSSPS AESQYLSSVD SFGSPPTAAA SQECAGLGEM
 51 PGSFVPTVTA ITTSQDLQWL VQPTLISSMA QSQGQPLASQ PPAVDPYDMP
101 GTSYSTPGLS AYSTGGASGS GGPSTSTTTS GPVSARPARA RPRRPREETL
151 TPEEEEKRRV RRERNKLAAA KCRNRRRELT DRLQAETDQL EEEKAELESE
201 IAELQKEKER LEFVLVAHKP GCKIPYEEGP GPGPLAEVRD LPGSTSAKED
251 GFGWLLPPPP PPPLPFQSSR DAPPNLTASL FTHSEVQVLG DPFPVVSPSY
301 TSSFVLTCPE VSAFAGAQRT SGSEQPSDPL NSPSLLAL (338)
```

**Domain structure**

161–179 Basic, DNA binding motif
183–211 Leucine zipper (underlined)
229–338 *Trans*-activation domain (see ***FOS***). All four members of the FOS family that transform established rodent fibroblast cell lines (FBR-v-FOS, FBJ-v-FOS, FOS and FOSB) contain an equivalent C-terminal domain[13,14]
243–282 TFIID binding site
304–310 TATA box binding protein binding site[15]

**Database accession numbers**

| | *PIR* | *SWISSPROT* | *EMBL/GENBANK* | *REFERENCES* |
|---|---|---|---|---|
| Mouse *Fosb* | S04108 | P13346 | X14897 | [1] |
| | | | M77748 | [8] |

***References***


[1] Zerial, M. et al. (1989) EMBO J. 8, 805–813.
[2] Nakabeppu, Y. et al. (1993) Mol. Cell. Biol. 13, 4157–4166.
[3] Gruda, M.C. et al. (1996) Oncogene 12, 2177–2185.
[4] Welter, J.F. and Eckert, R.L. (1995) Oncogene 11, 2681–2687.
[5] Kalousek, M.B. et al. (1994) J. Biol. Chem. 269, 6866–6873.
[6] Schuermann, M. et al. (1991) Oncogene 6, 567–576.
[7] Brown, J.R. et al. (1996) Cell 86, 297–309.
[8] Lazo, P.S. et al. (1992) Nucleic Acids Res. 20, 343–350.
[9] Nishina, H. et al. (1990) Proc. Natl Acad. Sci. USA 87, 3619–3623.
[10] Nakabeppu, Y. and Nathans, D. (1991) Cell 64, 751–759.

[11] Yen, J. et al. (1991) Proc. Natl Acad. Sci. USA 88, 5077–5081.
[12] Dobrzanski, P. et al. (1991) Mol. Cell. Biol. 11, 5470–5478.
[13] Wisdom, R. and Verma, I.M. (1993) Mol. Cell. Biol. 13, 2635–2643.
[14] Wisdom, R. and Verma, I.M. (1993) Mol. Cell. Biol. 13, 7429–7438.
[15] Metz, R. et al. (1994) EMBO J. 13, 3832–3842.

## Identification

v-*fps* (Fujinami-PRCII sarcoma) is the oncogene of the acutely transforming Fujinami sarcoma virus (FSV). Four other retroviruses also carry v-*fps*: PRCII and PRCIV (Poultry Research Centre II and IV), UR1 (University of Rochester) and 16L. *fes* (feline sarcoma virus) is the cognate gene of *fps* present in three retroviruses (Snyder–Theilen (ST)-FeSV, Gardner–Arnstein (GA)-FeSV and Hardy–Zuckerman 1 (HZ1)-FeSV)[1]. *FPS/FES* was detected by screening human lung carcinoma DNA with a v-*fes* probe[2].

## Related genes

*FPS/FES* is related to *FER*, human FES/FPS-related tyrosine kinase gene, *Flk* (fps/fes-like kinase, the rat homologue of human *FER*), murine *Fer*$^{T}$ tyrosine kinase and *Drosophila dfps*85D. Sequences homologous to *FPS/FES* also occur in the sea urchin genome and vesicular stomatitis virus L polymerase. The FPS/FES family have C-terminal homology with SRC (SH2 and tyrosine kinase domains). FER shares homology with TRK.

| | ***FPS/Fes*** | **v-*fps/fes*** |
|---|---|---|
| **Nucleotides (kb)** | 13 | 5.3 (FSV genome) |
| **Chromosome** | 15q25–q26 (*FES*)<br>5q21 (*FER*) | |
| **Mass (kDa): predicted** | 94 | |
| **expressed** | p92$^{FES}$<br>p94$^{FER}$<br>p98$^{fps}$ (chicken) | P130$^{gag\text{-}fps}$, P140$^{gag\text{-}fps}$ (FSV) |
| **Cellular location** | | Membrane associated |

FPS/FES: Human: nucleus, granules and some plasma membrane[3]. Avian: 60–90% cytosolic[4]. FER: Cytoplasm but also present in the nucleus associated with chromatin.

**Tissue distribution**

*FPS/FES* is expressed in immature and differentiated haematopoietic cells of the myeloid lineage and in leukaemic myeloid cells[5–7]. *FER* is ubiquitously expressed[8].

## Protein function

*FPS/FES* genes encode cytoplasmic tyrosine kinases that are autophosphorylated. The tissue distribution suggests that its normal role may be in the control of proliferation and differentiation of haematopoietic cells[9] and the introduction of human *FES* into cells can confer the capacity to undergo myeloid differentiation[10].

In human polymorphonuclear leukocytes GM-CSF causes the tyrosine phosphorylation of p93$^{FES}$, JAK2 and STAT1 and STAT3 and all of these factors become associated with the transducing β subunit of the GM-CSF receptor[11]. In the human erythroleukaemia cell line TF-1 GM-CSF or IL-3 also activate $p93^{FES}$, GM-CSF promoting FES association with the GM-CSF receptor. Over-expression of FPS/FES in a CSF1-dependent macrophage cell line has revealed 75 and 130 kDa substrates[12]. gp130, a member of the IL-6 family, activates FES and associates with $p92^{FES}$[13].

Expression of v-*fps* causes the phosphorylation of at least nine proteins that are also phosphorylated in response to v-*src*, v-*abl* or v-*fgr*[14] and disrupts gap junctional communication[15]. v-FPS also causes tyrosine phosphorylation of fibronectin receptor proteins and the GAP-associated proteins p62 and p190[16].

**Cancer**

*FES/FPS* expression is enhanced in some lung cancers and haematopoietic malignancies[17,18]. *FER* is highly expressed in cell lines derived from human kidney carcinomas and glioblastomas[19] and is frequently deleted from chromosome 5 in acute myeloid leukaemia and myelodysplastic syndromes[20].

**Transgenic animals**

Transgenic mice expressing v-*fps* have severe cardiac or neurological disorders and develop a variety of lymphoid or mesenchymal tumours[21,22]. Lymphoid tumours are monoclonal and appear between 2 and 12 months, indicating a requirement for other genetic changes in addition to the expression of v-*fps*.

The expression in mice of a gain-of-function mutant allele of human *FPS/FES* that encodes an N-terminally myristylated, membrane-associated protein causes widespread hypervascularity progressing to multifocal haemangiomas[23]. This protein has increased tyrosine kinase activity and weak transforming potential.

**In animals**

FSV induces fibrosarcomas or myxosarcomas[24,25].

***In vitro***

*Fps/Fes* over-expression transforms NIH 3T3 fibroblasts, inducing the expression of *Egr-1*, the *Ras*-related immediate-early gene *RhoB* and the transformation-related *9E3* gene and increasing the concentrations of PtdIns(3)*P*, PtdIns(3,4)$P_2$ and PtdIns(3,4,5)$P_3$[26–30]. Microinjection of anti-RAS antibody causes reversion of *FPS/FES*-transformed cells[31] and revertant cells derived from *Kras*-transformed cells are resistant to *FPS/FES* transformation. FSV transforms erythroid cells, osteoblasts[32,33] and quail myogenic cells[34] and FeSV transforms pre-B cells[35,36]. Rat-2 fibroblasts transfected with v-*fps* are tumorigenic and have metastatic potential[37].

## Gene structure

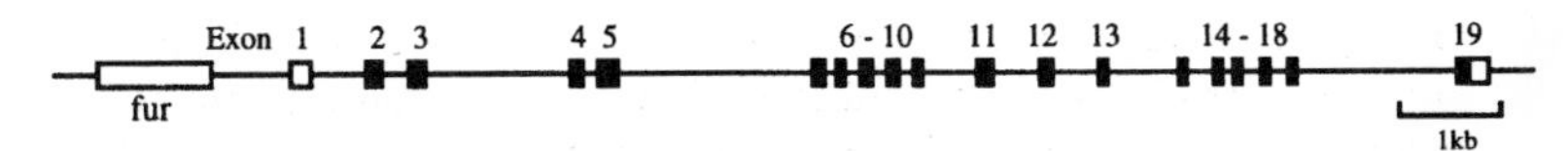


A 9 kb upstream region (*FUR*, fps/fes upstream region) encodes a putative 499 amino acid protein that is a member of the proprotein convertase gene family[38].

Human *FES* contains four distinct sites that bind myeloid nuclear proteins (–408 to –386, –293 to –254, –76 to –65 and –34 to +3): these include SP1, PU.1 and/or ELF1 binding sites and an FP4–3′-binding factor [39].

## Protein structure

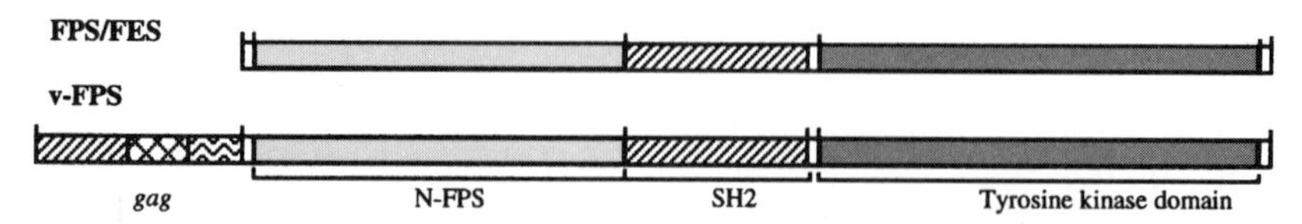


Human and chicken FPS share 70% amino acid identity. In chicken FPS and the derived oncogene P130$^{gag\text{-}fps}$ the N-terminal domain adjacent to SH2 is comprised of ~250 amino acids (N-FPS) that may determine the subcellular localization. FPS- and *gag-fps*-encoded proteins contain an N-terminal potential myristylation site (Gly2), in common with members of the *SRC* family, but, with the exception of the ST-FeSV and GA-FeSV proteins, do not appear to be myristylated or plasma membrane associated [40].

N-terminal *gag* substitution is sufficient to activate the oncogenic potential of *Fps/Fes* although the scattered mutations occurring in most of the forms of v-*fps* can do likewise. However, v-*fps* is linked to N-terminal *gag* sequences in all spontaneously arising, *fps/fes*-containing transforming viruses [41].

## Amino acid sequence of human FES/FPS

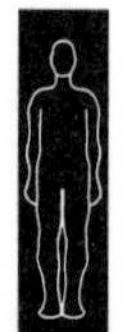

```
  1 MGFSSELCSP QGHGVLQQMQ EAELRLLEGM RKWMAQRVKS DREYAGLLHH
 51 MSLQDSGGQS RAISPDSPIS QSWAEITSQT EGLSRLLRQH AEDLNSGPLS
101 KLSLLIRERQ QLRKTYSEQW QQLQQELTKT HSQDIEKLKS QYRALARDSA
151 QAKRKYQEAS KDKDRDKAKD KYVRSLWKLF AHHNRYVLGV RAAQLHHQHH
201 HQLLLPGLLR SLQDLHEEMA CILKEILQEY LEISSLVQDE VVAIHREMAA
251 AAARIQPEAE YQGFLRQYGS APDVPPCVTF DESLLEEGEP LEPGELQLNE
301 LTVESVQHTL TSVTDELAVA TEMVFRRQEM VTQLQQELRN EEENTHPRER
351 VQLLGKRQVL QEALQGLQVA LCSQAKLQAQ QELLQTKLEH LGPGEPPPVL
401 LLQDDRHSTS SSEQEREGGR TPTLEILKSH ISGIFRPKFS LPPPLQLIPE
451 VQKPLHEQLW YHGAIPRAEV AELLVHSGDF LVRESQGKQE YVLSVLWDGL
501 PRHFIIQSLD NLYRLEGEGF PSIPLLIDHL LSTQQPLTKK SGVVLHRAVP
551 KDKWVLNHED LVLGEQIGRG NFGEVFSGRL RADNTLVAVK SCRETLPPDL
601 KAKFLQEARI LKQYSHPNIV RLIGVCTQKQ PIYIVMELVQ GGDFLTFLRT
651 EGARLRVKTL LQMVGDAAAG MEYLESKCCI HRDLAARNCL VTEKNVLKIS
701 DFGMSREEAD GVYAASGGLR QVPVKWTAPE ALNYGRYSSE SDVWSFGILL
751 WETFSLGASP YPNLSNQQTR EFVEKGGRLP CPELCPDAVF RLMEQCWAYE
801 PGQRPSFSTI YQELQSIRKR HR (822)
```

### Domain structure

| | |
|---|---|
| 567–575 and 590 | ATP binding region (Lys$^{590}$ is equivalent to SRC Lys$^{295}$) |
| 683 | Active site |
| 713 | Autophosphorylation site (equivalent to SRC Tyr$^{416}$). Lys$^{590}$ and Tyr$^{713}$ are essential for kinase and transforming activity [41,42] |

460–541 SH2 domain (underlined). FPS/FES has a group I SH2 domain (Phe or Tyr at the $\beta$D5 position) that recognizes phosphopeptides with the general motif P-Tyr-hydrophilic-hydrophilic-hydrophobic [43]

**Database accession numbers**

| | *PIR* | *SWISSPROT* | *EMBL/GENBANK* | *REFERENCES* |
|---|---|---|---|---|
| Human *FES/FPS* | A24673 | P07332 | X06292 | 2 |
| Human *FER* | | | J03358 | 44 |

***References***


1. **Hanafusa, H. (1988) In The Oncogene Handbook, Eds T. Curran, E.P. Reddy and A. Skalka, Elsevier, Amsterdam, pp. 39–57.**
2. Roebroek, A.J.M. et al. (1985) EMBO J. 4, 2897–2903.
3. Yates, K.E. (1995) Oncogene 10, 1239–1242.
4. Young, J.C. and Martin, G.S. (1984) J. Virol. 52, 913–918.
5. Feldman, R.A. et al. (1985) Proc. Natl Acad. Sci. USA 82, 2379–2383.
6. MacDonald, I. et al. (1985) Mol. Cell. Biol. 5, 2543–2551.
7. Samarut, J. et al. (1985) Mol. Cell. Biol. 5, 1067–1072.
8. Feller, S.M. and Wong, T.W. (1992) Biochemistry, 31, 3044–3051.
9. Carmier, J.F. and Samarut, J. (1986) Cell 44, 159–165.
10. Yu, G. et al. (1989) J. Biol. Chem. 264, 10276–10281.
11. Brizzi, M.F. et al. (1996) J. Biol. Chem. 271, 3562–3567.
12. Areces, L.B. et al. (1994) Mol. Cell. Biol. 14, 4606–4615.
13. Matsuda, T. et al. (1995) J. Biol. Chem. 270, 11037–11039.
14. Kamps, M.P. and Sefton, B.M. (1988) Oncogene 2, 305–315.
15. Kurata, W.E. and Lau, A.F. (1994) Oncogene 9, 329–335.
16. Moran, M.F. et al. (1990) Proc. Natl Acad. Sci. USA 87, 8622–8626.
17. Slamon, D.J. et al. (1984) Science 224, 256–262.
18. Jucker, M. et al. (1992) Oncogene 7, 943–952.
19. Hao, Q.L. et al. (1989) Mol. Cell. Biol. 9, 1587–1593.
20. Morris, C. et al. (1990) Cytogenet. Cell Genet. 53, 196–200.
21. Pawson, T. et al. (1989) Mol. Cell. Biol. 9, 5722–5725.
22. Chow, L.H. et al. (1991) Lab. Invest. 64, 457–462.
23. Greer, P. et al. (1994) Mol. Cell. Biol. 14, 6755–6763.
24. Hanafusa, T. et al. (1980) Proc. Natl Acad. Sci. USA 77, 3009–3013.
25. Sadowski, I. et al. (1988) Oncogene 2, 241–247.
26. Fukui, Y. et al. (1991) Oncogene 6, 407–411.
27. Barker, K. and Hanafusa, H. (1990) Mol. Cell. Biol. 10, 3813–3817.
28. Alexandropoulos, K. et al. (1991) J. Biol. Chem. 266, 15583–15586.
29. Spangler, R. et al. (1989) Proc. Natl Acad. Sci. USA 86, 7017–7021.
30. Feldman, R.A. et al. (1990) Oncogene Res. 5, 187–197.
31. Cogliano, A. et al. (1987) Bone 8, 299–304.
32. Birek, C. et al. (1988) Carcinogenesis 9, 1785–1791.
33. Falcone, G. et al. (1985) Proc. Natl Acad. Sci. USA 82, 426–430.
34. Kahn, P. et al. (1984) Proc. Natl Acad. Sci. USA 81, 7122–7126.
35. Pierce, J.H. and Aaronson, S.A. (1983) J. Virol. 46, 993–1002.
36. Dennis, J.W. et al. (1989) Oncogene 4, 853–860.
37. Smith, M.R. et al. (1986) Nature 320, 540–543.
38. Roebroek, A.J.M. et al. (1986) Mol. Biol. Rep. 11, 117–125.

[39] Heydemann, A. et al. (1996) Mol. Cell. Biol. 16, 1676–1686.
[40] Beemon, K. and Mattingly, B. (1986) Virology 155, 716–720.
[41] Foster, D.A. et al. (1985) Cell 42, 105–115.
[42] Weinmaster, G. et al. (1984) Cell 37, 559–568.
[43] Songyang, Z et al. (1994) Mol. Cell. Biol. 14, 2777–2785.
[44] Hao, Q.L. et al. (1989) Mol. Cell. Biol. 11, 1180–1183.

## Identification

*FRA1* and *FRA2* (*FOS*-related antigens) were detected by screening cDNA libraries with *FOS* DNA as a probe [1].

## Related genes

*FRA1* and *FRA2* are members of the helix-loop-helix/leucine zipper superfamily (see ***JUN***). Closely related genes are *FOS*, *FOSB*, Δ*FOSB* and r-*fos*.

| | ***FRA1*** | ***FRA2*** |
|---|---|---|
| **Mass (kDa): predicted** | 29 | 35 |
| **Cellular location** | Nucleus | Nucleus |

**Tissue distribution**

*FRA1* and *FRA2* expression is low in quiescent cells and induced rapidly (*FRA1* detectable after 1 h, *FRA2* after 2 h) and transiently following cell stimulation with a slight lag with respect to *FOS* [1–4]. After the initial activation period of serum-stimulated quiescent cells when FOS is the principal protein associated with JUN proteins (~3 h), FRA1 and FRA2 are the predominant partners in heterodimers with JUN proteins [5]. FRA1 is present in all layers of normal human epidermis except the basal layer: FRA2 is present in all layers of normal human epidermis but maximally in the upper spinous layer [6].

Murine *Fra-2* is expressed during late organogenesis in a pattern distinct from that of other *Fos*-related genes [7]. FRA2 shows circadian expression in the rat pineal gland [8].

## Protein function

FRA1 and FRA2 form heterodimeric transcription factors with JUN proteins (see ***FOS***). In Swiss 3T3 fibroblasts FRA1 and FRA2 are involved in asynchronous growth, rather than being required for the $G_0$ to $G_1$ transition [5]. FRA1 and FRA2 are phosphorylated *in vitro* by MAP kinase, cAMP-dependent kinase, protein kinase C and CDC2 and phosphorylation by MAP kinase increases DNA binding activity [9]

FRA2 occurs in a complex with the p50 subunit of NF-κB that binds to enhancer A of the major histocompatibility complex class I gene [10].

## Gene structure

The overall structure of the *Fra-1* and *Fra-2* genes is very similar to that of *Fos* (see ***FOS***). The chicken *Fra-2* promoter closely resembles that of *Fos* but includes two additional AP1 sites. FRA2/JUN and FOS/JUN dimers bind to this site [11].

Rat *Fra-1* is upregulated by FOS, FOSB, FRA1 or JUN, basal and AP1-regulated expression depending on sequences in the first intron comprising a consensus AP1 site two AP1-like elements [12].

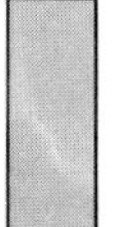

## Protein structure

The structures of FRA1 and FRA2 are closely related to each other and to that of FOS (see ***FOS***). Human FRA1 and FRA2 are 45% identical in sequence: FRA1 is 51% identical to FOS.

## Amino acid sequence of human FRA1

```
  1 MFRDFGEPGP SSGNGGGYGG PAQPPAAAQA AQQKFHLVPS INTMSGSQEL
 51 QWMVQPHFLG PSSYPRPLTY PQYSPPQPRP GVIRALGPPP GVRRRPCEQI
101 SPEEEERRRV RRERNKLAAA KCRNRRKELT DFLQAETDKL EDEKSGLQRE
151 IEELQKQKER LELVLEAHRP ICKIPEGAKE GDTGSTSGTS SPPAPCRPVP
201 CISLSPGPVL EPEALHTPTL MTTPSLTPFT PSLVFTYPST PEPCASAHRK
251 SSSSSGDPSS DPLGSPTLLA L (271)
```

**Domain structure**

111–129 Basic, DNA binding motif
133–161 Leucine zipper (underlined)

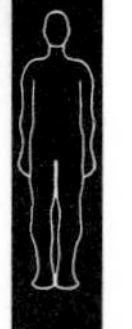

## Amino acid sequence of human FRA2

```
  1 MYQDYPGNFD TSSRGSSGSP AHAESYSSGG GGQQKFRVDM PGSGSAFIPT
 51 INAITTSQDL QWMVQPTVIT SMSNPYPRSH PYSPLPGLAS VPGHMALPRP
101 GVIKTIGTTV GRRRRDEQLS PEEEEKRRIR RERNKLAAAK CRNRRRELTE
151 KLQAETEELE EEKSGLQKEI AELQKEKEKL EFMLVAHGPV CKISPEERRS
201 PPAPGLQPMR SGGGSVGAVV VKQEPLEEDS PSSSSAGLDK AQRSVIKPIS
251 IAGGFYGEEP LHTPIVVTST PAVTPGTSNL VFTYPSVLEQ ESPASPSESC
301 SKAHRRSSSS GDQSSDSLNS PTLLAL (326)
```

**Domain structure**

130–148 Basic, DNA binding motif
152–180 Leucine zipper (underlined)

**Database accession numbers**

| | *PIR* | *SWISSPROT* | *EMBL/GENBANK* | *REFERENCES* |
|---|---|---|---|---|
| Human FRA1 | S08010, S15750 | P15407 | X16707 | 1 |
| Human FRA2 | S15749 | P15408 | X16706 | 1 |

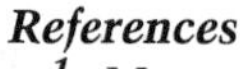

### *References*


1. Matsui, M. et al. (1990) Oncogene 5, 249–255.
2. Cohen, D.R. and Curran, T. (1988) Mol. Cell. Biol. 8, 2063–2069.
3. Nishina, H. et al. (1990) Proc. Natl Acad. Sci. USA 87, 3619–3623.
4. Yoshida, T. et al. (1993) Nucleic Acids Res. 21, 2715–2721.
5. Kovary, K. and Bravo, R. (1992) Mol. Cell. Biol. 12, 5015–5023.
6. Welter, J.F. and Eckert, R.L. (1995) Oncogene 11, 2681–2687.
7. Carrasco, D. and Bravo, R. (1995) Oncogene 10, 1069–1079.
8. Baler, R. and Klein, D.C. (1995) J. Biol. Chem. 270, 27319–27325.
9. Gruda, M.C. et al. (1994) Oncogene 9, 2537–2547.
10. Giuliani, C. et al. (1995) J. Biol. Chem. 270, 11453–11462.
11. Sonobe, M.H. et al. (1995) Oncogene 10, 689–696.
12. Bergers, G. et al. (1995) Mol. Cell. Biol. 15, 3748–3758.

## Identification

*FYN* (formerly *SYN* (src/yes-related novel gene) or *SLK* (src-like kinase)) was originally cloned from a SV40-transformed human fibroblast library [1,2]. There is no known naturally occurring *fyn*-containing retrovirus.

## Related genes

*FYN* is a member of the *SRC* tyrosine kinase family (*Blk, FGR, FYN, HCK, LCK/Tkl, LYN, SRC, YES*). FRK/RAK (FYN-related kinase) and its rat homologue Gtk (gastrointestinal-associated tyrosine kinase) comprise a subfamily of FYN-related proteins [3].

| | *FYN* |
|---|---|
| **Nucleotides** | Not fully mapped |
| **Chromosome** | 6q21 |
| **Mass (kDa): predicted** | 60 |
| **expressed** | 59 |
| **Cellular location** | Plasma membrane |

**Tissue distribution**
$p59^{fyn(B)}$ is mainly expressed in the brain and spinal cord but is detectable in most cell types other than epithelial cells [4,5]. $p59^{fyn(T)}$ is mainly expressed in T cells. An additional form, $p72^{fyn}$, has been detected in transformed T cells and in *in vitro* translation systems [6,7]. *FYN* expression is also activated in HL60 cells stimulated to differentiate by TPA [8] and in natural killer cells [9].

## Protein function

*FYN* encodes a tyrosine kinase implicated in the control of cell growth. Over-expression of $p59^{fyn(T)}$ enhances IL-2 secretion and DNA synthesis in response to anti-CD3 antibody. FYN expression is activated by stimulation of the TCR [10] and both FYN and the ZAP-70 tyrosine kinase associate with the tyrosine-based activation motif (TAM: 20–25 amino acids characterized by two YXXL/I motifs) in the CD3 $\varepsilon$ chain of the TCR [11,12], leading to the phosphorylation and activation of PtdIns-PLC$\gamma_1$ and subsequent cellular responses. Activation of the T cell receptor in Jurkat cells causes FYN to associate with phosphatidylinositol 3′-kinase and CBL.

The CD45 tyrosine phosphatase regulates the tyrosine kinase activity of FYN [13]. In $CD45^-$ cells, the TCR is uncoupled from protein tyrosine phosphorylation, PLC$\gamma_1$ regulation, inositol phosphates accumulation, $[Ca^{2+}]_i$ responses, diacylglycerol production and protein kinase C activation, suggesting that FYN plays a critical role in coupling the activated TCR to these responses. $p59^{fyn(T)}$ promotes greater increases in $[Ca^{2+}]_i$ in T cells than $p59^{fyn(B)}$ and this arises from the difference in the catalytic domain caused by the 52 amino acid region of SH2 and SH1 that differs between the two isoforms [14]. In T cells, FYN activates the *FOS* and *IL-2* promoters: both these activities are inhibited by the tyrosine kinase CSK [15]. Expression of the TCR appears to be regulated by the Ly-6A antigen and Ly-6A may also regulate FYN kinase activity [16].

In activated B cells, FYN and LCK and LYN all associate with membrane Ig and with CD20. However, FYN(T) also binds *via* its SH2 domain to a set of phosphoproteins distinct from those associating with LYN [17]. In pro-B cells IL-2 induces activation of FYN and its association with the IL-2R $\beta$ chain [18]. FYN also associates with CD2 and CD5 in T cells, with GAP in thrombin stimulated platelets, with $p125^{FAK}$ in chicken cells, with the PDGF receptor in PDGF-stimulated fibroblasts [19–23] and with the EPH family receptor tyrosine kinase SEK in keratinocytes [24].

FYN is expressed in oligodendrocytes and during the initial stages of myelination it is activated by associating with the large myelin-associated glycoprotein (MAG). Its importance for myelination is indicated by impairment of this process in *Fyn*-deficient mice [25].

The FYN SH3 domain binds specifically to the sequence XXXRPLPP(I/L)-PXX [26]. The SH3 domain of FYN (and also of LYN and HCK) interact with the B cell-specific Bruton's tyrosine kinase, BTK, which is probably a substrate for SRC family kinases [27].

## Cancer

*FYN* is over-expressed in some human tumour cell lines [1,2,4].

## Transgenic animals

Mice that do not express FYN show no overt phenotype but increases in thymocyte $[Ca^{2+}]_i$ and proliferation stimulated by activating the TCR or by Thy-1 are markedly reduced, whereas the proliferative response of peripheral T cells remains essentially unaltered [28,29]. However, disruption of *Fyn* (by the insertion of the $\beta$-galactosidase gene) has also been reported to give rise to homozygous *Fyn*-mutant neonates that die from a suckling problem [30]. T cells from $Fyn^{-/-}$ mice have elevated lifespans and reduced susceptibility to apoptosis compared with normal T cells, being less sensitive to activation of the Fas (Apo1/CD95) receptor with which $p59^{fynT}$ associates [31].

The over-expression of $p59^{fyn(T)}$ in transgenic mice gives rise to T cells that are hyperstimulable, showing greatly enhanced phosphotyrosine accumulation and a two-fold increase in $[Ca^{2+}]_i$ on activation by Con A or *via* CD3 or Thy-1 [32].

## In animals

Chickens inoculated with a recombinant avian retrovirus expressing *fyn* develop fibrosarcomas that may arise from mutations in the kinase or SH2 domains of FYN [33].

## *In vitro*

Over-expression of normal *Fyn* transforms NIH 3T3 cells [34]. Substitution of the N-terminal two-thirds of FYN by the corresponding v-FGR region activates the FYN tyrosine kinase and transforming potential [1]. *Fyn* is activated during the differentiation of murine keratinocytes, either by TPA or by calcium[35].

Over-expression in mouse T cells of FYN mutants with an activated kinase stimulates the guanine nucleotide exchange activity (mSOS) and complexes of FYN/SHC and SHC/GRB2 form in FYN-transformed cells[36]. Activation of the TCR also stimulates mSOS activity, indicating that this activation pathway for RAS may occur in response to activation of both FYN and the TCR.

In CHO cells stimulation by insulin promotes complex formation between the SH2 domain of $p59^{fyn}$ and the IRS-1 protein that mediates downstream signalling events from the activated insulin receptor[37].

In polyoma-transformed cells, FYN forms complexes with middle T antigen (as do SRC and YES), an interaction mediated by the C-terminus of FYN[1,38,39].

## Gene structure

*FYN* shares the common 12 exon organization of the *SRC* family.

## Protein structure

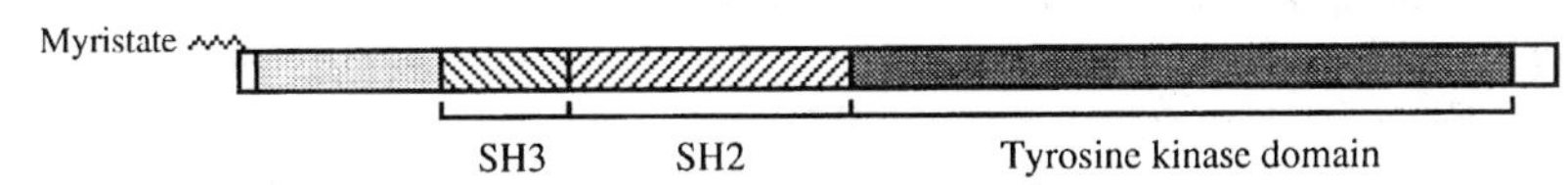


The two *Fyn* mRNAs (brain and thymus forms) arise by mutually exclusive splicing of alternative seventh exons: their products differ by 27 of the 55 amino acids encoded by the exon which is positioned at the beginning of the kinase domain and includes the presumptive nucleotide binding site[40].

Deletion of either the SH2 or the SH3 domain generates potent transforming agents, in contrast to the effects of the corresponding mutations in FYN[41].

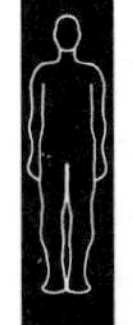

## Amino acid sequence of human FYN

```
  1 MGCVQCKDKE ATKLTEERDG SLNQSSGYRY GTDPTPQHYP SFGVTSIPNY
 51 NNFHAAGGQG LTVFGGVNSS SHTGTLRTRG GTGVTLFVAL YDYEARTEDD
101 LSFHKGEKFQ ILNSSEGDWW EARSLTTGET GYIPSNYVAP VDSIQAEEWY
151 FGKLGRKDAE RQLLSFGNPR GTFLIRESET TKGAYSLSIR DWDDMKGDHV
201 KHYKIRKLDN GGYYITTRAQ FETLQQLVQH YSERAAGLCC RLVVPCHKGM
251 PRLTDLSVKT KDVWEIPRES LQLIKRLGNG QFGEVWMGTW NGNTKVAIKT
301 LKPGTMSPES FLEEAQIMKK LKHDKLVQLY AVVSEEPIYI VTEYMNKGSL
351 LDFLKDGEGR ALKLPNLVDM AAQVAAGMAY IERMNYIHRD LRSANILVGN
401 GLICKIADFG LARLIEDNEY TARQGAKFPI KWTAPEAALY GRFTIKSDVW
451 SFGILLTELV TKGRVPYPGM NNREVLEQVE RGYRMPCPQD CPISLHELMI
501 HCWKKDPEER PTFEYLQSFL EDYFTATEPQ YQPGENL (537)
```

**Domain structure**

| | |
|---|---|
| 2 | Myristate attachment site |
| 12 | Protein kinase C phosphorylation site |
| 149–247 | SH2 domain |
| 233–287 | Region corresponding to exon 7 (italics): corresponding sequence of FYN(T): EKADGLCFNL TVVSSSCTPQ TSGLAKDAWE VARDSLFLEK KLGQGCFAEV WL |

| | |
|---|---|
| 277–285 and 299 | ATP binding site |
| 390 | Active site |
| 420 | Tyrosine autophosphorylation site |
| 531 | Phosphorylation site. Mutation of $Tyr^{531}$ to Phe does not activate the transforming capacity, in contrast to the effect of the equivalent mutation in SRC |
| 87–139 | SH3 domain (or A box: underlined). The crystal structure of this domain has been determined at 1.9 Å resolution [42]. |

Amino acids 83–537 are 80% homologous to YES and 77% to SRC and FGR. The C-terminal 191 amino acids are 86% homologous to SRC: the N-terminal 82 residues are 6% homologous.

**Database accession numbers**

| | *PIR* | *SWISSPROT* | *EMBL/GENBANK* | *REFERENCES* |
|---|---|---|---|---|
| Human *FYN* | A24314, A25389 | P06241 | M14333 | 2 |
| | | | M14676 | 1,43 |

***References***


1 Kawakami, T. et al. (1986) Mol. Cell. Biol. 6, 4195–4201.
2 Semba, K. et al. (1986) Proc. Natl Acad. Sci. USA 83, 5459–5463.
3 Sunitha, I. and Avigan, M.I. (1996) Oncogene 13, 547–559.
4 Kypta, R.M. et al. (1988) EMBO J. 7, 3837–3844.
5 Yagi, T. et al. (1993) Oncogene 8, 3343–3351.
6 Espino, P.C. et al. (1992) Oncogene 7, 317–322.
7 Da Silva, A.J. and Rudd, C.E. (1993) J. Biol. Chem. 268, 16537–16543.
8 Katagiri, K. et al. (1991) J. Immunol. 146, 701–707.
9 Biondi, A. et al. (1991) Eur. J. Immunol. 21, 843–846.
10 Katagiri, K. et al. (1989) Proc. Natl Acad. Sci. USA 86, 10064–10068.
11 Samelson, L.E. et al. (1990) Proc. Natl Acad. Sci. USA 87, 4358–4362.
12 Gauen, L.K.T. et al. (1994) Mol. Cell. Biol. 14, 3729–3741.
13 Shiroo, M. et al. (1992) EMBO J. 11, 4887–4897.
14 Davidson, D. et al. (1994) Mol. Cell. Biol. 14, 4554–4564.
15 Takeuchi, M. et al. 1994) J. Biol. Chem. 268, 27413–27419.
16 Lee, S.-K. et al. (1994) EMBO J. 13, 2167–2176.
17 Malek, S.N. and Desiderio, S. (1993) J. Biol. Chem. 268, 22557–22565.
18 Kobayashi, N. et al. (1993) Proc. Natl Acad. Sci. USA 90, 4201–4205.
19 Bell, G.M. et al. (1992) Mol. Cell. Biol. 12, 5548–5554.
20 Burgess, K.E. et al. (1992) Proc. Natl Acad. Sci. USA 89, 9311–9315.
21 Cichowski, K. et al. (1992) J. Biol. Chem. 267, 5025–5028.
22 Cobb, B.S. et al. (1994) Mol. Cell. Biol. 14, 147–155.
23 Twamley, G.M. et al. (1992) Oncogene 7, 1893–1901.
24 Ellis, C. et al. (1996) Oncogene 12, 1727–1736.
25 Umemori, H. et al. (1994) Nature 367, 572–576.
26 Rickles, R.J. et al. (1994) EMBO J. 13, 5598–5604.
27 Afar, D.E.H. et al. (1996) Mol. Cell. Biol. 16, 3465–3471.
28 Appleby, M.W. et al. (1992) Cell 70, 751–763.
29 Stein, P.L. et al. (1992) Cell 70, 741–750.
30 Yagi, T. et al. (1993) Nature 366, 742–745.
31 Atkinson, E.A. et al. (1996) J. Biol. Chem. 271, 5968–5971.

[32] Cooke, M.P. et al. (1991) Cell 65, 281–291.
[33] Semba, K. et al. (1990) Mol. Cell. Biol. 10, 3095–3104.
[34] Kawakami, T. et al. (1988) Proc. Natl Acad. Sci. USA 85, 3870–3874.
[35] Calautti, E. et al. (1995) Genes Dev. 9, 2279–2291.
[36] Li, B. et al. (1996) Proc. Natl Acad. Sci. USA 93, 1001–1005.
[37] Sun, X.J. et al. (1996) J. Biol. Chem. 271, 10583–10587.
[38] Cheng, S.H. et al. (1991) J. Virol. 65, 170–179.
[39] Pleiman, C.M. et al. (1993) Mol. Cell. Biol. 13, 5877–5887.
[40] Cooke, M.P. and Perlmutter, R.M. (1989) New Biol. 1, 66–74.
[41] Rivero-Lezcano, O.M. et al. (1995) Oncogene 11, 2675–2679.
[42] Noble, M.E.M. et al. (1993) EMBO J. 12, 2617–2624.
[43] Peters, D.J. et al. (1990) Oncogene 5, 1313–1319.

## Identification

*HCK* (haematopoietic cell kinase) encodes an *SRC*-family protein tyrosine kinase detected in human cells with v-*src* and murine *Lck* probes [1–3].

## Related genes

*HCK* is a member of the *SRC* tyrosine kinase family (*Blk, FGR, FYN, HCK, LCK/Tkl, LYN, SRC, YES*). *Hck-2* and *Hck-3* are the murine homologues of *LYN* and *LCK*, respectively.

| | ***HCK*** |
|---|---|
| **Nucleotides (kb)** | >30 |
| **Chromosome** | 20q11 |
| **Mass (kDa): predicted** | 57 |
| **expressed** | 56/59 |

**Cellular location**
Membrane associated: p56 and p59. Cytoplasm: p59 alone.

**Tissue distribution**
Expression of *HCK* is restricted to haematopoietic cells, predominantly of the myeloid and B-lymphoid lineages [2,3]. *HCK* is expressed when acute myeloid leukaemic cells are induced to differentiate *in vitro* to cells with monocytic characteristics [4]. HCK is also expressed in undifferentiated embryonic stem cells, expression declining when the cells are allowed to differentiate by withdrawing leukaemia inhibitory factor (LIF) [5].

## Protein function

The expression pattern suggests that HCK, in common with other members of the SRC family, has distinct cell lineage-specific functions and in particular may be involved in the differentiation of monocytes and granulocytes. LIF causes rapid, transient enhancement of HCK tyrosine kinase activity and HCK is associated with gp130, a signal transducing component of the LIF receptor [5].

HCK associates *in vitro via* its SH3 domain with the RAS-GTPase-activating protein GAP and GAP is phosphorylated by HCK [6].

HIV and simian immunodeficiency virus NEF proteins bind to the SH3 domains of HCK (and LYN) in a process that appears to be required for enhanced growth of Nef$^+$ viruses [7].

**Cancer**
*HCK* expression is detectable in several human leukaemia cell lines [2,3,8].

**Transgenic animals**
Transgenic male mice that are hemizygous for the *Hck* transgene are sterile, indicating that the gene may be important in spermatogenesis [9]. *Hck*-deficient

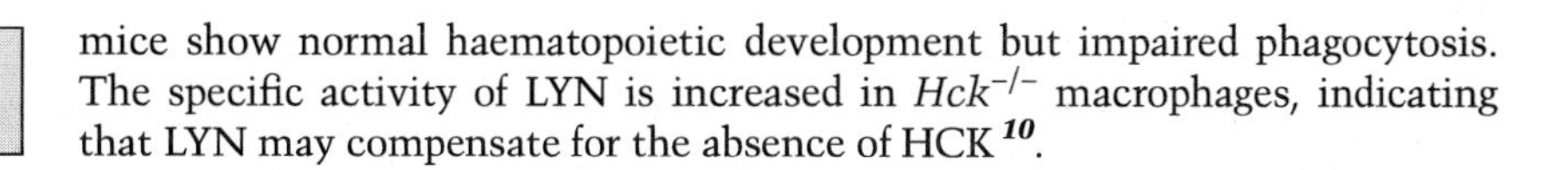


mice show normal haematopoietic development but impaired phagocytosis. The specific activity of LYN is increased in $Hck^{-/-}$ macrophages, indicating that LYN may compensate for the absence of HCK [10].

## Gene structure

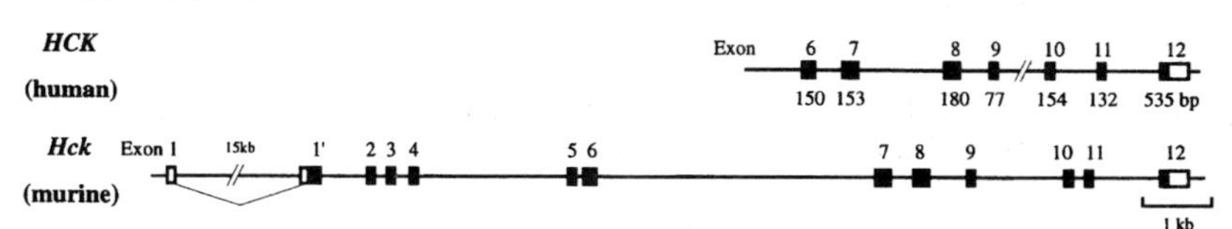


Human *HCK* exons 6–12 are almost identical in size to the corresponding exons of *FGR*, *LCK* and *SRC* [11].

## Transcriptional regulation

The murine *Hck* promoter has no TATA or CAAT elements but contains GC-rich regions, three SP1 and two AP2 consensus binding sites and an LPS-responsive element [12,13]. p59 is generated by translation from a CTG codon 5′ of the ATG used to generate p56 [14].

## Protein structure

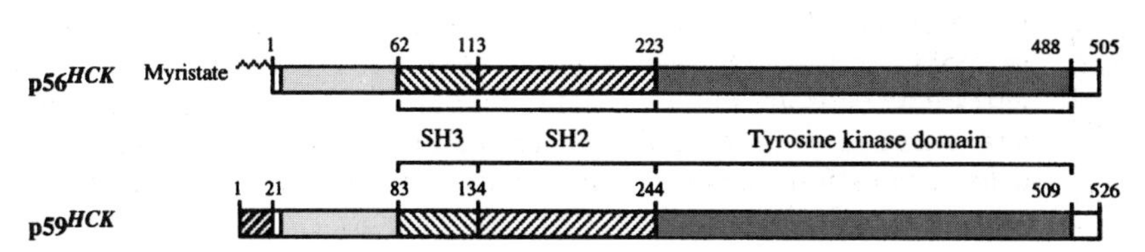


## Amino acid sequence of human HCK

```
  1 MGSMKSKFLQ VGGNTFSKTE TSASPHCPVY VPDPTSTIKP GPNSHNSNTP
 51 GIREAGSEDI IVVALYDYEA IHHEDLSFQK GDQMVVLEES GEWWKARSLA
101 TRKEGYIPSN YVARVDSLET EEWFFKGISR KDAERQLLAP GNMLGSFMIR
151 DSETTKGSYS LSVRDYDPRQ GDTVKHYKIR TLDNGGFYIS PRSTFSTLQE
201 LVDHYKKGND GLCQKLSVPC MSSKPQKPWE KDAWEIPRES LKLEKKLGAG
251 QFGEVWMATY NKHTKVAVKT MKPGSMSVEA FLAEANVMKT LQHDKLVKLH
301 AVVTKEPIYI ITEFMAKGSL LDFLKSDEGS KQPLPKLIDF SAQIAEGMAF
351 IEQRNYIHRD LRAANILVSA SLVCKIADFG LARVIEDNEY TAREGAKFPI
401 KWTAPEAINF GSFTIKSDVW SFGILLMEIV TYGRIPYPGM SNPEVIRALE
451 RGYRMPRPEN CPEELYNIMM RCWKNRPEER PTFEYIQSVL DDFYTATESQ
501 YQQQP (505)
```

### Domain structure

| | |
|---|---|
| −21–1 | 21 amino acid N-terminal extension present in the p59 isoform (MGGRSSCEDPGCPRDEERAPR) |
| 247–255 and 269 | ATP binding site |
| 360 | Active site |
| 390 | Autophosphorylation site |

| | |
|---|---|
| 62–113 | SH3 domain (A box: underlined) |
| 114–223 | SH2 domain (italics) |
| 224–488 | Catalytic domain |

**Database accession numbers**

| | *PIR* | *SWISSPROT* | *EMBL/GENBANK* | *REFERENCES* |
|---|---|---|---|---|
| Human *HCK* | A27812 | P08631 | M16591, M16592 | 2,3 |
| Human *HCK* | | | X59741, X59742, X59743 | 11 |

***References***


[1] Klemsz, M.J. et al. (1987) Nucleic Acids Res. 15, 9600.
[2] Quintrell, N. et al. (1987) Mol. Cell. Biol. 7, 2267–2275.
[3] Ziegler, S.F. et al. (1987) Mol. Cell. Biol. 7, 2276–2285.
[4] Willman, C.L. et al. (1991) Blood 77, 726–734.
[5] Ernst, M. et al. (1994). EMBO J. 13, 1574–1584.
[6] Briggs, S.D. et al. (1995) J. Biol. Chem. 270, 14718–14724.
[7] Lee, C.-H. et al. (1995) EMBO J. 14, 5006–5015.
[8] Perlmutter, R.M. et al. (1988) Biochim. Biophys. Acta 948, 245–262.
[9] Magram, J. and Bishop, J.M. (1991) Proc. Natl Acad. Sci. USA 88, 10327–10331.
[10] Lowell, C.A. et al. (1994) Genes Dev. 8, 387–398.
[11] Hradetzky, D. et al. (1992) Gene, 113, 275–280.
[12] Ziegler, S.F. et al. (1991) Oncogene 6, 283–288.
[13] Lichtenberg, U. et al. (1992) Oncogene 7, 849–858.
[14] Lock, P. et al. (1991) Mol. Cell. Biol. 11, 4363–4370.

# HER2/ErbB-2/Neu

## Identification

*Neu* was first identified by transfection with DNA from neuroglioblastomas that had arisen in rats exposed *in utero* to ethylnitrosourea [1]. *HER2* was identified by screening genomic and cDNA libraries [2–4].

## Related genes

*HER2/ErbB-2* are members of the *EGFR* family (*EGFR*, *HER3* and *HER4*). HER2 is ~80% homologous to EGFR and 80% identical to rat NEU.

| | ***HER2*** |
|---|---|
| **Nucleotides (kb)** | Not fully mapped |
| **Chromosome** | 17q21–q22 |
| **Mass (kDa): predicted** | 138 |
| **expressed** | gp185<br>100 (HER2 ECD) |
| **Cellular location** | Plasma membrane<br>HER2 ECD is a perinuclear cytoplasmic protein |

**Tissue distribution**

*HER2/ErbB-2/Neu* are widely distributed. The pattern of expression is closely similar to that of *EGFR* [1,2], although high levels of *HER2* mRNA occur in melanocytes in which *EGFR* is undetectable. *HER2 ECD* is an alternatively processed form of *HER2* produced in some human breast carcinoma cell lines [5]. HER2 protein is detectable throughout the myofibre membrane of rat skeletal muscle [6].

## Protein function

HER2 is a receptor-like tyrosine kinase activated by differentially spliced neuregulins, also called heregulins or neu differentiation factors (NRGs or NDFs: they include gp30, heregulin (HRG), NEU differentiation factor (NDF, the rat homologue of HRG), glial growth factor, sensory and motor neurone-derived factor and acetylcholine receptor-inducing factor [7]) or by NEU protein-specific activating factor (NAF) [8]. HER2 can form heterodimers with EGFR, HER3 or HER4 [9–11]. HER2 *trans*-phosphorylates HER3 but the converse does not occur [12]. Activation of HER2 phosphorylation by NDF/HRG requires the coexpression of either HER3 or HER4. In mammary carcinoma cells coexpression of HER3 appears essential [13]. A 30 kDa glycoprotein secreted from human breast cancer cells and a 25 kDa peptide secreted by activated macrophages also bind to HER2 [14].

Ligand binding promotes tyrosine phosphorylation of HER2 and activated HER2 phosphorylates and binds to SHC proteins and GRB2/SOS, as does activated EGFR, causing stimulation of the RAS-GTP pathway [15]. Other signalling proteins activated by ligand binding include phospholipase Cγ, phosphatidylinositol 3-kinase and PtdIns 4-kinase. HRD/NDF increases acetylcholine receptor α subunit mRNA levels in myotubes *via* tyrosine phosphorylation of both HER3 and HER2 [6].

**Heterodimerization in EGFR family signalling**

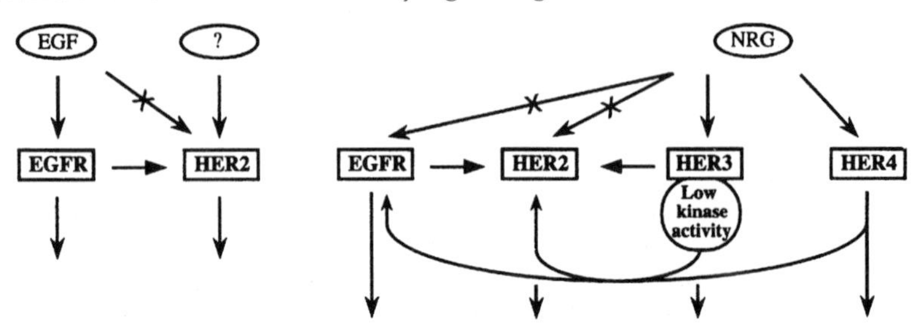


Oncogenic activation of *Neu* (by point mutation, over-expression or by truncation of non-catalytic sequences) results in its constitutive phosphorylation and in the tyrosine phosphorylation of PtdIns$P_2$-specific phospholipase Cγ, which is permanently associated with activated NEU [16]. Activated NEU also phosphorylates SHC [17], causes accumulation of GTP-RAS *via* GRB2/SOS and phosphorylation of GAP [18,19], activates PtdIns 3-kinase and causes phosphorylation of chromatin-associated p33 [20]. ETS, AP1 and NF-κB reporter genes are also activated by oncogenic NEU [21].

The truncated version of HER2, HER2 ECD, suppresses the growth-inhibitory effects of antibodies directed against HER2 [5].

TPA stimulates serine/threonine phosphorylation of normal and oncogenic NEU and inhibits the tyrosine kinase activity of oncogenic NEU: thus, like the EGFR oncogenic NEU may be negatively regulated by protein kinase C [22]. Kinase-deficient HER2 can suppress EGFR function [23].

**Proteins interacting with HER2**

| *Protein* | *Physical interaction* | *Tyrosine phosphorylation* | *Enzymatic activity* |
|---|---|---|---|
| NDF/HRG | + | + | + (+HER4 or HER3) |
| NAF | + | + | + |
| ASGP-2 | + | ? | + |
| AchR-inducing factor | + | | |
| EGFR | + | + | + |
| HER3 | + | + | + |
| HER4 | + | + | + |
| Oestrogen receptor | ? | + | + (activates progesterone receptor gene) |
| PLCγ | + | + | N.D. |
| GAP | + | + | N.D. |
| PtdIns 3-kinase | + | ? | + |
| PtdIns 4-kinase | + | N.D. | + |
| SHC | + | + | N.D. |
| GRB2/SOS | + | + | + |
| GRB7 | + | − | − |
| GRP94 | + | ? | |

| Protein | Physical interaction | Tyrosine phosphorylation | Enzymatic activity |
|---|---|---|---|
| SRC | + | N.D. | + |
| PTP 1C | + | N.D. | + |
| PTP 1D | + | N.D. | – |
| p56 | + | + | N.D. |
| p30 | + | ? | ? |
| p25 | + | ? | ? |

## Cancer

Unlike the rat *Neu* oncogene, *HER2* does not have an activated mutant form. However, *HER2* is over-expressed with high frequency in breast, stomach, ovarian and bladder cancers[24–26]. Increased expression of HER2 (and of EGFR, HER3 and HER4) has been detected in papillary thyroid carcinomas and its over-expression or amplification has been detected in a variety of other cancers[27–30]. In lymph node positive or node negative breast cancer patients there is a strong correlation between *HER2* amplification and poor prognosis[31] although the evidence that *HER2* promotes metastasis is controversial[32]. The extracellular region is proteolytically released from COS-7 cells in a form that activates $p185^{neu}$ and is similarly released as a 110 kDa protein into the sera of advanced-stage breast carcinoma patients[33]. Breast cancer cell lines that over-express *HER2* have GRB2/SOS complexes associated with HER2 and increased MAP kinase activity[34]. Breast tumours and breast cancer cell lines that over-express *HER2* also have a high frequency of over-expression of the SH2 domain protein GRB7, which binds tightly to tyrosine phosphorylated HER2[35]. The glucose-regulated chaperone protein GRP94 associates stably with HER2, and the complex is disrupted by geldanamycin[36].

There is an inverse correlation between *HER2* over-expression and oestrogen receptor (ER) expression in invasive cancer. The introduction of *HER2* cDNA into breast cancer cells expressing low levels of HER2 results in down-regulation of ER and oestrogen-independent growth that is insensitive to tamoxifen[37]. Activation of HER2 by heregulin (HRG) causes rapid tyrosine phosphorylation of the ER and interaction with oestrogen response elements in DNA followed by progesterone receptor synthesis. Thus oestrogen resistance may arise from long-term suppression of the ER by HER2-mediated pathways.

The administration of monoclonal anti-HER2 antibody to patients with HER2-overexpressing metastatic breast cancer has been shown to give varying degrees of remission in ~11% of cases[38].

## Transgenic animals

Transgenic mice carrying an activated *Neu* oncogene develop mammary adenocarcinomas and, occasionally, other tumours[19,39,40]. Double transgenic animals expressing both Neu and TGFα in mammary epithelium develop multifocal tumours with a rapidity exceeding that in mice expressing either one of these genes alone[41].

**In animals**

The expression of oncogenic NEU in mouse mammary epithelium can cause ductal carcinoma but may also induce epithelial abnormalities resembling those of human sclerosing adenosis and atypical hyperplasia that may be precursors of ductal carcinoma [42].

***In vitro***

Over-expression of either human *HER2* or rat *Neu* transforms NIH 3T3 fibroblasts and *Neu*-induced transformation can occur independently of the EGF receptor [43]. Transformation of NIH 3T3 fibroblasts by oncogenic *Neu* causes the cells to exhibit metastatic properties both *in vitro* and *in vivo* [44] that may be associated with the over-expression of CD44 [45]. MYC or pRb can suppress *Neu* transformation and adenovirus E1A protein suppresses *Neu*-induced metastasis of NIH 3T3 cells, although mechanisms additional to the suppression of *Neu* transcription are involved [46]. Nevertheless, liposome-mediated E1A gene transfer suppresses ovarian cancer cell tumorigenicity in nude mice [47]. Monoclonal antibodies directed against HER2 specifically inhibit the proliferation of human breast carcinoma cells [48]. The activation of HER3, HER4 and HER2 (*via* heterodimer formation) by NDF/HRG or by antibodies to HER2 stimulate expression of p53 and of the p53-inducible *WAF1* gene, hence the mechanism of HER2-mediated growth arrest of breast cancer cells appears to be through a p53-mediated pathway [49,50].

Microvilli of rat mammary adenocarcinoma cells contain actin-associated large glycoprotein complexes, designated signal transduction particles (STPs), which also include $p185^{neu}$, SRC, ABL and $p58^{gag}$, a retroviral-like cellular protein [51]. In NIH 3T3 cells ERBB-2 interacts with a 45 kDa protein (TOB: transducer of ErbB-2) to regulate negatively the growth suppressive effect of TOB [52].

The human mammary epithelial cell line MTSV1–7 expresses low levels of E-cadherin and $\alpha_2$ integrin: blockade of signalling from HER2 by antibody 4D5 restores expression of these genes, suggesting that HER2 may modulate metastatic events [53].

## Gene structure

There is strict conservation of exon–intron boundaries in the regions encoding the catalytic domains of HER2 and EGFR (see ***EGFR/ErbB-1***). The *HER2* promoter is similar to that of EGFR, having a high GC content, three SP1 binding sites (*HER2* only, one of high affinity) and multiple transcription initiation sites. *Neu* has no TATA box whereas *HER2* contains both TATA and CAAT boxes. A 100 bp region 5′ to the TATA box enhances transcription 200-fold *via* the binding of a heterodimeric nuclear binding protein (PBP) recognizing the core sequence TGGGAG [54]. *HER2* also has an ETS-binding consensus 38 bases downstream from the CAAT box [55]. In cells over-expressing HER2, promoter activity is enhanced as a result of the increased abundance of the transcription factor OB2-1 (comprised of AP2$\alpha$, AP2$\beta$ and AP2$\gamma$) [56]. Triplex formation (i.e. binding of a third strand of DNA or RNA) appears to downregulate selectively expression of HER2 [31].

## Protein structure

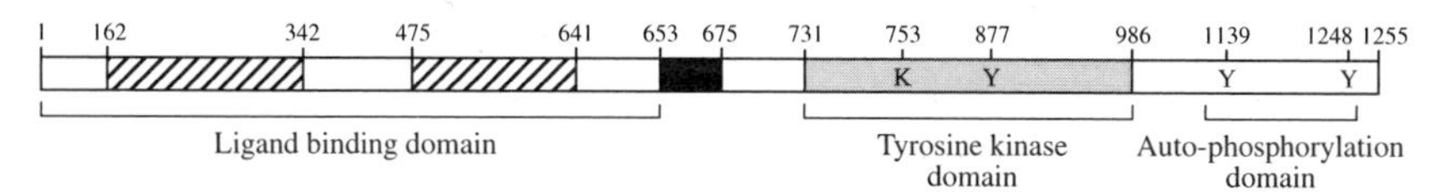


The dominant transforming *Neu* oncogene isolated from rat neuroblastoma DNA was oncogenically activated by a single point mutation (Val$^{664}$ to Glu) in the transmembrane domain[57]. The Glu$^{664}$ mutation confers high-affinity ligand binding on the receptor[58] and enhances tyrosine kinase activity and autophosphorylation of Tyr$^{1248}$ (the major autophosphorylation site). Mutation of Tyr$^{1248}$ to Phe lowers tyrosine kinase and transforming activities. Thus Tyr$^{1248}$ negatively regulates transformation, the effect being blocked by phosphorylation[59].

## Amino acid sequence of human HER2

```
   1 MELAALCRWG LLLALLPPGA ASTQVCTGTD MKLRLPASPE THLDMLRHLY
  51 QGCQVVQGNL ELTYLPTNAS LSFLQDIQEV QGYVLIAHNQ VRQVPLQRLR
 101 IVRGTQLFED NYALAVLDNG DPLNNTTPVT GASPGGLREL QLRSLTEILK
 151 GGVLIQRNPQ LCYQDTILWK DIFHKNNQLA LTLIDTNRSR ACHPCSPMCK
 201 GSRCWGESSE DCQSLTRTVC AGGCARCKGP LPTDCCHEQC AAGCTGPKHS
 251 DCLACLHFNH SGICELHCPA LVTYNTDTFE SMPNPEGRYT FGASCVTACP
 301 YNYLSTDVGS CTLVCPLHNQ EVTAEDGTQR CEKCSKPCAR VCYGLGMEHL
 351 REVRAVTSAN IQEFAGCKKI FGSLAFLPES FDGDPASNTA PLQPEQLQVF
 401 ETLEEITGYL YISAWPDSLP DLSVFQNLQV IRGRILHNGA YSLTLQGLGI
 451 SWLGLRSLRE LGSGLALIHH NTHLCFVHTV PWDQLFRNPH QALLHTANRP
 501 EDECVGEGLA CHQLCARGHC WGPGPTQCVN CSQFLRGQEC VEECRVLQGL
 551 PREYVNARHC LPCHPECQPQ NGSVTCFGPE ADQCVACAHY KDPPFCVARC
 601 PSGVKPDLSY MPIWKFPDEE GACQPCPINC THSCVDLDDK GCPAEQRASP
 651 LTSIISAVVG ILLVVVLGVV FGILIKRRQQ KIRKYTMRRL LQETELVEPL
 701 TPSGAMPNQA QMRILKETEL RKVKVLGSGA FGTVYKGIWI PDGENVKIPV
 751 AIKVLRENTS PKANKEILDE AYVMAGVGSP YVSRLLGICL TSTVQLVTQL
 801 MPYGCLLDHV RENRGRLGSQ DLLNWCMQIA KGMSYLEDVR LVHRDLAARN
 851 VLVKSPNHVK ITDFGLARLL DIDETEYHAD GGKVPIKWMA LESILRRRFT
 901 HQSDVWSYGV TVWELMTFGA KPYDGIPARE IPDLLEKGER LPQPPICTID
 951 VYMIMVKCWM IDSECRPRFR ELVSEFSRMA RDPQRFVVIQ NEDLGPASPL
1001 DSTFYRSLLE DDDMGDLVDA EEYLVPQQGF FCPDPAPGAG GMVHHRHRSS
1051 STRSGGGDLT LGLEPSEEEA PRSPLAPSEG AGSDVFDGDL GMGAAKGLQS
1101 LPTHDPSPLQ RYSEDPTVPL PSETDGYVAP LTCSPQPEYV NQPDVRPQPP
1151 SPREGPLPAA RPAGATLERP KTLSPGKNGV VKDVFAFGGA VENPEYLTPQ
1201 GGAAPQPHPP PAFSPAFDNL YYWDQDPPER GAPPSTFKGT PTAENPEYLG
1251 LDVPV (1255)
```

### Domain structure

| | |
|---|---|
| 1–21 | Signal sequence (italics) |
| 22–652 | Extracellular domain |
| 653–675 | Transmembrane domain (underlined) |
| 676–1255 | Cytoplasmic domain |
| 726–734 and 753 | ATP binding site |
| 1139, 1221, 1222, 1248 | Phosphorylation sites (underlined) |
| 68, 124, 187, 259, 530, 571, 629 | Putative carbohydrate attachment sites |

**Database accession numbers**

| | *PIR* | *SWISSPROT* | *EMBL/GENBANK* | *REFERENCES* |
|---|---|---|---|---|
| Human *HER2* | A25491 | P04626 | X03363 | 2 |
| | A25471 | | | 60 |

*References*


1 Shih, C. et al. (1981) Nature 290, 261–264.
2 Semba, K. et al. (1985) Proc. Natl Acad. Sci. USA 82, 6497–6501.
3 King, C.R. et al. (1985) Science 229, 974–976.
4 Coussens, L. et al. (1985) Science 230, 1132–1139.
5 Scott, G.K. et al. (1993) Mol. Cell. Biol. 13, 2247–2257.
6 Altiok, N. et al. (1995) EMBO J. 14, 4258–4266.
7 Marchionni, M.A. et al. (1993) Nature 362, 312–318.
8 Dobashi, K. et al. (1991) Proc. Natl Acad. Sci. USA 88, 8582–8586.
9 Karunagaran, D. et al. (1996) EMBO J. 15, 254–264.
10 **Carraway, K.L. and Cantley, L.C. (1994) Cell 78, 5–8.**
11 **Dougall, W.C. et al. (1994) Oncogene 9, 2109–2123.**
12 Wallasch, C. et al. (1995) EMBO J. 14, 4267–4275.
13 Chan, S.D.H. et al. (1995) J. Biol. Chem. 270, 22608–22613.
14 Tarakhovsky, A. et al. (1991) Oncogene 6, 2187–2196.
15 Segatto, O. et al. (1993) Oncogene 8, 2105–2112.
16 Peles, E. et al. (1991) EMBO J. 10, 2077–2086.
17 Xie, Y. et al. (1995) Oncogene 10, 2409–2413.
18 Satoh, T. et al. (1990) Proc. Natl Acad. Sci. USA 87, 7926–7929.
19 Fazioli, F. et al. (1993) Mol. Cell. Biol. 11, 2040–2048.
20 Samanta, A. and Greene, M.I. (1995) Proc. Natl Acad. Sci. USA 92, 6582–6586.
21 Galang, C.K. et al. (1996) J. Biol. Chem. 271, 7992–7998.
22 Cao, H. et al. (1991) Oncogene 6, 705–711.
23 Qian, X. et al. (1994) Oncogene 9, 1507–1514.
24 Pauletti, G. et al. (1996) Oncogene 13, 63–72.
25 Wright, C. et al. (1990) Br. J. Cancer 62, 764–765.
26 Underwood, M. et al. (1995) Cancer Res. 55, 2422–2430.
27 Saffari, B. et al. (1995) Cancer Res. 55, 5628–5631.
28 Yonemura, Y. et al. (1991) Cancer Res. 51, 1034–1038.
29 Riviere, A. et al. (1991) Cancer 67, 2142–2149.
30 Harpole, D.H. et al. (1995) Clin. Cancer Res. 1, 659–664.
31 Benz, C.C. et al. (1995) Gene 159, 3–7.
32 Yu, D. et al. (1994) Cancer Res. 54, 3260–3266.
33 Pupa, S.M. et al. (1993) Oncogene 8, 2917–2923.
34 Janes, P.W. et al. (1994) Oncogene 9, 3601–3608.
35 Stein, D. et al. (1994) EMBO J. 13, 1331–1340.
36 Chavany, C. et al. (1996) J. Biol. Chem. 271, 4974–4977.
37 Pietras, R.J. et al. (1995) Oncogene 10, 2435–2446.
38 Baselga, J. et al. (1996) J. Clin. Oncol. 14, 737–744.
39 Muller, W.J. et al. (1988) Cell 54, 105–115.
40 Bouchard, L. et al. (1989) Cell 57, 931–936.
41 Muller, W.J. et al. (1996) Mol. Cell. Biol. 16, 5726–5736.
42 Bradbury, J.M. et al. (1993) Oncogene 8, 1551–1558.
43 Chazin, V.R. et al. (1992) Oncogene 7, 1859–1866.

[44] Yu, D. and Hung, M.-C. (1991) Oncogene 6, 1991–1996.
[45] Zhu, D. and Bourguignon, L. (1996) Oncogene 12, 2309–2314.
[46] Yu, D. et al. (1993) Cancer Res. 53, 5784–5790.
[47] Yu, D. et al. (1995) Oncogene 11, 1383–1388.
[48] Carter, P. et al. (1992) Proc. Natl Acad. Sci. USA 89, 4285–4289.
[49] Peles, E. et al. (1992) Cell 69, 205–216.
[50] Bacus, S.S. et al. (1996) Oncogene 12, 2535–2547.
[51] Juang, S.-H. et al. (1996) Oncogene 12, 1033–1042.
[52] Matsuda, S. et al. (1996) Oncogene 12, 705–713.
[53] D'Souza, B. and Taylor-Papadimitriou, J. (1994) Proc. Natl Acad. Sci. USA 91, 7202–7206.
[54] Chen, Y. and Gill, G.N. (1996) J. Biol. Chem. 271, 5183–5188.
[55] Scott, G.K. et al. (1994) J. Biol. Chem. 269, 19848–19858.
[56] Bosher, J.M. et al. (1996) Oncogene 13, 1701–1707.
[57] Cao, H. et al. (1992) EMBO J. 11, 923–932.
[58] Ben-Levy, R. et al. (1992) J. Biol. Chem. 267, 17304–17313.
[59] Akiyama, T. et al. (1991) Mol. Cell. Biol. 11, 833–842.
[60] Yamamoto, T. et al. (1986) Nature 319, 230–234.

# HER3/ErbB-3

## Identification

*HER3/ErbB-3* was detected by screening genomic DNA using v-*erbB* as a probe and by PCR amplification of genomic DNA using degenerate oligonucleotide primers based on sequences encoding regions of the EGFR family catalytic domains [1].

## Related genes

*HER3* is a member of the *EGFR* family, distinct from *EGFR*, *HER2* and *HER4*.

| | ***HER3*** |
|---|---|
| **Nucleotides (kb)** | Not fully mapped |
| **Chromosome** | 12q13 |
| **Mass (kDa): predicted** | 146 |
| **expressed** | gp160 |
| **Cellular location** | Plasma membrane |

**Tissue distribution**

*HER3* is widely expressed with a pattern of expression closely similar to that of *EGFR* and *HER2*, although *HER3* is undetectable in fetal or adult skin fibroblasts. It is transcribed in term placenta, the respiratory and urinary tracts, stomach, lung, kidney and brain but not in skeletal muscle or lymphoid cells [2]. It is expressed in normal fetal liver, kidney and brain but not in the heart. HER3 protein is detectable at the motor endplates of rat skeletal muscle [3]. In C2 and L6 muscle cell lines HER3 and HER4 proteins are expressed in highly concentrated regions at neuromuscular synapses [4].

## Protein function

HER3 is a receptor tyrosine kinase. However, although HER3 has a tyrosine kinase domain that is highly homologous to those of EGFR and HER2, the C-terminal domain and a 29 amino acid region C-terminal to the ATP binding domain diverge markedly and the kinase activity of this receptor is impaired [5]. HER3 can form heterodimers with EGFR, HER2 or HER4.

The isoforms of Neu differentiation factors (NDF, heregulin or neuregulin) bind to HER3: these are members of the EGFR family of ligands and include gp30, heregulin (HRG), NEU differentiation factor, glial growth factor, sensory and motor neurone-derived factor and acetylcholine receptor-inducing factor. The major groups of isoforms are NDF-$\alpha$ and NDF-$\beta$, expressed mainly in mesenchymal and neuronal cells, respectively [6,7]. Both NDF-$\alpha$ and NDF-$\beta$ bind to ERBB-3/HER3 but mitogenic stimulation requires *trans*-activation by other members of the EGFR family, either ERBB-1/EGFR or ERBB-2/HER2. Both NDF-$\alpha$ and NDF-$\beta$ signal via ERBB-3/ERBB-2 heterodimers but only NDF-$\beta$ is mitogenically active via ERBB-3/ERBB-1 complexes [8,9].

Activated receptors associate with PtdIns 3-kinase, SHC and GRB2 but not with PLC$\gamma$ or GAP [10], although whether GRB2 binds is controversial [11]. HER3 contains seven repeats of Tyr-X-X-Met (the p85 SH2 recognition sequence) in its C-terminus, which are not present in the EGFR [12].

HRD/NDF increases acetylcholine receptor $\alpha$ subunit mRNA levels in myotubes via tyrosine phosphorylation of both HER3 and HER2 [3].

### Cancer

*HER3* is expressed in some carcinomas and, with low frequency, in sarcomas. It is detectable in cell lines derived from haematopoietic tumours but is over-expressed in a subset of mammary tumour-derived cell lines [13]. Increased expression of HER3 (and of EGFR, HER2 and HER4) has been detected in papillary thyroid carcinomas.

### *In vitro*

High coexpression of HER2 and HER3 transforms NIH 3T3 fibroblasts: heterodimers are formed with increased tyrosine phosphorylation of HER3 and recruitment of PtdIns 3-kinase. Similar interactions occur in human breast tumour cell lines [14].

## Gene structure

There is strict conservation of exon–intron boundaries in the regions encoding the catalytic domains of HER3 and EGFR (see ***EGFR/ErbB-1***). The *HER3* promoter is similar to that of EGFR, having a high GC content, no TATA box and multiple transcription initiation sites [15]. In cells over-expressing HER3, promoter activity is enhanced as a result of the increased abundance of the transcription factor OB2-1 [16].

## Protein structure

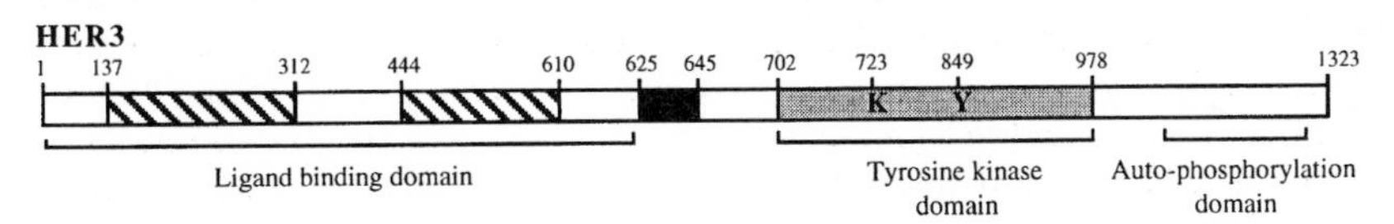


HER3 is ~44% identical in amino acid sequence to EGFR and HER2 in the extracellular domain and ~60% identical in the tyrosine kinase domain (see ***EGFR/ErbB-1***). The tyrosine kinase domain homology with EPH, MET, FMS and the insulin receptor is ~30%. There is no significant homology between the C-terminal 353 amino acids of HER3 and the corresponding regions of EGFR, HER2 and HER4 that contain the major tyrosine autophosphorylation sites.

## Amino acid sequence of human HER3

```
   1 MRANDALQVL GLLFSLARGS EVGNSQAVCP GTLNGLSVTG DAENQYQTLY
  51 KLYERCEVVM GNLEIVLTGH NADLSFLQWI REVTGYVLVA MNEFSTLPLP
 101 NLRVVRGTQV YDGKFAIFVM LNYNTNSSHA LRQLRLTQLT EILSGGVYIE
 151 KNDKLCHMDT IDWRDIVRDR DAEIVVKDNG RSCPPCHEVC KGRCWGPGSE
 201 DCQTLTKTIC APQCNGHCFG PNPNQCCHDE CAGGCSGPQD TDCFACRHFN
 251 DSGACVPRCP QPLVYNKLTF QLEPNPHTKY QYGGVCVASC PHNFVVDQTS
 301 CVRACPPDKM EVDKNGLKMC EPCGGLCPKA CEGTGSGSRF QTVDSSNIDG
 351 FVNCTKILGN LDFLITGLNG DPWHKIPALD PEKLNVFRTV REITGYLNIQ
 401 SWPPHMHNFS VFSNLTTIGG RSLYNRGFSL LIMKNLNVTS LGFRSLKEIS
 451 AGRIYISANR QLCYHHSLNW TKVLRGPTEE RLDIKHNRPR RDCVAEGKVC
 501 DPLCSSGGCW GPGPGQCLSC RNYSRGGVCV THCNFLNGEP REFAHEAECF
 551 SCHPECQPME GTATCNGSGS DTCAQCAHFR DGPHCVSSCP HGVLGAKGPI
 601 YKYPDVQNEC RPCHENCTQG CKGPELQDCL GQTLVLIGKT HLTMALTVIA
 651 GLVVIFMMLG GTFLYWRGRR IQNKRAMRRY LERGESIEPL DPSEKANKVL
 701 ARIFKETELR KLKVLGSGVF GTVHKGVWIP EGESIKIPVC IKVIEDKSGR
 751 QSFQAVTDHM LAIGSLDHAH IVRLLGLCPG SSLQLVTQYL PLGSLLDHVR
 801 QHRGALGPQL LLNWGVQIAK GMYYLEEHGM VHRNLAARNV LLKSPSQVQV
 851 ADFGVADLLP PDDKQLLYSE AKTPIKWMAL ESIHFGKYTH QSDVWSYGVT
 901 VWELMTFGAE PYAGLRLAEV PDLLEKGERL AQPQICTIDV YMVMVKCWMI
 951 DENIRPTFKE LANEFTRMAR DPPRYLVIKR ESGPGIAPGP EPHGLTNKKL
1001 EEVELEPELD LDLDLEAEED NLATTTLGSA LSLPVGTLNR PRGSQSLLSP
1051 SSGYMPMNQG NLGESCQESA VSGSSERCPR PVSLHPMPRG CLASESSEGH
1101 VTGSEAELQE KVSMCRSRSR SRSPRPRGDS AYHSQRHSLL TPVTPLSPPG
1151 LEEEDVNGYV MPDTHLKGTP SSREGTLSSV GLSSVLGTEE EDEDEEYEYM
1201 NRRRRHSPPH PPRPSSLEEL GYEYMDVGSD LSASLGSTQS CPLHPVPIMP
1251 TAGTTPDEDY EYMNRQRDGG GPGGDYAAMG ACPASEQGYE EMRAFQGPGH
1301 QAPHVHYARL KTLRSLEATD SAFDNPDYWH SRLFPKANAQ RT (1342)
```

### Domain structure

| | |
|---|---|
| 1–19 | Signal sequence (italics) |
| 20–643 | Extracellular domain |
| 644–664 | Transmembrane domain (underlined) |
| 665–1342 | Cytoplasmic domain |
| 715–723 and 742 | ATP binding site |
| 126, 250, 353, 408, 414, 437, 469, 522, 566, 616 | Putative carbohydrate attachment sites |
| 1328 | HER3 SHC binding site (Tyr$^{1328}$/Tyr$^{1309}$) (NPXY) [11]. |

### Database accession numbers

| | *PIR* | *SWISSPROT* | *EMBL/GENBANK* | *REFERENCES* |
|---|---|---|---|---|
| Human *HER3* | A36223 | P21860 | M29366 | 13 |
| | | | M34309 | 1 |

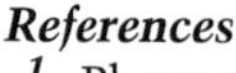

### *References*


1 Plowman, G.D. et al. (1990) Proc. Natl Acad. Sci. USA 87, 4905–4909.
2 Prigent, S.A. et al. (1992) Oncogene 7, 1273–1278.
3 Altiok, N. et al. (1995) EMBO J. 14, 4258–4266.
4 Zhu, X. et al. (1995) EMBO J. 14, 5842–5848.
5 Guy, P.M. et al. (1994) Proc. Natl Acad. Sci. USA 91, 8132–8136.
6 Carraway, K.L. et al. (1994) J. Biol. Chem. 269, 14303–14306.
7 **Carraway, K.L. and Cantley, L.C. (1994) Cell 78, 5–8.**

[8] Marikovsky, M. et al. (1995) Oncogene 10, 1403–1411.
[9] Pinkas-Kramarski, R. et al. (1996) J. Biol. Chem. 271, 19029–19032.
[10] Fedi, P. et al. (1994) Mol. Cell. Biol. 14, 492–500.
[11] Prigent, S.A. and Gullick, W.J. (1994) EMBO J. 13, 2831–2841.
[12] Soltoff, S.P. et al. (1994) Mol. Cell. Biol. 14, 3550–3558.
[13] Kraus, M.H. et al. (1989) Proc. Natl Acad. Sci. USA 86, 9193–9197.
[14] Alimandi, M. et al. (1995) Oncogene 10, 1813–1821.
[15] Suen, T.-C. and Hung, M.-C. et al. (1990) Mol. Cell Biol. 10, 6306–6315.
[16] Skinner, A. and Hurst, H.C. (1993) Oncogene 8, 3393–3401.

# HER4/ErbB-4

## Identification

*HER4/ErbB-4* was detected by PCR amplification of genomic DNA using degenerate oligonucleotide primers based on sequences encoding regions of the EGFR family catalytic domains [1].

## Related genes

*HER4* is a member of the *EGFR* family, distinct from *EGFR*, *HER2* and *HER3*.

| | ***HER4*** |
|---|---|
| **Nucleotides (kb)** | Not fully mapped |
| **Chromosome** | 2q33.3–34 |
| **Mass (kDa): predicted** | 144 |
| **expressed** | gp180 |
| **Cellular location** | Plasma membrane |

**Tissue distribution**
Maximum expression of *HER4* occurs in brain, heart and kidney but it is also expressed in parathyroid, cerebellum, pituitary, spleen, testis and breast with lower levels detectable in thymus, lung, salivary gland and pancreas. In C2 and L6 muscle cell lines HER4 and HER3 proteins are expressed in highly concentrated regions at neuromuscular synapses [2].

## Protein function

The intrinsic tyrosine kinase activity of HER4 is specifically stimulated by a heparin-binding growth factor related to heregulin that has no direct effect on EGFR, HER2 or HER3 [1] and also by heregulin [3] or betacellulin [4]. Heregulin $\beta$2 and recombinant heregulin-Fc fusion proteins also activate HER4 and cause phosphorylation of SHC and PtdIns 3-kinase [5]. HER4 can form heterodimers with EGFR, HER2 or HER3 and heregulin $\beta$2 can promote heterodimerization with EGFR/HER1.

**Cancer**
*HER4* is expressed in a variety of mammary adenocarcinoma and neuroblastoma cell lines [1]. Increased expression of HER4 (and of EGFR, HER2 and HER3) has been detected in papillary thyroid carcinomas.

**Transgenic animals**
*ErbB-4* null mice die during mid-embryogenesis as a consequence of aberrant cardiac muscle differentiation and central nervous system development. The effects of mutation of the neuregulin gene are consistent with its being the ERBB-4 ligand in the heart but suggest the existence of a novel ligand in the central nervous system [6].

*In vitro*

In NIH 3T3 cells phorbol ester-induced activation of protein kinase C causes proteolytic cleavage of ERBB-4 which negatively regulates signalling through this receptor [7].

## Protein structure

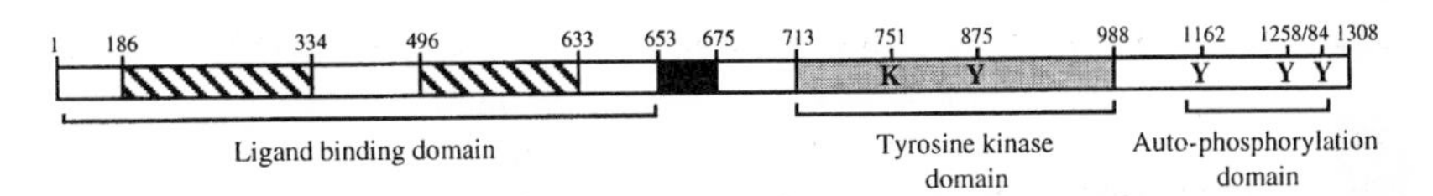


The extracellular domains are between 56% and 67% identical to the corresponding regions of HER3 and 43–51% and 34–46% identical to EGFR and HER2. The 50 conserved extracellular cysteines of EGFR, HER2 and HER3 are also conserved in HER4 except for the fourth cysteine in domain IV (496–633). The transmembrane 37 amino acids are 73% identical with those of EGFR and the 276 residue catalytic domain is 79%, 77% and 63% identical to those of EGFR, HER2 and HER3, respectively. In the C-terminus homology is much lower (19% (EGFR), 27% (HER2), respectively) but the major tyrosine autophosphorylation sites of EGFR are conserved.

## Amino acid sequence of human HER4

```
   1 MKPATGLWVW VSLLVAAGTV QPSDSQSVCA GTENKLSSLS DLEQQYRALR
  51 KYYENCEVVM GNLEITSIEH NRDLSFLRSV REVTGYVLVA LNQFRYLPLE
 101 NLRIIRGTKL YEDRYALAIF LNYRKDGNFG LQELGLKNLT EILNGGVYVD
 151 QNKFLCYADT IHWQDIVRNP WPSNLTLVST NGSSGCGRCH KSCTGRCWGP
 201 TENHCQTLTR TVCAEQCDGR CYGPYVSDCC HRECAGGCSG PKDTDCFACM
 251 NFNDSGACVT QCPQTFVYNP TTFQLEHNFN AKYTYGAFCV KKCPHNFVVD
 301 SSSCVRACPS SKMEVEENGI KMCKPCTDIC PKACDGIGTG SLMSAQTVDS
 351 SNIDKFINCT KINGNLIFLV TGIHGDPYNA IEAIDPEKLN VFRTVREITG
 401 FLNIQSWPPN MTDFSVFSNL VTIGGRVLYS GLSLLILKQQ GITSLQFQSL
 451 KEISAGNIYI TDNSNLCYYH TINWTTLFST INQRIVIRDN RKAENCTAEG
 501 MVCNHLCSSD GCWGPGPDQC LSCRRFSRGR ICIESCNLYD GEFREFENGS
 551 ICVECDPQCE KMEDGLLTCH GPGPDNCTKC SHFKDGPNCV EKCPDGLQGA
 601 NSFIFKYADP DRECHPCHPN CTQGCNGPTS HDCIYYPWTG HSTLPQHART
 651 PLIAAGVIGG LFILVIVGLT FAVYVRRKSI KKKRALRRFL ETELVEPLTP
 701 SGTAPNQAQL RILKETELKR VKVLGSGAFG TVYKGIWVPE GETVKIPVAI
 751 KILNETTGPK ANVEFMDEAL IMASMDHPHL VRLLGVCLSP TIQLVTQLMP
 801 HGCLLEYVHE HKDNIGSQLL LNWCVQIAKG MMYLEERRLV HRDLAARNVL
 851 VKSPNHVKIT DFGLARLLEG DEKEYNADGG KMPIKWMALE CIHYRKFTHQ
 901 SDVWSYGVTI WELMTFGGKP YDGIPTREIP DLLEKGERLP QPPICTIDVY
 951 MVMVKCWMID ADSRPKFKEL AAEFSRMARD PQRYLVIQGD DRMKLPSPND
1001 SKFFQNLLDE EDLEDMMDAE EYLVPQAFNI PPPIYTSRAR IDSNRSEIGH
1051 SPPPAYTPMS GNQFVYRDGG FAAEQGVSVP YRAPTSTIPE APVAQGATAE
1101 IFDDSCCNGT LRKPVAPHVQ EDSSTQRYSA DPTVFAPERS PRGELDEEGY
1151 MTPMRDKPKQ EYLNPVEENP FVSRRKNGDL QALDNPEYHN ASNGPPKAED
1201 EYVNEPLYLN TFANTLGKAE YLKNNILSMP EKAKKAFDNP DYWNHSLPPR
1251 STLQHPDYLQ EYSTKYFYKQ NGRIRPIVAE NPEYLSEFSL KPGTVLPPPP
1301 YRHRNTVV (1308)
```

**Domain structure**

| | |
|---|---|
| 1–25 | Signal sequence (italics) |
| 26–649 | Extracellular domain |
| 650–675 | Transmembrane domain (underlined) |
| 676–1308 | Cytoplasmic domain |
| 679 | Potential protein kinase C phosphorylation site (Ser$^{679}$) |
| 699 | Potential MAP kinase phosphorylation site (Thr$^{699}$) |
| 713–988 | Catalytic domain |
| 725–730 and 751 | ATP binding site |
| 1162, 1188, 1258 and 1284 | Phosphorylation sites (underlined) |
| 138, 174, 181, 253, 358, 410, 473, 495, 576, 620 | Putative carbohydrate attachment sites |

**Database accession numbers**

| | *PIR* | *SWISSPROT* | *EMBL/GENBANK* | *REFERENCES* |
|---|---|---|---|---|
| Human *HER4* | | | L07868 | 1 |

***References***

[1] Plowman, G.D. et al. (1993) Nature 366, 473–475.
[2] Zhu, X. et al. (1995) EMBO J. 14, 5842–5848.
[3] Plowman, G.D. et al. (1993) Proc. Natl Acad. Sci. USA 90, 1746–1750.
[4] Riese, D.J. et al. (1996) Oncogene 12, 345–353.
[5] Cohen, B.D. et al. (1996) J. Biol. Chem. 271, 4954–4960.
[6] Gassmann, M. et al. (1995) Nature 378, 390–394.
[7] Vecchi, M. et al. (1996) J. Biol. Chem. 271, 18989–18995.

# HSTF1/Hst-1, HST2/Hst-2

## Identification

*HSTF1* (heparan secretory transforming protein 1) is a human transforming gene originally detected by NIH 3T3 fibroblast transfection with DNA from human stomach tumours that has no homology with known viral oncogenes[1]. *HST2* is a close homologue of *HSTF1* that was cloned by cross-hybridization with *HSTF1* probes[2].

## Related genes

The *HSTF1/Hst-1* and *INT2/Int-2* oncogenes are members of the fibroblast growth factor (FGF) family[3–5]. The FGF family includes seven proteins: acidic fibroblast growth factor (*FGFA*, aFGF, HBGF-1 or FGF1), basic FGF (*FGFB*, bFGF, HBGF-2 or FGF2), *INT2* (FGF3), *HSTF1* (*Hst-1*/Kaposi-FGF/K-FGF or FGF4), *FGF5*, *HST2/Hst-2* (*FGF6*) and keratinocyte growth factor, KGF (FGF7). HSTF1 is homologous to FGFA (38%), FGFB (43%) and mouse INT2 (40%). Murine FGF-6 is 66% identical to murine HST-1/K-FGF and 39% identical to INT2.

| | ***HSTF1/Hst-1*** | ***HST2/Hst-2*** |
|---|---|---|
| **Nucleotides (kb)** | 11 | 8 |
| **Chromosome** | 11q13.3 | 12p13 |
| **Mass (kDa): predicted** | 22 | 14 |
| **expressed** | 22 | |

### Cellular location

HSTF1 is glycosylated and secreted, the secreted protein being stabilized by heparin[6]. HST2 also has the characteristics of a heparin-binding growth factor.

### Tissue distribution

*HSTF1/Hst-1* is expressed at a limited stage of embryonal development[7]. *HST2* is only detectable in leukaemic cell lines[8].

## Protein function

The members of the FGF family appear to act as paracrine or autocrine growth factors. FGFA and FGFB may function in both cell growth and in differentiation and have been implicated in cell transformation, angiogenesis and embryonic development. HST-1 and INT2 induce mesoderm formation in *Xenopus laevis* animal pole cells[9].

### Cancer

FGFs have oncogenic potential: they can induce blood vessel formation and are synthesized by many tumour cells[3–5]. *HSTF1*, together with *INT2* and *BCL1*, is amplified in up to 22% of human breast carcinomas[10,11]. *HSTF1*, *INT2* and anionic glutathione-S-transferase are co-amplified in ~30% of breast carcinomas [12]. However, there is no evidence that the HSTF1 or INT2 proteins are expressed in breast tumours. Co-amplification of *HSTF1/INT2* occurs in up to 47% of oesophageal carcinomas [13–15] and *HSTF1* is expressed together with *KIT* in some testicular germ cell tumours [16]. *HSTF1* has also been identified in human DNA from gastric cancers, hepatomas, colon carcinomas [17], melanoma [18], osteosarcoma [19] and in Kaposi's sarcoma [20].

The human FGF receptors *BEK* (bacterial expressed kinase) and *FLG* (Fms-like gene, also *Flt2*) are amplified in ~10% of human breast tumours. *BEK* expression may correlate with that of *MYC* and expression of *FGFR1/FLG* with *HSTF1/INT2/BCL1* [21]. K-*SAM*, the KATO-III cell-derived stomach cancer amplified gene, is also a member of the *BEK* family [22].

**In animals**

*Hst-1* may be activated by MMTV proviral insertion on either side of the gene: some insertions activate both *Hst-1* and the 17 kb distant *Int-2* gene [23]. *Hst-1* may be involved in tumour progression from a non-metastatic to a metastatic phenotype in the mouse mammary tumour system [24].

***In vitro***

*HSTF1* or *FGF5* genomic and cDNA sequences transfected into NIH 3T3 cells induce morphological transformation, anchorage-independent growth and tumorigenicity [2,20,25–28]. *HST2* transforms NIH 3T3 fibroblasts into cells that are tumorigenic in nude mice [8].

## Gene structure

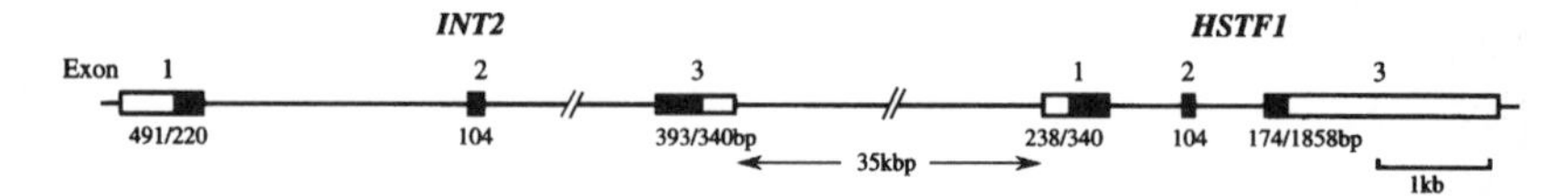


The FGF family has a common genomic organization and each member has a conserved 104 bp exon encoding the central 35 amino acids. *INT2/Int-2* and *HSTF1/Hst-2* are in the same transcriptional orientation.

## Transcriptional regulation

The *HSTF1* promoter contains a TATA box and three putative SP1 binding sites. The 5′ non-coding region and exon 1 are GC-rich regions, a characteristic of house-keeping genes [29]. Expression of *HSTF1* may be regulated by protein factor(s) binding to an enhancer located in the third exon of the gene [30].

## Protein structure

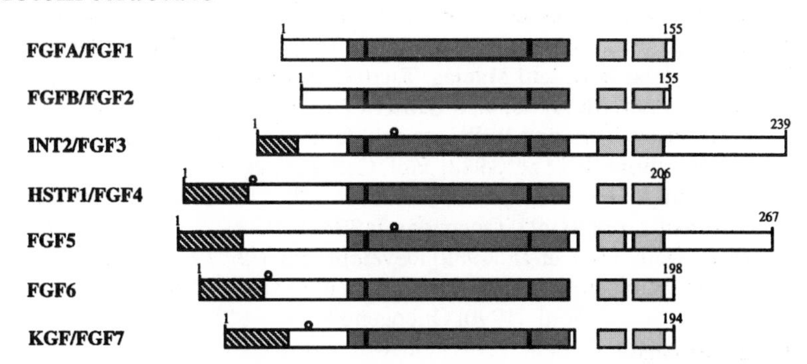


FGF family proteins contain two major regions of homology (heavy and light shading) and there are two absolutely conserved cysteine residues indicated by the vertical bars (see below). The cross-hatched boxes indicate N-terminal signal sequences, circles potential *N*-glycosylation sites. The human proteins are represented, with the exception of FGF6, for which the structure of the mouse protein is shown.

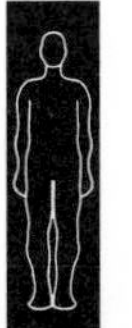

## Amino acid sequence of human HSTF1 (FGF4)

```
  1 MSGPGTAAVA LLPAVLLALL APWAGRGGAA APTAPNGTLE AELERRWESL
 51 VALSLARLPV AAQPKEAAVQ SGAGDYLLGI KRLRRLYCNV GIGFHLQALP
101 DGRIGGAHAD TRDSLLELSP VERGVVSIFG VASRFFVAMS SKGKLYGSPF
151 FTDECTFKEI LLPNNYNAYE SYKYPGMFIA LSKNGKTKKG NRVSPTMKVT
201 HFLPRL (206)
```

### Domain structure

9–20 Potential signal sequence (italics)
88, 155 Conserved cysteine residues (underlined)

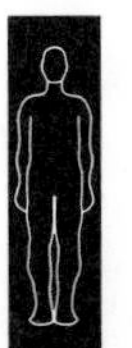

## Amino acid sequence of human HST2 (FGF6)

```
  1 MSRGAGRLQG TLWALVFLGI LVGMVVPSPA GTRANNTLLD SRGWGTLLSR
 51 SRAGLAGEIA GVNWESGYLV GIKRQRRLYC NVGIGFHLQV LPDGRISGTH
101 EENPYSLLEI STVERGVVSL FGVRSALFVA MNSKGRLYAT PSFQEECKFR
151 ETLLPNNYNA YESDLYQGTY IALSKYGRVK RGSKVSPIMT VTHFLPRI(198)
```

### Domain structure

12–26 Potential signal sequence (italics)
80, 147 Conserved cysteine residues (underlined)

### Database accession numbers

| | PIR | SWISSPROT | EMBL/GENBANK | REFERENCES |
|---|---|---|---|---|
| Human *HSTF1* (FGF4) (or *HST* or *KS3*) | A28417 | P08620 | J02986, M17446 | 20,25,29 |
| Human *HST2* (*FGF6*) | S04204 | P10767 | Hshst2, X63454, X14071–X14073 | 31 |

*References*


1. Sakamoto, H. et al. (1986) Proc. Natl Acad. Sci. USA 83, 3997–4001.
2. Sakamoto, H. et al. (1988) Biochem. Biophys. Res. Commun. 151, 965–972.
3. **Burgess, W.H. and Maciag, T. (1989) Annu. Rev. Biochem. 58, 575–606.**
4. **Goldfarb, M. (1990) Cell Growth Differ. 1, 439–445.**
5. **Johnson, D.E. and Williams, L.T. (1993) Adv. Cancer Res. 60, 1–41.**
6. Delli-Bovi, P. et al. (1988) Mol. Cell. Biol. 8, 2933–2941.
7. Suzuki, H.R. et al. (1992) Dev. Biol. 150, 219–222.
8. Iida, S. et al. (1992) Oncogene 7, 303–310.
9. Paterno, G.D. et al. (1989) Development 106, 79–83.
10. Tsuda, H. et al. (1989) Cancer Res. 49, 3104–3108.
11. Theillet, C. et al. (1990) Oncogene 5, 147–149.
12. Saint-Ruf, C. et al. (1991) Oncogene 6, 403–406.
13. Tsuda, T. et al. (1989) Cancer Res. 49, 5505–5508.
14. Kitagawa, Y. et al. (1991) Cancer Res. 51, 1504–1508.
15. Wagata, T. et al. (1991) Cancer Res. 51, 2113–2117.
16. Strohmeyer, T. et al. (1991) Cancer Res. 51, 1811–1816.
17. Yoshida, T. et al. (1991) Methods Enzymol. 198, 124–138.
18. Adelaide, J. et al. (1988) Oncogene 2, 413–416.
19. Zhan, X. et al. (1987) Oncogene 1, 369–376.
20. Delli-Bovi, P. et al. (1987) Cell 50, 729–737.
21. Adnane, J. et al. (1991) Oncogene 6, 659–663.
22. Hattori, Y. et al. (1990) Proc. Natl Acad. Sci. USA 87, 5983–5987.
23. Peters, G. et al. (1989) Proc. Natl Acad. Sci. USA 86, 5678–5682.
24. Murakami, A. et al. (1990) Cell Growth Differ. 1, 225–231.
25. Taira, M. et al. (1987) Proc. Natl Acad. Sci. USA 84, 2980–2984.
26. Wellstein, A. et al. (1990) Cell Growth Differ. 1, 63–71.
27. Fuller-Pace, F. et al. (1991) J. Cell. Biol. 115, 547–555.
28. Talarico, D. and Basilico, C. (1991) Mol. Cell. Biol. 11, 1138–1145.
29. Yoshida, T. et al. (1987) Proc. Natl Acad. Sci. USA 84, 7305–7309 [erratum Proc. Natl Acad. Sci. USA 85, 1967].
30. Sasaki, A. et al. (1991) Jpn J. Cancer Res. 82, 1191–1195.
31. Marics, I. et al. (1989) Oncogene 4, 335–340.

## Identification

Murine *Int-2* was initially identified as a frequent target for activation by proviral insertion of mouse mammary tumour virus [1]. Human *INT2* was detected by cross-hybridization with mouse *Int-2* probes [2].

## Related genes

*INT2/Int-2* and *HSTF1/Hst-1* are members of the fibroblast growth factor (FGF) family (see ***HSTF1/Hst***).

| | *INT2* |
|---|---|
| **Nucleotides (kb)** | 10 |
| **Chromosome** | 11q13 |
| **Mass (kDa): predicted** | 27 |
| **expressed** | 27.5–31.5 |

### Cellular location

NIH 3T3 cells transformed by mouse *Int-2* cDNA express a series of INT2-related proteins. The use of different initiation codons gives rise to proteins located in the endoplasmic reticulum, the Golgi apparatus or the nucleus [3]. In highly transformed clonal lines INT2 proteins undergo further post-translational processing and are secreted, becoming associated with the cell surface and the extracellular matrix [4,5].

### Tissue distribution

In addition to its activation in MMTV-induced tumours in mice, *Int-2* is expressed in embryos and in some teratocarcinomas [6]. Murine mRNA expression occurs in specific tissues during development and is rarely detected in adult tissues [7].

## Protein function

INT2 (and HST-1) induces mesoderm formation in *Xenopus laevis* animal pole cells [8]. The complex pattern of murine expression suggests multiple roles for INT2 during fetal development. INT2 secreted by highly transformed cells can be displaced from the cell surface by glycosaminoglycans and the addition of excess heparin causes the cells to revert to normal morphology [4,5].

### Cancer

*INT2*, together with *HSTF1* and *BCL1*, is amplified in up to 22% of human breast carcinomas [9,10] and one study has detected polymorphism of *INT2* in 61% of lymph-node positive patients [11]. However, there is no evidence that INT2 (or HSTF1) proteins are expressed in breast tumours. Co-amplification of *HSTF1/INT2* occurs in up to 47% of oesophageal carcinomas [12–14]. *INT2* is frequently amplified in squamous cell carcinomas (SCC) of the head and neck. Co-amplification of *INT2* and *EGFR* has been detected in a laryngeal SCC and an SCC metastatic to the neck[15].

**Transgenic animals**

In mice the *Int-2* transgene causes mammary hyperplasia and in some males prostatic hyperplasia although induction of tumours is rare[16,17]. Thus *Int-2* acts as a potent growth factor in these epithelial cells. In double transgenic mice *Int-2* and *Wnt-1* cooperate in the induction of mammary tumours (see ***WNT1***, ***WNT2***, ***WNT3***) and, in *Wnt-1* transgenic mice infected with MMTV, *Int-2* and *Hst-1* can cooperate with *Wnt-1*[18].

**In animals**

*Int-2* is frequently activated by MMTV proviral insertion. In spontaneously arising murine mammary tumours *Int-2* may be activated together with other *Int* genes, for example, *Wnt-1*[19].

***In vitro***

*Int-2* transforms NIH 3T3 cells with very low efficiency[20] but can substitute for FGFB in some cell lines[21].

## Gene structure

See ***HSTF1/Hst-1***, ***HST2/Hst-2***.

## Transcriptional regulation

In the mouse *Int-2* gene transcription can initiate at multiple sites within three separate promoter domains and terminate distal to one of two polyadenylation signals[22,23]. An alternative non-coding first exon (exon 1a) occurs in minor classes of *Int-2* mRNA. Homology between murine *Int-2* and human *INT2* extends throughout the promoter domains, although there is no evidence for the use of exon 1a in *INT2*[24].

## Protein structure

See ***HSTF1/Hst-1***, ***HST2/Hst-2***.

## Amino acid sequence of human INT2

```
  1 MGLIWLLLLS LLEPGWPAAG PGARLRRDAG GRGGVYEHLG GAPRRRKLYC
 51 ATKYHLQLHP SGRVNGSLEN SAYSILEITA VEVGIVAIRG LFSGRYLAMN
101 KRGRLYASEH YSAECEFVER IHELGYNTYA SRLYRTVSST PGARRQPSAE
151 RLWYVSVNGK GRPRRGFKTR RTQKSSLFLP RVLDHRDHEM VRQLQSGLPR
201 PPGKGVQPRR RRQKQSPDNL EPSHVQASRL GSQLEASAH (239)
```

**Domain structure**

1–17 Signal sequence (italics)

65 Potential glycosylation site

Human and mouse INT2 proteins are 89% identical up to the C-terminus in which 22 amino acids of human INT2 are replaced by 27 unrelated residues in the mouse protein[24].

**Database accession numbers**

| | *PIR* | *SWISSPROT* | *EMBL/GENBANK* | *REFERENCES* |
|---|---|---|---|---|
| Human *INT2* (FGF3) | S04742 | P11487 | X14445 | 24 |

***References***


1 **Peters, G. (1991) Semin. Virol. 2, 319–328.**
2 Casey, G. et al. (1986) Mol. Cell. Biol. 6, 502–510.
3 Acland, P. et al. (1990) Nature 343, 662–665.
4 Dixon, M. et al. (1989) Mol. Cell. Biol. 9, 4896–4902.
5 Kiefer, P. et al. (1991) Mol. Cell. Biol. 11, 5929–5936.
6 Jakobovits, A. et al. (1986) Proc. Natl Acad. Sci. USA 83, 7806–7810.
7 Wilkinson, D.G. et al. (1989) Development, 105, 131–136.
8 Paterno, G.D. et al. (1989) Development, 106, 79–83.
9 Tsuda, H. et al. (1989) Cancer Res. 49, 3104–3108.
10 Theillet, C. et al. (1990) Oncogene 5, 147–149.
11 Meyers, S.L. et al. (1990) Cancer Res. 50, 5911–5918.
12 Tsuda, T. et al. (1989) Cancer Res. 49, 5505–5508.
13 Kitagawa, Y. et al. (1991) Cancer Res. 51, 1504–1508.
14 Wagata, T. et al. (1991) Cancer Res. 51, 2113–2117.
15 Somers, K.D. et al. (1990) Oncogene 5, 915–920.
16 Muller, W.J. et al. (1990) EMBO J. 9, 907–913.
17 Ornitz, D.M. et al. (1991) Proc. Natl Acad. Sci. USA 88, 698–702.
18 Shackleford, G.M. et al. (1993) Proc. Natl Acad. Sci. USA 90, 740–744.
19 Peters, G. et al. (1986) Nature 320, 628–631.
20 Goldfarb, M. et al. (1991) Oncogene 6, 65–71.
21 Venesio, T. et al. (1992) Cell Growth Differ. 3, 63–71.
22 Smith, R. et al. (1988) EMBO J. 7, 1013–1022.
23 Mansour, S.L. and Martin, G.R. (1988) EMBO J. 7, 2035–2041.
24 Brookes, S. et al. (1989) Oncogene 4, 429–436.

## Identification

v-*jun* is the oncogene of avian sarcoma virus 17 (ASV 17) isolated from a spontaneous chicken sarcoma [1]. It is specifically responsible for the oncogenicity of ASV 17 (*ju-nana* is Japanese for the number 17). *JUN* was identified by screening a human genomic library with a v-*jun* probe [2].

## Related genes

*JUN* is a member of the *JUN* gene family (*JUN, JUNB, JUND* and murine *Jund-2*) and of the helix-loop-helix/leucine zipper superfamily [3,4] JUN shares 44% and 45% identity with JUNB and JUND, respectively. The human and mouse JUN and JUNB proteins are 98% identical and human and mouse JUND are 77% identical.

| | ***JUN*** | **v-*jun*** |
|---|---|---|
| **Nucleotides (kb)** | 3.622 | 0.93<br>3.5 (ASV 17) |
| **Chromosome** | 1p32–p31 | |
| **Mass (kDa): predicted** | 35.7 | 32 |
| **expressed** | 39 | $65^{gag\text{-}jun}$ |
| **Cellular location** | Nucleus | Nucleus |

**Tissue distribution**

*JUN* transcription, like that of *FOS*, is rapidly activated in mitogenically stimulated fibroblasts in which essentially all JUN protein exists as JUN/FOS heterodimers. Two other closely related genes, *JUNB* and *JUND*, are also members of the group of "early response genes" (see ***JUNB*** and ***JUND***) [5–7]. JUN is also activated in response to ionizing radiation [8]. *JUN* is expressed in human peripheral blood granulocytes [9]. In normal human epidermis JUN expression is limited to the granular layer [10]. *JUN, JUNB* and *JUND* mRNA is expressed in ovarian tumours as well as in normal ovaries [11].

mRNA expression in the mouse is maximal in the lung, ovary and heart and is very low in the intestine and liver [12]. *Jun* is also transiently expressed in IL-2- and IL-6-dependent mouse myeloma cell lines following withdrawal of growth factor and the onset of apoptosis [13].

## Protein function

JUN forms heterodimers with FOS and with FOS-related antigens (FRA1, FRA2) that bind with high affinity to the AP1 consensus site (the *cis*-acting TPA response element TRE, 5′-TGA$^{C}/_{G}$TCA-3′). Human JUN was originally termed AP1 (PEA 1 in mice) [2] and was first identified by its selective binding to enhancer elements in the *cis* control regions of SV40 and human metallothionein IIA. AP1 is a complex of polypeptides comprised predominantly of heterodimers of members of the JUN and FOS families[14]. The AP1 complex binds to the same consensus sequence as GCN4 and in artificial constructs both v-JUN and JUN activate transcription *via* consensus AP1 binding sites although they differ in their capacity to recognize AP1-like and CREB-like target sequences[15]. JUN (and FOS) also interact with the NF-$\kappa$B subunit p65 (RELA) through its REL homology domain to enhance binding to $\kappa$B and AP1 sites[16] and JUN dimers and JUN/FOS heterodimers also form complexes with the T-cell transcription factor $NFAT_p$[17] (see ***FOS***).

JUN homodimers also bind to AP1 sites with high affinity[18] and JUN *trans*-activates the human *MYB* promoter *via* an AP1-like sequence[19]. However, JUN affinity depends on the flanking sequences (ATGACTCAPy >> ATGACTCAPu: for the sequence CTGACTCAT more distant nucleotides may confer high or low affinity). Proteins of the FOS family enhance JUN binding to AP1 in the order FOSB > FRA1 > FOS, increasing the half-lives of the JUN/FOS/DNA complexes. JUN/JUN and JUN/FOS complexes also bind to CRE-containing nucleotides but with affinities that depend on the flanking sequences. The FOS/JUN and ATF/CREB families of transcription factors form selective cross-family heterodimers having distinct DNA binding specificities (see Chapter 1)[20], and heterodimerization of JUN with ATF2 targets JUN to several CRE-related sequences which control a subset of AP1 responsive genes. The latter includes the human urokinase (uPA) enhancer[21]. FOS inhibits the *in vitro* binding of JUN/ATF2 to the uPA 5′-TRE. JUN also forms a ternary complex with ERM, a member of the ETS transcription factor family, that synergistically activates transcription from an EBS/CRE site[22].

JUN acts as a negative regulator of insulin, MyoD and myogenin gene expression[23], of MHC class I genes in L cells[24] and of SPARC (secreted protein acidic and rich in cysteine) and thrombospondin genes in rat embryo fibroblasts[25]. In Lewis lung carcinoma (3LL), B16 melanoma and K1735 melanoma cells, however, expression of either *Fos* and/or *Jun* by transfection upregulates MHC H-2 whereas expression of *JunB* causes downregulation[26]. Reduced expression of MHC H-2 correlates with high metastatic capacity and *Fos/Jun* over-expression with reduced metastatic potential. In mouse papilloma cells, the expression of v-*jun* suppresses TPA-induced stromelysin gene expression and this also inhibits metastasis[27].

JUN and JUND but not JUNB also form complexes with OCT-1 that activate transcription from an element in the IL-2 promoter[28]. This mechanism of transcriptional activation occurs during T cell stimulation and is $Ca^{2+}$-sensitive and inhibited by cyclosporin A. JUN cooperates with STAT3$\beta$, a truncated form of STAT3, to activate the IL-6-responsive element of the rat $\alpha_2$-macroglobulin gene[29].

The IP-1 protein specifically blocks FOS/JUN binding to DNA when unphosphorylated but not after PKA-mediated phosphorylation. AP1 DNA binding is increased in cells treated with A23187 or dibutyryl cAMP, consistent with an inhibition of IP-1 activity by phosphorylation[30,31].

A protein distinct from IP-1 and designated JUN interacting factor 1 (JIF-1) binds to the leucine zipper of viral and cellular JUN, inhibiting DNA binding and *trans*-activation by JUN[32]. JIF-1 is homologous to the product of the putative tumour suppressor gene *QM* (see Section IV, ***WT1***).

JUN shares homology with GCN4 and can function as a transcriptional regulator in GCN4-deficient yeast strains. JUN, JUNB, JUND and GCN4 can each form homodimers that bind to DNA but the heterodimeric forms have greatly increased affinity and efficacy as transcriptional activators [14,33].

In HeLa and F9 cells the hepatitis B virus transactivator pX can increase the *trans*-activating capacity of JUN [34]. This action is RAS- and RAF-dependent and also requires the presence of the $Ser^{63}/Ser^{76}$ phosphorylation sites in the N-terminus of JUN.

### Cancer

*JUN* over-expression has been detected in small cell and non-small cell lung cancers [35,36].

### Transgenic animals

Transgenic mice that over-express *Jun* exhibit a normal phenotype but in *Jun/Fos* double transgenics the effect of *Jun* over-expression is to increase the frequency with which osteosarcomas develop relative to that observed in single *Fos* transgenics [37].

In mice carrying v-*jun* in the germline, wounding is a pre-requisite for tumorigenesis, following which ~25% of animals homozygous for the transgene develop dermal fibrosarcomas. v-*jun* transgenic cells require TNFα or IL-1 to become anchorage-independent for *in vitro* growth, indicating that these agents may act as cofactors in wound-induced carcinogenesis [38].

Murine *Jun* null embryos die at mid-gestation, exhibiting impaired hepatogenesis and fibroblasts from live mutant embryos have greatly reduced growth rates *in vitro* [39,40]. Many of the characteristics of RAS transformation are lost in $JUN^{-/-}$ cells, indicating that JUN is a crucial effector in RAS signalling pathways [41].

Bovine papillomavirus transgenic mice develop dermal fibrosarcoma and the enhanced expression of *Jun* and *Junb* but not of *Jund* or *Fos* correlates with the onset of an intermediate stage of tumour development [42].

### In animals

Viruses carrying v-*jun* or a recombinant between v-*jun* and *Jun* induce tumours when injected into the wing web of chicks [43]. Embryonic stem (ES) cells induce teratocarcinomas when subcutaneously injected into syngeneic mice: however, the tumorigenicity of ES cells is drastically reduced when both copies of the *Jun* gene have been inactivated by homologous recombination, indicating that JUN may be necessary for efficient tumour growth [44].

### *In vitro*

The cellular *Jun* gene, placed in an appropriate retroviral vector, is able by itself to transform immortalized rat fibroblasts and, in cooperation with *Hras*, primary rat fibroblasts [45]. *Jun* and *Hras* transforms more effectively than *Junb* and *Hras* and *Junb* inhibits *Jun/Ras* transformation, consistent with the differential DNA affinities of JUN and JUNB.

The transformation potential of v-*jun* in chick embryo fibroblasts is increased $10^3$-fold relative to *Jun* by deletion of the δ region and further increased by loss of the 3′ untranslated region [46,47]. v-*jun* cooperates with v-*erbB* to transform bone marrow cells and to enhance transformation of chick embryo fibroblasts caused by either oncogene alone [48].

Transformation by *Jun* of chick embryo fibroblasts is prevented (or reversed) by the expression of a synthetic *Fos* gene lacking a DNA binding domain [49], the mutant FOS protein forming heterodimers with JUN that are inactive as transcription factors.

In rodent fibroblasts transformation by *Raf* causes constitutive expression of *Jun*: dominant negative mutants of *Jun* inhibit growth and transformation [50]. This inhibition also occurs in cells transformed by other oncogenes (*Mos, Myc, Fos*, SV40 T Ag), suggesting that inhibition of JUN function may be a useful strategy for controlling cell growth.

## Gene structure

The single 3.4 kb exon (993 bp coding) has heterogeneous initiation and polyadenylation sites that give rise to multiple transcripts all encoding the same protein. *Jun* contains a 3′ AU-rich RNA-destabilizing element (ARE) made up of three distinct elements that do not contain the AUUUA motif but, like AUUUA-containing AREs, directs rapid shortening of the poly-A tail as a necessary first step for mRNA degradation [51].

## Transcriptional regulation

*JUN* expression is positively autoregulated by JUN binding to AP1 sites. *JUN* promoters also contain NF-JUN [52], SP1, CTF, RSRF and E1A 13S RNA translation product binding sites. The cAMP-responsive element binding protein can act as an inhibitor or an activator of *Jun* transcription, depending on its phosphorylation state [2,53]. In some cells (HeLa and U2OS) E1A represses the activity of JUN and JUNB by interacting with p300 which functions as a cofactor for JUN and JUNB [54]. The retinoblastoma gene product pRb stimulates *JUN* transcription *via* the SP1 binding site in the JUN promoter [55].

*Jun* transcription is transiently increased in rat aortic smooth muscle cells by angiotensin II and by hydrogen peroxide [56,57] and in hamster tracheal epithelial cells by asbestos [58].

## Protein structure

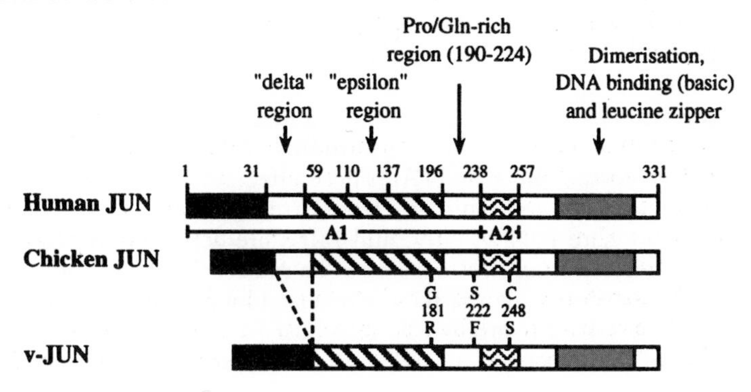

A1: activator domain: confers transcriptional activity when fused to a heterologous DNA binding domain (e.g. SP1 or GAL4). Residues 67–77 and 108–117 constitute the two homology box regions (HOB1 and HOB2) that are conserved in FOS and C/EBP (see ***FOS***).

A2: activation domain essential for *in vivo* activity. The $\delta$ domain (amino acids 28–54 in chicken JUN), missing in v-JUN, stabilizes the interaction of a cell-specific transcriptional inhibitor with A1 [59]. Expression of v-SRC or oncogenic RAS disrupts the JUN/inhibitor complex by interacting with the A1 domain, increasing transcriptional activity [60].

In v-JUN 220 residues encoding viral *gag* are joined in-frame to 296 Jun-encoded amino acids. There is a 27 amino acid deletion (the "$\delta$ region") in the N-terminal region of JUN and three non-conservative substitutions in the C-terminal half, two of which are in the DNA binding domain. A $Ser^{222}$ to Phe mutation prevents the phosphorylation *in vitro* of the negative regulatory site by glycogen synthase kinase-3 and the mutation $Cys^{248}$ to Ser may disrupt a regulatory mechanism involving the reversible oxidation of $Cys^{248}$ which can inactivate DNA binding to AP1 sites *in vitro* [61].

## Phosphorylation of JUN

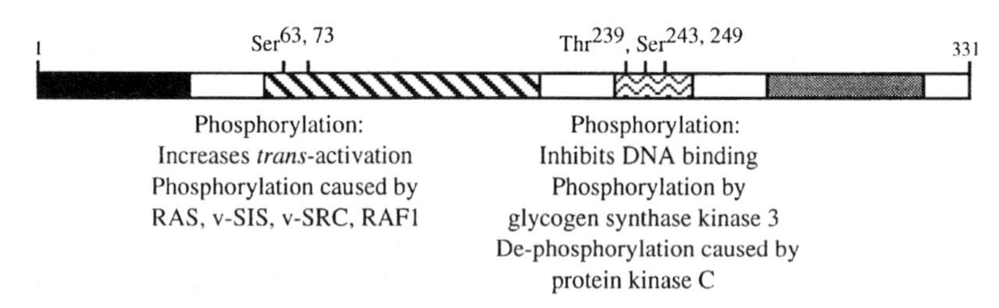


Transforming oncogenes, including *Hras*, stimulate AP1 activity: in rat embryo fibroblasts *Hras* increases JUN synthesis by 4.5-fold but causes a 35-fold increase in overall JUN phosphorylation [62] and similar increases in phosphorylation are caused by v-SIS, v-SRC and RAF-1 [63] and by angiotensin II or UV light, and the signals generated by each of these agents activate JUN N-terminal kinases (JNKs, MAP kinases also called stress-activated protein kinases (SAPKs)) [64–66]. Microinjection of Rho, Rac or Cdc42 GTPases stimulates fibroblast cell cycle progression through $G_1$/S: none of these activate the MAPK cascade but Rac and Cdc42 (not Rho) stimulate JNK [67]. JNK phosphorylated JUN binds to CBP, which mediates PKA-induced transcription by binding to the PKA-phosphorylated activation domain of CREB, and the interaction with CBP promotes the activity of JUN. This co-activation effect depends on JUN residues $Ser^{63/73}$ [68]. *Hras* expression causes hypophosphorylation of the C-terminus of JUN, promoting binding to DNA (similar to the effect of TPA). These effects may underlie the cooperativity between *Ras* and *Jun* in transforming these cells. Although ERKs phosphorylate JUN *in vitro*, and ERKs are activated by RAS, RAS-induced phosphorylation of JUN appears to be indirect [69]. Phosphorylation of sites in the N-terminal region of JUN is reversed by protein phosphatase 2A (PP2A), to which v-JUN is insensitive [70]. A possible PP2A-sensitive site lies in the $\delta$ domain, deleted from v-JUN, but other sites have not yet been mapped.

Mutation of $Ser^{243}$ to Phe blocks phosphorylation of all three glycogen synthase kinase-3 sites *in vitro* and increases the *trans*-activation capability of JUN by 10-fold. In v-JUN the corresponding serine residue is not present and v-JUN binds to the AP1 site to activate constitutive transcription of the gene [71].

Chicken JUN is efficiently phosphorylated *in vitro* by $p34^{cdc2}$, ERK-1 ($p44^{mapk}$), protein kinase C or casein kinase II [72]. The major sites of phosphorylation are serines 63, 73 and 246 but, in contrast to the observations summarized above, the phosphorylation state of these residues does not affect FOS/JUN dimerization, DNA binding or *in vitro* transcription activity.

In unstimulated cells JUN mainly exists as a phosphorylated, inactive form in equilibrium with a hypophosphorylated form that is competent to bind to DNA. The equilibrium is maintained by a specific protein kinase and a phosphatase and shifted to the right by agents stimulating JUN-mediated transcription from AP1 sites. The equilibrium is also shifted to the right in cells transfected so as to increase the number of AP1 binding sites, which raises the concentration of bound dimer and causes a net dephosphorylation of JUN [73].

Different protein kinases can distinguish between the monomeric, homodimeric and heterodimeric forms of JUN and FOS proteins, as well as between DNA bound and unbound forms [74]. Thus JUN/JUN dimers are efficiently phosphorylated by casein kinase II and JUN/FOS dimers are not, whereas phosphorylation of FOS by cAMP-dependent protein kinase and by $p34^{CDC2}$ is relatively insensitive to dimerization and DNA binding.

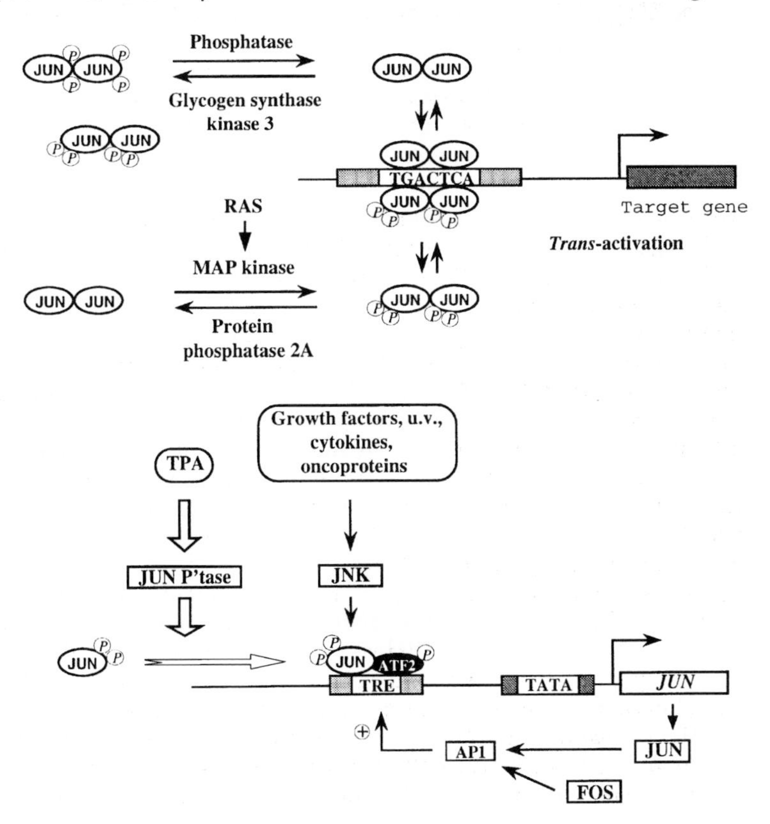

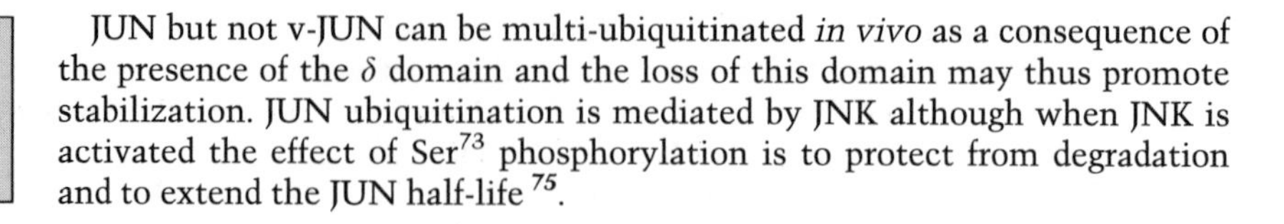

JUN but not v-JUN can be multi-ubiquitinated *in vivo* as a consequence of the presence of the $\delta$ domain and the loss of this domain may thus promote stabilization. JUN ubiquitination is mediated by JNK although when JNK is activated the effect of $Ser^{73}$ phosphorylation is to protect from degradation and to extend the JUN half-life [75].

## Amino acid sequence of human JUN

```
  1 MTAKMETTFY DDALNASFLP SESGPYGYSN PKILKQSMTL NLADPVGSLK
 51 PHLRAKNSDL LTSPDVGLLK LASPELERLI IQSSNGHITT TPTPTQFLCP
101 KNVTDEQEGF AEGFVRALAE LHSQNTLPSV TSAAQPVNGA GMVAPAVASV
151 AGGSGSGGFS ASLHSEPPVY ANLSNFNPGA LSSGGGAPSY GAAGLAFPAQ
201 PQQQQQPPHH LPQQMPVQHP RLQALKEEPQ TVPEMPGETP PLSPIDMESQ
251 ERIKAERKRM RNRIAASKCR KRKLERIARL EEKVKTLKAQ NSELASTANM
301 LREQVAQLKQ KVMNHVNSGC QLMLTQQLQT F (331)
```

**Domain structure**

| | |
|---|---|
| 1–89 | Region of bacterially expressed JUN binding to normal and oncogenic RAS [76] |
| 258–276 | Basic DNA binding region |
| 280–308 | Leucine zipper domain (underlined) |
| 31–47 | JNKs/SAPKs bind directly to this region [77] |
| 63 and 73 | Serine phosphorylation by MAP kinase, JUN N-terminal kinases [78]. $Ser^{73}$ phosphorylation directly enhances *trans*-activation function [79] |
| 91 and 93 | Phosphorylation of $Thr^{91}$ and/or $Thr^{93}$ promotes the dephosphorylation of the C-terminus and DNA binding [80] |
| 67–77 | HOB1 domain |
| 108–117 | HOB2 domain |
| 321 and 249 | Threonine and serine phosphorylation by casein kinase II |
| 239, 243 and 249 | Threonine and serine phosphorylation by glycogen synthase kinase-3 |

The substitution of $Glu^{293}$ and $Asn^{299}$ by glycines, $Ile^{264}$ by Leu and $Ser^{267}$ by Thr is responsible for the 10-fold lower DNA binding activity of JUNB [81].

The X-ray crystal structure of a heterodimer of the bZIP regions of FOS (139–200) and JUN (263–324) bound to DNA reveals that both subunits form continuous $\alpha$ helices. The C-terminal regions form an asymmetric coiled-coil: the N-terminal regions make base-specific contacts (FOS: $Asn^{147}$, $Ala^{150,151}$, $Cys^{154}$, $Arg^{155}$; JUN $Asn^{271}$, $Ala^{274,275}$, $Cys^{278}$, $Arg^{279}$) in the major groove of DNA [82].

**Database accession numbers**

| | *PIR* | *SWISSPROT* | *EMBL/GENBANK* | *REFERENCES* |
|---|---|---|---|---|
| Human *JUN* | A30009 | P05412 | J04111 | 2,83 |

***References***


1 Maki, Y. et al. (1987) Proc. Natl Acad. Sci. USA 84, 2848–2852.
2 Bohmann, D. et al. (1987) Science 238, 1386–1392.
3 **Busch, S.J. and Sassone-Corsi, P. (1990) Trends Genet. 6, 36–40.**

[4] **Kouzarides, T. and Ziff, E. (1989) Cancer Cells 1, 71–76.**
[5] Quantin, B. and Breathnach, R. (1988) Nature 334, 538–539.
[6] Ryder, K. and Nathans, D. (1988) Proc. Natl Acad. Sci. USA 85, 8464–8467.
[7] Ryseck, R.P. et al. (1988) Nature 334, 535–537.
[8] Hallahan, D.E. et al. (1995) J. Biol. Chem. 270, 30303–30309.
[9] Mollinedo, F. et al. (1991) Biochem. J. 273, 477–479.
[10] Welter, J.F. and Eckert, R.L. (1995) Oncogene 11, 2681–2687.
[11] Neyns, B. et al. (1996) Oncogene 12, 1247–1257.
[12] Kovary, K. and Bravo, R. (1991) Mol. Cell. Biol. 11, 2451–2459.
[13] Colotta, F. et al. (1992) J. Biol. Chem. 267, 18278–18283.
[14] Chiu, R. et al. (1988) Cell 54, 541–552.
[15] Hadman, M. et al. (1993) Oncogene 8, 1895–1903.
[16] Stein, B. et al. (1993) EMBO J. 12, 3879–3891.
[17] Jain, J. et al. (1993) Nature 365, 352–355.
[18] Ryseck, R.-P. and Bravo, R. (1991) Oncogene 6, 533–542.
[19] Nicolaides, N.C. et al. (1992) J. Biol. Chem. 267, 19665–19672.
[20] Chatton, B. et al. (1994) Oncogene 9, 375–385.
[21] De Cesare, D. et al. (1995) Oncogene 11, 365–376.
[22] Nakae, K. et al. (1995) J. Biol. Chem. 270, 23795–23800.
[23] Robinson, G.L.W.G. et al. (1995) Mol. Cell. Biol. 15, 1398–1404.
[24] Howcroft, T.K. et al. 1993) EMBO J. 12, 3163–3169.
[25] Mettouchi, A. et al. (1994) EMBO J. 13, 5668–5678.
[26] Yamit-Hezi, A. et al. (1994) Oncogene 9, 1065–1079.
[27] Tsang, T.C. et al. (1994) Cancer Res. 54, 882–886.
[28] Ullman, K.S. et al. (1993) Genes Dev. 7, 188–196.
[29] Schaefer, T.S. et al. (1995) Proc. Natl Acad. Sci. USA 92, 9097–9101.
[30] Auwerx, J. and Sassone-Corsi, P. (1992) Oncogene 7, 2271–2280.
[31] de Groot, R.P. and Sassone-Corsi, P. (1992) Oncogene 7, 2281–2286.
[32] Monteclaro, F.S. and Vogt, P.K. (1993) Proc. Natl Acad. Sci. USA 90, 6726–6730.
[33] Nakabeppu, Y. et al. (1988) Cell 55, 907–915.
[34] Natoli, G. et al. (1994) Oncogene 9, 2837–2843.
[35] Schutte, J. et al. (1988) Proc. Amer. Assoc. Cancer Res. Art. 1808, 455.
[36] Szabo, E. et al. (1996) Cancer Res. 56, 305–315.
[37] Wang, Z.Q. et al. (1995) Cancer Res. 55, 6244–6251.
[38] Vanhamme, L. et al. (1993) Cancer Res. 53, 615–621.
[39] Hilberg, F. et al. (1993) Nature 365, 179–181.
[40] Johnson, R.S. et al. (1993) Genes Dev. 7, 1309–1317.
[41] Johnson, R. et al. (1996) Mol. Cell. Biol. 16, 4504–4511.
[42] Bossy-Wetzel, E. et al. (1992) Genes Dev. 6, 2340–2351.
[43] Wong, W.-Y. et al. (1992) Oncogene 7, 2077–2080.
[44] Hilberg, F. and Wagner, E.F. (1992) Oncogene 7, 2371–2380.
[45] Schutte, J. et al. (1989) Proc. Natl Acad. Sci. USA 86, 2257–2261.
[46] Bos, T.J. et al. (1990) Genes Dev. 4, 1677–1687.
[47] **Vogt, P.K. and Bos, T.J. (1990) Adv. Cancer Res. 55, 1–35.**
[48] Garcia, M. and Samarut, J. (1993) Proc. Natl Acad. Sci. USA 90, 8837–8841.
[49] Okuno, H. et al. (1991) Oncogene 6, 1491–1497.
[50] Rapp, U.R. et al. (1994) Oncogene 9, 3493–3498.
[51] Peng, S.S.-Y. et al. (1996) Mol. Cell. Biol. 16, 1490–1499.
[52] Brach, M.A. et al. (1993) Mol. Cell. Biol. 13, 4284–4290.

[53] Angel, P. et al. (1988) Cell 55, 875–885.
[54] Lee, J.-S. et al. (1996) Mol. Cell. Biol. 16, 4312–4326.
[55] Chen, L.I. et al. (1994) Mol. Cell. Biol. 14, 4380–4389.
[56] Naftilan, A.J. et al. (1990) Mol. Cell. Biol. 10, 5536–5540.
[57] Rao, G.N. et al. (1993) Oncogene 8, 2759–2764.
[58] Heintz, N.H. et al. (1993) Proc. Natl Acad. Sci. USA 90, 3299–3303.
[59] Baichwal, V.R. and Tjian, R. (1990) Cell 63, 815–825.
[60] Baichwal, V.R. et al. (1991) Nature 352, 165–168.
[61] Frame, M.C. et al. (1991) Oncogene 6, 205–209.
[62] Binetruy, B. et al. (1991) Nature 351, 122–127.
[63] Smeal, T. et al. (1992) Mol. Cell. Biol. 12, 3507–3513.
[64] Gupta, S. et al. (1996) EMBO J. 15, 2760–2770.
[65] Zohn, I.E. et al. (1995) Mol. Cell. Biol. 15, 6160–6168;
[66] Pulverer, B.J. et al. (1993) Oncogene 8, 407–415.
[67] Olson, M.F. et al. (1995) Science 269, 1270–1272.
[68] Bannister, A.J. et al. (1995) Oncogene 11, 2509–2514.
[69] Westwick, J.K. et al. (1994) Proc. Natl Acad. Sci. USA 91, 6030–6034.
[70] Black, E.J. et al. (1991) Oncogene 6, 1949–1958.
[71] **Hunter, T. and Karin, M. (1992) Cell 70, 375–387.**
[72] Baker, S.J. et al. (1992) Mol. Cell. Biol. 12, 4694–4705.
[73] Papavassiliou, A.G. et al. (1992) Proc. Natl Acad. Sci. USA 89, 11562–11565.
[74] Abate, C. et al. (1993) Proc. Natl Acad. Sci. USA 90, 6766–6770.
[75] Fuchs, S.Y. et al. (1996) Oncogene 13, 1531–1535.
[76] Adler, V. et al. (1995) Proc. Natl Acad. Sci. USA 92, 10585–10589.
[77] Dai, T. et al. (1995) Oncogene 10, 849–855.
[78] Kyriakis, J.M. et al. (1994) Nature 369, 156–160.
[79] Smeal, T. et al. (1994) EMBO J. 13, 6006–6010.
[80] Papavassiliou, A.G. et al. (1995) EMBO J. 14, 2014–2091.
[81] Deng, T. and Karin, M. (1993) Genes Dev. 7, 479–490.
[82] Glover, J.N.M. and Harrison, S.C. (1995) Nature 373, 257–261.
[83] Hattori, K. et al. (1988) Proc. Natl Acad. Sci. USA 85, 9148–9152.

## Identification

*JUNB* was identified by screening a human genomic library with a v-*jun* probe [1].

## Related genes

*JUNB* is a member of the *JUN/Jun* gene family (see ***JUN***). JUNB is 44% identical to JUN.

| | ***JUNB*** |
|---|---|
| **Nucleotides (bp)** | 2136 |
| **Chromosome** | 19p13.2 |
| **Mass (kDa): predicted** | 36 |
| **expressed** | 42 |
| **Cellular location** | Nucleus |

**Tissue distribution**

*JUNB* is expressed in human peripheral blood granulocytes [2] and in normal ovarian tissue as well as in ovarian tumours [3]. Murine *Junb* expression is ubiquitous but is particularly high in the testis where *Jun* and *Jund* are barely detectable. JUNB is present in all layers of normal human epidermis [4].

## Protein function

*JUNB* is an "early response gene" induced by serum [5] but induction is attenuated in cells transformed by tyrosine kinase oncoproteins [6].

JUNB homodimers bind very weakly to AP1 sites to which JUN binds with high affinity [7]. Proteins of the FOS family confer significant affinity for DNA on JUNB, increasing the half-lives of the protein/DNA complexes. JUNB and other members of the FOS/JUN family (FRA1 and JUND) activate human involucrin promoter transcription in keratinocytes [8].

**Cancer**

In the cervical cancer cell line CC7T-a an abnormal *JUNB* transcript is expressed that results from insertion of two stretches of HPV16 DNA from the L1 open reading frame. A second integration event has also been detected in which the 3′ end of *JUNB* is replaced with an expressed sequence tag locus (EST182) from chromosome 15. In HeLa cells three-fold amplification and over-expression of JUNB occurs [9].

***In vitro***

*Junb* and *Hras* are less effective transforming agents than *Jun* and *Hras* and *Junb* inhibits *Jun/Ras* transformation [10], consistent with the differential DNA affinities of JUN and JUNB.

## Transcriptional regulation

The *Junb* locus contains nine flanking evolutionarily conserved sequences (FECS) that are between 72% and 91% identical in sequence in the human and mouse genes[11]. These include four DNAase I-hypersensitive sites (DHR1 (−100 to +250), DHR2 (around −1000), DHR3 (−1650) and DHR4 (+2040) that include serum response elements (SREs) at −1452 and +2091, a TRE at −949, two cAMP response elements (CREs) at +2071 and +2116, two ETS sites and two ETS-linked motifs[11–13]. The SREs and CREs regulate the *Junb* response to serum. −196 to −91 mediates the response to IL-6 and ciliary neurotrophic factor (CNTF) and IL-6-inducible complexes on this element contain STAT3 and a CRE-like site binding protein[14].

The E1A 13S RNA product directly activates *Fos*, *Jun* and *Junb* promoters, inducing JUN/AP1 binding to a TRE (TTACCTCA). E1A *trans*-activation is mediated by the JUN2 TRE sequence to which JUN/ATF-2 heterodimers bind[15]. E1A 12S RNA expression alone does not activate *via* TRE but, together with JUN does so: this activation is, in turn, blocked by expression of FOS[16]. Thus FOS modulates the dominance of E1A 12S or 13S products. *Junb* transcription is specifically stimulated by the expression of v-SRC, the tyrosine kinase activity of which modulates binding of factors to a 121bp region encompassing the CCAAT and TATAA elements[17].

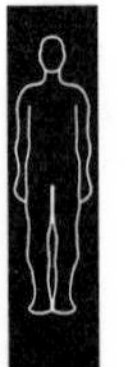

## Amino acid sequence of human JUNB

```
  1 MCTKMEQPFY HDDSYTATGY GRAPGGLSLH DYKLLKPSLA VNLADPYRSL
 51 KAPGARGPGP EGGGGGSYFS GQGSDTGASL KLASSELERL IVPNSNGVIT
101 TTPTPPGQYF YPRGGGSGGG AGGAGGGVTE EQEGFADGFV KALDDLHKMN
151 HVTPPNVSLG ATGGPPAGPG GVYAGPEPPP VYTNLSSYSP ASASSGGAGA
201 AVGTGSSYPT TTISYLPHAP PFAGGHPAQL GLGRGASTFK EEPQTVPEAR
251 SRDATPPVSP INMEDQERIK VERKRLRNRL AATKCRKRKL ERIARLEDKV
301 KTLKAENAGL SSTAGLLREQ VAQLKQKVMT HVSNGCQLLL GVKGHAF (347)
```

**Domain structure**

273–292 Basic DNA binding region
296–324 Leucine zipper domain (underlined)

**Database accession numbers**

| | *PIR* | *SWISSPROT* | *EMBL/GENBANK* | *REFERENCES* |
|---|---|---|---|---|
| Human *JUNB* | S10183 | P17275 | M29039 | 18 |
| | | | X51345 | 1 |

### *References*


1. Schutte, J. et al. (1989) Cell 59, 987–997.
2. Mollinedo, F. et al. (1991) Biochem. J., 273, 477–479.
3. Neyns, B. et al. (1996) Oncogene 12, 1247–1257.
4. Welter, J.F. and Eckert, R.L. (1995) Oncogene 11, 2681–2687.
5. Ryder, K. et al. (1988) Proc. Natl Acad. Sci. USA 85, 1487–1491.
6. Yu, C.-L. et al. (1993) Mol. Cell. Biol. 13, 2011–2019.
7. Ryseck, R.-P. and Bravo, R. (1991) Oncogene 6, 533–542.
8. **Welter, J.F. et al. (1995) J. Biol. Chem. 270, 12614–12622.**
9. Choo, K.B. et al. (1995) Cancer Lett. 93, 249–253.

[10] Schutte, J. et al. (1989) Proc. Natl Acad. Sci. USA 86, 2257–2261.
[11] Phinney, D.G. et al. (1996) Oncogene 13, 1875–1883.
[12] Perez-Albuerne et al. (1994) Proc. Natl Acad. Sci. USA 90, 11960–11964.
[13] Phinney, D.G. et al. (1994) Oncogene 9, 2353–2362.
[14] Kojima, H. et al. (1996) Oncogene 12, 547–554.
[15] van Dam, H. et al. (1993) EMBO J. 12, 479–487.
[16] de Groot, R. et al. (1991) Mol. Cell Biol. 11, 192–201.
[17] Apel, I. et al. (1992) Mol. Cell. Biol. 12, 3356–3364.
[18] Nomura, N. et al. (1990) Nucleic Acids Res. 18, 3047–3048.

## Identification

*JUND* was identified by screening a human genomic library with a v-*jun* probe [1].

## Related genes

*JUND* is a member of the *JUN*/*Jun* gene family (see ***JUN***). JUND is 45% identical to JUN.

| | ***JUND*** |
|---|---|
| **Nucleotides (bp)** | Not fully mapped |
| **Chromosome** | 19p13.2 |
| **Mass (kDa): predicted** | 35 |
| **expressed** | 40–50 |
| **Cellular location** | Nucleus |

**Tissue distribution**

*JUND* mRNA is 5- to 10-fold more abundant than *JUN* mRNA and occurs in most tissues, being maximal in the intestine and thymus and readily detectable in human peripheral blood granulocytes [2,3]. It is present in serum-starved cells and is not significantly induced by the addition of serum [4]. JUND is present in all layers of normal human epidermis [5]. *JunD*, together with *JunB* and *Fos*, is transcriptionally activated in mouse mammary epithelial cells after lactation. AP1 DNA binding activity (primarily FOS/JUND) is transiently induced and may mediate the apoptosis of mammary cells during involution [6]. *JUND* mRNA is expressed in ovarian tumours as well as in normal ovaries [7].

## Protein function

*JUND* is an "early response gene", as are *JUN* and *JUNB*, although *JUND* is significantly expressed in quiescent cells.

JUND (and JUNB) homodimers bind very weakly to AP1 sites to which JUN binds with high affinity [8]. JUND (and JUN) *trans*-activate the human *MYB* promoter *via* an AP1-like sequence [9]. In PC-12 cells stimulated by NGF JUND activates *nur77* transcription [10]. JUND and other members of the FOS/JUN family (FRA1 and JUNB) activate human involucrin promoter transcription in keratinocytes [11]. In *Kras*-transformed NIH 3T3 cells oxamflatin induces fibronectin expression and phenotypic reversion as a result of high levels of expression of JUND [12]. JUND DNA binding activity is enhanced by 1,25-dihydroxyvitamin $D_3$ during cell cycle arrest in human chronic myelogenous leukaemia cells [13].

**Cancer**

*JUND* loci may be involved in chromosome translocations 19;11 and 19;1 that occur in some cases of acute lymphocytic leukaemia (ALL) and acute non-lymphocytic leukaemia [14].

### *In vitro*

Mouse JUND negatively regulates fibroblast growth and antagonizes transformation by *Ras*[15]. The expression of *Jund* from a retroviral vector does not transform chick embryo fibroblasts but JUND is converted into a transforming protein in these cells by substitution of the 79 N-terminal amino acids of JUN[16] or by mutations in two of the N-terminal regions conserved within JUN proteins[17].

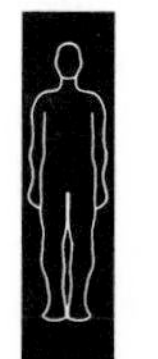

## Amino acid sequence of human JUND

```
  1 METPFYGDEA LSGLGGGASG SGGTFASPGR LFPGAPPTAA AGSMMKKDAL
 51 TLSLSEQVAA ALKPAPAPAS YPPAADGAPS AAPPDGLLAS PDLGLLKLAS
101 PELERLIIQS NGLVTTTPTS SQFLYPKVAA SEEQEFAEGF VKALEDLHKQ
151 NQLGAGRAAA AAAAAAGGPS GTATGSAPPG ELAPAAAAPE APVYANLSSY
201 AGGAGGAGGA ATVAFAAEPV PFPPPPPPGA LGPPRLAALK DEPQTVPDVP
251 SFGESPPLSP IDMDTQERIK AERKRLRNRI AASKCRKRKL ERISRLEEKV
301 KTLKSQNTEL ASTASLLREQ VAQLKQKVLS HVNSGCQLLP QHQVPAY (347)
```

### Domain structure

158–166 Alanine-rich region
273–292 Basic DNA binding region
296–324 Leucine zipper domain (underlined)

### Database accession numbers

| | *PIR* | *SWISSPROT* | *EMBL/GENBANK* | *REFERENCES* |
|---|---|---|---|---|
| Human *JUND* | S10184 | P17535 | X51346 | 1 |
| | | | X56681 | 18 |

### *References*


1 Nomura, N. et al. (1990) Nucleic Acids Res. 18, 3047–3048.
2 Mollinedo, F. et al. (1991) Biochem. J., 273, 477–479.
3 Ryder, K. et al. (1989) Proc. Natl Acad. Sci. USA 86, 1500–1503.
4 Hirai, S.-I. et al. (1989) EMBO J. 8, 1433–1439.
5 Welter, J.F. and Eckert, R.L. (1995) Oncogene 11, 2681–2687.
6 Marti, A. et al. (1994) Oncogene 9, 1213–1223.
7 Neyns, B. et al. (1996) Oncogene 12, 1247–1257.
8 Ryseck, R.-P. and Bravo, R. (1991) Oncogene 6, 533–542.
9 Nicolaides, N.C. et al. (1992) J. Biol. Chem. 267, 19665–19672.
10 Yoon, J.K. and Lau, L.F. (1994) Mol. Cell. Biol. 14, 7731–7743.
11 Welter, J.F. et al. (1995) J. Biol. Chem. 270, 12614–12622.
12 Sonoda, H. et al. (1996) Oncogene 13, 143–149.
13 Lasky, S. et al. (1995) J. Biol. Chem. 270, 19676–19679.
14 Mattei, M.G. et al. (1990) Oncogene 5, 151–156.
15 Pfarr, C.M. et al. (1994) Cell 76, 747–760.
16 Metivier, C. et al. (1993) Oncogene 8, 2311–2315.
17 Kameda, T. et al. (1993) Proc. Natl Acad. Sci. USA 90, 9369–9373.
18 Berger, I. and Shaul, Y. (1991) Oncogene 6, 561–566.

## Identification

v-*kit* is the oncogene of the Hardy–Zuckerman 4 strain of acutely transforming feline sarcoma virus (HZ4-FeSV)[1]. *KIT* was detected by screening cDNA with v-*kit* probes[2].

## Related genes

KIT is a member of receptor tyrosine kinase subclass III which also includes FLK2/FLT3, PDGFRA, PDGFRB, and CSF1R, each having five Ig-like loops in an extracellular domain and a cytoplasmic region containing a 60–100 residue tyrosine kinase insert region. *KIT* also shares homology with the tyrosine-specific protein kinase family (ABL, FES, FGR, FMS and SRC), with the EGF and insulin receptors and with mouse *Flk-1* (fetal liver kinase 1) and *Flk-2/Flt-3*.

| | *KIT* | v-*kit* |
|---|---|---|
| **Nucleotides (kb)** | 80 | 2.37 (HZ4-FeSV genome) |
| **Chromosome** | 4q11–q21 | |
| **Mass (kDa): predicted** | 110 | |
| **expressed** | gp124/gp160 | $p81^{gag\text{-}kit}$ |
| **Cellular location** | Plasma membrane | |

**Tissue distribution**

Expression of the KIT receptor and its ligand occurs in complementary tissues throughout the body from the early presomite stage to the mature adult. For example, the ligand (*steel*) is expressed in the follicular cells of the ovary and in Sertoli cells of the testes, the layers immediately surrounding the germ cells that express *Kit*[3–5]. *Kit* is also expressed in the interstitial cells of Cajal in the gut[6].

*KIT* mRNA and protein are present in normal human neonatal and adult melanocytes and in the early stages of erythroid and myeloid cell differentiation[7,8] but are not detectable in most human melanoma-derived cell lines[9,10].

## Protein function

*KIT* is a membrane tyrosine kinase receptor. In humans KIT is the receptor for stem cell factor (SCF, also called mast cell growth factor, MGF)[11–16].

The KIT receptor in mice is encoded by the dominant white spotting (*W*) locus. The ligand, KL (or SCF or MGF) is a growth factor encoded by the mouse *steel* (*Sl*) locus[17–19]. The phenotypes of *W* and *Sl* mutant mouse strains that bear germline loss of function mutations in *Kit* and its ligand, respectively, demonstrate roles for *Kit* in haematopoiesis, melanogenesis and germ cell development[20–23]. Furthermore, SCF acts synergistically with other cytokines to stimulate *in vitro* growth of a number of committed haematopoietic precursors[24–29]. These activities can be blocked by antibodies directed against KIT[30] or by anti-sense oligodeoxynucleotides that block *Kit* expression[31]. Anti-KIT antibodies also block melanoblast migration and prevent melanocyte activation in postnatal mice[32]. SCF promotes germ cell survival *in vitro*[33,34] and anti-KIT antibodies inhibit survival and/or proliferation of mature type-A spermatogonia[35].

The interaction of the ligand KL with its receptor causes receptor dimerization[36] that leads to enhanced KIT tyrosine autophosphorylation and association with PtdIns 3-kinase and PLC$\gamma$ though not detectably with GAP[37]. In megakaryocytes MATK tyrosine kinase associates with activated KIT[38]. KL causes protein kinase C-mediated phosphorylation of KIT, which inhibits tyrosine autophosphorylation of the receptor[39], and an increase in the cellular concentration of RAS-GTP[40] followed by the serine phosphorylation of RAF1[41]. KIT/KL complexes are rapidly internalized, undergoing ubiquitin-mediated degradation. Proteolytic cleavage of KIT also occurs, mediated by protein kinase C and independent of KIT kinase activity[42]. SCF causes the SRC-related TEC kinase to associate with KIT[43] and activated KIT also transiently associates with the tyrosine phosphatase haematopoietic cell phosphatase (HCP, SHP1 or PTP1C) that can dephosphorylate KIT *in vitro*[44] and negatively regulates KIT *in vivo*[45].

The phosphotyrosine phosphatase SYP associates with KIT following SCF activation and SYP also complexes with GRB2[46]. GRAP (GRB2-related adapter protein) also associates with activated KIT[47]. In the EPO-dependent erythroid progenitor cell line HCD57, SCF rapidly induces tyrosine phosphorylation of the EPO receptor and KIT associates with the cytoplasmic domain of the EPO receptor[48].

**Cancer**

*KIT* and its ligand are highly expressed in small cell lung cancer[49,50], on acute myeloblastic leukaemia blast cells[51] and in a significant proportion of testicular germ cell tumours[52]. The level of expression of *KIT* declines during the progression of cutaneous melanoma[53]. *KIT* expression is down-regulated in follicular thyroid carcinomas[54].

The Asp$^{816}$ → Val mutation in humans, which causes constitutive activation and phosphorylation of KIT *in vitro*, occurs in patients who have mastocytosis with an associated haematologic disorder and in urticaria pigmentosa[55].

**Transgenic animals**

Mice expressing the dominant negative $W^{42}$ mutation show effects on pigmentation and the number of tissue mast cells that are characteristic of some *W* phenotypes. Germ cell development and erythropoiesis are not affected[56].

Testicular tumours are induced in male transgenic mice expressing HPV16 E6/E7 in which the activation of *KIT* by its ligand is essential for tumorigenesis to develop[57].

**In animals**

Seven dominant mutations at the *W/Kit* locus affect pigmentation in mice ($W^{a}$, $W^{12}$, $W^{37}$, $W^{42}$, $W^{ei}$, $W^{v}$ (= $W^{55}$) and $W^{rio}$). The $W^{rio}$ mutation can be rescued by

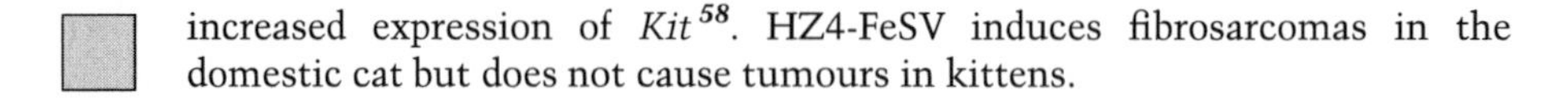

increased expression of *Kit*[58]. HZ4-FeSV induces fibrosarcomas in the domestic cat but does not cause tumours in kittens.

***In vitro***

v-*kit* transforms NIH 3T3 cells. Tumour cells bearing HZ4-FeSV transform feline embryo fibroblasts and CCL64 mink cells[1,59].

## Gene structure

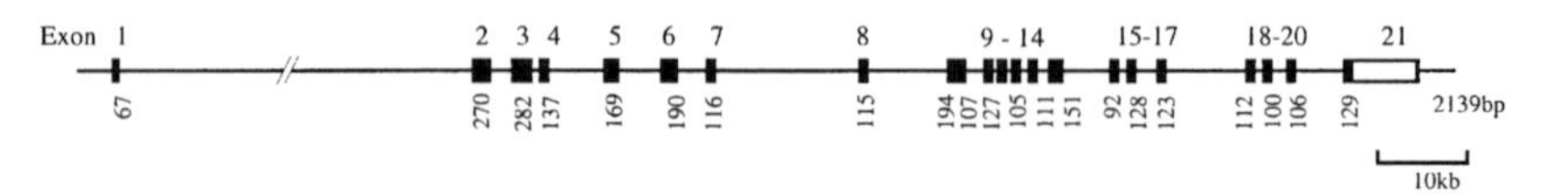


The overall structure closely resembles that of *FMS* including the large first intron and identical exon/intron boundaries in their two kinase domains. There are two *KIT* isoforms in both mice and humans distinguished by alternative splicing of the final 12 bp of exon 9, immediately upstream of the region encoding the transmembrane domain. The inclusion of this exon gives rise to *KitA*$^{+}$[60–62].

## Protein structure of mouse KIT

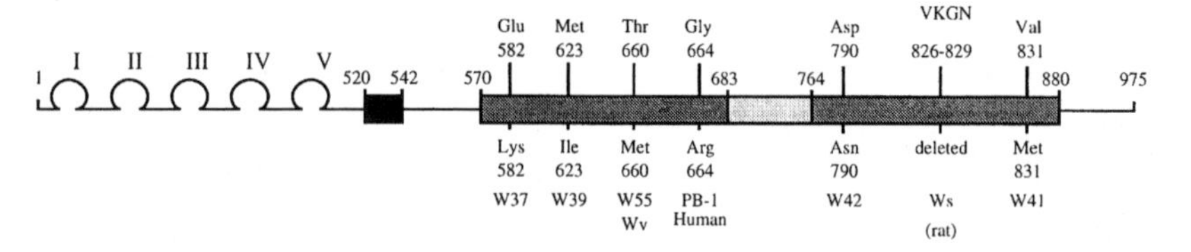


Point mutations and deletions occurring in mutated *Kit* alleles are shown below the diagram. The third immunoglobulin domain is essential for rodent SCF binding whereas the major binding determinant for human SCF is the second domain[63]. In the mouse, $W^{55}$ and $W^{v}$ arise from the same substitution in independent alleles giving identical phenotypes. The $W^{42}$ mutation, which is particularly severe in its effects in both the homozygous and heterozygous states, arises from a missense mutation[23]. The original *W* mutant bears a deletion of the transmembrane domain and part of the kinase domain[61]. The product of this gene is not expressed at the cell surface, kinase activity is lost and the homozygous mutation is lethal. For the other mutations indicated in the diagram immunoprecipitable KIT protein is synthesized: its kinase activity is reduced in $W^{39}$, $W^{55}$, $W^{v}$ and $W^{41}$ mutants and abolished in $W^{37}$ and $W^{42}$ mutants[21–23]. The $W^{v}$, $W^{41}$ and $W^{42}$ mutations also differentially affect the binding of p145$^{kit}$ to PLC$\gamma$, GAP and the p85 subunit of PtdIns 3-kinase[64]. $W^{f}$, $W^{Jic}$ and $W^{n}$ mutants arise from point mutations in the kinase-encoding region[65]. The $W^{ei}$ mutation occurs in the ATP binding domain which affects KIT kinase activity and gives rise to adult skin partly devoid of differentiated pigmented cells[66]. In *W*-sash ($W^{sh}$) and patch (*Ph*) mutant mice, *Kit* mis-expression affects early melanogenesis and is responsible for the pigment deficiency[67].

PB-1 refers to a substitution at $Gly^{664}$ in the human sequence detected in a case of piebaldism, a condition similar to the dominant white spotting (*W*) disorder in mice [68]. Truncation of KIT has also been detected in piebaldism [69]. *Ws* refers to a mutant rat strain having a four amino acid deletion in the conserved phosphotransferase region [70].

Human, mouse and feline KIT share >80% sequence identity. v-KIT differs in six scattered point mutations from the corresponding region of human KIT and its oncogenicity appears to derive from the removal of the extracellular and transmembrane domains, together with 50 amino acids from the C-terminus of KIT [71].

## Amino acid sequence of human KIT

1 *MRGARGAWDF LCVLLLLLRV QT*GSSQPSVS PGEPSPPSIH PGKSDLIVRV  
51 GDEIRLLCTD PGFVKWTFEI LDETNENKQN EWITEKAEAT NTGKYTCTNK  
101 HGLSNSIYVF VRDPAKLFLV DRSLYGKEDN DTLVRCPLTD PEVTNYSLKG  
151 CQGKPLPKDL RFIPDPKAGI MIKSVKRAYH RLCLHCSVDQ EGKSVLSEKF  
201 ILKVRPAFKA VPVVSVSKAS YLLREGEEFT VTCTIKDVSS SVYSTWKREN  
251 SQTKLQEKYN SWHHGDFNYE RQATLTISSA RVNDSGVFMC YANNTFGSAN  
301 VTTTLEVVDK GFINIFPMIN TTVFVNDGEN VDLIVEYEAF PKPEHQQWIY  
351 MNRTFTDKWE DYPKSENESN IRYVSELHLT RLKGTEGGTY TFLVSNSDVN  
401 AAIAFNVYVN TKPEILTYDR LVNGMLQCVA AGFPEPTIDW YFCPGTEQRC  
451 SASVLPVDVQ TLNSSGPPFG KLVVQSSIDS SAFKHNGTVE CKAYNDVGKT  
501 SAYFNFAFKG NNKEQIHPHT LFTPLLIGFV IVAGMMCIIV MILTYKYLQK  
551 PMYEVQWKVV EEINGNNYVY IDPTQLPYDH KWEFPRNRLS FGKTLGAGAF  
601 GKVVEATAYG LIKSDAAMTV AVKMLKPSAH LTEREALMSE LKVLSYLGNH  
651 MNIVNLLGAC TIGGPTLVIT EYCCYGDLLN FLRRKRDSFI CSKQEDHAEA  
701 ALYKNLLHSK ESSCSDSTNE YMDMKPGVSY VVPTKADKRR SVRIGSYIER  
751 DVTPAIMEDD ELALDLEDLL SFSYQVAKGM AFLASKNCIH RDLAARNILL  
801 THGRITKICD FGLARDIKND SNYVVKGNAR LPVKWMAPES IFNCVYTFES  
851 DVWSYGIFLW ELFSLGSSPY PGMPVDSKFY KMIKEGFRML SPEHAPAEMY  
901 DIMKTCWDAD PLKRPTFKQI VQLIEKQISE STNHIYSNLA NCSPNRQKPV  
951 VDHSVRINSV GSTASSSQPL LVHDDV (976)

**Domain structure**

| | |
|---|---|
| 1–22 | Signal sequence (italics) |
| 23–520 | Extracellular domain |
| 521–543 | Transmembrane domain (underlined) |
| 544–976 | Intracellular domain |
| 596–601 and 623 | ATP binding region |
| 719 | PtdIns 3-kinase (p85 subunit) binding site ($Tyr^{719}$) [72]. $Tyr^{719}$ is essential for a full KL-induced mitogenic response and for cell adhesion |
| 821 | $Tyr^{821}$ is essential for KIT-mediated mitogenesis and survival but not cell adhesion [73] |
| 792 | Active site |
| 823 | Autophosphorylation site |
| 130, 145, 283, 293, 300, 320, 352, 367, 463, 486 | Potential carbohydrate attachment sites |

The insertion in murine *KitA*$^{+}$ encodes Gly-Asn-Asn-Lys between amino acids 512 and 513 [62]: this insert is also present in human KIT.

## Database accession numbers

| | PIR | SWISSPROT | EMBL/GENBANK | REFERENCES |
|---|---|---|---|---|
| Human *KIT* | S01426 | P10721 | X06182 | 2,60,74 |

## *References*


[1] Besmer, P. et al. (1986) Nature 320, 415–421.
[2] Yarden, Y. et al. (1987) EMBO J. 6, 3341–3351.
[3] Keshet, E. et al. (1991) EMBO J. 10, 2425–2435.
[4] Matsui, Y. et al. (1990) Nature 347, 667–669.
[5] Motro, B. et al. (1991) Development 113, 1207–1222.
[6] Huizinga, J.D. et al. (1995) Nature 373, 347–349.
[7] Andre, C. et al. (1989) Oncogene 4, 1047–1049.
[8] Nocka, K. et al. (1989) Genes Dev. 3, 816–826.
[9] Lassam, N. and Bickford, S. (1992) Oncogene 7, 51–56.
[10] Zakut, R. et al. (1993) Oncogene 8, 2221–2229.
[11] Anderson, D.M. et al. (1990) Cell 63, 235–243.
[12] Huang, E. et al. (1990) Cell 63, 225–233.
[13] Martin, F.H. et al. (1990) Cell 63, 203–211.
[14] Nocka, K. et al. (1990) EMBO J. 9, 3287–3294.
[15] Williams, D.E. et al. (1990) Cell 63, 167–174.
[16] Zsebo, K.M. et al. (1990) Cell 63, 195–201.
[17] Copeland, N.G. et al. (1990) Cell 63, 175–183.
[18] Brannan, C.I. et al. (1991) Proc. Natl Acad. Sci. USA 88, 4671–4674.
[19] Zsebo, K.M. et al. (1990) Cell 63, 213–224.
[20] Silvers, W.K. (1979) In The Coat Colors of Mice: A Model for Mammalian Gene Action and Interaction. Springer-Verlag, New York, pp. 206–241.
[21] Nocka, K. et al. (1990) EMBO J. 9, 1805–1813.
[22] Reith, A.D. et al. (1990) Genes Dev. 4, 390–400.
[23] Tan, J.C. et al. (1990) Science 247, 209–212.
[24] Bernstein, I.D. et al. (1991) Blood 77, 2316–2321.
[25] Briddell, R.A. et al. (1991) Blood 78, 2854–2859.
[26] Broxmeyer, H.E. et al. (1991) Blood 77, 2142–2149.
[27] McNeice, I.K. et al. (1991) Exp. Hematol., 19, 226–231.
[28] Metcalf D. and Nicola, N.A. (1991) Proc. Natl Acad. Sci. USA 88, 6239–6243.
[29] Migliaccio, G. et al. (1991) Proc. Natl Acad. Sci. USA 88, 7420–7424.
[30] Ogawa, M. et al. (1991) J. Exp. Med. 174, 63–71.
[31] Ratajczak, M.Z. et al. (1992) Proc. Natl Acad. Sci. USA 89, 1710–1714.
[32] Nishikawa, S. et al. (1991) EMBO J. 10, 2111–2118.
[33] Dolci, S. et al. (1991) Nature 352, 809–811.
[34] Godin, I. et al. (1991) Nature 352, 807–809.
[35] Yoshinaga, K. et al. (1991) Development 113, 689–699.
[36] Lev, S. et al. (1992) J. Biol. Chem. 267, 15970–15977.
[37] Shearman, M.S. et al. (1993) EMBO J. 12, 3817–3826.
[38] Jhun, B.H. et al. (1995) J. Biol. Chem. 270, 9661–9666.
[39] Blume-Jensen, P. et al. (1993) EMBO J. 12, 4199–4209.
[40] Duronio, V. et al. (1992) Proc. Natl Acad. Sci. USA 89, 1587–1591.
[41] Lev, S. et al. (1991) EMBO J. 10, 647–654.
[42] Yee, N.S. et al. (1994) J. Biol. Chem. 269, 31991–31998.
[43] Tang, B. et al. (1994) Mol. Cell. Biol. 14, 8432–8437.

[44] Yi, T. and Ihle, J.N. (1993) Mol. Cell. Biol. 13, 3350–3358.
[45] Paulson, R.F. et al. (1996) Nature Genet. 13, 309–323.
[46] Tauchi, T. et al. (1994) J. Biol. Chem. 269, 25206–25211.
[47] Feng, G.-S. et al. (1996) J. Biol. Chem. 271, 12129–12132.
[48] Wu, H. et al. (1995) Nature 377, 242–246.
[49] Hibi, K. et al. (1991) Oncogene 6, 2291–2296.
[50] Sekido, Y. et al. (1991) Cancer Res. 51, 2416–2419.
[51] Buhring, H.-J. et al. (1993) Cancer Res. 53, 4424–4431.
[52] Strohmeyer, T. et al. (1991) Cancer Res. 51, 1811–1816.
[53] Natali, P.G. et al. (1992) Int. J. Cancer 52, 197–201.
[54] Natali, P.G. et al. (1995) Cancer Res. 55, 1787–1791.
[55] Longley, B.J. et al. (1996) Nature Genet. 12, 312–314.
[56] Ray, P. et al. (1991) Genes Dev. 5, 2265–2273.
[57] Kondoh, G. et al. (1995) Oncogene 10, 341–347.
[58] De Sepulveda, P. et al. (1995) Mol. Cell. Biol. 15, 5898–5905.
[59] Hampe, A. et al. (1984) Proc. Natl Acad. Sci. USA 81, 85–89.
[60] Vandenbark, G.R. et al. (1992) Oncogene 7, 1259–1266.
[61] Hayashi, S.-I. et al. (1991) Nucleic Acids Res. 19, 1267–1271.
[62] Reith, A.D. et al. (1991) EMBO J. 10, 2451–2459.
[63] Lev, S. et al. (1993) Mol. Cell. Biol. 13, 2224–2234.
[64] Herbst, R. et al. (1992) J. Biol. Chem. 267, 13210–13216.
[65] Koshimizu, U. et al. (1994) Oncogene 9, 157–162.
[66] De Sepulveda, P. et al. (1994) Oncogene 9, 2655–2661.
[67] Duttlinger, R. et al. (1995) Proc. Natl Acad. Sci. USA 92, 3754–3758.
[68] Giebel, L.B. and Spritz, R.A. (1991) Proc. Natl Acad. Sci. USA 88, 8696–8699.
[69] Spritz, R. A. et al. (1993) Hum. Mol. Genet. 2, 1499–1500.
[70] Tsujimura, T. et al. (1991) Blood 78, 1942–1946.
[71] Qiu, F. et al. (1988) EMBO J. 7, 1003–1011.
[72] Serve, H. et al. (1994) J. Biol. Chem. 269, 6026–6030.
[73] Serve, H. et al. (1995) EMBO J. 14, 473–483.
[74] Giebel, L.B. et al. (1992) Oncogene 7, 2207–2217.

## Identification

*Lck* encodes a cellular tyrosine kinase (formerly *lsk*$^{T}$ or *tck*), detected originally in the LSTRA murine lymphoma cell line derived from MuLV-induced thymomas [1].

## Related genes

Member of the *SRC* tyrosine kinase family (*Blk, FGR, FYN, HCK, LCK/Tkl, LYN, SRC, YES*). Murine BMK has 70% homology with murine LCK. A *Lck*-related gene is expressed in murine eggs. *Tkl* is the avian cellular tyrosine kinase homologue of Lck.

| | *LCK* |
|---|---|
| **Nucleotides (kb)** | 2.2 |
| **Chromosome** | 1p35–p32 |
| **Mass (kDa): predicted** | 58 |
| **expressed** | 56 |
| **Cellular location** | Plasma membrane |

**Tissue distribution**
*LCK, FGR, LYN* and *HCK* are expressed only in haematopoietic cells and *LCK* is found specifically only in lymphoid cells, T cells, NK cells and some B cells. Chicken *Tkl* is abundantly expressed in the thymus with lesser expression in the spleen [2].

## Protein function

LCK is a lymphocyte-specific tyrosine kinase associated with the cytoplasmic domains of CD4 and CD8$\alpha$ (CD8 can exist as $\alpha\alpha$ or $\alpha\beta$ dimers) that bind to class II and class I MHC molecules respectively [3]. LCK co-precipitates with CD20, PtdIns 3-kinase, PtdIns 4-kinase, the IL-2R $\beta$ chain [4], CD5 (which it also phosphorylates) [5,6], CD2 [7] and CD44 [8], and also with GAP and immunoglobulin receptors in LSTRA cells and B cells, respectively [9–13].

Stimulation of the T cell receptor (TCR) activates LCK but it is FYN, rather than LCK, that co-precipitates with the TCR [14]. Furthermore, T cells from LCK-deficient mice proliferate in response to activation of the TCR or to IL-2 [15]. Cross-linking CD4 enhances the autophosphorylation of $p56^{lck}$, which may cause the phosphorylation of the $\zeta$ chain of the CD3/TCR complex with which LCK associates [16]. RAF1 and MAP kinase are substrates for LCK [17,18]. Thus MAP kinases may be directly regulated by LCK (and other members of the SRC family) and participate in cell signalling cascades that may activate substrates such as ribosomal S6 kinases (RSKs).

An alternative role for LCK may be in regulating T cell maturation and differentiation: it is expressed before CD4/8 in developing T cells and is present in complexes with CD4 and CD8 in immature $CD4^{+}/CD8^{+}$ cells.

The expression in transgenic mice of a catalytically inactive form of LCK that functions in a dominant negative manner indicates that normal LCK plays a critical role in $CD4^+8^+$ thymocyte development for which FYN cannot substitute[19,20]. However, this role may be independent of association between LCK and CD4 or CD8.

The intracellular domain of CD4 negatively regulates the replicative rate of HIV-1 in T cells and its association with LCK is necessary for this effect[21]. The product of the HIV *nef* gene, which plays an unresolved role in viral replication, promotes downregulation of CD4 and interacts with LCK to depress its kinase activity[22]. In herpes saimiri-transformed human T cells, LCK associates with TIP protein, the product of the viral *orf1* gene[23]. LCK kinase activity is strongly activated and TIP undergoes phosphorylation[24].

### Cancer

Greatly reduced expression of LCK has been detected in T cells infiltrating renal cell carcinoma, correlating with reduced anti-tumour activity[25]. *LCK* is expressed in B cells from patients with chronic lymphocytic leukaemia (CLL) though not in normal B cells[26]. A translocation breakpoint within the first *LCK* intron has been detected in a T cell acute lymphoblastic leukaemia (T cell ALL) cell line with a t(1;7)(p34;q34). This separates the two *LCK* promoters (see below) and juxtaposes the constant region of the TCR $\beta$ chain to the proximal promoter and protein-coding region of the *LCK* gene[27].

Mutations in the translocated *LCK* gene detected in the HSB2 leukaemia cell line increase the kinase activity by 10-fold, although LCK protein expression is elevated only two-fold[28].

### Transgenic animals

Transgenic mice that do not express LCK show considerable thymic atrophy with a marked decrease in the number of $CD4^+/CD8^+$ thymocytes[15]. Transgenic mice expressing high levels of a catalytically inactive form of LCK are defective in the production of virtually all T lymphocytes[19].

### In animals

A significant proportion (14%) of primary tumours induced in rats by Moloney murine leukaemia virus have a proviral insertion upstream of *Lck* that increases *Lck* transcription, generating three different hybrid transcripts[29].

### *In vitro*

The phenotype of LSTRA murine lymphoma cells appears to be caused by the enhanced expression of *Lck* caused by promoter insertion[30]. Substitution of Phe for $Tyr^{505}$ increases the apparent kinase activity and this mutant form of LCK transforms NIH 3T3 cells[31]. The over-expression of *Lck* promotes tumorigenesis in otherwise normal thymocytes[32].

## Gene structure

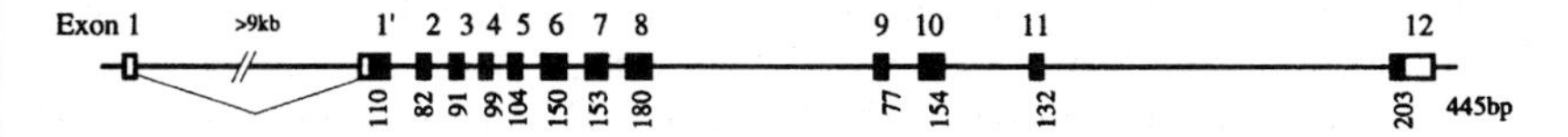

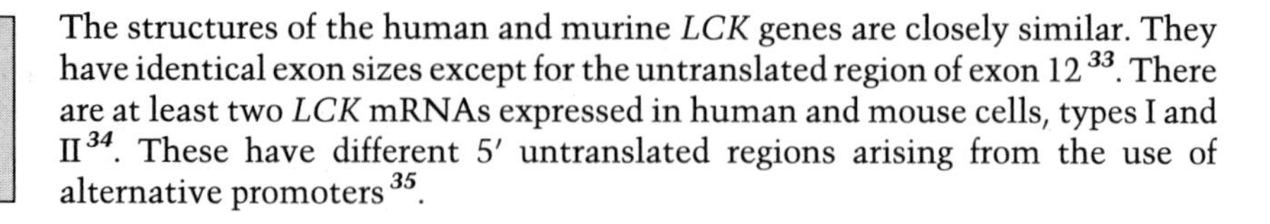

The structures of the human and murine *LCK* genes are closely similar. They have identical exon sizes except for the untranslated region of exon 12 [33]. There are at least two *LCK* mRNAs expressed in human and mouse cells, types I and II [34]. These have different 5′ untranslated regions arising from the use of alternative promoters [35].

## Transcriptional regulation

The type I promoter in Jurkat cells and in the colon carcinoma cell line SW620 contains an ETS-binding element that is essential for its activity [36]. Three MYB binding sites act synergistically with ETS-related factors to activate the human type I promoter.

Type II *LCK* transcripts initiate from a promoter ~9 kb from the downstream promoter and have an alternative 5′ UTR. The sequence between −584 and +37 with respect to the proximal promoter transcription start site directs tissue-specific and temporally appropriate transcription of *LCK* [37]. The type II promoter is used in both normal, mature T cells and in transformed T cells and an alternatively spliced transcript utilizing this promoter is also expressed in both types of cell [38]. In this type IIB mRNA the deletion of exon 1′ results in the use of a different AUG codon in exon 1. In the type IIB translation product 10 residues encoded by exon 1 replace the first 35 residues encoded by exon 1′.

## Protein structure

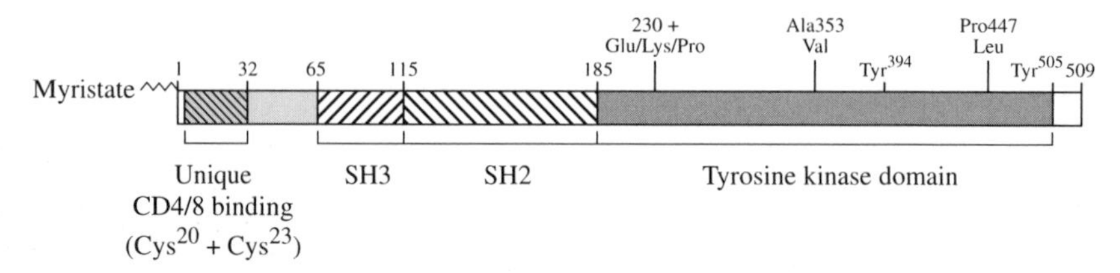


LCK has 65% sequence homology to SRC in the C-terminal 450 amino acids: the N-termini are unrelated. Deletion of the SH2 or SH3 domains increases LCK tyrosine kinase activity and phosphorylation of the autophosphorylation site $Tyr^{394}$ [39]. Phosphorylation of $Tyr^{192}$ in the SH2 domain negatively regulates signalling from the T cell receptor *via* LCK [40] and there is evidence that LCK proteins may dimerize following T cell receptor activation [41]. The structure of the SH2 domain of LCK complexed with an 11 amino acid peptide derived from hamster polyoma middle-T antigen has been analysed by high-resolution crystallography [42], as has the crystal structure of the SH2 and SH3 domains complexed with a phosphotyrosyl peptide containing the sequence of the C-terminal regulatory tail [43].

The regulatory site, phosphorylated in resting T cells, is $Tyr^{505}$ (equivalent to $Tyr^{527}$ in SRC). Site-directed mutagenesis of $Tyr^{505}$ activates the oncogenic potential of LCK, as does deletion of the C-terminal region containing $Tyr^{505}$ [44]. The SH2 but not the SH3 domain is required for full oncogenic activity in LCK from which $Tyr^{505}$ has been removed [45] and the SH2 domain binds *in vitro* to several TCR-regulated tyrosine phosphorylation substrates[46]. $Tyr^{505}$ is specifically phosphorylated by the human $p50^{CSK}$ tyrosine kinase which thus negatively regulates LCK kinase activity. However, $p50^{CSK}$ also phosphorylates the CD45 tyrosine phosphatase which promotes its binding to LCK *via* the SH2 domain of the kinase and increases the activity of the phosphatase[47]. The non-transmembrane phosphatase SH-PTP1, expressed primarily in haematopoietic cells including T cells, undergoes tyrosyl phosphorylation in response to LCK activation *via* CD4 or CD8[48]. A phosphotyrosine-independent ligand of the LCK SH2 domain has also been cloned[49].

$Tyr^{505}$ is dephosphorylated by CD45 when cells are activated *via* surface receptors, thus permitting the phosphorylation of $Tyr^{394}$ (equivalent to $Tyr^{416}$ in SRC) which represents the switch to the activated enzyme form. Physical association between LCK and CD45 occurs independently of the TCR[50] and such complexes can also contain GAP, GRB2 and SOS, indicating that CD45 may serve as a docking site for components of the RAS signalling pathway[51]. However, there is no simple relationship between CD45 loss, C-terminal tyrosine phosphorylation and enzymatic activity as in some cell lines hyperphosphorylation of the negative regulatory C-terminal tyrosine correlates with increased activity[52].

Oxidative reagents induce phosphorylation of $Tyr^{394}$ and $Tyr^{505}$ in LCK, indicating that a redox-sensitive signalling mechanism in T cells is in part mediated by LCK[53]. In human leukaemia T cells, hydrogen peroxide treatment causes phosphorylation of $Tyr^{394}$ and increases LCK catalytic activity. Even mutant forms of LCK that are catalytically inactive can be reactivated by this means[54].

## Amino acid sequence of human LCK

```
  1 MGCGCSSHPE DDWMENIDVC ENCHYPIVPL DGKGTLLIRN GSEVRDPLVT
 51 YEGSNPPASP LQDNLVIALH SYEPSHDGDL GFEKGEQLRI LEQSGEWWKA
101 QSLTTGQEGF IPFNFVAKAN SLEPEPWFFK NLSRKDAERQ LLAPGNTHGS
151 FLIRESESTA GSFSLSVRDF DQNQGEVVKH YKIRNLDNGG FYISPRITFP
201 GLHELVRHYT NASDGLCTRL SRPCQTQKPQ KPWWEDEWEV PRETLKLVER
251 LGAGQFGEVW MGYYNGHTKV AVKSLKQGSM SPDAFLAEAN LMKQLQHQRL
301 VRLYAVVTQE PIYIITEYME NGSLVDFLKT PSGIKLTINK LLDMAAQIAE
351 GMAFIEERNY IHRDLRAANI LVSDTLSCKI ADFGLARLIE DNEYTAREGA
401 KFPIKWTAPE AINYGTFTIK SDVWSFGILL TEIVTHGRIP YPGMTNPEVI
451 QNLERGYRMV RPDNCPEELY QLMRLCWKER PEDRPTFDYL RSVLEDFFTA
501 TEGQYQPQP (509)
```

**Domain structure**

| | |
|---|---|
| 2 | Myristate attachment site |
| 3, 5 | Palmitoylation of $Cys^3$ or $Cys^5$ is required for membrane binding and biological activity[55] |
| 42, 59 | Major sites for TPA-induced phosphorylation. $Ser^{59}$ phosphorylation by MAP kinase *in vitro* reduces LCK activity[56] |
| 34–150 | Recombinant CD45 binding site[57] |
| 232–493 | Tyrosine kinase domain |
| 251–259 and 273 | ATP binding region |

364 Active site
394 Autophosphorylation site
505 Phosphorylation site
66–117 SH3 domain (underlined)

Mutations detected in human HSB2 leukaemia cells: Glu, Lys, Pro insertion after 230 (sequence thus reads QKPQKPQKPWW); $Val^{28} \rightarrow$ Leu; $Ala^{353} \rightarrow$ Val; $Pro^{447} \rightarrow$ Leu [28].

## Database accession numbers

| | *PIR* | *SWISSPROT* | *EMBL/GENBANK* | *REFERENCES* |
|---|---|---|---|---|
| Human *LCK* | JQ0152 | P06239, P07100 | X13529 | 58 |
| | | | X04476 | 59 |
| | | | X14055 | 33 |
| | | | X06369 | 60 |
| | | | X05027, M21510 | 61 |

### *References*


1 **Sefton, B.M. (1991) Oncogene 6, 683–686.**
2 Chow, L.M.L. et al. (1992) Mol. Cell. Biol. 12, 1226–1233.
3 Thompson, P.A. et al. (1992) Oncogene 7, 719–725.
4 Hatakeyama, M. et al. (1991) Science 252, 1523–1528.
5 Burgess, K.E. et al. (1992) Proc. Natl Acad. Sci. USA 89, 9311–9315.
6 Raab, M. et al. (1994) Mol. Cell. Biol. 14, 2862–2870.
7 Bell, G.M. et al. (1992) Mol. Cell. Biol. 12, 5548–5554.
8 Taher, T.E.I. et al. (1996) J. Biol. Chem. 271, 2863–2867.
9 Ellis, C. et al. (1991) Oncogene 6, 895–901.
10 Amrein, K.E. et al. (1992) Proc. Natl Acad. Sci. USA 89, 3343–3346.
11 Campbell, M.A. and Sefton, B.M. (1992) Mol. Cell. Biol. 12, 2315–2321.
12 Taieb, J. et al. (1993) J. Biol. Chem. 268, 9169–9171.
13 Prasad, K.V.S. et al. (1993) Mol. Cell. Biol. 13, 7708–7717.
14 Samelson, L.E. et al. (1990) Proc. Natl Acad. Sci. USA 87, 4358–4362.
15 Molina, T.J. et al. (1992) Nature 357, 161–164.
16 August, A. and Dupont, B. (1996) J. Biol. Chem. 271, 10054–10059.
17 Thompson, P.A. et al. (1991) Cell Growth Differ. 2, 609–617.
18 Prasad, K.V.S. and Rudd, C.E. (1992) Mol. Cell. Biol. 12, 5260–5267.
19 Levin, S.D. et al. (1993) EMBO J. 12, 1671–1680.
20 Anderson, S.J. et al. (1993) Nature 365, 552–554.
21 Tremblay, M. et al. (1994) EMBO J. 13, 774–783.
22 Collette, Y. et al. (1996) J. Biol. Chem. 271, 6333–6341.
23 Jung, J.U. et al. (1995) J. Biol. Chem. 270, 20660–20667.
24 Wiese, N. et al. (1996) J. Biol. Chem. 271, 847–852.
25 Finke, J.H. et al. (1993) Cancer Res. 53, 5613–5616.
26 Abts, H. et al. (1991) Leuk. Res. 15, 987–997.
27 Burnett, R.C. et al. (1991) Genes Chromosom. Cancer 3, 461–467.
28 Wright, D.D. et al. (1994) Mol. Cell. Biol. 14, 2429–2437.
29 Shin, S. and Steffen, D.L. (1993) Oncogene 8, 141–149.
30 Marth, J.D. et al. (1985) Cell 43, 393–404.
31 Marth, J.D. et al. (1988) Mol. Cell. Biol. 8, 540–550.
32 Abraham, K.M. et al. (1991) Proc. Natl Acad. Sci. USA 88, 3977–3981.

[33] Rouer, E. et al. (1989) Gene 84, 105–113.
[34] Garvin, A.M. et al. (1988) Mol. Cell. Biol. 8, 3058–3064.
[35] Adler, H.T. et al. (1988) J. Virol. 62, 4113–4122.
[36] McCracken, S. et al. (1994) Oncogene 9, 3609–3615.
[37] Allen, J.M. et al. (1992) Mol. Cell. Biol. 12, 2758–2768.
[38] Rouer, E. and Benarous, R. (1992) Oncogene 7, 2535–2538.
[39] Reynolds, P.J. et al. (1992) Oncogene 7, 1949–1955.
[40] Couture, C. et al. (1996) J. Biol. Chem. 271, 24880–24884.
[41] Lee-Fruman, K.K. et al. (1996) J. Biol. Chem. 271, 25003–25010.
[42] Eck, M.J. et al. (1993) Nature 362, 87–91.
[43] Eck, M.J. et al. (1994) Nature 368, 764–769.
[44] Adler, H.T. and Sefton, B.M. (1992) Oncogene 7, 1191–1199.
[45] Veillette, A. et al. (1992) Oncogene 7, 971–980.
[46] Peri, K.G. et al. (1993) Oncogene 8, 2765–2772.
[47] Autero, M. et al. (1994) Mol. Cell. Biol. 14, 1308–1321.
[48] Lorenz, U. et al. (1994) Mol. Cell. Biol. 14, 1824–1834.
[49] Joung, I. et al. (1996) Proc. Natl Acad. Sci. USA 93, 5991–5995.
[50] Koretzky, G.A. et al. (1993) J. Biol. Chem. 268, 8958–8964.
[51] Lee, J.M. et al. (1996) Oncogene 12, 253–263.
[52] Burns, C.M. et al. (1994) J. Biol. Chem. 269, 13594–13600.
[53] Nakamura , K. et al. (1993) Oncogene 8, 3133–3139.
[54] Yurchak, L.K. et al. (1996) J. Biol. Chem. 271, 12549–12554.
[55] Yurchak, L.K. and Sefton, B.M. (1995) Mol. Cell. Biol. 15, 6914–6922.
[56] Winkler, D.G. et al. (1993) Proc. Natl Acad. Sci. USA 90, 5176–5180.
[57] Ng, D.H.W. et al. (1996) J. Biol. Chem. 271, 1295–1300.
[58] Perlmutter, R.M. et al. (1988) J. Cell. Biochem. 38, 117–126.
[59] Koga, Y. et al. (1986) Eur. J. Immunol. 16, 1643–1646.
[60] Trevillyan, J.M. et al. (1986) Biochim. Biophys. Acta 888, 286–295.
[61] Veillette, A. et al. (1987) Oncogene Res. 1, 357–374.

## Identification

*LYN* (LCK/YES-related novel tyrosine kinase, formerly *SYN*)[1] was initially cloned from a placental cDNA library using a v-*yes* probe.

## Related genes

*LYN* is a member of the *SRC* tyrosine kinase family (*Blk, FGR, FYN, HCK, LCK/Tkl, LYN, SRC, YES*). It is highly homologous in the kinase domain to LCK and YES.

| | ***LYN*** |
|---|---|
| **Nucleotides (bp)** | Not fully mapped |
| **Chromosome** | 8q13–qter |
| **Mass (kDa): predicted** | 58.5 |
| **expressed** | 53/56 |
| **Cellular location** | Cytoplasm |

**Tissue distribution**
*LYN* is expressed in brain and in B lymphocytes. Related forms of *LYN* have been detected in the spleen[2].

## Protein function

LYN proteins are tyrosine kinases. $p53^{LYN}$ and $p56^{LYN}$ are physically associated with IgM on B cells and cross-linking IgM increases the kinase activity of LYN. Cross-linking of surface Ig on immature B cells causes growth arrest, apoptosis, clonal anergy and clonal deletion, in contrast to the stimulation of proliferation of mature B cells. The use of anti-sense oligonucleotides has indicated that activation of LYN is necessary for anti-Ig-mediated cell cycle arrest but not for apoptosis[3]. LYN directly activates SYK which phosphorylates SHC, promoting its association with GRB2, following B cell receptor activation[4]. Chemoattractant receptors may also activate both LYN and SHC, presumably acting *via* G proteins[5]. In immature B cells LYN is a substrate for CD45 protein tyrosine phosphatase[6].

LYN also associates with CD20 on B cells and LYN kinase activity is stimulated by IL-2 or IL-3. LYN activates PtdIns 3-kinase (binding *via* its SH3 domain that specifically recognizes the sequence RXXRPLPPLPXP[7]) and PLC$\gamma_2$, MAP kinase and GAP *via* its N-terminal 27 amino acids[8,9]. LYN may modulate receptor-mediated changes in $[Ca^{2+}]_i$[10]. LYN is also physically associated with the CD19 receptor on human B-lineage lymphoid cells. B43-Gen is a specific immunoconjugate containing genistein that activates CD19 and triggers apoptosis in highly radiation-resistant $p53^{-}Bax^{-}$ Ramos-BT B-lineage lymphoma cells[11].

In the rat basophilic leukaemia cell line RBL-2H3 the kinase activity of both LYN and SRC is increased after cellular stimulation *via* the high-affinity IgE receptor [12]. LYN immunoprecipitates with both the activated IgE and Thy-1 receptors [13] and may therefore be responsible for the tyrosine phosphorylation of the $\beta$ and $\gamma$ subunits of the activated receptor [14]. Association of LYN, together with YES and FYN, with a major signalling receptor also occurs in human platelets and in some cell lines (see ***YES***).

In HL60 cells and in B-lineage lymphoid cells LYN is activated by ionizing radiation [15] and arrest at the $G_2$/M checkpoint appears to be caused by LYN-mediated tyrosine phosphorylation of $p34^{CDC2}$ [16]. In human B cell precursors LYN kinase activity has been reported to be stimulated by exposure to 1 gauss, 60 Hz electromagnetic fields [17]. In FDC-P1 cells $p53^{LYN}$ and $p56^{LYN}$ associate constitutively with the N-terminal unique domain of TEC protein tyrosine kinase [18].

In U-937 cells the inhibitor of DNA replication 1-$\beta$-D-arabinofuranosyl-cytosine (araC) inhibits CDK2 activity and promotes colocalization of CDK2 and LYN in the nucleus [19].

Coexpression of LYN and BTK (Bruton's tyrosine kinase) causes *trans*-phosphorylation of BTK in an EBV-transformed B cell line. Activated BTK is membrane-associated and may serve to integrate distinct receptor signals [20].

**Transgenic animals**

*Lyn*$^{-/-}$ mice have immune system defects with impaired T-dependent and T-independent responses as well as reduced numbers of circulating B cells, although the animals are IgM hyperglobulinaemic [21].

## Gene structure

Common splicing patterns indicate that *LYN*, *SRC*, *FGR*, *FYN* and *YES* are derived from a common gene but each differs in C-terminus and tissue-specific expression and thus, presumably, in function.

## Transcriptional regulation

The *LYN* promoter lacks TATA or CAAT boxes but contains four GC-rich regions, a cAMP-responsive element, an octamer-binding motif, PEA3-like motifs and an NF-$\kappa$B-binding sequence. In T cells transcription is induced by the HTLV-I-encoded $p40^{tax}$ protein [22].

## Protein structure

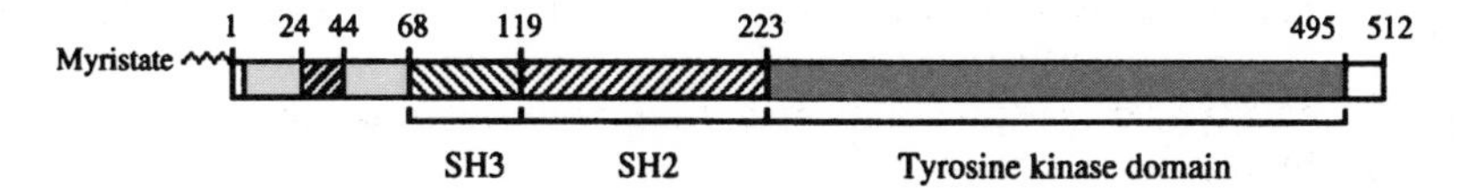

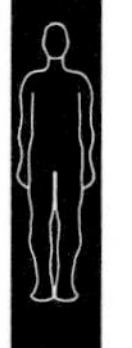

## Amino acid sequence of human LYN

```
  1 MGCIKSKGKD SLSDDGVDLK TQPVRNTERT IYVRDPTSNK QQRPVPESQL
 51 LPGQRFQTKD PEEQGDIVVA LYPYDGIHPD DLSFKKGEKM KVLEEHGEWW
101 KAKSLLTKKE GFIPSNYVAK LNTLETEEWF FKDITRKDAE RQLLAPGNSA
151 GAFLIRESET LKGSFSLSVR DFDPVHGDVI KHYKIRSLDN GGYYISPRIT
201 FPCISDMIKH YQKQADGLCR RLEKACISPK PQKPWDKDAW EIPRESIKLV
251 KRLGAGQFGE VWMGYYNNST KVAVKTLKPG TMSVQAFLEE ANLMKTLQHD
301 KLVRLYAVVT REEPIYIITE YMAKGSLLDF LKSDEGGKVL LPKLIDFSAQ
351 IAEGMAYIER KNYIHRDLRA ANVLVSESLM CKIADFGLAR VIEDNEYTAR
401 EGAKFPIKWT APEAINFGCF TIKSDVWSFG ILLYEIVTYG KIPYPGRTNA
451 DVMTALSQGY RMPRVENCPD ELYDIMKMCW KEKAEERPTF DYLQSVLDDF
501 YTATEGQYQQ QP (512)
```

### Domain structure

| | |
|---|---|
| 2 | Myristate attachment site |
| 24–44 | Region deleted in the alternative form of LYN [23,24] |
| 253–261 and 275 | ATP binding site |
| 367 | Active site |
| 397, 508 | Autophosphorylation site |
| 68–119 | SH3 domain (or A box: italics) |

### Database accession numbers

| | *PIR* | *SWISSPROT* | *EMBL/GENBANK* | *REFERENCES* |
|---|---|---|---|---|
| Human *LYN* | A26719 | P07948 | M16038 | 25 |

### *References*


1. Semba, K. et al. (1986) Proc. Natl Acad. Sci. USA 83, 5459–5463.
2. Brunati, A.M. et al. (1991) Biochim. Biophys. Acta 1091, 123–126.
3. Scheuermann, R.H. et al. (1994) Proc. Natl Acad. Sci. USA 91, 4048–4052.
4. Nagai, K. et al. (1995) J. Biol. Chem. 270, 6824–6829.
5. Ptasznik, A. et al. (1995) J. Biol. Chem. 270, 19969–19973.
6. Katagiri, T. et al. (1995) J. Biol. Chem. 270, 27987–27990.
7. Rickles, R.J. et al. (1994) EMBO J. 13, 5598–5604.
8. Yamanashi, Y. et al. (1992) Proc. Natl. Acad. Sci USA 89, 1118–1122.
9. Pleiman, C.M. et al. (1993) Mol. Cell. Biol. 13, 5877–5887.
10. Takata, M. et al. (1994) EMBO J. 13, 1341–1349.
11. Myers, D.E. et al. (1995) Proc. Natl Acad. Sci. USA 92, 9575–9579.
12. Eiseman, E. and Bolen, J.B. (1992) Nature 355, 78–80.
13. Draberova, L. and Draber, P. (1993) Proc. Natl Acad. Sci. USA 90, 3611–3615.
14. Jouvin, M.-H.E. et al. (1994) J. Biol. Chem. 269, 5918–5925.
15. Kharbanda, S. et al. (1994) J. Biol. Chem. 269, 20739–20743.
16. Uckun, F. et al. (1996) J. Biol. Chem. 271, 6389–6397.
17. Uckun, F. et al. (1995) J. Biol. Chem. 270, 27666–27670.
18. Mano, H. et al. (1994) Oncogene 9, 3205–3211.
19. Yuan, Z.-M. et al. (1996) Oncogene 13, 939–946.
20. Rawlings, D.J. et al. (1996) Science 271, 822–825.

[21] Hibbs, M.L. et al. (1995) Cell 83, 301–311.
[22] Uchiumi, F. et al. (1992) Mol. Cell. Biol. 12, 3784–3795.
[23] Stanley, E. et al. (1991) Mol. Cell. Biol. 11, 3399–3406.
[24] Yi, T. et al. (1991) Mol. Cell. Biol. 11, 2391–2398.
[25] Yamanashi, Y. et al. (1987) Mol. Cell. Biol. 7, 237–243.

## Identification

*MAS* is a human transforming gene having no homology with known viral oncogenes that was originally detected by NIH 3T3 fibroblast transfection of DNA from an epidermoid carcinoma [1].

## Related genes

*MAS* is related to *Rta* (rat thoracic aorta) and *MRG* (human mas-related gene). MAS is a member of the seven transmembrane-spanning receptor family that includes the visual opsins, $\alpha_2$-, $\beta_1$- and $\beta_2$-adrenergic receptors, M1 and M2 muscarinic acetylcholine receptors and substance K receptor.

| | ***MAS*** |
|---|---|
| **Nucleotides (kb)** | Not fully mapped |
| **Chromosome** | 6q24–q27 |
| **Mass (kDa): predicted** | 37 |
| **expressed** | 45 |
| **Cellular location** | Transmembrane cell surface protein |

**Tissue distribution**

Rat *Mas* is strongly expressed in the hippocampus and cerebral cortex [2,3]. The related RTA protein is significantly expressed in the gut, vas deferens, uterus and aorta [4]. *MRG* expression has not been detected [5].

In the mouse *Mas* is genomically imprinted, being expressed exclusively from the paternal allele during early development with selective relaxation of imprinting occurring by 13.5 days of gestation [6].

## Protein function

MAS is the receptor for neuronal type angiotensin III. Its action is mediated by G proteins that activate PtdIns$P_2$ hydrolysis causing increase in $[Ca^{2+}]_i$ [7]. *MAS* is activated oncogenically by rearrangement of 5′ non-coding regions in which human centromeric $\alpha$ satellite repeat DNA (alphoid sequences) are juxtaposed upstream of *MAS* [8].

**In animals**

*MAS* renders NIH 3T3 fibroblasts tumorigenic in nude mice and has a weak focus-inducing activity in NIH 3T3 cells [1].

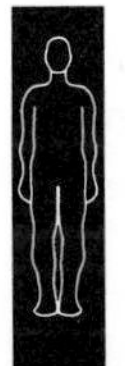

## Amino acid sequence of human MAS

```
  1 MDGSNVTSFV VEEPTNISTG RNASVGNAHR QIPIVHWVIM SISPVGFVEN
 51 GILLWFLCFR MRRNPFTVYI THLSIADISL LFCIFILSID YALDYELSSG
101 HYYTIVTLSV TFLFGYNTGL YLLTAISVER CLSVLYPIWY RCHRPKYQSA
151 LVCALLWALS CLVTTMEYVM CIDREEESHS RNDCRAVIIF IAILSFLVFT
201 PLMLVSSTIL VVKIRKNTWA SHSSKLYIVI MVTIIIFLIF AMPMRLLYLL
251 YYEYWSTFGN LHHISLLFST INSSANPFIY FFVGSSKKKR FKESLKVVLT
301 RAFKDEMQPR RQKDNCNTVT VETVV (325)
```

### Domain structure

| | |
|---|---|
| 1–30, 98–104, 173–185, 251–257 | Extracellular domains |
| 31–61, 66–97, 105–135, 150–172, 186–214, 225–250, 258–286 | Seven transmembrane domains (underlined) |
| 62–65, 136–149, 215–224, 287–325 | Cytoplasmic domains |
| 5, 16, 22 | Potential carbohydrate attachment sites |

### Database accession numbers

| | *PIR* | *SWISSPROT* | *EMBL/GENBANK* | *REFERENCES* |
|---|---|---|---|---|
| Human *MAS* | A01375 | P04201 | M13150 | 1 |

### *References*


[1] Young, D. et al. (1986) Cell 45, 711–719.
[2] Young, D. et al. (1988) Proc. Natl Acad. Sci. USA 85, 5339–5342.
[3] Bunnemann, B. et al. (1990) Neurosci. Lett. 114, 147–153.
[4] Ross, P.C. et al. (1990) Proc. Natl Acad. Sci. USA 87, 3052–3056.
[5] Monnot, C. et al. (1991) Mol. Endocrinol. 5, 1477–1487.
[6] Villar, A.J. and Pedersen, R.A. (1994) Nature Genet. 8, 373–379.
[7] Jackson, T.R. et al. (1988) Nature 335, 437–440.
[8] van't Veer, L.J. et al. (1993) Oncogene 8, 2673–2681.

## Identification

MAX was detected by screening a cDNA expression library with the C-terminus of human MYC [1].

## Related genes

*MAX* is a member of the helix-loop-helix/leucine zipper superfamily. ΔMAX is a truncated form. The murine homologue is *Myn*. Variant forms of *Xmax* occur in *Xenopus laevis*.

| | ***MAX*** |
|---|---|
| **Nucleotides** | Not fully mapped |
| **Chromosome** | 14q22–q24 |
| **Mass (kDa): predicted** | 18 |
| **expressed** | 21/22<br>16.5 (ΔMAX) |
| **Cellular location** | Nucleus |

Translocation of MAX and ΔMAX from the cytosol is dependent on association with MYC. The steady-state concentration of MAX is constant throughout the cell cycle [2] and intracellular MYC is always complexed with MAX.

**Tissue distribution**

MAX and MYN, its murine homologue to which human MYCN also binds, occur in many cell types of diverse origin: MAX has been detected in NIH 3T3 fibroblasts, HeLa cells and neuroblastoma-derived cell lines. The ratio of MAX to ΔMAX is cell line-specific [3].

The nerve growth factor-responsive PC12 cell line expresses a mutant form of MAX incapable of homo- or heterodimerization and thus can divide, differentiate or apoptose by MAX/MYC-independent pathways [4].

## Protein function

MAX is a helix-loop-helix protein that forms sequence-specific DNA binding homodimers and also heterodimerizes with MYC [1,5]. The core consensus sequence is CACGTG but the complete 12 bp consensus binding sites are RACCACGTGGTY (MYC/MAX) and RANCACGTGNTY (MAX/MAX) where R represents purine and Y pyrimidine [6,7]. MYC/MAX does not bind when the core is flanked by a 5′T or 3′A whereas MAX/MAX readily binds such sequences. Thus the flanking motif can strongly inhibit MYC/MAX *trans*-activation. The leucine zipper domain is critical to the formation of heterodimers with MAX [8]. Unlike MYC, MAX can homodimerize efficiently to bind to the same DNA sequence as the MYC/MAX heterodimer, the homodimers acting as transcriptional repressors [9].

MYC/MAX heterodimers are required for the induction of both cell cycle progression and apoptosis [10]. Phosphorylation of the N-terminus of MAX by casein kinase II inhibits DNA binding of MAX homodimers *in vitro* but does not affect that of MYC/MAX heterodimers [11]. MAX is also phosphorylated *in vitro* and *in vivo* by MXI2, a mitogen-activated protein kinase that has a helix-loop-helix region [12]. ΔMAX retains the capacity to form heterodimers with MYC that bind to the CACGTG motif but lacks the putative regulatory domain of MAX[2,3,13].

The basic/helix-loop-helix/leucine zipper proteins MAD1 and MXI1 [14,15] form heterodimers with MAX that bind efficiently to the MYC/MAX consensus sequence to repress transcription. MAD1 protein is rapidly induced upon differentiation of cells of the myeloid and epidermal lineages and MAD1/MAX replaces the MYC/MAX complexes found in undifferentiated cells. Thus the relative abundance of these binding partners may determine the extent of formation of the transcriptionally active MYC/MAX complex. MAD1 inhibits *Myc*, mutant p53, adenovirus E1A or HPV type 16 transformation of rat embryo cells in cooperation with activated *Ras*. Thus MAD1 may have a MYC-independent function in repression of transformation in addition to its role in modulating MYC activity [16]. MAD3 and MAD4 are MAX-interacting transcriptional repressors that suppress MYC-dependent transformation and are expressed during neural and epidermal differentiation [17].

mSin3A and mSin3B, homologous to the yeast transcriptional repressor Sin3, bind specifically to MAD1 and MXI1 [18]. MAD1/MAX and mSIN3 form ternary complexes in solution specifically recognizing the MAD1/MAX E box binding site (CACGTG). Point mutations in the first 25 residues of MAD1 that contains a putative amphipathic α helical region (mSin interaction domain, SID) eliminate interaction with SIN3 proteins and block MAD1 transcriptional repression. The region including SID is necessary and sufficient for MAD-mediated transcriptional repression [19]. Two murine *Mxi1* mRNAs arise from alternative splicing: the gene products differ by a 36 amino acid N-terminal extension, homologous to the α helical repression domain of MAD1, which mediates association with SIN3 and is essential for anti-oncogene activity [20].

## Transcriptional regulation

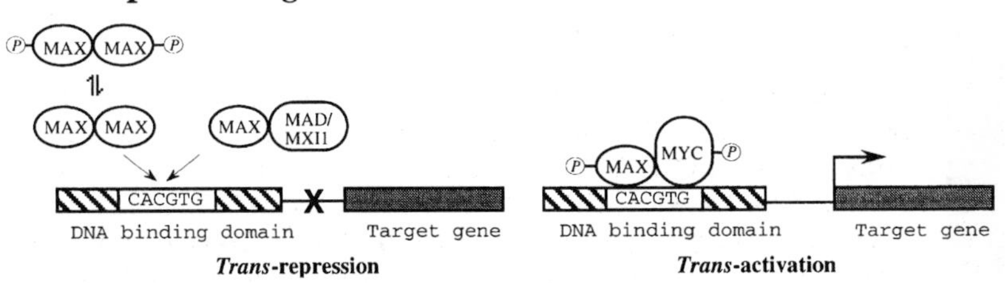


MAX homodimers phosphorylated by casein kinase II do not interact with binding domains. Unphosphorylated MAX homodimers repress transcription of genes that are targets for MYC, for example, in quiescent cells in which the concentration of MYC is low. Phosphorylated MAX/MYC heterodimers *trans*-activate target genes, for example, in proliferating cells [11].

### Cancer

Abnormal expression of MAX has not been detected but LOH, inactivating mutations and missense mutations in the *MXI1* gene have been detected in primary prostate cancers [21] and LOH occurs with high frequency in glioblastomas [22].

### Transgenic animals

Over-expressed *Max* is not oncogenic and attenuates *Myc*-induced lymphoproliferation and lymphomagenesis in transgenic mice, presumably by inhibiting the function of MYC [23].

## Protein structure

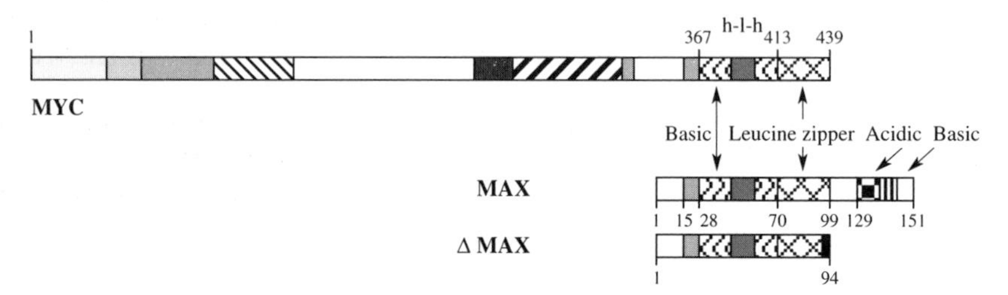


The helix-loop-helix/leucine zipper domain mediates the interaction between MYC and MAX and these regions are critical for transforming potential, autoregulation of *MYC* expression and inhibition of differentiation [24,25]. Artificial deletion of the N-terminus generates a dominant-negative mutant of MAX [26]. X-ray structural analysis of the MAX homodimer has revealed a symmetrical, parallel, left-handed, four-helix bundle in which each monomer contributes two α helical segments separated by a loop [27].

The suppressive activity of MAX requires the C-terminal acidic and basic regions and the leucine zipper. Replacement of the end of the zipper with the ΔMAX-specific sequence is responsible for the enhancement of transformation by ΔMAX [28].

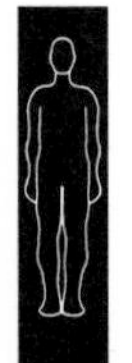

## Amino acid sequence of human MAX

```
  1 MSDNDDIEVE SDADKRAHHN ALERKRRDHI KDSFHSLRDS VPSLQGEKAS
 51 RAQILDKATE YIQYMRRKNH THQQDIDDLK RQNALLEQQV RALGKARSSA
101 QLQTNYPSSD NSLYTNAKGS TISAFDGGSD SSSESEPEEP QSRKKLRMEA
151 S (151)
```

### Domain structure

| | |
|---|---|
| 15–28 | Basic motif |
| 13–21 | The alternative forms of MAX and ΔMAX contain a nine amino acid insertion (EEQPRFQSA) after Asp[12] |
| 30–41 and 48–66 | Helix I and helix II of the basic helix-loop-helix region (underlined) |
| 79–93 | Three leucines spaced seven residues apart (underlined) |
| 90–151 | C-terminal 62 amino acids of MAX replaced in ΔMAX by GESES |

| | |
|---|---|
| 90–151 | An alternative 36 residue C-terminus is predicted to be encoded by the 5′ region of the first intron of *MAX* (GEHPSSWGSW PCCAPARSGF GTWACRVRAS HGVCAQ)[29] |
| 2, 11, 20, 140, 142, 144 | Potential casein kinase II phosphorylation sites |
| 129–139 | Acidic region |
| 140–147 | Nuclear localization signal (italics)[13] |

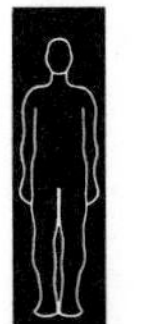

## Amino acid sequence of human MXI1

```
  1 MSQERPTFYR QELNKTIWEV PERYQNLSPV GSGAYGSVCA AFDTKTGLRV
 51 AVKKLSRPFQ SIIHAKRTYR ELRLLKHMKH ENVIGLLDVF TPARSLEEFN
101 DVYLVTHLMG ADLNNIVKCQ KLTDDHVQFL IYQILRGLKY IHSADIIHRD
151 LKPSNLAVNE DCELKILDFG LARHTDDEMT GYVATRWYRA PEIMLNWMHY
201 NQTVDIWSVG CIMAELLTGR TLFPGTDHID QLKLILRLVG TPGAELLKKI
251 SSESARNYIQ SLTQMPKMNF ANVFIGANPL GKLTIYPHLM DIELVMI (297)
```

Mutations detected in prostate cancer: nucleotide deletion codon 141 or 142 (the frameshift results in loss of the zipper terminal leucine and all downstream residues), T → C transition adjacent to codon 117 changes invariant GT splice donor to GC; $Glu^{152}$ → Ala[20].

### Database accession numbers

| | PIR | SWISSPROT | EMBL/GENBANK | REFERENCES |
|---|---|---|---|---|
| Human *MAX* | | P25912 | M64240 | 1 |
| Human Δ*MAX* | | | X60287 | 3 |
| Human Δmax (3.5 kb mRNA) | | | X66867 | 29 |
| Human *MXI1* | | | HS197751, U19775 | 30 |
| | | | U32512–U32515 | 31 |

### *References*


1 Blackwood, E.M. and Eisenman, R.N. (1991) Science 251, 1211–1217.
2 Blackwood, E.M. et al. (1992) Genes Dev. 6, 71–80.
3 Makela, T.P. et al. (1992) Science 256, 373–377.
4 Hopewell, R. and Ziff, E.B. (1995) Mol. Cell. Biol. 15, 3470–3478.
5 **Cole, M.D. et al. (1991) Cell 65, 715–716.**
6 Solomon, D.L.C. et al. (1993) Nucleic Acids Res. 21, 5372–5376.
7 Fisher, F. et al. (1993) EMBO J. 12, 5075–5082.
8 Reddy, C.D. et al. (1992) Oncogene 7, 2085–2092.
9 Kato, G.J. et al. (1992) Genes Dev. 6, 81–92.
10 Amati, B. et al. (1993) EMBO J. 12, 5083–5087.
11 Berberich, S.J. and Cole, M.D. (1992) Genes Dev. 6, 166–176.
12 Zervos, A.S. et al. (1995) Proc. Natl Acad. Sci. USA 92, 10531–10534.
13 Prendergast, G.C. et al. (1992) Genes Dev. 6, 2429–2439.
14 Ayer, D.E. and Eisenman, R.N. (1993) Genes Dev. 7, 2110–2119.
15 Larsson, L.-G. et al. (1994) Oncogene 9, 1247–1252.
16 Cerni, C. et al. (1995) Oncogene 11, 587–596.
17 Hurlin, P.J. et al. (1995) EMBO J. 14, 5646–5659.
18 Kasten, M.M. et al. (1996) Mol. Cell. Biol. 16, 4215–4221.
19 Ayer, D.E. et al. (1996) Mol. Cell. Biol. 16, 5772–5781.
20 Schreiber-Agus, N. et al. (1995) Cell 80, 777–786.

[21] Eagle, L.R. et al. (1995) Nature Genet. 9, 249–255.
[22] Albarosa, R. et al. (1995) Human Genet. 95, 709–711.
[23] Lindeman, G.J. et al. (1995) Oncogene 10, 1013–1017.
[24] Penn, L.J.Z. et al. (1990) Mol. Cell. Biol. 10, 4961–4966.
[25] Crouch, D.H. et al. (1990) Oncogene 5, 683–689.
[26] Billaud, M. et al. (1993) Proc. Natl Acad. Sci. USA 90, 2739–2743.
[27] Ferre-D'Amare, A.R. et al. (1993) Nature 363, 38–45.
[28] Västrik, I. et al. (1995) Oncogene 11, 553–560.
[29] Västrik, I. et al. (1993) Oncogene 8, 503–507.
[30] Zervos, A.S. et al. (1995) Proc. Natl Acad. Sci. USA 92, 10531–10534.
[31] Wechsler, D.S. et al. (1996) Genomics 32, 466–470.

## Identification

Identified as an activated oncogene in an *N*-methyl-*N′*-nitro-*N*-nitroso-guanidine (MNNG)-treated human osteosarcoma cell line (MNNG-HOS)[1,2] and in MNNG-treated human xeroderma pigmentosum cells[3]. *MET* was activated as an oncogene by the formation of a chimeric gene generated by chromosomal rearrangement fusing the *TPR* gene (translocated promoter region) to the N-terminally truncated *MET* kinase domain[4].

## Related genes

MET has homology with SEA and RON and in the extracellular domain with SEX, SEP, OCT and NOV.

*TPR* has weak homology in α helical regions to tropomyosin, spectrin, laminin B1, myosin heavy chain and *Drosophila glued* protein and to vimentin. The C-terminus of the larger TPR protein (TPR-L) is homologous to *Drosophila engrailed* protein, *E. coli* RNA polymerase subunit and nucleolin.

| | ***MET/Met*** | ***TPR/Tpr*** | ***TPR-MET*** |
|---|---|---|---|
| **Nucleotides** | Not fully mapped | Not fully mapped | |
| **Chromosome** | 7q31 | 1 | |
| **Mass (kDa): predicted** | 155.5 | 84 | |
| **expressed** | gp190 ($p50^{\alpha}/p145^{\beta}$disulfide-linked, heterodimer) gp140 ($p50^{\alpha}/p85^{\beta}$) /gp130 ($p50^{\alpha}/p75^{\beta}$) (alternative splicing) | | $p65^{TPR\text{-}MET}$ |

### Cellular location

MET β subunit (p145): transmembrane.
MET α subunit (p50): extracellular.
Multimeric forms of the α/β heterodimer occur suggesting that the receptors exist as patches on the cell surface[5].
TPR-MET: probably cytoplasmic[6].

### Tissue distribution

*MET* mRNA is expressed at high levels in liver, gastrointestinal tract, thyroid and kidney and gp140 is present in the microglial cells of the human CNS[7]. Low levels are expressed in normal colorectal mucosa[8,9].

*TPR* is expressed in tumour cell lines of epithelial and mesenchymal origin and in T and B cell neoplasia[2]. The alternatively spliced *Tpr-l* is expressed in rat testis, lung, thymus and spleen[10].

## Protein function

MET is a tyrosine kinase receptor that binds hepatocyte growth factor (HGF, also called scatter factor)[11–13]. The activity of MET may be greatly enhanced by autophosphorylation of the cytoplasmic domain[14]. HGF causes rapid tyrosine phosphorylation of the $\beta$ subunit of MET[11,15]. HGF is a potent mitogen for hepatocytes *in vitro* and is an hepatotrophic factor possibly involved in liver regeneration[16]. In epithelial cells HGF stimulates the RAS guanine nucleotide exchanger and increases the proportion of RAS-GTP[17]. HGF can activate RAS both directly by recruiting GRB2 to the receptor or indirectly *via* SHC, which also binds to the activated MET receptor[18]. The GRB2-binding protein GAB1 interacts with activated MET with a specificity that is high relative to that of its association with other tyrosine kinase receptors[19]. GAB1/MET binding is dependent on the kinase activity of the receptor and over-expression of GAB1 activates the MAPK kinase pathway, indicating the critical role of this protein in signalling from MET. Pathways downstream of SHC and GRB2 are essential for transformation, whereas PtdIns 3′-kinase, phospholipase C$\gamma$ and SHPTP2/SYP are not[20]. In oral squamous carcinoma cells HGF stimulates tyrosine phosphorylation of MET and of $p125^{FAK}$ and promotes cellular migration and invasion[21].

TPR is a protein of unknown function but it activates the oncogenic potential of both *MET*, *RAF* and *TRK*[22]. The *TPR-MET* oncogene appears to be activated by an insertion of chromosome 1 (*TPR*) DNA into the *MET* locus on chromosome 7, both upstream and downstream portions of *MET* being conserved. This is similar to the mechanism of activation of *RAF* by *TPR*[23]. Both *TPR-MET* and *TPR-RAF* rearrangements occur within introns but there is no sequence homology between the sites.

### Cancer

*MET* mRNA and protein concentrations are increased in some human carcinomas, including colorectal cancers, and in epithelial tumour cell lines[9,24,25]. In thyroid papillary carcinomas MET protein concentration is increased by 100-fold[26] but it is not detectably over-expressed in breast carcinomas. *MET* is amplified and over-expressed in cell lines from human tumours of non-haematopoietic origin, particularly gastric tumours[4], and from rhabdomyosarcomas, although *MET* is repressed in normal, differentiated striated muscle[27].

MET over-expression has been detected in 60% of osteosarcomas examined and coexpression of HGF and MET, operating an autocrine or paracrine mechanism, may contribute to the aggressive behaviour of these tumours[28]. Enhanced coexpression of *MET* and HGF has been detected in pancreatic cancer[29]. The steady state levels of the 8 kb *MET* mRNA are strongly increased by IL-1$\alpha$, IL-6, TNF$\alpha$, TGF$\beta$, EGF, HGF or the steroid hormones oestrogen, progesterone, tamoxifen and dexamethasone in human carcinoma cell lines[30]. MET protein is present at high levels in primary ovarian tumours.

Sarcomas in individuals with Li–Fraumeni syndrome and in p53-deficient mice express high levels of MET[31].

### Transgenic animals

Mutation of two MET C-terminal tyrosine residues (1349 and 1356) in the mouse genome causes embryonic death. Abrogation of GRB2 binding by

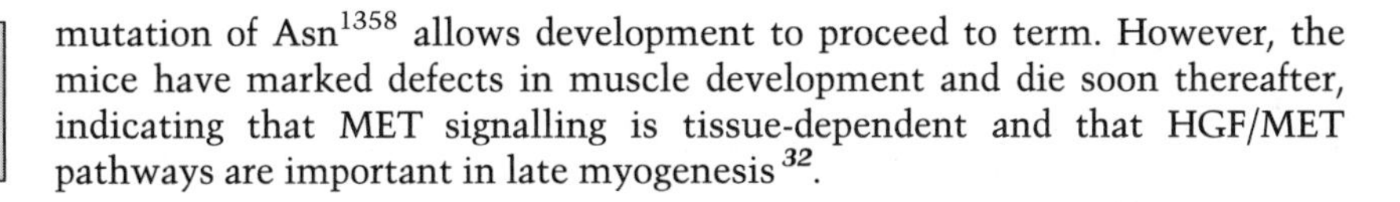

mutation of Asn$^{1358}$ allows development to proceed to term. However, the mice have marked defects in muscle development and die soon thereafter, indicating that MET signalling is tissue-dependent and that HGF/MET pathways are important in late myogenesis[32].

**In animals**

NIH 3T3 fibroblasts that over-express the human *MET* proto-oncogene are only weakly tumorigenic in nude mice but cells co-transfected with *MET* and *HGF* are highly tumorigenic, indicating that an autocrine transformation mechanism occurs[33].

***In vitro***

HGF was originally discovered as an activity that causes MDCK epithelial cells to change shape and become motile and invasive in *in vitro* assays[34]. HGF exerts similar effects on NIH 3T3 cells transfected to express p190$^{MET}$[35], which are also metastatic *in vivo*[36]. Responses to HGF are cell-dependent and it can exert mitogenic, motogenic or morphogenic effects[16]. HGF is a co-mitogen for normal human melanocytes, acting synergistically with basic fibroblast growth factor (bFGF) or mast cell growth factor (MGF). Melanocytes express the MAP2 kinases ERK1 and ERK2, the latter being phosphorylated in response to HGF[37].

Over-expression of *MET* transforms NIH 3T3 fibroblasts[1] and HGF transforms immortalized mouse liver epithelial cells[38].

A second *MET* allele is rearranged in the chemically treated human cell line MNNG-HOS. der(7)t(1;7)(q23;q32) represents a deletion of the N-terminus of the MET extracellular ligand binding domain but the rearranged allele also includes sequences derived from chromosome 2[22].

## Transcriptional regulation

The human *MET* promoter includes two PAX3 consensus binding sequences at 257 bp and 114 bp (opposite orientation) upstream of the transcription start site[39].

## Protein structure

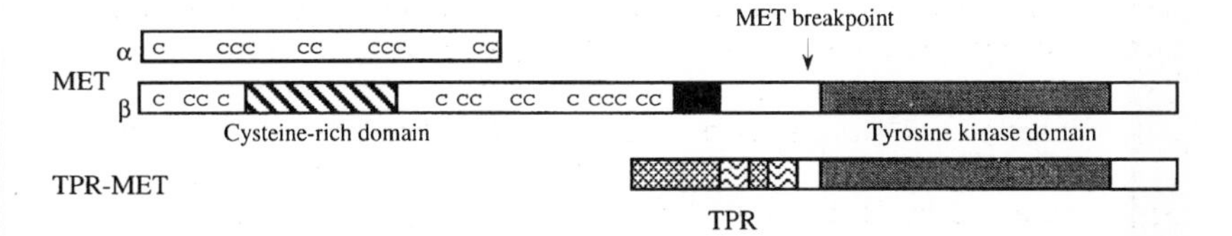


The 307 N-terminal amino acids of the MET precursor are cleaved to release the $\alpha$ subunit which then associates with the membrane bound $\beta$ subunit. The alternatively spliced isoform of MET is also expressed as a membrane tyrosine kinase but does not undergo cleavage to yield $\alpha$ and $\beta$ subunits[40]. In TPR/MET the tyrosine kinase domain of MET is fused to 142 N-terminal amino acids of TPR. This region of TPR includes two leucine zipper motifs that mediate dimerization of TPR/MET and are essential for transforming

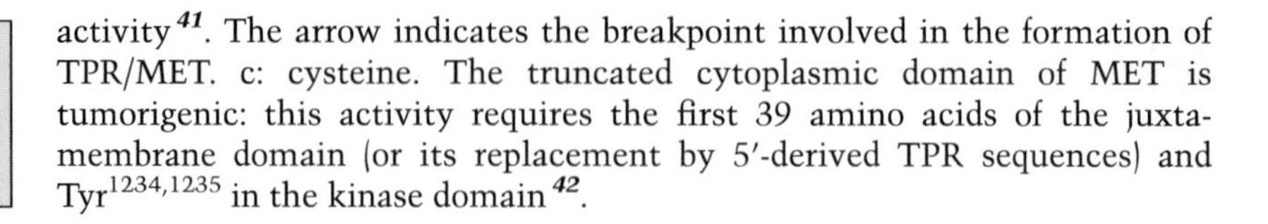

activity[41]. The arrow indicates the breakpoint involved in the formation of TPR/MET. c: cysteine. The truncated cytoplasmic domain of MET is tumorigenic: this activity requires the first 39 amino acids of the juxtamembrane domain (or its replacement by 5′-derived TPR sequences) and $Tyr^{1234,1235}$ in the kinase domain[42].

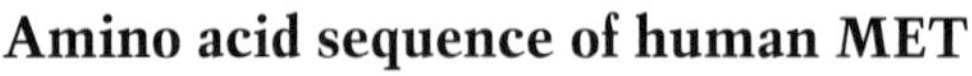

## Amino acid sequence of human MET

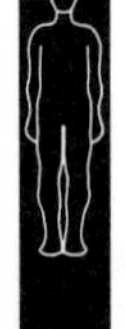

```
   1 MKAPAVLAPG ILVLLFTLVQ RSNGECKEAL AKSEMNVNMK YQLPNFTAET
  51 PIQNVILHEH HIFLGATNYI YVLNEEDLQK VAEYKTGPVL EHPDCFPCQD
 101 CSSKANLSGG VWKDNINMAL VVDTYYDDQL ISCGSVNRGT CQRHVFPHNH
 151 TADIQSEVHC IFSPQIEEPS QCPDCVVSAL GAKVLSSVKD RFINFFVGNT
 201 INSSYFPDHP LHSISVRRLK ETKDGFMFLT DQSYIDVLPE FRDSYPIKYV
 251 HAFESNNFIY FLTVQRETLD AQTFHTRIIR FCSINSGLHS YMEMPLECIL
 301 TEKRKKRSTK KEVFNILQAA YVSKPGAQLA RQIGASLNDD ILFGVFAQSK
 351 PDSAEPMDRS AMCAFPIKYV NDFFNKIVNK NNVRCLQHFY GPNHEHCFNR
 401 TLLRNSSGCE ARRDEYRTEF TTALQRVDLF MGQFSEVLLT SISTFIKGDL
 451 TIANLGTSEG RFMQVVVSRS GPSTPHVNFL LDSHPVSPEV IVEHTLNQNG
 501 YTLVITGKKI TKIPLNGLGC RHFQSCSQCL SAPPFVQCGW CHDKCVRSEE
 551 CLSGTWTQQI CLPAIYKVFP NSAPLEGGTR LTICGWDFGF RRNNKFDLKK
 601 TRVLLGNESC TLTLSESTMN TLKCTVGPAM NKHFNMSIII SNGHGTTQYS
 651 TFSYVDPVIT SISPKYGPMA GGTLLTLTGN YLNSGNSRHI SIGGKTCTLK
 701 SVSNSILECY TPAQTISTEF AVKLKIDLAN RETSIFSYRE DPIVYEIHPT
 751 KSFISGGSTI TGVGKNLNSV SVPRMVINVH EAGRNFTVAC QHRSNSEIIC
 801 CTTPSLQQLN LQLPLKTKAF FMLDGILSKY FDLIYVHNPV FKPFEKPVMI
 851 SMGNENVLEI KGNDIDPEAV KGEVLKVGNK SCENIHLHSE AVLCTVPNDL
 901 LKLNSELNIE WKQAISSTVL GKVIVQPDQN FTGLIAGVVS ISTALLLLLG
 951 FFLWLKKRKQ IKDLGSELVR YDARVHTPHL DRLVSARSVS PTTEMVSNES
1001 VDYRATFPED QFPNSSQNGS CRQVQYPLTD MSPILTSGDS DISSPLLQNT
1051 VHIDLSALNP ELVQAVQHVV IGPSSLIVHF NEVIGRGHFG CVYHGTLLDN
1101 DGKKIHCAVK SLNRITDIGE VSQFLTEGII MKDFSHPNVL SLLGICLRSE
1151 GSPLVVLPYM KHGDLRNFIR NETHNPTVKD LIGFGLQVAK GMKYLASKKF
1201 VHRDLAARNC MLDEKFTVKV ADFGLARDMY DKEYYSVHNK TGAKLPVKWM
1251 ALESLQTQKF TTKSDVWSFG VVLWELMTRG APPYPDVNTF DITVYLLQGR
1301 RLLQPEYCPD PLYEVMLKCW HPKAEMRPSF SELVSRISAI FSTFIGEHYV
1351 HVNATYVNVK CVAPYPSLLS SEDNADDEVD TRPASFWETS (1390)
```

### Domain structure

| | |
|---:|---|
| 1–24 | Signal sequence (italics) |
| 25–932 | Extracellular domain |
| 933–955 | Transmembrane domain (underlined) |
| 956–1390 | Intracellular domain |
| 1009–1010 | Breakpoint for the translocation to form TPR/MET |
| 1085–1330 | Tyrosine kinase domain |
| 1084–1092 and 1110 | ATP binding site |
| 1204 | Active site |
| 307–308 | Potential cleavage site |
| 1234–1235 | Autophosphorylation sites critical for activation of *MET*[43] (equivalent to $Tyr^{365}$, $Tyr^{366}$ of TPR/MET)[44] |

| | |
|---|---|
| 1356 | Tyr$^{1356}$ essential for signal transduction [45] |
| 45, 106, 149, 202, 399, 405, 607, 635, 785, 879, 930 | Potential carbohydrate attachment sites |
| 755 | S → STWWKEPLNIVSFLFCFAS (in ref. 46) |

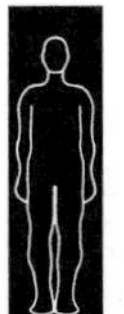

## Amino acid sequence of human TPR

```
  1 MAAVLQQVLE RTELNKLPKS VQNKLEKFLA DQQSEIDGLK GRHEKFKVES
 51 EQQYFEIEKR LSHSQERLVN ETRECQSLRL ELEKLNNQLK ALTEKNKELE
101 IAQDRNIAIQ SQFTRTKEEL EAEKRDLIRT NERLSQELEY LTEDVKRLNE
151 KLKESNTTKG ELQLKLDELQ ASDVSVKYRE KRLEQEKELL HSQNTWLNTE
201 LKTKTDELLA LGREKGNEIL ELKCNLENKK EEVSRLEEQM NGLKTSNEHL
251 QKHVEDLLTK LKEAKEQQAS MEEKFHNELN AHIKLSNLYK SAADDSEAKS
301 NELTRAVEEL HKLLKEAGEA NKAIQDHLLE VEQSKDQMEK EMLEKIGRLE
351 KELENANDLL SATKRKGAIL SEEELAAMSP TAAAVAKIVK PGMKLTELYN
401 AYVETQDQLL LEKLENKRIN KYLDEIVKEV EAKAPILKRQ REEYERAQKA
451 VASLSVKLEQ AMKEIQRLQE DTDKANKQSS VLERDNRRME IQVKDLSQQI
501 RVLLMELEEA RGNHVIRDEE VSSADISSSS EVISQHLVSY RNIEELQQQN
551 QRLLVALREL GETREREEQE TTSSKITELQ LKLESALTEL EQLRKSRQHQ
601 MQLVDSIVRQ RDMYRILLSQ TTGVAIPLHA SSLDDVSLAS TPKRPSTSQT
651 VSTPAPVPVI ESTEAIEAKA ALKQLQEIFE NYKKEKAENE KIQNEQLEKL
701 QEQVTDLRSQ NTKISTQLDF ASKRYL (726)
                                   EMLQD NVEGYRREIT SLHERNQKLT
751 ATTQKQEQII NTMTQDLRGA NEKLAVAEVR AENLKKEKEM LKLSEVRLSQ
801 QRESLLAEQR (810)
```

### Domain structure

| | |
|---|---|
| 75–99 and 117–141 | Leucine zipper domains (underlined) |
| 726–810 | The 30 bp deletion that extends the ORF to 810 amino acids causes the deletion of the C-terminal leucine from p84$^{TPR}$ and the substitution of 85 amino acids |

### Database accession numbers

| | *PIR* | *SWISSPROT* | *EMBL/GENBANK* | *REFERENCES* |
|---|---|---|---|---|
| Human *MET* | A40175 | P08581 | J02958, X54559 | 46–48 |
| Human *TPR-S* | | | X63105 | 49 |
| Human *TPR-L* | | | M15326, X66397 | 10 |

### *References*


1 Cooper, C.S. et al. (1984) Nature 311, 29–33.
2 Park, M. et al. (1986) Cell 45, 895–904.
3 Michelin, S. et al. (1993) Oncogene 8, 1983–1991.
4 Soman, N.R. et al. (1991) Proc. Natl Acad. Sci. USA 88, 4892–4896.
5 Faletto, D.L. et al. (1992) Oncogene 7, 1149–1157.
6 **Cooper, C.S. (1992) Oncogene 7, 3–7.**
7 Di Renzo, M.F. et al. (1993) Oncogene 8, 219–222.
8 Iyer, A. et al. (1990) Cell Growth Differ. 1, 87–95.
9 Liu, C. et al. (1992) Oncogene 7, 181–185.
10 Mitchell, P.J. and Cooper, C.S. (1992) Oncogene 7, 2329–2333.
11 Bottaro, D.P. et al. (1991) Science 251, 802–804.

[12] Naldini, L. et al. (1991) EMBO J. 10, 2867–2878.
[13] Hartmann, G. et al. (1992) Proc. Natl Acad. Sci. USA 89, 11574–11578.
[14] Naldini, L. et al. (1991) Mol. Cell. Biol. 11, 1793–1803.
[15] Naldini, L. et al. (1991) Oncogene 6, 501–504.
[16] **Vande Woude, G. (1992) Jpn J. Cancer Res. 83, cover article.**
[17] Graziani, A. et al. (1993) J. Biol. Chem. 268, 9165–9168.
[18] Pelicci, G. et al. (1995) Oncogene 10, 1631–1638.
[19] Weidner, K.M. et al. (1996) Nature 384, 173–176.
[20] Fixman, E.D. et al. (1996) J. Biol. Chem. 271, 13116–13122.
[21] Matsumoto, K. et al. (1994) J. Biol. Chem. 269, 31807–31813.
[22] Testa, J.R. et al. (1990) Oncogene 5, 1565–1571.
[23] Ishikawa, F. et al. (1987) Mol. Cell. Biol. 7, 1226–1232.
[24] Di Renzo, M.F. et al. (1991) Oncogene 6, 1997–2003.
[25] Di Renzo, M.F. et al. (1995) Clin. Cancer Res. 1, 147–154
[26] Di Renzo, M.F. et al. (1992) Oncogene 7, 2549–2553.
[27] Ferracini, R. et al. (1996) Oncogene 12, 1697–1705.
[28] Ferracini, R. et al. (1995) Oncogene 10, 739–749.
[29] Ebert, M. et al. (1994) Cancer Res. 54, 5775–5778.
[30] Moghul, A. et al. (1994) Oncogene 9, 2045–2052.
[31] Rong, S. et al. (1995) Cancer Res. 55, 1963–1970.
[32] Maina, F. et al. (1996) Cell 87, 531–542.
[33] Rong, S. et al. (1992) Mol. Cell. Biol. 12, 5152–5158.
[34] Rosen, E.M. et al. (1991) Cell Growth Differ. 2, 603–607.
[35] Giordano, S. et al. (1993) Proc. Natl Acad. Sci. USA 90, 649–653.
[36] Rong, S. et al. (1994) Proc. Natl Acad. Sci. USA 91, 4731–4735.
[37] Halaban, R. et al. (1992) Oncogene 7, 2195–2206.
[38] Kanda, H. et al. (1993) Oncogene 8, 3047–3053.
[39] Epstein, J.A. et al. (1996) Proc. Natl Acad. Sci. USA 93, 4213–4218.
[40] Rodrigues, G.A. et al. (1991) Mol. Cell. Biol. 11, 2962–2970.
[41] Rodrigues, G.A. and Park, M. (1993) Mol. Cell. Biol. 13, 6711–6722.
[42] Zhen, Z. et al. (1994) Oncogene 9, 1691–1697.
[43] Longati, P. et al. (1994) Oncogene, 9, 49–57.
[44] Rodrigues, G.A. and Park, M. (1994) Oncogene 9, 2019–2027.
[45] Zhu, H. et al. (1994) J. Biol. Chem. 269, 29943–29948.
[46] Park, M. et al. (1987) Proc. Natl Acad. Sci. USA 84, 6379–6383.
[47] Chan, A.M.L. et al. (1987) Oncogene 1, 229–233.
[48] Dean, M. et al. (1985) Nature 318, 385–388.
[49] Mitchell, P.J. and Cooper, C.S. (1992) Oncogene 7, 383–388.

## Identification

v-*mil* (also v-*mht*) is the oncogene of avian retrovirus Mill-Hill-2 (MH2), which also carries v-*myc* [1–4]. v-*Rmil* is the oncogene of the IC10 and IC11 retroviruses generated during *in vitro* passaging of RAV-1 in chicken NR cells [5]. *Mil* was detected by screening chicken DNA with a v-*mil* probe [6].

## Related genes

*Mil* is the avian homologue of mammalian *RAF1*. v-*Rmil* is the avian homologue of human *RAFB1*. *Mil* has homology with *Src, Fes, Fms, Mos, Yes, Fps, ErbB*, the catalytic subunit of cAMP-dependent protein kinase and protein kinase C.

| | ***Mil* (chicken)** | ***Rmil*** | **v-*mil*** |
|---|---|---|---|
| **Nucleotides (bp)** | Not fully mapped | Not fully mapped | 1154 (MH2 genome) |
| **Mass (kDa): predicted** | 73 | 89 | 42.8 |
| **expressed** | 71/73 | 93.5/95 | $P100^{gag\text{-}mil}$ |
| **Cellular location** | Unknown | Unknown | Cytoplasm |

**Tissue distribution**

In most chicken tissues *Mil* mRNA lacking exon 7a is expressed: mRNA containing exon 7a occurs only in heart, skeletal muscle and brain [7]. *Rmil* is expressed at much higher levels in neural cells, neuroretinas and brain than in other embryonic tissues.

## Protein function

MIL is a serine/threonine kinase belonging to the RAF-MOS subfamily [8]. MIL is phosphorylated *in vivo* [9] and itself directly phosphorylates the N-terminus of JUN ($Ser^{63,73}$), independently of MAPKs [10]. $P100^{gag\text{-}mil}$ binds RNA and DNA *in vitro* [11].

**In animals**

MH2 induces monocytic leukaemias and liver tumours in chickens [12].

***In vitro***

MH2 rapidly transforms chick haematopoietic cells (macrophages) and fibroblasts [12]. In contrast to its mammalian homologue v-*raf*, v-*mil* alone does not fully transform avian primary fibroblasts, probably because v-*mht*/*mil* contains an extra 5′ segment relative to v-*raf* that affects substrate recognition [13].

Viral and activated forms of MIL/RAF1 proteins associate with a 34 kDa phosphoprotein in interphase but not during mitosis [14]. GAG/MIL, like

RAF1, is hyperphosphorylated and activated during mitosis, suggesting a role in the $G_2$ to M phase transition, although the activators and substrates of MIL/RAF1 at this stage of the cell cycle are unknown [15].

## Gene structure

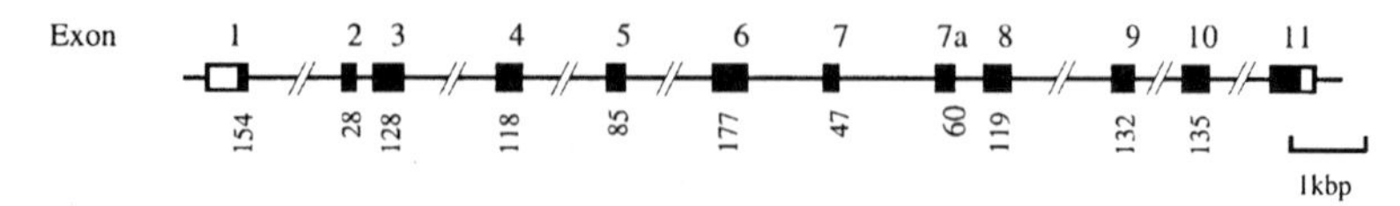


## Protein structure

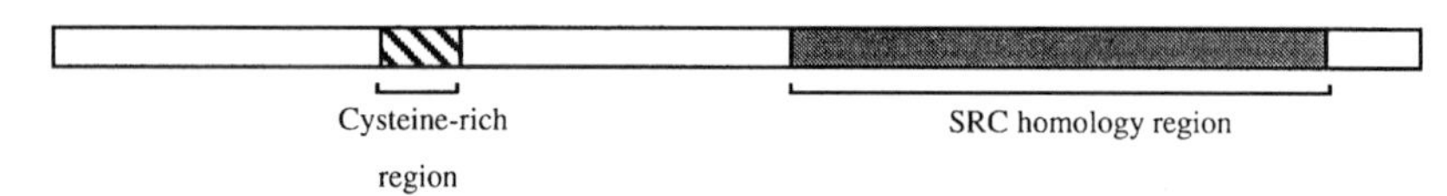

## Amino acid sequence of chicken MIL

```
  1 MEHIQGAWKT ISNGFGLKDS VFDGPNCISP TIVQQFGYQR RASDDGKISD
 51 TSKTSNTIRV FLPNKQRTVV NVRNGMTLHD CLMKALKVRG LQPECCAVFR
101 LVTEPKGKKV RLDWNTDAAS LIGEELQVDF LDHVPLTTHN FARKTFLKLA
151 FCDICQKFLL NGFRCQTCGY KFHEHCSTKV PTMCVDWSNI RQLLLFPNSN
201 ISDSGVPALP PLTMRRMRES VSRIPVSSQH RYSTPHVFTF NTSNPSSEGT
251 LSQRQRSTST PNVHMVSTTM PVDSRIIEDA IRNHSESASP SALSGSPNNM
301 SPTGWSQPKT PVPAQRERAP GTNTQEKNKI RPRGQRDSSY YWEIEASEVM
351 LSTRIGSGSF GTVYKGKWHG DVAVKILKVV DPTPEQFQAF RNEVAVLRKT
401 RHVNILLFMG YMTKDNLAIV TQWCEGSSLY KHLHVQETKF QMFQLIDIAR
451 QTAQGMDYLH AKNIIHRDMK SNNIFLHEGL TVKIGDFGLA TVKSRWSGSQ
501 QVEQPTGSIL WMAPEVIRMQ DSNPFSFQSD VYSYGIVLYE LMTGELPYSH
551 INNRDQIIFM VGRGYASPDL SKLYKNCPKA MKRLVADCLK KVREERPLFP
601 QILSSIELLQ HSLPKINRSA SEPSLHRASH TEDINSCTLT STRLPVF (647)
```

### Domain structure

| | |
|---|---|
| 139–184 | Phorbol ester and diacylglycerol binding region (underlined) |
| 355–363 and 375 | ATP binding site |
| 468 | Active site |
| 269–647 | v-MIL (379 amino acids) differs from this region in two point mutations ($S^{293}$ and $L^{350}$ (italics)) |

### Database accession numbers

| | PIR | SWISSPROT | EMBL/GENBANK | REFERENCES |
|---|---|---|---|---|
| Chicken *Mil* | S00644 | P05625 | X07017 | 6 |
| Chicken *Mil* exon 7a | | | X55430 | 7 |

### *References*


1 Saule, S. et al. (1983) EMBO J. 2, 805–809.
2 Kan, N.C. et al. (1983) Proc. Natl Acad. Sci. USA 80, 6566–6570.
3 Coll, J. et al. (1983) EMBO J. 2, 2189–2194.

[4] Jansen, H.W. et al. (1983) EMBO J. 2, 1969–1975.
[5] Felder, M.-P. et al. (1991) J. Virol. 65, 3633–3640.
[6] Koenen, M. et al. (1988) Oncogene 2, 179–185.
[7] Dozier, C. et al. (1991) Oncogene 6, 1307–1311.
[8] Moelling, K. et al. (1984) Nature 312, 558–561.
[9] Patschinsky, T. et al. (1986) Mol. Cell. Biol. 6, 739–744.
[10] Radziwill, G. et al. (1995) Proc. Natl Acad. Sci. USA 92, 1421–1425.
[11] Bunte, T. et al. (1983) EMBO J. 2, 1087–1092.
[12] Graf, T. et al. (1986) Cell 45, 357–364.
[13] Kan, N.C. et al. (1991) Avian Dis. 35, 941–949.
[14] Lovric, J. et al. (1996) Oncogene 12, 1145–1151.
[15] Lovric, J. and Moelling, K. (1996) Oncogene 12, 1109–1116.

## Identification

v-*mos* is the oncogene of the acutely transforming murine Moloney sarcoma virus (Mo-MuSV). Isolated from a rhabdosarcoma in BALB/c mice infected with Moloney murine leukaemia virus [1]. *MOS* was detected by screening placental DNA with a v-*mos* probe [2].

## Related genes

*MOS* is a member of the *RAF-MOS* subfamily and has homology with *SRC* but is not a tyrosine kinase.

| | ***MOS/Mos*** | **v-*mos*** |
|---|---|---|
| **Nucleotides (kb)** | 1.2 | >1.2 |
| **Chromosome** | 8q11 | |
| **Mass (kDa): predicted** | 37.8 | |
| **expressed** | 39<br>24, 29, 42 and 44 in transformed cells | $P37/39^{env\text{-}mos}$<br>$P85^{gag\text{-}mos}$ |
| **Cellular location** | Cytoplasm | Nucleus (mainly) |

**Tissue distribution**

*Mos* mRNA is only expressed at relatively high levels in germ cells in testes [3] and ovaries [4,5]. Human *MOS* is expressed at low levels in normal T and B lymphocytes and in neuroblastoma and cervical carcinoma cell lines [6].

## Protein function

MOS is a serine/threonine kinase and the protein from *Xenopus* eggs exhibits autophosphorylation activity *in vitro* [7]. *Mos* expression is sufficient to cause meiosis I [8] and is required for meiosis II [9,10]. MOS is an active component of cytostatic factor (CSF) [11,12], an activity responsible for arrest in metaphase at the end of meiosis II [13]. The inactivation of maturation promoting factor (MPF ($p34^{CDC2}$ and cyclin)) is required to allow oocytes to complete nuclear division, thus, MOS directly or indirectly activates and/or stabilizes MPF. Proteins other than MOS contribute to CSF and are not required for meiosis I [8]. MOS is an upstream activator of mitogen-activated protein (MAP) kinase, directly phosphorylating MAP kinase kinase during *Xenopus* oocyte entry into meiosis [14,15], and MAP kinase is required for Cdc2 activation [16]. MAPKK activity is necessary for MOS-induced metaphase arrest [17].

In unfertilized eggs and transformed cells MOS associates with tubulin and the $p35^{CDK}$ isoform of $p34^{CDC2}$. MOS phosphorylates tubulin *in vitro* [18–20] and may promote the reorganization of microtubules that leads to meiotic spindle formation. The transforming capacity of MOS probably derives from expression of its M phase activity during interphase [8,11,21].

v-MOS, v-SRC or *Hras* expression enhances the *trans*-activating capacity of ELK3/NET, an ETS-related factor that has sequence similarity to three regions of ELK1 and SAP1 [22].

**Cancer**

*MOS* and the flanking regions of its gene are mutated in some benign pleomorphic adenomas of the salivary glands [23].

**Transgenic animals**

Transgenic mice expressing *Mos* develop pheochromocytomas and medullary C cell carcinomas of the thyroid resembling the human syndrome multiple endocrine neoplasia type 2 [24].

$Mos^{-/-}$ mice are viable but only males are fully fertile, females displaying reduced fertility as a result of the parthenogenetic activation of mature oocytes [25,26]. Thus MOS is not required for male gametogenesis or the development and growth of somatic lineages but is necessary for metaphase arrest of oocytes at meiosis II. *Mos*-deficient female mice develop ovarian cysts and teratomas.

**In animals**

MOS causes fibrosarcomas in mice following Mo-MuSV infection and can also cause osteosarcomas in other species [27,28]. Activation of *Mos* by insertion of an endogenous intracisternal A-particle has been detected in some mouse plasmacytomas [29].

***In vitro***

*Mos*/LTR or v-*mos*/LTR hybrid genes transform NIH 3T3 fibroblasts [30] and this is dependent on MAP kinase activation [31]. The expression of v-MOS in 3T3 fibroblasts blocks PDGF-mediated signalling by inhibiting PDGFR$\beta$ autophosphorylation [32].

## Gene structure

Human, mouse, rat and chicken *MOS* genes contain a single contiguous ORF of ~1050 bp.

## Transcriptional regulation

Transcription of *Mos* in mouse oocytes is directed by a simple promoter (consensus PyPyCAPyPyPyPyPy) comprised of sequences within 20 bp of the transcription start site [33]. A transcription start site ~1580 bp upstream from the ORF of the $G_2$ transcript regulates the very low level of expression detected in mouse somatic cells [34]. A negative regulatory region (NRE) between 392 and 502 bp upstream from the *Mos* ATG inhibits transcription in somatic cells. A repressor protein binds to one of the three NRE boxes in this region, the binding site being conserved in other germ cell-specific genes [35]. In rat and mouse a 200 bp region upstream of the *Mos* exon has *cis*-inhibitory activity (e.g. can block the transforming activity of v-*mos*). The region contains two polyadenylation signals and when located downstream of a gene causes termination of transcription [36].

Mouse and *Xenopus* oocyte maturation is dependent on cytoplasmic polyadenylation of *Mos* mRNA (anti-sense ODN to *Mos* prevents progress to meiosis II after emission of the first polar body): polyadenylation requires three *cis* elements in the 3′ UTR (the hexanucleotide AAUAAA and two U-rich cytoplasmic polyadenylation elements (CPEs)) and meiotic maturation is prevented by selective amputation of polyadenylation signals from *Mos* mRNA[37,38].

## Protein structure

v-MOS from the earliest Mo-MuSV isolated (HT1-Mo-MuSV) is identical in amino acid sequence to murine MOS[39]. A number of other MuSV isolates show some sequence changes in transduced v-*mos*[40]. Human and mouse MOS are 77% identical in sequence.

## Amino acid sequence of human MOS

```
  1 MPSPLALRPY LRSEFSPSVD ARPCSSPSEL PAKLLLGATL PRAPRLPRRL
 51 AWCSIDWEQV CLLQRLGAGG FGSVYKATYR GVPVAIKQVN KCTKNRLASR
101 RSFWAELNVA RLRHDNIVRV VAASTRTPAG SNSLGTIIME FGGNVTLHQV
151 IYGAAGHPEG DAGEPHCRTG GQLSLGKCLK YSLDVVNGLL FLHSQSIVHL
201 DLKPANILIS EQDVCKISDF GCSEKLEDLL CFQTPSYPLG GTYTHRAPEL
251 LKGEGVTPKA DIYSFAITLW QMTTKQAPYS GERQHILYAV VAYDLRPSLS
301 AAVFEDSLPG QRLGDVIQRC WRPSAAQRPS ARLLLVDLTS LKAELG (346)
```

### Domain structure

66–74 and 87 ATP binding site (underlined)
201 Active site

### Database accession numbers

| | *PIR* | *SWISSPROT* | *EMBL/GENBANK* | *REFERENCES* |
|---|---|---|---|---|
| Human *MOS* | A00649 | P00540 | J00119 | 2 |

### *References*


1. Moloney, J.B. (1966) Natl. Cancer Inst. Monogr. 22, 139–142.
2. Watson, R. (1982) Proc. Natl Acad. Sci. USA 79, 4078–4082.
3. Propst, F. and Vande Woude, G.F. (1985) Nature 315, 516–518.
4. Goldman, D.S. et al. (1988) Proc. Natl Acad. Sci. USA 84, 4509–4513.
5. Keshet, E. et al. (1987) Oncogene 2, 234–240.
6. Li, C.-C.H. et al. (1993) Oncogene 8, 1685–1691.
7. Watanabe, N. et al. (1989) Nature 342, 505–511.
8. Yew, N. et al. (1992) Nature 355, 649–652.
9. Kanki, J.P. and Donoghue, D.J. (1991) Proc. Natl Acad. Sci. USA 88, 5794–5798.
10. Daar, I. et al. (1991) J. Cell Biol. 114, 329–335.
11. Sagata, N. et al. (1989) Science 245, 643–646.
12. Sagata, N. et al. (1989) Nature 342, 512–518.
13. Lorca, T. et al. (1991) EMBO J. 10, 2087–2093.
14. Posada, J. et al. (1993) Mol. Cell. Biol. 13, 2546–2553.
15. Nebreda, A.R. and Hunt, T. (1993) EMBO J. 12, 1979–1986.

[16] Huang, C.F. and Ferrell, J.E. (1996) EMBO J. 15, 2169–2173.
[17] Kosako, H. et al. (1994) J. Biol. Chem. 269, 28354–28358.
[18] Zhou, R. et al. (1992) Mol. Cell. Biol. 12, 3583–3589.
[19] Bai, W. et al. (1992) Oncogene 7, 493–500.
[20] Bai, W. et al. (1992) Oncogene 7, 1757–1763.
[21] Daar, I. et al. (1991) Science 253, 74–76.
[22] Giovane, A. et al. (1994) Genes Dev. 8, 1502–1513.
[23] Stenman, G. et al. (1991) Oncogene 6, 1105–1108.
[24] Schulz, N. et al. (1992) Cancer Res. 52, 450–455.
[25] Colledge, W.H. et al. (1994) Nature 370, 65–68.
[26] Hashimoto, N. et al. (1994) Nature 370, 68–71.
[27] Fefer, A. et al. (1967) Cancer Res. 27, 1626–1631.
[28] Fujinaga, S. et al. (1970) Cancer Res. 30, 1698–1708.
[29] Horowitz, M. et al. (1984) EMBO J. 3, 2937–2941.
[30] Freeman, R.S. et al. (1989) Proc. Natl Acad. Sci. USA 86, 5805–5809.
[31] Okazaki, K. and Sagata, N. (1995) Oncogene 10, 1149–1157.
[32] Faller, D.V. et al. (1994) J. Biol. Chem. 269, 5022–5029.
[33] Pal, S.K. et al. (1991) Mol. Cell. Biol. 11, 5190–5196.
[34] Gao, C. et al. (1996) Oncogene 12, 1571–1576.
[35] Xu, W. and Cooper, G.M. (1995) Mol. Cell. Biol. 15, 5369–5375.
[36] McGeady, M.L. et al. (1986) DNA 5, 289–298.
[37] Gebauer, F. et al. (1994) EMBO J. 13, 5712–5720.
[38] Sheets, M.D. et al. (1995) Nature 374, 511–516.
[39] Seth, A. and Vande-Woude, G.F. (1985) J. Virol. 56, 144–152.
[40] Brow, M.A. et al. (1984) J. Virol. 49, 579–582.

## Identification

v-*myb* is the oncogene of the acutely transforming avian myeloblastosis virus (AMV) and E26 leukaemia virus[1]. *MYB* was identified by screening a cDNA library with a v-*myb* probe.

## Related genes

Related genes are human *MYBA* and *MYBB* and *MYB*-like genes *MYBL1* and *MYBL2*, and chicken B-*myb*. MYB related proteins also occur in *Schizosaccharomyces pombe* (*cdc5*$^+$), *Xenopus laevis*, *Drosophila melanogaster*, yeast (BAS1, REB1), *D. discoideum*, *Zea mays*, barley, potato and *Arabidopsis thaliana*[2].

The "telobox" is a MYB-related motif found in a number of yeast, plant and human proteins[3].

| | ***MYB/Myb*** | **v-*myb*** |
|---|---|---|
| **Nucleotides (kb)** | >25 | 7.14 (AMV genome)<br>5.7 (E26 genome) |
| **Chromosome** | 6q22–q23 | |
| **Mass (kDa): predicted** | 72.5 | $43^{v\text{-}myb}$<br>$75^{gag\text{-}myb\text{-}ets}$ |
| **expressed** | 75<br>90 | $48^{v\text{-}myb}$ (AMV)<br>$P135^{gag\text{-}myb\text{-}ets}$ (E26) |
| **Cellular location** | Nucleus | Nucleus |

**Tissue distribution**

*MYB* is expressed in immature cells of the lymphoid, erythroid and myeloid lineages[4–6]. It is also strongly expressed in CD4$^+$ thymocytes, induced in T lymphocytes and fibroblasts by mitogenic stimulation[7] and is detectable in vascular smooth muscle cells[8,9].

Chicken *Myb* is coexpressed with *Mim-1* during granulopoiesis in the pancreas and spleen[10].

Murine *Myb* is expressed at high levels in the developing thymus and liver but is not confined to haematopoietic cells, being expressed in the neural retina and respiratory tract epithelia. *Myb* (and *Myc*) is high in T cells in regenerating mouse spleen[11].

Murine *Myba* is expressed in immature neuronal cells, in sperm cell precursors and during B cell development and in adult mice is predominantly expressed in the testis[12]. In contrast to *Myb* and *Myba*, *Mybb* is widely expressed in embryogenesis. A testis-specific form of *Mybb* has been detected[12]. *Myba* and *Mybb* show cell- and stage-specific expression in murine testis development and MYBA appears to be necessary for progression through the first meiotic prophase whilst MYBB may be involved in the proliferation and differentiation of gonocytes and spermatogonia[13]. *Mybb* is strongly expressed only during the late $G_1$ and S phases of the cell cycle[14] and appears to undergo cyclin-mediated phosphorylation as cells enter S phase[15].

## Protein function

MYB contains DNA binding, transcriptional activation and negative regulatory domains and binds directly to double-stranded DNA, inducing DNA bending[16–20]. The consensus binding site is YAAC$^{G}/_{T}$G, most commonly CCTAACTG[21] or YAAC$^{T/}$(C)$_{/G}$GYCA[22], from which intact MYB or v-MYB activates transcription. Binding of v-MYB and MYB is decreased by CpG methylation of the binding motif[23]. The C-terminal EVES domain can associate with the N-terminal 192 amino acids of MYB and also with the ubiquitously expressed transcriptional coactivator p100[24]. This domain can be phosphorylated and this may regulate intra-molecular interactions that control association of MYB with the transcriptional apparatus.

High *MYB* expression is generally associated with immature cells of haematopoietic lineage and *MYB* is essential for normal haematopoiesis. In differentiated cells *Myb* expression is associated with cell proliferation and in lymphoid cells the appearance of *MYB* mRNA correlates with activation by IL-1α[25], IL-2[26–31] or IL-3[32–34]. Transformed cells increase *Myb* expression as the cells enter the cycle: avian thymocytes express high levels of *Myb* in $G_0$ or $G_1$ and throughout the cycle. Differentiation (stimulated, for example, by erythropoietin, retinoic acid, TPA, HMBA or DMSO) is accompanied by suppression of *Myb* expression but this does not seem to be necessary for the initiation of differentiation[4,35–43]. In Friend erythroleukaemia cells and M1 myeloblastic leukaemia cells sustained expression of *Myb* blocks differentiation[44–46] although ectopic expression of *Mbm2* (the alternatively spliced form of *Myb*) accelerates differentiation[47]. Although sustained *Myb* (or *Fos*) expression blocks erythropoietin-induced differentiation, it only partially blocks the effect of DMSO. Thus chemically and erythropoietin-induced differentiation mechanisms may differ[48].

Thus *MYB* is probably an important regulator of cell differentiation and the generation in normal cells of multiple isoforms of *MYB* by alternative splicing and/or alternative initiation indicates that the protein products may have differing, tissue-specific roles[49]. However, transgenic mice that do not express *Myb* develop normally until day 13 but then rapidly become anaemic[50]. Embryonic erythropoiesis is unaffected but adult erythropoiesis, which first occurs in the liver, is greatly diminished, suggesting that the function of *Myb* is unique to haematopoiesis.

There is a dramatic increase in *MYB* expression (30- to 60-fold) associated with generalized autoimmune diseases, which appears to occur in the greatly expanded population of $CD4^{-}8^{-}$ cells. Studies in MRL-*lpr*/*lpr* mice, which carry the same defect, indicate the existence of specific nuclear DNA binding proteins that regulate *Myb* expression[51]. *lpr* cells lack mature T cell surface markers (Lyt-2 and L3T4), express high levels of *Myb* mRNA and are generally unresponsive to T cell mitogens. Induction of differentiation (by TPA + A23187) depresses *Myb* and activates IL-2R transcription[52].

*MYB* may be critically involved in the proliferation of smooth muscle cells. The expression of anti-sense phosphorothiolate oligodeoxynucleotides (ODNs) against *MYB* or non-muscle myosin heavy chain reversibly suppresses growth of these cells [9,53]. ODNs directed against *MYB* also inhibit proliferation of lymphoid cells [54–57] and of human colon cancer cell lines [58] and interfere with haematopoiesis [59,60]. Anti-sense *MYB* oligonucleotides also inhibit the two-fold increase in the concentration of intracellular free calcium that normally occurs at the $G_1$/S phase interface in rat vascular smooth muscle cells [61]. The MYB-dependent rise in $[Ca^{2+}]_i$ is dependent on extracellular $Ca^{2+}$ but is not mediated by L type channels (i.e. is nifedipine-insensitive) or T type channels (which are not normally present in vascular smooth muscle cells). The antiproliferative activity of *Myc* and *Myb* anti-sense oligonucleotides in smooth muscle cells appears not to be caused by a hybridization-dependent mechanism but arises from the action of a stretch of four contiguous guanosine residues [62].

In Friend erythroleukaemia cells the endoplasmic reticulum $Ca^{2+}$ pump inhibitors thapsigargin and cyclopiazonic acid elevate $[Ca^{2+}]_i$, which causes transient suppression of *Myb* expression and terminal erythroid differentiation [63].

The coexpression of MYB prevents the rapid growth arrest but accelerates apoptosis induced by TGF$\beta_1$ in M1 myeloid leukaemia cells [64]. MYB can also promote apoptosis when coexpressed with p53 by a mechanism in which MYB (but not MYBB) indirectly increases transcription of the cell death-associated *BAX* gene [65]. In E26-transformed myeloid cells, MYB/ETS functions as a survival factor by directly inducing the expression of the anti-apoptotic protein BCL2 [66]. A dominant mutant form of MYB that functions as a repressor has the converse effect in T cells, directly downregulating transcription of *Bcl-2* and inducing apoptosis [67].

**Regulation of gene expression by MYB**

Human MYB autoregulates transcription of its own gene (see below) and may exert a positive effect during proliferation and/or differentiation. The following genes have also been shown to be *trans*-activated by MYB:

Adenosine deaminase (ADA) [68]
*Bcl-2* [67]
*CD13*/aminopeptidase N (APN) [69]
*CD34* [70]
*CD4* [71]
*CDC2* [72]
DNA polymerase $\alpha$ [73]
Epstein–Barr virus (EBV) transcription factor Z (BZLF1) [74]
*GATA-1* [75]
*HER2* (human) [76]
HIV-1 [77]
HPV16 promoter [78]
HTLV-I LTRs [79, 80]
Human HSP70 [81,82]
Insulin-like growth factor I (IGF-I) [83,84]

*Mim-1* (Myb-induced myeloid protein 1)[85,86]
*MYC*[87]
*PAX6*[88]
Ribonuclease A-related gene[89]
SV40 enhancer[90–92]
T cell receptor $\delta$ enhancer[93]
Murine neutrophil elastase promoter[94]

A large number of genes having potential MYB binding sites have also been identified, including cyclins B and D1, *Src, Bcr, Ets, Jun, Kit, P53* and IL-2[95].

**Cancer**
Amplification of *MYB* has been detected in acute myelogenous leukaemia (AML), chronic myelogenous leukaemia (CML), acute lymphocytic leukaemia (ALL), T cell leukaemias, colon carcinomas and melanomas[58,96–102]. The stability of *MYB* and *MYC* mRNAs is increased in the cells of some AML patients[103]. Malignant haematopoietic colony forming units can be removed from the cells of CML patients by exposure to *MYB* anti-sense oligodeoxy-nucleotides[104] and proliferation of a variety of malignant cell lines is inhibited by *MYB* anti-sense ODNs[57,58]. Abnormal MYB expression has also been detected in some ovarian and breast carcinomas[105,106] and in a high proportion of cervical carcinomas examined[107], and *MYB* mRNA is detectable in human neuroblastoma[39] and teratocarcinoma cell lines[108]. Only oestrogen receptor-positive breast tumour cell lines show significant MYB expression, transcription being induced by addition of oestrogen[109]. *MYBA* and *MYBB* are expressed in carcinoma- and sarcoma-derived cell lines[110].

**In animals**
AMV induces acute myeloid leukaemia in chickens[111]. The recombinant avian leukosis virus (ALV) EU-8, injected into chicken embryos, induces a high incidence of B cell lymphomas[112], caused by proviral integration of EU-8 in the *Myb* locus. A similar metastatic lymphoma develops when chicken embryos are infected with the RAV-1 isolate of ALV[113]. The induction of lymphomas following infection of embryos, rather than the classic lymphoid leukosis caused by ALV in adult animals, implies that the target cells in which *Myb* is activated occur only in embryos. Chemically induced rat colon tumours have frequent rearrangements, insertions or deletions in the *Myb* and *Hras* loci[114]. Over-expression of MYB lacking the C-terminal 214 residues induces fibrosarcomas in chickens[115].

**Transgenic animals**
Homozygous *Myb* mutant mice appear normal at day 13 of gestation but by day 15 are severely anaemic. Embryonic erythropoiesis is not impaired but adult-type erythropoiesis is greatly diminished[50]. This indicates that *Myb* is not essential for early development but may be required to maintain the proliferative state of haematopoietic progenitor cells. T cell specific expression of the v-*myb* transgene in mice gives rise to elevated levels of $CD4^+$ cells and animals are predisposed to the development of T lymphomas[116].

***In vitro***

AMV transforms macrophage precursors (monoblasts) *in vivo* and *in vitro*, chicken bone marrow cells and avian yolk-sac cells. It does not transform fibroblasts and may be a unique oncogene in this respect. Retrovirally driven expression of full length *Myb* leads to density-dependent transformation of murine haemopoietic cells [117].

E26 transforms fibroblasts (quail) and erythroid or myeloid cells. Erythroid cell transformation by E26 is caused by v-*ets* which has a cooperative effect with v-*myb* in this lineage, although neither the DNA binding domain nor the *trans*-activating domain of v-MYB is required [118,119]. E26 stimulates the proliferation of chicken neuroretina cells, as does AMV in the presence of basic fibroblast growth factor [120]. v-MYB appears to block differentiation [121] and, in normal or v-*myc* transformed macrophages, AMV or E26 cause "de-differentiation", inducing changes characteristic of immature cells. Blockade of differentiation of haematopoietic progenitors by E26 MYB/ETS involves inhibition of both retinoic acid receptor and ERBA transcriptional activation *via* specific hormone response elements [122].

The *in vitro* survival and proliferation of *myb*-transformed myeloid cells requires haematopoietic growth factors, for example, IL-3 or cMGF (chicken myeloid growth factor) [32]. The requirement for an exogenous growth factor is relieved by the expression of v-*src*, v-*fps*, v-*yes*, v-*ros*, v-*mil* or v-*erbB* [123,124]. These oncogenes cause autocrine growth stimulation *via* the action of a myeloid cell-specific transcription factor. The expression of E26-derived *gag-myb-ets* in IL-3-dependent murine haematopoietic cell lines indicates that protection from apoptosis and changes in growth factor dependence are critical for the induction of erythroid differentiation [125]. In addition to *Bcl-2*, *Rem-1*, a member of the recoverin family, is a cellular target of MYB/ETS [126].

*MYB* expression is repressed by WT1 in T and B cell lines [127].

## Gene structure

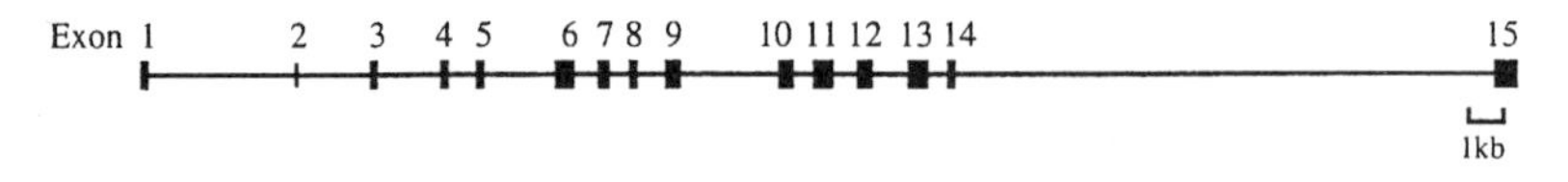


The organization of *MYB* sequences is not completely resolved. The complete chicken and human *MYB* genes are expressed as the result of an intermolecular recombination process involving coding sequences from transcription units on different chromosomes and a putative splicing factor (PR264) is encoded by the opposite strand of the *Myb trans*-spliced $E_T$ exon [128]. As $E_T$ bears 85% homology to the equivalent human and mouse exons and human $E_T$ maps to a different chromosome to that carrying the remainder of *MYB*, it seems probable that the complex organization of the chicken gene reflects a general property of mammalian *MYB*.

## Transcriptional regulation

The region of the human *MYB* gene between nucleotides –616 and –575 upstream from the cap site contains putative MYB binding sites that confer

MYB-inducible expression when linked to a reporter gene [129]. Mutation of the putative binding sites inhibits *trans*-activation by MYB and also reduces the binding of MYB protein to the sites. Human MYB contains a region to which CMAT (c-*myb* in activated T cells) protein binds (−784 to −758) to induce expression during T cell activation [130].

The promoters of chicken and mouse *Myb* contain GC-rich upstream regions with no TATA box (mouse) or with a TATA box that is not associated with a CAAT box (chicken). In murine *Myb* a positive intragenic regulatory mechanism operates *via* two tandem repeats of AP1 sites in the first intron [131] (see also ***FOS*** and ***MYC***). Thus the decrease in the level of *Myb* mRNA that accompanies differentiation of mouse erythroleukaemic cells correlates with a decrease in sequence-specific protein binding to this region. Transcriptional attenuation is also the major mechanism of regulation of human *MYB* during retinoic acid- or vitamin $D_3$-induced differentiation of HL60 cells [132]. However, DMSO or phorbol ester regulate *MYB* expression by an additional, post-translational mechanism that, for DMSO, requires continuous transcription.

Members of the NF-κB family (p50+p65 and p65+REL) *trans*-activate murine *Myb* reporter constructs *via* regions of *Myb* intron 1 that comprise NF-κB binding sites [133]. MZF1 negatively regulates the *Myb* promoter in haematopoietic and non-haematopoietic cells [134].

*B-Myb* transcription is maximal during the $G_1$ and S phases of the cell cycle. Transcription is regulated by p107/E2F [135] and HPV16 E7 activates *B-Myb* transcription by directly interfering with this complex. p107 represses human *MYBB* promoter activity and over-expression of MYBB reverses p107 growth arrest [136]. In the M1 murine myeloid leukaemia cell line, *Mybb* prevents growth arrest and its expression is necessary for completion of IL-6-induced monocyte/macrophage differentiation [137].

## Protein structure

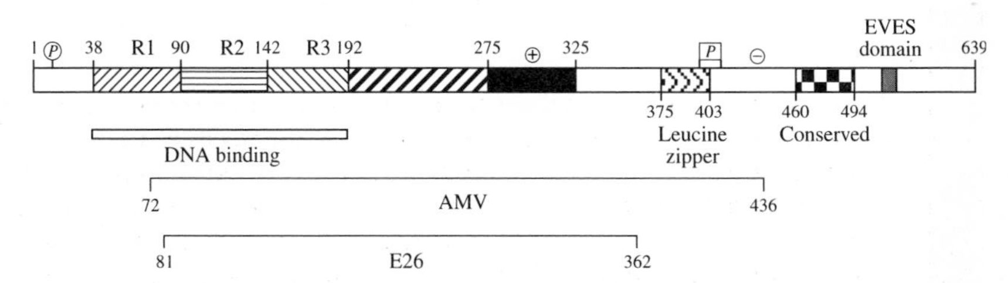


The overall identity between human and chicken MYB is 82%. AMV p45$^{v\text{-}myb}$ is derived from chicken *Myb via* extensive 5′ and 3′ deletions, generating *myb*$^{A}$ (lacking 71 N-terminal amino acids and 198 C-terminal amino acids of MYB, and with 11 point mutations). v-*myb* replaces 26 codons of the 3′ end of *pol* and most of *env*. There are six *gag* encoded amino acids and 33 bp of *env* give rise to 11 C-terminal amino acids, up to TAG [138]. E26 v-*myb* (*myb*$^{E}$) lacks 80 N-terminal and 278 C-terminal amino acids of p75$^{myb}$, has one point mutation and is expressed as p135$^{gag\text{-}myb\text{-}ets}$.

## Amino acid sequence of human MYB

```
  1 MARRPRHSIY SSDEDDEDFE MCDHDYDGLL PKSGKRHLGK TRWTREEDEK
 51 LKKLVEQNGT DDWKVIANYL PNRTDVQCQH RWQKVLNPEL IKGPWTKEED
101 QRVIELVQKY GPKRWSVIAK HLKGRIGKQC RERWHNHLNP EVKKTSWTEE
151 EDRIIYQAHK RLGNRWAEIA KLLPGRTDNA IKNHWNSTMR RKVEQEGYLQ
201 ESSKASQPAV ATSFQKNSHL MGFAQAPPTA QLPATGQPTV NNDYSYYHIS
251 EAQNVSSHVP YPVALHVNIV NVPQPAAAAI QRHYNDEDPE KEKRIKELEL
301 LLMSTENELK GQQVLPTQNH TCSYPGWHST TIADHTRPHG DSAPVSCLGE
351 HHSTPSLPAD PGSLPEESAS PARCMIVHQG TILDNVKNLL EFAETLQFID
401 SFLNTSSNHE NSDLEMPSLT STPLIGHKLT VTTPFHRDQT VKTQKENTVF
451 RTPAIKRSIL ESSPRTPTPF KHALAAQEIK YGPLKMLPQT PSHLVEDLQD
501 VIKQESDESG FVAEFQENGP PLLKKIKQEV ESPTDKSGNF FCSHHWEGDS
551 LNTQLFTQTS PVRDAPNILT SSVLMAPASE DEDNVLKAFT VPKNRSLASP
601 LQPCSSTWEP ASCGKMEEQM TSSSQARKYV NAFSARTLVM (640)
```

### Domain structure

| | |
|---|---|
| 1–200 | Dispersed nuclear localization signal |
| 11–12 | Casein kinase II phosphorylation sites which, when phosphorylated, reduce MYB binding to DNA [139] |
| 34–86, 87–138, 139–189 | Repeat regions R1 and R2 (essential for MYB binding to the MYB recognition element (YAACNG or YAACGN) and R3 [140,141]. Deletion of R1 diminishes DNA binding and activates transformation of myeloid cells *in vitro* [142] |
| 66–454 | Homologous to the region of chicken MYB incorporated in AMV v-MYB (388 amino acids: underlined) |
| 186–363 | Homologous to chicken MYB expressed in P135$^{gag\text{-}myb\text{-}ets}$ with the substitution of $Met^{152}$ for $Arg^{337}$ |
| 43, 63, 82, 95, 115, 134, 147, 166, 185 | Conserved tryptophan residues in the R1, R2 and R3 domains (bold) |
| 130 | $Cys^{130}$ is conserved in all MYB-related proteins. Its reduction is essential for MYB to bind to DNA and for transformation of myeloid cells [143,144]. The equivalent amino acid in v-MYB ($Cys^{65}$) is essential for the transcription factor activity of v-MYB |
| 275–325 | Transcription activation domain that interacts with the MYB response element in the promoters of *Myc* and *Mim-1* [145,146]. CBP (CREB-binding protein), a transcriptional coactivator of CREB protein, coactivates MYB by binding to the *trans*-activation domain in a phosphorylation-independent manner [147] |

| | |
|---|---|
| 383, 389, 396, 403 | Leucine zipper [148] domain that interacts with cellular proteins, including the nuclear proteins p67 and p160 [149]. Mutations in this region increase both the *trans*-activating and transforming capacities of MYB |
| 512–566 | *cis*-acting negative regulatory domain, removal of which increases *trans*-activation by MYB and confers transforming capacity *in vitro* [150]. This region is highly conserved and the corresponding domain is deleted from chicken MYB in the AMV and E26 forms. |
| 529–532 | EVES domain [151] |

**Database accession numbers**

| | *PIR* | *SWISSPROT* | *EMBL/GENBANK* | *REFERENCES* |
|---|---|---|---|---|
| Human *MYB* | A26661 | P10242 | M15024 | 152 |
| | | | M13665, M13666 | 153 |
| Human *MYBA* | S03423 | P10243 | X66087 | 92 |
| Human *MYBB* | S01991 | P10244 | X13293 | 110 |
| Human *MBM2* | | | X52125, X52126 | 154 |

***References***


1 **Shen-Ong, G.L.C. (1990) Biochim. Biophys. Acta 1032, 39–52.**
2 Lipsick, J.S. (1996) Oncogene 13, 223–235.
3 Bilaud, T. et al. (1996) Nucleic Acids Res. 24, 1294–1303.
4 Gonda, T.J. and Metcalf, D. (1984) Nature 310, 249–251.
5 Sheiness, D. and Gardinier, M. (1984) Mol. Cell. Biol. 4, 1206–1212.
6 Duprey, S.P. and Boettiger, D. (1985) Proc. Natl Acad. Sci. USA 82, 6937–6941.
7 Thompson, C.B. et al. (1986) Nature 319, 374–380.
8 Brown, K.E. et al. (1992) J. Biol. Chem. 267, 4625–4630.
9 Simons, M. and Rosenberg, R.D. (1992) Circ. Res. 70, 835–843.
10 Queva, C. et al. (1992) Development 114, 125–133.
11 Sihvola, M. et al. (1989) Biochem. Biophys. Res. Commun. 160, 181–188.
12 Sitzmann, J. et al. (1996) Oncogene 12, 1889–1894.
13 Latham, K.E. et al. (1996) Oncogene 13, 1161–1168.
14 Marhamati, D.J. and Sonenshein, G.E. (1996) J. Biol. Chem. 271, 3359–3365.
15 Robinson, C. et al. (1996) Oncogene 12, 1855–1864.
16 **Lüscher, B. and Eisenman, R.N. (1990) Genes Dev. 4, 2235–2241.**
17 **Graf, T. (1992) Curr. Opin. Genet. Dev. 2, 249–255.**
18 **Thompson, M.A. and Ramsay, R.G. (1995) BioEssays 17, 341–350.**
19 Sakura, H. et al. (1989) Proc. Natl Acad. Sci. USA 86, 5758–5762.
20 Saikumar, P. et al. (1994) Oncogene 9, 1279–1287.
21 Biedenkapp, H. et al. (1988) Nature 335, 835–837.
22 Weston, K. (1992) Nucleic Acids Res. 20, 3043–3049.
23 Klempnauer, K.-H. (1993) Oncogene 8, 111–115.
24 Dash, A.B. et al. (1996) Genes Dev. 10, 1858–1869.
25 Zubiaga, A.M. et al. (1991) J. Immunol. 146, 3849–3856.
26 Stern, J.B. and Smith, K.A. (1986) Science 233, 203–206.

[27] Pauza, C.D. (1987) Mol. Cell. Biol. 7, 342–348.
[28] Reed, J.C. et al. (1987) Oncogene 1, 223–228.
[29] Kelly, K. and Siebenlist, U. (1988) J. Biol. Chem. 263, 4828–4831.
[30] Churilla, A.M. et al. (1989) J. Exp. Med. 170, 105–121.
[31] Bohjanen, P.R. et al. (1990) Proc. Natl Acad. Sci. USA 87, 5283–5287.
[32] Weinstein, Y. et al. (1986) Proc. Natl Acad. Sci. USA 83, 5010–5014.
[33] Avanzi, G.C. et al. (1991) Cancer Res. 51, 1741–1743.
[34] Dautry, F. et al. (1988) J. Biol. Chem. 263, 17615–17620.
[35] Westin, E.H. et al. (1982) Proc. Natl Acad. Sci. USA 79, 2194–2198.
[36] Ramsay, R.G. et al. (1986) Proc. Natl Acad. Sci. USA 83, 6849–6853.
[37] Fukuda, M. et al. (1987) Biochem. Int. 15, 73–79.
[38] Lockett, T.J. and Sleigh, M.J. (1987) Exp. Cell Res. 173, 370–378.
[39] Thiele, C.J. et al. (1988) Mol. Cell. Biol. 8, 1677–1683.
[40] McClinton, D. et al. (1990) Mol. Cell. Biol. 10, 705–710.
[41] Makover, D. et al. (1991) Oncogene 6, 455–460.
[42] Danish, R. et al. (1992) Oncogene 7, 901–907.
[43] Smarda, J. et al. (1995) Mol. Cell. Biol. 15, 2474–2481.
[44] Clarke, M.F. et al. (1988) Mol. Cell. Biol. 8, 884–892.
[45] Hoffman-Liebermann, B. and Liebermann, D.A. (1991) Mol. Cell. Biol. 11, 2375–2381.
[46] Selvakumaran, M. et al. (1992) Mol. Cell. Biol. 12, 2493–2500.
[47] Weber, B.L. et al. (1990) Science 249, 1291–1293.
[48] Todokoro, K. et al. (1988) Proc. Natl Acad. Sci. USA 85, 8900–8904.
[49] Ramsay, R.G. et al. (1989) Oncogene Res. 4, 259–269.
[50] Mucenski, M.L. et al. (1991) Cell 65, 677–689.
[51] Mountz, J.D. and Steinberg, A.D. (1989) J. Immunol. 142, 328–335.
[52] Yokota, S. et al. (1987) J. Immunol. 139, 2810–2817.
[53] Simons, M. et al. (1992) Nature 359, 67–70.
[54] Anfossi, G. et al. (1989) Proc. Natl Acad. Sci. USA 86, 3379–3383.
[55] Gewirtz, A. M. et al. (1989) Science 245, 180–183.
[56] Furukawa, Y. et al. (1990) Science 250, 805–808.
[57] Citro, G. et al. (1992) Proc. Natl Acad. Sci. USA 89, 7031–7035.
[58] Melani, C. et al. (1991) Cancer Res. 51, 2897–2901.
[59] Gewirtz, A.M. and Calabretta, B. (1988) Science 243, 1303–1306.
[60] Valtieri, M. et al. (1991) Blood 77, 1181–1190.
[61] Simons, M. et al. (1993) J. Biol. Chem. 268, 627–632.
[62] Burgess, T.L. et al. (1995) Proc. Natl Acad. Sci. USA 92, 4051–4055.
[63] Schaefer, A. et al. (1994) J. Biol. Chem. 269, 8786–8791.
[64] Selvakumaran, M. et al. (1994) Mol. Cell. Biol. 14, 2352–2360.
[65] Sala, A. et al. (1996) Cancer Res. 56, 1991–1996.
[66] Frampton, J. et al. (1996) Genes Dev. 10, 2720–2731.
[67] Taylor, D. et al. (1996) Genes Dev. 10, 2732–2744.
[68] Ess, K.C. (1995) Mol. Cell. Biol. 15, 5707–5715.
[69] Shapiro, L.H. (1995) J. Biol. Chem. 270, 8763–8771.
[70] Melotti, P. and Calabretta, B. (1994) J. Biol. Chem. 269, 25303–25309.
[71] Siu, G. et al. (1992) Mol. Cell. Biol. 12, 1592–1604.
[72] Ku, D.-H. et al. (1993) J. Biol. Chem. 268, 2255–2259.
[73] Sudo, T. et al. (1992) Oncogene 7, 1999–2006.
[74] Kenney, S.C. et al. (1992) Mol. Cell. Biol. 12, 136–146.
[75] Aurigemma, R.E. et al. (1992) J. Virol. 66, 3056–3061.

[76] Mizuguchi, G. et al. (1995) J. Biol. Chem. 270, 9384–9389.
[77] Dasgupta, P. et al. (1990) Proc. Natl Acad. Sci. USA 87, 8090–8094.
[78] Nürnberg, W. et al. (1995) Cancer Res. 55, 4432–4437.
[79] Bosselut, R. et al. (1992) Virology 186, 764–769.
[80] Dasgupta, P. et al. (1992) J. Virol. 66, 270–276.
[81] Foos, G. et al. (1993) Oncogene 8, 1775–1782.
[82] Kanei-Ishii, C. et al. (1994) J. Biol. Chem. 269, 15768–15775.
[83] Reiss, K. et al. (1991) Cancer Res. 51, 5997–6000.
[84] Travali, S. et al. (1991) Mol. Cell. Biol. 11, 731–736.
[85] Ness, S.A. et al. (1993) Genes Dev. 7, 749–759.
[86] Mink, S. et al. (1996) Mol. Cell. Biol. 16, 1316–1325.
[87] Nakagoshi, H. et al. (1992) Oncogene 7, 1233–1239.
[88] Plaza, S. et al. (1995) Oncogene 10, 329–340.
[89] Nakano, T. and Graf, T. (1992) Oncogene 7, 527–534.
[90] Nishina, Y. et al. (1989) Nucleic Acids Res. 17, 107–117.
[91] Mizuguchi, G. et al. (1990) J. Biol. Chem. 265, 9280–9284.
[92] Golay, J. et al. (1994) Oncogene 9, 2469–2479.
[93] Hernandez-Munain, C. and Krangel, M.S. (1995) Mol. Cell. Biol. 15, 3090–3099.
[94] Oelgeschläger, M. et al. (1996) Mol. Cell. Biol. 16, 4717–4725.
[95] Deng, Q.-L. et al. (1996) Nucleic Acids Res. 24, 766–774.
[96] Alitalo, K. et al. (1984) Proc. Natl Acad. Sci. USA 81, 4534–4538.
[97] Balaban, G.B. et al. (1984) Cancer Genet. Cytogenet. 11, 429–439.
[98] Barletta, C. et al. (1987) Science 235, 1064–1067.
[99] Griffin, C.A. and Baylin, S.B. (1985) Cancer Res. 45, 272–275.
[100] Pellici, P.G. et al. (1984) Science 224, 1117–1121.
[101] Slamon, D.J. et al. (1984) Science 224, 256–262.
[102] Tesch, H. et al. (1992) Leuk. Res. 16, 265–274.
[103] Baer, M.R. et al. (1992) Blood 79, 1319–1326.
[104] Ratajczak, M.Z. et al. (1992) Blood 79, 1956–1961.
[105] Guerin, M. et al. (1990) Oncogene 5, 131–135.
[106] Barletta, C. et al. (1992) Eur. J. Gynaecol. Oncol. 13, 53–59.
[107] Nürnberg, W. et al. (1995) Cancer Res. 55, 4432–4437.
[108] Janssen, J.W.G. et al. (1986) Cytogenet. Cell. Genet. 41, 129–135.
[109] Gudas, J.M. et al. (1995) Clin. Cancer Res. 1, 235–243.
[110] Nomura, N. et al. (1988) Nucleic Acids Res. 16, 11075–11090.
[111] Baluda, M.A. and Goetz, I.E. (1961) Virology 15, 185–199.
[112] Kanter, M.R. et al. (1988) J. Virol. 62, 1423–1432.
[113] Pizer, E. and Humphries, E.H. (1989) J. Virol. 63, 1630–1640.
[114] Alexander, R.J. et al. (1992) Am. J. Med. Sci., 303, 16–24.
[115] Press, R.D. et al. (1994) Mol. Cell. Biol. 14, 2278–2290.
[116] Badiani, P.A. et al. (1996) Oncogene 13, 2205–2212.
[117] Ferrao, P. et al. (1995) Oncogene 11, 1631–1638.
[118] Domenget, C. et al. (1992) Oncogene 7, 2231–2241.
[119] Metz, T. and Graf, T. (1991) Genes Dev. 5, 369–380.
[120] Garrido, C. et al. (1992) J. Virol. 66, 160–166.
[121] Patel, G. et al. (1993) Mol. Cell. Biol. 13, 2269–2276.
[122] Rascle, A. et al. (1996) Mol. Cell. Biol. 16, 6338–6351.
[123] Adkins, B. et al. (1984) Cell 39, 439–445.
[124] Sterneck, E. et al. (1992) EMBO J. 11, 115–126.

[125] Athanasiou, M. et al. (1996) Oncogene 12, 337–344.
[126] Kraut, N. et al. (1995) Oncogene 10, 1027–1036.
[127] McCann, S. et al. (1995) J. Biol. Chem. 270, 23785–23789.
[128] Vellard, M. et al. (1992) Proc. Natl Acad. Sci. USA 89, 2511–2515.
[129] Nicolaides, N.C. et al. (1991) Mol. Cell. Biol. 11, 6166–6176.
[130] Phan, S.-C. et al. (1996) Mol. Cell. Biol. 16, 2387–2393.
[131] Reddy, C.D. and Reddy, E.P. (1989) Proc. Natl Acad. Sci. USA 86, 7326–7330.
[132] Boise, L.H. et al. (1992) Oncogene 7, 1817–1825.
[133] Toth, C.R. et al. (1995) J. Biol. Chem. 270, 7661–7671.
[134] Perrotti, D. et al. (1995) Mol. Cell. Biol. 15, 6075–6087.
[135] Liu, N. et al. (1996) Nucleic Acids Res. 24, 2905–2910.
[136] Sala, A. et al. (1996) J. Biol. Chem. 271, 9363–9367.
[137] Bies, J. et al. (1996) Oncogene 12, 355–363.
[138] Klempnauer, K.-H. et al. (1983) Cell 33, 345–355.
[139] Oelgeschläger, M. et al. (1995) Mol. Cell. Biol. 15, 5966–5974.
[140] Ording, E. et al. (1996) Oncogene 13, 1043–1051.
[141] Oehler, T. et al. (1990) Nucleic Acids Res. 18, 1703–1710.
[142] Dini, P.W. and Lipsick, J.S. (1993) Mol. Cell. Biol. 13, 7334–7348.
[143] Grasser, F.A. et al. (1992) Oncogene 7, 1005–1009.
[144] Guehmann, S. et al. (1992) Nucleic Acids Res. 20, 2279–2286.
[145] Evans, J.L. et al. (1990) Mol. Cell. Biol. 10, 5747–5752.
[146] Zobel, A. et al. (1991) Oncogene 6, 1397–1407.
[147] Dai, P. et al. (1996) Genes Dev. 10, 528–540.
[148] Kanei-Ishii, C. et al. (1992) Proc. Natl Acad. Sci. USA 89, 3088–3092.
[149] Favier, D. and Gonda, T. (1994) Oncogene 9, 305–311.
[150] Vorbrueggen, G. et al. (1994) Nucleic Acids Res. 22, 2466–2475.
[151] Dash, A.B. et al. (1996) Genes Dev. 10, 1858–1869.
[152] Majello B. et al. (1986) Proc. Natl Acad. Sci. USA 83, 9636–9640.
[153] Slamon, D.J. et al. (1986) Science 233, 347–351.
[154] Westin, E.H. et al. (1990) Oncogene 5, 1117–1124.

## Identification

The v-*myc* oncogene was first detected in avian myelocytomatosis virus MC29[1]. *Myc* has also been transduced by the acutely transforming avian retroviruses CMII, OK10, MH2 and FH3. *MYC* was detected by screening human DNA with an MC29 *myc* probe[2].

## Related genes

MYC is a member of the helix–loop–helix/leucine zipper superfamily. The *MYC* gene family contains at least seven closely related genes, *MYC, MYCN, MYCL, PMYC, RMYC, SMYC* and *BMYC*, together with *LMYCΨ*, an inactive pseudogene. pAv-*myc* is expressed in the testis of the northern sea star *Asterias vulgaris*[3] and the maize *Lc* gene has homology with *MYC*[4].

| | *MYC* | *MYCN* | *MYCL* | v-*myc* (MC29) |
|---|---|---|---|---|
| **Nucleotides (kb)** | 6–7 | 6–7 | 6–7 | 5.7 |
| **Chromosome** | 8q24 | 2p24.1 | 1p32 (*MYCL1*) 7p15 (*MYCLK1*) | |
| **Mass (kDa): predicted** | 49 | 49.5 | 40 | 96 |
| **expressed** | 64/67 | 66 | 60/66/68 | $P110^{\Delta gag\text{-}myc}$ |
| **Cellular location** | Nucleus | Nucleus | Nucleus | Nucleus |

MYC is detectable in the cytoplasm of colorectal tumour cells and serum-starved fibroblasts. It can colocalize with α-tubulin and polymerized microtubules in HL60 cells[5].

### Tissue distribution

*MYC* is expressed during proliferation in a wide variety of adult tissues and at all stages of embryonal development. *MYCN* and *MYCL* expression is generally restricted to embryonic brain, kidney and lung, suggesting their possible involvement in differentiation[6], although *Lmyc* is not essential for normal development and its deletion is not accompanied by compensatory changes in the expression of *Myc* or *Nmyc*.

MYC protein concentration: undetectable in quiescent Swiss 3T3 fibroblasts; ~$10^5$ molecules/cell in Burkitt's lymphoma and other tumour cells[7,8].

## Protein function

MYC, MYCN and MYCL are helix-loop-helix/leucine zipper (HLH/LZ) proteins that form heterodimers with MAX that have a high affinity for the core consensus sequence CACGTG (E box MYC sites (EMS)) and lower affinity for various non-canonical DNA sequences [9,10] (see ***MAX***). MYCN also binds to asymmetric (CATGTG) sequences [11]. Both the HLH and LZ domains of MYC contribute to specific heterodimer formation [12] and mutation of the LZ domain can modulate binding to nucleosomal DNA [13]. The binding of MYC/MAX dimers causes a change in the conformation of DNA [14,15] and His$^{336}$ of avian MYC contacts or is close to the thymine 5-methyl group at position 2 of the DNA half site [16]. The homologous residue in MAX also recognizes the same site. The members of the MYC family possess common functional elements: thus, *trans*-activation incompetent mutants of one member can act in *trans* to suppress dominantly the cotransformation activities of all three MYC proteins [17]. Truncated MYC that retains the basic, HLH and LZ domains binds to the sequence GGGCAC$^{G}/_{A}$TGCCC [10]. MYC proteins do not appear to dimerize with other HLH/LZ proteins (e.g. FOS, JUN, MyoD or E12), although a point mutation in the basic domain of MyoD confers the capacity to bind to a *Myc* DNA site with high affinity [18].

In addition to MAX, MYC and MYCN also interact with NMI and BIN1. NMI has homology with a coiled-coil heptad repeat in the *C. elegans* protein CEF59 and with the interferon-induced leucine zipper protein, IFP 35 [19]. BIN1 (box-dependent myc-interacting protein 1) is a nuclear protein that binds to the MYC box regions of MYC (MB1 and MB2) to block transformation. The N-terminal region of BIN1 is strongly homologous to those of amphiphysin, a breast cancer-associated autoimmune antigen, and RVS167, a negative regulator of the cell cycle in yeast, and BIN1 may therefore be a tumour suppressor gene [20]. MYCN, but not MYC or MYCL, interacts with a brain-specific protein [19]. MYC also interacts directly with SSRP1, an HMG-box protein, in a bacterial expression screen [21].

**Regulation of gene expression by MYC**

Expression of the following genes has been reported to be modulated by highly expressed MYC or MYCN:

***Activated***

- Adenovirus E4 (*via* the E1A activation region) [22]
- *CDC25A* [23]
- Carbamoyl-phosphate synthase (glutamine hydrolysing)/aspartate carbamoyltransferase/dihydroorotase (*cad*) [24]
- Cyclins A and E [25]
- *Cdc2* [26]
- Dihydrofolate reductase [27]
- *ECA39* [28]
- Eukaryotic initiation factor 4E (eIF4E) mRNA cap-binding protein [29]
- Human heat shock protein 70 [30]
- *MrDb* (RNA helicase of the DEAD box family) [31]
- Ornithine decarboxylase [32–34]
- Plasminogen activator inhibitor 1 (PAI-1/*Mr1*) and *Mr2* [35]
- α-Prothymosin and serum-inducible genes [36–38].

Two translational forms of MYC (MYC1 and MYC2, see below) and also of MYCN and MYCL occur in all species examined and the non-AUG-initiated form (MYC1) strongly and specifically activates transcription of CCAAT/enhancer binding protein (C/EBP) sequences within the EFII enhancer[39].

***Repressed***

- Adenovirus-2 major late promoter (MLP)[22]
- Albumin[40]
- $\beta$1 integrin[41]
- C/EBP$\alpha$[42]
- Collagen genes[43]
- Cyclin D1[25,44]
- $\lambda$5 and TdT (terminal deoxynucleotidyl transferase) initiators[45]
- Lymphocyte function-related antigen 1 (LFA-1)[46,47]
- MHC class I antigens[48]
- *Mim-1* (by v-MYC in myelomonocytic cells)[49]
- Mouse metallothionein I promoter[50]
- MT-1[51]
- Neural cell adhesion molecule (NCAM)[52]

For the most part the considerable number of genes that have been shown to be responsive to MYC have been identified through the expression of exogenous *MYC* and it is partly for this reason that the critical physiological targets remain to be resolved. The ubiquitous requirement for ornithine decarboxylase activation as an early event in cell proliferation suggests that the regulation of *ODC1* by MYC may important, together with that of several cyclin genes and dihydrofolate reductase. It is noteworthy that the *trans*-activation of *ODC1* by MYC may be inhibited by the putative tumour suppressor BIN1.

Repression by MYC depends on the presence of initiator (INR) elements in the basal promoters of susceptible genes, for example, those of cyclin D1 and the adenovirus late promoter[22]. The zinc finger protein YY1 can either activate or repress transcription *via* INRs. Yeast two hybrid screening has indicated that YY1 can interact with MYC[53]. Thus MYC could influence both facets of YY1 activity. Thus YY1 may be involved in negative regulation of *Myc* expression in terminally differentiated B cells[54,55]. In primary myoblasts, however, YY1 activates the expression of *Myc* whilst suppressing skeletal $\alpha$-actin transcription, and may therefore function to drive proliferation and suppress terminal differentiation in these cells[56]. MYC can also heterodimerize with the transcription initiation factor TFII-I and with the TATA binding protein (TBP[57]; see ***RB1***). Association with TFII-I correlates with inhibition of complex formation with TBP and of transcription initiation[58].

**Cell cycle regulation**

*MYC* is implicated in the control of normal proliferation, differentiation and neoplastic transformation. Expression of *MYC* in untransformed cells is growth factor dependent and essential for progression through the cell cycle[59]. High levels of expression accelerate growth. Downregulation of *MYC* expression usually correlates with the onset of differentiation and constitutive expression interferes with normal differentiation. *MYCN* expression correlates with metastatic potential[60], consistent with the repressive effects of MYC on genes coding for MHC class I antigens, $\beta$1 integrin subunits and NCAM.

Transcription of CDC25, the phosphatase that mediates the activation of cyclin-dependent kinases, is directly activated by MYC/MAX [23]. Like MYC, CDC25A (or CDC25B) cooperates with *Hras* to transform primary fibroblasts and CDC25A also induces apoptosis in the absence of growth factors. Thus CDC25A may be a critical target of MYC in regulating both cell cycle progression and apoptosis. The induction of MYC expression in $G_0$ arrested fibroblasts promotes hyperphosphorylation of pRb and activation of both cyclin D1/ and cyclin E/CDKs. This activation does not require increased cyclin/CDK synthesis but may require, in addition to activation of CDC25 phosphatase [61], the dissociation of an inhibitory complex present in starved cells. This inhibitory complex does not appear to include WAF1 but may contain KIP1.

The retinoblastoma protein pRb associates with the MYC and NMYC *trans*-activation domains although this association is not essential for *trans*-activation by NMYC [62]. Nevertheless, pRb has been shown to stimulate GAL4/Myc-mediated transcription in a cell-specific manner [63]. Microinjection of pRb reversibly arrests cell cycle progression in $G_1$ and this effect is antagonized by the coinjection of MYC [64]. The pRb-related protein p107 also forms a specific complex with the N-terminal *trans*-activation domain of MYC [65]. This permits phosphorylation of $Ser^{62}$ by a p107/cyclin A/CDK2 complex and subsequent phosphorylation of $Thr^{58}$. Phosphorylation of $Ser^{62}$ and $Thr^{58}$ has been reported to attenuate MYC function leading to reduced *trans*-activation [66]. Mutations in Burkitt's lymphoma and AIDS-related lymphoma cluster in this region and the corresponding absence of phosphorylation may confer a growth advantage on lymphoma cells. However, there is conflicting evidence indicating that point mutation of $Thr^{58}$ does not affect suppression of *trans*-activation by p107 and that suppression may result from interaction with other mutation sites that are scattered throughout the *trans*-activation domain in Burkitt's lymphoma cell lines [67]. As the p107/pRb binding site overlaps that of TBP, it is possible that p107 may disrupt association of MYC with the TBP.

**Apoptosis**

In haematopoietic cells sustained expression of *MYC* accelerates apoptosis [68–71]; see also Chapter 5 and ***BCL2***). Thus, in the IL-3-dependent myeloid cell line 32D, withdrawal of IL-3 inhibits *Myc* transcription and causes the cells to arrest in $G_1$. Cells constitutively expressing *Myc*, however, do not arrest in $G_1$ but within 6 h show morphological changes characteristic of apoptosis. In immature T cells and in some T cell hybridomas, activation of the T cell receptor causes apoptosis that is dependent on the sustained expression of *Myc* [70]. In Rat-1 fibroblasts constitutive *Myc* expression also causes apoptosis when proliferation is inhibited by the absence of serum [71] and increases sensitivity to tumour necrosis factor $\alpha$ (TNF$\alpha$) cytotoxicity [72]. MYC is induced during vitamin $K_3$-stimulated apoptosis of nasopharyngeal carcinoma cells [73] and the high rate of apoptosis in Burkitt lymphoma cells is suppressed when MYC protein levels are lowered [74]. In immature WEHI 231 B lymphoma cells, however, a decrease in *Myc* expression appears to precede induction of cell death [34].

Genes that are known to cooperate with *MYC* in transformation, e.g. *Pim-1*, *BCL2*, *HRAS* and v-*raf*, may maintain cell viability without being directly mitogenic: in the absence of any of these gene products or of cytokines such as PDGF or insulin-like growth factors [75] MYC may accelerate programmed cell death but in *P53*$^{-/-}$ fibroblasts, activation of *Myc* induces cell cycle arrest but not apoptosis, suggesting that p53 mediates apoptosis [76]. However, in epithelial cells MYC can promote apoptosis by mechanisms either dependent or independent of p53 [77]. Furthermore, in E$\mu$-*myc*/*P53*$^{+/-}$ transgenic mice, the loss of wild-type p53 cooperates with deregulated *Myc* expression to enhance proliferation during lymphomagenesis without inhibition of apoptosis, loss of the wild-type p53 allele being necessary for lymphoid transformation [78,79]. However, loss of p53 does not accelerate mammary tumour formation in mice carrying a *Myc* transgene.

MYC-induced apoptosis is associated with elevated cyclin A expression and may be mediated by ornithine decarboxylase [32]: protection by BCL2 correlates with increased cyclin C, D1 and E transcription [80]. In *Myc*-transformed chick embryo fibroblasts undergoing apoptosis on serum removal, proteolysis of the focal adhesion kinase pp125$^{FAK}$ occurs prior to commitment to cell death, indicating that disruption of integrin signalling pathways is important in apoptosis [81].

**Cancer**

*MYC* may be activated to become a transforming gene by (i) proviral insertion, (ii) chromosomal translocation, or (iii) gene amplification. In general the result is to elevate expression of *MYC*, rather than to change the structure of the protein itself although point mutations that enhance pathogenicity may arise [82]. Deregulated *MYC* expression correlates with the occurrence of many types of human tumours, particularly small cell lung carcinoma (SCLC) [83], breast [84–86] and cervical carcinomas [87,88]. In Ewing's sarcoma upregulation of *MYC* arises from the *trans*-activation of the *MYC* promoter by EWS/FLI1 [89]. Enhanced transcription arising from translocation of *MYC* occurs in some but not all Burkitt's lymphomas, following translocation of the gene to the vicinity of the immunoglobulin enhancer [90–92]. In many such rearrangements the first (non-coding) exon is lost from the gene. Point mutations in the coding region of MYC may also occur, particularly within the MYC boxes (MB1 and MB2 within the *trans*-activation domain), and have been detected in Burkitt's lymphoma and also in transformed follicular lymphomas without MYC rearrangement, in aggressive lymphomas arising in the acquired immunodeficiency syndrome (AIDS) [93] and in some carcinomas [94,95]. Deletions 3′ of the *MYC* gene have been detected in *MYC* amplicons in some human tumour cell lines [96].

*MYCN* is frequently amplified in neuroblastomas [97], retinoblastomas, astrocytomas, gliomas and SCLC. Expression correlates with appearance of the more severe forms of cervical intraepithelial carcinoma (CIN types II and III) and with increased metastasis in the advanced stages of neuroblastoma. A chimeric *RLF-MYCL1* mRNA has been detected in some SCLC cell lines.

*MYCL1* displays two allele polymorphism that can give rise to three genotypes (LL, LS and SS). Loss of heterozygosity at the *MYCL1* locus has been detected in breast and colon cancers and the SS genotype may contribute to the progression of colorectal cancer [98].

**Transgenic animals**

*Myc* can cause mammary carcinomas in transgenic mice [99,100]. Nevertheless excessive expression of *Myc* in transgenic mice does not prevent normal development and mammary tumours and lymphomas develop in a stochastic manner, indicating that over-expression of *Myc* is necessary but not sufficient for tumorigenesis.

In E$\mu$ transgenics B cell tumour formation is accelerated by retroviral infection with Moloney murine leukaemia virus. Identification of loci occupied by the integrated provirus (*Pim-1, Pim-2, Bmi-1, Pal-1, Bla-1* and *Emi-1*, see Table II, page 74) shows that *Bmi-1* and *Pim-1* cooperate with *Myc* in tumour formation [101–104].

These observations are generally consistent with *in vitro* data indicating that alteration in the *Myc* content of cells, rather than mutation, activates its oncogenic potential and that transformation probably requires cooperation between MYC and other oncogene products, including RAS.

Targeting individual oncogenes (*Myc, Hras, Neu, Int-2*) to breast epithelium gives rise to morphologically distinct types of tumour. This indicates that although further genetic alterations may be essential for full tumour development, the initiating event determines the type of tumour that will evolve. This is confirmed by the finding that a specific set of genes induced by *Neu* or *Hras* initiation are not induced by *Myc* or *Int-2* nor do they occur in tumours induced by *Myc* or *Int-2* that are subsequently transfected with *Neu* or *Hras* [105]. Thus specific oncogenes appear to target specific cell types.

Homozygous null mutation of *Myc* results in embryonic lethality: female mice that are heterozygous for the mutation have reduced fertility [106]. Homozygous deletion of *Nmyc* results in embryonic lethality [107,108]. In these animals the lung airway epithelium is underdeveloped and death results from inability to oxygenate their blood. Homozygous null *Lmyc* mice are viable and reproductively competent [109].

**In animals**

Viruses expressing *Myc* cause myeloid leukaemias, sarcomas and carcinomas [110,111]. Proviral integration of non-defective retroviruses may cause activation of *Myc*. Thus avian retroviruses may induce B cell lymphomas, adenocarcinoma or T lymphoma in chickens that are associated with activation of *Myc* [112–114]. *Myc* may also be activated by murine and feline leukaemia viruses that cause T lymphomas [115–119], by intracisternal A-particle [120] and by retroposon insertion [11]. *Nmyc* is frequently activated by proviral insertion of murine leukaemia viruses [122,123].

***In vitro***

Primary fibroblasts are immortalized by *Myc, Nmyc* or v-*myc* but are rendered tumorigenic only when activated *Ras* is also expressed [124]. Fibroblast cell lines, however, show reduced growth factor requirements and are tumorigenic when transfected with *Myc* alone. *Lmyc* also co-transforms primary cells with an activated *Ras* gene but is <10% as effective as *Myc*, a difference that reflects the relative potencies of the activation domains of the MYC proteins [125].

Low frequency (~60 Hz) electromagnetic radiation of human leukaemic (HL60) cells has been reported to stimulate *MYC* transcription but this finding has proved irreproducible [126,127].

## Gene structure

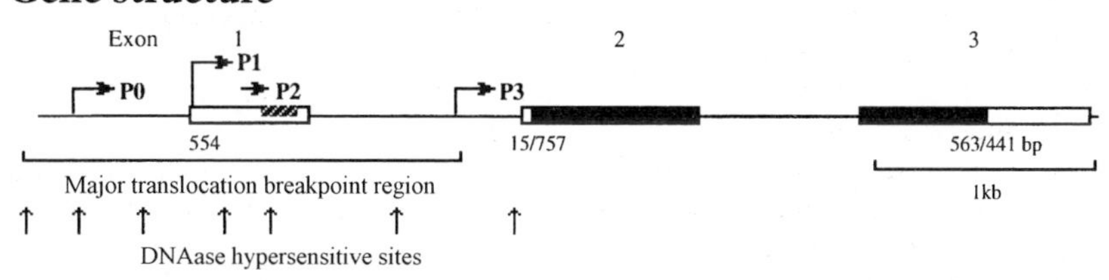


Exons 2 and 3 are between 70% and 90% identical between species and exon 1 is 70% conserved between the human and mouse genes. The three exon gene organization is similar in *MYCN* and *MYCL*. The vertical arrows indicate DNAase hypersensitive sites: additional sites have been detected within exon 2 and in the 3′ UTR. The 3′ region contains constitutive and tissue-specific DNAase I hypersensitive sites, some of which colocalize with *cis*-acting enhancer elements [128,129].

The 5′ end of exon 1 (non-coding) includes two major transcriptional initiation sites (TATAA boxes), P1 and P2. Two minor promoters, P0 (upstream of P1) and P3 (near the 3′ end of intron 1) lack TATAA boxes. Cross-hatched bar: sequence downstream of the P2 initiation site that is the site of the conditional block to transcriptional elongation [130,131]. Two MYC proteins having different N-termini (p64/p67) arise from alternative translation initiation between AUG in exon 2 and a CUG codon near the 3′ end of exon 1.

## Transcriptional regulation

Factors shown to bind within the human *MYC* promoter are nuclear factor 1 (NF1) [132], PuF/NM23-H2/NDPK-B [133], FUSE-binding protein (FBP) that activates the far upstream element (FUSE) at −1500 relative to the P1 promoter [134], MSSP-1/MSSP-2 [135,136], pur1 and cellular nucleic acid binding protein (CNBP), heterogeneous nuclear ribonucleoprotein K [137,138], TGF$\beta_1$ and pRb [139], E2F and p55 [140], v-ABL [141], *MYC* binding protein 1 (MBP1) [142], v-MYB and MYB [143] (see ***MYB***), SP1, Sp3, NF1 and CCAAT binding protein (CBP) [144] and the terminator binding factor TBF I [145].

Human CTCF represses *MYC* and binds to two regions (+5 to +45 immediately downstream of the P2 initiation site and to another GC-rich region immediately downstream of the P1 initiation site) of the promoter [146]. E2F-1 induces transcription from the P2 promoter that is repressed by pRb [147]. However, over-expression of pRb may increase or decrease *MYC* transcription depending on cell type [148]. The activation of *MYC* by cyclin A is enhanced by the coexpression of E2F-1. Cyclin D1 expression generates E2F binding site-dependent activation of *MYC* that is inhibited by pRb. In fibroblast and epithelial cell lines the human P2 promoter is *trans*-activated by SV40 T Ag by a mechanism dependent on the E2F binding sites [148].

Regulatory elements mapped in the 5′ untranslated regions of murine *Myc* are AP1 and AP2 sites, the P2 promoter elements ME1a2, E2F and ME1a1 [149], two NF-κB sites [150] and binding sites for MYC-associated zinc finger protein (MAZ) [151], MYC-PRF and MYC-CF1 (common factor 1, also identical to the zinc finger protein Yin-yang 1 (YY1) [152]. In addition to these sites regulatory elements upstream from −3500 (with respect to the mouse P1 promoter) and 1500 bp 3′ of the polyadenylation sites are required for correct transcription *in vivo* [153].

The rapid increase in fibroblast mRNA from very low levels in $G_0$ cells that occurs on stimulation with EGF is due to relief of a block to transcriptional elongation, as evidenced by the high concentration of RNA polymerase II in the exon 1 region in $G_0$ cells. *In vitro* studies have defined a 95 bp 5′ region of exon 1 of the human *MYC* gene that specifies premature termination and the deletion of the first exon/intron region of *Myc* genes elevates the levels of mRNA and oncogenic activity[154,155]. The efficiency of premature termination sites declines markedly when they are placed >~400 bp from the start site[131]. A transcriptional inhibitor binds within intron 1[156] and point mutations that abolish binding occur in some Burkitt's lymphomas. Intragenic regulatory regions also occur in the *Fos* and *Myb* genes but the sequence in *MYC* is unrelated to the FIRE sequence of *Fos*[157].

Transcription of *MYC* and *MYCN* occurs in both the sense and anti-sense direction although the role of anti-sense transcription is unknown[158–160].

## RNA stability

Two regions within *MYC* mRNA regulate the half-life of the molecule: one is in the 3′ untranslated region (three copies of AUUUA at the 3′ end of exon 3) and the other is the C-terminal portion of the coding region (see also ***FOS***). The 3′ UTR is the stronger destabilizing element. Two proteins (37 kDa and 40 kDa) bind to AU-rich regions in the 3′ UTR of *MYC* and other unstable mRNAs and increase the *in vitro* rate of RNA degradation[161]. The C-terminal coding region determinant (CRD) of stability requires translation for its efficient action[162]. A 70 kDa protein (CRD-BP) binds to this region of *MYC* mRNA and appears to confer stability on the transcript[163]. In transgenic mice deletion of both the 5′ and 3′ non-coding sequences of the *MYC* gene do not affect MYC expression, indicating that in this system sequences of exon 2 and/or exon 3 are critical in post-transcriptional regulation[161,164].

In contrast to the effect of EGF, serum stimulation (which causes up to 40-fold elevation of *Myc* mRNA) increases initiation by stabilizing mRNA[165]. A human T cell leukaemia-derived cell line carries a translocation 24 nucleotides 5′ of the first poly-A addition signal of *MYC*. This replaces a 61 bp AU-rich region with sequences derived from chromosome 2 and causes a five-fold increase in *MYC* expression due to enhanced mRNA stability. The hybrid gene transforms rat fibroblasts to a tumorigenic phenotype[166]. This *MYC* rearrangement contrasts with those occurring in Burkitt's lymphoma and other cancers in that it does not involve T cell receptor or immunoglobulin loci. In human colon cancer cells, thymidylate synthase binds to the 3′ coding region of MYC mRNA and inhibits MYC translational efficiency[167].

## Protein structure

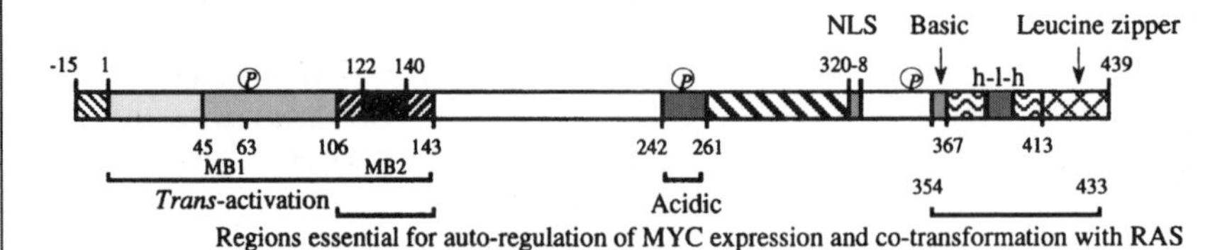

Cross-hatched box (−15 to 1): additional N-terminal region of p64 translated from a CUG codon in exon 1. Circles: phosphorylation sites. The major site for casein kinase II (CKII) phosphorylation, hyperphosphorylated during mitosis, is in the central acidic domain. The acidic domain resembles those of other transcriptional activators. Deletions in this region have little effect on MYC/RAS cotransformation but in v-MYC influence the transforming host range[168].

The conserved sequences MYC box 1 and 2 (MB1 and MB2) lie within the *trans*-activation domain[169]. The regions of MYC protein that are essential for apoptosis are identical to those required for cotransformation, autosuppression and inhibition of differentiation, namely part of the N-terminus (amino acids 7–91 and 106–143), the helix-loop-helix region (371–412) and the leucine zipper (414–433).

The v-MYC protein encoded by the CMII virus is identical to chicken MYC. Other viral forms contain scattered mutations, including that of $Thr^{61}$ (corresponding to human $Thr^{58}$) to a non-phosphorylatable residue in the MC29, MH2 and OK10 v-MYC proteins, which enhances oncogenicity[170].

## Amino acid sequence of human MYC

```
  1 MPLNVSFTNR NYDLDYDSVQ PYFYCDEEEN FYQQQQQSEL QPPAPSEDIW
 51 KKFELLPTPP LSPSRRSGLC SPSYVAVTPF SLRGDNDGGG GSFSTADQLE
101 MVTELLGGDM VNQSFICDPD DETFIKNIII QDCMWSGFSA AAKLVSEKLA
151 SYQAARKDSG SPNPARGHSV CSTSSLYLQD LSAAASECID PSVVFPYPLN
201 DSSSPKSCAS QDSSAFSPSS DSLLSSTESS PQGSPEPLVL HEETPPTTSS
251 DSEEEQEDEE EIDVVSVEKR QAPGKRSESG SPSAGGHSKP PHSPLVLKRC
301 HVSTHQHNYA APPSTRKDYP AAKRVKLDSV RVLRQISNNR KCTSPRSSDT
351 EENVKRRTHN VLERQRRNEL KRSFFALRDQ IPELENNEKA PKVVILKKAT
401 AYILSVQAEE QKLISEEDLL RKKREQLKHK LEQLRNSCA (439)
```

**Domain structure**

| | |
|---|---|
| −15–1 | 15 amino acid N-terminal extension in p67 (MDFFRVVENQQPPAT) |
| 1–143 | Transactivation domain |
| 45–63 | MB1 (underlined) |
| 122–140 | MB2 (underlined) |
| 58 | MAP kinase and glycogen synthase kinase-3 phosphorylation site[171]. $Thr^{58}$ is frequently mutated in Burkitt's lymphoma[94,172]. Phosphorylation of $Thr^{58}$ and $Ser^{62}$ is necessary for high levels of *trans*-activation by MYC[173] |
| 39, 62, 138 | $Glu^{39}$, $Ser^{62}$ and $Phe^{138}$ are additional mutational hot spots in Burkitt's and AIDS-related lymphoma |
| 33–245 | Region over which point mutations occur in Burkitt's lymphoma[174] |
| 62 | $Ser^{62}$ is phosphorylated by MAP kinases (ERK, ERT (EGF receptor $Thr^{699}$) and MAP2 protein kinase). $Ser^{62}$ is within a $p34^{CDC2}$ kinase recognition motif but MYC is not phosphorylated by $p34^{CDC2}$ *in vitro*[175] |
| 1–252 | α-tubulin binding domain |

320–328 Major nuclear localization signal [176]
355–367 Basic region (underlined)
413–437 Leucine zipper (underlined)
290–318 Non-specific DNA binding region
359 His$^{359}$ corresponds to avian His$^{336}$ that contacts or is close to the thymine 5-methyl group at position 2 of the DNA half site [16]: the homologous residue in MAX recognizes the same site
364–374 Incomplete nuclear localization signal: highly conserved between MYC, MYCN and MYCL and essential for oncogenicity
106–143 and 354–433 Essential for *Ras* complementation in transforming normal rat embryo cells: these domains are conserved in MYCN and MYCL (MYC box proteins). Deletion of 106–143 dominantly inhibits the cooperation of normal MYC with oncogenic RAS to transform rat embryo fibroblasts

Mutations detected in transformed follicular lymphomas: Asn$^{11}$ → Ser, Val$^{170}$ → Ile; non-AIDS aggressive lymphomas: Asp$^{127}$ → Asp, Ile$^{129}$ → Leu; AIDS-associated lymphomas: Thr$^{58}$ → Ala/Asp, Ser$^{71}$ → Trp, Leu$^{82}$ → His, Asp$^{87}$ → Gly, Ile$^{129}$ → Leu, Leu$^{56}$ → Val, Pro$^{57}$ → Thr, Leu$^{176}$ → Met [93].

## Amino acid sequence of human MYCL

```
  1 MDYDSYQHYF YDYDCGEDFY RSTAPSEDIW KKFELVPSPP TSPPWGLGPG
 51 AGDPAPGIGP PEPWPGGCTG DEAESRGHSK GWGRNYASII RRDCMWSGFS
101 ARERLERAVS DRLAPGAPRG NPPKASAAPD CTPSLEAGNP APAAPCPLGE
151 PKTQACSGSE SPSDSENEEI DVVTVEKRQS LGIRKPVTIT VRADPLDPCM
201 KHFHISIHQQ QHNYAARFPP ESCSQEEASE RGPQEEVLER DAAGEKEDEE
251 DEEIVSPPPV ESEAAQSCHP KPVSSDTEDV TKRKNHNFLE RKRRNDLRSR
301 FLALRDQVPT LASCSKAPKV VILSKALEYL QALVGAEKRM ATEKRQLRCR
351 QQQLQKRIAY LSGY (364)
```

**Domain structure**

343–361 Leucine zipper (underlined)

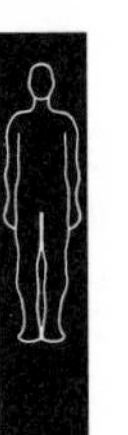

## Amino acid sequence of human MYCN

```
  1 MPSCSTSTMP GMICKNPDLE FDSLQPCFYP DEDDFYFGGP DSTPPGEDIW
 51 KKFELLPTPP LSPSRGFAEH SSEPPSWVTE MLLENELWGS PAEEDAFGLG
101 GLGGLTPNPV ILQDCMWSGF SAREKLERAV SEKLQHGRGP PTAGSTAQSP
151 GAGAASPAGR GHGGAAGAGR AGAALPAELA HPAAECVDPA VVFPFPVNKR
201 EPAPVPAAPA SAPAAGPAVA SGAGIAAPAG APGVAPPRPG GRQTSGGDHK
251 ALSTSGEDTL SDSDDEDDEE EDEEEEIDVV TVEKRRSSSN TKAVTTFTIT
301 VRPKNAALGP GRAQSSELIL KRCLPIHQQH NYAAPSPYVE SEDAPPQKKI
351 KSEASPRPLK SVIPPKAKSL SPRNSDSEDS ERRRNHNILE RQRRNDLRSS
401 FLTLRDHVPE LVKNEKAAKV VILKKATEYV HSLQAEEHQL LLEKEKLQAR
451 QQQLLKKIEH ARTC (464)
```

**Domain structure**

| | |
|---|---|
| 36–55 | MB1 (underlined) |
| 102–115 | MB2 (underlined) |
| 262–278 | Asp/Glu-rich (acidic) |
| 433–454 | Leucine zipper (underlined) |
| 261, 263 | Phosphorylation by casein kinase II |
| 58, 62 | MAP kinase and glycogen synthase kinase-3 phosphorylation site |
| 255, 261, 263 | Casein kinase I phosphorylation site (underlined) |
| 375, 377, 380 | Casein kinase II phosphorylation site (underlined) |

**Database accession numbers**

| | *PIR* | *SWISSPROT* | *EMBL/GENBANK* | *REFERENCES* |
|---|---|---|---|---|
| Human *MYC* | A10349, A10350 | P01106, P01107 | X00196, X00364, V00568 | 177–182 |
| Human *MYCL1* | A27675 | P12524 | M19720, | 183 |
| | | | X07262, X07263 | 184 |
| Human *MYCL2* | A30146 | P12525 | J03069 | 185 |
| Human *MYCN* | A01355 | P04198 | M13228, M13241 | 186 |
| | A22937 | | X03294, X03295 | 187 |
| | A25744 | | X02363 | 188 |
| | S02249 | | Y00664 | 189 |
| | | | M32092, X02363 | 190 |
| Human *NMI* | | | U32849 | 19 |
| AMV-MC 29 v-*myc* | A01353 | P01110 | V01173 | 191 |
| | | | V01174 | 192 |

***References***


[1] Sheiness, D. and Bishop, M.J. (1979) J. Virol. 31, 514–521.
[2] Eva, A. et al. (1982) Nature 295, 116–119.
[3] Walker, C.W. et al. (1992) Oncogene 7, 2007–2012.
[4] Ludwig, S.R. et al. (1989) Proc. Natl Acad. Sci. USA 86, 7092–7096.
[5] Alexandrova, N. et al. (1995) Mol. Cell. Biol. 15, 5188–5195.
[6] Zimmerman, K.A. et al. (1986) Nature 319, 780–783.
[7] Moore, J.P. et al. (1987) Oncogene Res. 2, 65–80.
[8] Waters, C.M. et al. (1991) Oncogene 6, 797–805.
[9] Blackwood, E.M. and Eisenman, R.N. (1991) Science 251, 1211–1217.
[10] Kato, G.J. et al. (1992) Genes Dev. 6, 81–92.
[11] Ma, A. et al. (1993) Oncogene 8, 1093–1098.
[12] Davis, L.J. and Halazonetis, T.D. (1993) Oncogene 8, 125–132.
[13] Wechsler, D.S. et al. (1994) Mol. Cell. Biol. 14, 4097–4107.
[14] Wechsler, D.S. and Dang, C.V. (1992) Proc. Natl Acad. Sci. USA 89, 7635–7639.
[15] Fisher, D.E. et al. (1992) Proc. Natl Acad. Sci. USA 89, 11779–11783.
[16] Dong, Q. et al. (1994) EMBO J. 13, 200–204.
[17] Mukherjee, B. et al. (1992) Genes Dev. 6, 1480–1492.
[18] van Antwerp, M.E. et al. (1992) Proc. Natl Acad. Sci. USA 89, 9010–9014.
[19] Bao, J. and Zervos, A.S. (1996) Oncogene 12, 2171–2176.
[20] Sakamuro, D. et al. (1996) Nature Genet. 14, 69–77.
[21] Bunker, C.A. and Kingston, R.E. (1995) Nucleic Acids Res. 23, 269–276.
[22] Li, L. et al. (1994) EMBO J. 13, 4070–4079.
[23] Galaktionov, K. et al. (1996) Nature 382, 511–517.

[24] Miltenberger, R.J. et al. (1995) Mol. Cell. Biol. 15, 2527–2535.
[25] Jansen-Durr, P. et al. (1993) Proc. Natl Acad. Sci. USA 90, 3685–3689.
[26] Born, T.L. et al. (1994) Mol. Cell. Biol. 14, 5710–5718.
[27] Mai, S. and Jalava, A. (1994) Nucleic Acids Res. 22, 2264–2273.
[28] Ben-Yosef, T. et al. (1996) Oncogene 13, 1859–1866.
[29] Jones, R.M. et al. (1996) Mol. Cell. Biol. 16, 4754–4764.
[30] Kaddurah-Daouk, R. et al. (1987) Genes Dev. 1, 347–357.
[31] Grandori, C. et al. (1996) EMBO J. 15, 4344–4357.
[32] Packham, G. and Cleveland, J.L. (1994) Mol. Cell. Biol. 14, 5741–5747.
[33] Tobias, K.T. et al. (1995) Oncogene 11, 1721–1727.
[34] Wu, S. et al. (1996) Oncogene 12, 621–629.
[35] Prendergast, G.C. et al. (1990) Mol. Cell. Biol. 10, 1265–1269.
[36] Eilers, M. et al. (1991) EMBO J. 10, 133–141.
[37] Mol, P.C. et al. (1995) Mol. Cell. Biol. 15, 6999–7009.
[38] Lutz, W. et al. (1996) Oncogene 13, 803–812.
[39] Hann, S.R. et al. (1994) Genes Dev. 8, 2441–2452.
[40] Gorski, K. et al. (1986) Cell 47, 767–776.
[41] Judware, R. and Culp, L.A. (1997) Oncogene 14, 1341–1350.
[42] Christy, R. et al. (1991) Proc. Natl Acad. Sci. USA 88, 2593–2597.
[43] Yang, B.-S. et al. (1991) Mol. Cell. Biol. 11, 2291–2295.
[44] Philipp, A. et al. (1994) Mol. Cell. Biol. 14, 4032–4043.
[45] Mai, S. and Mårtensson, I.-L. (1995) Nucleic Acids Res. 23, 1–9.
[46] Versteeg, R. et al. (1989) J. Exp. Med. 170, 621–635.
[47] Inghirami, G. et al. (1990) Science 250, 682–686.
[48] **Schrier, P.I. and Peltenburg, L.T.C. (1993) Adv. Cancer Res. 60, 181–246.**
[49] Mink, S. et al. (1996) Proc. Natl Acad. Sci. USA 93, 6635–6640.
[50] Jin, P. and Ringertz, N.R. (1990) J. Biol. Chem. 265, 14061–14064.
[51] Stuart, G.W. et al. (1984) Proc. Natl Acad. Sci. USA 81, 7318–7322.
[52] Akeson, R. and Bernards, R. (1990) Mol. Cell. Biol. 10, 2012–2016.
[53] Shrivastava, A. et al. (1993) Science 262, 889–1892.
[54] Kakkis, E. et al. (1989) Nature 339, 718–721.
[55] Riggs, K.J. et al. (1993) Mol. Cell. Biol. 13, 7487–7495.
[56] Lee, T.C. et al. (1994) Oncogene 9, 1047–1052.
[57] Maheswaran, S. et al. (1994) Mol. Cell. Biol. 14, 1147–1152.
[58] Roy, A.L. et al. (1993) Nature 365, 359–361.
[59] Seth, A. et al. (1993) Mol. Cell. Biol. 13, 4125–4136.
[60] Bernards, R. et al. (1986) Cell 47, 667–674.
[61] Steiner, P. et al. (1995) EMBO J. 14, 4814–4826.
[62] Cziepluch, C. et al. (1993) Oncogene, 8, 2833–2838.
[63] Adnane, J. and Robbins, P.D. (1995) Oncogene 10, 381–387.
[64] Goodrich, D.W. and Lee, W.-H. (1992) Nature 360, 177–179.
[65] Beijersbergen, R. et al. (1994) EMBO J. 13, 4080–4086.
[66] Hoang, A.T. et al. (1995) Mol. Cell. Biol. 15, 4031–4042.
[67] Smith-Sørensen, B. et al. (1996) J. Biol. Chem. 271, 5513–5518.
[68] Askew, D.S. et al. (1991) Oncogene 6, 1915–1922.
[69] **Williams, G.T. (1991) Cell 65, 1097–1098.**
[70] Shi, Y. et al. (1992) Science 257, 212–214.
[71] Evan, G.I. et al. (1992) Cell 69, 119–128.
[72] Klefstrom, J. et al. (1994) EMBO J. 13, 5442–5450.
[73] Wu, F.Y.-H. et al. (1993) Oncogene 8, 2237–2244.

[74] Milner, A.E. et al. (1993) Oncogene 8, 3385–3391.
[75] Harrington, E.A. et al. (1994) EMBO J. 13, 3286–3295.
[76] Hermeking, H. and Eick, D. (1994) Science 265, 2091–2093.
[77] Sakamuro, D. et al. (1995) Oncogene 11, 2411–2418.
[78] Hsu, B. et al. (1995) Oncogene 11, 175–179.
[79] Elson, A. et al. (1995) Oncogene 11, 181–190.
[80] Hoang, A.T. et al. (1994) Proc. Natl Acad. Sci. USA 91, 6875–6879.
[81] Crouch, D.H. et al. (1996) Oncogene 12, 2689–2696.
[82] Symonds, G. et al. (1989) Oncogene 4, 285–294.
[83] Gazdar, A.F. et al. (1985) Cancer Res. 45, 2924–2930.
[84] Guerin, M. et al. (1988) Oncogene Res. 3, 21–31.
[85] Tsuda, H. et al. (1989) Cancer Res. 49, 3104–3108.
[86] Mariani-Costantini, R. et al. (1988) Cancer Res. 48, 199–205.
[87] Ocadiz, R. et al. (1987) Cancer Res. 47, 4173–4177.
[88] **Dang, C.V. and Lee, L.A. (1995) c-Myc Function in Neoplasia. R.G. Landes Co., Austin, Texas.**
[89] Bailly, R.-A. et al. (1994) Mol. Cell. Biol. 14, 3230–3241.
[90] **DePinho, R.A. et al. (1991) Adv. Cancer Res. 57, 1–46.**
[91] **MaGrath, I. (1990) Adv. Cancer Res. 55, 133–270.**
[92] Polack, A. et al. (1993) EMBO J. 12, 3913–3920.
[93] Clark, H.M. et al. (1994) Cancer Res. 54, 3383–3386.
[94] Albert, T. et al. (1994) Oncogene 9, 759–763.
[95] Yano, T. et al. (1993) Oncogene 8, 2741–2748.
[96] Feo, S. et al. (1994) Oncogene 9, 955–961.
[97] Schwab, M. et al. (1983) Nature 305, 245–248.
[98] Young, J. et al. (1994) Oncogene 9, 1053–1056.
[99] Sinn, E. et al. (1987) Cell 49, 465–475.
[100] Schoenenberger, C.A. et al. (1988) EMBO J. 7, 169–175.
[101] van Lohuizen, M. et al. (1991) Cell 65, 737–752.
[102] Haupt, Y. et al. (1991) Cell 65, 753–763.
[103] Verbeek, S. et al. (1991) Mol. Cell. Biol. 11, 1176–1179.
[104] Moroy, T. et al. (1991) Oncogene 6, 1941–1948.
[105] **Morrison, B.W. and Leder, P. (1994) Oncogene 9, 3417–3426.**
[106] Davis, A.C. et al. (1993) Genes Dev. 7, 671–682.
[107] Sawai, S. et al. (1991) New Biol. 3, 861–869.
[108] Moens, C.B. et al. (1992) Genes Dev. 6, 691–704.
[109] Hatton, K.S. et al. (1996) Mol. Cell. Biol. 16, 1794–1804.
[110] **Bister, K. and Jansen, H.W. (1986) Adv. Cancer Res. 47, 99–188.**
[111] Chen, C. et al. (1989) J. Virol. 63, 5092–5100.
[112] Hayward, W.S. et al. (1981) Nature 290, 475–480.
[113] Payne, G.S. et al. (1982) Nature 295, 209–214.
[114] Swift, R.A. et al. (1987) J. Virol. 61, 2084–2090.
[115] Corcoran, L.M. et al. (1984) Cell 37, 113–122.
[116] Selten, G. et al. (1984) EMBO J. 3, 3215–3222.
[117] O'Donnell, P.V. et al. (1985) J. Virol. 55, 500–503.
[118] Mucenski, M.L. et al. (1987) Oncogene Res. 2, 33–48.
[119] Forrest, D. et al. (1987) Virology 158, 194–205.
[120] Greenberg, R. et al. (1985) Mol. Cell. Biol. 5, 3625–3628.
[121] Katzir, N. et al. (1985) Proc. Natl Acad. Sci. USA 82, 1054–1058.
[122] Dolcetti, R. et al. (1989) Oncogene 4, 1009–1014.

[123] van Lohuizen, M. et al. (1989) EMBO J. 8, 133–136.
[124] Schreiber-Agus, N. et al. (1993) Mol. Cell. Biol. 13, 2765–2775.
[125] Barrett, J. et al. (1992) Mol. Cell. Biol. 12, 3130–3137.
[126] Lacy-Hulbert, A. et al. (1995) Nature 375, 22–23.
[127] Lacy-Hulbert, A. et al. (1995) Radiat. Res. 144, 9–17.
[128] Mautner, J. et al. (1995) Nucleic Acids Res. 23, 72–80.
[129] Mautner, J. et al. (1995) Oncogene 12, 1299–1307.
[130] Krumm, A. et al. (1992) Genes Dev. 6, 2201–2213.
[131] Roberts, S. and Bentley, D.L. (1992) EMBO J. 11, 1085–1093.
[132] Siebenlist, U. et al. (1984) Cell 37, 381–391.
[133] Berberich, S.J. and Postel, E.H. (1995) Oncogene 10, 2297–2305.
[134] Duncan, R. et al. (1994) Genes Dev. 8, 465–480.
[135] Negishi, Y. et al. (1994) Oncogene 9, 1133–1143.
[136] Takai, T. et al. (1994) Nucleic Acids Res. 22, 5576–5581.
[137] Michelotti, G.A. et al. (1996) Mol. Cell. Biol. 16, 2656–2669.
[138] **Spencer, C.A. and Groudine, M. (1990) Adv. Cancer Res. 56, 1–48.**
[139] Pietenpol, J.A. et al. (1991) Proc. Natl Acad. Sci. USA 88, 10227–10231.
[140] Parkin, N.T. and Sonenberg, N. (1989) Oncogene 4, 815–822.
[141] Wong, K.-K. et al. (1995) Mol. Cell. Biol. 15, 6535–6544.
[142] Ray, R. et al. (1995) Cancer Res. 55, 3747–3751.
[143] Cogswell, J.P. et al. (1993) Mol. Cell. Biol. 13, 2858–2869.
[144] Lang, J.C. et al. (1991) Oncogene 6, 2067–2075.
[145] Roberts, S. et al. (1992) Genes Dev. 6, 1562–1574.
[146] Filippova, G.N. et al. (1996) Mol. Cell. Biol. 16, 2802–2813.
[147] Oswald, F. et al. (1994) Oncogene 9, 2029–2036.
[148] Batsché, E. et al. (1994) Oncogene 9, 2235–2243.
[149] Dufort, D. et al. (1993) Oncogene 8, 165–171.
[150] Kessler, D.J. et al. (1992) Oncogene 7, 2447–2453.
[151] Bossone, S.A. et al. (1992) Proc. Natl Acad. Sci. USA 89, 7452–7456.
[152] Lee, T.C. et al. (1994) Oncogene 9, 1047–1052.
[153] Lavenu, A. et al. (1994) Oncogene 9, 527–536.
[154] Xu, L. et al. (1993) Oncogene 8, 2547–2553.
[155] Strobl, L.J. et al. (1993) Oncogene 8, 1437–1447.
[156] Reinhold, W. et al. (1995) Mol. Cell. Biol. 15, 3041–3048.
[157] Tourkine, N. et al. (1989) Oncogene 4, 973–978.
[158] Nepveu, A. and Marcu, K.B. (1986) EMBO J. 5, 2859–2865.
[159] Krystal, G.W. et al. (1990) Mol. Cell. Biol. 10, 4180–4191.
[160] Celano, P. et al. (1992) J. Biol. Chem. 267, 15092–15096.
[161] Morello, D. et al. (1993) Oncogene 8, 1921–1929.
[162] Herrick, D.J. and Ross, J. (1994) Mol. Cell. Biol. 14, 2119–2128.
[163] Prokipcak, R.D. et al. (1994) J. Biol. Chem. 269, 9261–9262.
[164] Yeilding, N.M. et al. (1996) Mol. Cell. Biol. 16, 3511–3522.
[165] Nepveu, A. et al. (1987) Oncogene 1, 243–250.
[166] Aghib, D.F. and Bishop, M.J. (1991) Oncogene 6, 2371–2375.
[167] Chu, E. et al. (1995) Mol. Cell. Biol. 15, 179–185.
[168] Heaney, M.L. et al. (1986) J. Virol. 60, 167–176.
[169] Kato G.J. et al. (1990) Mol. Cell. Biol. 10, 5914–5920.
[170] Frykberg, L. et al. (1987) Oncogene 1, 415–421.
[171] Henriksson, M. et al. (1993) Oncogene 8, 3199–3209.
[172] Pulverer, B.J. et al. (1994) Oncogene 9, 59–70.

[173] Gupta, S. et al. (1993) Proc. Natl Acad. Sci. USA 90, 3216–3220.
[174] Bhatia, K. et al. (1993) Nature Genet. 5, 56–61.
[175] Luscher, B. and Eisenman, R.N. (1992) J. Cell Biol. 118, 775–784.
[176] Dang, C.V. and Lee, W.M.F. (1988) Mol. Cell. Biol. 8, 4048–4054.
[177] Colby, W.W. et al. (1983) Nature 301, 722–725.
[178] Saito, H. et al. (1983) Proc. Natl Acad. Sci. USA 80, 7476–7480.
[179] Watt, R. et al. (1983) Nature 303, 725–728.
[180] Bernard, O. et al. (1983) EMBO J. 2, 2375–2383.
[181] Rabbitts, T.H. et al. (1983) Nature 306, 760–765.
[182] Gazin, C. et al. (1984) EMBO J. 3, 383–387.
[183] Kaye, F. J. et al. (1988) Mol. Cell. Biol. 8, 186–195.
[184] DePinho, R.A. et al. (1987) Genes Dev. 1, 1311–1326.
[185] Morton, C.C. et al. (1989) Genomics 4, 367–375.
[186] Slamon, D.J. et al. (1986) Science 232, 768–772.
[187] Stanton, L.W. et al. (1986) Proc. Natl Acad. Sci. USA 83, 1772–1776.
[188] Kohl, N.E. et al. (1986) Nature 319, 73–77.
[189] Ibson, J.M. and Rabbitts, P.H. (1988) Oncogene 2, 399–402.
[190] Michitsch, R.W. et al. (1985) Nucleic Acids Res. 13, 2545–2558.
[191] Alitalo, K. et al. (1983) Proc. Natl Acad. Sci. USA 80, 100–104.
[192] Reddy, E.P. et al. (1983) Proc. Natl Acad. Sci. USA 80, 2500–2504.

## Identification

v-*sis* is the acutely transforming oncogene of simian sarcoma virus (SSV) isolated from woolly monkey sarcoma[1] and derived from the platelet-derived growth factor B chain gene (*Pdgfb*).

## Related genes

The 3′ untranslated regions of *PDGFB* share sequence homology with the corresponding regions of human *IL-2*, human IFN-$\beta$1, human and mouse NGF$\beta$ and proenkephalin. PDGFB is 60% similar to PDGFA.

| | ***PDGFB/Pdgfb*** | **v-*sis*** |
|---|---|---|
| **Nucleotides (kb)** | 24 | 5.1 (SSV) |
| **Chromosome** | 22q12.3–q13.1 | |
| **Mass (kDa): predicted** | 27 | 33 |
| **expressed** | 26 | 28 |
| **Cellular location** | Secreted | PDGFR-associated |

**Tissue distribution**

PDGF is a disulfide-bonded dimer of two chains (A and B) that occurs in three forms (AA, BB and AB). PDGFB is generated by proteolytic cleavage of a precursor. PDGF B chains are secreted as part of 30 kDa AB or BB dimers: 24 kDa BB dimers remain cell associated and are degraded in lysosomes. PDGF is released from the $\alpha$ granules of platelets during blood clotting but is also synthesized by many other types of cell, including vascular endothelial cells, macrophages, activated monocytes and bone marrow megakaryocytes, and all three forms occur in both normal and transformed cells[2].

v-SIS is detectable in SSV-transformed NRK (normal rat kidney) cells[3]. It is retained at the cell surface by virtue of a hydrophilic membrane retention domain located at the C-terminus and functions by interaction with the PDGF receptor[4].

## Protein function

PDGF is the major growth factor in human serum and *in vitro* it is a potent mitogen and chemoattractant for normal and neoplastic cells.

**Cancer**

*PDGF* and *PDGF* receptor genes are coexpressed in primary human astrocytomas[5] and *PDGFB* transcription occurs in fibrosarcomas and glioblastomas. Malignant mesothelioma cell lines express primarily PDGFB receptors (and PDGFB) whereas the equivalent normal cells express only PDGFA receptors[6] and synthesize only PDGFA. This suggests that both types of cell may undergo autocrine stimulation, PDGF-AA acting *via* the α-receptor and PDGF-BB *via* the β-receptor [7].

*PDGFB* is translocated to chromosome 9 in chronic myelogenous leukaemia (i.e. the reciprocal translocation to that undergone by *ABL*). There is no evidence that the translocated gene is transcribed, however, and *PDGFB* probably does not play a role in CML.

A subgroup of chronic myelogenous leukaemia has the balanced t(5;12)(q33;p13) translocation which results in the expression of a fusion transcript encoding the tyrosine kinase domain of the PDGFβ receptor coupled to the product of the *ETS*-related *TEL* (translocation, ETS, leukemia) gene [8].

**In animals**

SSV induces fibrosarcomas and glioblastomas in monkeys [9]. In human WM9 melanoma cells expression of PDGF-BB promotes vascularization of the tumours that arise when the cells are injected into mice.

***In vitro***

SSV transforms rat kidney cells into a fibroblastic morphology unusual for cells infected by acute transforming viruses and also transforms fibroblasts. Suramin, a polyanionic drug used clinically for parasitic infections, disrupts PDGF ligand–receptor binding and induces reversion of fibroblasts transformed by v-*sis* [10].

Murine cell lines over-expressing v-*sis*/*Pdgfb* are highly tumorigenic: *in vitro* their growth is independent of the presence of PDGF and they show constitutively high expression of *Myc*, *Fos*, *Jun* and *Jund* but not of other early response genes (*Fra-1*, *Fosb*, *Junb* and *Krox20*) [11], emphasizing the probable importance of *Myc*, *Fos*, *Jun* and *Jund* in the regulation of growth. Transformation of NIH 3T3 cells by v-*sis* is inhibited by expression of the transcription factor *Egr-1* and by WT1 [12].

**Transgenic animals**

PDGF-B null mice die perinatally [13].

## Gene structure of human *PDGFB*

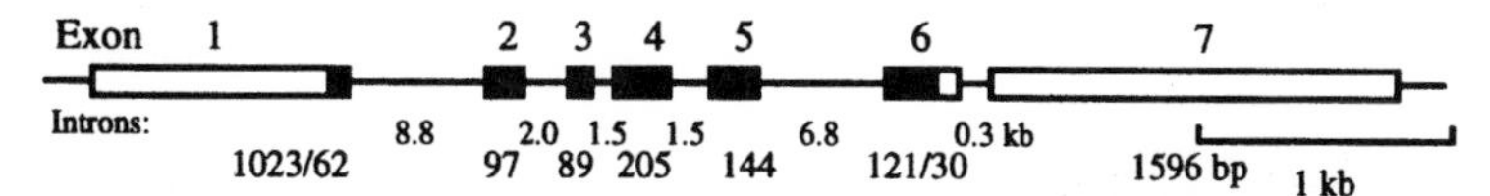


The first intron contains an enhancer-like element responsible for the high level of activity of the PDGFB promoter in choriocarcinoma cell lines. A sequence flanking the TATA box (enhancer-dependent *cis* coactivator (EDC)) is also necessary for enhancer-mediated transcription [14]. There are two HTLV-I TAX responsive elements at −64 to −45 (TRE1 or SPE (sis proximal element)) and −34 to −15 (TATA box region). The transcription factors SP1, SP3 and EGR1 bind to TRE1 [15]. SPE is essential for transcription in K562 and U2-OS cells and contains the motif CCACCC to which SP1 and SP3 bind [16].

## Protein structure

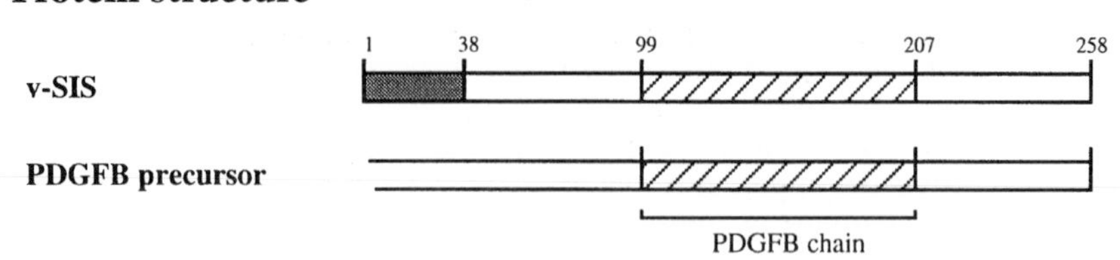


The PDGFB precursor undergoes proteolytic cleavage to generate a final form in which the N- and C-termini correspond to amino acids 99 and 207 of v-SIS, respectively[17]. Residues 99 to 207 of v-SIS (220 amino acids) differ in only four positions from the 108 residues of PDGFB.

## Amino acid sequence of human PDGFB precursor

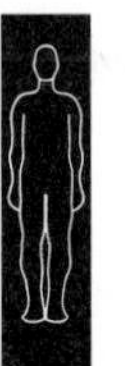

```
  1 MNRCWALFLS LCCYLRLVSA EGDPIPEELY EMLSDHSIRS FDDLQRLLHG
 51 DPGEEDGAEL DLNMTRSHSG GELESLARGR RSLGSLTIAE PAMIAECKTR
101 TEVFEISRRL IDRTNANFLV WPPCVEVQRC SGCCNNRNVQ CRPTQVQLRP
151 VQVRKIEIVR KKPIFKKATV TLEDHLACKC ETVAAARPVT RSPGGSQEQR
201 AKTPQTRVTI RTVRVRRPPK GKHRKFKHTH DKTALKETLG A (241)
```

### Domain structure

| | |
|---|---|
| 1–20 | Signal sequence (italics) |
| 21–81 and 191–241 | Propeptide |
| 82–190 | PDGFB (underlined) |
| 108 and 111 | Involved in receptor binding |
| 97–141, 130–178, 134–180 | Disulfide bonds |
| 124–133 | Interchain disulfide bond |

### Database accession numbers

| | PIR | SWISSPROT | EMBL/GENBANK | REFERENCES |
|---|---|---|---|---|
| Human *PDGFB* (*PDGF2* or *SIS*) | A94276 | P01127 | M12783 | 18,19 |
| | | | K01913–K01916 | 20,21 |
| | | | M16288 | 22 |
| | | | X03702 | 23 |
| | | | X00556, X00559–X00562 | 24 |
| | | | HSPDGFA, X06374 | 25 |
| Human *SIS* (v-*sis* homologous region) | | | V00504 | 26 |
| Human *SIS* (3′ flank) | | | M32009 | 27 |
| Human *SIS* (clone pSM-1) | | X02744 | | 23 |

### *References*


[1] Theilen, G.P. et al. (1971) J. Natl Cancer Inst. 47, 881–889.
[2] **Heldin, C.-H. and Westermark, B. (1990) J. Cell Sci. 96, 193–196.**
[3] Devare, S.G. et al. (1983) Proc. Natl Acad. Sci. USA 80, 731–735.
[4] LaRochelle, W.J. et al. (1991) Genes Dev. 5, 1191–1199.

[5] Maxwell, M. et al. (1990) J. Clin. Invest. 86, 131–140.
[6] Versnel, M.A. et al. (1991) Oncogene 6, 2005–2011.
[7] **Heldin, C.-H. (1992) EMBO J. 11, 4251–4259.**
[8] Golub, T.R. et al. (1994) Cell 77, 307–316.
[9] Wolfe, L.G. et al. (1972) J. Natl. Cancer Inst. 48, 1905–1907.
[10] Fleming, T.P. et al. (1989) Proc. Natl Acad. Sci. USA 86, 8063–8067.
[11] Sonobe, M.H. et al. (1991) Oncogene 6, 1531–1537.
[12] Huang, R.-P. et al. (1994) Oncogene 9, 1367–1377.
[13] Levéen, P. et al. (1994) Genes Dev. 8, 1875–1887.
[14] Franklin, G.C. et al. (1995) Oncogene 11, 1873–1884.
[15] Trejo, S.R. et al. (1996) J. Biol. Chem. 271, 14584–14590.
[16] Liang, Y. et al. (1996) Oncogene 13, 863–871.
[17] Robbins, K.C. et al. (1983) Nature 305, 605–608.
[18] Rao, C.D. et al. (1986) Proc. Natl Acad. Sci. USA 83, 2392–2396.
[19] Antoniades, H.N. and Hunkapiller, M.W. (1983) Science 220, 963–965.
[20] Chiu, I.-M. et al. (1984) Cell 37, 123–129.
[21] Collins, T. et al. (1985) Nature 316, 748–750.
[22] Josephs, S.F. et al. (1984) Science 225, 636–639.
[23] Ratner, L. et al. (1985) Nucleic Acids Res. 13, 5007–5018.
[24] Waterfield, M.D. et al. (1983) Nature 304, 35–39.
[25] Weich, H.A. et al. (1986) FEBS Lett. 198, 344–348 [erratum in FEBS Lett. (1986) 201, 180].
[26] Josephs, S.F. et al. (1983) Science 219, 503–505.
[27] Tong, B.D. et al. (1986) Mol. Cell. Biol. 6, 3018–3022.

## Identification

*Pim-1* was first identified as a common proviral integration site in MuLV-induced murine T cell lymphomas [1–3].

## Related genes

PIM1 has extensive homology with the protein kinase gene family, high homology with the $\gamma$ subunit of phosphorylase kinase, C-terminal homology with ABL and N-terminal homology with MOS. Murine PIM1 is 53% identical to PIM2.

| | *PIM1* |
|---|---|
| **Nucleotides (kb)** | ~5 |
| **Chromosome** | 6p21 |
| **Mass (kDa): predicted** | 35.7 |
| **expressed** | 35 |
| **Cellular location** | Cytoplasm |

**Tissue distribution**

*PIM1* is expressed at high concentrations in haematopoietic tissues, testis and ovaries and in embryonic stem cells [4,5]. Alternative gene products arise in rodents from variable polyadenylation and the use of different translational initiation codons [6,7].

## Protein function

PIM1 is a serine/threonine protein kinase that may be involved in early B and T cell lymphomagenesis [8]. However, physiological substrates have not been identified and *Pim-1*-deficient mice are apparently normal in all respects other than in having a decreased erythrocyte mean cell volume [9].

**Cancer**

Enhanced *PIM1* transcription occurs in some acute myeloid and lymphoid leukaemias [2,4] although the 6;9 translocation is not the direct cause [10]. Thus, although the *PIM1* human chromosome site (6p21) is fragile, elevated PIM1 protein synthesis occurs in many human leukaemias by mechanisms other than translocation or amplification.

**Transgenic animals**

*Pim-1* is the integration locus of Mo-MuLV in 35% of B cell lymphomas generated in E$\mu$-*myc* transgenic mice [11]. Transgenic animals over-expressing *Pim-1* in lymphoid cells show a low frequency of predisposition to lymphomagenesis but have a greatly increased susceptibility to tumour induction by MuLV or by *N*-ethyl-nitrosourea [12]. Tumours over-expressing *Pim-1* also show activation of *Myc*, *Nmyc* and *Pal-1* (see ***MYC***: Transgenic animals).

**In animals**

*Pim-1* and *Myc* are the most frequently occupied insertion sites in MuLV-induced tumours and both may be activated within the same cell lineage[13,14]. Activation of *Pim-1* also occurs in B cell lymphomas[15] and in murine thymomas induced by NMU[16].

***In vitro***

*Pim-1* alone does not transform 3T3 fibroblasts but causes transformation in cooperation with *Myc* or *Ras*. *Pim-1* does not appear to be necessary for proliferation or differentiation of embryonic stem cells *in vitro*[17].

## Gene structure

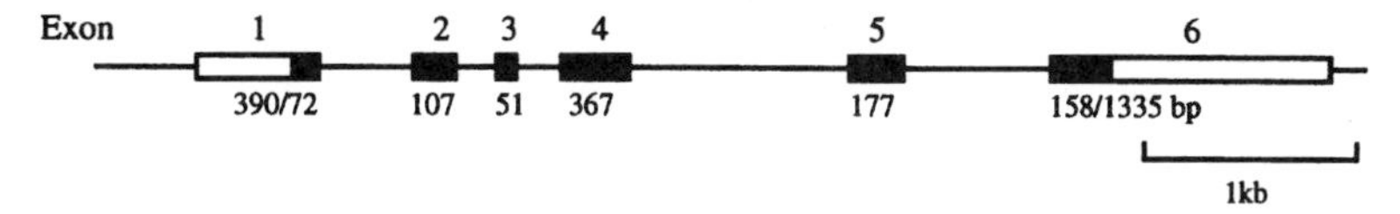


The human and murine genes share >80% identity and the proteins are 94% identical.

## Transcriptional regulation

The *PIM1* promoter does not contain TATA or CAAT box sequences but has consensus binding sites for AP1, AP2, SP1, NF-κB, NF-A2 (Oct2) and *PIM1* promoter factor 348.

Proviral activation of murine *Pim-1* involves elevated transcription by enhancer insertion and, usually, removal of 3′ untranslated $(ATTT)_5$ sequences that destabilize mRNAs. The protein coding domain is unaffected by insertions and the transforming effects of *Pim-1* are thus due to abnormally high expression of the gene. In most lymphomas the provirus is integrated within the *Pim-1* gene and has duplicated or triplicated enhancer regions within their LTRs. Integrations within the gene (in the 3′ untranslated region) or 3′ of the gene are all in the same transcriptional orientation as *Pim-1*. The concentration of *Pim-1* mRNA is higher in such tumours than when integration is outside the transcription unit: transcription is terminated at the polyadenylation signal in the 5′ LTR, generating truncated *Pim-1* transcripts lacking up to 1300 bases. Proviruses integrated upstream of *Pim-1* and in the same transcriptional orientation have intact 5′ and 3′ LTRs but major internal deletions. Such integrations do not provide a promoter for *Pim-1*, nor do they alter the transcript size: thus transcriptional enhancement appears to be the mechanism of activation for integrations outside coding regions of *Pim-1*.

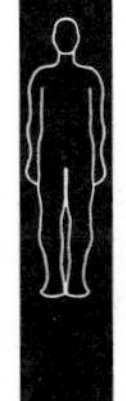

## Amino acid sequence of human PIM

```
  1 MLLSKINSLA HLRAAPCNDL HATKLAPGKE KEPLESQYQV GPLLGSGGFG
 51 SVYSGIRVSD NLPVAIKHVE KDRISDWGEL PNGTRVPMEV VLLKKVSSGF
101 SGVIRLLDWF ERPDSFVLIL ERPEPVQDLF DFITERGALQ EELARSFFWQ
151 VLEAVRHCHN CGVLHRDIKD ENILIDLNRG ELKLIDFGSG ALLKDTVYTD
201 FDGTRVYSPP EWIRYHRYHG RSAAVWSLGI LLYDMVCGDI PFEHDEEIIR
251 GQVFFRQRVS SECQHLIRWC LALRPSDRPT FEEIQNHPWM QDVLLPQETA
301 EIHLHSLSPG PSK (313)
```

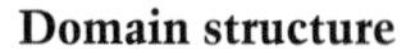

**Domain structure**

44–52 and 67 ATP binding site (underlined)
167 Active site

**Database accession numbers**

| | *PIR* | *SWISSPROT* | *EMBL/GENBANK* | *REFERENCES* |
|---|---|---|---|---|
| Human *PIM1* | A27476, JU0327 | P11309 | M27903 | 18 |
| | | | M16750 | 19 |
| | | | M54915 | 20 |

***References***


[1] Cuypers, H.T. et al. (1986) Human Genet. 72, 262–265.
[2] Nagarajan, L. et al. (1986) Proc. Natl Acad. Sci. USA 83, 2556–2560.
[3] Wirschubsky, Z. et al. (1986) Int. J. Cancer 38, 739–745
[4] Amson, R. et al. (1989) Proc. Natl Acad. Sci. USA 86, 8857–8861.
[5] Meeker, T.C. et al. (1990) Mol. Cell. Biol. 10, 1680–1688.
[6] Saris, C.J.M. et al. (1991) EMBO J. 10, 655–664.
[7] Wingett, D. et al. (1992) Nucleic Acids Res. 20, 3183–3189.
[8] Dautry, F. et al. (1988) J. Biol. Chem. 263, 17615–17620.
[9] Laird, P.W. et al. (1993) Nucleic Acids Res. 21, 4750–4755.
[10] von Lindern, M. et al. (1989) Oncogene 4, 75–79.
[11] van Lohuizen, M. et al. (1991) Cell 65, 737–752.
[12] Breuer, M. et al. (1991) Cancer Res. 51, 958–963.
[13] Selten, G. et al. (1984) EMBO J. 3, 3215–3222.
[14] O'Donnell, P.V. et al. (1985) J. Virol. 55, 500–503.
[15] Mucenski, M.L. et al. (1987) Oncogene Res. 2, 33–48.
[16] Warren, W. et al. (1987) Carcinogenesis 8, 163–172.
[17] te Riel, H. et al. (1990) Nature 348, 649–651.
[18] Reeves, R. et al. (1990) Gene 90, 303–307.
[19] Zakut-Houri, R. et al. (1987) Gene 54, 105–111.
[20] Domen, J. et al. (1987) Oncogene Res. 1, 103–112.

## Identification

v-*raf*-is the oncogene of murine transforming retrovirus (MuSV) 3611 [1,2]. 3611-MuSV arose after transforming gene rescue in culture with MuLV followed by infection of a mouse treated with butylnitrosourea [3]. This mouse developed histiocytic lymphoma and lung adenocarcinoma.

The mammalian *RAF* family (*RAF1*, *RAFA1* and *RAFB1* (murine *Raf-1*, *Araf-1* and *Braf-1*)) were detected by screening genomic libraries with v-*raf* probes [4–6].

## Related genes

*RAF1*, *RAFA1* and *RAFB1* are members of the *RAF-MOS* subfamily and the *SRC* superfamily of protein kinases [7]. *RAF1P1/RAF2*, *ARAF2* and *BRAF-2* are pseudogenes. The avian homologue of *Raf* is v-*mil* (see ***Mil***). Other RAF1-related genes are: *PKS*, Xe-*raf* (*Xenopus laevis*) D-*raf*-1, D-*raf*-2 (*Drosophila melanogaster*) and *Elegans raf*-1 (*Caenorhabditis elegans*).

| | ***RAFA1/Araf-1*** | ***RAFB1/Braf-1*** | ***RAF1/Raf-1*** | **v-*raf*** |
|---|---|---|---|---|
| **Nucleotides (kb)** | 11 | Not fully mapped | >100 | 1.1<br>7.6<br>(3611-MuSV) |
| **Chromosome** | Xp11.2 | 7q33–36 | 3p25 | |
| **Mass (kDa): predicted** | 67.5 | 84/72.5 | 73 | 37 (v-*raf*) |
| **expressed** | 68 | 73/95 (multiple) | 70–74 (pp74–pp78) | $gp90^{gag\text{-}raf}$<br>$P75^{gag\text{-}raf}$ |
| **Cellular location** | Cytosolic | Cytosolic | Cytosolic | Cytosolic |
| **Tissue distribution** | Predominantly urogenital tissues | Cerebrum, testes, haematopoietic cell lines | Ubiquitous | |

## Protein function

*RAF* genes encode serine/threonine protein kinases. RAF1 is positively regulated by serine/tyrosine phosphorylation [8] and has no autophosphorylating protein kinase activity (unlike Δ*gag*-v-*raf* and Δ*gag*-v-*mil*) [9]. The ubiquitous distribution of RAF1 suggests that it may have a basic regulatory function. *Xenopus* RAF-1 appears to mediate the developmental effects of basic fibroblast growth factor during mesoderm induction [10].

RAF1 exists in a complex with the heat shock protein HSP90 and this association is essential for RAF1 stability and cellular localization [11]. Serine phosphorylated RAF1 also associates with the SH2 domains of FYN and SRC [12], with the CD3 $\delta$ and $\gamma$ chains of the T cell receptor in unstimulated mouse T cells [13] and with two members of the 14-3-3 protein family (14-3-3$\beta$ and 14-3-3$\zeta$). The latter naturally dimeric proteins co-segregate with RAF to the plasma membrane[14,15] and are able to promote RAF1 oligomerization which *per se* promotes RAF1 activation (as indicated by phosphorylation of ERKs) *via* a RAS-dependent mechanism[16,17]. In *Xenopus* oocytes 14-3-3 proteins promote Raf-1-dependent maturation[18]. $Cys^{165}$, $Cys^{168}$ and $Ser^{259}$ of RAF1 interact with 14-3-3 and 14-3-3ζ which may maintain RAF1 in an inactive conformation through binding to both its N-terminus and C-terminus. Activated RAS binds to RAF1, inducing conformational changes and displacement of 14-3-3ζ from the N-terminus[19]. The inactivation of RAF1 by purified membrane-associated phosphatases (either tyrosine or serine/threonine) is blocked by either HSP90 or 14-3-3ζ protein[20]. However, RAF1 can be activated independently of 14-3-3 in a RAS-independent manner[21] and formation of ternary RAS/RAF/14-3-3 complexes in NIH 3T3 cells has been shown to be without effect on RAF activity[22].

Two-hybrid screening in yeast has identified a specific interaction between RAF-1 (amino acids 136–239) and the $G\beta_2$ subunit of heterotrimeric G proteins[23]. RAF1 also binds to and phosphorylates CDC25B[24].

In NIH 3T3 cells RAF-1 undergoes rapid phosphorylation, mainly on serine and threonine, in response to PDGF, EGF, insulin, thrombin, endothelin, acidic FGF, CSF1 or TPA, and also in response to the oncoproteins of v-*fms*, v-*src*, v-*sis*, *Hras* or polyoma middle T antigen[25–30]. RAF1 is also activated in mitogenically stimulated epithelial and lymphoid cells[31–35]. The RAF1 signalling pathway is activated in B lymphocytes by cross-linking surface IgM[36] and RAF1 is activated by prolactin in the prolactin-dependent T cell line Nb2[37], by TNF-α in human myeloid or monocytic leukaemia cells and lung fibroblasts[38] and by IL-13 in the human monocytic progenitor cell line U937[39].

Viral and activated forms of MIL/RAF1 proteins associate with a 34 kDa phosphoprotein in interphase but not during mitosis[40]. RAF1 is also hyperphosphorylated and activated during mitosis, suggesting a role in the $G_2$ to M phase transition, although the activators and substrates of RAF1 at this stage of the cell cycle are unknown[41,42].

## RAF1 signalling pathways

RAF1 forms complexes with the activated PDGF β receptor and with RAS-GTP in intact, stimulated cells[43,44]. The activated enzyme is translocated to the perinuclear region and the nucleus. RAF1 activates MAP kinase kinase (MAPKK or MEK) which in turn stimulates the mitogen-activated protein kinases (MAPKs) ERK1 and ERK2, resulting in phosphorylation of $p62^{TCR}$ (see ***FOS***) and JUN (see ***JUN***). MAPKK, ERK1 and ERK2 are constitutively active in v-*raf*-transformed cells[45–49] and RAF1 is itself an *in vitro* substrate for MAPKK[45]. Oncogenic *Raf* also activates p70 ribosomal S6 kinase by a pathway that is independent of MAPK[50]. Reconstitution of RAF1-mediated signal transduction in a baculovirus expression system indicates that RAS/RAF1/ERK1 activation causes hyperphosphorylation of JUN whereas FOS phosphorylation is stimulated by both ERK1-dependent and ERK1-independent pathways[51]. Mutants of MEK1 in which the RAF1/MAPKKK-dependent regulatory phosphorylation sites are substituted with aspartic acid residues render fibroblasts both transformed and tumorigenic[52]. EGF and TPA can also activate MAPK by a RAF1-independent pathway[53].

In neutrophils wortmannin inhibits RAF1, BRAF and MAPK activation in response to IL-8, indicating that PtdIns 3-kinase is involved in the activation of RAF1 in this G protein-coupled response[54].

## RAF1 and transcriptional activation

RAF oncoproteins *trans*-activate expression of genes driven by AP1, ETS and NF-$\kappa$B binding motifs[55–57]. Activation of oncogenic RAF, BRAF or RAS induces the expression and secretion of heparin binding epidermal growth factor (HB-EGF) in NIH 3T3 fibroblasts[58]. Inhibition of RAF increases collagen gene expression in hepatic stellate cells: inhibition of both RAF and MAPK decreases collagen expression, indicating that pathways diverging at RAF coordinately regulate expression of this gene[59].

Transfection of a v-*raf* expression vector into NIH 3T3 cells induces transcription of the growth factor-regulated early response genes *Egr-1*, *Fos* and $\beta$-actin[60–63]. The expression of v-*raf* or *RAF1* activates transcription of the *MDR1* multidrug resistance genes in human and rodent cells[64]. In myeloid cells transformation by v-*raf* is associated with suppression of apoptosis *via* BCL2 and a reduction in the requirement for IL-3 for proliferation[65].

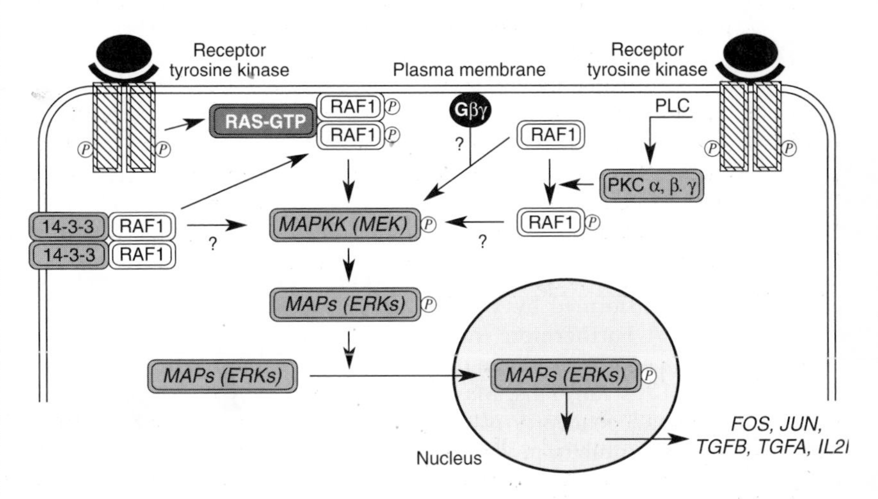


The generalized scheme summarizes the above data, indicating that RAF1 kinase may be activated by a wide variety of receptors with intrinsic tyrosine kinase activity or by receptors that interact with tyrosine kinases (e.g. CD4). RAS may mediate coupling to RAF1 of pathways that either require or are independent of protein kinase C.

### Cancer

Oncogenic *RAF1* has been detected by transfection of DNA from a primary stomach cancer[66], laryngeal[67], lung and other carcinomas and sarcomas[68] and a glioblastoma cell line[69]. High levels of *RAF1* expression occur in many small cell lung cancers (SCLC) and derived cell lines[70]. *RAF1* is also amplified in some non-small cell lung cancers[71].

### In animals

The oncogenically active v-*raf*, derived from mouse *Raf-1*, induces a defined spectrum of tumours *in vivo*[72]. The differences in tumour spectra induced by *Raf*- and *Myc*-expressing retroviruses suggest that the preferred *Myc* targets (lymphoid) differ from the preferred *Raf* targets (erythroid, fibroblast) in the rate-limiting pathways through which their growth is normally controlled.

### *In vitro*

*Raf-1* preferentially transforms erythroid cells and also transforms fibroblasts and epithelial cells[73]. *Raf* and *Myc* act synergistically to transform cells of all haematopoietic lineages: their coexpression in lymphoid and erythroid cells induces differentiation to a myeloid form[74]. In fibroblasts *Raf* cooperates with *Hras* to promote transformation: $G\alpha_{12}$, the $\alpha$ subunit of the heterotrimeric G protein $G_{12}$, is an even more potent cotransforming agent with *Raf* that may activate RHO-dependent pathways[75].

Expression of the kinase domains of *Araf-1* or *Braf-1* in 3T3 cells indicates that *Braf-1* is the more potent transforming agent and that *Braf-1* causes stronger and more rapid activation of both MEK1 and MAPKs[76]. In HeLa cells RAFA1 is activated in response to EGF and it in turn phosphorylates and activates MEK1 but not MEK2[77]. In PC-12 cells BRAF-1 also activates MEK1 in response to EGF, PDGF or NGF[78]. In addition to MEK1, BRAF-1 can interact directly with RAS, MEK2 and three members ($\eta$, $\theta$ and $\zeta$) of the 14-3-3 family[79]. In addition to RAS, RAP1 is also involved in the activation of BRAF[80].

In PC-12 cells oncogenic RAF1 activates both promoters of the transforming growth factor $\beta_1$ (TGF$\beta_1$) gene and appears to mediate the effects of NGF, FGF and EGF on these cells[81].

Revertant cell lines that suppress transformation by v-*ras* and by v-*fes* or v-*src* are transformed by v-*fms*, v-*mos* or v-*sis* and by 3611-MuSV or *Araf*-MuSV[72,82]. Furthermore, neither microinjection of anti-RAS antibody nor block of RAS activity by mutation affects growth in cells transformed by *Raf* or *Mos*. Thus *Raf*, *Fms*, *Mos* and *Sis* are the only oncogenes that appear to function independently of *Ras*. In NIH 3T3 cells expression of *Raf-1* antisense RNA inhibits proliferation, causes reversion of *Raf*-transformed cells and blocks transformation by *Kras* or *Hras*[83]. Thus RAF-1 functions downstream of *Ras*.

*Ras* and *Raf-1* cooperate to transform NIH 3T3 cells and dominant negative mutants of *Ras* inhibit the activation of RAF-1 kinase in NIH 3T3 cells stimulated by serum or TPA and of both RAF-1 and BRAF-1 in NGF-stimulated PC-12 cells[84], indicating that RAS may control the coupling of growth factor receptors and protein kinase C to cytosolic RAF kinases. This is consistent with the finding that N-terminally truncated, oncogenic RAF activates the MAP kinase ERK2 independently of RAS function[47] and that the kinase activity of RAF-1 is enhanced as a result of direct phosphorylation by protein kinase C$\alpha$, $\beta$ or $\gamma$[85,86]. Fusion of the membrane localization signal of KRAS(4B) to the C-terminus of RAF confers constitutive activity that can be enhanced by EGF, independently of RAS[87]. RAS acts as a regulated, membrane bound anchor to recruit RAF-1 to the plasma membrane: this permits tyrosine phosphorylation and further activation of RAF-1, for example on $Tyr^{340/341}$ by oncogenic SRC [88].

*Araf-1* transforms NIH 3T3 fibroblasts after incorporation into the genome of murine leukaemia virus and expression of the *gag-raf* gene product [89].

## Gene structure

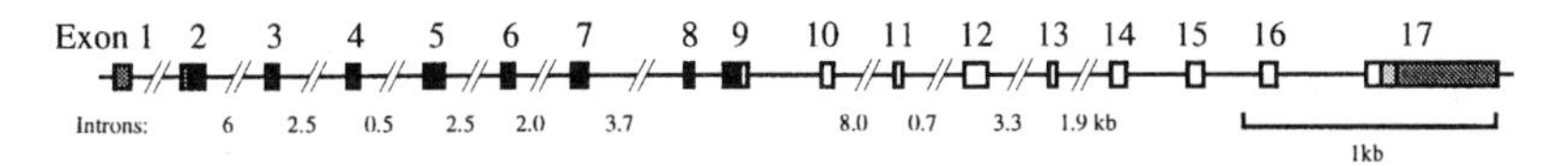


Open boxes: exons homologous to those in the mouse genome that are transduced in v-*raf*.

## Transcriptional regulation

The human *RAF1* promoter is located in an HTF-island (CpG-rich, non-methylated) and lacks TATA and CAAT boxes [90]. The 5′ untranslated exon 1 is located at least 55 kb upstream of the body of the gene which spans 45 kb [4]. The last *RAF1* exon (17) contains 905 bp of 3′ untranslated sequence. The large size of the gene may account for the relatively high frequency with which truncation-activated oncogenic versions of *RAF1* have been obtained.

The *RAFA1* promoter includes three glucocorticoid response elements (at −17, −34 and −168), two of which are targets for the glucocorticoid receptor [91].

## Protein structure

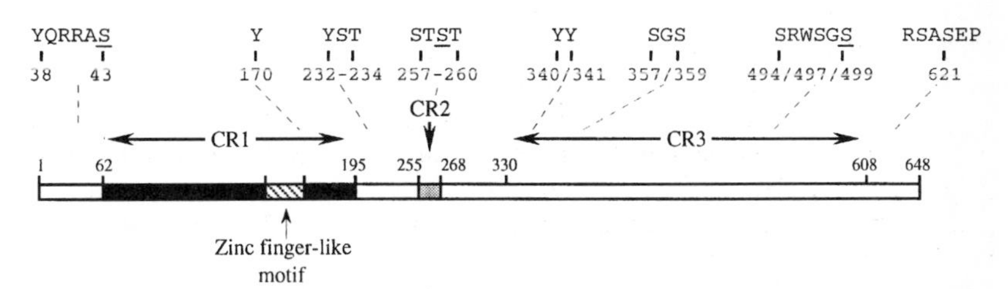


Conserved region 1 (CR1) contains a zinc finger-like motif; CR2 is conserved in virtually all forms of RAF and interacts directly with RAS proteins, binding GTP-RAS in preference to GDP-RAS and inhibiting RAS-GAP activity [43,44]. CR3 comprises the minimal transforming element [92]. CR1, CR2 and CR3 are between 61% and 100% homologous in RAF1, RAFA1 and RAFB1. Potential phosphorylation sites are shown above the figure. RAF1 is phosphorylated at multiple sites in the N-terminus that are deleted in the transforming protein. Underlined: major phosphorylation sites.

Deletion of residues 245–261 in CR2 causes oncogenic activation [92,93] but mutation of $Ser^{259}$ alone is not sufficient to activate transforming potential: the additional deletion of the region 283–309 activates weak transforming power but the complete removal of the N-terminal 303 residues (that includes nine serines) is required for maximal activity [94].

The crystal structure of the RAS binding domain (RBD) of RAF1 in complex with RAP1A and a GTP analogue reveals that RBD has a ubiquitin $\alpha/\beta$ roll: protein interaction is mediated by a central anti-parallel $\beta$ sheet formed by strands B1–B2 of RBD and $\beta2$–$\beta3$ of RAP1A [95].

The structure of the cysteine-rich domain of RAF1 differs from the related domains in protein kinase C in that it does not bind diacylglycerol but does interact with phosphatidylserine and with RAS [96].

## Amino acid sequence of human RAF1

```
  1 MEHIQGAWKT ISNGFGFKDA VFDGSSCISP TIVQQFGYQR RASDDGKLTD
 51 PSKTSNTIRV FLPNKQRTVV NVRNGMSLHD CLMKALKVRG LQPECCAVFR
101 LLHEHKGKKA RLDWNTDAAS LIGEELQVDF LDHVPLTTHN FARKTFLKLA
151 FCDICQKFLL NGFRCQTCGY KFHEHCSTKV PTMCVDWSNI RQLLLFPNST
201 IGDSGVPALP SLTMRRMRES VSRMPVSSQH RYSTPHAFTF NTSSPSSEGS
251 LSQRQRSTST PNVHMVSTTL PVDSRMIEDA IRSHSESASP SALSSSPNNL
301 SPTGWSQPKT PVPAQRERAP VSGTQEKNKI RPRGQRDSSY YWEIEASEVM
351 LSTRIGSGSF GTVYKGKWHG DVAVKILKVV DPTPEQFQAF RNEVAVLRKT
401 RHVNILLFMG YMTKDNLAIV TQWCEGSSLY KHLHVQETKF QMFQLIDIAR
451 QTAQGMDYLH AKNIIHRDMK SNNIFLHEGL TVKIGDFGLA TVKSRWSGSQ
501 QVEQPTGSVL WMAPEVIRMQ DNNPFSFQSD VYSYGIVLYE LMTGELPYSH
551 INNRDQIIFM VGRGYASPDL SKLYKNCPKA MKRLVADCVK KVKEERPLFP
601 QILSSIELLQ HSLPKINRSA SEPSLHRAAH TEDINACTLT TSPRLPVF(648)
```

**Domain structure**

| | |
|---|---|
| 89 | Arg[89]: essential for RAS binding and RAS-mediated activation [97] |
| 51–131 | Minimal RAS binding region |
| 80–103 | Primary region of interaction with RAS |
| 64–67, 120–125 | Secondary interactions with RAS [98] |
| 132–149 | High-affinity binding to RAS [99] may be a cryptic binding site unmasked by RAS interaction with the minimal site [100] |
| 139–184 | Cysteine-rich domain |
| 139–184 | Phorbol ester and DAG binding |
| 259, 499 | Serine residues phosphorylated by protein kinase C$\alpha$ |
| 355–363 and 375 | ATP binding |
| 340, 341 | Major tyrosine phosphorylation sites. The introduction of negative charge at these sites appears essential for enzymatic activation of RAF1 by tyrosine kinases [101] |
| 378 | Mutation of corresponding v-RAF residue (Lys[53]) eliminates transforming capacity |
| 468 | Active site |
| 621 | Ser[621] is phosphorylated by the cAMP-dependent protein kinase (PKA) and is essential for RAF1 catalytic activity [102] |
| 165, 168, 259 | Cys[165], Cys[168] and Ser[259] interact with 14-3-3 [103] |

RAF proteins can be oncogenically activated by N-terminal fusion, truncation or point mutations [68,92,104]. The generation of an activated oncoprotein by

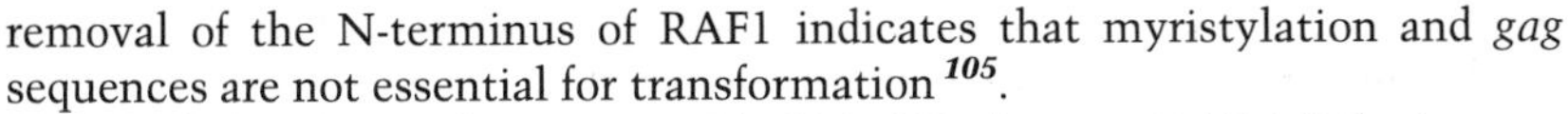


removal of the N-terminus of RAF1 indicates that myristylation and *gag* sequences are not essential for transformation[105].

The v-*raf* gene encodes amino acids 326–648 of mouse RAF-1 linked to *gag* sequences. The sequence of v-RAF is identical to residues 326–648 of human RAF1 with the substitutions $EI^{343/344} \rightarrow KM$; $F^{387} \rightarrow L$; $T^{543} \rightarrow A$; $S^{549} \rightarrow A$; $K^{572} \rightarrow R$; $M^{581} \rightarrow I$; $S^{621} \rightarrow P$.

## Amino acid sequence of human RAFA1

```
  1 MEPPRGPPAN GAEPSRAVGT VKVYLPNKQR TVVTVRDGMS VYDSLDKALK
 51 VRGLNQDCCV VYRLIKGRKT VTAWDTAIAP LDGEELIVEV LEDVPLTMHN
101 FVRKTFFSLA FCDFCLKFLF HGFRCQTCGY KFHQHCSSKV PTVCVDMSTN
151 RQQFYHSVQD LSGGSRQHEA PSNRPLNELL TPQGPSPRTQ HCDPEHFPFP
201 APANAPLQRI RSTSTPNVHM VSTTAPMDSN LIQLTGQSFS TDAAGSRGGS
251 DGTPRGSPSP ASVSSGRKSP HSKSPAEQRE RKSLADDKKK VKNLGYRXSG
301 YYWEVPPSEV QLLKRIGTGS FGTVFRGRWH GDVAVKVLKV SQPTAEQAQA
351 FKNEMQVLRK TRHVNILLFM GFMTRPGFAI ITQWCEGSSL YHHLHVADTR
401 FDMVQLIDVA RQTAQGMDYL HAKNIIHRDL KSNNIFLHEG LTVKIGDFGL
451 ATVKTRWSGA QPLEQPSGSV LWMAAEVIRM QDPNPYSFQS DVYAYGVVLY
501 ELMTGSLPYS HIGCRDQIIF MVGRGYLSPD LSKISSNCPK AMRRLLSDCL
551 KFQREERPLF PQILATIELL QRSLPKIERS ASEPSLHRTQ ADELPACLLS
601 AARLVP (606)
```

**Domain structure**

99–144 Phorbol ester and DAG binding
316–324 and 336 ATP binding
429 Active site

## Amino acid sequence of human RAFB1

```
  1 MDTVTSSSSS SLSVLPSSLS VFQNPTDVAR SNPKSPQKPI VRVFLPNKQR
 51 TVVPARCGVT VRDSLKKALM MRGLIPECCA VYRIQDGEKK PIGWDTDISW
101 LTGEELHVEV LENVPLTTHN FVRKTFFTLA FCDFCRKLLF QGFRCQTCGY
151 KFHQRCSTEV PLMCVNYDQL DLLFVSKFFE HHPIPQEEAS LAETALTSGS
201 SPSAPASDSI GPQILTSPSP SKSIPIPQPF RPADEDHRNQ FGQRDRSSSA
251 PNVHINTIEP VNIDDLIRDQ GFRGDGGSTT GLSATPPASL PGSLTNVKAL
301 QKSPGPQRER KSSSSSEDRN RMKTLGRRDS SDDWEIPDGQ ITVGQRIGSG
351 SFGTVYKGKW HGDVAVKMLN VTAPTPQQLQ AFKNEVGVLR KTRHVNILLF
401 MGYSTKPQLA IVTQWCEGSS LYHHLHIIET KFEMIKLIDI ARQTAQGMDY
451 LHAKSIIHRD LKSNNIFLHE DLTVKIGDFG LATVKSRWSG SHQFEQLSGS
501 ILWMAPEVIR MQDKNPYSFQ SDVYAFGIVL YELMTGQLPY SNINNRDQII
551 FMVGRGYLSP DLSKVRSNCP KAMKRLMAEC LKKKRDERPL FPQILASIEL
601 LARSLPKIHR SASEPSLNRA GFQTEDFSLY ACASPKTPIQ AGGYGAFPVH
                                                    (650)
```

**Domain structure**

6–11 Poly serine
119–164 Phorbol ester and DAG binding
347–355 and 367 ATP binding
460 Active site

## Database accession numbers

| | PIR | SWISSPROT | EMBL/GENBANK | REFERENCES |
|---|---|---|---|---|
| Human *RAF1* | A00637 | P04049 | X03484 | 4,106 |
| Human *RAFA1* | A26439 | P10398 | X04790 | 5 |
| Human *RAFB1* | A31850, S13798 | P15056 | M95712, M95721, X54072, M21001 | 6,107–109 |

## *References*


1 Rapp, U.R. et al. (1983) Proc. Natl Acad. Sci. USA 80, 4218–4222.
2 **Rapp, U.R. (1991) Oncogene 6, 495–500.**
3 Rapp, U.R. and Todaro, G.J. (1978) Science 201, 821–824.
4 Bonner, T.I. et al. (1985) Mol. Cell. Biol. 5, 1400–1407.
5 Beck, T.W. et al. (1987) Nucleic Acids Res. 15, 595–609.
6 Ikawa, S. et al. (1988) Mol. Cell. Biol. 8, 2651–2654.
7 **Hanks, S.K. et al. (1988) Science 241, 42–52.**
8 **Heidecker, G. et al. (1992) Adv. Cancer Res. 58, 53–73.**
9 Schultz, A.M. et al. (1988) Oncogene 2, 187–193.
10 MacNicol, A.M. et al. (1993) Cell 73, 571–583.
11 Schulte, T.W. et al. (1995) J. Biol. Chem. 270, 24585–24588.
12 Cleghon, V. and Morrison, D.K. (1994) J. Biol. Chem. 269, 17749–17755.
13 Loh, C. et al. (1994) J. Biol. Chem. 269, 8817–8825.
14 Freed, E. et al. (1994) Science 265, 1713–1716.
15 Li, S. et al. (1995) EMBO J. 14, 685–696.
16 Luo, Z. et al. (1996) Nature 383, 181–185.
17 Farrar, M.A. et al. (1996) Nature 383, 178–181.
18 Fanti, W.J. et al. (1994) Nature 371, 612–614.
19 Rommel, C. et al. (1996) Oncogene 12, 609–619.
20 Dent, P. et al. (1995) Science 268, 1902–1906.
21 Michaud, N.R. et al. (1995) Mol. Cell. Biol. 15, 3390–3397.
22 Suen, K.-L. et al. (1995) Oncogene 11, 825–831.
23 Pumiglia, K. et al. (1995) J. Biol. Chem. 270, 14251–14254.
24 Galaktionov, K. et al. (1995) Science 269, 1575–1577.
25 Morrison, D.K. et al. (1989) Cell 58, 649–657.
26 Baccarini, M. et al. (1990) EMBO J. 9, 3649–3657.
27 Kovacina, K.S. et al. (1990) J. Biol. Chem. 265, 12115–12118.
28 Blackshear, P.J. et al. (1990) J. Biol. Chem. 265, 12131–12134.
29 App, H. et al. (1991) Mol. Cell. Biol. 11, 913–919.
30 Kolch, W. et al. (1993) Nature 364, 249–252.
31 Turner, B. et al. (1991) Proc. Natl Acad. Sci. USA 88, 1227–1231.
32 Maslinski, W. et al. (1992) J. Biol. Chem. 267, 15281–15284.
33 Carroll, M.P. et al. (1991) J. Biol. Chem. 266, 14964–14969.
34 Thompson, P.A. et al. (1991) Cell Growth Differ. 2, 609–617.
35 Siegel, J.N. et al. (1990) J. Biol. Chem. 265, 18472–18480.
36 Tordai et al. (1994) J. Biol. Chem. 269, 7538–7543.
37 Clevenger, C.V. et al. (1994) J. Biol. Chem. 269, 5559–5565.
38 Belka, C. et al. (1995) EMBO J. 14, 1156–1165.
39 Adunyah, S.E. et al. (1995) Biochem. Biophys. Res. Commun. 206, 103–111.
40 Lovric, J. et al. (1996) Oncogene 12, 1145–1151.
41 Laird, A.D. et al. (1995) J. Biol. Chem. 270, 26742–26745.

[42] Lovric, J. and Moelling, K. (1996) Oncogene 12, 1109–1116.
[43] Warne, P.H. et al. (1993) Nature 364, 352–355.
[44] Zhang, X. et al. (1993) Nature 364, 308–313.
[45] Kyriakis, J.M. et al. (1993) J. Biol. Chem. 268, 16009–16019.
[46] Dent, P. et al. (1992) Science 257, 1404–1407.
[47] Howe, L.R. et al. (1992) Cell 71, 335–342.
[48] Muslin, A.J. et al. (1993) Mol. Cell. Biol. 13, 4197–4202.
[49] Huang, W. et al. (1993) Proc. Natl Acad. Sci. USA 90, 10947–10951.
[50] Lenormand, P. et al. (1996) J. Biol. Chem. 271, 15762–15768.
[51] Agarwal, S. et al. (1995) Oncogene 11, 427–438.
[52] Brunet, A. et al. (1994) Oncogene 9, 3379–3387.
[53] Chao et al. (1994) J. Biol. Chem. 269, 7337–7341.
[54] Knall, C. et al. (1996) J. Biol. Chem. 271, 2832–2838.
[55] Wasylyk, C. et al. (1989) Mol. Cell. Biol. 9, 2247–2250.
[56] Bruder, J.T. et al. (1993) Nucleic Acids Res. 21, 5229–5234.
[57] Li, S. and Sedivy, J.M. (1993) Proc. Natl Acad. Sci. USA 90, 9247–9251.
[58] McCarthy, S.A. et al. (1995) Genes Dev. 9, 1953–1964.
[59] Davis, B.H. et al. (1996) J. Biol. Chem. 271, 11039–11042.
[60] Kaibuchi, K. et al. (1989) J. Biol. Chem. 264, 20855–20858.
[61] Jamal, S. and Ziff, E. (1990) Nature 344, 463–466.
[62] Qureshi, S.A. et al. (1991) J. Biol. Chem. 266, 20594–20597.
[63] Rim, M. et al. (1992) Oncogene 7, 2065–2068.
[64] Cornwell, M.M. and Smith, D.E. (1993) J. Biol. Chem. 268, 15347–15350.
[65] Cleveland, J.L. et al. (1994) Oncogene 9, 2217–2226.
[66] Shimizu, K. et al. (1985) Proc. Natl Acad. Sci. USA 82, 5641–5645.
[67] Kasid, U. et al. (1987) Science 237, 1039–1041.
[68] Stanton, V.P. and Cooper, G.M. (1987) Mol. Cell. Biol. 7, 1171–1179.
[69] Fukui, M. et al. (1985) Proc. Natl Acad. Sci. USA 82, 5954–5958.
[70] Graziano, S.L. et al. (1987) Cancer Res. 48, 2148–2155.
[71] Hajj, C. et al. (1990) Cancer 66, 733–739.
[72] **Rapp, U.R. et al. (1988) In The Oncogene Handbook, eds T. Curran, E.P. Reddy and A. Skalka. Elsevier, Amsterdam, pp. 213–253.**
[73] Keski-Oja, A. et al. (1982) J. Cell. Biochem. 20, 139–148.
[74] Klinken, S.P. et al. (1989) J. Virol. 63, 1489–1492.
[75] Zhang, Y. et al. (1996) Oncogene 12, 2377–2383.
[76] Pritchard, C.A. et al. (1995) Mol. Cell. Biol. 15, 6430–6442.
[77] Wu, X. et al. (1996) J. Biol. Chem. 271, 3265–3271.
[78] Vaillancourt, R.R. et al. (1994) Mol. Cell. Biol. 14, 6522–6530.
[79] Papin, C. et al. (1996) Oncogene 12, 2213–2221.
[80] Ohtsuka, T. et al. (1996) J. Biol. Chem. 271, 1258–1261.
[81] Cosgaya, J.M. and Aranda, A. (1996) Oncogene 12, 2651–2660.
[82] Noda, M. et al. (1983) Proc. Natl Acad. Sci. USA 80, 5602–5606.
[83] Kolch, W. et al. (1991) Nature 349, 426–428.
[84] Troppmair, J. et al. (1992) Oncogene 7, 1867–1873.
[85] Sozeri, O. et al. (1992) Oncogene 7, 2259–2262.
[86] Kolch, W. et al. (1993) Oncogene 8, 361–370.
[87] Leevers, S.J. et al. (1994) Nature 369, 411–414.
[88] Marais, R. et al. (1995) EMBO J. 14, 3136–3145.
[89] Huleihel, M. et al. (1986) Mol. Cell. Biol. 6, 2655–2662.
[90] Beck, T.W. et al. (1990) Mol. Cell. Biol. 10, 3325–3333.

[91] Lee, J.-E. et al. (1996) Oncogene 12, 1669–1677.
[92] Heidecker, G. et al. (1990) Mol. Cell. Biol. 10, 2503–2512.
[93] Ishikawa, F. et al. (1988) Oncogene 3, 635–658.
[94] McGrew, B.R. et al. (1992) Oncogene 7, 33–42.
[95] Nassar, N. et al. (1995) Nature 375, 554–560.
[96] Mott, H.R. et al. (1996) Proc. Natl Acad. Sci. USA 93, 8312–8317.
[97] Fabian et al. (1994) Proc. Natl Acad. Sci. USA 91, 5982–5986.
[98] Barnard, D. et al. (1995) Oncogene 10, 1283–1290.
[99] Chuang, E. et al. (1994) Mol. Cell. Biol. 14, 5318–5325.
[100] Drugan, J.K. et al. (1996) J. Biol. Chem. 271, 233–237.
[101] Fabian, J.R. et al. (1993) Mol. Cell. Biol. 13, 7170–7179.
[102] Mischak, H. et al. (1996) Mol. Cell. Biol. 16, 5409–5418.
[103] Michaud, N.R. et al. (1995) Mol. Cell. Biol. 15, 3390–3397.
[104] Ishikawa, F. et al. (1987) Mol. Cell. Biol. 7, 1226–1232.
[105] Schultz, A.M. et al. (1985) Virology 146, 78–89.
[106] Bonner, T.I. et al. (1986) Nucleic Acids Res. 14, 1009–1015.
[107] Sithanandam, G. et al. (1990) Oncogene 5, 1775–1780.
[108] Eychene, A. et al. (1992) Oncogene 7, 1657–1660.
[109] Stephens, R.M. et al. (1992) Mol. Cell. Biol. 12, 3733–3742.

## Identification

There are three forms of *Ras*: *Hras* (the oncogene of Harvey murine sarcoma virus, Ha-MuSV), *Kras* (oncogene of Kirsten murine sarcoma virus, Ki-MuSV) and *Nras* (detected in tumours but not in retroviruses). Ha-MuSV and Ki-MuSV were first isolated by inoculating rats with the corresponding mouse leukaemia viruses (Mo-MuLV and Ki-MuLV): following the induction of leukaemia, plasma from these animals was injected into BALB/c mice which rapidly developed solid tumours (<u>ra</u>t <u>s</u>arcomas) from the effects of v-*ras*$^{H}$ or v-*ras*$^{K}$ [1–3]. BALB-MuSV, AF-1 and Rasheed-MuSV are additional murine sarcoma viruses in which acquired cellular *Ras* genes have been identified. The human homologues are *HRAS1*, *KRAS2* and *NRAS*: *HRAS2* and *KRAS1* are inactive pseudogenes.

## Related genes

The *Ras* superfamily comprises ~50 currently known *Ras*-related genes encoding GTP-binding proteins (G proteins). These include *GEM, NRASL1, NRASL2* and *NRASL3, RRAS, RhoA, RhoB* and *RhoC, Rac1* and *Rac2, Ral, Rap1A* (also called *Krev-1* or *Smg*-p21A), *RAB2*, yeast *YPT1*, human and mouse *MEL*, the *Drosophila Dras* genes, *C. elegans let-60* and *Saccharomyces cerevisiae RAS1* and *RAS2*. RAS family proteins share regions of strong homology with the α subunits of heterotrimeric G proteins and with the *E. coli* elongation factor EF-Tu.

Based on sequence homology, RAS-like proteins can be grouped in three main families: RAS proteins, RHO/RAC proteins and RAB proteins. RHO/RAC proteins are involved in the organization of the cytoskeleton and RAB proteins regulate intracellular vesicular transport. Further subgroups are typified by RAN (nuclear GTPases) and ARF (ADP-ribosylation factor)[4].

| | ***NRAS/Nras*** | ***HRAS1/Hras*** | ***KRAS2/Kras-2*** |
|---|---|---|---|
| **Nucleotides (kb)** | 32 | 4.5 | 50 |
| **Chromosome** | 1p13 | 11p15.5 | 12p12.1 |
| **Mass (kDa): expressed** | 21 | 21 | 21 |
| **Cellular location** | | Inner surface of the plasma membrane | |
| **Tissue distribution** | | Ubiquitous | |

NIH 3T3 fibroblasts contain ~500 and 1.3 fmol of GDP-RAS and GTP-RAS, respectively; cells over-expressing wild-type *Hras* contain ~700 and 20 fmol (<0.3% of RAS in GTP-RAS form); cells over-expressing activated *Hras* contain ~5000 and ~2000 fmol, respectively (29% of RAS in GTP-RAS form).

These figures correspond to ~$5 \times 10^4$ and ~$5 \times 10^5$ RAS molecules per cell for normal and HRAS over-producing cells, respectively[5–7].

## Protein function

RAS proteins are membrane bound GTPases. Normal $p21^{ras}$ (RAS) hydrolyses GTP at rates comparable with those reached by purified G proteins and exists in an equilibrium between an active (GTP-RAS) and an inactive (GDP-RAS) state. The rates of GDP release and GTP hydrolysis are increased by the actions of three classes of regulatory proteins: GTPase-activating proteins (GAPs) that increase the rate of hydrolysis of GTP, guanine nucleotide release proteins (GNRPs, also called guanine nucleotide exchange factors (GEFs) or guanine nucleotide dissociation stimulators (GDSs)) that catalyse the release of bound GDP[8] and guanine nucleotide dissociation inhibitors (GDIs) that inhibit the replacement of GDP by GTP and may also inhibit the action of GAPs[4,9–12].

Normal RAS proteins are involved in the control of cell growth and differentiation but any one of many single amino acid mutations can give rise to highly oncogenic proteins (activating point mutations in natural *RAS* oncogenes occur at codons 12, 13, 59 or 61). The action of a variety of growth factors increases the concentration of GTP-RAS in normal cells and the conformational change induced by GTP binding activates RAS, enabling it to interact with cellular target ("effector") proteins. In mammalian cells RAS activates a cascade of serine/threonine protein kinases that includes RAF1, mitogen-activated protein kinase kinase (MAPKK) and extracellular signal-regulated kinases (ERKs or MAP kinases)[13,14]. The RAS-RAF pathway is also activated by hepatitis B virus HBx protein[15]. MAPK phosphorylates and activates ELK1 and JUN and the CREB kinase/RSK2 which is a member of the $pp90^{RSK}$ family[16].

Oncogenic RAS stimulates the activity of protein kinase C and the $Na^+/H^+$ exchange protein, phospholipid metabolism and in various types of cell has been reported to activate transcription of many genes, including ornithine decarboxylase (*ODC1*), *FOS*, *JUN*, *JUNB*, *MDR1*, *MYC*, transin, heparin-binding epidermal growth factor, p9Ka/42A, *TGFα* and *TGFβ*. RAS represses transcription of the *MYOD1*, *MYOH*, Myf5, MRF4, myogenin, PDGF receptor and fibronectin genes[17].

Two families of transcription factors, AP1/ATF and ETS, have been identified as key nuclear mediators of RAS action. *Hras* expression enhances the *trans*-activating capacity of NET, an ETS-related factor that has sequence similarity to three regions of ELK1 and SAP1[18]. Sustained expression of oncogenic *Ras* can cause apoptosis when protein kinase C is downregulated: RAS-dependent cell death is prevented by the action of BCL2 which is phosphorylated in a RAS-dependent manner and associates with RAS[19].

The yeast *RAS1* and *RAS2* genes encode proteins with GTPase activity that have strong homology with human RAS proteins. The yeast proteins activate adenylate cyclase in a manner analogous to the action of $G_s$ in mammalian plasma membranes. However, there is no evidence that RAS proteins regulate adenylate cyclase in vertebrate cells, although they can substitute for RAS1 and RAS2 in yeast[20].

**RAS–protein interactions**

RAS can interact directly with RAF1, BRAF, RALGDS (RAL guanine nucleotide dissociation stimulator), RLF (RALGDS-like factor)[21], the catalytic subunit of phosphatidylinositol-3-OH kinase[22] and with protein kinase Cζ[23]. Dominant negative RAS mutants inhibit growth factor-induced production of 3′ phosphorylated phosphoinositides and transfection of *Ras*, but not *Raf*, into COS cells causes a large elevation in the level of these lipids. RAS also interacts directly with mitogen-activated protein kinase kinase kinase (MEKK1) which is the yeast counterpart of RAF[24]. Human RIN1 interacts directly with yeast Ras2p and competes with RAF1 for binding to HRAS *in vitro*[25]. RIN1 is primarily a plasma membrane protein and may be an effector or regulator of RAS. The binding of MEK1 to RAS requires RAF1 and MEK2 does not associate with RAS/RAF1 complexes[26].

The herpesvirus saimiri oncogene product STP-C488 forms a complex with RAS and increases levels of GTP-RAS, causing constitutive activation of the MAPK pathway[27].

**Regulation of RAS by GAPs**

Most cells express two GAPs, type I p120$^{GAP}$ and NF1-GAP (see ***NF1***), with similar activities. p120$^{GAP}$ acts catalytically on normal, but not transforming, RAS proteins to stimulate their relatively weak hydrolytic activity for bound GTP by 100-fold[28]. In general, GAP appears to function as an upstream regulator of normal RAS, maintaining it in an inactive, GDP-bound state. However, GAP may also be involved in coupling RAS to downstream effector proteins. Thus mutant forms of RAS lacking effector function still associate with GAP but remain bound to GTP, although such complexes are not oncogenic[29] and either RAS or GAP proteins inhibit the coupling of muscarinic receptors to atrial potassium channels[30]. p120$^{GAP}$ possesses SRC homology domains SH2 and SH3 through which it can associate with tyrosine phosphorylated proteins whereas NF1-GAP does not, indicating that different activated RAS complexes may have distinct cellular targets[31]. Three point mutations in the SH2 domain of GAP have been detected in human basal cell carcinomas: mutations in this region may therefore play a role in tumorigenesis[32].

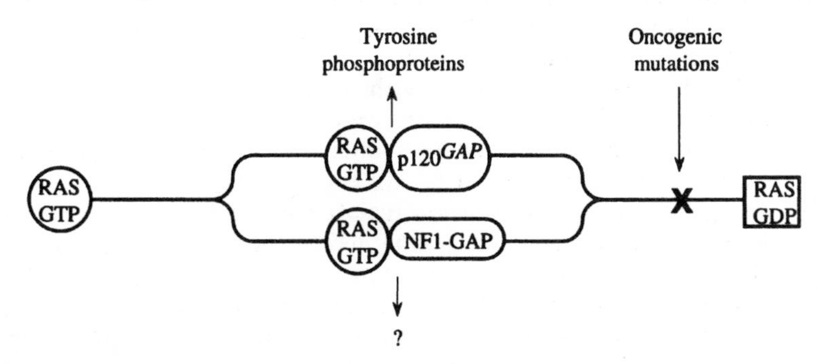


**Conserved RAS signalling pathways**

Homologues of RAS act as crucial signal transducing elements in all eukaryotic organisms that have been examined. Homologous components of pathways that couple to RAS in *C. elegans*, *Drosophila* and mammals are shown below. In mammals, phosphorylated tyrosine residues on activated tyrosine kinase receptors associate with the SH2 domain of growth factor receptor bound protein 2 (GRB2). GRB2 binds to SOS1 *via* its SH3 domains. SOS1 is thus recruited to the plasma membrane where it activates RAS[33–39].

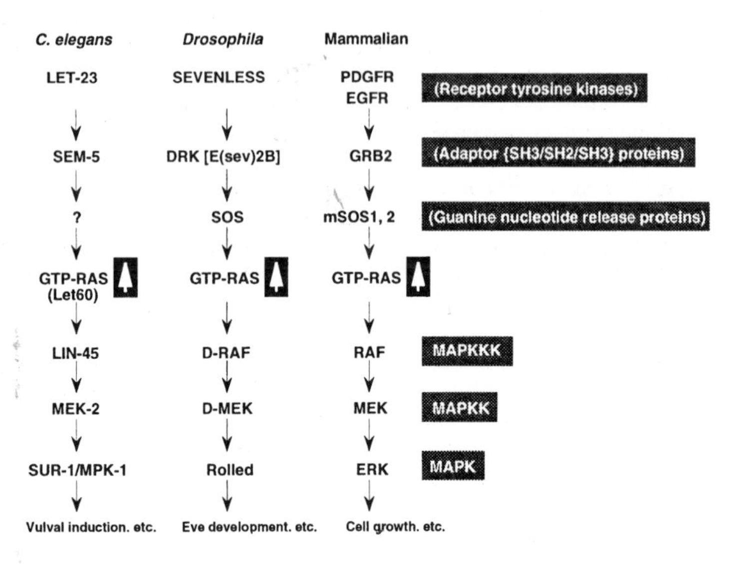


A kinase suppressor of RAS (KSR-1) has been detected in *Drosophila* and in *C. elegans* that acts either downstream or in parallel with RAS. Mutations in KSR-1 synergize with defects in RAF and MAPK homologues[40].

**RAS in cell proliferation**

The evidence indicates that, except for oncoproteins located in the cytosol (e.g. MOS), the stimulation of proliferation involves RAS. Thus, insulin, together with bombesin, PDGF or EGF form potent comitogenic combinations for quiescent cells *in vitro*. PDGF or EGF cause activation of RAS (i.e. increase the cellular GTP-RAS concentration) and the tyrosine phosphorylation of GAP. PDGF causes ~10% of total cellular GAP to associate with the PDGFR[41,42], together with PLC$\gamma$, PtdIns 3-kinase and RAF1. Ligand-activated association of GAP with receptors may thus be a mechanism for switching on the mitogenic pathway.

The question of whether RAS functions before or after PtdIns 3-kinase in growth factor-dependent pathways is controversial. In CHO cells, PtdIns 3-kinase appears to function upstream of both RAF1 and RAS. Thus, insulin stimulation of the serum response element (SRE) of the *Fos* promoter is inhibited by expression of dominant negative *Ras*- or *Raf*-encoding plasmids[43]. Furthermore, PDGF receptors mutated in the PtdIns 3-kinase binding sites ($Tyr^{708}$, $Tyr^{719}$) are unable to stimulate RAS whereas GAP binding site mutants ($Tyr^{739}$) do so, suggesting an important role for PtdIns 3-kinase or a protein binding to the same site in PDGF-stimulated RAS activation[44]. However, other studies indicate that PDGFRs that bind PtdIns 3-kinase or PtdIns 3-kinase and SHPTP2 can fully activate PtdIns 3-kinase, whereas receptors that associate with PtdIns 3-kinase and either RAS-GAP or PLC$\gamma$ are defective in their capacity to cause accumulation of PtdIns(3,4)$P_2$ and PtdIns(3,4,5)$P_3$ [45]. A mutant PDGFR lacking its SHPTP2 binding site displays reduced GRB2 binding and hence RAS activation [46] and there is a strong correlation between the activation of PtdIns 3-kinase and that of RAS. Thus activation of PtdIns 3-kinase by the PDGR requires both binding of the enzyme to the receptor and accumulation of RAS-GTP and this activation is negatively modulated by PLC$\gamma$ and RAS-GAP.

In fibroblastic cells an important target for PtdIns 3-kinase appears to be the serine/theonine protein kinases AKT and AKT2 which are activated by PDGF *via* a pathway dependent on the PDGFR$\beta$ PtdIns 3-kinase binding tyrosines (Tyr$^{740}$ and Tyr$^{751}$) and blocked by the PtdIns 3-kinase-specific inhibitor wortmannin and the dominant inhibitory Asn$^{17}$Ras [47].

Neither insulin nor bombesin elevate the GTP-RAS concentration in normal cells, although insulin causes tyrosine phosphorylation of the SRC tyrosine kinase substrate SHC which then binds to GRB2/SOS. Furthermore, in Swiss 3T3 fibroblasts, microinjection of anti-RAS antibody inhibits DNA synthesis stimulated by PDGF or EGF and also by insulin and bombesin, suggesting that RAS is involved in the insulin/bombesin-stimulated pathway. Oncogenic RAS also activates expression of cyclin D1, indicating a link between external signals transduced *via* RAS and the cell cycle machinery that regulates progression through $G_1$ [48].

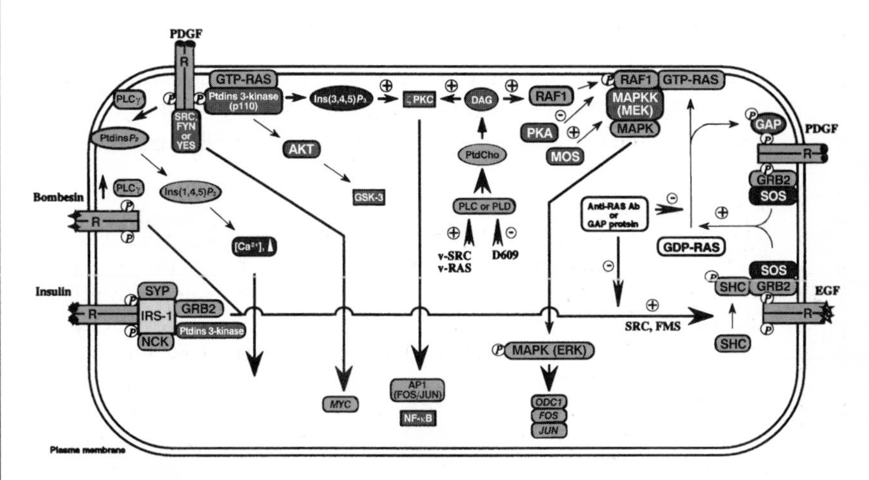


PtdIns$P_2$: phosphatidylinositol 4,5-bisphosphate; PtdCho: phosphatidylcholine; PLC$\gamma$: phospholipase C$\gamma$; PKA: cAMP-dependent protein kinase; PKC: protein kinase C; DAG: diacylglycerol; GSK-3: glycogen synthase kinase 3; $[Ca^{2+}]_i$: intracellular free $Ca^{2+}$ concentration.

## Cancer

Activating mutations in *RAS* oncogenes occur in a wide variety of human tumours. The overall incidence is only between 10% and 15% but is as high as 95% in pancreatic carcinomas [49–51]. Individual *RAS* genes are commonly associated with specific tumours, for example, *KRAS2* with cancers of the lung, in which *HRAS* mutations are rare [52,53], the colon or pancreas, *NRAS* with acute myelogenous leukaemia (AML). However, there is no specificity in thyroid tumours and in thyroid adenomas and carcinomas mutations in all three genes (*HRAS, KRAS2* and *NRAS*) may occur within one tumour. Simultaneous mutations in *KRAS2* and *NRAS* have also been detected in multiple myeloma [54]. The highly polymorphic *HRAS1* minisatellite locus (unstable repetitive DNA sequences) just downstream from the *HRAS1* gene consists of four common progenitor alleles and several dozen rare alleles, which apparently derive from mutations of the progenitors. Mutant alleles of the *HRAS1* minisatellite locus represent a major risk factor for common types of cancer (~10% of breast, colorectum, and bladder cancers) [55]. Mutations in *KRAS2* have been detected in ~40% of pancreatic cancers and these mutations appear to correlate strongly with the presence of microsatellite instability [56]. In non-small cell lung cancer hypomethylation of CCGG sites in the 3′ region and allelic loss of *HRAS* can occur, even in the absence of mutations at codons 12, 13 or 61 [57].

These observations indicate the probable importance of *RAS* oncogenes in neoplasia although there is as yet no discernible pattern to the expression of activated *RAS* genes and they occur in benign as well as malignant tumours.

Over-expression of GRB2 has been detected in human breast cancer cell lines [58].

**Transgenic animals**

The expression of *Nras*, v-*Hras* or human *HRAS1* causes non-neoplastic proliferation and malignant tumours in a tissue-specific manner. Coexpression of v-*Hras* and *Myc* causes a synergistic increase in the initiation of tumours, indicating that the expression of an activated *Ras* gene alone is not sufficient to transform differentiated cells *in vivo* [59–61].

Deletion of p120$^{GAP}$ in mice stimulates aberrant RAS signalling of downstream pathways, affecting the ability of endothelial cells to organize into a highly vascularized network and causing extensive neuronal cell death. Loss of GAP synergizes with the deletion of *Nf-1*, indicating that RAS-GAP and neurofibromin act together to regulate RAS activity during embryonic development [62].

**In animals**

In tumours chemically (e.g. by nitrosomethylurea (NMU), dimethylbenzanthracene (DMBA) or *N*-methyl-*N'*-nitro-*N*-nitrosoguanidine (MNNG)) or physically (e.g. by X-ray treatment) induced in rodents the frequency of *Ras* mutations is usually ~70%, commonly arising from point mutations at codons 12 or 61 [63,64]. The expression of oncogenic *Hras in vivo* can rapidly confer metastatic potential on benign rat mammary cells [65].

***In vitro***

Normal fibroblasts are not transformed either by cellular or retroviral *Ras* oncogenes [66,67] and mutant *RAS* from human tumours only transforms transfected primary fibroblasts when supplemented with immortalizing oncogenes such as *Myc* [68], v-*myc*, *Nmyc* [69], adenovirus E1A [70] mutant *P53* or polyoma large T antigen. E1B or polyoma middle T antigen can replace *Ras*.

However, NIH 3T3 fibroblasts are transformed by over-expression of normal RAS proteins [71–73]. Transformation of NIH 3T3 cells by *Hras* is greatly enhanced by the expression of *Raf-1* and RAF-1 kinase appears to be rate limiting for *Ras* transformation [74]. *Hras* transformation is inhibited by the expression of normal pRb [75]. Sustained over-expression of ornithine decarboxylase (ODC) can transform NIH 3T3 cells and rat fibroblasts and transformation by ODC, *Hras* or v-*src* causes tyrosine phosphorylation of the CRK-associated substrate $pp130^{CAS}$ which may thus act as a downstream tyrosine kinase activated *via* ODC [76]. *Ras* oncogenes efficiently transform erythroid [77], myeloid [78] and murine mast cells [79].

Transformation by either *Kras* or *Hras* oncogenes of a variety of rodent cells increases TGF$\alpha$ and TGF$\beta$ secretion. TGF$\alpha$ may function as an autocrine growth factor for *Ras*-transformed malignant cells and TGF$\beta$ may promote metastasis [80]. In PC-12 cells oncogenic RAS activates both promoters of the TGF$\beta_1$ gene and appears to mediate the effects of NGF, FGF and EGF on these cells [81].

Oncogenic *Hras* or *Kras* (or *Raf*) markedly upregulate VEGF/VPF expression in transformed cells, suggesting that RAS may contribute to the growth of solid tumours by indirectly promoting angiogenesis [82].

The evidence from cellular studies of *Ras* is consistent with the pathology and epidemiology of spontaneously arising cancer in humans, suggesting that transformation is at least a two-stage process and that one oncogene may be needed for immortalization and another for transformation. However, massive over-expression of a single gene (e.g. v-*myc* or *Ras*) can probably over-ride this distinction.

*Rap1A/Krev-1* encodes a RAS-related protein that suppresses the activity of v-*Kras-2* and causes transformed NIH 3T3 cells to revert to a normal phenotype [83]. RAP1 has a higher affinity for GAP than RAS and may also interact specifically with effector molecules. Reversion of the transformed phenotype has also been demonstrated on expression of tropomyosin 1 [84] or of a mutated form of gelsolin, the actin regulatory protein, which may inhibit PtdIns$P_2$ hydrolysis [85]. Reversion of *Ras* transformation of fibroblasts also occurs following the over-expression of serum response factor (SRF) [86]. Trichostatin A (TSA) inhibits both oncogenic *Ras*- and NGF-induced neurite outgrowth of PC-12 cells and causes reversion of transformed NIH 3T3 cells to a normal morphology [87]. Reversion of an adreno-cortical tumour cell line by $p120^{GAP}$ or by a *trans*-dominant negative mutant of *Jun* is correlated with an increase in levels of DNA methylation and DNA methyltransferase (MeTase) activity, indicating that RAS and AP1 regulate the expression of the DNA MeTase promoter [88].

The GTPase Rac1, a key component in the reorganization of the actin cytoskeleton induced by growth factors or oncogenic RAS, is essential for RAS transformation [89]. Thus, the expression of *Rac1* together with *RhoA* enhances *Ras* transformation and dominant negative mutants of RAC1 ($Asn^{17}$) or RHOA ($Asn^{19}$) inhibit transformation by oncogenic RAS but not by RAF targeted to the plasma membrane [90]. Thus, RAS activation of RAC1 and RHOA, together with activation of the RAF/MAPK pathway, appears to be necessary for full RAS transformation in some cell types, although in epithelial cells, for example, RAF-independent RAS-mediated pathways are sufficient to cause transformation [91].

RHOB appears necessary for transformation of 3T3 fibroblasts by oncogenic *Ras* but not *Raf*[92]. The observation that p190 has selective affinity for RHO rather than RAC proteins and regulates RHO activity *in vitro* suggests the pathways shown below for RAS signal transduction.

In rat aortic smooth muscle cells, angiotensin II stimulates the formation of GTP-RAS and the tyrosine phosphorylation of p120 RAS-GAP and p190 RHO-GAP by a mechanism that appears to require SRC[93]. In human aortic smooth muscle cells lactosylceramide stimulates the formation of GTP-RAS[94].

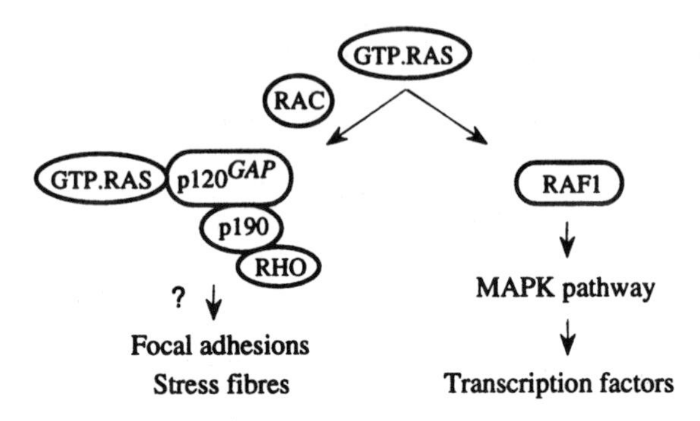


*NRAS* transformed human PA-1 teratocarcinoma cells have six-fold enhanced AP2 mRNA levels which results in self-interference of AP2-mediated *trans*-activation[95].

Human cell lines transformed by the presence of activated *NRAS* or *KRAS2* revert to a largely normal *in vitro* morphology upon loss of the mutant gene. However, these cells remain tumorigenic in nude mice, suggesting that, although *RAS* mutations can activate the transformed phenotype, their continued presence is not necessary for tumour development[96].

## Gene structure of *NRAS*

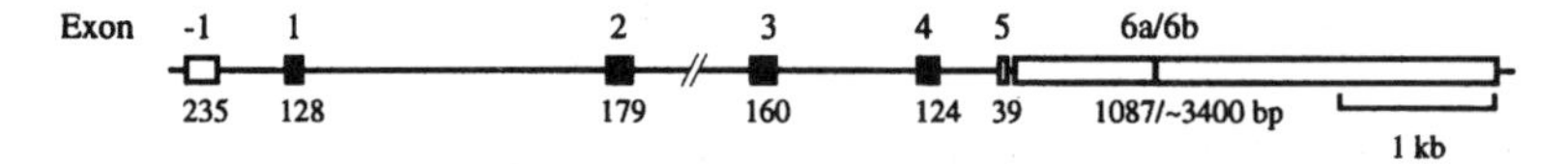


Exons 1–4 encode 37, 60, 53 and 39 amino acids, respectively. The sizes of the corresponding exons in *HRAS* are 110, 230, 160 and 117 bp and in *KRAS2* 122, 179, 160 and 124 bp, respectively. *NRAS* contains two 3′ non-coding exons (5 and 6) not present in *HRAS1* or *KRAS2*. *KRAS2* has alternative fourth exons (4A and 4B) separated by ~5600 bp that encode 39 and 38 amino acids, respectively. Codons 12 and 13 of wild-type *Hras* constitute a strong pause site for *in vitro* DNA synthesis catalysed by DNA polymerase α[97]. Pausing at these codons is abolished by mutation to an activated oncogene.

## Transcriptional regulation

The promoter regions of *RAS* genes do not contain TATA or CAT boxes but have multiple GC boxes and utilize more than one transcription start site[98]. The human *NRAS* promoter contains consensus binding sites for CREB/ATF, AP1, AP2, MYB, E4TF1, SP1 and MLTF/MYC[99].

## Protein structure

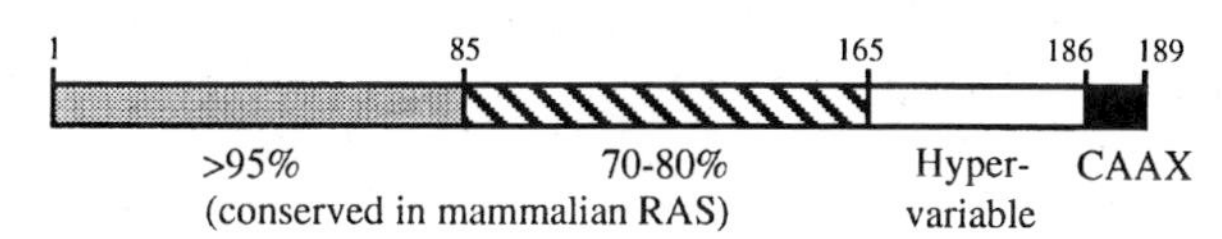


Several different wild-type and oncogenic RAS complexes have been crystallized to provide the first atomic descriptions of proto-oncogenes and oncogenes. RAS comprises a central, six-stranded $\beta$ sheet and five helices, two of which ($\alpha2$ and $\alpha3$) lie below the plane defined by the $\beta$ sheet. Of the ten loops in the protein, L1 contains $Gly^{12}$ (the most frequent site of mutation in human tumours), L2 includes the residues believed to interact with the effector and L4 contains $Gln^{61}$ [100].

The sequence of Ha-MuSV HRAS differs from human HRAS1 only at positions 12 (lysine for glycine) and 143 (lysine for glutamate): that of Ki-MuSV KRAS differs from human KRAS 2A in five positions ($Gly^{12} \rightarrow$ Ser, $Glu^{37} \rightarrow$ Gln, $Ala^{59} \rightarrow$ Thr, $Asp^{132} \rightarrow$ Glu, $Ile^{187} \rightarrow$ Val).

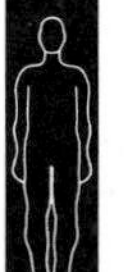

## Amino acid sequence of human NRAS

1 MTEYKLVVVG AGGVGKSALT IQLIQNHFVD EYDPTIEDSY RKQVVIDGET
51 CLLDILDTAG QEEYSAMRDQ YMRTGEGFLC VFAINNSKSF ADINLYREQI
101 KRVKDSDDVP MVLVGNKCDL PTRTVDTKQA HELAKSYGIP FIETSAKTRQ
151 GVEDAFYTLV REIRQYRMKK LNSSDDGTQG CMGLPCVVM (189)

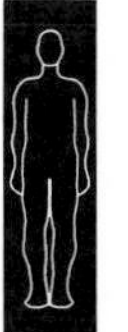

## Amino acid sequence of human HRAS1

1 MTEYKLVVVG AGGVGKSALT IQLIQNHFVD EYDPTIEDSY RKQVVIDGET
51 CLLDILDTAG QEEYSAMRDQ YMRTGEGFLC VFAINNTKSF EDIHQYREQI
101 KRVKDSDDVP MVLVGNKCDL AARTVESRQA QDLARSYGIP YIETSAKTRQ
151 GVEDAFYTLV REIRQHKLRK LNPPDESGPG CMSCKCVLS (189)

Underlined residues differ from NRAS.

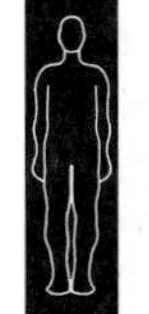

## Amino acid sequence of human KRAS 2A

1 MTEYKLVVVG AGGVGKSALT IQLIQNHFVD EYDPTIEDSY RKQVVIDGET
51 CLLDILDTAG QEEYSAMRDQ YMRTGEGFLC VFAINNTKSF EDIHHYREQI
101 KRVKDSEDVP MVLVGNKCDL PSRTVDTKQA QDLARSYGIP FIETSAKTRQ
151 RVEDAFYTLV REIRQYRLKK ISKEEKTPGC VKIKKCIIM (189)

Underlined residues differ from NRAS.

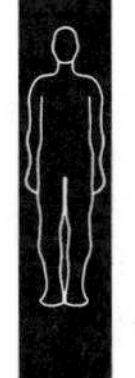

## Amino acid sequence of human KRAS 2B

1 MTEYKLVVVG AGGVGKSALT IQLIQNHFVD EYDPTIEDSY RKQVVIDGET
51 CLLDILDTAG QEEYSAMRDQ YMRTGEGFLC VFAINNTKSF EDIHHYREQI
101 KRVKDSEDVP MVLVGNKCDL PSRTVDTKQA QDLARSYGIP FIETSAKTRQ
151 GVDDAFYTLV REIRKHKEKM SKDGKKKKKK SKTKCVIM (188)

Underlined residues differ from KRAS 2A.
175–180 Polybasic region involved in plasma membrane targeting.

### Domain structure

| Residues | Description |
|---|---|
| 5–63, 77–92, 109–123, 139–165 and 186–189 | Non-contiguous domains essential for RAS transforming activity |
| 10–16, 57–62 and 116–119 | Highly conserved nucleotide binding regions |
| 32–40 | "Effector domain" or Switch I region: substitutions in this region reduce the biological effect of RAS proteins in both mammalian and yeast cells but do not affect GTP binding or hydrolysis. Essential for stimulation of GTPase activity by GAP. Mutations in this region reduce the direct interaction that occurs between RAS and RAF1 [101,102]. Specifically, mutations at amino acids 31, 34, 35, 38, 57 and 59 block RAF1 binding and mutations at 26, 29, 39, 40, 41, 44, 45, 56 and 58 reduce the interaction [103] |
| 61–65 | Confer RAS-GAP sensitivity on RAS protein |
| 60–72 | Switch II region: together with Switch I forms the two domains that undergo large conformational changes upon exchange of bound GDP for GTP. Point mutations at 64 or 65 combined with mutation at 75 or 71 prevents interaction of nucleotide-free RAS with GNRPs: mutations at 75 and 71 increase by two-fold the affinity for BRAF and thus for MEK1: mutations at 64 with mutations at either 65 or 71 block interaction with PtdIns 3-kinase or neurofibromin, though not with BRAF and RAF1 [104] |
| 96–110 | Region mediating association with JUN and JNK [105] |
| 115–126 | Region mediating asociation with JNK [105] |
| 165–184 | Hypervariable region |
| 186–189 | CAAX box: $Cys^{186}$ essential for transforming activity [106] |
| 12, 13, 59, 61 | Sites of naturally occurring activating point mutations that inhibit GTP hydrolysis. Oncogenic mutations therefore enable the GTP-RAS complex to remain in an active form that is presumed to stimulate a growth-promoting event |
| 63, 116, 117, 119 and 146 | Activating point mutations created by *in vitro* mutagenesis |
| 164 | Mutagenesis causes loss of GTP binding |
| 177 | HRAS1 serine phosphorylation site |
| 186 | Farnesyl alkylation of NRAS, KRAS 2A, HRAS1: palmitoylation of Cys residues (HRAS1 $Cys^{184}$) in the hypervariable region |
| 185 | Farnesyl alkylation of KRAS 2B. Inhibition of farnesyltransferase activity selectively blocks *Ras*-dependent transformation *in vitro* [107,108] |

## Database accession numbers

| | PIR | SWISSPROT | EMBL/GENBANK | REFERENCES |
|---|---|---|---|---|
| Human *NRAS* | A01359 | P01111 | X02751 | 109 |
| | A21700 | | X00642–X00645 | 110 |
| | | | L00040–L00043 | 111 |
| | | | M10055 | 112 |
| | | | K03211 | 113 |
| Human *KRAS2A* | A01364 | P01116 | L00045–L00049 | 114 |
| Human *KRAS2B* | A01367 | P01118 | K01519 | 115 |
| | | | K01520 | 116 |
| | | | X01669 | 117 |
| | | | X02825 | 118 |
| | | | K03209 | 119 |
| | | | K03210 | 120 |
| Human *HRAS1* | A01360 | P01112 | V00574 | 121–124 |

## *References*


[1] Harvey, J.J. and East, J. (1971) Int. Rev. Exp. Pathol. 10, 265–360.
[2] Kirsten, W.H. and Mayer, C.A. (1967) J. Natl. Cancer Inst. 39, 311–335.
[3] Harvey, J.J. (1964) Nature 204, 1104–1105.
[4] **Boguski, M.S. and McCormick, F. (1993) Nature 366, 643–654.**
[5] Scheele, J.S. et al. (1995) Proc. Natl Acad. Sci. USA 92, 1097–1100.
[6] Hand, P.H. et al. (1987) Biochim. Biophys. Acta 908, 131–142.
[7] Miller, A.C. et al. (1993) Mol. Cell. Biol. 13, 4416–4422.
[8] Schweighoffer, F. et al. (1993) Oncogene 8, 1477–1485.
[9] **Bourne, H.R. et al. (1990) Nature 348, 125–132.**
[10] **Downward, J. (1992) BioEssays 14, 177–184.**
[11] **Santos, E. and Nebreda, A.R. (1989) FASEB J. 3, 2151–2163.**
[12] Polakis, P. and McCormick, F. (1993) J. Biol. Chem. 268, 9157–9160.
[13] De Vries-Smits, A.M.M. et al. (1992) Nature 357, 602–604.
[14] Cook, S.J. et al. (1993) EMBO J. 12, 3475–3485.
[15] Doria, M. et al. (1995) EMBO J. 14, 4747–4757.
[16] Xing, J. et al. (1996) Science 273, 959–963.
[17] Sistonen, L. et al. (1989) EMBO J. 8, 815–822.
[18] Giovane, A. et al. (1994) Genes Dev. 8, 1502–1513.
[19] Chen, C.Y. and Faller, D.V. (1995) Oncogene 11, 1487–1498.
[20] **Broach, J.R. and Deschenes, R.J. (1990) Adv. Cancer Res. 54, 79–139.**
[21] Wolthuis, R.M.F. et al. (1996) Oncogene 13, 353–362.
[22] Rodriguez-Viciana, P. et al. (1994) Nature 370, 527–532.
[23] Diaz-Merco, M.T. et al. (1994) J. Biol. Chem. 269, 31706–31710.
[24] Russell, M. et al. (1995) J. Biol. Chem. 270, 11757–11760.
[25] Han, L. and Colicelli, J. (1995) Mol. Cell. Biol. 15, 1318–1323.
[26] Jelinek, T. et al. (1994) Mol. Cell. Biol. 14, 8212–8218.
[27] Jung, J.U. and Desrosiers, R.C. (1995) Mol. Cell. Biol. 15, 6506–6512.
[28] Marshall, M.S. et al. (1989) EMBO J. 1989, 8, 1105–1110.
[29] Krengel, U. et al. (1990) Cell 62, 539–548.
[30] Yatani, A. et al. (1990) Cell 61, 769–776.
[31] McCormick, F. et al. (1992) Philos. Trans. R. Soc. Lond. (Biol.) 336, 43–48.
[32] Friedman, E. et al. (1993) Nature Genet. 5, 242–247.
[33] Lowenstein, E.J. et al. (1992) Cell 70, 431–442.
[34] Gale, N.W. et al. (1993) Nature 363, 88–92.
[35] Rozakis-Adcock, M. et al. (1993) Nature 363, 83–88.

[36] Liu, B.X. et al. (1993) Oncogene 8, 3081–3084.
[37] Burgering, B.M.Th. et al. (1993) EMBO J. 12, 4211–4220.
[38] Egan, S.E. et al. (1993) Nature 363, 45–51.
[39] Skolnik, E.Y. et al. (1993) EMBO J. 12, 1929–1936.
[40] **Downward, J. (1995) Cell 83, 831–834.**
[41] Molloy, C.J. et al. (1989) Nature 342, 711–714.
[42] Kazlauskas, A. et al. (1990) Science 247, 1578–1581.
[43] Yamauchi, K. et al. (1993) J. Biol. Chem. 268, 14597–14600.
[44] Satoh, T. et al. (1993) Mol. Cell. Biol. 13, 3706–3713.
[45] Klinghoffer, R.A. et al. (1996) Mol. Cell. Biol. 16, 5905–5914.
[46] Bennett, A.M. et al. (1994) Proc. Natl Acad. Sci. USA 91, 7335–7339.
[47] Franke, T.F. et al. (1995) Cell 81, 727–736.
[48] Winston, J.T. et al. (1996) Oncogene 12, 127–134.
[49] **Bos, J.L. (1988) Mutat. Res. 195, 255–271.**
[50] Almoguera, C. et al. (1988) Cell 53, 549–554.
[51] Rochlitz, C.F. et al. (1989) Cancer Res. 49, 357–360.
[52] Vachtenheim, J. et al. (1994) Clin. Cancer Res. 1, 359–365.
[53] Rodenhuis, S. and Slebos, R.J.C. (1992) Cancer Res. 52, 2665s–2669s.
[54] Portier, M. et al. (1992) Oncogene 7, 2539–2543.
[55] Krontiris, T.G. et al. (1993) N. Engl. J. Med. 329, 517–523.
[56] Brentnall, T.A. et al. (1995) Cancer Res. 55, 4264–4267.
[57] Vachtenheim, J. et al. (1994) Cancer Res. 54, 1145–1148.
[58] Daly, R.J. et al. (1994) Oncogene 9, 2723–2727.
[59] Sinn, E. et al. (1987) Cell 49, 465–475.
[60] Andres, A.-C. et al. (1987) Proc. Natl Acad. Sci. USA 84, 1299–1303.
[61] Mangues, R. et al. (1992) Oncogene 7, 2073–2076.
[62] Henkemeyer, M. et al. (1995) Nature 377, 695–701.
[63] **Barbacid, M. (1987) Annu. Rev. Biochem. 56, 779–827.**
[64] Nakazawa, H. et al. (1992) Oncogene 7, 2295–2301.
[65] Nicolson, G.L. et al. (1992) Oncogene 7, 1127–1135.
[66] Newbold, R.F. and Overell, R.W. (1983) Nature 304, 648–651.
[67] Sager, R. et al. (1983) Proc. Natl Acad. Sci. USA 87, 7601–7605.
[68] Land, H. et al. (1983) Nature 304, 596–602.
[69] Schwab, M. et al. (1985) Nature 316, 160–162.
[70] Ruley, H.E. (1983) Nature 304, 602–606.
[71] Ricketts, M.H. and Levinson, A.D. (1988) Mol. Cell. Biol. 8, 1460–1468.
[72] Reynolds, V.L. et al. (1987) Oncogene 1, 323–330.
[73] Doppler, W. et al. (1987) Gene, 54, 147–153.
[74] Cuadrado, A. et al. (1993) Oncogene 8, 2443–2448.
[75] Kivinen, L. et al. (1993) Oncogene 8, 2703–2711.
[76] Auvinen, M. et al. (1995) Mol. Cell. Biol. 15, 6513–6525.
[77] Hankins, W.D. and Scolnick, E.M. (1981) Cell 26, 91–97.
[78] Pierce, J.H. and Aaronson, S.A. (1985) Mol. Cell. Biol. 5, 667–674.
[79] Rein, A. et al. (1985) Mol. Cell. Biol. 5, 2257–2264.
[80] Colletta, G. et al. (1991) Oncogene 6, 583–587.
[81] Cosgaya, J.M. and Aranda, A. (1996) Oncogene 12, 2651–2660.
[82] Rak, J. et al. (1995) Cancer Res. 55, 4575–4580.
[83] Beranger, F. et al. (1991) Proc. Natl Acad. Sci. USA 88, 1606–1610.
[84] Braverman, R.H. et al (1996) Oncogene 13, 537–545.
[85] Mullauer, L. et al. (1993) Oncogene 8, 2531–2536.

[86] Kim et al. (1994) J. Biol. Chem. 269, 13740–13743.
[87] Futamura, M. et al. (1995) Oncogene 10, 1119–1123.
[88] MacLeod, A.R. et al. (1995) J. Biol. Chem. 270, 11327–11337.
[89] Qui, R.-G. et al. (1995) Nature 374, 457–459.
[90] Khosravi-Far, R. (1995) Mol. Cell. Biol. 15, 6443–6453.
[91] Khosravi-Far, R. (1996) Mol. Cell. Biol. 16, 3923–3933.
[92] Prendergast et al. (1995) Oncogene 10, 2289–2296.
[93] Schieffer, B. et al. (1996) J. Biol. Chem. 271, 10329–10333.
[94] Bhunia, A.K. et al. (1996) J. Biol. Chem. 271, 10660–10666.
[95] Kannan, P. et al. (1994) Genes Dev. 8, 1258–1269.
[96] Plattner, R. et al. (1996) Proc. Natl Acad. Sci. USA 93, 6665–6670.
[97] Hoffmann, J.-S. et al. (1993) Cancer Res. 53, 2895–2900.
[98] Hoffman, E.K. et al. (1987) Mol. Cell. Biol. 7, 2592–2596.
[99] Thorn, J.T. et al. (1991) Oncogene 6, 1843–1850.
[100] **Wittinghofer, F. (1992) Semin. Cancer Biol. 3, 189–198.**
[101] Warne, P.H. et al. (1993) Nature 364, 352–355.
[102] Zhang, X. et al. (1993) Nature 364, 308–313.
[103] Shirouzu, M. et al. (1994) Oncogene 9, 2153–2157.
[104] Moodie, S.A. et al. (1995) Oncogene 11, 447–454.
[105] Adler, V. et al. (1995) Proc. Natl Acad. Sci. USA 92, 10585–10589.
[106] Willumsen, B.M. et al. (1996) Oncogene 13, 1901–1909.
[107] Kohl, N.E. et al. (1993) Science 260, 1934–1937.
[108] James, G.L. et al. (1993) Science 260, 1937–1942.
[109] Taparowsky, E. et al. (1983) Cell 34, 581–586.
[110] Brown, R. et al. (1984) EMBO J. 3, 1321–1326.
[111] Yuasa, Y. et al. (1984) Proc. Natl Acad. Sci. USA 81, 3670–3674
[112] Gambke, C. et al. (1985) Proc. Natl Acad. Sci. USA 82, 879–882.
[113] Hall, A. and Brown, R. (1985) Nucleic Acids Res. 13, 5255–5268.
[114] McGrath, J.P. et al. (1983) Nature 304, 501–506.
[115] Shimizu, K. et al. (1983) Nature 304, 497–500.
[116] Capon, D.J. et al. (1983) Nature 304, 507–513.
[117] McCoy, M.S. et al. (1984) Mol. Cell. Biol. 4, 1577–1582.
[118] Nakano, H. et al. (1984) Proc. Natl Acad. Sci. USA 81, 71–75.
[119] Hirai, H. et al. (1985) Biochem. Biophys. Res. Commun. 127, 68–174.
[120] Yamamoto, F. and Perucho, M. (1988) Oncogene Res. 3, 123–138.
[121] Capon, D.J. et al. (1983) Nature 302, 33–37.
[122] Reddy, E.P. (1983) Science 220, 1061–1063.
[123] Tabin, C.J. et al. (1982) Nature 300, 143–149.
[124] Sekiya, T. et al. (1984) Proc. Natl Acad. Sci. USA 81, 4771–4775.

## Identification

v-*rel* is the oncogene of avian reticuloendotheliosis virus strain T (REV-T), originally isolated from turkeys that had contracted lymphoid leukosis [1,2]. *REL* was detected by screening DNA with v-*rel* probes [3].

## Related genes

*REL* genes are members of the nuclear factor κB (NF-κB) transcription factor family [4–6]. REL family proteins have a conserved N-terminal REL homology domain of ~300 amino acids. This NF-κB/REL/dorsal (NRD) domain represents the minimal region for DNA binding and dimerization. The 30 N-terminal amino acids confer differential DNA binding affinities between family members.

### Nomenclature of REL family members [7]

| *Protein* | *Human gene* | *Non-human homologue* | *Former names* | *References* |
|---|---|---|---|---|
| NF-κB complex (p50/p65) | *NFKB1/RELA* | | | |
| **Class I:** | | | | |
| NF-κB1 (p105/p50) | *NFKB1* | *Nfkb-1* | Human κ binding factor (KBF1), p110, EBP-1 | 8 |
| NF-κB2 (p100/p52) | *NFKB2* | *Nfkb-2* | p100, p97, p98/p55, p50B, p49, NF-κB, H2TF1, LYT10 | 9 |
| **Class II:** | | | | |
| REL | *REL* | *Rel* | | |
| RELA/p65 | *RELA* | *RelA* | p65 NF-κB | |
| RELB | *RELB* | *RelB* | *I-Rel* | 10 |
| Dorsal | | *Dorsal* | | |
| Dif, Cif | | *Dif* | | |
| IκB-α | *IKBA* | *Ikba* | p37, MAD-3, pp40, RL/IF-1, ECL-6 | 11,12 |
| IκB-β | *IKBB* | *Ikbb* | p43 | |
| IκB-γ | *NFKB1* | *Nfkb-1* | p70, p105/pdI, C-terminal portion of p105 | 13 |
| BCL3 | *BCL3* | *Bcl-3* | 46–56 kDa | |
| Cactus | | *Cactus* | | |

Class I REL family proteins are synthesized as precursor proteins (p105 and p100 for NF-κ1 and NF-κB2, respectively) that undergo proteolytic cleavage to yield mature DNA binding proteins containing essentially the N-terminal NRD domain and C-terminal inhibitory factors that contain multiple copies of the ankyrin motif also found in the IκB family.

Class II REL proteins include REL, RELA/p65, RELB and the *dorsal* gene product of *Drosophila melanogaster*. They are not post-translationally processed and contain C-terminal *trans*-activation domains [14].

| | *REL/Rel* | v-*rel* |
|---|---|---|
| **Nucleotides (kb)** | >24 (chicken, turkey) | 1.4 (REV-T) |
| **Chromosome** | 2p13–p12 | |
| **Mass (kDa): predicted** | 65 | 56 |
| **expressed** | 68 | pp59$^{v\text{-}rel}$ |

**Cellular location**
REL is a transcription factor the activity of which is regulated by its subcellular location[15]. Thus REL may be detected in either the cytoplasm or nucleus depending on the cell type and state of the cell. NF-κB1 is ubiquitously expressed and occurs in both homodimeric form and in heterodimers with other REL proteins. In chick embryo fibroblasts (CEFs) REL is cytoplasmic whereas v-REL is nuclear in non-transformed CEFs[16]. In transformed avian and murine cells v-REL is primarily cytoplasmic and can form complexes with REL, p50/p105, p52/p100, IκBα and NF-κB[10,17–19]. However, v-REL with an added SV40 T antigen nuclear localization signal also transforms these cells, when the protein is nuclear. In transformed lymphoid cells the majority of v-REL (90%) is cytoplasmic and complexed with IκB, the minority is nuclear and associated with the p50 precursor encoded by NF-κB1. In human lymphoid cells complexes of REL and NF-κB1 p105 are detectable in both the cytoplasm and the nucleus[20] and these findings suggest that complex formation may unmask a p105 nuclear localization signal.

**Tissue distribution**
In humans high concentrations of *REL* mRNA occur in relatively mature lymphocytes[21]. In chickens *Rel* mRNA is mainly in haematopoietic cells[22] but there is ubiquitous, low level embryonic expression[23]. *Rel* mRNA expression is depressed in immature thymocytes and may therefore play a role in lymphocyte differentiation, in contrast to the evidence for *Myb* and *Ets*[21].

## Protein function

The complex originally described as NF-κB is a transcription factor composed of two DNA-binding subunits (NF-κB1 and RELA) which contact specific decameric sequences (κB sequences: NGGNN$^{A}/_{T}$TTCC) as a heterodimer. NF-κB appears to be ubiquitously expressed, is activated in many different cell types in response to a variety of primary or secondary pathogenic stimuli and controls gene expression in a cell-specific manner (acute phase proteins in liver, cytokines and cell surface receptors in lymphoid cells). It is probable therefore that NF-κB acts synergistically with cell-specific transcription factors. NF-κB is generally present in an inactive, cytoplasm form bound to inhibitory IκB proteins. Cell stimulation (by cytokines, mitogenic lectins, phorbol esters or some viral gene products) causes dissociation of NF-κB from IκB and translocation to the nucleus. Activation is rapid, e.g. NF-κB

DNA binding activity can be detected in the nucleus of HL60 cells within 1 min of their exposure to tumour necrosis factor[24]. REL is tyrosine-phosphorylated in T cells following PHA or TPA treatment and in neutrophils in response to granulocyte colony-stimulating factor (G-CSF)[25]. The activity of IκB-α is regulated by phosphorylation, hypophosphorylated IκB-α acting *via* RELA/p65 to prevent NF-κB translocation to the nucleus[26]. Phosphorylation of the N-terminus of IκB-α (e.g. by protein kinase C) disrupts its interaction with NF-κB[27–29] and targets it for ubiquitin-mediated degradation[30]. IκB-β is structurally related to IκB-α and both appear to interact with the same REL family members. However, IκB-α transcription is regulated by NF-κB, whereas that of IκB-β is not. Furthermore, although in T lymphocytes at least, unphosphorylated IκB-α associates with REL, IκB-β/REL complex formation requires the phosphorylation of the IκB-β PEST domain at consensus casein kinase II sites[31].

REL family homodimers and heterodimers show distinct DNA-binding specificities and affinities. Homodimers have a higher affinity for symmetrical sequences (e.g. those found in the MHC class I gene promoter) whereas heterodimers preferentially bind to asymmetrical sequences[32].

Class I REL proteins appear to have two functions: as homodimers they are located in the nucleus of many cell types and act as transcriptional repressors (p50 dimers do not appear to activate transcription *in vivo* and may downregulate *trans*-activation by NF-κB). As heterodimers with RELA or other class II REL proteins, they bind to DNA with increased affinity, permitting the strong *trans*-activating capacity of the class II protein to operate. The C-terminus of the class I NF-κB1 precursor protein comprises the IκB-γ protein, and cleavage of p105 thus generates a regulator of the subcellular location of REL proteins.

## REL and transcriptional activation

The activation regions of both REL and v-REL proteins interact with the TATA binding protein (TBP) and transcription factor IIB (TFIIB), suggesting that the transcription regulating activities of REL proteins may be mediated by interaction with basal transcription factors[33]. REL family proteins also interact with SP1 to enhance transcription from SP1 site-containing promoters[34]. v-REL contains a complete NRD domain and can therefore homodimerize and bind to κB sites[35]. It also has weak *trans*-activating, C-terminal domains[36–38]. v-REL appears to compete with endogenous proteins of the *REL* family, the expression and activity of which is cell specific and probably transforms cells by acting as a dominant negative version of REL[6].

IκB-α inhibits both DNA binding and nuclear translocation of dimeric REL complexes containing either REL or RELA. RELA activates the CD28 response element in the IL-2 promoter[39] and causes *trans*-repression of the progesterone receptor, which itself represses RELA-mediated transcription[40].

REL homodimers activate transcription of IL-6, IL-2 receptor, IFN-β, IFN-γ and NF-κB[38,41–43] and the heterodimer of REL/NF-κB subunit p65 binds to the phorbol ester (TPA) responsive sequence 5′-GGGAAAGTAC-3′ in the 5′ flanking region of the human urokinase gene[44]. REL proteins also modulate gene expression by acting *via* the HIV LTR NF-κB sites[45] and induce κB-site-dependent stimulation of polyomavirus replication[36]. Mutant

NF-κB (p50) that is unable to bind to DNA but can form homo- or heterodimers prevents transcriptional activation *via* the HIV LTR or the MHC class I H-2K$^b$ promoter[46]. The HTLV-I TAX protein promotes degradation of IκB-α and IκB-β: it binds to the REL homology domain of REL, NF-κB1 and RELA and enhances transcription synergistically[47].

In most cells v-REL represses gene expression, including that of *Myc*[45,48,49] but in some cells (e.g. undifferentiated F9 cells, rat fibroblasts, chicken cells) it acts as a κB-specific transcriptional activator rather than as a repressor[50] and can stimulate transcription of IκB-α[51], NF-κB1[52], Jun[53], IL-2Rα[54], MIP-1β[55], p75[56], DM-GRASP (see ***In vitro***), high mobility group protein 14b and the MHC class I and II gene clusters[57,58].

**Cancer**

Rearrangement or amplification of the *REL* locus occurs in some lymphomas including non-Hodgkin lymphomas[59]. A cell line derived from a diffuse large cell lymphoma expresses a *REL* fusion mRNA (*NRG*, non-rel gene) that contains 284 REL N-terminal amino acids and 166 C-terminal NRG residues[60].

A B cell lymphoma-associated chromosomal translocation, t(10;14)(q24;q32), translocates the immunoglobulin $C\alpha_1$ locus into that of *NFKB2*: the fusion gene product includes the REL homology domain and binds κB sequences *in vitro*[61,62]. Truncated forms of *NFKB2* are expressed in the T cell lymphoma line HUT78 and in some patients with cutaneous T cell leukaemia (CTCL) in which the C-terminal coding regions are deleted[63,64].

These rearrangements in human cancers appear to resemble functionally the effect of transduction of *Rel* to form v-*rel* in that the C-terminally truncated gene product localizes to the nucleus and activates transcription. However, their oncogenicity probably requires cooperation with other transcription factors.

Over-expression of the p50 subunit of NF-κB has been detected in non-small cell lung carcinoma[65] and over-expression of both p50 and its p105 precursor occurs in some colon and ovarian carcinoma cell lines. This may act to neutralize the activity of IκB.

High levels of expression of p52 precursor p100 (NF-κB2) have been detected in some breast cancer cell lines and primary tumours, relative to normal breast epithelial cells. High levels of p100 promote association with p50/p65 heterodimers in the cytosol[66]. Hence over-expressed p100 may sequester active NF-κB complexes in a manner functionally resembling the effect of IκB.

**Transgenic animals**

*Rel*$^{-/-}$ mice show impaired humoral immunity with mature B and T cells being unresponsive to most mitogenic stimuli. Exogenous IL-2 restores T cell but not B cell proliferation in these animals[67].

p65/RELA deficiency causes embryonic lethality in mice accompanied by massive liver apoptosis[68]. In fibroblasts from these animals TNF-induced expression of IκB-α and GM-CSF is inhibited although basal transcription levels of these genes are unaffected.

v-REL expressed in thymocytes gives rise to aggressive T cell leukaemias and lymphomas *in vivo*. This transforming activity is due to v-REL homodimers as it also occurs in p50 null mice[58].

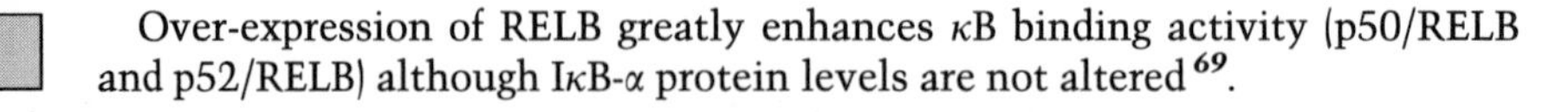

Over-expression of RELB greatly enhances κB binding activity (p50/RELB and p52/RELB) although IκB-α protein levels are not altered [69].

### In animals

In REV-T infection p59$^{v\text{-}rel}$ causes acute neoplasia in birds that is rapidly fatal [22,70].

### *In vitro*

Over-expression of REL transforms primary avian fibroblasts but in bone marrow cells induces programmed cell death [23]. v-*rel* primarily transforms lymphoid cells [8,19,71–74] but may also transform erythroid and myeloid cells and only partially transforms CEFs [22,75]. Immunoprecipitates of v-REL have an associated serine/threonine protein kinase activity [76,77]. Chicken spleen cells transformed *in vitro* by REV-T are tumorigenic on transplantation [78].

In HeLa cells TNF-α causes the rapid formation or translocation to the nucleus of multiple κB binding complexes, including p50/p65, p50/REL, p65/REL, p52/REL and p52/p65 [79]. In HT1080 fibrosarcoma cells TNF and some other agents that strongly activate NF-κB confer protection from apoptosis. Inhibition of NF-κB translocation to the nucleus by expression of IκB-α enhances apoptosis induced by these agents [80,81].

In chicken spleen cells v-REL blocks a normal pathway of apoptosis by a mechanism that may involve inhibition of IκB-α (p40) proteolysis [82]. In chick embryonic fibroblasts over-expression of REL or v-REL induces expression of the Ig superfamily adhesion molecule DM-GRASP [58,83]. In P19 embryonal carcinoma cells either REL or v-REL induce differentiation [84].

## Gene structure

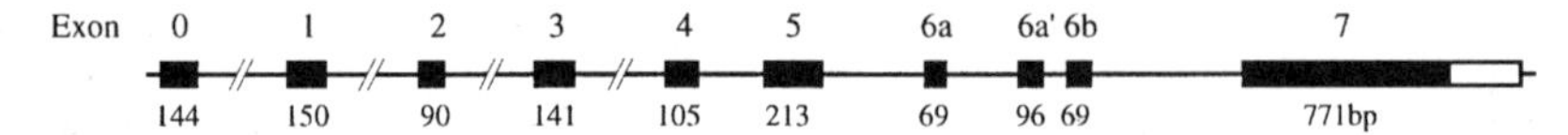


Exon 6a′ is a portion of an inverted *Alu* repeat and is not present in turkey *Rel*, from which v-*rel* was transduced [3].

## Transcriptional regulation

The promoter region of the normal chicken *Rel* gene is GC-rich, contains an NF-κB consensus binding sequence and lacks a TATA box. *In vitro* v-*rel* expression suppresses transcription from the *Rel* promoter by a mechanism that does not involve the NF-κB site [85].

## Protein structure

Within the REL homology domain (dark cross-hatched) all REL proteins have a nuclear localization signal, although transformation of spleen cells appears independent of whether v-REL is nuclear or cytoplasmic [16]. REL (and *dorsal*) contain C-terminal sequences that are important for cytoplasmic retention and transcriptional activation that are deleted in v-REL [48,86]. p50 contains a unique ~40 amino acid insert in the REL domain. The p50 precursor is

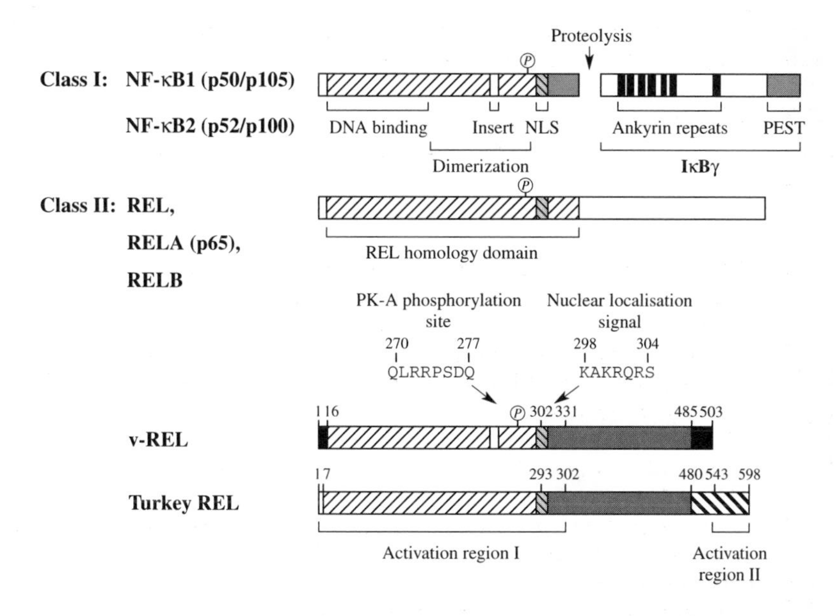


cleaved to release the REL domain, containing the DNA binding subunit. The C-terminal region of NF-κB1 contains ~six ankyrin repeats and constitutes IκB-γ [4,8,13,87]. IκB-γ shares homology with IκB-α, IκB-β and IκB-ε and all possess a C-terminal PEST sequence.

## Amino acid sequence of human REL

```
  1 MASGAYNPYI EIIEQPRQRG MRFRYKCEGR SAGSIPGEHS TDNNRTYPSI
 51 NIMNYYGKGK VRITLVTKND PYKPHPHDLV GKDCRDGYYE AEFGNERRPL
101 FFQNLGIQCV KKKEVKEAII TRIKAGINPF NVPEKQLNDI EDCDLNVVRL
151 CFQVFLPDEH GNLTTALPPV VSNPIYDNRA PNTAELRICR VNKNCGSVRG
201 GDEIFLLCDK VQKDDIEVRF VLNDWEAKGI FSQADVHRQV AIVFKTPPYC
251 KAITEPVTVK MQLRRPSDQE VSESMDFRYL PDEKDTYGNK AKKQKTTLIF
301 QKLCQDHVET GFRHVDQDGL ELLTSGDPPT LASQSAGITV NFPERPRPGL
351 LGSIGEGRYF KKEPNLFSHD AVVREMPTGV GVQAESYYPS PGPISSGLSH
401 HASMAPLPSS SWSSVAHPTP RSGNTNPLSS FSTRTLPSNS QGIPPFLRIP
451 VGNDLNASNA CIYNNADDIV GMEASSMPSA DLYGISDPNM LSNCSVNMMT
501 TSSDSMGETD NPRLLSMNLE NPSCNSVLDP RDLRQLHQMS SSSMSAGANS
551 NTTVFVSQSD AFEGSDFSCA DNSMINESGP SNSTNPNSHG FVQDSQYSGI
601 GSMQNEQLSD SFPYEFFQV (619)
```

**Domain structure**

1–300 REL homology domain
19–27 DNA binding motif (RXXRXRXXC) conserved in all REL proteins (underlined)
264–267 Putative cAMP-dependent protein kinase phosphorylation site (underlined)
290–295 Nuclear localization signal (underlined)
308–340 *Alu* sequence encoded by exon 6a′ (italics)

v-*rel* has transduced the turkey *Rel* sequence encoding the first 478 amino acids of REL. In v-REL this is flanked by 11 N-terminal amino acids and 18 C-terminal amino acids derived from *env*[88,89]. The loss of the C-terminus of REL causes nuclear localization and is principally responsible for the powerful oncogenic activity of v-REL[90]. The percentage homology (identical plus conserved residues) between turkey and human REL is 87–100% (exons 0–6a), 70% (exon 6b) and 49% (exon 7).

**Database accession numbers**

| | *PIR* | *SWISSPROT* | *EMBL/GENBANK* | *REFERENCES* |
|---|---|---|---|---|
| Human *REL* | A60646 | Q04864 | X75042 | 3 |
| Human *NFKB1* | | KBF1_HUMAN P19838 | | 8 |
| Human *NFKB2* | | KBF2_HUMAN Q00653 | | 91 |
| Human *NFKB2* (p49) | | KBF3_HUMAN Q04860 | | 92 |
| Human *RELA* (p65) | | TF65_HUMAN Q04206 | | 93 |
| Turkey *Rel* | A01377 | P01125 | X03508, X03616–X03623, K02447 | 88 |

***References***


1. Hoelzer, J.D. et al. (1979) Virology 93, 20–30.
2. Robinson, F.R. and Twiehaus, M.J. (1974) Avian Dis. 18, 278–288.
3. Brownell, E. et al. (1989) Oncogene 4, 935–942.
4. **Rushlow, C. and Warrior, R. (1992) BioEssays 14, 89–95.**
5. **Hannink, M. and Temin, H.M. (1991) Crit. Rev. Oncog. 2, 293–309.**
6. **Gilmore, T.D. et al. (1996) Oncogene 13, 1367–1378.**
7. **Nabel, G.J. and Verma, I.M. (1993) Genes Dev. 7, 2063.**
8. Kieran, M. et al. (1990) Cell 62, 1007–1018.
9. Bours, V. et al. (1990) Nature 348, 76–80.
10. Bours, V. et al. (1994) Oncogene 9, 1699–1702.
11. Davis, N. et al. (1991) Science 253, 1268–1271.
12. Haskill, S. et al. (1991) Cell 65, 1281–1289.
13. Inoue, J.-I. et al. (1992) Cell 68, 1109–1120.
14. Nolan, G.P. et al. (1991) Cell 64, 961–969.
15. Nakayama, K. et al. (1992) Mol. Cell. Biol. 12, 1736–1746.
16. Gilmore, T.D. and Temin, H.M. (1988) J. Virol. 62, 703–714.
17. Simek, S. and Rice, N.R. (1988) J. Virol. 62, 4730–4736.
18. Davis, J.N. et al. (1990) Oncogene 5, 1109–1115.
19. Capobianco, A.J. et al. (1992) J. Virol. 66, 3758–3767.
20. Neumann, M. et al. (1992) Oncogene 7, 2095–2104.
21. Brownell, E. et al. (1987) Mol. Cell. Biol. 7, 1304–1309.
22. Moore, B.E. and Bose, H.R. Jr. (1989) Oncogene 4, 845–852.
23. Abbadie, C. et al. (1993) Cell 75, 899–912.
24. Hohmann, H.P. et al. (1990) J. Biol. Chem. 265, 15183–15188.
25. Druker, B.J. et al. (1994) J. Biol. Chem. 269, 5387–5390.
26. Chen, Z.J. et al. (1996) Cell 84, 853–862.
27. Shirakawa, F. and Mizel, S.B. (1989) Mol. Cell. Biol. 9, 2424–2430.
28. Ghosh, S. and Baltimore, D. (1990) Nature 344, 678–682.
29. **Baeuerle, P.A. and Baltimore, D. (1996) Cell 87, 13–20.**
30. Baldi, L. et al. (1996) J. Biol. Chem. 271, 376–379.

[31] Chu, Z.-L. et al. (1996) Mol. Cell. Biol. 16, 5974–5984.
[32] Grimm, S. and Bauerle, M.K.A. (1994) Oncogene 9, 2391–2398.
[33] Xu, X. et al. (1993) Mol. Cell. Biol. 13, 6733–6741.
[34] Sif, S. and Gilmore, T.D. (1994) J. Virol. 68, 7131–7138
[35] Mosialos, G. and Gilmore, T.D. (1993) Oncogene 8, 721–730.
[36] Sarkar, S. and Gilmore, T.D. (1993) Oncogene 8, 2245–2252.
[37] Ishikawa, H. et al. (1993) Oncogene 8, 2889–2896.
[38] Smardova, J. et al. (1995) Oncogene 10, 2017–2026.
[39] Lai, J.-H. et al. (1995) Mol. Cell. Biol. 15, 4260–4271.
[40] Kalkhoven, E. et al. (1996) J. Biol. Chem. 271, 6217–6224.
[41] Muchardt, C. et al. (1992) J. Virol. 66, 244–250.
[42] Sica, A. et al. (1992) Proc. Natl Acad. Sci. USA 89, 1740–1744.
[43] Cogswell, P.C. et al. (1993) J. Immunol. 150, 2794–2804.
[44] Hansen, S.K. et al. (1992) EMBO J. 11, 205–213.
[45] McDonnell, P.C. et al. (1992) Oncogene 7, 163–170.
[46] Logeat, F. et al. (1991) EMBO J. 10, 1827–1832.
[47] Suzuki, T. et al. (1994) Oncogene 9, 3099–3105.
[48] Richardson, P.M. and Gilmore, T.D. (1991) J. Virol. 65, 3122–3130.
[49] Ballard, D.W. et al. (1992) Proc. Natl Acad. Sci. USA 89, 1875–1879.
[50] Walker, W.H. et al. (1992) J. Virol. 66, 5018–5029.
[51] Schatzle, J.D. et al. (1995) J. Virol. 69, 5383–5390.
[52] Walker, A.K. and Enrietto, P.J. (1996) Oncogene 12, 2515–2525.
[53] Fujii, M. et al. (1996) Oncogene 12, 2193–2202.
[54] Hrdlickova, R. et al. (1994) J. Virol. 68, 308–319.
[55] Petrenko, O. et al. (1995) Gene 160, 305–306.
[56] Zhang, G. and Humphries, E.H. (1996) Oncogene 12, 1153–1157.
[57] Boehmelt, G. et al. (1992) EMBO J. 11, 4641–4652.
[58] Carrasco, D. et al. (1996) EMBO J. 15, 3640–3650.
[59] Houldsworth, J. et al. (1996) Blood 87, 25–29.
[60] Lu, D. et al. (1991) Oncogene 6, 1235–1241.
[61] Neri, A. et al. (1991) Cell 67, 1075–1087.
[62] Chang, C.-C. et al. (1995) Mol. Cell. Biol. 15, 5180–5187.
[63] Zhang, J. et al. (1994) Oncogene 9, 1931–1937.
[64] Thakur, S. et al. (1994) Oncogene 9, 2335–2344.
[65] Mukhopadhyay, T. et al. (1995) Oncogene 11, 999–1003.
[66] Dejardin, E. et al. (1995) Oncogene 11, 1835–1841.
[67] Köntgen, F. et al. (1995) Genes Dev. 9, 1965–1977.
[68] Beg, A.A. et al. (1995) Nature 376, 167–170.
[69] Weih, F. et al. (1996) Oncogene 12, 445–449.
[70] Moore, B.E. and Bose, H.R. (1988) Virology 162, 377–387.
[71] White, D.W. et al. (1996) Mol. Cell. Biol. 16, 1169–1178.
[72] Kamens, J. and Brent, R. (1991) New Biol. 3, 1005–1013.
[73] Kochel, T. and Rice, N.R. (1992) Oncogene 7, 567–572.
[74] Kunsch, C. et al. (1992) Mol. Cell. Biol. 12, 4412–4421.
[75] Morrison, L.E. et al. (1991) Oncogene 6, 1657–1666.
[76] Rice, N.R. et al. (1986) Virology 149, 217–229.
[77] Walro, D.S. et al. (1987) Virology 160, 433–444.
[78] Lewis, R.B. et al. (1981) Cell 25, 421–431.
[79] Beg, A.A. and Baldwin, A.S. (1994) Oncogene 9, 1487–1492.
[80] Wang, C.-Y. et al. (1996) Science 274, 784–787

[81] Van Antwerp, D.J. et al. (1996) Science 274, 787–789.
[82] White, D.W. et al. (1995) Oncogene 10, 857–868.
[83] Zhang, G. et al. (1995) Mol. Cell. Biol. 15, 1489–1498.
[84] Inuzuka, M. et al. (1994) Oncogene 9, 133–140.
[85] Capobianco, A.J. and Gilmore, T.D. (1991) Oncogene 6, 2203–2210.
[86] Kamens, J. et al. (1990) Mol. Cell Biol. 10, 2840–2847.
[87] Ghosh, S. et al. (1990) Cell 62, 1019–1029.
[88] Wilhelmsen, K.C. et al. (1984) J. Virol. 52, 172–182.
[89] Capobianco, A.J. et al. (1990) Oncogene 5, 257–265.
[90] Hrdlickova, R. et al. (1995) J. Virol. 69, 3369–3380.
[91] Schmid R.M. et al. (1991) Nature 352, 733–736.
[92] Bours V. et al. (1992) Mol. Cell. Biol. 12, 685–695.
[93] Ruben, S.M. et al. (1991) Science 251, 1490–1493.

# RET

## Identification

*RET* is a human transforming gene with no homology with known viral oncogenes, originally detected by NIH 3T3 fibroblast transfection with DNA from a T cell lymphoma [1].

## Related genes

*RET* has homology with *Tek* receptor tyrosine kinase and the $Ca^{2+}$ binding sites of cadherins [2]. $RET^{TPC}$ and $RET^{PTC}$ (papillary thyroid carcinoma) are synonymous notations for the oncoprotein.

| | ***RET*** |
|---|---|
| **Nucleotides (kb)** | 55 |
| **Chromosome** | 10q11.2 |
| **Mass (kDa): predicted** | 91 (RET)<br>57–60 (RET/PTC1) |
| **expressed** | gp150/gp170<br>p96/p100 (in *RET*-transformed cells)<br>$p57^{RET\text{-}PTC1}/p64^{RET\text{-}PTC1}$<br>$p76^{RET\text{-}PTC2}/p81^{RET\text{-}PTC2}$<br>$p76^{RET\text{-}PTC3}/p81^{RET\text{-}PTC3}$<br>$p80^{RET\text{-}PTC4}/p85^{RET\text{-}PTC4}$ |
| **Cellular location** | Plasma membrane<br>p96 and oncoproteins: cytoplasm |

**Tissue distribution**

RET is expressed in the developing central and peripheral nervous systems and the excretory systems of mice. Expression is undetectable or low in most adult rat or mouse normal tissues but it is present in the rat placenta during the mid-term of gestation and in dopaminergic neurones of the substantia nigra [3].

## Protein function

RET encodes a receptor tyrosine kinase for which the ligand is glial cell line-derived neurotrophic factor (GDNF) when GDNF is bound to a GPI-anchored cell surface protein GDNFR-α [4,5]. The high levels of expression of *RET* in the central nervous system suggest that this receptor may mediate the neurotrophic effects of GDNF on motor and dopaminergic neurones [6,7].

The major dominant oncogenic forms of RET (RET/PTC1, RET/PTC2 and RET/PTC3) are constitutively phosphorylated on tyrosine and have autophosphorylation activity [8,9]. $p76^{RET\text{-}PTC2}$ and $p81^{RET\text{-}PTC2}$ form homo- and heterodimers with each other [10]. RET oncoproteins bind to SHC proteins that are phosphorylated on tyrosine and bound to GRB2 [11] and also directly to GRB7 and GRB10 [12]. RET/PTC1 and RET/PTC2 bind to phospholipase Cγ and this interaction is essential for the full oncogenic activity of RET/PTC2 [13].

However, mitogenic stimulation by RET/PTC2 is independent of PLC$\gamma$ or GRB10 binding but does require association of the LIM protein enigma with $Tyr^{586}$ [14].

Activation of RET in EGFR/RET chimeric receptors is associated with phosphorylation of paxillin and a 23 kDa protein [15].

*RET* may be involved in neuronal differentiation [16,17] and activated forms of *RET* (*RET/PTC1* and *RET/PTC3*) induce the specific expression in PC-12 cells of immediate early and delayed response genes [18]. The same activated forms of *RET* induce meiotic maturation of *Xenopus* oocytes, as does TPR/MET, *via* a pathway dependent on endogenous RAS [19].

**Cancer**

*RET* is the first gene detected in which dominantly acting point mutations initiate human hereditary neoplasia. The forms so far detected in papillary thyroid carcinomas are RET/PTC1, RET/PTC2, RET/PTC3, RET/ΔPTC3, RET/ΔELE1 and RET/PTC4. Germline mutations in *RET* cause multiple endocrine neoplasia type 2A (MEN2A), MEN2B and familial medullary thyroid carcinoma (FMTC) [20–23] and are associated with Hirschsprung's disease (HSCR) or congenital colonic aganglionosis [16,24,25].

An activated form of *RET* (*RET/PTC1*) has been found with high frequency (11–33%) in papillary thyroid carcinomas [26,27] and in the TPC-1 human papillary thyroid carcinoma cell line [28]. *RET/PTC1* activation is an early event in thryoid cell transformation [29]. There is one report of *RET/PTC1* activation in follicular adenomas and adenomatous goitres [30]. A second type of *RET* oncogenic rearrangement, *RET/PTC2*, occurs with lower frequency in papillary thyroid carcinomas [10]. These tumours are usually sporadic. RET/ELE1 fusion oncoproteins have also been detected. The latter, together with *RET/PTC3* and *RET/PTC4* oncogenes, occur with high frequency in thyroid tumours of children exposed to fallout from the Chernobyl reactor accident [31,32].

**Transgenic animals**

Mice carrying the metallothionein/*Ret* (MT/*Ret*) fusion gene develop melanosis and melanocytic tumours [33]. In MMTV/*Ret* transgenic mice mammary and salivary gland adenocarcinomas develop in a stochastic manner [34]. Mice homozygous for the deletion of *Ret* die soon after birth and reveal that *Ret* is essential for renal organogenesis and enteric neurogenesis [35].

Mice carrying the *RET/PTC1* oncogene with thyroid-specific expression develop slowly progressing thyroid tumours that mimic aspects of human papillary carcinomas [36]. When *RET/PTC1* is placed under the control of the *H4* promoter the transgene is expressed in several tissues but tumour development is more restricted, with mammary adenocarcinoma and cutaneous gland tumours arising [37]. Thus, although tumour development is restricted, the transforming capacity of *RET/PTC1* is not restricted to the thyroid epithelium.

***In vitro***

RET proteins carrying MEN2A ($Cys^{634} \rightarrow Arg$) and MEN2B ($Met^{918} \rightarrow Thr$) mutations (see **Mutations detected in *RET***) are constitutively phosphorylated on tyrosine and activated as kinases. These proteins transform NIH 3T3 cells and cause PC-12 cells to differentiate[38]. The MEN2A mutation induces ligand-independent homodimerization: the MEN2B mutation alters catalytic activity without dimerization although dimerization of the MEN2B mutant form still appears to be necessary for the realization of full enzymatic activity.

## Gene structure

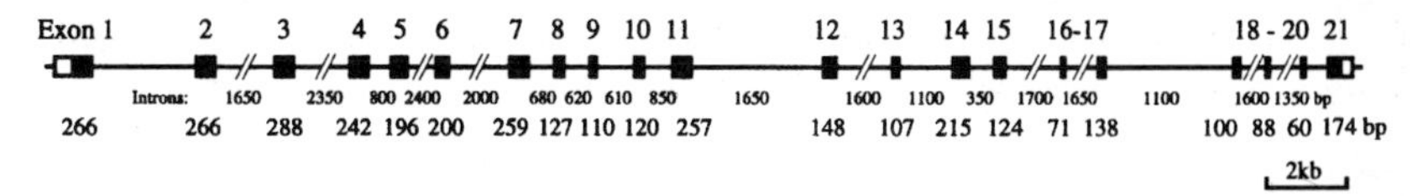


## Transcriptional regulation

The promoter region of proto-*RET* contains a GC-rich region without a TATA box. Putative binding motifs for SP1, AP2, epidermal growth factor receptor-specific transcription factor (ETF) and the transcription suppressor GC factor (GCF) occur in this repeated GC region[39–41].

## Protein structure

The *RET* transforming gene isolated by Takahashi *et al.*[1] had been activated *in vitro* during the transfection assay in a rearrangement that juxtaposed two unlinked human DNA segments, the *RET* proto-oncogene and the putative zinc finger-containing *RFP* gene (ret finger protein)[42].

RET/PTC1 is a different fusion protein of a 5′ non-*RET* region (*D10S170*) and the kinase domain encoded by *RET*. This is a somatic, tumour-specific event, in contrast to the recombination between *RET* and *RFP*. Alternative splicing of proto-*RET* gives rise to differing C-termini and corresponding *PTC* cDNAs have been isolated.

A second rearrangement (RET/PTC2) has been detected in papillary thyroid carcinoma in which the C-terminus of normal RET, including the tyrosine

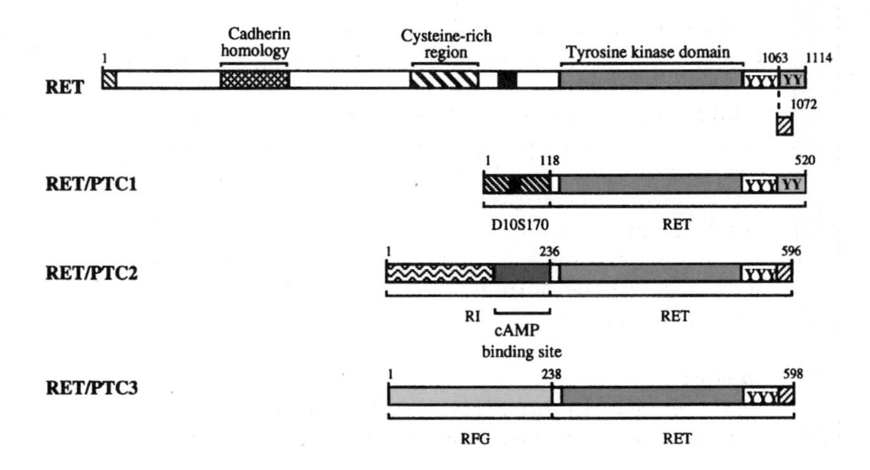

kinase domain, is fused with part of the RIα regulatory subunit of protein kinase A[9,10].

A third rearrangement (*RET/PTC3*) has also been detected in papillary thyroid carcinomas in which the C-terminal 360 amino acids of RET are linked to 238 N-terminal residues of ELE1/RFG (RET fused gene)[43,44].

## Amino acid sequence of human RET

```
   1 MAKATSGAAG LRLLLLLLLP LLGKVALGLY FSRDAYWEKL YVDQAAGTPL
  51 LYVHALRDAP EEVPSFRLGQ HLYGTYRTRL HENNWICIQE DTGLLYLNRS
 101 LDHSSWEKLS VRNRGFPLLT VYLKVFLSPT SLREGECQWP GCARVYFSFF
 151 NTSFPACSSL KPRELCFPET RPSFRIRENR PPGTFHQFRL LPVQFLCPNI
 201 SVAYRLLEGE GLPFRCAPDS LEVSTRWALD REQREKYELV AVCTVHAGAR
 251 EEVVMVPFPV TVYDEDDSAP TFPAGVDTAS AVVEFKRKED TVVATLRVFD
 301 ADVVPASGEL VRRYTSTLLP GDTWAQQTFR VEHWPNETSV QANGSFVRAT
 351 VHDYRLVLNR NLSISENRTM QLAVLVNDSD FQGPGAGVLL LHFNVSVLPV
 401 SLHLPSTYSL SVSRRARRFA QIGKVCVENC QAFSGINVQY KLHSSGANCS
 451 TLGVVTSAED TSGILFVNDT KALRRPKCAE LHYMVVATDQ QTSRQAQAQL
 501 LVTVEGSYVA EEAGCPLSCA VSKRRLECEE CGGLGSPTGR CEWRQGDGKG
 551 ITRNFSTCSP STKTCPDGHC DVVETQDINI CPQDCLRGSI VGGHEPGEPR
 601 GIKAGYGTCN CFPEEEKCFC EPEDIQDPLC DELCRTVIAA AVLFSFIVSV
 651 LLSAFCIHCY HKFAHKPPIS SAEMTFRRPA QAFPVSYSSS GARRPSLDSM
 701 ENQVSVDAFK ILEDPKWEFP RKNLVLGKTL GEGEFGKVVK ATAFHLKGRA
 751 GYTTVAVKML KENASPSELR DLLSEFNVLK QVNHPHVIKL YGACSQDGPL
 801 LLIVEYAKYG SLRGFLRESR KVGPGYLGSG GSRNSSSLDH PDERALTMGD
 851 LISFAWQISQ GMQYLAEMKL VHRDLAARNI LVAEGRKMKI SDFGLSRDVY
 901 EEDSYVKRSQ GRIPVKWMAI ESLFDHIYTT QSDVWSFGVL LWEIVTLGGN
 951 PYPGIPPERL FNLLKTGHRM ERPDNCSEEM YRLMLQCWKQ EPDKRPVFAD
1001 ISKDLEKMMV KRRDYLDLAA STPSDSLIYD DGLSEEETPL VDCNNAPLPR
1051 ALPSTWIENK LYGMSDPNWP GESPVPLTRA DGTNTGFPRY PNDSVYANWM
1101 LSPSAAKLMD TFDS (1114)
```

Alternative C-termini:

9 amino acid: GRISHAFTRF

43 amino acid: DAQHSSSLVGAAFGKSQQLFWLCCQHCNFAEKSRITKTLPALQT

The RET sequence is that encoded by one of the species of mRNA in the human monocytic leukaemia cell line THP-1[45]. 3.9, 4.5 and possibly 7.0 kb species encode a 1072 amino acid protein and a 6.0 kb and possibly a minor 4.6 kb form encode a 1114 amino acid protein[46]. The italicized 51 C-terminal amino acids represent one isoform of RET. Alternative 9 and 43 amino acid C-termini occur, the latter by splicing of exon 19 to exon 21[47].

Two further isoforms are predicted to encode transmembrane receptors with truncated ligand binding domains (minus exon 3 (deletes amino acids 113–209) and minus exons 3, 4 and 5 (deletes amino acids 113–354)) and an additional form encodes a soluble, secreted N-terminal form of the receptor which lacks exons 3 and 4 and translates exon 5 in another frame to include a stop codon and a unique 63 amino acid C-terminus[48].

### Domain structure

1–587 RET sequence replaced by RFP in the chimeric protein generated *in vitro* (underlined)

| Residues | Feature |
|---|---|
| 1–712 | RET sequence replaced by D10S170 sequence in the PTC1 fusion protein |
| 29–635 | Extracellular domain |
| 636–657 | Transmembrane region (underlined) |
| 722–999 | Tyrosine protein kinase domain |
| 730–738, 758 | ATP binding site |
| 874 | Active site |
| 687, 826, 1015, 1029, 1062 and 1096 | Six tyrosine phosphorylation sites. The MEN2B mutant lacks phosphorylation at $Tyr^{1096}$ [49] |
| 905 | $Tyr^{905}$ is essential for the transforming activity of MEN2A RET but not MEN2B, whereas the activity of MEN2B RET but not MEN2A is significantly reduced by mutation of $Tyr^{864}$ or $Tyr^{952}$ [50] |
| 1015 | Phospholipase Cγ docking site (in seq. KRRDYLDLAA): corresponds to $Tyr^{539}$ of RET/PTC2 [51] |
| 98, 151, 199, 336, 343, 361, 367, 377, 394 ,448, 468, 554 | Potential carbohydrate attachment sites |
| 1064–1114 | 51 C-terminal amino acids of one isoform of RET (italics). The alternative nine amino acid terminus is shown in the RET/PTC2 sequence below [43] |

**Mutations detected in *RET***

The majority of germline mutations in MEN2A [52,53] and FMTC occur in exons 10 and 11 (codons 609, 611, 618, 620 and 634) and result in the substitution of a cysteine by another amino acid [54,55]. The point mutations that occur in MEN2A affect $Cys^{634}$ in 95% of cases [52]. In the remainder $Cys^{618}$ is mutated. The $Cys^{634}$ mutation causes constitutive RET dimerization and hence activation. The Cys residues disrupted by MEN mutations may normally be involved intracellular disulfide bond formation and mutation of one may release the other to form activated RET homodimers. In MEN2B (and also MTC and spontaneous pheochromocytoma) [56] a point mutation at $Met^{918}$ in exon 16 is involved and in sporadic MTC there is a low frequency of somatic mutations at codons 768 and 883. The $Met^{918} \rightarrow Thr$ mutation alters the catalytic properties of RET both quantitatively and in terms of target substrates, leading to the MEN2A phenotype [57]. In the dominant form of Hirschsprung's disease (HSCR) or congenital colonic aganglionosis heterozygous deletions of *RET* or frameshift, nonsense or missense mutations are scattered throughout the entire RET coding sequence [24,25,58].

Mutations in MEN2A, 2B, FMTC and HSCR are shown in bold underlined in the above sequence [59–63]. Two maternally derived missense mutations in the RET tyrosine kinase domain have also been detected in MEN2B [64].

Somatic mutations can occur in sporadic MTC ($Met^{918} \rightarrow Thr$) [23,65,66] and in pheochromocytomas and neuroblastoma cells after induction of differentiation [17,67].

The mutation $Cys^{609} \rightarrow Trp$ has been detected in HSCR. Although the mutations $Cys^{618} \rightarrow Arg$ and $Cys^{620} \rightarrow Arg$ result in MEN2A or FMTC they can also predispose to HSCR with low penetrance [68]. The overall frequency of *RET* mutations in familial HSCR is ~50%. In sporadic HSCR mutations are scattered throughout the *RET* gene and there is currently no obvious relationship between *RET* genotype and HSCR phenotype. Thus, additional modifier genes may be involved in familial HSCR [69,70].

The HSCR mutations $Ser^{289} \rightarrow Pro$, $Arg^{421} \rightarrow Gln$ and $Arg^{496} \rightarrow Gly$ inactivate RET/PTC2 fibroblast mitogenicity [71]. Three HSCR mutations ($Ser^{765} \rightarrow Pro$, $Arg^{897} \rightarrow Gln$, $Arg^{972} \rightarrow Gly$) when introduced into RET/PTC2 abolish biological activity and two others ($Ser^{32} \rightarrow Leu$, $Phe^{393} \rightarrow Leu$) block maturation and transport to the cell surface of the receptor [72]. The rare polymorphism in exon 18 ($Arg^{982} \rightarrow Cys$) does not affect the transforming capability of RET/PTC2 or its tyrosine phosphorylation [73]. Thus loss of function HSCR mutations may act through a dominant negative mechanism.

Artificial mutation of $Asp^{300} \rightarrow Lys$ prevents translocation to the plasma membrane [74].

## Amino acid sequence of human RFP

```
  1 MASGSVAECL QQETTCPVCL QYFAEPMMLD CGHNICCACL ARCWGTAETN
 51 VSCPQCRETF PQRHMRPNRH LANVTQLVKQ LRTERPSGPG GEMGVCEKHR
101 EPLKLYCEED QMPICVVCDR SREHRGHSVL PLEEAVEGFK EQIQNQLDHL
151 KRVKDLKKRR RAQGEQARAE LLSLTQMERE KIVWEFEQLY HSLKEHEYRL
201 LARLEELDLA IYNSINGAIT QFSCNISHLS SLIAQLEEKQ QQPTRELLQD
251 IGDTLSRAER IRIPEPWITP PDLQEKIHIF AQKCLFLTES LKQFTEKMQS
301 DMEKIQELRE AQLYSVDVTL DPDTAYPSLI LSDNLRQVRY SYLQQDLPDN
351 PERFNLFPCV LGSPCFIAGR HYWEVEVGDK AKWTIGVCED SVCRKGGVTS
401 APQNGFWAVS LWYGKEYWAL TSPMTALPLR TPLQRVGIFL DYDAGEVSFY
451 NVTERCHTFT FSHATFCGPV RPYFSLSYSG GKSAAPLIIC PMSGIDGFSG
501 HVGNHGHSME TSP (513)
```

The chimeric gene product was generated *in vitro* by the fusion of the RET C-terminus to the first 315 amino acids of the 513 amino acid protein RFP [41]. The N-terminus of RET and the N-terminus of RFP that replaces it in the chimeric protein are underlined.

## Amino acid sequence of human H4/D10S170 (RET/PTC1)

```
  1 MADSASESDT DGAGGNSSSS AAMQSSCSST SGGGGGGGGG GGGGKSGGIV
 51 ISPFRLEELT NRLASLQQEN KVLKIELETY KLKCKALQEE NRDLRKASVT
101 IQARAEQEEE FISNTLFKKI QALQKEKETL AVNYEKEEEF LTNELSRKLM
151 QLQHEKGELE QHLEQEQEFQ VNKLMKKIKK LENDTISKQL TLEQLRREKI
201 DLENTLEQEQ EALVNRLWKR MDKLEAETRI LQEKLDQPVS APPSPRDISM
251 EIDSPENMMR HIRFLKNEVE RLKKQLRAAQ LQHSEKMAQY LEEERHMREE
301 NLRLQRKLQR EMERREALCR QLSESESSLE MDDERYFNEM SAQGLRPRTV
351 SSPIPYTPSP SSSRPISPGL SYASHTVGFT PPTSLTRAGM SYYNSPGLHV
401 QHMGTSHGIT RPSPRRSNSP DKFKRPTPPP SPNTQTPVQP PPPPPPPPMQ
451 PTVPSGSHLA AYSFATFGAH LLPALMHELS LNFKLGLIQW SRLLNAKGSF
501 SGIFGYDLFA LRLSRLHYPL CCKCLSEMQP VLWVYNTNQT TFSISVLLES
551 SCTSIPWLEP SLFGIWYFSS SVQFLLGPEL HSPGF (585)
```

In the RET/PTC1 fusion protein amino acids 1–339 (underlined) replace 1–712 of RET[75].

## Amino acid sequence of human RIα/RET (RET/PTC2)

```
  1 MQSGSTAASQ QARSLRQCQL YVEKHNIEAL LKDSIVQLCT ARPERPMAFL
 51 REYFERLEKE EAKQIQNLQK AGTRTDSRED EISPPPPNPV VKGRRRRGAI
101 SAEVYTEEDA ASYVRKVIPK DYKTMAALAK AIEKNVLFSH LDDNERSDIF
151 DAMFSVSFIA GETVIQQGDE GDNFYVIDQG ETDVYVNNEW ATSVGEGGSF
                                              RIα/RET
201 GELALIYGTP RAATVKAKTN VKLWGIDRDS YRRILMEDPK WEFPRKNLVL
251 GKTLGEGEFG KVVKATAFHL KGRAGYTTVA VKMLKENASP SELRDLLSEF
301 NVLKQVNHPH VIKLYGACSQ DGPLLLIVEY AKYGSLRGFL RESRKVGPGY
351 LGSGGSRNSS SLDHPDERAL TMGDLISFAW QISQGMQYLA EMKLVHRDLA
401 ARNILVAEGR KMKISDFGLS RDVYEEDPYV KRSQGRIPVK WMAIESLFDH
451 IYTTQSDVWS FGVLLWEIVT LGGNPYPGIP PERLFNLLKT GHRMERPDNC
501 SEEMYRLMLQ CWKQEPDKRP VFADISKDLE KMMVKRRDYL DLAASTPSDS
551 LIYDDGLSEE ETPLVDCNNA PLPRALPSTW IENKLYGRIS HAFTRF (596)
```

**Domain structure**

| | |
|---|---|
| 1–236 | Sequence of RIα (underlined) |
| 236/237 | RIα/RET fusion point (the same as in D10S170-RET) |
| 588–596 | Nine amino acid form of C-terminus (italics) |
| 586 | Enigma binding site[14] |

The underlined sequence is that of RIα and the RIα/RET fusion point, which is the same as in D10S170-RET, is shown.

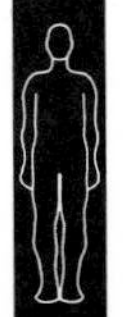

## Amino acid sequence of human ELE1/RFG (RET/PTC3)

```
  1 MNTFQDQSGS SSNREPLLRC SDARRDLELA IGGVLRAEQQ IKDNLREVKA
 51 QIHSCISRHL ECLRSREVWL YEQVDLIYQL KEETLQQQAQ QLYSLLGQFN
101 CLTHQLECTQ NKDLANQVSV CLERLGSLTL KPEDSTVLLF EADTITLRQT
151 ITTFGSLKTI QIPEHLMAHA SSANIGPFLE KRGCISMPEQ KSASGIVAVP
201 FSEWLLGSKP ASGYQAPYIP STDPQDWLTQ KQTLENSQTS SRACNFFNNV
251 GGNLKGLENW LLKSEKSSYQ KCNSHSTTSS FSIEMEKVGD QELPDQDEMD
301 LSDWLVTPQE SHKLRNAENG SRETSEKFKL LFQSYNVNDW LVKTDSCTNC
351 QGNQPKGVEI ENLANLKCLN DHLEAKKPLS TPSMVTEDWL VQNHQDPCKV
401 EEVCRANEPC TSFAECVCDE NCEKEALYKW LLKKEGKDKN GMPVEPKPEP
451 EKHKDSLNMW LCPRKEVIEQ TKAPKAMTPS RIADSFQVIK NSPLSEWLIR
501 PPYKEGSPKE VPGTEDRAGK QKFKSPMNTS WCSFNTADWV LPGKKMGNLS
551 QLSSGEDKWL LRKKAQEVLL NSPLQEEHNF PPDHYGLPAV CDLFACMQLK
601 VDKEKWLYRT PLQM (614)
```

The N-terminal 238 amino acids of ELE1/RFG (614 amino acids) are fused in RET/PTC3 to 360 amino acids of RET[44].

**Database accession numbers**

| | PIR | SWISSPROT | EMBL/GENBANK | REFERENCES |
|---|---|---|---|---|
| Human *RET* | A27203 | P07949 | Hstykret M16029 | 16 |
| | | | Hsret5 X15262, | |
| | | | Hsretpon X12949 | 45 |
| | | | U11504–U11540, | 76 |
| | | | U11546 | 47 |
| Human *RET/PTC2* | | | Hsretri L03357 | 10 |
| Human *RET/PTC3* | | | Hsptcaa M31213 | 77 |
| Human *ELE1/RFG* | | | Hsrfg1 X77548 | 43 |
| | | | Hsele1 X71413 | 44 |
| Human *RFP* | A28101 | P14373 | J03407, M16029 | 42 |
| Human *D10S170* | | S72869 | | 75 |

***References***


[1] Takahashi, M. et al. (1985) Cell 42, 581–588.
[2] Iwamoto, T. et al. (1993) Oncogene 8, 1087–1091.
[3] Szentirmay, Z. et al. (1990) Oncogene 5, 701–705.
[4] Treanor, J. et al. (1996) Nature 382, 80–83.
[5] Jing, S. et al. (1996) Cell 85, 1113–1124.
[6] Trupp, M. et al. (1996) Nature 381, 785–789;
[7] Durbec, P. et al. (1996) Nature 381, 789–793.
[8] Ishizaka, Y. et al. (1992) Oncogene 7, 1441–1444.
[9] Lanzi, C. et al. (1992) Oncogene 7, 2189–2194.
[10] Bongarzone, I. et al. (1993) Mol. Cell. Biol. 13, 358–366.
[11] Borrello, M.G. et al. (1994) Oncogene 9, 1661–1668.
[12] Pandey, A. et al. (1996) J. Biol. Chem. 271, 10607–10610.
[13] Borrello, M.G. et al. (1996) Mol. Cell. Biol. 16, 2151–2163.
[14] Durick, K. et al. (1996) J. Biol. Chem. 271, 12691–12694.
[15] Romano, A. et al. (1994) Oncogene 9, 2923–2933.
[16] Takahashi, M. and Cooper, G.M. (1987) Mol. Cell. Biol. 7, 1378–1385.
[17] Tahira, T. et al. (1991) Oncogene 6, 2333–2338.
[18] Califano, D. et al. (1995) Oncogene 11, 107–112.
[19] Grieco, D. et al. (1995) Oncogene 11, 113–117.
[20] Mulligan, L.M. et al. (1994) Hum. Mol. Genet. 3, 1007–1008 and 2163–2167.
[21] Hofstra, R.M.W. et al. (1994) Nature 367, 375–376.
[22] Donis-Keller, H. et al. (1993) Hum. Mol. Genet. 2, 851–856.
[23] Zedenius, J. et al. (1994) Hum. Mol. Genet. 3, 1259–1262.
[24] Edery, P. et al. (1994) Nature 367, 378–380.
[25] Romeo, G. et al. (1994) Nature 367, 377–378.
[26] Jhiang, S.M. et al. (1992) Oncogene 7, 1331–1337.
[27] Santoro, M. et al. (1992) J. Clin. Invest. 89, 1517–1522.
[28] Ishizaka, Y. et al. (1990) Biochem. Biophys. Res. Commun. 168, 402–408.
[29] Viglietto, G. et al. (1995) Oncogene 11, 1207–1210.
[30] Ishizaka, Y. et al. (1991) Oncogene 6, 1667–1672.
[31] Fugazzola, L. et al. (1996) Oncogene 13, 1093–1097.
[32] Klugbauer, S. et al. (1996) Oncogene 13, 1099–1102.
[33] Taniguchi, M. et al. (1992) Oncogene 7, 1491–1496.
[35] Schuchardt, A. et al. (1994) Nature 367, 380–383.
[36] Santoro, M. et al. (1996) Oncogene 12, 1821–1826.
[37] Portella, G. et al. (1996) Oncogene 13, 2021–2026.

[38] Borrello, M.G. et al. (1995) Oncogene 11, 2419–2427.
[39] Itoh, F. et al. (1992) Oncogene 7, 1201–1205.
[40] Kwok, J.B.J. et al. (1993) Oncogene 8, 2575–2582.
[41] Pasini, B. et al. (1995) Oncogene 11, 1737–1743.
[42] Takahashi, M. et al. (1988) Mol. Cell. Biol. 8, 1853–1856.
[43] Santoro, M. et al. (1994) Oncogene 9, 509–516.
[44] Bongarzone, I. et al. (1994) Cancer Res. 54, 2979–2985.
[45] Takahashi, M. et al. (1988) Oncogene 3, 571–578.
[46] Tahira, T. et al. (1990) Oncogene 5, 97–102.
[47] Myers, S.M. et al. (1995) Oncogene 11, 2039–2045.
[48] Lorenzo, M.J. et al. (1995) Oncogene 10, 1377–1383.
[49] Liu, X. et al. (1996) J. Biol. Chem. 271, 5309–5312.
[50] Iwashita, T. et al. (1996) Oncogene 12, 481–487.
[51] Borrello, M.G. et al. (1996) Mol. Cell. Biol. 16, 2151–2163.
[52] Mulligan, L. et al. (1993) Nature 363, 458–460.
[53] Bugalho, M.J.M. et al. (1994) Hum. Mol. Genet. 3, 2263.
[54] Gardner, E. et al. (1994) Hum. Mol. Genet. 3, 1771–1774.
[55] Bolino, A. et al. (1995) Oncogene 10, 2415–2419.
[56] Eng, C. et al. (1994) Hum. Mol. Genet. 3, 237–241, 686.
[57] Santoro, M. et al. (1995) Science 267, 381–383.
[58] Attie, T. et al. (1994) Hum. Mol. Genet. 3, 1439–1440.
[59] Carlson, K.M. et al. (1994) Proc. Natl Acad. Sci. USA 91, 1579–1583.
[60] Xue, F. et al. (1994) Hum. Mol. Genet. 3, 635–638.
[61] McMahon, R. et al. (1994) Hum. Mol. Genet. 3, 643–646.
[62] Eng, C. et al. (1995) Oncogene 10, 509–513.
[63] Blaugrund, J.E. et al. (1994) Hum. Mol. Genet. 3, 1895–1897.
[64] Kitamura, Y. et al. (1995) Hum. Mol. Genet. 4, 1987–1988.
[65] Schuffenecker, I. et al. (1994) Hum. Mol. Genet. 3, 1939–1943.
[66] Marsh, D.J. et al. (1996) Cancer Res. 56, 1241–1243.
[67] Santoro, M. et al. (1990) Oncogene 5, 1595–1598.
[68] Mulligan, L.M. et al. (1994) Hum. Mol. Genet. 3, 2163–2167.
[69] Angrist, M. et al. (1995) Hum. Mol. Genet. 4, 821–830.
[70] Attie, T. et al. (1995) Hum. Mol. Genet. 4, 1381–1386.
[71] Durick, K. et al. (1995) J. Biol. Chem. 270, 24642–24645.
[72] Carlomagno, F. et al. (1996) EMBO J. 15, 2717–2725.
[73] Pasini, B. et al. (1995) Nature Genet. 10, 35–40.
[74] Asai, N. et al. (1995) Mol. Cell. Biol. 15, 1613–1619.
[75] Grieco, M. et al. (1994) Oncogene 9, 2531–2535.
[76] Ceccherini, I. et al. (1994) Oncogene 9, 3025–3029.
[77] Grieco, M. et al. (1990) Cell 60, 557–563.

## Identification

v-*ros* is the oncogene of the acutely transforming avian sarcoma virus UR2 (University of Rochester 2)[1], so designated because it is unrelated to any other ASV gene[2]. *MCF3* is an activated form of *ROS1* detected by transfection of cDNA derived from a human mammary carcinoma cell line (MCF-7) into NIH 3T3 cells and injection of these cells into nude mice. The *MCF3* gene was expressed in some of the tumours generated[3]. *ROS1* was detected by screening cDNA with an *MCF3* probe[4].

## Related genes

ROS1 has homology with SRC tyrosine kinases and with receptor-type tyrosine kinases including the insulin and EGF receptors. Close similarity in overall structure and sequence exists between vertebrate ROS and *Drosophila sevenless*.

| | ***ROS1/Ros-1*** | **v-*ros*** |
|---|---|---|
| **Nucleotides** | 32 kb | 1273 bp |
| **Chromosome** | 6q21–q22 | |
| **Mass (kDa): predicted** | 256 | 61 |
| **expressed** | gp260 | P68$^{gag\text{-}ros}$ |
| **Cellular location** | Plasma membrane | |

**Tissue distribution**

*Ros-1* mRNA expression in the mouse occurs transiently during the development of the kidney, intestine and lung and coincides with major morphogenetic and differentiation events[5]. Chicken *Ros-1* is significantly expressed in the kidney and intestine with low expression in the gonads, thymus, bursa and brain[6].

## Protein function

ROS1 is a receptor-like tyrosine kinase with autophosphorylation capacity. p68$^{gag\text{-}ros}$ co-precipitates with phosphatidylinositol kinase and in UR2-transformed cells the levels of PtdIns4*P*, PtdIns(4,5)$P_2$ and Ins(1,4,5)$P_3$ are increased[7]. The pattern of expression during the differentiation of normal tissues is unusual for a tyrosine kinase receptor and indicates a specific role for *Ros-1* during development[5,8].

**Cancer**

ROS1, undetectable in normal brain tissue, is expressed in some primary human gliomas and derived cell lines[9,10].

**Transgenic animals**

Male $Ros^{-/-}$ mice are infertile whereas the fertility of female $Ros^{-/-}$ mice is unaffected. The defect is manifested in sperm maturation in the epididymis[11].

### In animals

Injected UR2 induces tumours in chickens. UR2-transformed rat cells induce fatal fibrosarcomas on injection into rats[12]. *MCF3*-transformed NIH 3T3 cells form tumours in nude mice.

### *In vitro*

Infected cells from chicken tumours transform chick embryo fibroblasts and infected CEFs transform rat-1 cell lines.

## Gene structure

The genomic structure of *Ros1* has not been completely characterized.

## Protein structure

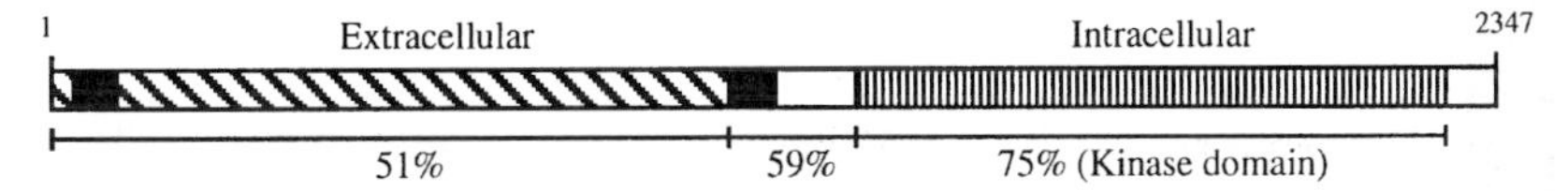


The figures indicate % identity with chicken ROS1. v-ROS is an N-terminally truncated version of chicken ROS1 with an alternative 12 amino acid C-terminus. Rat ROS1 is related to *Drosophila sev* protein and both have a unique insert in the tyrosine kinase domain. ROS proteins contain a 20 residue hydrophobic sequence 50 amino acids downstream of the initiating Met residue and may form a loop structure from the plasma membrane. In MCF3, an activated form of ROS1, all but eight amino acids of the ROS1 extracellular domain are replaced by sequences of unknown origin.

## Amino acid sequence of human ROS1

```
   1 MKNIYCLIPK LVNFATLGCL WISVVQCTVL NSCLKSCVTN LGQQLDLGTP
  51 HNLSEPCIQG CHFWNSVDQK NCALKCRESC EVGCSSAEGA YEEEVLENAD
 101 LPTAPFASSI GSHNMTLRWK SANFSGVKYI IQWKYAQLLG SWTYTKTVSR
 151 PSYVVKPLHP FTEYIFRVVW IFTAQLQLYS PPSPSYRTHP HGVPETAPLI
 201 RNIESSSPDT VEVSWDPPQF PGGPILGYNL RLISKNQKLD AGTQRTSFQF
 251 YSTLPNTIYR FSIAAVNEVG EGPEAESSIT TSSSAVQQEE QWLFLSRKTS
 301 LRKRSLKHLV DEAHCLRLDA IYHNITGISV DVHQQIVYFS EGTLIWAKKA
 351 ANMSDVSDLR IFYRGSGLIS SISIDWLYQR MYFIMDELVC VCDLENCSNI
 401 EEITPPSISA PQKIVADSYN GYVFYLLRDG IYRADLPVPS GRCAEAVRIV
 451 ESCTLKDFAI KPQAKRIIYF NDTAQVFMST FLDGSASHLI LPRIPFADVK
 501 SFACENNDFL VTDGKVIFQQ DALSFNEFIV GCDLSHIEEF GFGNLVIFGS
 551 SSQLHPLPGR PQELSVLFGS HQALVQWKPP ALAIGANVIL ISDIIELFEL
 601 GPSAWQNWTY EVKVSTQDPP EVTHIFLNIS GTMLNVPELQ SAMKYKVSVR
 651 ASSPKRPGPW SEPSVGTTLV PASEPPFIMA VKEDGLWSKP LNSFGPGEFL
 701 SSDIGNVSDM DWYNNSLYYS DTKGDVFVWL LNGTDISENY HLPSIAGAGA
 751 LAFEWLGHFL YWAGKTYVIQ RQSVLTGHTD IVTHVKLLVN DMVVDSVGGY
 801 LYWTTLYSVE STRLNGESSL VLQTQPWFSG KKVIALTLDL SDGLLYWLVQ
 851 DSQCIHLYTA VLRGQSTGDT TITEFAAWST SEISQNALMY YSGRLFWING
 901 FRIITTQEIG QKTSVSVLEP ARFNQFTIIQ TSLKPLPGNF SFTPKVIPDS
 951 VQESSFRIEG NASSFQILWN GPPAVDWGVV FYSVEFSAHS KFLASEQHSL
1001 PVFTVEGLEP YALFNLSVTP YTYWGKGPKT SLSLRAPETV PSAPENPRIF
```

1051 ILPSGKCCNK NEVVVEFRWN KPKHENGVLT KFEIFYNISN QSITNKTCED
1101 WIAVNVTPSV MSFQLEGMSP RCFIAFQVRA FTSKGPGPYA DVVKSTTSEI
1151 NPFPHLITLL GNKIVFLDMD QNQVVWTFSA ERVISAVCYT ADNEMGYYAE
1201 GDSLFLLHLH NRSSSELFQD SLVFDITVIT IDWISRHLYF ALKESQNGMQ
1251 VFDVDLEHKV KYPREVKIHN RNSTIISFSV YPLLSRLYWT EVSNFGYQMF
1301 YYSIISHTLH RILQPTATNQ QNKRNQCSCN VTEFELSGAM AIDTSNLEKP
1351 LIYFAKAQEI WAMDLEGCQC WRVITVPAML AGKTLVSLTV DGDLIYWIIT
1401 AKDSTQIYQA KKGNGAIVSQ VKALRSRHIL AYSSVMQPFP DKAFLSLASD
1451 TVEPTILNAT NTSLTIRLPL AKTNLTWYGI TSPTPTYLVY YAEVNDRKNS
1501 SDLKYRILEF QDSIALIEDL QPFSTYMIQI AVKNYYSDPL EHLPPGKEIW
1551 GKTKNGVPEA VQLINTTVRS DTSLIISWRE SHKPNGPKES VRYQLAISHL
1601 ALIPETPLRQ SEFPNGRLTL LVTRLSGGNI YVLKVLACHS EEMWCTESHP
1651 VTVEMFNTPE KPYSLVPENT SLQFNWKAPL NVNLIRFWVE LQKWKYNEFY
1701 HVKTSCSQGP AYVCNITNLQ PYTSYNVRVV VVYKTGENST SLPESFKTKA
1751 GVPNKPGIPK LLEGSKNSIQ WEKAEDNGCR ITYYILEIRK STSNNLQNQN
1801 LRWKMTFNGS CSSVCTWKSK NLKGIFQFRV VAANNLGFGE YSGISENIIL
1851 VGDDFWIPET SFILTIIVGI FLVVTIPLTF VWHRRLKNQK SAKEGVTVLI
1901 NEDKELAELR GLAAGVGLAN ACYAIHTLPT QEEIENLPAF PREKLTLRLL
1951 L*GSGAFG*EVY EGTAVDILGV GSGEIKVAV*K* TLKKGSTDQE KIEFLKEAHL
2001 MSKFNHPNIL KQLGVCLLNE PQYIILELME GGDLLTYLRK ARMATFYGPL
2051 LTLVDLVDLC VDISKGCVYL ERMHFIHRDL AARNCLVSVK DYTSPRIVKI
2101 GDFGLARDIY KNDYYRKRGE GLLPVRWMAP ESLMDGIFTT QSDVWSFGIL
2151 IWEILTLGHQ PYPAHSNLDV LNYVQTGGRL EPPRNCPDDL WNLMTQCWAQ
2201 EPDQRPTFHR IQNQLQLFRN FFLNSIYQCR DEANNSGVIN ESFEGEDGDV
2251 ICLNSDDIMP VVLMETKNRE GLNYMVLATE CGQGEEKSEG PLGSQESESC
2301 GLRKEEKEPH ADKDFCQEKQ VAYCPSGKPE GLNYACLTHS GYGDGSD
(2347)

## Domain structure

11–42 and 1860–1883 Hydrophobic regions (underlined)
1952–1957 and 1980 ATP binding site (italics)

## Database accession numbers

| | *PIR* | *SWISSPROT* | *EMBL/GENBANK* | *REFERENCES* |
|---|---|---|---|---|
| Human *ROS1* | | | M34353 | [4] |

## *References*


1 Balduzzi, P.C. et al. (1981) J. Virol. 40, 268–275.
2 Wang, L.-H. et al. (1982) J. Virol. 41, 833–841.
3 Birchmeier, C. et al. (1986) Mol. Cell. Biol. 6, 3109–3116.
4 Birchmeier, C. et al. (1990) Proc. Natl Acad. Sci. USA 87, 4799–4803.
5 Sonnenberg, E. et al. (1991) EMBO J. 10, 3693–3702.
6 Neckameyer, W.S. et al. (1986) Mol. Cell. Biol. 6, 1478–1486.
7 Macara, I.G. et al. (1984) Proc. Natl Acad. Sci. USA 81, 2728–2732.
8 Chen, J. et al. (1994) Oncogene 9, 773–780.
9 Maxwell, M. et al. (1996) Int. J. Oncol. 8, 713–718.
10 Sharma, S. et al. (1989) Oncogene Res. 5, 91–100.
11 Sonnenberg-Rietmacher, E. et al. (1996) Genes Dev. 10, 1184–1193.
12 Neckameyer, W.S. and Wang, L-H. (1985) J. Virol. 53, 879–884.

## Identification

v-*sea* (sarcoma, erythroblastosis and anaemia) is the oncogene of the acutely transforming virus AEV-S13 [1]. Chicken *Sea* was detected by low stringency hybridization screening of cDNA with a v-*src* probe [2].

## Related genes

SEA is a member of the MET/hepatocyte growth factor/scatter factor family of receptor protein tyrosine kinases with strong homology to both MET, RON/STK and the insulin receptor family and in the extracellular domain with SEX, SEP, OCT and NOV.

| | ***SEA/Sea*** | **v-*sea*** |
|---|---|---|
| **Nucleotides (kb)** | Not fully mapped | 8.5 (S13) |
| **Chromosome** | 11q13 | |
| **Mass (kDa): predicted** | 190 | 42 (v-SEA) |
| **expressed** | 160, 180 | $gp155^{env\text{-}sea}$<br>$gp85^{env}/gp70^{env\text{-}sea}$ |
| **Cellular location** | Plasma membrane<br>35 kDa $\alpha$ chain: 160 kDa $\beta$ chain | Plasma membrane |

### Tissue distribution

Chicken *Sea* is expressed in most tissues with highest levels in peripheral white blood cells and the intestine [2]. The protein is expressed at low levels in kidney, intestine, liver, stomach, white blood cells and allantochorion [3]. *Sea* mRNA is elevated five-fold in chicken embryo cells transformed by v-*src*.

## Protein function

SEA is a growth factor receptor tyrosine protein kinase. Transformation by *env-sea* correlates with tyrosine phosphorylation of SHC proteins [4].

### Cancer

Expressed at low frequency (~1%) in breast carcinomas together with *BCL1*, *HSTF1* and *INT2* [5].

### In animals

Injection of AEV-S13 into young chickens causes sarcomas, erythroblastosis and anaemias [1]. Fibroblasts transformed by v-*sea* are only weakly oncogenic when injected into chicks: cells expressing a retrovirus carrying both v-*sea* and v-*ski* are highly malignant [6].

### *In vitro*

v-*sea* transforms fibroblasts and erythroblasts [7] but does not transform avian myeloid cells [8]. v-*sea* induces the synthesis of chicken myeloid growth factor (cMGF) and causes autocrine growth in myeloid cells transformed by v-*myb* or v-*myc* [9]. The differentiation of v-*sea*-transformed erythroid cells into erythrocytes is blocked by the expression of v-*ski* [6]. This effect of v-*ski* is similar to that of v-*erbA* but is associated with different effects on gene expression.

## Gene structure

The *env* sequence is 96% homologous with that of the RAV2 strain of ALV. The ~1085 bp insert between *env* and the viral 3′ non-coding region encodes the *sea* ORF. A single base deletion 14 bp upstream of the normal *env* termination codon permits read-through into the *sea* gene. The termination codon lies within the 3′ viral non-coding region.

The S13 virus encodes normal *gag* and *gag-pol* proteins and an abnormal *env* glycoprotein (gp155) that is cleaved to gp85$^{env}$ and gp70$^{env\text{-}sea}$. gp155$^{env\text{-}sea}$ retains the entire extracellular and transmembrane domains of *env*. Replacement of the entire *env* sequence by the myristylation target signal of pp60$^{v\text{-}src}$ does not affect the capacity of v-*sea* to transform fibroblasts [7]. The uncleaved but fully glycosylated gp155$^{env\text{-}sea}$ retains the capacity to transform chicken embryo fibroblasts [10].

## Protein structure

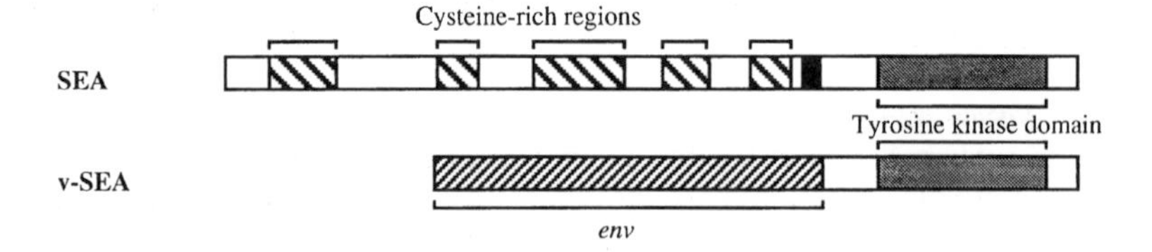


The extracellular and transmembrane regions of v-SEA are derived from the viral *env* gene.

## Amino acid sequence of chicken SEA

```
  1 MGPRCLVCLL LLLAPSLLQA GAWQCRRIPF SSTRNFSVPY TLPSLDAGSP
 51 VQNIAVFPDP PTVFVAVRNR ILVVDPELRL RSVLVTGPTG SAPCEICRLC
101 PAAVDAPGPE DVDNVLLLLD PVEPWLYSCG TARRGLCYLH QLDVRGSEVT
151 IASTRCLYSA AANSPVNCPD CVASPLGSTA TVVADRYTAS FYLGSTVNSS
201 VAARYSPRSV SVRRLKGTRD GFADPFHSLT VLPHYQDVYP IHYVHSFTDG
251 DHVYLVTVQP EFPGSSTFHT RLVRLSAHEP ELRRYREIVL DCRYESKRRR
301 RRRGAEEETE RDVAYNVLQA AHAARPGARL ARDLGIDGTE TVLFGAFAES
351 HPESRAPQHN SAVCAFPLRL LNQAIREGMD KCCGTGTQTL KRGLAFFQPQ
401 QYCPHSVNLS APVTNTSCWD QPTLVPAASH KVDLFNGRLS GTLLTSIFVT
451 VLQNVTVAHL GTAQGRVLQM VLQRSSSYVV ALTNFSLGEP GLVQHATGLQ
501 GHSLLFAAGT KVWRVNVTGP GCRHFSTCDR CLRAERFMGC GWCGNGCTRH
551 HECAGPWVQD SCPPVLTDFH PRSAPLRGQT RVTLCGMTFH SPPDPTAHHS
601 LPGPYRVAVG GRSCTVLLDE SESYRPLPTF RRKDFVDVLV CVLEPGEPAV
651 AAGPADVVLN VTESAGTSRF RVQGSSTLSG FVFVEPHIST LHPSFGPQGG
701 GTLMSLYGTH LSAGSSWRVT INGSECLLDG QPSEGDGEIR CTAPAATSLG
751 AAPVALWIDG EEFLAPLPFE YRPDPSVLTV VPNCSYGGST LTLIGTHLDS
```

```
 801 VYRAKIQFQG GGGGKTEATE CEGPQSPNWL LCRSPAFPIE IKPVPGNLSV
 851 LLDGAADRWL FRLRYFPQPQ MFSFGQQGER YQLKPGDNEI KVNQLGLDSV
 901 AGCMNITMTV GGRDCHPNVL KNEVTCRVPR DVDLTPAGAP VQICVNGDCQ
 951 ALGLVLPASS LDMAASLALG TGVTFLVCCV LAAVLLRWRW RKRRGLENLE
1001 LLVHPPRIEH PITIQRPNVD YREVQVLPVA DSPGLARPHA HFASAGADAA
1051 GGGSPVPLLR TTSCCLEDLR PELLEEVKDI LIPEERLITH RSRVIGRGHF
1101 GSVYHGTYMD PLLGNLHCAV KSLHRITDLE EVEEFLREGI LMKSFHHPQV
1151 LSLLGVCLPR HGLPLVVLPY MRHGDLRHFI RAQERSPTVK ELIGFGLQVA
1201 LGMEYLAQKK FVHRDLAARN CMLDETLTVK VADFGLARDV FGKEYYSIRQ
1251 HRHAKLPVKW MALESLQTQK FTTKSDVWSF GVLMWELLTR GASPYPEVDP
1301 YDMARYLLRG RRLPQPQPCP DTLYGVMLSC WAPTPEERPS FSGLVCELER
1351 VLASLEGERY VNLAVTYVNL ESGPPFPPAP RGQLPDSEDE EDEEDEEDED
1401 AAVR (1404)
```

**Domain structure**

| | |
|---|---|
| 1–22 | Signal sequence (italics) |
| 297–303 | Proteolytic processing site |
| 964–986 | Transmembrane domain (underlined) |
| 1096–1357 | Tyrosine kinase domain |
| 1118–1121 | ATP binding site |
| 35, 408, 415, 454, 484, 516, 660, 722, 847, 905 | Potential *N*-linked glycosylation sites |

**Database accession numbers**

| | *PIR* | *SWISSPROT* | *EMBL/GENBANK* | *REFERENCES* |
|---|---|---|---|---|
| Chicken *Sea* | | | L12024 | 2 |

***References***


[1] Stubbs, E.L. and Furth, J. (1935) J. Exp. Med. 61, 593–616.
[2] Huff, J.L. et al. (1993) Proc. Natl. Acad. Sci. USA 90, 6140–6144.
[3] Huff, J.L. et al. (1996) Oncogene 12, 299–307.
[4] Crowe, A.J. et al. (1994) Oncogene 9, 537–544.
[5] Theillet, C. et al. (1990) Oncogene 5, 147–149.
[6] Larsen, J. et al. (1992) Oncogene 7, 1903–1911.
[7] Crowe, A.J. and Hayman, M.J. (1991) J. Virol. 65, 2533–2538.
[8] Beug, H. and Graf, T. (1989) Eur. J. Clin. Invest. 19, 491–502.
[9] Adkins, B. et al. (1984) Cell 39, 439–445.
[10] Crowe, A.J. and Hayman, M.J. (1993) Oncogene 8, 181–189.

## Identification

v-*ski* is the common oncogene of Sloan–Kettering viruses (SKVs), a group of acutely transforming chicken retroviruses. It was first detected in chick embryo fibroblasts infected with an originally non-transforming ALV strain, from which three isolates (SKV770, SKV780, SKV790) were prepared[1]. SKI was detected by screening cDNA with a v-*ski* probe[2].

## Related genes

*SNOA*, *SNOI* and *SNON* (SKI-related novel) are produced by alternative splicing of the same gene (giving different C-termini) and *SNOI* utilizes an alternative third exon[2,3]. *SKI* and *SNO* are closely related but show no marked sequence homology to other oncogenes. SKI proteins contain an extensive C-terminal helical domain that has homology with domains present in myosin, intermediate filaments and lamins.

| | ***SKI*** | **v-*ski*** |
|---|---|---|
| **Nucleotides (kb)** | >70 (chicken) | 3.0–8.9 (SKV-derived genomes) |
| **Chromosome** | 1q22–q24 | |
| **Mass (kDa): predicted** | 80 | $p49^{v\text{-}ski}$ |
| **expressed (chicken)** | p90 (7 exons) p50 (lacking exon 7), p60 (lacking exon 6) | $p125^{\Delta gag\text{-}ski}$; $p110^{\Delta gag\text{-}ski\text{-}pol}$; $p45^{\Delta gag\text{-}ski}$ |
| **Cellular location** | Nucleus | Nucleus |

**Tissue distribution**
Detectable at low levels in all chicken and quail tissues[4]. *Xenopus Ski* RNA accumulates in developing oocytes: following fertilization, the level declines during the mid-blastula transition. In *Xenopus* adult tissues *Ski* expression is high in the lungs and ovaries[5].

## Protein function

Unknown. v-*ski* induces MyoD and myogenin expression and myogenesis in non-muscle cells[6]. These genes are also induced by a transformation-defective v-*ski* mutant that does not induce myotube formation. However, in *Ski* transgenic mice the levels of MyoD and myogenin are not affected[7]. The effect of wild-type v-*ski* is the opposite to that of v-*jun* which inhibits myogenic differentiation.

The combination of v-*ski* and v-*sea* is highly malignant, indicating that v-*ski* can cooperate with a tyrosine kinase oncogene, as does v-*erbA* with v-*erbB*[8].

The C-terminal helical domain, deleted in v-SKI, may permit the formation of homodimers or interaction with other proteins. Phosphorylation in the

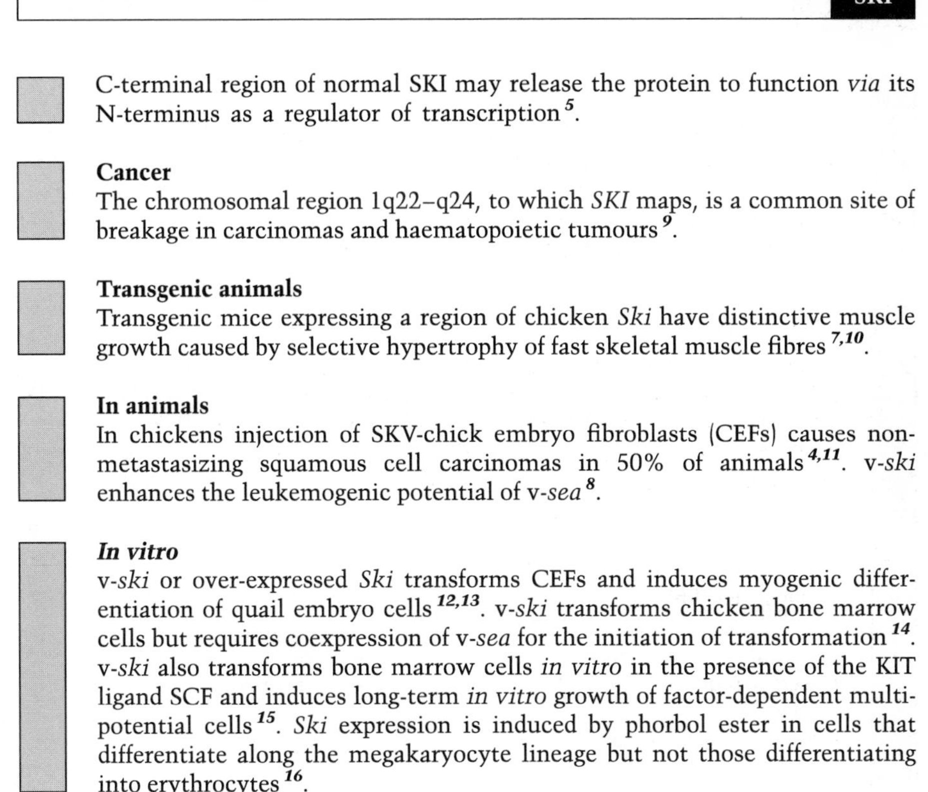

C-terminal region of normal SKI may release the protein to function *via* its N-terminus as a regulator of transcription[5].

**Cancer**
The chromosomal region 1q22–q24, to which *SKI* maps, is a common site of breakage in carcinomas and haematopoietic tumours[9].

**Transgenic animals**
Transgenic mice expressing a region of chicken *Ski* have distinctive muscle growth caused by selective hypertrophy of fast skeletal muscle fibres[7,10].

**In animals**
In chickens injection of SKV-chick embryo fibroblasts (CEFs) causes non-metastasizing squamous cell carcinomas in 50% of animals[4,11]. v-*ski* enhances the leukemogenic potential of v-*sea*[8].

***In vitro***
v-*ski* or over-expressed *Ski* transforms CEFs and induces myogenic differentiation of quail embryo cells[12,13]. v-*ski* transforms chicken bone marrow cells but requires coexpression of v-*sea* for the initiation of transformation[14]. v-*ski* also transforms bone marrow cells *in vitro* in the presence of the KIT ligand SCF and induces long-term *in vitro* growth of factor-dependent multipotential cells[15]. *Ski* expression is induced by phorbol ester in cells that differentiate along the megakaryocyte lineage but not those differentiating into erythrocytes[16].

## Partial structure of the chicken *Ski* gene

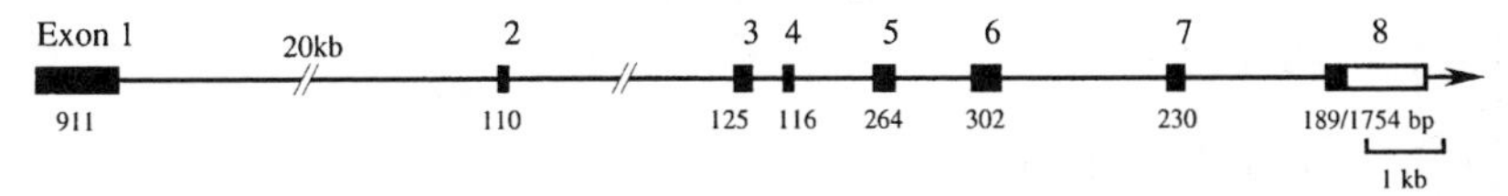


The region between exons 2 and 3 has not been fully mapped. The open box indicates the untranslated region of exon 8 and the arrow indicates the 3′ end has not been defined[17].

## Protein structure

v-SKI is highly homologous (91%) with human SKI except for a 37 amino acid insertion in v-SKI (280–316) that corresponds to exon 2 of chicken *Ski* from which it is derived and a 15 amino acid insertion in SKI (55–69).

## Amino acid sequence of human SKI

```
  1 MEAAAGGRGC FQPHPGLQKT LEQFHLSSMS SLGGPAAFSA RWAQEAYKKE
 51 SAKEAGAAAV PAPVPAATEP PPVLHLPAIQ PPPPVLPGPF FMPSDRSTER
101 CETVLEGETI SCFVVGGEKR LCLPQILNSV LRDFSLQQIN AVCDELHIYC
151 SRCTADQLEI LKVMGILPFS APSCGLITKT DAERLCNALL YGGAYPPPCK
201 KELAASLALG LELSERSVRV YHECFGKCKG LLVPELYSSP SAACIQCLDC
251 RLMYPPHKFV VHSHKALENR TCHWGFDSAN WRAYILLSQD YTGKEEQARL
```

```
301 GRCLDDVKEK FDYGNKYKRR VPRVSSEPPA SIRPKTDDTS SQSPAPSEKD
351 KPSSWLRTLA GSSNKSLGCV HPRQRLSAFR PWSPAVSASE KELSPHLPAL
401 IRDSFYSYKS FETAVAPNVA LAPPAQQKVV SSPPCAAAVS RAPEPLATCT
451 QPRKRKLTVD TPGAPETLAP VAAPEEDKDS EAEVEVESRE EFTSSLSSLS
501 SPSFTSSSSA KDLGSPGARA LPSAVPDAAA PADAPSGLEA ELEHLRQALE
551 GGLDTKEAKE KFLHEVVKMR VKQEEKLSAA LQAKRSLHQE LEFLRVAKKE
601 KLREATEAKR NLRKEIERLR AENEKKMKEA NESRLRLKRE LEQARQARVC
651 DKGCEAGRLR AKYSAQIEDL QVKLQHAEAD REQLRADLLR EREAREHLEK
701 VVKELQEQLW PRARPEAAGS EGAAELEP (728)
```

## Domain structure

| | |
|---|---|
| 112, 122, 143, 150, 153, 174, 187, 224, 228, 244, 247, 250 and 272 | 13 cysteine residues conserved among all known SKI/SNO proteins (underlined italics) |
| 538–561, 562–586, 587–611, 612–636 | Four major contiguous 25-mer repeated elements defined by the regular position of five hydrophobic, one acidic and two basic residues (underlined)[5]. |

## Database accession numbers

| | *PIR* | *SWISSPROT* | *EMBL/GENBANK* | *REFERENCES* |
|---|---|---|---|---|
| Human *SKI* | S06053 | P12755 | X15218 | 2 |
| Human *SNOA* | | P12756 | S06054, X15217 | 3 |
| Human *SNON* | | P12757 | X15219 | 2 |
| | | | S06052 | 3 |
| Human *SNOI* | | | Z19588 | 3 |

## *References*


1. Stavnezer, E. et al. (1981) J. Virol. 39, 920–934.
2. Nomura, N. et al. (1989) Nucleic Acids Res. 17, 5489–5500.
3. Pearson-White, S. et al. (1993) Nucleic Acids Res. 21, 4632–4638.
4. Stavnezer, E. et al. (1989) Mol. Cell. Biol. 9, 4038–4045.
5. Sleeman, J.P. and Laskey, R.A. (1993) Oncogene 8, 67–77.
6. Colmenares, C. et al. (1991) Mol. Cell. Biol. 11, 1167–1170.
7. Sutrave, P. et al. (1992) In Neuromuscular Development and Disease, Molecular and Cellular Biology, vol. 2, Eds A.M.Kelly and H.M.Blau. Raven Press, New York, pp. 107–114.
8. Larsen, J. et al. (1992) Oncogene 7, 1903–1911.
9. Koduru, P.R.K. et al. (1987) Blood 69, 97–102.
10. Sutrave, P. et al. (1990) Genes Dev. 4, 1462–1473.
11. **Sutrave, P. and Hughes, S.H. et al. (1991) Oncogene 6, 353–356.**
12. Colmenares, C. and Stavnezer, E. (1989) Cell 59, 293–303.
13. Colmenares, C. et al. (1991) J. Virol. 65, 4929–4935.
14. Larsen, J. et al. (1993) Oncogene 8, 3221–3228.
15. Beug, H. et al. (1995) Oncogene 11, 59–72.
16. Namciu, S. et al. (1994) Oncogene 9, 1407–1416.
17. Grimes, H.L. et al. (1992) Nucleic Acids Res. 20, 1511–1516.

## Identification

v-*src* is the transforming gene of the chicken Rous sarcoma virus (RSV)[1]. *SRC* was identified by screening human genomic DNA with a v-*src* probe[2,3].

## Related genes

*SRC*-like genes have been found in all species that have been examined. They are members of the superfamily of kinases[4]. In vertebrates eight genes of the *SRC* family have been identified (*SRC, Blk, FGR* (*SRC-2*), *FYN, HCK, LCK/Tkl, LYN, YES*) of which *SRC, FGR* and *YES* have viral homologues. All the proteins contain $Gly^2$ that becomes myristylated and contributes to membrane anchoring of the proteins; all share sequences throughout the SRC homology domains 1 (the tyrosine kinase catalytic region), 2 and 3 (SH1, SH2, SH3)[5] and all have the capacity to be regulated by phosphorylation of a common C-terminal tyrosine residue.

SH2 and/or SH3 domains also occur in the ABL/ARG family of non-receptor tyrosine kinases, the FPS/FES family, the growth factor receptor coupling proteins (SHCs and GRBs), v-AKT, the PTP1 and PTP2 tyrosine phosphatases, ATK, HTK-16, ITK, SYK, TEC, TYK and ZAP-70 tyrosine kinases, RAS-GAP, $p47^{gag\text{-}crk}$, NCK, the transcription factor ISGF3α, phospholipase $C\gamma_1$, PtdIns 3-kinase, ASH, yeast actin binding protein ABP1p, myosin-I, tensin and α-spectrin.

**Percentage sequence identity with SRC in domains of the SRC family proteins**

| | *Myristylation and unique domains* | *SH2 and SH3 domains* | *Catalytic domain* |
|---|---|---|---|
| YES | 22 | 74 | 89 |
| FYN | 20 | 67 | 81 |
| FGR | 11 | 57 | 78 |
| LCK | 4 | 50 | 67 |
| HCK | 17 | 56 | 69 |
| LYN | 11 | 52 | 66 |
| BLK | 18 | 53 | 67 |

| | ***SRC/Src*** | **v-*src*** |
|---|---|---|
| **Nucleotides (kb)** | ~60 | 7–9 (RSVs) |
| **Chromosome** | 20q13.3 | |
| **Mass (kDa): predicted** | 60 | 59 |
| **expressed** | 60 ($pp60^{SRC}$) | $pp60^{v\text{-}src}$ |

**Cellular location** Plasma membrane associated[6]
v-SRC stably associates with the detergent-insoluble cytoskeletal matrix (SH2 domain mediated); SRC does not[7].

**Tissue distribution**

SRC ($pp60^{SRC}$) is expressed in most avian and mammalian cells although it is barely detectable in lymphocytes. The highest concentrations of protein and tyrosine kinase activity occur in neuronal tissues[8] and in platelets where SRC comprises 0.2–0.4% of total protein. SRC is activated *via* G protein-coupled receptors (thrombin, $\alpha_2$-adrenergic and M1 muscarinic) in lung fibroblasts[9].

## Protein function

SRC is the prototype of the SRC family of membrane-associated protein tyrosine kinases[10]. The activity of v-SRC greatly exceeds that of SRC which is normally inhibited *in vivo* by nearly stoichiometric phosphorylation of $Tyr^{527}$. The specificity of SRC is remarkably broad (see Table), including L-amino acids and D-amino acids as well as aromatic and aliphatic alcohols[11].

SRC and/or other members of the SRC family (e.g. YES and FYN) are rapidly activated by the stimulation of a variety of transmembrane signalling receptors[12], including the PDGFR with which phosphorylated SRC associates. The SH3 domain is essential for signalling from the activated PDGF or EGF receptors and for the former but not the latter SRC binding *via* its SH2 domain is a prerequisite for activation[13]. Activation of the PDGFR subsequently leads to the release of activated SRC from the plasma membrane caused by cAMP-dependent phosphorylation of the SRC N-terminus[14]. Dominant negative forms of SRC can block PDGF-induced DNA synthesis, a block that is reversible by over-expression of *Myc* but not of *Fos* and/or *Jun*, suggesting that SRC kinases may control *Myc* transcription[15]. SRC is also phosphorylated during mitosis by $p34^{CDC2}$ and v-SRC can accelerate meiotic maturation. Phosphorylation may therefore sensitize SRC $Tyr^{527}$ to phosphatase action or desensitize it to a kinase and there is evidence that sustained expression of protein tyrosine phosphatase PTP$\alpha$ causes SRC activation, cell transformation and tumorigenesis[16–18]. However, the ubiquitous protein phosphatase 2A (PP2A) undergoes tyrosine phosphorylation with reduction in its activity in cells transformed by v-SRC[19]. The activation of $\alpha_2$-adrenergic receptors involves the G$\beta\gamma$ subunit-mediated formation of SHC/SRC complexes that are early events in RAS-dependent activation of the MAP kinase pathway[20].

SRC forms complexes with polyoma middle T antigen in polyoma-transformed cells. This augments many-fold the tyrosine kinase activity of SRC and is necessary but not sufficient for transformation by polyomavirus. Activation of SRC is partly caused by the enzyme being locked in a conformation that prevents its being negatively regulated by $Tyr^{527}$ phosphorylation[21–24].

v-SRC induces the cell type-specific activation of MAP kinase kinase[25] and the sustained transcription of the *Egr-1*, *Junb*, *TIS10* and *CEF-4/9E3* genes that are transiently activated by exposure of fibroblasts to serum[26,27]. v-SRC also activates the prostaglandin synthase-2 gene[28]. v-SRC does not activate *Fos* transcription in murine fibroblasts, although it causes detectable activation in chick embryo fibroblasts[29]. v-SRC causes multiple alterations in the metabolism of phosphatidylinositol and its derivatives, stimulating PtdIns$P_2$ hydrolysis and the accumulation of inositol trisphosphate, enhancing the activity of Ins(1,4,5)$P_3$ 3-kinase and activating PtdIns 3-kinase[30,31]. Transformation of fibroblasts by v-SRC also activates S6 kinase II, although it is not a substrate for v-SRC *in vitro*[32,33]. Transformation by v-SRC also causes the tyrosine phosphorylation of STAT3 (signal transducer and activator of transcription 3) which is accompanied by induction of DNA binding activity[34].

v-SRC, but not SRC, associates with the 90 kDa heat shock protein (HSP90) during or immediately after synthesis. This interaction decreases v-SRC kinase activity during transfer to the plasma membrane and modulates its specificity[35]. v-SRC also activates the HIV-1 LTR *via* its two NF-κB binding sites[36]. v-SRC, v-MOS or *Hras* expression enhances the *trans*-activating capacity of NET, an ETS-related factor that has sequence similarity to three regions of ELK1 and SAP1[37]. In avian cells transformed by the Rous sarcoma virus v-SRC activates the *BCL2*-related gene *NR-13*[38].

**SH2 and SH3 domains**

SH2 domains regulate protein interactions by binding directly to tyrosine phosphorylated proteins and the SH2 and SH3 domains of SRC regulate substrate specificity in a host-dependent manner and are thus important for transformation. PtdIns 3-kinase, which possesses two SH2 domains, binds to v-SRC and also to SRC in cells transformed by polyomavirus. v-SRC causes tyrosine phosphorylation of the SH2-containing SHC proteins and promotes their association with the GRB2 SH2 domain: GRB2 can in turn couple *via* SOS to RAS activation. RAS-GAP binds to the SH2 domain of SRC but is tyrosine phosphorylated only in complexes with v-SRC from transformed cells[39]. SRC SH3 (and also that of FYN and LYN) binds SHC, p62 (which can also bind to RAS) and heterogeneous nuclear ribonuclear protein K (a pre-mRNA binding protein)[40]. The RNA binding protein p68 (closely related to the RAS-GAP binding protein p62) is tyrosine phosphorylated and associates with SRC during mitosis *via* SRC SH2 and SH3 domains. SRC may therefore regulate processing or translation of RNA in a cell cycle-dependent fashion.[41,42]. The SRC SH3 domain binds specifically to the sequence Ψ(aliphatic) XXRPLPXLP[43].

**SRC substrates**

The only known function of SRC is as a tyrosine kinase and it is generally believed that the loss of anchorage dependence and growth control, together with the changes in metabolite transport and organization of the cytoskeleton that occur in v-SRC-transformed cells derive from the phosphorylation of specific target proteins[44]. The activation of SRC correlates with the phosphorylation on tyrosine residues of a wide range of substrates, many of which are similarly phosphorylated during the mitogenesis of normal cells and in transformed cells. However, in intact cells essential targets of SRC involved in regulating cell proliferation have not been identified, although the activation by phosphorylation of focal adhesion kinase ($p125^{FAK}$), a possible mediator of integrin signalling that contains a high affinity binding site for the SRC SH2 domain[27], may represent a common point of convergence of growth factor- and SRC-activated pathways.

***Proteins phosphorylated by* SRC**

p42, p50, p60, p75, p80/85*EMS1* [45], p130 [46], p120 [47], gp130, p62 [40]
α-Fodrin, α- and β-tubulin and the microtubule-associated proteins MAP2 and tau [48–50]
Annexin II (p36, calpactin I or lipocortin II) [51]
Calmodulin and clathrin [52]
Catenin
CD3-ζ [53]
Connexin 43 [54]
Cortactin (p80/85) [55]
Cytoskeletal proteins p110 (actin filament-associated protein (AFAP-110) [46]
Enolase, lactate dehydrogenase, phosphoglycerate mutase [56]
G protein α subunits [57]
IGF-I receptor [58]
Integrin [59]
NCK [60]
p125*FAK* [61–64]
p34*CDC2* [65,66]
p68 RNA binding protein [41,42]
Paxillin [67]
Platelet fibrinogen receptor (gpIIb/gpIIIa) [68,69]
Protein tyrosine phosphatases PTP1C, PTP1D/SYP [70]
Low molecular weight phosphotyrosine-protein phosphatase [71]
RAS-GAP [72–74]
Ribonuclear protein K [40]
SHC [40]
SIN (SRC interacting or signal integrating protein) [75]
STAT3 [76]
Synaptophysin [48,49,77]
SYP (SH-PTP2) [78]
Talin [79]
Vinculin [80–82]
YRP [83]

**Regulation of SRC tyrosine kinase activity**

(a) In normal, unstimulated cells SRC $Tyr^{527}$ is phosphorylated and its interaction with the SH2 domain results in very low kinase activity [84].

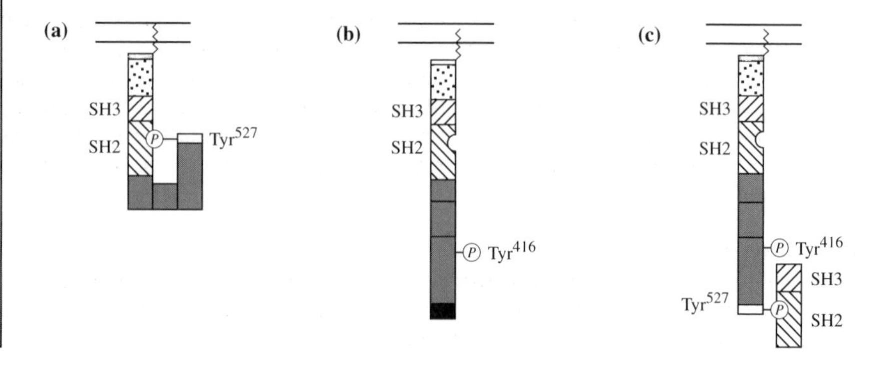

Treatment with phosphatase or an antibody directed against the C-terminal region of SRC generates a kinase activity comparable to that of v-SRC [85]. The substitution of $Tyr^{527}$ by Phe produces an oncogenic mutant that has high kinase activity throughout the cell cycle [86–89].
(b) v-SRC in which the absence of $Tyr^{527}$ permits constitutive activation of the kinase.
(c) Hypothetical model for the activation of SRC by a cellular SH2 domain protein associating with $Tyr^{527}$.

**Cancer**
SRC protein kinase activity is enhanced in some human colon cancers [90,91], skin tumours [92] and breast carcinomas. Association of SRC with EGFR and also with HER2 and p89 has been detected in human breast carcinoma cell lines [93]. SRC expression and kinase activity are increased in some human neuroblastoma-derived cells lines.

TGF$\beta$ downregulates the expression and activity of SRC (and of $pp53/56^{lyn}$) in the human prostate carcinoma cell line PC-3 [94]. There is a concomitant increase in the level of unphosphorylated SHC protein, suggesting that this pathway may negatively regulate RAS activity. In a variety of human cell lines hypoxia increases levels of *SRC* mRNA and activates SRC kinase, a target of which is the activation of transcription of vascular endothelial growth factor (VEGF) which encodes a powerful angiogenic protein [95].

**Transgenic animals**
Mice homozygous for a null mutation in *Src* have impaired osteoclast function, are deficient in bone remodelling and develop osteopetrosis [96–98]. Cells from such animals express reduced levels of osteopontin (OPN) compared with those from $Src^{+/-}$ or $Src^{+/+}$ animals, suggesting that SRC may regulate *OPN* expression [99].

Rapidly metastasizing tumours that arise in transgenic mice expressing polyomavirus middle T antigen are rarely detected when *Src* is inactivated [100]. Transgenic mice expressing high levels of v-SRC kinase activity die by midgestation with the formation of twin or multiple embryos [101].

Fibroblasts from $Src^{-/-}$ mice have decreased rates of spreading on fibronectin: the normal rate is restored by the expression of *Src* but the effect is independent of kinase activity, being mediated by the SH2 and SH3 domains [102].

CSK is a negative regulator of SRC family kinases and several cytoskeletal proteins are hyperphosphorylated on tyrosine residues in $Csk^{-}$ cells: regulation of cortactin and tensin hyperphosphorylation is SRC-dependent whereas FAK and paxillin hyperphosphorylation is partly dependent on both SRC and FYN [103]. The $Src^{-}$ mutation can restore normal cortactin distribution and partially correct the filamentous actin organization in $Csk^{-}$ cells.

**In animals**
The original RSV strain is tumorigenic in only a few strains of chicken: later variants are tumorigenic in a range of avian species. Some *src* variants induce cellular anti-tumour immunity [104].

***In vitro***

Infection with RSV transforms many types of cell. Expression of v-SRC is sufficient to initiate and maintain cellular transformation of chicken or mammalian fibroblasts. Chicken SRC is non-transforming, even when expressed at high levels [105–107], although over-expressed mouse SRC can induce tumorigenicity in NIH 3T3 fibroblasts [108]. Over-expression of *Src* in mouse fibroblasts potentiates the tumorigenic effect of over-expressed EGFR [109].

Transformation by v-SRC downregulates transcription of *Nov* in chicken embryo fibroblasts and of the *QR1* gene in the embryonic avian neuroretina [110]. In rat fetal lung cells SRC is activated and translocated to the cytoskeletal fraction by mechanical strain [111]. In granulocytes and the colonic cell line HT29, SRC associates with the adhesion molecule biliary glycoprotein (CD66a) [112].

## Gene structure

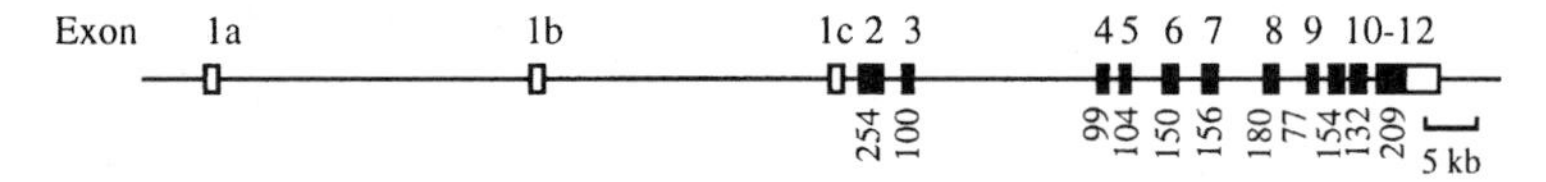


An additional fourth 5′ non-coding exon is present in chicken *Src*. Multiple mRNAs are generated by the differential splicing of the human or chicken non-coding exons and by the use of distinct initiation sites: all have the potential to encode SRC, as their 5′ exons are all eventually joined to exon 2.

An alternative splicing mechanism occurs in normal avian and rodent neurones and in some human neuroblastomas that yields a variant ($p60^{src+}$) with 6, 11 or 17 additional amino acids [113].

## Transcriptional regulation

The human *SRC* promoter has a high GC content and several SP1 and AP2 sites but no TATA or CAAT boxes [114].

## Protein structure

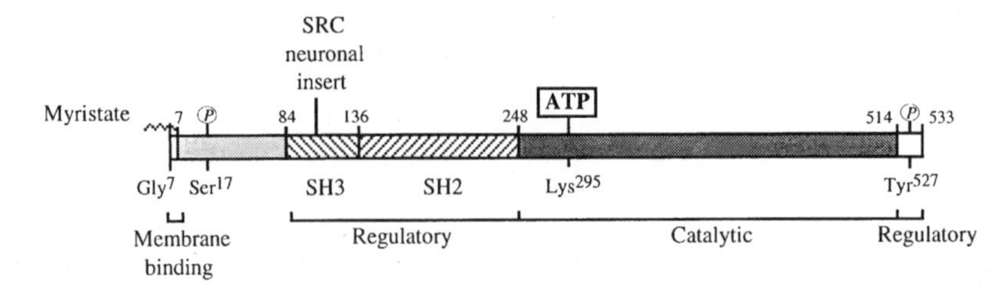


Chicken SRC (shown above) and human SRC are 94.6% identical. v-*src* is derived from chicken *Src*. The C-terminal regions (12 amino acids in v-SRC and 19 in SRC) are unrelated in sequence.

SRC can be converted to a transforming protein by various amino acid substitutions[115,116], by replacement or truncation of the C-terminus[89,117] or by dephosphorylation of Tyr$^{527}$[21]. Mutations of Lys$^{295}$ or in the vicinity of the consensus sequence block kinase activity and transformation[118,119]. The site of SRC *trans*-phosphorylation, Tyr$^{416}$, is highly conserved in tyrosine kinases and is the major phosphorylation site in v-SRC (Tyr$^{527}$ is deleted). All transforming SRC mutants have increased kinase activity that correlates with Tyr$^{416}$ *trans*-phosphorylation[120,121].

The three-dimensional structures of SRC SH2 and SH3 domains have been determined[122,123].

## Amino acid sequence of human SRC

1 MGSNKSKPKD ASQRRRSLEP AENVHGAGGG AFPASQTPSK PASADGHRGP
51 SAAFAPAAAE PKLFGGFNSS DTVTSPQRAG PLAGGVTTFV ALYDYESRTE
101 TDLSFKKGER LQIVNNTEGD WWLAHSLSTG QTGYIPSNYV APSDSIQAEE
151 *WYFGKITRRE SERLLLNAEN PRGTFLVRES ETTKGAYCLS VSDFDNAKGL*
201 *NVKHYKIRKL DSGGFYITSR TQFNSLQQLV AYYSKHADGL CHRLTT*VCPT
251 SKPQTQGLAK DAWEIPRESL RLEVKLGQGC FGEVWMGTWN GTTRVAIKTL
301 KPGTMSPEAF LQEAQVMKKL RHEKLVQLYA VVSEEPIYIV TEYMSKGSLL
351 DFLKGETGKY LRLPQLVDMA AQIASGMAYV ERMNYVHRDL RAANILVGEN
401 LVCKVADFGL ARLIEDNEYT ARQGAKFPIK WTAPEAALYG RFTIKSDVWS
451 FGILLTELTT KGRVPYPGMV NREVLDQVER GYRMPCPPEC PESLHDLMCQ
501 CWRKEPEERP TFEYLQAFLE DYFTSTEPQY QPGENL (536)

Underlined: differences between human and chicken SRC.

**Domain structure**

| | |
|---|---|
| 1–7 | Myristylation domain[64] |
| 2 | Myristate attachment site |
| 5, 7, 9, 14–16 | Six basic N-terminal residues (Lys$^{5,7,9}$, Arg$^{14-16}$) essential for high affinity binding to membranes for transformation by v-SRC[124] |
| 8–87 | Unique domain |
| 276–284 and 298 | ATP binding |
| 385 | Arg$^{385}$ is essential for high kinase activity and transformation by v-SRC[125] |
| 389 | Active site |
| 420 | Autophosphorylation site |
| 530 | Phosphorylation site |
| 84–145 | SH3 domain |
| 151–248 | SH2 domain (italics) |
| 249–517 | Catalytic domain |
| 518–536 | Regulatory domain |
| 93, 95, 134, 139 and 152 | Potential tyrosine phosphorylation sites: the phosphorylation of each of these amino acids does not directly affect kinase activity although, as the total kinase activity is dependent on the overall phosphorylation state of this region, it presumably influences the structure of SRC[126]. |

**Database accession numbers**

| | PIR | SWISSPROT | EMBL/GENBANK | REFERENCES |
|---|---|---|---|---|
| Human *SRC1* | A26891 | P12931 | M16243–M16245 | 2 |
| | | | K03212–K03218 | 127 |
| | | | X03995–X04000 | 3 |
| RSV (Prague C) v-*src* | A00632 | P00526 | V01197 | 128,129 |
| RSV v-*src* | A00631 | P00524 | X13745 | 130 |
| | | | V01169 | 131 |
| Chicken *src* | | | Ggsvc J00844 | 132 |

***References***


1 Radke, K. et al. (1980) Cell 21, 821–828.
2 Anderson, S.K. et al. (1985) Cell. Biol. 5, 1112–1129.
3 Parker, R.C. et al. (1985) Mol. Cell. Biol. 5, 831–838.
4 **Hanks, S.K. et al. (1988) Science 241, 42–52.**
5 **Pawson, T. and Gish, G.D. (1992) Cell 71, 359–362.**
6 Anand, R. et al. (1993) Oncogene 8, 3013–3020.
7 Okamura, H. and Resh, M.D. (1994) Oncogene 9, 2293–2303.
8 Brugge, J. et al. (1987) Genes Dev. 1, 287–296.
9 Chen, Y. et al. (1994) J. Biol. Chem. 269, 27372–27377.
10 **Jove, R. and Hanafusa, H. (1987) Annu. Rev. Cell Biol. 3, 31–56.**
11 Lee, T.R. et al. (1995) J. Biol. Chem. 270, 5375–5380.
12 **Eiseman, E. and Bolen, J.B. (1990) Cancer Cells 2, 303–310.**
13 Broome, M.A. and Hunter, T. (1996) J. Biol. Chem. 271, 16798–16806.
14 Walker, F. et al. (1993) J. Biol. Chem. 268, 19552–19558
15 Barone, M.V. and Courtneidge, S.A. (1995) Nature 378, 509–512.
16 Zheng, X.M. et al. (1992) Nature 359, 336–339.
17 den Hertog, J. et al. (1993) EMBO J. 12, 3789–3798.
18 Cobb, B.S. and Parsons, J.T. (1993) Oncogene 8, 2897–2903.
19 Chen, J. et al. (1994) J. Biol. Chem. 269, 7957–7962.
20 Luttrell, L.M. et al. (1996) J. Biol. Chem. 271, 19443–19450.
21 Courtneidge, S.A. (1985) EMBO J. 4, 1471–1477.
22 Bolen, J.B. et al. (1984) Cell 38, 767–777.
23 Cartwright, C.A. et al. (1985) Mol. Cell. Biol. 5, 2647–2652.
24 **Cooper, J.A. and Howell, B. (1993) Cell 73, 1051–1054.**
25 Gardner, A.M. et al. (1993) J. Biol. Chem. 268, 17896–17901.
26 Dehbi, M. et al. (1992) Mol. Cell. Biol. 12, 1490–1499.
27 Apel, I. et al. (1992) Mol. Cell. Biol. 12, 3356–3364.
28 Xie, W. and Herschman, H.R. (1995) J. Biol. Chem. 270, 27622–27628.
29 Catling, A.D. et al. (1993) Oncogene 8, 1875–1886.
30 Fukui, Y. et al. (1991) Oncogene 6, 407–411.
31 Ruggiero, M. et al. (1991) FEBS Lett. 291, 203–207.
32 Sweet, L.J. et al. (1990) Mol. Cell. Biol. 10, 2413–2417.
33 Chung, J. et al. (1991) Mol. Cell. Biol. 11, 1868–1874.
34 Yu, C.-L. et al. (1995) Science 269, 81–83.
35 Xu, Y. and Lindquist, S. (1993) Proc. Natl Acad. Sci. USA 90, 7074–7078.
36 Dehbi, M. et al. (1994) Oncogene 9, 2399–2403.
37 Giovane, A. et al. (1994) Genes Dev. 8, 1502–1513.
38 Gillet, G. et al. (1995) EMBO J. 14, 1372–1381.
39 Brott, B.K. et al. (1991) Proc. Natl Acad. Sci. USA 88, 755–759.
40 Weng, Z. et al. (1994) Mol. Cell. Biol. 14, 4509–4521.

[41] Fumagalli, S. et al. (1994) Nature 368, 871–874.
[42] Taylor, S.J. et al. (1995) J. Biol. Chem. 270, 10120–10124.
[43] Sparks, A.B. et al. (1996) Proc. Natl Acad. Sci. USA 93, 1540–1544.
[44] **Kellie, S. et al. (1991) J. Cell Sci. 99(2), 207–211.**
[45] Patel, A.M. et al. (1996) Oncogene 12, 31–35.
[46] Flynn, D.C. et al. (1993) Mol. Cell. Biol. 13, 7892–7900.
[47] Reynolds, A.B. et al. (1992) Oncogene 7, 2439–2445.
[48] Cheng, N. and Sahyoun, N. (1988) J. Biol. Chem. 263, 3935–3942.
[49] Matten, W.T. et al. (1990) J. Cell Biol. 111, 1959–1970.
[50] Akiyama, T. et al. (1986) J. Biol. Chem. 261, 14797–14803.
[51] Ozaki, T. and Sakiyama, S. (1993) Oncogene 8, 1707–1710.
[52] Fukami, Y. et al. (1986) Proc. Natl Acad. Sci. USA 83, 4190–4193.
[53] O'Shea, J.J. et al. (1991) Proc. Natl Acad. Sci. USA 88, 1741–1745.
[54] Loo, L.W.M. et al. (1992) J. Biol. Chem. 270, 12751–12761.
[55] Okamura and Resh, (1995) J. Biol. Chem. 270, 26613–26618.
[56] Cooper, J.A. et al. (1983) Nature 302, 218–223.
[57] Hausdorff, W.P. et al. (1992) Proc. Natl Acad. Sci. USA 89, 5720–5724.
[58] Peterson, J.E. et al. (1996) J. Biol. Chem. 271, 31562–31571.
[59] Aneskievich, B.J. et al. (1991) Oncogene 6, 1381–1390.
[60] Meisenhelder, J. and Hunter, T. (1992) Mol. Cell. Biol. 12, 5843–5856.
[61] Cobb, B.S. et al. (1994) Mol. Cell. Biol. 14, 147–155.
[62] Schaller, M.D. et al. (1994) Mol. Cell. Biol. 14, 1680–1688.
[63] Eide, B.L. et al. (1995) Mol. Cell. Biol. 15, 2819–2827.
[64] Calalb, M.B. et al. (1995) Cell. Biol. 15, 954–963.
[65] Ferris, D.K. et al. (1991) Cell Growth Differ. 2, 343–349.
[66] Cheng, H.-C. et al. (1991) J. Biol. Chem. 266, 17919–17925.
[67] Weng, Z. et al. (1993) J. Biol. Chem. 268, 14956–14963.
[68] Findik, D. et al. (1990) FEBS Lett. 262, 1–4.
[69] Elmore, M.A. et al. (1990) FEBS Lett. 269, 283–287.
[70] Møller, N.P.H. et al. (1994) Proc. Natl Acad. Sci. USA 91, 7477–7481.
[71] Rigacci, S. et al. (1996) J. Biol. Chem. 271, 1278–1281.
[72] Brott, B.K. et al. (1991) Mol. Cell. Biol. 11, 5059–5067.
[73] Brott, B.K. et al. (1991) Proc. Natl Acad. Sci. USA 88, 755–759.
[74] DeClue, J.E. et al. (1991) Mol. Cell. Biol. 11, 2819–2825.
[75] Alexandropoulos, K. and Baltimore, D. (1996) Genes Dev. 10, 1341–1355.
[76] Cao, X. et al. (1996) Mol. Cell. Biol. 16, 1595–1603.
[77] Barnekow, A. et al. (1990) Oncogene 5, 1019–1024.
[78] Peng, Z.-Y. and Cartwright, C.A. (1995) Oncogene 11, 1955–1962.
[79] Volberg, T. et al. (1991) Cell. Regul. 2, 105–120.
[80] Sefton, B.M. et al. (1981) Cell 24, 165–174.
[81] Sefton, B.M. et al. (1982) Cold Spring Harbor Symp. Quant. Biol. 46, 939–951.
[82] Felice, G. et al. (1990) Biochem. Soc. Trans. 18, 69–72.
[83] Scholz, G. et al. (1995) Proc. Natl Acad. Sci. USA 89, 2592–2596.
[84] Liu, X. et al. (1993) Oncogene 8, 1119–1126.
[85] Cooper, J.A. and King, C.S. (1986) Mol. Cell. Biol. 6, 4467–4477.
[86] Cartwright, C.A. et al. (1987) Cell 49, 83–91.
[87] Kmiecik, T.E. and Shalloway, D. (1987) Cell 49, 65–73.
[88] Piwnica-Worms, H. et al. (1987) Cell 49, 75–82.
[89] Reynolds, A.B. et al. (1987) EMBO J. 6, 2359–2364.

90 Garcia, R. et al. (1991) Oncogene 6, 1983–1989.
91 Park, J. and Cartwright, C.A. (1995) Mol. Cell. Biol. 15, 2374–2382.
92 Barnekow, A. et al. (1987) Cancer Res. 47, 235–240.
93 Muthuswamy, S.K. and Muller, W.J. (1995) Oncogene 11, 1801–1810.
94 Atfi, A. et al. (1994) J. Biol. Chem. 269, 30688–30693.
95 Mukhopadhyay, D. et al. (1995) Nature 375, 577–581.
96 Soriano, P. et al. (1991) Cell 64, 693–702.
97 Lowe, C. et al. (1993) Proc. Natl Acad. Sci. USA 90, 4485–4489.
98 Lowell, C.A. and Soriano, P. (1996) Genes Dev. 10, 1845–1857.
99 Chackalaparampil, I. et al. (1996) Oncogene 12, 1457–1467.
100 Guy, C.T. et al. (1994) Genes Dev. 8, 23–32.
101 Boulter, C.A. et al. (1991) Development 111, 357–366.
102 Kaplan, K.B. (1995) Genes Dev. 9, 1505–1517.
103 Thomas, S.M. et al. (1995) Nature 376, 267–271.
104 Gelman, I.H. et al. (1993) Oncogene 8, 2995–3004.
105 Iba, H. et al. (1984) Proc. Natl Acad. Sci. USA 81, 4424–4428.
106 Parker, R.C. et al. (1984) Cell 37, 131–139.
107 Shalloway, D. et al. (1984) Proc. Natl Acad. Sci. USA 81, 7071–7075.
108 Lin, P.-H. et al. (1995) Oncogene 10, 401–405.
109 Maa, M.-C. et al. (1995) Proc. Natl Acad. Sci. USA 92, 6981–6985.
110 Scholz, G. et al. (1996) Mol. Cell. Biol. 16, 481–486.
111 Liu, M. et al. (1996) J. Biol. Chem. 271, 7066–7071.
112 Brümmer, J. et al. (1995) Oncogene 11, 1649–1655.
113 Wiestler, O.D. and Walter, G. (1988) Mol. Cell. Biol. 8, 502–504.
114 Bonham, K. and Fujita, D.J. (1993) Oncogene 8, 1973–1981.
115 Kato, J.-Y. et al. (1986) Mol. Cell. Biol. 6, 4155–4160.
116 Levy, J.B. et al. (1986) Proc. Natl Acad. Sci. USA 83, 4228–4232.
117 Yaciuk, P. et al. (1989) Mol. Cell. Biol. 9, 2453–2463.
118 Kamps, M.P. and Sefton, B.M. (1986) Mol. Cell. Biol. 6, 751–757.
119 DeClue, J.E. and Martin, G.S. (1989) J. Virol. 63, 542–554.
120 Jove, R. et al. (1989) Oncogene Res. 5, 49–60.
121 Sato, M. et al. (1989) J. Virol. 63, 683–688.
122 Waksman, G. et al. (1993) Cell 72, 779–790.
123 Yu, H. et al. (1992) Science 258, 1665–1668.
124 Sigal, C.T. et al. (1994) Proc. Natl Acad. Sci. USA 91, 12253–12257.
125 Senften, M. et al. (1995) Oncogene 10, 199–203.
126 Espino, P.C. et al. (1990) Oncogene 5, 283–93.
127 Tanaka, A. et al. (1987) Mol. Cell. Biol. 7, 1978–1983.
128 Schwartz, D.E. et al. (1983) Cell 32, 853–869.
129 Neil, J.C. et al. (1981) Nature 291, 675–677.
130 Barnier, J.V. et al. (1989) Nucleic Acids Res. 17, 1252.
131 Czernilofsky, A.P. et al. (1983) Nature 301, 736–738.
132 Takeya, T. and Hanafusa, H. (1983) Cell 32, 881–890 (and correction: Cell 34, 319, 1983).

## Identification

*TAL1* (*T* cell *a*cute *l*eukemia *1*, also called *SCL* (*s*tem *c*ell *l*eukemia) or *TCL5*) was identified in a chromosome translocation in a stem cell leukaemia [1].

## Related genes

TAL1 is a member of the basic helix-loop-helix (bHLH) family closely related to TAL2 and LYL1.

| | ***TAL1*** |
|---|---|
| **Nucleotides (kb)** | 16 |
| **Chromosome** | 1p32 |
| **Mass (kDa): predicted** | 34 |
| **expressed** | 22/42 |
| **Cellular location** | Nucleus |

**Tissue distribution**

Expressed in developing brain, normal bone marrow and mast cells, mast cell lines, leukaemic T cell, megakaryocytic and erythroleukaemic cell lines [2] and in endothelial cells [3] but not in normal T cells.

Complexes of TAL1 and RBTN2 occur in the nucleus of erythroid cells and TAL1/RBTN1 complexes have been detected in a T cell acute leukaemia cell line [4].

## Protein function

TAL1 is a transcription factor that forms heterodimers with any of the known class A bHLH proteins (E12, E47, E2-2 and HEB) that bind to the E box (CANNTG) eukaryotic enhancer element, as does MyoD [5]. In mice TAL1 is essential for haematopoiesis. In erythroid cell differentiation *TAL1* expression is regulated by the erythroid transcription factor GATA1 [6], although levels of TAL1 protein may fall during differentiation as mRNA levels increase [7]. In the early myeloid cell line 416B, GATA1 but not TAL1 induces differentiation into megakaryocytes [8]. Transcription of *TAL1* and phosphorylation of TAL1 is stimulated by erythropoietin [9]. TAL1 may be involved in neural and endothelial cell differentiation [2,3].

Ectopic expression of *TAL1* blocks induction of myotube differentiation by myogenin in fibroblasts [10]. TAL1 also associates *via* its HLH domain with the GTP binding protein DRG [11]. Depending on the availability of negative regulatory proteins such as Id, TAL1 can potentially either activate or suppress transcription [12,13].

The N-terminal domain of TAL1, ectopically activated in T cell acute leukaemias after chromosomal abnormalities caused by V-D-J recombinase error, has specific *trans*-activating capacity [14].

**Cancer**

*TAL1* rearrangements occur in ~25% of childhood T cell acute lymphoblastic leukaemias (T-ALLs). 3% of these rearrangements are the result of the translocation (1;14)(p32;q11) that transposes *TAL1* into the T cell receptor $\delta$ gene, resulting in elevated expression of *TAL1* mRNA. The remainder of the rearrangements are submicroscopic and involve an almost precise 90 kb deletion of *TAL1*. The incidence of *TAL1* rearrangements in adult ALL is low [15]. The 90 kb deletion ($tal^{d}$ or $tal^{d1}$) occurs upstream from one allele of the *TAL1* locus, probably due to aberrant Ig recombinase activity [16,17]. A second specific deletion ($tal^{d2}$) occurs in 6% of T-ALLs. The (1;14) translocations and both $tal^{d}$ deletions disrupt the 5′ end of the *TAL1* gene so that its expression is controlled by the regulatory elements of the TCR$\delta$ or *SIL* genes that are both normally expressed in T cell ontogeny [18].

In ~20% of T-ALL *SIL* (SCL interrupting locus, chromosome 1p33), a highly conserved mammalian gene, fuses with *TAL1*, even though neither encodes TCR genes that normally rearrange during T cell ontogeny and are common sites for chromosomal translocation in T cell leukaemias. The rearrangement is similar in effect to the (1;14)(p32;q11) translocation. There is also one example in which recombination with TCR-$\delta$ affects the 3′ side of *TAL1*, transcription being initiated from a promoter in the fourth exon [19] and one in which the gene is not disrupted, the breakpoint occurring 25 kb downstream of *TAL1* [20].

*TCTA* (T-cell leukaemia translocation-associated gene) undergoes translocation (t(1;3)(p34;p21)) to place it head-to-head with *TAL1* [21]. Generation of a fusion protein is therefore unlikely.

**In animals**

*TAL1* cooperates with v-ABL in promoting tumorigenesis when *TAL1* is expressed from a synthetic retrovirus in v-*abl*-transformed T cells injected into mice [22].

**Transgenic animals**

*Tal-1*$^{-/-}$ embryos die at the yolk sac stage and the absence of TAL1 prevents the generation of red blood cells, myeloid cells, megakaryocytes, mast cells, T cells and B cells [23].

The N-terminal domain of TAL1, ectopically activated in T cell acute leukaemias after chromosomal abnormalities caused by V-D-J recombinase error, has specific *trans*-activating capacity [24]. Nevertheless, transgenic mice over-expressing the CD2-*Tal-1* transgene do not develop tumours or show enhanced susceptibility to Moloney murine leukaemia virus [25]. However, expression of *Tal-1* driven by the *Lck* promoter does promote clonal T cell lymphoblastic leukaemia/lymphoma [26]. Furthermore, expression of *Tal-1* synergizes with that of either *Lmo2/Rbtn2* or casein kinase II$\alpha$ in accelerating leukaemogenesis [26,27].

## Gene structure

*TAL1* type A: exons 1a–4–5/6; type B: exons 1b–2–3–4–5/6. Alternative splicing can link exon 1a to 2 or 1b to 4 or utilize an alternative intron to

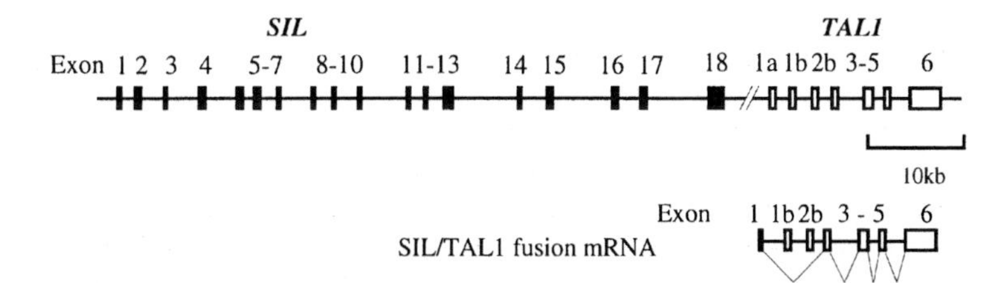


split exons 5 and 6. In *SIL/TAL1* an interstitial deletion removes most of the *SIL* gene, splicing *SIL* exon 1 to *TAL1* exon 3 and the 3′ *TAL1* exons (4, 5 and 6). The breakpoint regions in the *SIL* and *TAL1* genes (sildb and taldb1) are demethylated. Complete demethylation of sildb occurs in all T-ALL; complete demethylation of taldb1 occurs in most TCR$\alpha\beta^{+}$ and partial demethylation of taldb1 occurs in most TCR$\gamma\delta^{+}$ T-ALL[28]. The 5′ non-coding regions of both mouse and human TAL1 contain two GATA motifs (at −33 and −65) and one SP1 binding site (−59)[29].

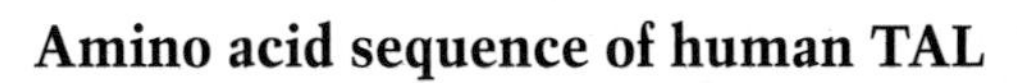

## Amino acid sequence of human TAL

```
  1 MTERPPSEAA RSDPQLEGRD AAEASMAPPH LVLLNGVAKE TSRAAAAEPP
 51 VIELGARGGP GGGPAGGGGA ARDLKGRDAA TAEARHRVPT TELCRPPGPA
101 PAPAPASVTA ELPGDGRMVQ LSPPALAAPA APGRALLYSL SQPLASLGSG
151 FFGEPDAFPM FTTNNRVKRR PSPYEMEITD GPHTKVVRRI FTNSRERWRQ
201 QNVNGAFAEL RKLIPTHPPD KKLSKNEILR LAMKYINFLA KLLNDQEEEG
251 TQRAKTGKDP VVGAGGGGGG GGGGAPPDDL LQDVLSPNSS CGSSLDGAAS
301 PDSYTEEPAP KHTARSLHPA MLPAADGAGP R (331)
```

### Domain structure

122 MAP kinase phosphorylation site
1–166 *Trans*-activation domain[30]
176–331 Sequence of p22$^{TAL1}$
188–239 Helix-loop-helix motif (underlined)
263–274 Polyglycine
122 Ser$^{122}$ is phosphorylated *in vitro* by ERK1 protein kinase and in intact cells after stimulation by EGF[31].
270 A G insertion at codon 270 modifies residues 272–278 (GGRGRRAPR), terminating at codon 279. The mutant form promotes premature apoptosis following medium depletion or serum reduction of Jurkat T cells[32].

### Database accession numbers

| | *PIR* | *SWISSPROT* | *EMBL/GENBANK* | *REFERENCES* |
|---|---|---|---|---|
| Human *TAL1* | A34519, A36358 | P17542 | M29038, M61103–M61105, M63572, M63576, M63584, M63589 | 16 |

### *References*


1 Begley, C.G. et al. (1989) Proc. Natl Acad. Sci. USA 86, 2031–2035.
2 Green, A.R. et al. (1992) Oncogene 7, 653–660.

[3] Hwang, L.-Y. et al. (1993) Oncogene 8, 3043–3046.
[4] Valge-Archer, V.E. et al. (1994) Proc. Natl Acad. Sci. USA 91, 8617–8621.
[5] Hsu, H.-L. et al. (1994) Mol. Cell. Biol. 14, 1256–1265.
[6] Aplan, P.D. et al. (1992) EMBO J. 11, 4073–4081.
[7] Murrell, A.M. et al. (1995) Oncogene 11, 131–139.
[8] Visvader, J.E. et al. (1992) EMBO J. 11, 4557–4564.
[9] Prasad, K.S.S. et al. (1995) J. Biol. Chem. 270, 11603–11611.
[10] Hofmann, T.J. and Cole, M.D. (1996) Oncogene 13, 617–624.
[11] Mahajan, M.A. et al. (1996) Oncogene 12, 2343–2350.
[12] Hsu, H.-L. et al. (1994) Proc. Natl Acad. Sci. USA 91, 5947–5951.
[13] Voronova, A.F. and Lee, F (1994) Proc. Natl Acad. Sci. USA 91, 5952–5956.
[14] Sánchez-García, I. and Rabbitts, T.H. (1994) Proc. Natl Acad. Sci. USA 91, 7869–7873.
[15] Stock, W. et al. (1995) Clin. Cancer Res. 1, 459–463.
[16] Aplan, P.D. et al. (1990) Mol. Cell. Biol. 10, 6426–6435.
[17] Brown, L. et al. (1990) EMBO J. 9, 3343–3351.
[18] Bernard, O. et al. (1991) Oncogene 6, 1477–1488.
[19] Bernard, O. et al. (1992) J. Exp. Med. 176, 919–925.
[20] Xia, Y. et al. (1992) Genes, Chromosom. Cancer 4, 211–216.
[21] Aplan, P.D. et al. (1995) Cancer Res. 55, 1917–1921.
[22] Elwood, N.J. et al. (1993) Oncogene 8, 3093–3101.
[23] Porcher, C. et al. (1996) Cell 86, 47–57.
[24] Sánchez-Garcia, I. and Rabbitts, T.H. (1994) Proc. Natl Acad. Sci. USA 91, 7869–7873.
[25] Robb, L. et al. (1995) Oncogene 10, 205–209.
[26] Kelliher, M.A. et al. (1996) EMBO J. 15, 5160–5166.
[27] Larson, R.C. et al. (1996) EMBO J. 15, 1021–1027.
[28] Breit, T.M. et al. (1994) Oncogene 9, 1847–1853.
[29] Lecointe, N. et al. (1994) Oncogene 9, 2623–2632.
[30] Wadman, I.A. et al. (1994) Oncogene 9, 3713–3716.
[31] Cheng, J.-T. et al. (1993) Mol. Cell. Biol. 13, 801–808.
[32] Leroy-Viard, K. et al. (1995) EMBO J. 14, 2341–2349.

## Identification

v-*erbA* and v-*erbB* are the oncogenes carried by the AEV-ES4 strain of avian erythroblastosis virus[1], detected by hybridization with cDNA probes directed against AEV DNA[2]. *THRA1/ErbA-1* encodes thyroid hormone receptor α. *THRB/ErbA-2* encodes thyroid hormone receptor β. Both are receptors for triiodothyronine ($T_3$), detected by screening cDNA libraries with a v-*erbA* probe[3–5].

## Related genes

*THRA1* and *THRB* are members of a superfamily numbering nearly 30 that includes the receptors for steroids, retinoic acid and vitamin $D_3$. The subfamily of thyroid hormone receptors (THRs) and retinoic acid receptors (RARs) comprises three RARs (α, β and γ), three retinoid X receptors (RXRα, β, γ), two THRs (α and β) and several "orphan receptors" for which ligands have yet to be identified.

Other *THRA1*-related genes are *EAR2* (erbA-related/*ERBAL2*), *EAR3* (*ERBAL3*) and *EAR3* (or chicken ovalbumin upstream promoter, COUP) which is closely similar to *Drosophila seven-up*.

| | ***THRA*** | ***THRB*** | **v-*erbA*** |
|---|---|---|---|
| **Nucleotides (kb)** | 27 | 60 | 5.7 (AEV-ES4) |
| **Chromosome** | 17q11.2–q12 | 3p24.1–p22 | |
| **Mass (kDa): predicted** | 47 (THRA1)/ 55 (THRA2) | 52 | 72 |
| **expressed** | 48/55 | 52/55 | $P75^{gag\text{-}erbA}$ |
| **Cellular location** | Nucleus | Nucleus | Nucleus |

**Tissue distribution**

*THRA1* is ubiquitous, *THRB* is restricted. Rat THRB is strongly expressed in the liver, thyroid, adrenal and anterior and posterior pituitary glands[6]. In chick brain ontogenesis *ErbA-1* is expressed from the early embryonic stages: *ErbA-2* is rapidly induced after embryonic day 19 and is substantially expressed in brain, lung, kidney, eye and yolk sac but is undetectable in haematopoietic tissues[7].

## Protein function

THRA1 and THRB are transcription factors possessing zinc finger domains. Human THRA1 and THRB are the high-affinity receptors ($K_d = 0.2$ nM) for $T_3$. THRA1 binds to thyroid hormone response elements (TREs) usually comprised of directly repeated half-sites of consensus sequence AGGTCA separated by 4 nt, repressing activity as a promoter[8]. $T_3$ binding to THRA1 activates transcription[9]. THRA2 does not bind $T_3$ and has been proposed to act as a dominant negative regulator of thyroid hormone receptors. There is a second functional form of human THRB (*THRB2/ErbA-2*).

Thyroid hormone receptor uncoupling protein (TRUP) interacts with the hinge region and the N-terminal portion of the ligand binding domain of THR in a hormone-independent manner to inhibit *trans*-activation by the THR [10].

Over-expression of THRB1 in murine muscle cells indicates that it is involved in triggering muscle terminal differentiation [11]. Thus, it blocks the activity of the myogenesis inhibitor AP1, increases $T_3$-induced *MyoD* expression and causes $T_3$ to stimulate growth arrest and terminal differentiation in the presence of serum factors acting, like SKI, as a positive regulator of muscle differentiation.

v-ERBA does not bind hormone or *trans*-activate and it acts as a constitutive repressor and an antagonist of thyroid hormone and retinoic acid receptors [12]. Thus v-ERBA blocks apoptosis of normal early erythrocyte progenitor cells induced by $T_3$ or retinoic acid in the absence of differentiation-inducing agents [13]. v-*erbA* cooperates with v-*erbB* and related sarcoma-inducing oncogenes (and with *Hras*) to block erythroid cell differentiation (into erythrocytes) and to promote transformation. v-ERBA suppresses transcription of avian erythrocyte anion transport (band III), carbonic anhydrase II and $\beta$-globin genes. The capacity of v-*erbA* to block differentiation correlates with transcriptional arrest of erythrocyte-specific genes [14]. Transcriptional activation *via* the AP1-induced TRE promoter by v-ERBA in rat thyroid epithelial cells inhibits iodide uptake and reduces thyrotropin growth dependence [15].

Rous sarcoma virus LTR contains a unique hormone responsive element (RSV-T3RE) that in the absence of $T_3$ mediates strong activation by THRA or v-ERBA homodimers or by THRA/RXR heterodimers, but not THRB. Activation is reversed by $T_3$ [16].

## Cancer

Loss of heterozygosity at the *THRA1* locus has been detected in sporadic breast cancers and in a breast cancer cell line (BT474) *THRA1* undergoes fusion to *BTR*, the resultant truncated form of *THRA1* resembling v-*erbA* [17]. In some colon carcinomas expression of the larger transcript (6 kb) of *THRB* is suppressed: expression of *THRA1* and *THRA2* is unaffected [18].

## Transgenic animals

Mice expressing v-*erbA* have breeding disorders, abnormal behaviour, reduced adipose tissue, hypothyroidism with inappropriate TSH response and enlarged seminal vesicles: males develop hepatocellular carcinoma [19]. These findings are consistent with the model of v-ERBA functioning as a dominant negative receptor.

## In animals

v-*erbA* alone is not tumorigenic but cooperates with v-*ets* to cause avian erythroleukaemia [20].

***In vitro***
v-*erbA* alone can transform erythrocytic progenitor cells and stimulate chick embryo fibroblast (CEF) growth, although it does not cause tumours. However, v-*erbA* enhances the tumorigenicity of v-*erbB*-transformed CEFs[21]. v-*erbA* affects transformed erythroblasts in two ways: (i) it increases the pH range within which the cells will grow and (ii) it blocks the differentiation of erythroblasts into erythrocytes. The potent repressor function of v-ERBA requires the formation of heterodimeric complexes with retinoid X receptor (RXR$\alpha$) and C-terminal mutations in v-ERBA that abolish heterodimer formation also block v-ERBA repressor function[22]. Despite being a constitutive repressor in animal cells, v-ERBA is a hormone-activated transcriptional activator in yeast. The functional domains of the protein required for activation of gene expression in yeast and for transformation of avian cells are closely similar[23].

However, v-*erbA* stimulates the differentiation of quail myoblasts by a mechanism that is independent of $T_3$. The mutation $Gly^{61} \rightarrow$ Ser in v-ERBA compared with ERBA is involved in the enhancement of myogenesis by v-ERBA[24].

## Gene structure of human *THRA1*

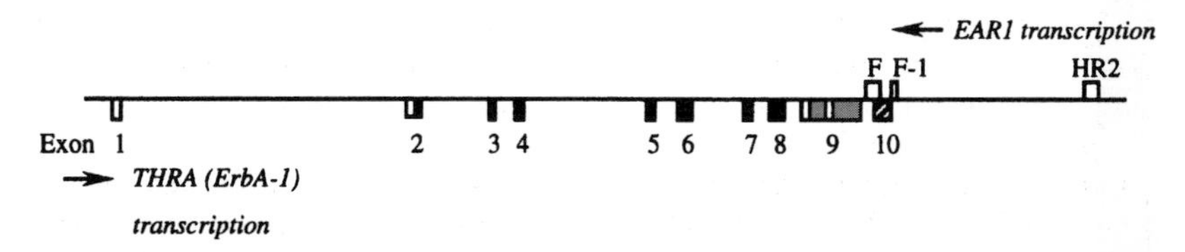


Boxes below the line represent *THRA1* (exons 1–9): open boxes above the line represent the exons of the *THRA*-related *EAR-1* gene (F: final, F-1: adjacent to final, HR2: homologue of rat exon 2) that is transcribed in the reverse direction[25]. The use of an alternative splice site in *THRA1* exon 9 generates *THRA2* (exons 1–10).

## Transcriptional regulation

The human *THRA* promoter lacks TATA elements but is very GC-rich and contains many SP1 sites as well as hormone responsive elements[26].

## Protein structure

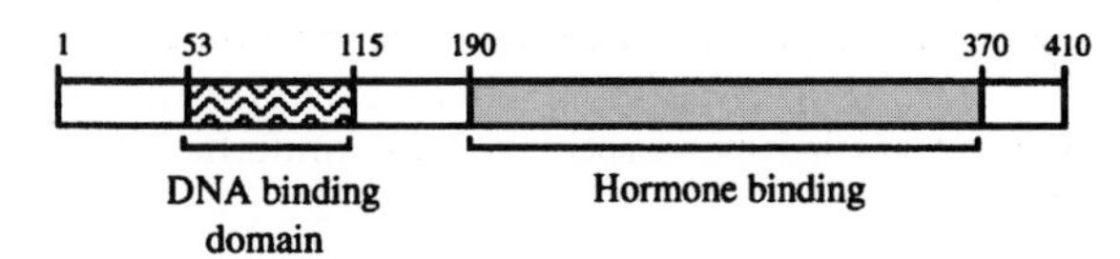


Human THRA1 differs from chicken ERBA-1, from which v-ERBA is derived, in having two additional amino acids in the N-terminal region and 41 scattered substitutions. $P75^{v\text{-}gag\text{-}erbA}$ is a highly mutated version of ERBA-1 with two point mutations in the DNA binding domain and 11 others in the hormone binding domain, a nine amino acid C-terminal deletion in the hormone binding domain and an N-terminal third encoded by *gag*. These changes result in the loss of $T_3$ binding capacity (though the hormone binding region is retained) but the retention of sequence-specific DNA binding. v-ERBA acts as a constitutive repressor of $T_3$-regulated genes. In the AEV-ES4 viral genome both the *ErbA* and *ErbB* genes are inserted between *gag* and *env* and are independently expressed, remaining colinear with the coding domains of their cellular progenitors.

The zinc finger motif contains a P-box helix that contacts the major groove of DNA. An additional α helix, the A-box, at the extreme C-terminus of the zinc finger motif makes minor groove contacts with bases at the 5′ end of the half-site. The differences in DNA recognition between ERBA and v-ERBA are principally determined by the N-terminal region in which a Cys substituted in v-ERBA for Tyr in the avian ERBA sequence is particularly important [27].

The N-terminal region (21–30) of chicken THRA contains five basic amino acids necessary for *trans*-activation and for interaction with the transcription factor TFIIB [28].

## Amino acid sequence of human THRA1

```
  1 MEQKPSKVEC GSDPEENSAR SPDGKRKRKN GQCSLKTSMS GYIPSYLDKD
 51 EQCVVCGDKA TGYHYRCITC EGCKGFFRRT IQKNLHPTYS CKYDSCCVID
101 KITRNQCQLC RFKKCIAVGM AMDLVLDDSK RVAKRKLIEQ NRERRRKEEM
151 IRSLQQRPEP TPEEWDLIHI ATEAHRSTNA QGSHWKQRRK FLPDDIGQSP
201 IVSMPDGDKV DLEAFSEFTK IITPAITRVV DFAKKLPMFS ELPCEDQIIL
251 LKGCCMEIMS LRAAVRYDPE SDTLTLSGEM AVKREQLKNG GLGVVSDAIF
301 ELGKSLSAFN LDDTEVALLQ AVLLMSTDRS GLLCVDKIEK SQEAYLLAFE
351 HYVNHRKHNI PHFWPKLLMK VTDLRMIGAC HASRFLHMKV ECPTELFPPL
401 FLEVFEDQEV (410)
```

**Domain structure**

| | |
|---|---|
| 1–52 | Modulating domain |
| 53–73 and 91–115 | Two C4-type zinc fingers |
| 158–179 | Putative *trans*-repression domain (underlined). Mutation of $Pro^{160}$ (or of the corresponding v-ERBA residue), between the DNA and hormone binding domains, abolishes the capacity to suppress basal transcription but does not prevent hormone-induced *trans*-activation by THRA1 [29]. v-ERBA may therefore function by *trans*-repression, rather than as a dominant negative inhibitor of THR or RAR activation |
| 190–370 | Hormone binding |
| 371–410 | C-terminal region (underlined) that differs from THRA2 (THRA1 and THRA2 are identical up to amino acid 370) |

**C-terminus of THRA2**

```
371 EREVQSSILY KGAAAEGRPG GSLGVHPEGQ QLLGMHVVQG PQVRQLEQQL
421 GEAGSLQGPV LQHQSPKSPQ QRLLELLHRS GILHARAVCG EDDSSEADSP
471 SSSEEEPEVC EDLAGNAASP (490)
```

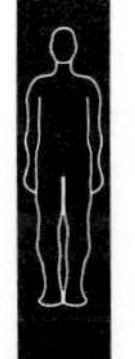

## Amino acid sequence of human THRB

```
  1 MTENGLTAWD KPKHCPDREH DWKLVGMSEA CLHRKSHSER RSTLKNEQSS
 51 PHLIQTTWTS SIFHLDHDDV NDQSVSSAQT FQTEEKKCKG YIPSYLDKDE
101 LCVVCGDKAT GYHYRCITCE GCKGFFRRTI QKNLHPSYSC KYEGKCVIDK
151 VTRNQCQECR FKKCIYVGMA TDLVLDDSKR LAKRKLIEEN REKRRREELQ
201 KSIGHKPEPT DEEWELIKTV TEAHVATNAQ GSHWKQKPKF LPEDIGQAPI
251 VNAPEGGKVD LEAFSHFTKI ITPAITRVVD FAKKLPMFCE LPCEDQIILL
301 KGCCMEIMSL RAAVRYDPES ETLTLNGEMA VIRGQLKNGG LGVVSDAIFD
351 LGMSLSSFNL DDTEVALLQA VLLMSSDRPG LACVERIEKY QDSFLLAFEH
401 YINYRKHHVT HFWPKLLMKV TDLRMIGACH ASRFLHMKVE CPTELLPPLF
451 LEVFED (456)
```

### Domain structure

| | |
|---|---|
| 1–101 | Modulating domain |
| 102–122 and 140–164 | Two C4-type zinc fingers |
| 239–456 | Hormone binding |

### Database accession numbers

| | *PIR* | *SWISSPROT* | *EMBL/GENBANK* | *REFERENCES* |
|---|---|---|---|---|
| Human *THRA1* | A30893, A40917, S06247, B32286 | P21205 | X55004–X55005, X55068–X55074, Y00479 | 4 |
| Human *THRA2* | A30893, S06247, B32286 | P10827 | J03239, X55004 | 3 |
| | | | X55066, X55069 | 25 |
| | | | X55071–X55074 | 30 |
| Human *THRB* | A25237 | P10828 | X04707 | 5,31 |

### *References*


1 Rothe Meyer, A. and Engelbreth-Holm, J. (1933) Acta Pathol. Microbiol. Scand. 10, 380–427.
2 Debuire, B. et al. (1984) Science 224, 1456–1459.
3 Pfahl, M. and Benbrook, D. (1987) Nucleic Acids Res. 15, 9613.
4 Nakai, A. et al. (1988) Proc. Natl Acad. Sci. USA 85, 2781–2785.
5 Weinberger, C. et al. (1986) Nature 324, 641–646.
6 Macchia, E. et al. (1990) Endocrinology 126, 3232–3239.
7 Forrest, D. et al. (1991) EMBO J. 10, 269–275.
8 Harbers, M. et al. (1996) Nucleic Acids Res. 24, 2252–2259.
9 Damm, K. et al. (1989) Nature 339, 593–597.
10 Burris, T.P. et al. (1995) Proc. Natl Acad. Sci. USA 92, 9525–9529.
11 Carnac, G. et al. (1993) Oncogene 8, 3103–3110.
12 Sande, S. et al. (1993) J. Virol. 67, 1067–1074.
13 Gandrillon, O. et al. (1994) Oncogene, 9, 749–758.
14 Zenke, M. et al. (1988) Cell 52, 107–119.
15 Trapasso, F. et al. (1996) Oncogene 12, 1879–1888.
16 Saatcioglu, F. et al. (1994) Cell 75, 1095–1105.
17 Futreal, P.A. et al. (1994) Cancer Res. 54, 1791–1794.
18 Markowitz, S. et al. (1989) J. Clin. Invest. 84, 1683–1687.
19 Barlow, C. et al. (1994) EMBO J. 13, 4241–4250.
20 Metz, T. and Graf, T. (1992) Oncogene 7, 597–605.
21 Zenke, M. et al. (1990) Cell 61, 1035–1049.
22 Yen, P.M. et al. (1994) J. Biol. Chem. 269, 903–909.

[23] Smit-McBride, Z. and Privalsky, M.L. (1993) Oncogene 8, 1465–1475.
[24] Cassar-Malek, I. et al. (1994) Oncogene 9, 2197–2206.
[25] Miyajima, N. et al. (1989) Cell 57, 31–39.
[26] Laudet, V. et al. (1993) Oncogene 8, 975–982.
[27] Jackson, C. and Privalsky, M.L. (1996) J. Biol. Chem. 271, 10800–10805.
[28] Hadzic, E. et al. (1995) Mol. Cell. Biol. 15, 4507–4517.
[29] Damm, K. and Evans, R.M. (1993) Proc. Natl Acad. Sci. USA 90, 10668–10672.
[30] Laudet, V. et al. (1991) Nucleic Acids Res. 19, 1105–1112.
[31] Sakurai, A. et al. (1990) Mol. Cell. Endocrinol. 71, 83–91.

# TIAM1

## Identification

*Tiam1* was cloned from a mouse T lymphoma cell line by proviral tagging in combination with *in vitro* selection of invasive T lymphoma variants [1]. Proviral insertion generated 5′ and 3′ truncated transcripts. The N-terminal and C-terminal truncated products induced invasiveness when transfected into non-invasive T lymphoma cells.

## Related genes

TIAM1 is a member of the class of GDP dissociation stimulators (GDS) for GTPases of the RAS superfamily.

| | ***TIAM1*** |
|---|---|
| **Nucleotides** | Not fully mapped |
| **Chromosome** | 21q22 [2] |

### Tissue distribution

*TIAM1* is expressed in a wide variety of human tumour cell lines including B and T lymphomas, neuroblastomas, melanomas and carcinomas [1].

## Protein function

TIAM1 activates the RHO-like GTPase RAC1 [3]. It contains a 240 amino acid DBL homology (DH) domain and two pleckstrin homology domains, in common with proteins that are thought to regulate the activity of RHO-like small GTPases (e.g. BCR, DBL, VAV, CDC24, ECT2, LBC, TIM and OST). TIAM1 carries a putative N-terminal myristylation signal and PEST sequences. Between the N-terminal PH domain and the DH domain, TIAM1 contains a discs-large homologous region (DHR), first identified in the *Drosophila* tumour suppressor gene *discs-large*.

TIAM1 or truncated forms of the protein confer invasiveness on normal cells.

### Transgenic animals

*Tiam-1* is activated by proviral insertion in tumours from double *Myc/Pim-1* transgenics and may therefore cooperate with the products of these genes.

## Amino acid sequence of human TIAM1

```
  1 MGNAESQHVE HEFYGEKHAS LGRNDTSRSL RLSHKTRRTR HASSGKVIHR
 51 NSEVSTRSSS TPSIPQSLAE NGLEPFSQDG TLEDFGSPIW VDRVDMGLRP
101 VSYTDSSVTP SVDSSIVLTA ASVQSMPDTE ESRLYGDDAT YLAEGGRRQH
151 SYTSNGPTFM ETASFKKKRS KSADIWREDS LEFSLSDLSQ EHLTSNEEIL
201 GSAEEKDCEE ARGMETRASP RQLSTCQRAN SLGDLYAQKN SGVTANMGPG
251 SKFAGYCRNL VSDIPNLANH KMPPAAAEET PPYSNYNTLP CRKSHCLSEG
301 ATNPQISHSN SMQGRRAKTT QDVNAGEGSE FADSGIEGAT TDTDLLSRRS
351 NATNSSYSPT TGRAFVGSDS GSSSTGDAAR QGVYENFRRE LEMSTTNSES
401 LEEAGSAHSD EQSSGTLSSP GQSDILLTAA QGTVRKAGAL AVKNFLVHKK
```

```
 451 NKKVESATRR KWKHYWVSLK GCTLFFYESD GRSGIDHNSI PKHAVWVENS
 501 IVQAVPEHPK KDFVFCLSNS LGDAFLFQTT SQTELENWIT AIHSACATAV
 551 ARHHHKEDTL RLLKSEIKKL EQKIDMDEKM KKMGEMQLSS VTDSKKKKTI
 601 LDQIFVWEQN LEQFQMDLFR FRCYLASLQG GELPNPKRLL AFASRPTKVA
 651 MGRLGIFSVS SFHALVAART GETGVRRRTQ AMSRSASKRR SRFSSLWGLD
 701 TTSKKKQGRP SINQVFGEGT EAVKKSLEGI FDDIVPDGKR EKEVVLPNVH
 751 QHNPDCDIWV HEYFTPSWFC LPNNQPALTV VRPGDTARDT LELICKTHQL
 801 DHSAHYLRLK FLIENKMQLY VPQPEEDIYE LLYKEIEICP KVTHSIHIEK
 851 SDTAADTYGF SLSSVEEDGI RRLYVNSVKE TGLASKKGLK AGDEILEINN
 901 RAADALNSSM LKDFLSQPSL GLLVRTYPEL EEGVELLESP PHRVDGPADL
 951 DESPLAFLTS NPGHSLCSEQ GSSAETAPEE TEGPDLESSD ETDHSSKSTE
1001 QVAAFCRSLH EMNPSDQNPS PQDSTGPQLA TMRQLSDADN VRKVICELLE
1051 TERTYVKDLN CLMERYLKPL QKETFLTQDE LDVLFGNLTE MVEFQVEFLK
1101 TLEDGVRLVP DLEKLEKVDQ FKKVLFSLGG SFLYYADRFK LYSAFCAIHT
1151 KVPKVLVKAK TDTAFKAFLD AQNPKQQHSS TLESYLIKPI QRILKYPLLL
1201 RELFALTDAE SEEHYHLDVA IKTMNKVASH INEMQKIHEE FGAVFDQLIA
1251 EQTGEKKEVA DLSMGDLLLH TTVIWLNPPA SLGKWKKEPE LAAFVFKTAV
1301 VLVYKDGSKQ KKKLVGSHRL SIYEDWDPFR FRHMIPTEAL QVRALASADA
1351 EANAVCEIVH VKSESEGRPE RVFHLCCSSP ESRKDFLKAV HSILRDKHRR
1401 QLLKTESLPS SQQYVPFGGK RLCALKGARP AMSRAVSAPS KSLGRRRRRL
1451 ARNRFTIDSD AVSASSPEKE SQQPPGGGDT DRWVEEQFDL AQYEEQDDIK
1501 ETDILSDDDE FCESVKGASV DRDLQERLQA TSISQRERGR KTLDSHASRM
1551 AQLKKQAALS GINGGLESAS EEVIWVRRED FAPSRKLNTE I (1591)
```

## Database accession numbers

| | PIR | SWISSPROT | EMBL/GENBANK | REFERENCES |
|---|---|---|---|---|
| Human *TIAM1* | | U16296 | HS162961 | 1 |
| | | | X86351, X86350, X86349 | 2 |
| Mouse *Tiam-1* | | | MM05245, U05245 | 4 |

## *References*


[1] Habets, G.G. et al. (1995) Oncogene 10, 1371–1376.
[2] Chen, H. and Antonarakis, S.E. (1995) Genomics 30, 123–127.
[3] van Leeuwen, F.N. et al. (1995) Oncogene 11, 2215–2221.
[4] Habets, G.G.M. et al. (1994) Cell 77, 537–549.

## Identification

*TRK* (tropomyosin-receptor-kinase) was originally detected by NIH 3T3 fibroblast transfection with genomic DNA from a colon carcinoma [1,2]. Additional *TRK* oncogenes have been isolated from thyroid tumours and over 40 *TRK* oncogenes have been generated *in vitro* [3].

## Related genes

*TRK* is a human transforming gene having no homology with known viral oncogenes. The extracellular regions of TRK proteins contain three leucine-rich motifs (LRMs) flanked by conserved cysteine residues, a characteristic of the LRM superfamily that includes human platelet von Willebrand factor receptor, ribonuclease/angiogenin inhibitor, cell adhesion proteins and extracellular matrix proteins. The extracellular domains also contain two $C_2$ Ig-like loops similar to those present in neural cell adhesion molecules and in the receptors for fibroblast growth factors, PDGF and CSF1, in KIT and in *Drosophila Dtrk* and *toll*. TRK also has homology with the receptor tyrosine kinase domains of EGFR, the insulin receptor family and the SRC family and with ROR1 and ROR2 and *Drosophila Dror*. A muscle-specific receptor in *Torpedo californica* contains a TRK-related kinase domain but also has a kringle domain close to the transmembrane region [4]. Human TRKA is 49% homologous to TRKB; TRKB and TRKC are 55% homologous [5]. TRK (human), TRKB (mouse) and TRKC (pig) share 67% overall amino acid homology. The catalytic domain of TRKE is 41% identical to that of TRKA (extracellular region 16%).

| | ***TRKA/TrkA*** | ***TRKB/TrkB*** | ***TRKC/TrkC*** |
|---|---|---|---|
| **Nucleotides (kb)** | 20 | >100 | Not fully mapped |
| **Chromosome** | 1q23–1q24 (*NTRK1*) | 9q22.1 (*NTRK2*) | 1q23–q31 (*NTRK3*) |
| **Mass (kDa): predicted** | 87<br>70 (TRK1)/49 (TRK2)<br>66 (TRK-T3) | 95 | 90 |
| **expressed** | gp140<br>70 (tropomyosin-TRK)<br>55 (TRK-T1)<br>68 (TRK-T3) | gp95/gp145 | $gp145^{trkC}$ K1<br>$gp145^{trkC}$ K2<br>$gp145^{trkC}$ K3 |
| **Cellular location** | Plasma membrane<br>Cytoplasm (TRK-T3) | Plasma membrane | Plasma membrane |

**Tissue distribution**

*TRKA*, *TRKB* and *TRKC* are primarily expressed in the nervous system [6]. In humans and rats the *TRKAII* isoform occurs in neuronal tissues and *TRKAI* is expressed mainly in non-neuronal tissues [7]. TRKC protein is expressed in pancreatic $\beta$ cells [8]. *TRKE* is the first member of the family found widely and abundantly expressed in human tissues (brain, placenta, lung, skeletal muscle, kidney, pancreas but not liver) [9,10].

In human neuroblastoma cell lines expression of both *TRKA* and *TRKB* mRNAs is increased by retinoic acid: interferon $\gamma$ selectively increases *TRKA* mRNA [11].

## Protein function

*TRK* genes (*TRK* (or *TRKA*), *TRKB*, *TRKC* and *TRKE*) encode transmembrane tyrosine kinases that are receptors for the nerve growth factor (NGF) family of neurotrophins (NGF, brain-derived neurotrophic factor (BDNF), NT-3 and NT-4, also called NT-5). TRK proteins probably mediate specific cell adhesion events during neuronal cell development [12]. TRK oncoproteins bind to SHC proteins that are phosphorylated on tyrosine and bound to GRB2/SOS, leading to activation of RAS: in neuronal cells the multimeric complex formed after NGF activation of TRKA includes not only SHC but CRK and $PLC\gamma_1$ [13]. Transphosphorylation has been detected between TRKB and TRKC but not with TRKA [14]. Some of the multiple *TrkB* and *TrkC* transcripts encode truncated, non-catalytic receptors [3,15].

The transmembrane glycoprotein $gp75^{LNGFR}$ constitutes a second class of specific neurotrophin receptor that binds all neurotrophins with equal, low affinity ($K_d$ ~1 nM). Targeted mutation of $gp75^{LNGFR}$ in mice indicates that this neurotrophin receptor is involved in the development of sensory neurones [16]. In Schwann cells NGF binding to $p75^{LNGFR}$ activates NF-$\kappa$B. $p75^{LNGFR}$ can promote the conversion of sphingomyelin to ceramide during apoptosis [17]. Two positively charged residues ($Arg^{31}$ and $His^{33}$) mediate binding to TRKA and TRKB but not to TRKC [18]. In PC-12 cells $p75^{NGFR}$ and TrkA receptors collaborate to activate rapidly a $p75^{NGFR}$-associated protein kinase [19].

**$gp140^{trkA}$**: High-affinity ($K_d$ ~30 pM) receptor for NGF also activated by NT-5 [20–22]. NGF binding stimulates the tyrosine kinase activity and rapid tyrosine autophosphorylation of $p140^{trkA}$ [23]. The activation of chimeric receptors comprised of the EGFR extracellular ligand binding domain and TRK transmembrane and intracellular sequences promotes association of phosphorylated $PLC\gamma$, RAS-GTP, SHC and the non-catalytic subunit of PtdIns 3-kinase (p85) with TRK. p85/PtdIns 3-kinase function appears to be non-essential for neuronal differentiation signal transmission [24]. In PC-12 cells the activation of $gp140^{trkA}$ by NGF stimulates RAF-1 and mitogen-activated protein (MAP) kinases and results in differentiation into neuronal-type cells [25]. NGF also causes RAS-dependent activation of a protein kinase that phosphorylates CREB which in turn contributes to the induction of *Fos* transcription [26]. Other transcription factors including NGFI-A that binds to G(S)G and proteins binding to CCAAT elements are also activated and contribute to the RAS-dependent transcriptional induction of *Vgf* [27]. The action of NGF on PC-12 cells is potentiated by ganglioside GM1 which binds to TRK and enhances NGF-induced autophosphorylation [28].

Mutations in *TRKA* have been detected in patients with congenital insensitivity to pain with anhydrosis (CIPA), suggesting that TRKA is involved in nociceptive function and thermoregulation *via* sweating in humans [29].
**gp145$^{trkB}$**: Receptor for BDNF, NT-4 and NT-5. *TrkB* mediates the survival and proliferation of NIH 3T3 cells in response to BDNF and the coexpression of gp145$^{trkB}$ and either BDNF or NT-4 transforms these cells. BDNF or NT-4 induces PC-12 cells expressing gp145$^{trkB}$ to differentiate [21,30–33]. In neuronal cells a neural-specific member of the SHC family, N-SHC, binds to BDNF-activated TRKB together with GRB2/SOS [34].
**gp95$^{trkB}$**: Function unknown but the differential pattern of expression with respect to gp145$^{trkB}$ suggests that this non-catalytic form may transport gp145$^{trkB}$ ligands within or to the brain.
**gp145$^{trkC}$**: All three isoforms are high-affinity receptors for NT-3 and, when expressed in NIH 3T3 fibroblasts, undergo rapid tyrosine phosphorylation in response to NT-3. Activated gp145$^{trkC}$ K1 phosphorylates PLC$\gamma_1$ and PtdIns 3-kinase but gp145$^{trkC}$ K2 and gp145$^{trkC}$ K3 do not [3,15,35]. In pancreatic $\beta$ cells NT-3 causes phosphorylation of TRKC on tyrosine residues and the activation of TRKC promotes a transient increase in intracellular calcium [8].
**TRKE**: Probable receptor for NGF.

## Cancer

Transforming *TRK* (*NTRK1*) alleles are found in human colon carcinomas and with high frequency in thyroid papillary carcinomas [2,36]. *TRKA* expression is high in many early-stage neuroblastomas but undetectable in most advanced tumours and appears to correlate inversely with amplification of *MYCN*. *TRKB* and its ligand BDNF are expressed in the *MYCN* amplified cell line SMS-KCN [37]. *TRKC* expression has been detected in 25% of a sample of neuroblastomas, mainly in lower stage tumours [38].

## Transgenic animals

Mice carrying a germline mutation that eliminates *TrkA* have severe sensory and sympathetic neuropathies and most die within one month of birth [39]. Most of the tissues known to express *TrkA* in both the peripheral and central nervous systems are affected, in contrast to the effects of *TrkB* or *TrkC* disruption in which only a limited subset of expressing cells are affected. Mice lacking *TrkB* have multiple central and peripheral nervous system deficiencies and die soon after birth [40]. Mice defective in *TrkC* lack Ia muscle afferent projections to spinal motor neurones and display abnormal movements and postures [41]. Thus NGF signalling *via Trk*A appears essential for the development of both the peripheral and central nervous systems.

## *In vitro*

In transfected NIH 3T3 fibroblasts expressing gp140$^{trkA}$ NGF elevates the concentration of free intracellular calcium [42] and activates *Fos* transcription, DNA synthesis and morphological transformation [43] and in *Xenopus* oocytes expressing gp140$^{trkA}$ NGF induces meiotic maturation [44]. Monoamine-activated $\alpha_2$-macroglobulin inhibits NGF-stimulated neurite outgrowth by binding to gp140$^{trkA}$ [45]. NIH 3T3 fibroblasts are also transformed by co-transfection of plasmids expressing *TrkB* and BDNF or NT-3 or by coexpression of *TrkC* and NT-3 [26,46,47]. The human oncogene *TRK-T1* transforms NIH 3T3 cells [48].

## Protein structure

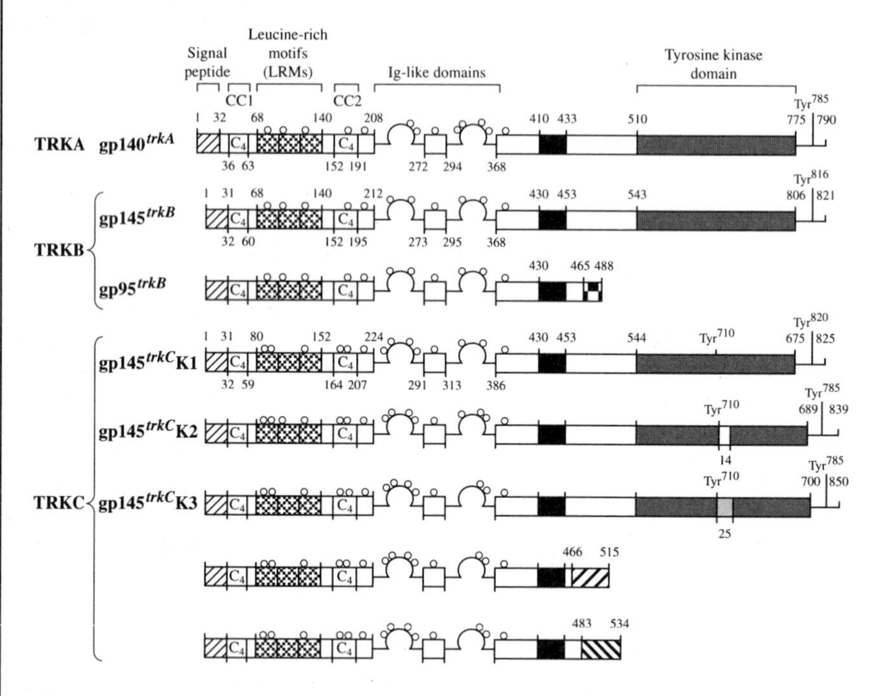


The N-terminal region of the extracellular domains contains a signal peptide, two cysteine clusters ($C_4$) each including four of the 12 cysteine residues conserved in all known TRK proteins, three tandem LRM repeats and two $C_2$ Ig-like domains. Circles (○) denote conserved potential *N*-glycosylation sites [49]. gp145$^{trkC}$ K2 contains an insert of 14 amino acids in the sequence of gp145$^{trkC}$ K1 and gp145$^{trkC}$ K3 has a 25 amino acid insert and a further form has a 39 residue insert. Two putative non-catalytic isoforms have also been identified. The 14, 25 and 39 amino acid tyrosine kinase insert variants have low intrinsic kinase activity and block the capacity of TRKC to mediate NT-3 stimulated *Fos* and *Myc* transcription, evidently by blocking TRKC-SHC association [50,51].

## *TRK* oncoproteins

The *TRK* oncogene first isolated from a colon carcinoma was generated by rearrangement of a non-muscle tropomyosin gene (*TPM3*) and *TRK* [2]. Tropomyosin has 27 residues deleted at the C-terminus and the N-terminal 360 amino acids of TRK are truncated but the transmembrane region and the tyrosine kinase domain are unmodified ((b) below). Thus activation of *TRK* to an oncogene is the result of fusion of the tropomyosin sequences [52].

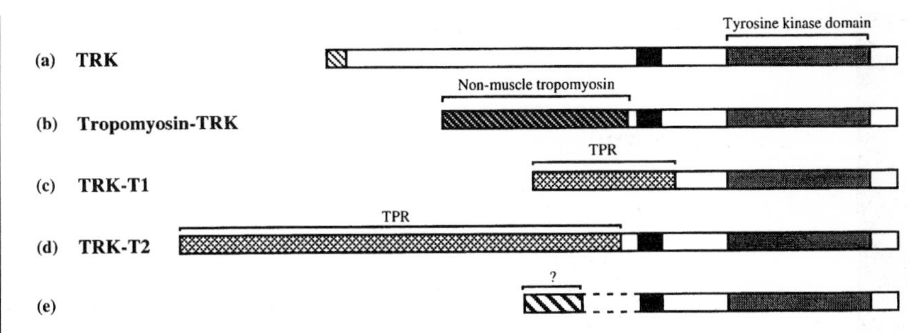


The tropomyosin-activated *TRK* oncogene has also been isolated from human thyroid papillary carcinomas but other chimeric *TRK* oncogenes have been isolated from these tumours that contain different activating sequences[36,48]. TRK-T1 (c) and TRK-T2 (d) contain TPR sequences. *TRK-T3* is an *NTRK1*-derived thyroid oncogene containing 1412 nt of *NTRK1* (398 amino acids) preceded by 598 nt (193 amino acids) of *TFG* (TRK-*f*used gene)[53]. The TFG region includes a coiled-coil motif that may promote complex formation. A further isolate contains uncharacterized sequences ((e) above).

Activation of *TRK* can also occur by recombination *in vitro* and over 40 such *TRK* oncogenes have been detected[54–57]. The products of these *TRK* oncogenes retain the parental tyrosine kinase activity and have an intact C-terminus. However, the N-termini acquired may generate non-glycosylated cytoplasmic molecules or transmembrane glycoproteins.

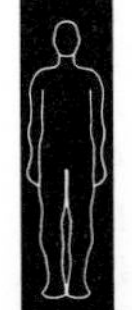

## Amino acid sequence of human TRKAI

```
  1 MLRGGRRGQL GWHSWAAGPG SLLAWLILAS AGAAPCPDAC CPHGSSGLRC
 51 TRDGALDSLH HLPGAENLTE LYIENQQHLQ HLELRDLRGL GELRNLTIVK
101 SGLRFVAPDA FHFTPRLSRL NLSFNALESL SWKTVQGLSL QELVLSGNPL
151 HCSCALRWLQ RWEEEGLGGV PEQKLQCHGQ GPLAHMPNAS CGVPTLKVQV
201 PNASVDVGDD VLLRCQVEGR GLEQAGWILT ELEQSATVMK SGGLPSLGLT
251 LANVTSDLNR KNLTCWAEND VGRAEVSVQV NVSFPASVQL HTAVEMHHWS
301 IPFSVDGQPA PSLRWLFNGS VLNETSFIFT EFLEPAANET VRHGCLRLNQ
351 PTHVNNGNYT LLAANPFGQA SASIMAAFMD NPFEFNPEDP IPDTNSTSGD
401 PVEKKDETPF GVSVAVGLAV FACLFLSTLL LVLNKCGRRN KFGINRPAVL
451 APEDGLAMSL HFMTLGGSSL SPTEGKGSGL QGHIIENPQY FSDACVHHIK
501 RRDIVLKWEL GEGAFGKVFL AECHNLLPEQ DKMLVAVKAL KEASESARQD
551 FQREAELLTM LQHQHIVRFF GVCTEGRPLL MVFEYMRHGD LNRFLRSHGP
601 DAKLLAGGED VAPGPLGLGQ LLAVASQVAA GMVYLAGLHF VHRDLATRNC
651 LVGQGLVVKI GDFGMSRDIY STDYYRVGGR TMLPIRWMPP ESILYRKFTT
701 ESDVWSFGVV LWEIFTYGKQ PWYQLSNTEA IDCITQGREL ERPRACPPEV
751 YAIMRGCWQR EPQQRHSIKD VHARLQALAQ APPVYLDVLG (790)
```

**Domain structure**

| | |
|---|---|
| 1–32 | Signal sequence (italics) |
| 36–63 and 152–191 | Conserved cysteine domains 1 and 2 |
| 68–92, 93–116 and 117–140 | LRM domains |
| 97–120 | High-affinity binding site (underlined) for NGF[58] |

| | |
|---|---|
| 208–272 and 294–268 | Ig-like domains |
| 33–409 | Extracellular domain |
| 418–433 | Transmembrane domain (underlined) |
| 434–790 | Cytoplasmic domain |
| 510–518 and 538 | ATP binding site |
| 644 | Active site |
| 674 | Autophosphorylation site |
| 393–394 | Translocation breakpoint for tropomyosin/TRK oncoprotein formation |
| 510–775 | Tyrosine kinase domain |
| 490 | SHC binding site ($Tyr^{490}$) |
| 785 | Phospholipase Cγ binding site. Mutation of $Tyr^{785}$ blocks NGF-promoted expression of the neurone-specific intermediate filament protein peripherin [59] |
| 751 | PtdIns 3-kinase (p85) binding site ($Tyr^{751}$) [10] |
| 67, 95, 121, 188, 202, 253, 262, 281, 318, 323, 338, 358 and 395 | Potential carbohydrate attachment sites |

The human TRK sequence shown is for the TRKAI isoform; human and rat TRKAII contain a six amino acid insert (VSFSPV) after $Pro^{392}$ (human) [7].

## Amino acid sequence of human TRKB

```
  1 MSSWIRWHGP AMARLWGFCW LVVGFWRAAF ACPTSCKCSA SRIWCSDPSP
 51 GIVAFPRLEP NSVDPENITE IFIANQKRLE IINEDDVEAY VGLRNLTIVD
101 SGLKFVAHKA FLKNSNLQHI NFTRNKLTSL SRKHFRHLDL SELILVGNPF
151 TCSCDIMWIK TLQEAKSSPD TQDLYCLNES SKNIPLANLQ IPNCGLPSAN
201 LAAPNLTVEE GKSITLSCSV AGDPVPNMYW DVGNLVSKHM NETSHTQGSL
251 RITNISSDDS GKQISCVAEN LVGEDQDSVN LTVHFAPTIT FLESPTSDHH
301 WCIPFTVKGN PKPALQWFYN GAILNESKYI CTKIHVTNHT EYHGCLQLDN
351 PTHMNNGDYT LIAKNEYGKD EKQISAHFMG WPGIDDGANP NYPDVIYEDY
401 GTAANDIGDT TNRSNEIPST DVTDKTGREH LSVYAVVVIA SVVGFCLLVM
451 LFLLKLARHS KFGMKGPASV ISNDDDSASP LHHISNGSNT PSSSEGGPDA
501 VIIGMTKIPV IENPQYFGIT NSQLKPDTFV QHIKRHNIVL KRELGEGAFG
551 KVFLAECYNL CPEQDKILVA VKTLKDASDN ARKDFHREAE LLTNLQHEHI
601 VKFYGVCVEG DPLIMVFEYM KHGDLNKFLR AHGPDAVLMA EGNPPTELTQ
651 SQMLHIAQQI AAGMVYLASQ HFVHRDLATR NCLVGENLLV KIGDFGMSRD
701 VYSTDYYRVG GHTMLPIRWM PPESIMYRKF TTESDVWSLG VVLWEIFTYG
751 KQPWYQLSNN EVIECITQGR VLQRPRTCPQ EVYELMLGCW QREPHMRKNI
801 KGIHTLLQNL AKASPVYLDI LG (822)
```

### Database accession numbers

| | *PIR* | *SWISSPROT* | *EMBL/GENBANK* | *REFERENCES* |
|---|---|---|---|---|
| Human *TRK* | A30124 | P04629 | M23102 | 2,60 |
| Human *TRK-T1/TRK-T2* | | | | 48 |
| Human *TRK-T3* | X85960 | | | 53 |
| Human *TRKB* | | | S76473 | 5 |
| Human *TRKE* | | | HSTRKE X74979 | 9 |
| Mouse *TrkB* | | P15209 | X17647 | 61 |
| Pig *TrkC* | | P24786 | M80800 | 47 |

### *References*


1. Pulciani, S. et al. (1982) Nature 300, 539–542.
2. Martin-Zanca, D. et al. (1986) Nature 319, 743–748.
3. **Barbacid, M. et al. (1993) Oncogene 8, 2033–2042.**
4. Jennings, C.G.B. et al. (1993) Proc. Natl Acad. Sci. USA 90, 2895–2899.
5. Nakagawara, A. et al. (1995) Genomics 25, 538–546.
6. **Barbacid, M. et al. (1991) Biochim. Biophys. Acta 1072, 115–127.**
7. Barker, P.A. et al. (1993) J. Biol. Chem. 268, 15150–15157.
8. Tazi, A. et al. (1996) J. Biol. Chem. 271, 10154–10160.
9. Di Marco, E. et al. (1993) J. Biol. Chem. 268, 24290–24295.
10. Valent, A. et al. (1996) Hum. Genet. 98, 2–15.
11. Lucarelli, E. et al. (1995) J. Biol. Chem. 270, 24725–24731.
12. McDonald, N.Q. and Chao, M.V. (1995) J. Biol. Chem. 270, 19669–19672.
13. Torres, M. and Bogenmann, E. (1996) Oncogene 12, 77–86.
14. Canossa, M. et al. (1996) J. Biol. Chem. 271, 5812–5818.
15. **Barbacid, M. et al. (1993) In Molecular Genetics of Nervous System Tumors. Wiley-Liss, New York, pp. 123–135.**
16. Lee, K.-F. et al. (1992) Cell 69, 737–749.
17. Carter, B.D. et al. (1996) Science 272, 542–545.
18. Rydén, M. and Ibáñez, C.F. (1996) J. Biol. Chem. 271, 5623–5627.
19. Canossa, M. et al. (1996) EMBO J. 15, 3369–3376.
20. Klein, R. et al. (1991) Cell 65, 189–197.
21. Berkemeier, L.R. et al. (1991) Neuron 7, 857–866.
22. Kaplan, D.R. et al. (1991) Science 252, 554–558.
23. Kaplan, D.R. et al. (1991) Nature 350, 158–160.
24. Dikic, I. et al. (1995) J. Biol. Chem. 270, 15125–15129.
25. Ohmichi, M. et al. (1992) J. Biol. Chem. 267, 14604–14610.
26. Ginty, D.D. et al. (1994) Cell 77, 713–725.
27. D'Arcangelo, G. et al. (1996) Mol. Cell. Biol. 16, 4621–4631.
28. Mutoh, T. et al. (1995) Proc. Natl Acad. Sci. USA 92, 5087–5091.
29. Indo, Y. et al. (1996) Nature Genet. 13, 485–488.
30. Klein, R. et al. (1992) Neuron 8, 947–956.
31. Soppet, D. et al. (1991) Cell 65, 895–903.
32. Squinto, S.P. et al. (1991) Cell 65, 886–893.
33. Ip, N.Y. et al. (1993) Neuron 10, 137–149.
34. Nakamura, T. et al. (1996) Oncogene 13, 1111–1121.
35. Lamballe, F. et al. (1993) EMBO J. 12, 3083–3094.
36. Bongarzone, I. et al. (1989) Oncogene 4, 1457–1462.
37. Nakagawara, A. et al. (1994) Mol. Cell. Biol. 14, 759–767.
38. Yamashiro, D.J. et al. (1996) Oncogene 12, 37–41.
39. Klein, R. et al. (1994) Nature 368, 246–249.
40. Klein, R. et al. (1993) Cell 75, 113–122.
41. Smeyne, R.J. et al. (1994) Nature 368, 249–251.
42. De Bernardi, M.A. et al. (1996) J. Biol. Chem. 271, 6092–6098.
43. Cordon-Cardo, C. et al. (1991) Cell 66, 173–183.
44. Nebreda, A.R. et al. (1991) Science 252, 558–561.
45. Koo, P.H. and Qiu, W.-S. (1994) J. Biol. Chem. 269, 5369–5376.
46. Klein, R. et al. (1991) Cell 66, 395–403.
47. Lamballe, F. et al. (1991) Cell 66, 967–979.

48 Greco, A. et al. (1992) Oncogene 7, 237–242.
49 Schneider, R. and Schweiger, M. (1991) Oncogene 6, 1807–1811.
50 Guiton, M. et al. (1995) J. Biol. Chem. 270, 20384–20390.
51 Tsoulfas, P. (1996) J. Biol. Chem. 271, 5691–5697.
52 Butti, M.G. et al. (1995) Genomics 28, 15–24.
53 Greco, A. et al. (1995) Mol. Cell. Biol. 15, 6118–6127.
54 Kozma, S.C. et al. (1988) EMBO J. 7, 147–154.
55 Oskam, R. et al. (1988) Proc. Natl Acad. Sci. USA 85, 2964–2968.
56 Coulier, F. et al. (1990) Mol. Cell. Biol. 10, 4202–4210.
57 Albor, A. et al. (1996) Oncogene 13, 1755–1763.
58 Windisch, J.M. et al. (1995) J. Biol. Chem. 270, 28133–28138.
59 Loeb, D.M. et al. (1994) J. Biol. Chem. 269, 8901–8910.
60 Martin-Zanca, D. et al. (1989) Mol. Cell. Biol. 9, 24–33.
61 Klein, R. et al. (1989) EMBO J. 8, 3701–3709.

## Identification

*VAV* (sixth letter of the Hebrew alphabet) was first identified by *in vitro* replacement of 67 N-terminal proto-VAV amino acids by 19 Tn5 transposase residues [1].

## Related genes

The central domain (198–434) shares homology with DBL, BCR, murine CDC25MM, CDC24 and CDC25 and (see ***ABL***). Human and mouse VAV are 95% homologous. Murine *Vav2* is 63% and 55% identical at nucleic acid and amino acid level to *Vav* [2].

| | ***VAV*** |
|---|---|
| **Nucleotides (kb)** | >35 |
| **Chromosome** | 19p13.2 |
| **Mass (kDa): predicted** | 96 |
| **expressed** | p95 (oncogenic form ~p85) |
| **Cellular location** | Cytoplasm and nucleus |

**Tissue distribution**
Normal human *VAV* appears to be specifically expressed in cells of haematopoietic origin (erythroid, lymphoid, myeloid) regardless of their differentiation lineage.

## Protein function

VAV is guanine nucleotide exchange factor for RHO family GTP-binding proteins (RHOA, RAC1 and CDC42). It undergoes rapid and transient tyrosine phosphorylation following activation of the T cell receptor or IgM antigen receptors on B cells [3] or in response to EGF or PDGF in transfected NIH 3T3 cells [4,5] and in human haematopoietic cells after stimulation by interferon $\alpha$ [6], IL-3 or GM-CSF or stem cell factor (see ***KIT***) [7]. IL-3 or erythropoietin cause VAV to associate with TEC kinase in murine haematopoietic progenitor cells and in these cells VAV is constitutively associated with GRB2 [8]. In activated haematopoietic cells VAV also interacts with GRB2/mSOS1 and with SHC [9], the tyrosine phosphatase PTP1C [10] and ZAP-70 [11]. Thus PTP1C may modulate RAS signalling pathways in haematopoietic cells. VAV also associates with GRB3-3, an isoform of GRB2 that can induce fibroblast apoptosis [12], and interacts with ENX-1 and may thus act as a regulator of homeobox genes in haematopoietic cells [13].

In addition to activation *via* the TCR, in haematopoietic cells VAV RAS exchange activity can be stimulated by insulin with transient VAV tyrosine phosphorylation in myeloma cells [14] or without tyrosine phosphorylation by TPA or 1,2-diacylglycerol [15]. However, in fibroblasts VAV does not appear to function either as a RAS guanine nucleotide exchange factor or as an upstream regulatory element of RAS although it can cooperate with RAS to transform these cells [16]. In activated B cells VAV rapidly and transiently associates *via* its SH2 domain with a 70 kDa tyrosine-phosphorylated protein, VAP1 and also with the GRB2-related adaptor protein CRK [17]. VAV may therefore participate in proliferation-associated signalling processes. In Jurkat cells VAV associates with ZAP-70 and tubulin [18] and with the heterogeneous ribonucleoprotein K [19] and the nuclear protein Ku-70 [20] and over-expression of VAV activates nuclear factors, including NFAT, involved in IL-2 expression [21]. The expression of BCR/ABL causes ligand-independent phosphorylation of VAV and proliferation [22]. VAV may be activated in these myeloid cells by JAK2 tyrosine kinase.

Following stimulation of T cells with prolactin, VAV is transiently associated with the prolactin receptor and there is an increase in VAV-associated guanine nucleotide exchange factor activity, following which VAV translocates to the nucleus [23].

Transformation by VAV (and also by DBL) involves activation of mitogen-activated protein kinases (MAPKs) independently of RAS activation [24]. Oncogenic VAV also activates the JNK/SAPK pathway *via* the Rho family member Rac-1 [25].

### Cancer

The region of chromosome 19 to which VAV maps is involved in karyotypic abnormalities in a variety of malignancies including melanomas and leukaemias [26].

### *In vitro*

VAV is a weak NIH 3T3 cell transforming agent; transformation is greatly enhanced by truncation of the N-terminal helix-loop-helix/leucine zipper region [27]. Dominant negative mutants of *Vav* inhibit *Ras*- and *Raf*-induced transformation of fibroblasts and dominant negative mutants of *Myc* inhibit transformation by *Vav* and *Ras*, suggesting that *Vav* and *Ras* signalling pathways overlap and that both require *Myc* activation [28].

### Transgenic animals

Transgenic mouse and *in vitro* studies indicate that *Vav* is essential for the early developmental steps that precede the onset of haematopoiesis [29] and its absence gives rise to defects in the antigen receptor-mediated proliferation of B and T cells [30,31].

## Protein structure

The C-terminal region of VAV contains one SH2 domain and two SH3 domains (see ***DBL***). Point mutations in the SH2 domain inhibit the transforming potential of oncogenic VAV but do not activate that of normal VAV [32]. VAV proteins also contain a pleckstrin homology domain and a cysteine-rich domain that includes two putative metal binding regions, Cys-$X_2$-Cys-$X_{13}$-Cys-$X_2$-Cys and His-$X_2$-Cys-$X_6$-Cys-$X_2$-His. Mutations in these regions can completely abolish transforming activity [33].

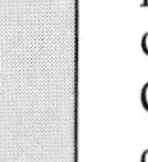

## Amino acid sequence of human VAV (see also *DBL*)

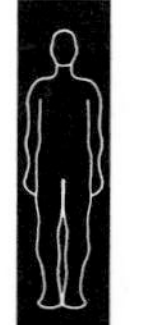

```
  1 MELWRQCTHW LIQCRVLPPS HRVTWDGAQV CELAQALRDG VLLCQLLNNL
 51 LPHAINLREV NLRPQMSQFL CLKNIRTFLS TCCEKFGLKR SELFEAFDLF
101 DVQDFGKVIY TLSALSWTPI AQNRGIMPFP TEEESVGDED IYSGLSDQID
151 DTVEEDEDLY DCVENEEAEG DEIYEDLMRS EPVSMPPKMT EYDKRCCCLR
201 EIQQTEEKYT DTLGSIQQHF LKPLQRFLKP QDIEIIFINI EDLLRVHTHF
250 LKEMKEALGT PGAPNLYQVF IKYKERFLVY GRYCSQVESA SKHLDRVAAA
301 REDVQMKLEE CSQRANNGRF TARPADGAYA ASSQISPPSP GAGETHAGGD
351 GARKLRLALD AMRDLAQCVN EVKRDNETLR QITNFQLSIE NLDQSLAHYG
401 RPKIDGELKI TSVERRSKMD RYAFLLDKAL LICKRRGDSY DLKDFVNLHS
451 FQVRDDSSGD RDNKKWSHMF LLIEDQGAQG YELFFKTREL KKKWMEQFEM
501 AISNIYPENA TANGHDFQMF SFEETTSCKA CQMLLRGTFY QGYRCHRCRA
551 SAHKECLGRV PPCGRHGQDF PGTMKKDKLH RRAQDKKRNE LGLPKMEVFQ
601 EYYGLPPPPG AIGPFLRLNP GDIVELTKAE AEQNWWEGRN TSTNEIGWFP
651 CNRVKPYVHG PPQDLSVHLW YAGPMERAGA ESILANRSDG TFLVRQRVKD
701 AAEFAISIKY NVEVKHTVKI MTAEGLYRIT EKKAFRGLTE LVEFYQQNSL
751 KDCFKSLDTT LQFPFKEPEK RTISRPAVGS TKYFGTAKAR YDFCARDRSE
801 LSLKEGDIIK ILNKKGQQGW WRGEIYGRVG WFPANYVEED YSEYC (845)
```

### Domain structure

| | |
|---|---|
| 515–563 | Phorbol ester and diacylglycerol binding |
| 671–761 | SH2. The VAV SH2 domain has Thr at the $\beta$D5 position and recognizes phosphopeptides with the general motif *P*-Tyr-Met-Glu-Pro [34] |
| 782–842 | SH3. The C-terminal SH3 domain binds to heterogeneous nuclear ribonucleoprotein K (hnRNPK), a poly(rC)-specific RNA-binding protein [35], and to the focal adhesion protein zyxin [13] |
| 486–493 and 575–582 | Putative nuclear localization signals (italics) |
| 541–548 and 553–566 | Zinc finger domains (underlined) |

### Database accession numbers

| | PIR | SWISSPROT | EMBL/GENBANK | REFERENCES |
|---|---|---|---|---|
| Human *VAV* | S05382 | P15498 | X16316 | 1 |
| | B39576 | | M59834 | 27 |
| Mouse *Vav* | | | Mmvavpo M59833 | 27 |

### *References*


1 Katzav, S. et al. (1989) EMBO J. 8, 2283–2290.
1 Schuebel, K.E. et al. (1996) Oncogene 13, 363–371.
3 Bustelo, X.R. and Barbacid, M. (1992) Science 256, 1196–1199.
4 Bustelo, X.R. et al. (1992) Nature 356, 68–71.
5 Margolis, B. et al. (1992) Nature 356, 71–74.
6 Platanias, L.C. and Sweet, M.E. (1994) J. Biol. Chem. 269, 3143–3146.
7 Alai, M. et al. (1992) J. Biol. Chem. 267, 18021–18025.

[8] Machide, M. et al. (1995) Oncogene 11, 619–625.
[9] Ramos-Morales, F. et al. (1994) Oncogene 9, 1917–1923.
[10] Kon-Kozlowski, M. et al. (1996) J. Biol. Chem. 271, 3856–3862.
[11] Katzav, S. et al. (1994) J. Biol. Chem. 269, 32579–32585.
[12] Ramos-Morales, F. et al. (1995) Oncogene 11, 1665–1669.
[13] Hobert, O. et al. (1996) Mol. Cell. Biol. 16, 3066–3073.
[14] Uddin, S. et al. (1995) J. Biol. Chem. 270, 7712–7716.
[15] Gulbins, E. et al. (1994) Mol. Cell. Biol. 14, 4749–4758.
[16] Bustelo, X.R. et al. (1994) Oncogene 9, 2405–2413.
[17] Smit, L. et al. (1996) J. Biol. Chem. 271, 8564–8569.
[18] Huby, R.D.J. et al. (1995) J. Biol. Chem. 270, 30241–30244.
[19] Hobert, O. et al. (1994) J. Biol. Chem. 269, 20225–20228.
[20] Romero, F. et al. (1996) Mol. Cell. Biol. 16, 37–44.
[21] Wu, J. et al. (1995) Mol. Cell. Biol. 15, 4337–4346.
[22] Matsuguchi, T. et al. (1994) EMBO J. 124, 257–265.
[23] Clevenger, C.V. et al. (1995) J. Biol. Chem. 270, 13246–13253.
[24] Khosravi-Far, R. et al. (1994) Mol. Cell. Biol. 14, 6848–6857.
[25] Crespo, P. et al. (1996) Oncogene 13, 455–460.
[26] Martinerie, C. et al. (1990) Hum. Genet. 86, 65–68.
[27] Katzav, S. et al. (1991) Mol. Cell. Biol. 11, 1912–1920.
[28] Katzav, S. et al. (1995) Oncogene 11, 1079–1088.
[29] Zmuidzinas, A. et al. (1995) EMBO J. 14, 1–11.
[30] Tarakhovsky, A. et al. (1995) Nature 374, 467–470.
[31] Zhang, R. et al. (1995) Nature 374, 470–473.
[32] Katzav, S. (1993) Oncogene 8, 1757–1763.
[33] Coppola, J. et al. (1991) Cell Growth Differ. 2, 95–105.
[34] Songyang, Z. et al. (1994) Mol. Cell. Biol. 14, 2777–2785.
[35] Bustelo, X.R. et al. (1995) Mol. Cell. Biol. 15, 1324–1332.

# WNT1, WNT2, WNT3

## Identification

*Wnt-1/Int-1* was originally identified as a frequent target for MMTV insertion in mammary carcinomas [1,2]. *Wnt-3* is related to *Wnt-1* and is also activated by proviral insertion [3]. *WNT1* and *WNT3* were detected by screening genomic DNA with mouse *Int-1* and *Wnt-3* probes. *WNT2* was detected by isolation of a methylation-free CpG island from a human lung cDNA library.

## Related genes

There are at least 12 genes in the mouse *Wnt-1/Int-1* family [4,5]. On the basis of their similarity to the *Drosophila melanogaster wingless* gene product, *Int-1* and related genes have been reclassified as *Wnt* (*wingless-type* MMTV integration site). At least 12 *Xenopus laevis Xwnt* genes have been detected and there are *Caenorhabditis elegans* homologues of *Wnt-1*, *Wnt-2* and *Wnt-5B*.

| | ***WNT1/Wnt-1 (INT1/Int-1)*** | ***WNT2/Wnt-2 (IRP/Irp)*** | ***WNT3/Wnt-3 (INT4/Int-4)*** |
|---|---|---|---|
| **Nucleotides (kb)** | 30 | Not fully mapped | >50 |
| **Chromosome** | 12q13 | 7q31 | 17q21–q22 |
| **Mass (kDa): predicted** | 41 | 38 | 40 |
| **expressed** | gp40/gp42 | gp35 | |

**Cellular location**

WNT1 and WNT2: Secreted. Tightly associated with extracellular matrix and cell surface [6,7].

**Tissue distribution**

Murine *Wnt* genes are expressed in a variety of embryonic and adult tissues, particularly in brain and lung. *Wnt-1* is also expressed in the adult testis [8] and *Wnt-2* during the ductal phase of mouse mammary gland development [9]. *WNT2* is expressed in human fetal lung fibroblast cell lines [10].

## Protein function

WNT1 is probably a growth factor but its receptor is unknown. Normal function is in embryogenesis: in the mouse it is required for development of the mid-brain and anterior hindbrain. Secretion of WNT1 is inefficient and substantial amounts are present in the endoplasmic reticulum associated with Bip [11,12]. Expression of *Wnt-1* in embryos can reproduce the effects of lithium treatment, suggesting that WNT1 may modulate the way cells respond to inducing agents by suppressing the concentration of phosphoinositides available to generate second messenger signals [13,14]. In PC-12 cells *Wnt-1* expression enhances cell–cell adhesion and correlates with increased expression of E-cadherin and plakoglobulin and decreased expression of NCAM and in murine cells *Wnt-1* expression promotes catenin accumulation and stabilizes the formation of catenin/APC and catenin/cadherin complexes[15]. *Xwnt-3A* may participate in patterning the central nervous system during early *Xenopus* development[16].

These observations are generally consistent with the evidence that *Drosophila* and *Xenopus wnt* genes are involved in the generation of the central nervous system[17].

**Cancer**

*WNT1* is amplified in some primary retinoblastoma tumours[18].

Expression of *WNT3, WNT4* and *WNT7b* has been detected in human breast cell lines and *WNT2, WNT3, WNT4* and *WNT7b* are detectable in breast tissue[19]. No amplification or rearrangement of *WNT3* has been detected in human breast tumours[20]. Over-expression of *WNT2* has been detected in human colorectal carcinoma, gastric and oesophageal cancers[21]. *WNT2* and *WNT4* expression is elevated in fibroadenomas with respect to normal or malignant tissue and in 10% of tumours *WNT7b* expression is 30-fold higher than in normal or benign breast tissues. *WNT5A* over-expression has been detected in primary human breast, lung and prostate carcinomas and in melanomas[22].

Amplification of *Wnt-2* has been detected in mouse mammary tumours[23].

**Transgenic animals**

*Wnt-1* causes hyperplasia in the mammary glands of male and female mice which can progress to mammary and salivary adenocarcinomas[24,25]. Bi-transgenic mice carrying both *Wnt-1* and *Int-2* transgenes regulated by the MMTV LTR develop mammary carcinomas more rapidly and with higher frequency than occurs when either gene is expressed alone[26]. This indicates that *Wnt-1* and *Int-2* cooperate in mammary tumorigenesis, although amplification or rearrangement of *WNT1* has not been detected in human breast carcinomas.

In *Wnt-1* transgenic mice homozygous for the deletion of *P53*, mammary tumour development is accelerated, $P53^-$ tumours showing enhanced chromosomal instability[27].

**In animals**

*Wnt-1* and *Int-2* are the most frequent targets in MMTV-induced tumours. Either *Wnt-1* or *Int-2* or both may be activated[28] and insertion at each site has been detected in both pre-malignant lesions and in malignant tumours[29]. The frequency with which the *Wnt-1* and *Int-2* loci are rearranged by MMTV insertional mutagenesis is a function of the host genetic background[30–32], indicating that strains inbred for a high incidence of mammary tumours have acquired host mutations that complement the activity of specific *Wnt/Int* genes.

***In vitro***

*Wnt-1* transforms fibroblasts with very low efficiency but expression of the gene by fibroblasts causes morphological transformation of co-cultured mammary epithelial cells[33]. Thus WNT1 protein may participate in a paracrine mechanism. *Wnt-1* partially transforms the C57 mammary epithelial cell line[34] and renders the RAC mammary cell line tumorigenic[35]. Constitutive expression of *Wnt-1* in PC-12 cells modulates the activation of MAP kinase by several growth factors. NGF stimulation (which causes differentiation) is decreased whereas FGF and EGF stimulation is enhanced and EGF is converted from a proliferation to a differentiation factor[36]. *Wnt-1* also induces expression of *Ret*[37]. *Wnt-2, Wnt-3, Wnt-3A, Wnt-4, Wnt-7A, Wnt-7B* and *Wnt-11* transform C57MG mouse mammary tumour cells[38].

## Gene structure of *WNT1*

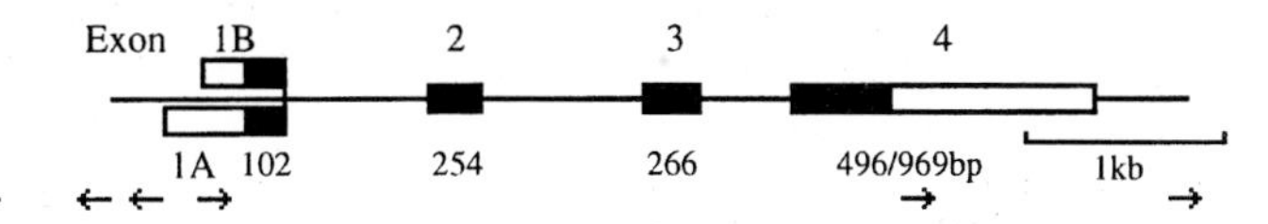


The sequence and organization of the *WNT1* gene is highly conserved in organisms ranging from man to *Drosophila*. The conservation between human *WNT1* and mouse *Wnt-1* includes extensive intronic regions as well as 5′ and 3′ non-translated regions[39].

Arrows indicate integration sites and orientations in *Wnt-1* of MMTV proviruses in a variety tumours, the majority being oriented away from *Wnt-1*. For *Wnt-1, Int-2, Hst-1* and *Wnt-4* proviral integration of MMTV does not perturb the DNA encoding these genes[3,40].

The structure of *Wnt-3* is similar to that of *Wnt-1*. All introns are at homologous positions compared with *Wnt-1* except for a unique intron before a fifth exon that is completely non-coding[3].

## Transcriptional regulation

The 5′ exon has two forms (1A and 1B) with identical 3′ ends but different 5′ ends[41]. There are two TATA boxes upstream of the 1B start site at –35 and –25. The region upstream of 1A contains no TATA boxes but is very GC-rich and includes at least two SP1 binding sites.

## Protein structure

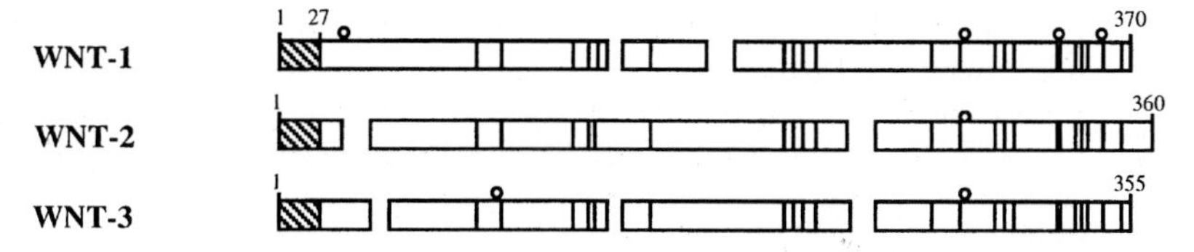


The vertical bars represent the 22 cysteine residues conserved throughout the WNT family. Circles indicate potential *N*-glycosylation sites. The cross-hatched box indicates the signal sequence.

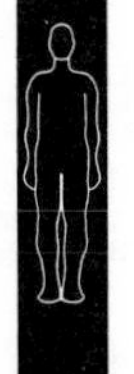

## Amino acid sequence of human WNT1 (INT1)

```
  1 MGLWALLPGW VSATLLLALA ALPAALAANS SGRWWGIVNV ASSTNLLTDS
 51 KSLQLVLEPS LQLLSRKQRR LIRQNPGILH SVSGGLQSAV RECKWQFRNR
101 RWNCPTAPGP HLFGKIVNRG CRETAFIFAI TSAGVTHSVA RSCSEGSIES
151 CTCDYRRRGP GGPDWHWGGC SDNIDFGRLF GREFVDSGEK GRDLRFLMNL
201 HNNEAGRTTV FSEMRQECKC HGMSGSCTVR TCWMRLPTLR AVGDVLRDRF
251 DGASRVLYGN RGSNRASRAE LLRLEPEDPA HKPPSPHDLV YFEKSPNFCT
301 YSGRLGTAGT AGRACNSSSP ALDGCELLCC GRGHRTRTQR VTERCNCTFH
351 WCCHVSCRNC THTRVLHECL (370)
```

**Domain structure**

1–27 Signal sequence (italics)
29, 316, 346, 359 Potential carbohydrate attachment sites
332–337 Basic region (RGHRTR) required for accumulation of WNT1 at the cell surface [42].

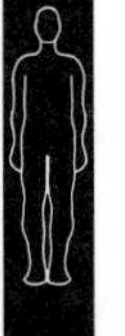

## Amino acid sequence of human WNT2

```
  1 MNAPLGGIWL WLPLLLTWLT PEVNSSWWYM RATGGSSRVM CDNVPGLVSS
 51 QRQLCHRHPD VMRAISQGVA EWTAECQHQF RQHRWNCNTL DRDHSLFGRV
101 LLRSSRESAF VYAISSAGVV FAITRACSQG EVKSCSCDPK KMGSAKDSKG
151 IFDWGGCSDN IDYGIKFARA FVDAKERKGK DARALMNLHN NRAGRKAVKR
201 FLKQECKCHG VSGSCTLRTC WLAMADFRKT GDYLWRKYNG AIQVVMNQDG
251 TGFTVANERF KKPTKNDLVY FENSPDYCIR DREAGSLGTA GRVCNLTSRG
301 MDSCEVMCCG RGYDTSHVTR MTKCGCKFHW CCAVRCQDCL EALDVHTCKA
351 PKNADWTTAT (360)
```

**Domain structure**

1–21 or 25 Signal sequence
24, 295 Potential carbohydrate attachment sites

## Amino acid sequence of mouse WNT3

```
  1 MEPHLLGLLL GLLLSGTRVL AGYPIWWSLA LGQQYTSLAS QPLLCGSIPG
 51 LVPKQLRFCR NYIEIMPSVA EGVKLGIQEC QHQFRGRRWN CTTIDDSLAI
101 FGPVLDKATR ESAFVHAIAS AGVAFAVTRS CAEGTSTICG CDSHHKGPPG
151 EGWKWGGCSE DADFGVLVSR EFADARENRP DARSAMNKHN NEAGRTTILD
201 HMHLKCKCHG LSGSCEVKTC WWAQPDFRAI GDFLKDKYDS ASEMVVEKHR
251 ESRGWVETLR AKYALFKPPT ERDLVYYENS PNFCEPNPET GSFGTRDRTC
301 NVTSHGIDGC DLLCCGRGHN TRTEKRKEKC HCVFHWCCYV SCQECIRIYD
351 VHTCK (355)
```

**Domain structure**

1–27 Signal sequence (italics)
90, 301 Potential carbohydrate attachment sites

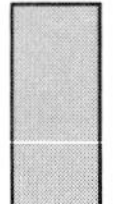

**Database accession numbers**

| | *PIR* | *SWISSPROT* | *EMBL/GENBANK* | *REFERENCES* |
|---|---|---|---|---|
| Human *WNT1 (INT1)* | A24674 | P04628 | X03072 | 39,41 |
| Human *WNT2* | S00834 | P09544 | X07876 | 43 |
| Mouse *Wnt-3* | A35503 | P17553 | M32502 | 3 |

### *References*


1. **Peters, G. (1991) Semin. Virol. 2, 319–328.**
2. Nusse, R. and Varmus, H.E. (1982) Cell 31, 99–109.
3. Roelink, H. et al. (1990) Proc. Natl Acad. Sci. USA 87, 4519–4523.
4. **Nusse, R. and Varmus, H.E. (1992) Cell 69, 1073–1087.**
5. Katoh, M. et al. (1996) Oncogene 13, 873–876.
6. Papkoff, J. (1994) Oncogene 9, 313–317.
7. Parkin, N.T. et al. (1993) Genes Dev. 7, 2181–2193.
8. Jakobovits, A. et al. (1986) Proc. Natl Acad. Sci. USA 83, 7806–7810.
9. Buhler, T. A. et al. (1993) Dev. Biol. 155, 87–96.
10. Levay-Young, B.K. and Navre, M. (1992) Am. J. Physiol. 262, L672–683.
11. van Ooyen, A. and Nusse, R. (1984) Cell 39, 233–240.
12. Kitajewski, J. et al. (1992) Mol. Cell. Biol. 12, 784–790.
13. Sokol, S. et al. (1991) Cell 67, 741–752.
14. Olson, D.J. et al. (1991) Science 252, 1173–1176.
15. Papkoff, J. et al. (1996) Mol. Cell. Biol. 16, 2128–2134.
16. Wolda, S.L. et al. (1993) Dev. Biol. 155, 46–57.
17. Christian, J.L. et al. (1991) Dev. Biol. 143, 230–234.
18. Arheden, K. et al. (1988) Cytogen. Cell Genet. 48, 174–177.
19. Huguet, E.L. et al. (1994) Cancer Res. 54, 2615–2621.
20. Roelink, H.J. et al. (1993) Genomics 17, 790–792.
21. Vider, B.-Z. et al. (1996) Oncogene 12, 153–158.
22. Iozzo, R.V. et al. (1995) Cancer Res. 55, 3495–3499.
23. Roelink, H. et al. (1992) Oncogene 7, 487–492.
24. Tsukamoto, A.S. et al. (1988) Cell 55, 619–625.
25. Edwards, P.A.W. et al. (1992) Oncogene 7, 2041–2051.
26. Kwan, H. et al. (1992) Mol. Cell. Biol. 12, 147–154.
27. Donehower, L.A. et al. (1995) Genes Dev. 9, 882–895.
28. Gray, D.A. et al. (1986) Virology 154, 271–278.
29. Morris, D.W. et al. (1990) J. Virol. 64, 1794–1802.
30. Escot, C. et al. (1986) J. Virol. 58, 619–625.
31. Etkind, P. (1989) J. Virol. 63, 4972–4975.
32. Marchetti., A. et al. (1991) J. Virol. 65, 4550–4554.
33. Jue, S.F. et al. (1992) Mol. Cell. Biol. 12, 321–328.
34. Blasband, A. et al. (1992) Oncogene 7, 153–161.
35. Rijsewijk, F. et al. (1987) EMBO J. 6, 127–131.
36. Pan, M.-G. et al. (1995) Oncogene 11, 2005–2012.
37. Zheng, S. et al. (1996) Oncogene 12, 547–554.
38. Christiansen, J.H. (1996) Oncogene 12, 2705–2711.
39. van Ooyen, A. et al. (1985) EMBO J. 4, 2905–2909.
40. Peters, G. et al. (1989) Proc. Natl Acad. Sci. USA 86, 5678–5682.
41. Nusse, R. et al. (1990) Mol. Cell. Biol. 10, 4170–4179.
42. Schryver, B. et al. (1996) Oncogene 13, 333–342.
43. Wainwright, B.J. et al. (1988) EMBO J. 7, 1743–1748.

## Identification

v-*yes* (originally *yas*) is the oncogene of two avian sarcoma viruses, Y73 and Esh sarcoma virus (ESV). Y73 avian sarcoma virus was isolated from a transplantable tumour arising in a chicken of the same strain on a farm in Yamaguchi Prefecture[1]. Esh was isolated from a tumour arising in a White leghorn owned by Mr Esh of Pennsylvania[2]. *YES* was detected by screening DNA with a v-*yes* probe[3].

## Related genes

*YES* is a member of the *SRC* tyrosine kinase family (*Blk, FGR, FYN, HCK, LCK/Tkl, LYN, SRC, YES*). Chicken *YRK* (Yes-related kinase) is 72.4% identical to chicken YES[4]. *YESP/YES2* is a human pseudogene.

| | ***YES1/Yes-1*** | **v-*yes*** |
|---|---|---|
| **Nucleotides** | >30 kb | 3718 bp (Y73) |
| **Chromosome** | 18q21.3 | |
| **Mass (kDa): predicted** | 61 | |
| **expressed** | 62 | $P90^{gag\text{-}yes}$ (Y73)<br>$P80^{gag\text{-}yes}$ (ESV) |

**Cellular location**
YES1: Plasma membrane and attached to the cytoskeleton.
v-YES: More diffusely distributed but concentrated in cell junction and adhesion plaques, similar to SRC.

**Tissue distribution**
*YES* mRNA is widely distributed. In humans expression is high in the brain and kidney and low in spleen, muscle and thymus[5–7]. YES kinase activity is high in platelets, peripheral blood T cells and natural killer cells but low in monocytes and B lymphocytes[8,9]. Expression is stimulated during lymphocyte mitogenesis[6].

## Protein function

*YES* genes encode membrane-associated protein tyrosine kinases, functionally related to SRC. As for SRC, there is evidence implicating YES in the tyrosine phosphorylation of multiple substrates and in the modulation of PtdIns metabolism that occurs during the activation of many normal cell types and in transformed cells.

In normal quiescent fibroblasts stimulated by PDGF a small proportion of YES associates with the PDGF receptor and PtdIns kinase[10]. This complex formation is correlated with a transient increase in the activity of the YES kinase. For transforming mutants of polyomavirus middle T antigen the extent of association of middle T antigen with YES and PtdIns kinase correlates with transforming capacity[11].

In a variety of cell types, activated YES associates with stimulated plasma membrane signalling receptors. Other SRC family kinases (FYN, LYN, SRC) may be simultaneously or alternatively activated, indicating that the SRC family kinases are involved in a cell-specific manner in generating the patterns of protein tyrosine phosphorylation that occur after cell stimulation[12,13]. In keratinocytes and other types of cell, YES is inactivated by a $Ca^{2+}$-dependent association with proteins that over-rides its activation by tyrosine dephosphorylation[14].

### Cancer

Moderate to strong expression of *YES1* has been detected at relatively low frequency in a variety of cancers including fibrosarcoma, malignant lymphoma, malignant melanoma, .glioblastoma, breast cancer, colorectal cancer, head and neck cancer, renal cancers, lung cancers, gastric carcinoma and stomach cancer[15–18].

### Transgenic animals

*Src*$^-$/*Yes*$^-$ (*Src*$^-$/*Fyn*$^-$) double mutants die perinatally[19]. However, some *Fyn*$^-$/*Yes*$^-$ double mutants are viable although they undergo degenerative renal changes leading to diffuse segmental glomerulosclerosis. Thus in some cells at least, members of the *Src* family of kinases can compensate for the loss of others.

### In animals

Y73 and ESV induce sarcomas in chickens[20].

### *In vitro*

Y73 and ESV transform fibroblasts[21]. In Y73-transformed cells vinculin and integrin are phosphorylated on tyrosine residues and the concentrations of PtdIns 3*P*, PtdIns(3,4)$P_2$ and PtdIns(3,4,5)$P_3$ are increased[22]. In chick embryo fibroblasts YES associates *via* its SH3 domain with a proline-rich phosphoprotein YAP65, which associates through a WW domain (two tryptophans) with WBP-1 and WBP-2[23].

## Transcriptional regulation

The *YES* promoter contains six GC box-like sequences but no TATA box[24]. Four of the GC boxes immediately 5′ of the gene bind SP1 and affect *YES* transcription.

## Protein structure

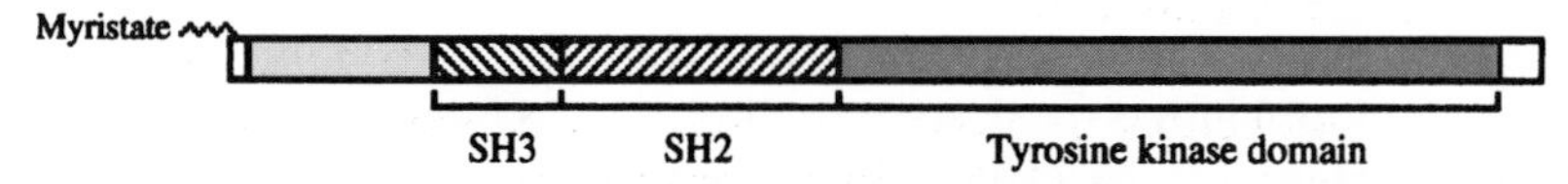


Human YES is 92% identical to chicken YES. v-YES differs from chicken YES by six substitutions and the replacement of eight amino acids at the extreme of the C-terminus of YES by three *env*-encoded residues. The latter modification may be the major feature responsible for the transforming properties of v-*yes* (see ***SRC***).

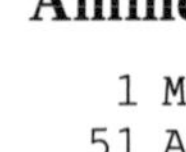

## Amino acid sequence of human YES

```
  1 MGCIKSKENK SPAIKYRPEN TPEPVSTSVS HYGAEPTTVS PCPSSSAKGT
 51 AVNFSSLSMT PFGGSSGVTP FGGASSSFSV VPSSYPAGLT GGVTIFVALY
101 DYEARTTEDL SFKKGERFQI INNTEGDWWE ARSIATGKNG YIPSNYVAPA
151 DSIQAEEWYF GKMGRKDAER LLLNPGNQRG IFLVRESETT KGAYSLSIRD
201 WDEIRGDNVK HYKIRKLDNG GYYITTRAQF DTLQKLVKHY TEHADGLCHK
251 LTTVCPTVKP QTQGLAKDAW EIPRESLRLE VKLGQGCFGE VWMGTWNGTT
301 KVAIKTLKPG TMMPEAFLQE AQIMKKLRHD KLVPLYAVVS EEPIYIVTEF
351 MSKGSLLDFL KEGDGKYLKL PQLVDMAAQI ADGMAYIERM NYIHRDLRAA
401 NILVGENLVC KIADFGLARL IEDNEYTARQ GAKFPIKWTA PEAALYGRFT
451 IKSDVWSFGI LQTELVTKGR VPYPGMVNRE VLEQVERGYR MPCPQGCPES
501 LHELMNLCWK KDPDERPTFE YIQSFLEDYF TATEPQYQPG ENL (543)
```

### Domain structure

| | |
|---|---|
| 1–7 | Necessary and sufficient for myristylation (Gly$^2$) |
| 8–96 | Unique domain variable between non-membrane receptor tyrosine kinases |
| 96–148 | SH3 domain (underlined): selects Ψ(aliphatic)XXRPLPXLP |
| 148–256 | SH2 domain |
| 256–524 | Catalytic domain, which shares sequence homology with other tyrosine kinases |
| 283–291 and 305 | ATP binding site surrounding Lys$^{302}$ |
| 396 | Active site |
| 426 | Autophosphorylation site |
| 525–543 | Regulatory domain |

### Database accession numbers

| | *PIR* | *SWISSPROT* | *EMBL/GENBANK* | *REFERENCES* |
|---|---|---|---|---|
| Human *YES1* | A26714 | P07947 | M15990 | 3 |

### *References*


1 Iothara, S. et al. (1978) Jpn J. Cancer Res. (Gann) 69, 825–830.
2 Wallbank, A.M. et al. (1966) Nature 209, 1265.
3 Sukegawa, J. et al. (1987) Mol. Cell. Biol. 7, 41–47.
4 Sudol, M. et al. (1993) Oncogene 8, 823–831.
5 Semba, K. et al. (1985) Science 227, 1038–1040.
6 Reed, J.C. et al. (1986) Proc. Natl Acad. Sci. USA 83, 3982–3986.
7 Sukegawa, J. et al. (1990) Oncogene 5, 611–614.
8 **Eiseman, E. and Bolen, J.B. et al. (1990) Cancer Cells 2, 303–310.**
9 Zhao, Y.-H. et al. (1990) Oncogene 5, 1629–1635.
10 Kypta, R.M. et al. (1990) Cell 62, 481–492.
11 Kornbluth, S. et al. (1990) J. Virol. 64, 1584–1589.
12 Huang, M.-M. et al. (1991) Proc. Natl Acad. Sci. USA 88, 7844–7848.
13 Eiseman, E. and Bolen, J.B. (1992) Nature 355, 78–80.
14 Zhao, Y. et al. (1993) Mol. Cell. Biol. 13, 7507–7514.
15 Sugawara, K. et al. (1991) Br. J. Cancer 63, 508–513.

[16] Loganzo, F. et al. (1993) Oncogene 8, 2637–2644.
[17] Park, J. and Cartwright, C.A. (1995) Mol. Cell. Biol. 15, 2374–2382.
[18] Seki, T. et al. (1985) Jpn J. Cancer Res. (Gann) 76, 907–910.
[19] Stein, P.L. et al. (1994) Genes Dev. 8, 1999–2007.
[20] Kawai, S. et al. (1980) Proc. Natl Acad. Sci. USA 77, 6199–6203.
[21] Ghysdael, J. et al. (1981) Virology 111, 386–400.
[22] Fukui, Y. et al. (1991) Oncogene 6, 407–411.
[23] Chen, H.I. and Sudol, M. (1995) Proc. Natl Acad. Sci. USA 92, 7819–7823.
[24] Matsuzawa, Y. et al. (1991) Oncogene 6, 1561–1567.

# Section IV

# TUMOUR SUPPRESSOR GENES

## Identification

*APC* was identified from the familial adenomatosis polyposis (FAP) locus by screening a YAC library with 5q21 markers[1–3]. The *MCC* (mutated in colorectal cancer) gene, ~180 kb from *APC*, was also isolated by positional cloning[4,5].

## Related genes

There are no closely related genes but both *APC* and *MCC* contain heptad repeats that give predicted coil-coiled domains, similar to those in myosins and keratins and the *SKI* family.

| | *APC* | *MCC* |
|---|---|---|
| **Nucleotides (kb)** | 120 | 170 |
| **Chromosome** | 5q21 | 5q21 |
| **Mass (kDa): predicted** | 312 | 93 |
| **expressed** | ~300 | 100 |
| **Cellular location** | Cytoplasm | Cytoplasm |

APC is concentrated in the basolateral portion of the crypt epithelial cells[6]. Wild-type but not mutant APC binds to the microtubule cytoskeleton and promotes its assembly *in vitro*[7,8]. In the mouse APC co-localizes with α-catenin in the lateral cytoplasm of intestinal epithelial cells but also occurs in microvilli and in the apical cytoplasm where α-catenin is not detected[9].

MCC is associated with the plasma membrane and membrane organelles[10].

**Tissue distribution**

Lymphoblastoid cell lines; both forms (see below) probably ubiquitous.

## Protein function

APC forms dimers *via* the leucine zipper region and associates with the E-cadherin binding proteins α- and β-catenin[11,12]. Catenin/cadherin complexes mediate adhesion, cytoskeletal anchoring and signalling. β-catenin is upregulated by WNT1 and catenin/APC complexes may play a role in regulating cell growth. APC is phosphorylated by glycogen synthase kinase 3β (GSK3β) which regulates the interaction of APC with β-catenin[13]. GSK3β, APC and β-catenin are homologues of components of the WINGLESS (*Drosophila*) and WNT (mice) signalling pathways (see scheme below).

APC also interacts with DLG, the human homologue of *Drosophila discs large* (*dlg*) tumour suppressor and progenitor of the membrane-associated guanylate kinase (MAGUK) protein family[14], and with EB1, a 280 amino acid protein lacking homology with any known protein[15].

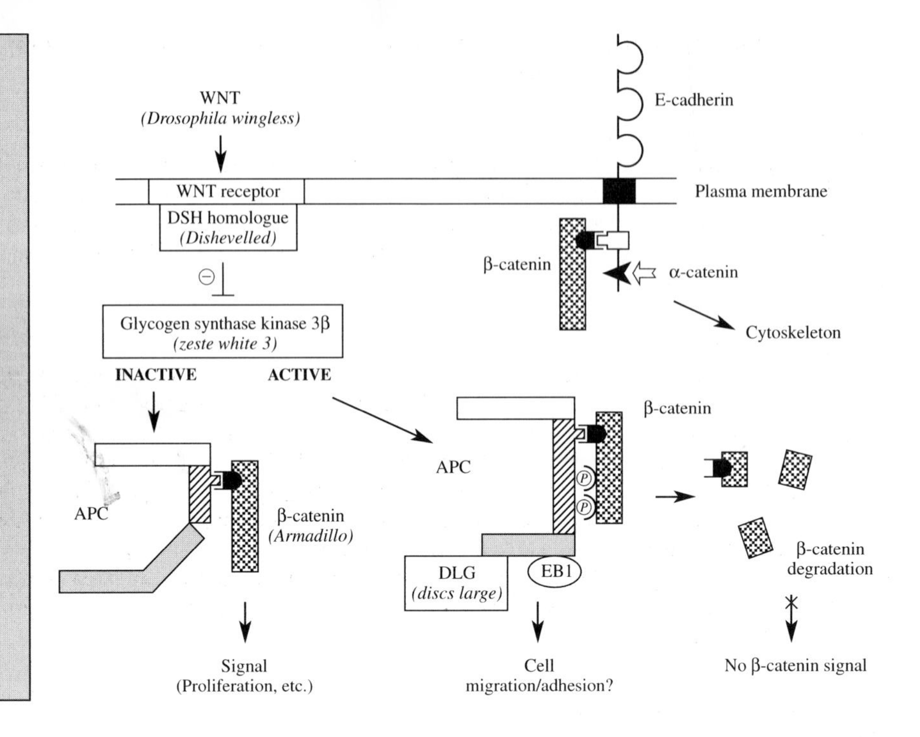


## Cancer

Familial adenomatosis polyposis (FAP) arises from the inheritance of one abnormal *APC* (adenomatous polyposis coli) allele. The incidence is 1 in 8000 and it causes the development of hundreds of colonic polyps in early life and leads, in untreated individuals, to colorectal cancer. *APC* is mutated in the germline of FAP patients, virtually all mutations inactivating the gene[16]. Variant alleles of *APC* are also involved in the attenuated form of familial adenomatosis polyposis (AAPC). The *APC* locus is involved in other forms of colorectal cancers, APC mutations occurring in ~80% of colon tumours studied, although in these cases the incidence of allelic losses from chromosome 5q is very variable[17], and in gastric carcinomas[18]. Mutations in *APC* show a negative correlation with microsatellite instability, as do mutations in *P53* and *KRAS*[19]. FAP kindreds show increased risk of hepatoblastoma and are associated with inactivation of the *APC* gene[20].

Prophylactic colectomy is effective in preventing large bowel cancer in FAP patients but extracolonic cancers, notably periampullary cancer, may subsequently be fatal. KRAS mutations at codon 12 occur frequently in these cancers, although somatic *APC* mutations are rare[21].

Identical allelic loss of APC has been detected in Barrett's oesophagus, dysplasia and invasive adenocarcinoma[22].

*MCC* is not mutated in the germline but undergoes somatic mutations in FAP and colon cancer[23], although it may not function as an independent tumour suppressor gene in colorectal cancer[24]. Deletions in the 5q21 region also occur in ~25% of lung cancers[25]. FAP is inconsistently associated with characteristic patches of congenital hypertrophy of the retinal pigment epithelium (CHRPE), the extent of which correlates with the location of mutation in the *APC* gene [26].

**Transgenic animals**
Transgenic mice homozygous for the null *APC* allele develop normally, indicating that in mice this gene is not essential for proliferation and differentiation. Animals with germline mutations of *APC* have a phenotype similar to that of FAP [27]. Expression of activated *Ras* in these mice causes colon carcinomas [28].

***In vitro***
Over-expression of APC in NIH 3T3 cells blocks cell cycle progression from $G_0/G_1$ to S phase and causes significant reduction in the activity of CDK2 [29]. Expression of exogenous APC in colorectal cancer cells containing only mutant APC reduces total β-catenin levels [30]. The central region of APC (1342–2075), typically deleted or truncated in tumours, is responsible for this downregulation. Transfection of full-length *APC* suppresses the tumorigenicity of some human colon carcinoma cell lines [31].

MCC phosphorylation increases as cells pass the $G_1$ to S phase transition and over-expression of the normal but not mutant protein blocks the serum induction of this transition [10].

## Gene structure of *APC*

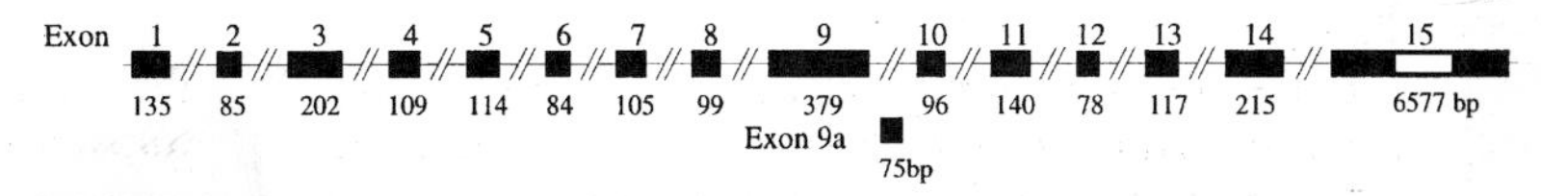


Three exons 5′ to exon 1 have recently been identified (exons 0.3, 0.1 and 0.2). Alternatively spliced forms including exons 0.3 + 1 + 2, 0.3 + 2, 0.1 + 0.2 + 1 + 2 and 0.1 + 1 + 2 [32].

Two alternative forms having the deletions associated with AAPC are (i) loss of exons 1–4 and the first 16 bases of exon 5 and (ii) loss of exons 2–4 and the first 16 bases of exon 5 [33].

## Protein structure

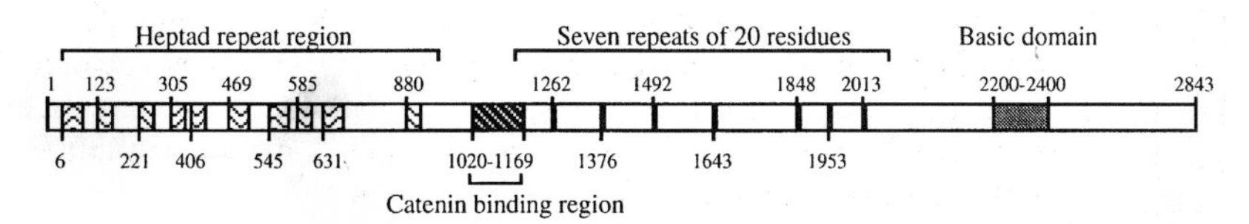


The alternatively spliced form (exon 9a) has a deletion of bases 934–1236 that removes 101 amino acids. Isoforms lacking exon 7 have been detected in human and mouse cell lines [34]. At least five isoforms arise from alternative splicing of 5′ non-coding sequences [35]. The initial amino acids of the 10 heptad repeats and the seven 20 residue repeats (consensus F-VE-TP-CFSR-SSLSSLS) are numbered. The last 14 amino acids of the fourth and the first seven amino acids of the fifth heptad repeats are present only in the longer gene product (containing exon 9).

## Amino acid sequence of human APC

1 MAAASYDQLL KQVEALKMEN SNLRQELEDN SNHLTKLETE ASNMKEVLKQ
51 LQGSIEDEAM ASSGQIDLLE RLKELNLDSS NFPGVKLRSK MSLRSYGSRE
101 GSVSSRSGEC SPVPMGSFPR RGFVNGSRES TGYLEELEKE RSLLLADLDK
151 EEKEKDWYYA QLQNLTKRID SLPLTENFSL QTDMTRRQLE YEARQIRVAM
201 EEQLGTCQDM EKRAQRRIAR IQQIEKDILR IRQLLQSQAT EAERSSQNKH
251 ETGSHDAERQ NEGQGVGEIN MATSGNGQGS TTRMDHETAS VLSSSSTHSA
301 PRRLTSHLGT *KVEMVYSLLS MLGTHDKDDM SRTLLAMSSS QDSCISMRQS*
351 *GCLPLLIQLL HGNDKDSVLL GNSRGSKEAR ARASAALHNI IHSQPDDKRG*
401 *RREIRVLHLL EQ*IRAYCETC WEWQEAHEPG MDQDKNPMPA PVEHQICPAV
451 CVLMKLSFDE EHRHAMNELG GLQAIAELLQ VDCEMYGLTN DHYSITLRRY
501 AGMALTNLTF GDVANKATLC SMKGCMRALV AQLKSESEDL QQVIASVLRN
551 LSWRADVNSK KTLREVGSVK ALMECALEVK KESTLKSVLS ALWNLSAHCT
601 ENKADICAVD GALAFLVGTL TYRSQTNTLA IIESGGGILR NVSSLIATNE
651 DHRQILRENN CLQTLLQHLK SHSLTIVSNA CGTLWNLSAR NPKDQEALWD
701 MGAVSMLKNL IHSKHKMIAM GSAAALRNLM ANRPAKYKDA NIMSPGSSLP
751 SLHVRKQKAL EAELDAQHLS ETFDNIDNLS PKASHRSKQR HKQSLYGDYV
801 FDTNRHDDNR SDNFNTGNMT VLSPYLNTTV LPSSSSSRGS LDSSRSEKDR
851 SLERERGIGL GNYHPATENP GTSSKRGLQI STTAAQIAKV MEEVSAIHTS
901 QEDRSSGSTT ELHCVTDERN ALRRSSAAHT HSNTYNFTKS ENSNRTCSMP
951 YAKLEYKRSS NDSLNSVSSS DGYGKRGQMK PSIESYSEDD ESKFCSYGQY
1001 PADLAHKIHS ANHMDDNDG*E LDTPINYSLK YSDE*QLNSGR QSPSQNERWA
1051 RPKHIIEDEI KQSEQRQSRN QSTTYPVYTE STDDKHLKFQ PHFGQQECVS
1101 PYRSRGANGS ETNRVGSNHG INQNVSQSLC QEDDY*EDDKP TNYSERYSEE*
1151 EQHE*EEERPT NYSIKYNEE*K RHVDQPIDYS LKYATDIPSS QKQSFSFSKS
1201 SSGQSSKTEH MSSSSENTST PSSNAKRQNQ LHPSSAQSRS GQPQKAATCK
1251 VSSINQETIQ TYCVEDTPIC FSRCSSLSSL SSAEDEIGCN QTTQEADSAN
1301 TLQIAEIKEK IGTRSAEDPV SEVPAVSQHP RTKSSRLQGS SLSSESARHK
1351 AVEFSSGAKS PSKSGAQTPK SPPEHYVQET PLMFSRCTSV SSLDSFESRS
1401 IASSVQSEPC SGMVSGIISP SDLPDSPGQT MPPSRSKTPP PPPQTAQTKR
1451 EVPKNKAPTA EKRESGPKQA AVNAAVQRVQ VLPDADTLLH FATESTPDGF
1501 SCSSSLSALS LDEPFIQKDV ELRIMPPVQE NDNGNETESE QPKESNENQE
1551 KEAEKTIDSE KDLLDDSDDD DIEILEECII SAMPTKSSRK AKKPAQTASK
1601 LPPPVARKPS QLPVYKLLPS QNRLQPQKHV SFTPGDDMPR VYCVEGTPIN
1651 FSTATSLSDL TIESPPNELA AGEGVRGGAQ SGEFEKRDTI PTEGRSTDEA
1701 QGGKTSSVTI PELDDNKAEE GDILAECINS AMPKGKSHKP FRVKKIMDQV
1751 QQASASSSAP NKNQLDGKKK KPTSPVKPIP QNTEYRTRVR KNADSKNNLN
1801 AERVFSDNKD SKKQNLKNNS KDFNDKLPNN EDRVRGSFAF DSPHHYTPIE
1851 GTPYCFSRND SLSSLDFDDD DVDLSREKAE LRKAKENKES EAKVTSHTEL
1901 TSNQQSANKT QAIAKQPINR GQPKPILQKQ STFPQSSKDI PDRGAATDEK
1951 LQNFAIENTP VCFSHNSSLS SLSDIDQENN NKENEPIKET EPPDSQGEPS
2001 KPQASGYAPK SFHVEDTPVC FSRNSSLSSL SIDSEDDLLQ ECISSAMPKK
2051 KKPSRLKGDN EKHSPRNMGG ILGEDLTLDL KDIQRPDSEH GLSPDSENFD
2101 WKAIQEGANS IVSSLHQAAA AACLSRQASS DSDSILSLKS GISLGSPFHL
2151 TPDQEEKPFT SNKGPRILKP GEKSTLETKK IESESKGIKG GKKVYKSLIT
2201 GKVRSNSEIS GQMKQPLQAN MPSISRGRTM IHIPGVRNSS SSTSPVSKKG
2251 PPLKTPASKS PSEGQTATTS PRGAKPSVKS ELSPVARQTS QIGGSSKAPS
2301 RSGSRDSTPS RPAQQPLSRP IQSPGRNSIS PGRNGISPPN KLSQLPRTSS
2351 PSTASTKSSG SGKMSYTSPG RQMSQQNLTK QTGLSKNASS IPRSESASKG

```
2401 LNQMNNGNGA NKKVELSRMS STKSSGSESD RSERPVLVRQ STFIKEAPSP
2451 TLRRKLEESA SFESLSPSSR PASPTRSQAQ TPVLSPSLPD MSLSTHSSVQ
2501 AGGWRKLPPN LSPTIEYNDG RPAKRHDIAR SHSESPSRLP INRSGTWKRE
2551 HSKHSSSLPR VSTWRRTGSS SSILSASSES SEKAKSEDEK HVNSISGTKQ
2601 SKENQVSAKG TWRKIKENEF SPTNSTSQTV SSGATNGAES KTLIYQMAPA
2651 VSKTEDVWVR IEDCPINNPR SGRSPTGNTP PVIDSVSEKA NPNIKDSKDN
2701 QAKQNVGNGS VPMRTVGLEN RLNSFIQVDA PDQKGTEIKP GQNNPVPVSE
2751 TNESSIVERT PFSSSSSSKH SSPSGTVAAR VTPFNYNPSP RKSSADSTSA
2801 RPSQIPTPVN NNTKKRDSKT DSTESSGTQS PKRHSGSYLV TSV (2843)
```

**Domain structure**

| | |
|---|---|
| 1–55 | Region sufficient for APC homodimerization |
| 1–730 | Leucine-rich region |
| 7–72 and 185–227 | Potential coil |
| 312–412 | 101 amino acids deleted in the alternatively spliced form (italics) |
| 6–57, 123–150, 221–245, 305–326, 406–426, 469–517, 545–579, 585–619, 631–687, 880–897 | 10 heptad repeat regions (underlined) |
| 731–2832 | Serine-rich region |
| 1020–1034, 1136–1150 and 1155–1169 | Three imperfect 15 amino acid repeats, any one of which is sufficient to bind catenins (underlined italics) |
| 1131–1156 and 1558–1577 | Aspartate/glutamate-rich (acidic) region |
| 1342–2075 | Responsible for downregulation of $\beta$-catenin |
| 1866–1893 | Highly charged region |

APC contains multiple serine phosphorylation, glycosylation and myristylation sites.

**Mutations detected in *APC***

*APC* germline mutations are usually point mutations or small deletions, insertions or splicing mutations leading to downstream termination, the most N-terminal being at codon 168. These include a single base pair change that introduces a translational stop at codon 168 and a deletion at codons 169–170. The most common mutation is a 5 bp deletion at codon 1309: numerous other mutations have been detected[36–41]. APC mutations occur in replication error (RER) and non-RER colorectal tumours but the frequency of frameshift mutations is greater in RER cases[42]. Multiple somatic mutations have also been detected in *APC* (together with *P53* mutations) following loss of *MSH2* in HNPCC[43]. Mutations in *APC* also occur in gastroduodenal tumours[44,45], in hepatoblastomas[46] and both somatic and germline mutations occur in desmoid tumours[47]. A mutation has been detected at intron 31 splice junction[48]. Mutations in exons 1–6 are rare in patients with familial colorectal cancer who do not have many colorectal polyps[49].

Mutations detected in AAPC are similar but are grouped nearer the 5′ end of *APC* (within the first 157 codons). These include (i) a 4 bp deletion in exon 3 that generates an 83 amino acid protein the first 77 residues of which correspond to normal APC; (ii) deletion of exon 3 caused by a point mutation in intron 3: this generates a 97 amino acid protein the first 73 residues of which correspond to normal APC; (iii) a nonsense mutation at codon 157 giving a truncated protein of 156 amino acids; (iv) a 2 bp deletion in exon 4 generating a 145 amino acid protein the first 142 residues of which correspond to normal APC [50].

Mutations between 1285 and 1465 correlate with profuse polyposis.

In CHRPE ocular lesions are present if *APC* is truncated after exon 9 but almost always absent if the mutation occurs before this exon [26]. Thus CHRPE mutations occur in codons 463–1397, not 136–302 or 1445–1578 (the latter give rise to desmoid tumours) [51].

## Amino acid sequence of human MCC

```
  1 MNSGVAMKYG NDSSAELSEL HSAALASLKG DIVELNKRLQ QTERERDLLE
 51 KKLAKAQCEQ SHLMREHEDV QERTTLRYEE RITELHSVIA ELNKKIDRLQ
101 GTTIREEDEY SELRSELSQS QHEVNEDSRS MDQDQTSVSI PENQSTMVTA
151 DMDNCSDLNS ELQRVLTGLE NVVCGRKKSS CSLSVAEVDR HIEQLTTASE
201 HCDLAIKTVE EIEGVLGRDL YPNLAEERSR WEKELAGLRE ENESLTAMLC
251 SKEEELNRTK ATMNAIREER DRLRRRVREL QTRLQSVQAT GPSSPGRLTS
301 TNRPINPSTG ELSTSSSSND IPIAKIAERV KLSKTRSESS SSDRPVLGSE
351 ISSIGVSSSV AEHLAHSLQD CSNIQEIFQT LYSHGSAISE SKIREFEVET
401 ERLNSRIEHL KSQNDLLTIT LEECKSNAER MSMLVGKYES NATALRLALQ
451 YSEQCIEAYE LLLALAESEQ SLILGQFRAA GVGSSPGDQS GDENITQMLK
501 RAHDCRKTAE NAAKALLMKL DGSCGGAFAV AGCSVQPWES LSSNSHTSTT
551 SSTASSCDTE FTKEDEQRLK DYIQQLKNDR AAVKLTMLEL ESIHIDPLSY
601 DVKPRGDSQR LDLENAVLMQ ELMAMKEEMA ELKAQLYLLE KEKKALELKL
651 STREAQEQAY LVHIEHLKSE VEEQKEQRMR SLSSTSSGSK DKPGKECADA
701 ASPALSLAEL RTTCSENELA AEFTNAIRRE KKLKARVQEL VSALERLTKS
751 SEIRHQQSAE FVNDLKRANS NLVAAYEKAK KKHQNKLKKL ESQMMAMVER
801 HETQVRMLKQ RIALLEEENS RPHTNETSL (829)
```

### Domain structure

220–243 Similarity to amino acids 249–272 of the G protein-coupled M3 muscarinic acetylcholine receptor (italics)

### Mutations detected in *MCC*

Mutations in colorectal cancer: $Arg^{267} \rightarrow$ Leu, $Pro^{486} \rightarrow$ Leu, $Ser^{490} \rightarrow$ Leu, $Arg^{506} \rightarrow$ Gln, $Ala^{698} \rightarrow$ Val [24].

### Database accession numbers

| | *PIR* | *SWISSPROT* | *EMBL/GENBANK* | *REFERENCES* |
|---|---|---|---|---|
| Human *APC* | A37261 | APC_HUMAN P25054 | Hsfapapc M74088 | 4 |
| Human *MCC* | A33166, A38434 | CRCM_HUMAN P23508 | Hscrcmut M62397 | 5 |
| Rat *Apc* | D38629 | | | 52 |

### *References*


[1] Kinzler, K.W. et al. (1991) Science 253, 661–665.
[2] Nishisho, I. et al. (1991) Science 253, 665–669.
[3] Groden, J. et al. (1991) Cell 66, 589–600.
[4] Joslyn, G. et al. (1991) Cell 66, 601–613.

[5] Kinzler, K.W. et al. (1991) Science 251, 1366–1370.
[6] Smith, K.J. et al. (1993) Proc. Natl Acad. Sci. USA 90, 2846–2850.
[7] Smith, K.J. et al. (1994) Cancer Res. 54, 3672–3675.
[8] Munemitsu, S. et al. (1994) Cancer Res. 54, 3676–3681.
[9] Miyashiro, I. et al. (1995) Oncogene 11, 89–96.
[10] Matsumine, A. et al. (1996) J. Biol. Chem. 271, 10341–10346.
[11] **Nagase, H. and Nakamura, Y. (1993) Hum. Mutations 2, 425–434.**
[12] Su, L.-K. et al. (1993) Science 262, 1734–1737.
[13] Rubinfeld, B. et al. (1996) Science 272, 1023–1026.
[14] Matsumine, A. et al. (1996) Science 272, 1020–1023.
[15] Su, L.-K. et al. (1995) Cancer Res. 55, 2972–2977.
[16] Miyoshi, Y. et al. (1992) Proc. Natl Acad. Sci. USA 89, 4452–4456.
[17] **Fearon, E.R. and Vogelstein, B. (1990) Cell 61, 759–767.**
[18] Nakatsuru, S. et al. (1993) Hum. Mol. Genet. 2, 1463–1465.
[19] Heinen, C.D. et al. (1995) Cancer Res. 55, 4797–4799.
[20] Kurahashi, H. et al. (1995) Cancer Res. 55, 5007–5011.
[21] Gallinger, S. et al. (1995) Oncogene 10, 1875–1878.
[22] Zhuang, Z. et al. (1996) Cancer Res. 56, 1961–1964.
[23] **Bourne, H.R. (1991) Nature 353, 696–698.**
[24] Curtis, L.J. et al. (1994) Hum. Mol. Genet. 3, 443–446.
[25] Ashton-Rickardt, P.G. et al. (1991) Oncogene 6, 1881–1886.
[26] Olschwang, S. et al. (1993) Cell 75, 959–968.
[27] Su, L.-K. et al. (1992) Science 256, 668–670.
[28] D'Abaco, G.M. et al. (1996) Mol. Cell. Biol. 16, 884–891.
[29] Baeg, G.-H. et al. (1995) EMBO J. 14, 5618–5625.
[30] Munemitsu, S. et al. (1995) Proc. Natl Acad. Sci. USA 92, 3046–3050.
[31] Groden, J. et al. (1995) Cancer Res. 55, 1531–1539.
[32] Thliveris, A. et al. (1994) Cancer Res. 54, 2991–2995.
[33] Samowitz, W.S. et al. (1995) Cancer Res. 55, 3732–3734.
[34] Oshima, M. et al. (1993) Cancer Res. 53, 5589–5591.
[35] Horii, A. et al. (1993) Hum. Mol. Genet. 2, 283–287.
[36] Gayther, S.A. et al. (1994) Hum. Mol. Genet. 3, 53–56.
[37] Stella, A. et al. (1994) Hum. Mol. Genet. 3, 1687–1688, 1918.
[38] Paffenholz, R. et al. (1994) Hum. Mol. Genet. 3, 1703–1704.
[39] Gebert, J.F et al. (1994) Hum. Mol. Genet. 3, 1167–1168
[40] Hamzehloei, T. et al. (1994) Hum. Mol. Genet. 3, 1023–1024.
[41] Mandl, M. et al. (1994) Hum. Mol. Genet. 3, 1009–1011.
[42] Huang, J. et al. (1996) Proc. Natl Acad. Sci. USA 93, 9049–9054.
[43] Lazar, V. et al. (1994) Hum. Mol. Genet. 3, 2257–2260.
[44] Tamura, G. et al. (1994) Cancer Res. 54, 1149–1151.
[45] Toyooka, M. et al. (1995) Cancer Res. 55, 3165–3170.
[46] Oda, H. et al. (1996) Cancer Res. 56, 3320–3323.
[47] Palmirotta, R. et al. (1995) Hum. Mol. Genet. 4, 1979–1981.
[48] Ainsworth, P. et al. (1991) Oncogene 6, 1881–1886.
[49] Joyce, J.A. et al. (1995) Clin. Genet. 48, 299–303.
[50] Spiro, L. et al. (1993) Cell 75, 951–957.
[51] Caspari, R. et al. (1995) Hum. Mol. Genet. 4, 337–340.
[52] Kakiuchi, H. et al. (1995) Proc. Natl Acad. Sci. USA 92, 910–914.

# BRCA1, BRCA2

## Identification

*BRCA1* was identified by positional cloning [1,2]. *BRCA2* was identified by a genomic linkage and search using breast cancer families unlinked to the *BRCA1* locus [3,4].

## Related genes

*BRCA1* and *BRCA2* are remarkably similar. Both have AT-rich coding sequences and a large exon 11 (3426 and 4932 bp, respectively). Both proteins are highly charged and contain a granin protein family consensus sequence.

| | ***BRCA1*** | ***BRCA2*** |
|---|---|---|
| **Nucleotides (kb)** | 100 | ~70 |
| **Chromosome** | 17q21 | 13q12–13 |
| **Mass (kDa): predicted** | 208 | >400 |
| **expressed** | 220 [5,6] | |
| **Cellular location** | Nucleus (Cytoplasmic in a high proportion of breast and ovarian cancers) [7] | |

**Tissue distribution**

*BRCA1* is most abundant in testis and thymus but also expressed in breast and ovary [2]. Mouse *Brca1* is widely expressed during development and also in adult tissues where it is associated with rapidly growing epithelial cells undergoing differentiation [8] and with meiosis and spermiogenesis but not mitosis of male germ cells [9]. In the mammary gland *Brca1* is induced during puberty, pregnancy and following treatment of ovariectomized animals with 17$\beta$-oestradiol and progesterone [10].

*BRCA2* is detectable in breast, testis and thymus and at lower levels in lung, ovary and spleen [11].

## Protein function

BRCA1 and BRCA2 are putative transcription factors undergoing cell cycle-dependent phosphorylation maximal in S and M phases [5]. The N-terminal region contains a $C_3HC_4$ zinc finger similar to those of many DNA binding proteins. Multiple alternative splicing patterns occur [2]. Deletion of exons 5 and 6 in mouse *Brca1* drastically decreases embryonic growth and BRCA1 may therefore be a positive regulator of cell proliferation during early embryogenesis [12].

**Cancer**

*BRCA1* and *BRCA2* are genes in which mutations predispose to breast cancer and together are probably involved in about two-thirds of familial breast cancer, approximately 5% of all cases. *BRCA1* is probably responsible for about one-third of families having multiple cases of breast cancer alone but 80% of families in which there is both breast cancer and epithelial ovarian cancer [13]. Inheritance of a *BRCA1* mutation in an extensive family carries a 60% risk of breast cancer for that individual by age 50 and a 90% lifetime risk. In some families the risk of ovarian cancer is similar but in others it is less, indicating allelic heterogeneity or the existence of an adjacent gene [14]. *BRCA1* mutations have also been detected at low frequency in apparently sporadic breast and ovarian cancers but mutations of this gene do not appear to be critical in the development of the majority of breast and ovarian cancers that arise in the absence of germline mutations. Thus *BRCA1* mutations may be relatively unimportant in sporadic tumours. However, anomalous location of BRCA1 in the cytoplasm appears to be associated with a high proportion of breast and ovarian cancers [7]. Homozygous mutation (for the $AA_{2800}$ deletion) has been detected in one woman [15].

Individuals with rare alleles of the *HRAS1* variable number of tandem repeats (VNTR) polymorphism have an increased risk of non-hereditary breast cancer and also of hereditary ovarian cancer but not of hereditary breast cancer [16].

Mutations in *BRCA2* confer a high risk of female breast cancer, accounting for ~35% of all inherited breast cancers, together with a more moderate risk of ovarian cancer. They also increase the risk of male breast cancer, in contrast to mutations in *BRCA1*. Both alleles appear to be lost in cancers [17]. Allelic imbalance in the *BRCA2* region has been detected in primary sporadic breast tumours [18]. LOH on 13q12–13 suggests that *BRCA2* may also be involved in cancer of the pancreas, prostate, cervix, colon and ureter [19].

**Transgenic animals**

*Brca1*$^{-/-}$ mice die *in utero* between 10 and 13 days of gestation with neuro-epithelial abnormalities [20], indicating that the gene is essential for embryonic development in mice, even though the homozygous mutation mentioned above indicates that *BRCA1* is dispensable for human development. Deletion of *Brca1* exons 5 and 6 or exon 11 is also lethal although mice with one mutated allele appear normal [21]. *Brca1*$^{5\text{-}6}$ mutant embryos have decreased expression of cyclin E and *Mdm-2* and increased levels of *Waf-1* [12].

***In vitro***

The expression of anti-sense RNA to *Brca1* accelerates the growth of NIH 3T3 cells [5]. Expression of wild-type BRCA1 inhibits growth of a range of breast and ovarian cancer cell lines but not of colon or lung cancer cells [22]. Mutant BRCA1 has no effect on breast cancer cell growth. Mutations in the 3′ (e.g. giving a 1835 amino acid protein) but not the 5′ end of the gene inhibit ovarian cancer cell growth.

Human breast cancer or mouse fibroblast cell lines expressing BRCA1 undergo apoptosis when deprived of serum or treated with calcium ionophore [23].

## Gene structure of *BRCA1*

The *BCRA1* region of the genome contains a tandem duplication of ~30 kb that has generated two copies of exons 1A and 1B and exon 2, together with two exons of the adjacent *1A1-3B* gene [24]. A splice variant lacking exon 11 is coexpressed with full length *BRCA1* in most tissues.

## Gene structure of *BRCA2*

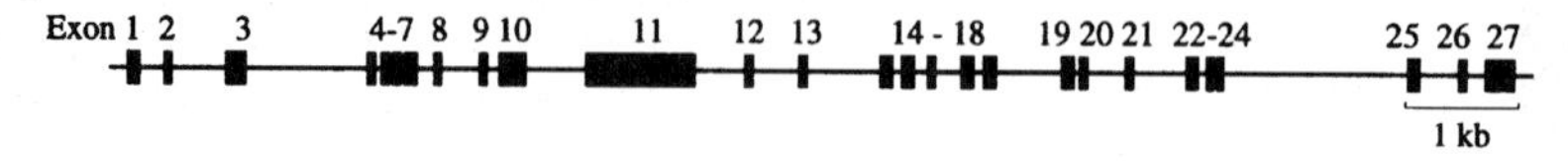


*BRCA2* contains a CpG-rich region at the 5′ end[11].

## Protein structure

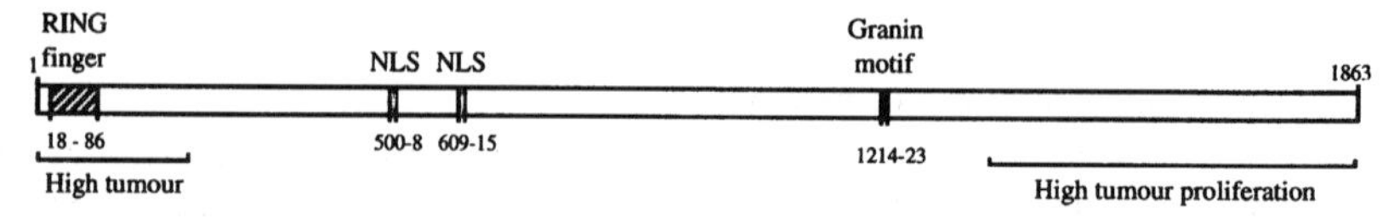


The N-terminal "zinc ring", "RING finger" or A-box homology domain includes the zinc ring $C_3HC_4$ sequence found in several viral proteins, proto-oncoproteins and transcription factors[25,26].

## Amino acid sequence of human BRCA1

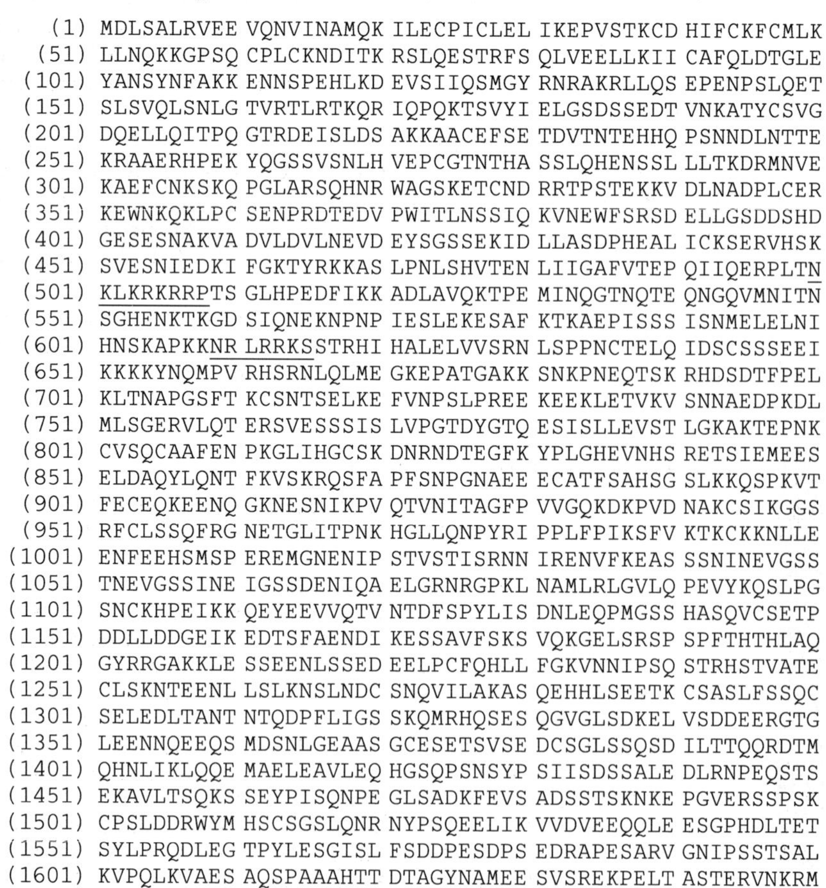

```
   (1) MDLSALRVEE VQNVINAMQK ILECPICLEL IKEPVSTKCD HIFCKFCMLK
  (51) LLNQKKGPSQ CPLCKNDITK RSLQESTRFS QLVEELLKII CAFQLDTGLE
 (101) YANSYNFAKK ENNSPEHLKD EVSIIQSMGY RNRAKRLLQS EPENPSLQET
 (151) SLSVQLSNLG TVRTLRTKQR IQPQKTSVYI ELGSDSSEDT VNKATYCSVG
 (201) DQELLQITPQ GTRDEISLDS AKKAACEFSE TDVTNTEHHQ PSNNDLNTTE
 (251) KRAAERHPEK YQGSSVSNLH VEPCGTNTHA SSLQHENSSL LLTKDRMNVE
 (301) KAEFCNKSKQ PGLARSQHNR WAGSKETCND RRTPSTEKKV DLNADPLCER
 (351) KEWNKQKLPC SENPRDTEDV PWITLNSSIQ KVNEWFSRSD ELLGSDDSHD
 (401) GESESNAKVA DVLDVLNEVD EYSGSSEKID LLASDPHEAL ICKSERVHSK
 (451) SVESNIEDKI FGKTYRKKAS LPNLSHVTEN LIIGAFVTEP QIIQERPLTN
 (501) KLKRKRRPTS GLHPEDFIKK ADLAVQKTPE MINQGTNQTE QNGQVMNITN
 (551) SGHENKTKGD SIQNEKNPNP IESLEKESAF KTKAEPISSS ISNMELELNI
 (601) HNSKAPKKNR LRRKSSTRHI HALELVVSRN LSPPNCTELQ IDSCSSSEEI
 (651) KKKKYNQMPV RHSRNLQLME GKEPATGAKK SNKPNEQTSK RHDSDTFPEL
 (701) KLTNAPGSFT KCSNTSELKE FVNPSLPREE KEEKLETVKV SNNAEDPKDL
 (751) MLSGERVLQT ERSVESSSIS LVPGTDYGTQ ESISLLEVST LGKAKTEPNK
 (801) CVSQCAAFEN PKGLIHGCSK DNRNDTEGFK YPLGHEVNHS RETSIEMEES
 (851) ELDAQYLQNT FKVSKRQSFA PFSNPGNAEE ECATFSAHSG SLKKQSPKVT
 (901) FECEQKEENQ GKNESNIKPV QTVNITAGFP VVGQKDKPVD NAKCSIKGGS
 (951) RFCLSSQFRG NETGLITPNK HGLLQNPYRI PPLFPIKSFV KTKCKKNLLE
(1001) ENFEEHSMSP EREMGNENIP STVSTISRNN IRENVFKEAS SSNINEVGSS
(1051) TNEVGSSINE IGSSDENIQA ELGRNRGPKL NAMLRLGVLQ PEVYKQSLPG
(1101) SNCKHPEIKK QEYEEVVQTV NTDFSPYLIS DNLEQPMGSS HASQVCSETP
(1151) DDLLDDGEIK EDTSFAENDI KESSAVFSKS VQKGELSRSP SPFTHTHLAQ
(1201) GYRRGAKKLE SSEENLSSED EELPCFQHLL FGKVNNIPSQ STRHSTVATE
(1251) CLSKNTEENL LSLKNSLNDC SNQVILAKAS QEHHLSEETK CSASLFSSQC
(1301) SELEDLTANT NTQDPFLIGS SKQMRHQSES QGVGLSDKEL VSDDEERGTG
(1351) LEENNQEEQS MDSNLGEAAS GCESETSVSE DCSGLSSQSD ILTTQQRDTM
(1401) QHNLIKLQQE MAELEAVLEQ HGSQPSNSYP SIISDSSALE DLRNPEQSTS
(1451) EKAVLTSQKS SEYPISQNPE GLSADKFEVS ADSSTSKNKE PGVERSSPSK
(1501) CPSLDDRWYM HSCSGSLQNR NYPSQEELIK VVDVEEQQLE ESGPHDLTET
(1551) SYLPRQDLEG TPYLESGISL FSDDPESDPS EDRAPESARV GNIPSSTSAL
(1601) KVPQLKVAES AQSPAAAHTT DTAGYNAMEE SVSREKPELT ASTERVNKRM
```

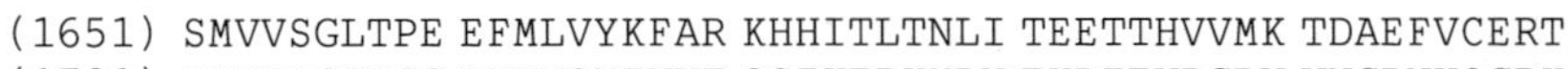


```
(1651) SMVVSGLTPE EFMLVYKFAR KHHITLTNLI TEETTHVVMK TDAEFVCERT
(1701) LKYFLGIAGG KWVVSYFWVT QSIKERKMLN EHDFEVRGDV VNGRNHQGPK
(1751) RARESQDRKI FRGLEICCYG PFTNMPTDQL EWMVQLCGAS VVKELSSFTL
(1801) GTGVHPIVVV PDAWTEDNGF HAIGQMCEAP VVTREWVLDS VALYQCQEL
(1851) DTYLIPQIPH SHY (1863)
```

**Domain structure**

500–508 and 609–615 Putative nuclear localization signals (underlined)
1214–1223 Granin motif

**Mutations detected in *BRCA1***

Over 100 mutations in *BRCA1* have been detected. The majority of small deletion and insertion mutations occur at single nucelotide repeats: substitutions often occur both at homonucleotides and at CpG/CpNpG motifs. The latter are methylated, which may contribute to mutagenesis[27]. Mutations are distributed throughout the gene but there are hotspots in exons 2 and 20. The majority are predicted to result in truncation of BRCA1 to a functionally inactive form. Mutations detected in breast/ovarian cancer kindreds include a stop codon, a frameshift and a missense resulting from a Met by Arg substitution[28–42]. Loss of expression of one allele has also been detected. There is a significant correlation between the location of the mutation and the ratio of breast to ovarian cancer within individual families[28].

Germline mutations in *BRCA1* in breast cancer appear to correlate with tumour proliferation. Thus clusters of mutations in the N- or C-termini broadly correlate with high rates of tumour proliferation (codons 23, 43, 47, 1756, 1759 and 1809, together with 757 and 1355. Mutations in the central region correlate with lower rates (codons 223, 374, 494, 785, 855, 1161 and 1355)[43].

Several germline mutations arise specifically in Japanese families with breast or ovarian cancer[44,45] and the 185delAG mutation may occur with a frequency as high as 1% in Ashkenazi Jewish women[46].

## Amino acid sequence of human BRCA2

```
  1 MPIGSKERPT FFEIFKTRCN KADLGPISLN WFEELSSEAP PYNSEPAEES
 51 EHKNNNYEPN LFKTPQRKPS YNQLASTPII FKEQGLTLPL YQSPVKELDK
101 FKLDLGRNVP NSRHKSLRTV KTKMDQADDV SCPLLNSCLS ESPVVLQCTH
151 VTPQRDKSVV CGSLFHTPKF VKGRQTPKHI SESLGAEVDP DMSWSSSLAT
201 PPTLSSTVLI VRNEEASETV FPHDTTANVK SYFSNHDESL KKNDRFIASV
251 TDSENTNQRE AASHGFGKTS GNSFKVNSCK DHIGKSMPNV LEDEVYETVV
301 DTSEEDSFSL CFSKCRTKNL QKVRTSKTRK KIFHEANADE CEKSKNQVKE
351 KYSFVSEVEP NDTDPLDSNV AHQKPFESGS DKISKEVVPS LACEWSQLTL
401 SGLNGAQMEK IPLLHISSCD QNISEKDLLD TENKRKKDFL TSENSLPRIS
451 SLPKSEKPLN EETVVNKRDE EQHLESHTDC ILAVKQAISG TSPVASSFQG
501 IKKSIFRIRE SPKETFNASF SGHMTDPNFK KETEASESGL EIHTVCSQKE
551 DSLCPNLIDN GSWPATTTQN SVALKNAGLI STLKKKTNKF IYAIHDETFY
601 KGKKIPKDQK SELINCSAQF EANAFEAPLT FANADSGLLH SSVKRSCSQN
651 DSEEPTLSLT SSFGTILRKC SRNETCSNNT VISQDLDYKE AKCNKEKLQL
701 FITPEADSLS CLQEGQCEND PKSKKVSDIK EEVLAACHP VQHSKVEYSD
```

```
 751 TDFQSQKSLL YDHENASTLI LTPTSKDVLS NLVMISRGKE SYKMSDKLKG
 801 NNYESDVELT KNIPMEKNQD VCALNENYKN VELLPPEKYM RVASPSRKVQ
 851 FNQNTNLRVI QKNQEETTSI SKITVNPDSE ELFSDNENNF VFQVANERNN
 901 LALGNTKELH ETDLTCVNEP IFKNSTMVLY GDTGDKQATQ VSIKKDLVYV
 951 LAEENKNSVK QHIKMTLGQD LKSDISLNID KIPEKNNDYM NKWAGLLGPI
1001 SNHSFGGSFR TASNKEIKLS EHNIKKSKMF FKDIEEQYPT SLACVEIVNT
1051 LALDNQKKLS KPQSINTVSA HLQSSVVVSD CKNSHITPQM LFSKQDFNSN
1101 HNLTPSQKAE ITELSTILEE SGSQFEFTQF RKPSYILQKS TFEVPENQMT
1151 ILKTTSEECR DADLHVIMNA PSIGQVDSSK QFEGTVEIKR KFAGLLKNDC
1201 NKSASGYLTD ENEVGFRGFY SAHGTKLNVS TEALQKAVKL FSDIENISEE
1251 TSAEVHPISL SSSKCHDSVV SMFKIENHND KTVSEKNNKC QLILQNNIEM
1301 TTGTFVEEIT ENYKRNTENE DNKYTAASRN SHNLEFDGSD SSKNDTVCIH
1351 KDETDLLFTD QHNICLKLSG QFMKEGNTQI KEDLSDLTFL EVAKAQEACH
1401 GNTSNKEQLT ATKTEQNIKD FETSDTFFQT ASGKNISVAK ESFNKIVNFF
1451 DQKPEELHNF SLNSELHSDI RKNKMDILSY EETDIVKHKI LKESVPVGTG
1501 NQLVTFQGQP ERDEKIKEPT LLGFHTASGK KVKIAKESLD KVKNLFDEKE
1551 QGTSEITSFS HQWAKTLKYR EACKDLELAC ETIEITAAPK CKEMQNSLNN
1601 DKNLVSIETV VPPKLLSDNL CRQTENLKTS KSIFLKVKVH ENVEKETAKS
1651 PATCYTNQSP YSVIENSALA FYTSCSRKTS VSQTSLLEAK KWLREGIFDG
1701 QPERINTADY VGNYLYENNS NSTIAENDKN HLSEKQDTYL SNSSMSNSYS
1751 YHSDEVYNDS GYLSKNKLDS GIEPVLKNVE DQKNTSFSKV ISNVKDANAY
1801 PQTVNEDICV EELVTSSSPC KNKNAAIKLS ISNSNNFEVG PPAFRIASGK
1851 IVCVSHETIK KVKDIFTDSF SKVIKENNEN KSKICQTKIM AGCYEALDDS
1901 EDILHNSLDN DECSTHSHKV FADIQSEEIL QHNQNMSGLE KVSKISPCDV
1951 SLETSDICKC SIGKLHKSVS SANTCGIFST ASGKSVQVSD ASLQNARQVF
2001 SEIEDSTKQV FSKVLFKSNE HSDQLTREEN TAIRTPEHLI SQKGFSYNVV
2051 NSSAFSGFST ASGKQVSILE SSLHKVKGVL EEFDLIRTEH SLHYSPTSRQ
2101 NVSKILPRVD KRNPEHCVNS EMEKTCSKEF KLSNNLNVEG GSSENNHSIK
2151 VSPYLSQFQQ DKQQLVLGTK VSLVENIHVL GKEQASPKNV KMEIGKTETF
2201 SDVPVKTNIE VCSTYSKDSE NYFETEAVEI AKAFMEDDEL TDSKLPSHAT
2251 HSLFTCPENE EMVLSNSRIG KRRGEPLILV GEPSIKRNLL NEFDRIIENQ
2301 EKSLKASKST PDGTIKDRRL FMHHVSLEPI TCVPFRTTKE RQEIQNPNFT
2351 APGQEFLSKS HLYEHLTLEK SSSNLAVSGH PFYQVSATRN EKMRHLITTG
2401 RPTKVFVPPF KTKSHFHRVE QCVRNINLEE NRQKQNIDGH GSDDSKNKIN
2451 DNEIHQFNKN NSNQAAAVTF TKCEEEPLDL ITSLQNARDI QDMRIKKKQR
2501 QRVFPQPGSL YLAKTSTLPR ISLKAAVGGQ VPSACSHKQL YTYGVSKHCI
2551 KINSKNAESF QFHTEDYFGK ESLWTGKGIQ LADGGWLIPS NDGKAGKEEF
2601 YRALCDTPGV DPKLISRIWV YNHYRWIIWK LAAMECAFPK EFANRCLSPE
2651 RVLLQLKYRY DTEIDRSRRS AIKKIMERDD TAAKTLVLCV SDIISLSANI
2701 SETSSNKTSS ADTQKVAIIE LTDGWYAVKA QLDPPLLAVL KNGRLTVGQK
2751 IILHGAELVG SPDACTPLEA PESLMLKISA NSTRPARWYT KLGFFPDPRP
2801 FPLPLSSLFS DGGNVGCVDV IIQRAYPIQW MEKTSSGLYI FRNEREEEKE
2851 AAKYVEAQQK RLEALFTKIQ EEFEEHEENT TKPYLPSRAL TRQQVRALQD
2901 GAELYEAVKN AADPAYLEGY FSEEQLRALN NHRQMLNDKK QAQIQLEIRK
2951 AMESAEQKEQ GLSRDVTTVW KLRIVSYSKK EKDSVILSIW RPSSDLYSLL
3001 TEGKRYRIYH LATSKSKSKS ERANIQLAAT KKTQYQQLPV SDEILFQIYQ
3051 PREPLHFSKF LDPDFQPSCS EVDLIGFVVS VVKKTGLAPF VYLSDECYNL
3101 LAIKFWIDLN EDIIKPHMLI AASNLQWRPE SKSGLLTLFA GDFSVFSASP
3151 KEGHFQETFN KMKNTVENID ILCNEAENKL MHILHANDPK WSTPTKDCTS
3201 GPYTAQIIPG TGNKLLMSSP NCEIYYQSPL SLCMAKRKSV STPVSAQMTS
3251 KSCKGEKEID DQKNCKKRRA LDFLSRLPLP PPVSPICTFV SPAAQKAFQP
3301 PRSCGTKYET PIKKKELNSP QMTPFKKFNE ISLLESNSIA DEELALINTQ
3351 ALLSGSTGEK QFISVSESTR TAPTSSEDYL RLKRRCTTSL IKEQESSQAS
3401 TEECEKNKQD TITTKKYI (3418)
```

### Domain structure

| | |
|---|---|
| 1783–1863 | Weak homology with BRCA1 (1394–1474) |
| 3334–3344 | Granin motif |
| 987–1068, 1198–1292, 1407–1497, 1501–1588, 1649–1733, 1822–1913, 1955–2035, 2036–2112 | Eight repetitive units (BRC repeats)[47] |

### Mutations detected in *BRCA2*

Mutations in breast cancer families have been detected in 6 (exons 2, 9, 10, 11, 16, 18, 20, 22 and 23) of the 26 coding exons of *BRCA2*[11,16,48]. Germline and somatic mutations have been detected in primary breast cancers, including male breast cancer[4,48,49] and also in astrocytomas, bladder, lung, renal, prostate, pancreatic and ovarian cancers[50,51] and in melanoma[52].

The mutation 6174delT (frameshift) in *BRCA2*, together with 185delAG in *BRCA1*, are the two most common mutations associated with hereditary breast cancer in the Ashkenazim[53,54].

A number of polymorphisms are common in *BRCA2* and a polymorphism has been detected that is predicted to cause the loss of the C-terminal 93 amino acids[55].

### Database accession numbers

| | *PIR* | *SWISSPROT* | *EMBL/GENBANK* | *REFERENCES* |
|---|---|---|---|---|
| Human *BRCA1* | | P38398, BRC1_HUMAN | HS14680, U14680 | [2] |
| Human *BRCA2* | | U43746 | HSU43746, HSBRCA211, X95161 | [11] |
| Mouse *Brca-1* | | | MM35641, U35641 | [56,57] |

### *References*


1 Hall, J.M. et al. (1990) Science 250, 684–1689
2 Miki, Y. et al. (1994) Science 266, 66–71.
3 Wooster, R. et al. (1994) Science 265, 2088–2090.
4 Wooster, R. et al. (1995) Nature 378, 789–792.
5 Rao, V.N. et al. (1996) Oncogene 12, 523–528.
6 Chen, Y. et al. (1996) Cancer Res. 56, 3168–3172.
7 Chen, Y. et al. (1995) Science 270, 789–791.
8 Lane, T.F. et al. (1995) Genes Dev. 9, 2712–2722.
9 Zabludoff, S.D. et al. (1996) Oncogene 13, 649–653.
10 Marquis, S.T. et al. (1995) Nature Genet. 11, 17–26.
11 Tavtigian, S.V. et al. (1996) Nature Genet. 12, 333–337.
12 Hakem, R. et al. (1996) Cell 85, 1009–1023.
13 Easton, D.F. (1993) Am. J. Hum. Genet. 52, 678–701.
14 Ford, D. et al. (1994) Lancet 343, 692–695.
15 Boyd, M. et al. (1995) Nature 375, 541–542.
16 Phelan, C.M. et al. (1996) Nature Genet. 12, 309–311.
17 Collins, N. et al. (1995) Oncogene 10, 1673–1675.
18 Hamann, U. et al. (1996) Cancer Res. 56, 1988–1990.
19 Gudmundsson, J. et al. (1995) Cancer Res. 55, 4830–4832.
20 Gowen, L.C. et al. (1996) Nature Genet. 12, 191–194.

[21] Liu, C.-Y. et al. (1996) Genes Dev. 10, 1835–1843.
[22] Holt, J.T. et al. (1996) Nature Genet. 12, 298–302.
[23] Shao, N. et al. (1996) Oncogene 13, 1–7.
[24] Brown, M.A. et al. (1996) Oncogene 12, 2507–2513.
[25] Bienstock, R.J. et al. (1996) Cancer Res. 56, 2539–2545.
[26] **Stratton, M.R. and Wooster, R. (1996) Curr. Opin. Genet. Dev. 6, 93–97.**
[27] Rodenhiser, D. et al. (1996) Oncogene 12, 2623–2629.
[28] Gayther, S.A. et al. (1995) Nature Genet. 11, 428–433.
[29] Castilla, L.H. et al. (1994) Nature Genet. 8, 387–391.
[30] Friedman, L.S. et al. (1994) Nature Genet. 8, 399–404.
[31] Futreal, P.A. et al. (1994) Science 266, 120–122.
[32] Simard, J. et al. (1994) Nature Genet. 8, 392–398.
[33] Hogervorst, F.B.L. et al. (1995) Nature Genet. 10, 208–212.
[34] Merajver, S.D. et al. (1995) Clin. Cancer Res. 1, 539–544.
[35] Kainu, T. et al. (1996) Cancer Res. 56, 2912–2915.
[36] Caligo, M.A. et al. (1996) Oncogene 13, 1483–1488.
[37] De Benedetti, V.M.G. et al. (1996) Oncogene 13, 1353–1357.
[38] Plummer, S.J. et al. (1995) Hum. Mol. Genet. 4, 1989–1991.
[39] Takahashi, H. et al. (1995) Cancer Res. 55, 2998–3002.
[40] Holt, J.T. et al. (1996) Nature Genet. 12, 298–302.
[41] Durocher, F. et al. (1996). J. Med. Genet. 33, 814–819.
[42] Garvin, A.M. et al. (1996) J. Med. Genet. 33, 721–725.
[43] Sobol, H. et al. (1996) Cancer Res. 56, 3216–3219.
[44] Inoue, R. et al. (1995) Cancer Res. 55, 3521–3524.
[45] Matsushima, M. et al. (1995) Hum. Mol. Genet. 4, 1953–1956.
[46] Struewing, J.P. et al. (1995) Nature Genet. 11, 198–200.
[47] Bork, P. et al. (1996) Nature Genet. 13, 22–23.
[48] Miki, Y. et al. (1996) Nature Genet. 13, 245–247.
[49] Couch, F.J. et al. (1996) Nature Genet. 13, 123–125.
[50] Thorlacius, S. et al. (1996) Nature Genet. 13, 117–119.
[51] Takahashi, H. et al. (1996) Cancer Res. 56, 2738–2741.
[52] Teng, D.H.-F. et al. (1996) Nature Genet. 13, 241–244.
[53] Roa, B.B. et al. (1996) Nature Genet. 14, 185–187.
[54] Oddoux, C. et al. (1996) Nature Genet. 14, 188–190.
[55] Puget, N. et al. (1996) Nature Genet. 14, 253–254.
[56] Abel, K.J. et al. (1995) Hum. Mol. Genet. 4, 2265–2273.
[57] Sharan, S.K. et al. (1995) Hum. Mol. Genet. 4, 2275–2278.

## Identification

The putative tumour suppressor gene, *DCC* (deleted in colorectal carcinomas) was identified as a heterozygous deletion in ~70% of colon carcinomas [1].

## Related genes

*DCC* is a member of the immunoglobulin gene superfamily, homologous to neural cell adhesion molecules (NCAMs). DCC shares 53% identity with chick neogenin.

| | ***DCC*** |
|---|---|
| **Nucleotides (kb)** | 1400 |
| **Chromosome** | 18q21–qter |
| **Mass (kDa): predicted** | 153 |
| **expressed** | gp175–200 |
| **Cellular location** | Type I transmembrane glycoprotein |

**Tissue distribution**

DCC is expressed in axons of the central and peripheral nervous system and in differentiated cell types of the intestine [2]. Expression is absent in most cancer cell lines. An isoform occurs in reticuloendothelial cells of the thymus, tonsil and lymph nodes that is not present in colonic epithelium [3]. Reduced levels of DCC occur in ~50% of brain tumours as a result of allelic loss, aberrant splicing or allele-specific loss of transcripts [4]. Murine *Dcc* is ubiquitously expressed at low levels with alternative transcripts in some tissues [5]: expression is maximal in the developing embryo and there is an embryo-specific alternatively spliced form [6].

## Protein function

DCC appears to be involved in the differentiation of a variety of types of epithelial cell and also in neuronal cell development [2,6].

**Cancer**

Allelic loss of chromosomes 17p or 18q occurs in 70% of colorectal carcinomas and with high frequency in ovarian adenocarcinomas [7,8]. The 17p region contains *P53*; the 18q region contains *DCC* and also *BCL2* and *YES1*. Reduced expression or allellic loss of *DCC* in colorectal, gastric, pancreatic, oesophageal, breast, haematological and glial malignancies [9] and in human male germ cell tumours (GCTs) [10] may reflect alterations in cellular attachment; anti-sense RNA to *DCC* inhibits cell adhesion *in vitro* [11]. In colorectal cancers, however, there is conflicting evidence indicating that DCC expression is sustained during tumorigenesis [12]. In addition to mutations in *P53*, *DCC*, *APC* and *MCC*, colorectal carcinomas also accumulate mutations in *RAS*. Mutations in *DCC* also occur in breast carcinomas[13]. At least four scrambled transcripts are present at low concentrations in normal and neoplastic cells in which *DCC* exons are joined accurately at consensus splice sites but in a different order to that in the primary transcript[14].

***In vitro***

*DCC* expression is reduced in HPV-18 immortalized human keratinocytes transformed to tumorigenicity by NMU[15]. Expression of full-length, but not truncated, DCC in NMU-transformed tumorigenic HPV-immortalized human epithelial cells suppresses tumorigenicity[16].

## Gene structure

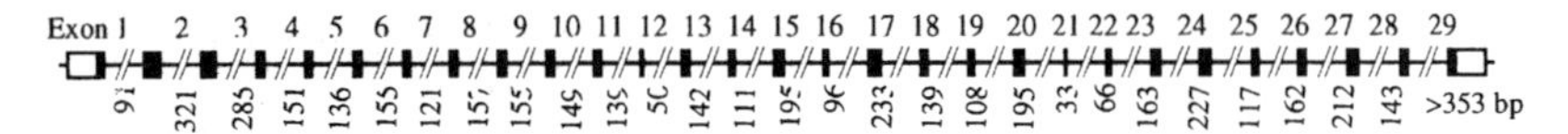


Two additional exons (1A and 28A) have been detected in brain tumours[4,17]. An in-frame insertion of 123 bp from exon 1A, a deletion of exon 9 (resulting in predicted premature termination), an in-frame deletion of exons 25 and 26, in-frame deletion of sequences from exons 17–22 and insertion of 71 bp from exon 28A (replacing the C-terminal 25 amino acids with 21 novel residues) occur. Transcripts containing exon 28A are predicted to yield DCC with the final 29 amino acids (exon 29) replaced by 25 amino acids encoded mostly by exon 28A.

## Amino acid sequence of human DCC

```
   (1) MENSLRCVWV PKLAFVLFGA SLLSAHLQVT GFQIKAFTAL RFLSEPSDAV
  (51) TMRGGNVLLD CSAESDRGVP VIKWKKDGIH LALGMDERKQ QLSNGSLLIQ
 (101) NILHSRHHKP DEGLYQCEAS LGDSGSIISR TAKVAVAGPL RFLSQTESVT
 (151) AFMGDTVLLK CEVIGEPMPT IHWQKNQQDL TPIPGDSRVV VLPSGALQIS
 (201) RLQPGDIGIY RCSARNPASS RTGNEAEVRI LSDPGLHRQL YFLQRPSNVV
 (251) AIEGKDAVLE CCVSGYPPPS FTWLRGEEVI QLRSKKYSLL GGSNLLISNV
 (301) TDDDSGMYTC VVTYKNENIS ASAELTVLVP PWFLNHPSNL YAYESMDIEF
 (351) ECTVSGKPVP TVNWMKNGDV VIPSDYFQIV GGSNLRILGV VKSDEGFYQC
 (401) VAENEAGNAQ TSAQLIVPKP AIPSSSVLPS APRDVVPVLV SSRFVRLSWR
 (451) PPAEAKGNIQ TFTVFFSREG DNRERALNTT QPGSLQLTVG NLKPEAMYTF
 (501) RVVAYNEWGP GESSQPIKVA TQPELQVPGP VENLQAVSTS PTSILITWEP
 (551) PAYANGPVQG YRLFCTEVST GKEQNIEVDG LSYKLEGLKK FTEYSLRFLA
 (601) YNRYGPGVST DDITVVTLSD VPSAPPQNVS LEVVNSRSIK VSWLPPPSGT
 (651) QNGFITGYKI RHRKTTRRGE METLEPNNLW YLFTGLEKGS QYSFQVSAMT
 (701) VNGTGPPSNW YTAETPENDL DESQVPDQPS SLHVRPQTNC IIMSWTPPLN
 (751) PNIVVRGYII GYGVGSPYAE TVRVDSKQRY YSIERLESSS HYVISLKAFN
 (801) NAGEGVPLYE SATTRSITDP TDPVDYYPLL DDFPTSVPDL STPMLPPVGV
 (851) QAVALTHDAV RVSWADNSVP KNQKTSEVRL YTVRWRTSFS ASAKYKSEDT
 (901) TSLSYTATGL KPNTMYEFSV MVTKNRRSST WSMTAHATTY EAAPTSAPKD
 (951) FTVITREGKP RAVIVSWQPP LEANGKITAY ILFYTLDKNI PIDDWIMETI
(1001) SGDRLTHQIM DLNLDTMYYF RIQARNSKGV GPLSDPILFR TLKVEHPDKM
(1051) ANDQGRHGDG GYWPVDTNLI DRSTLNEPPI GQMHPPHGSV TPQKNSNLLV
(1101) IIVVTVGVIT VLVVVIVAVI CTRRSSAQQR KKRATHSAGK RKGSQKDLRP
```

```
(1151) PDLWIHHEEM EMKNIEKPSG TDPAGRDSPI QSCQDLTPVS HSQSETQLGS
(1201) KSTSHSGQDT EEAGSSMSTL ERSLAARRAP RAKLMIPMDA QSNNPAVVSA
(1251) IPVPTLESAQ YPGILPSPTC GYPHPQFTLR PVPFPTLSVD RGFGAGRSQS
(1301) VSEGPTTQQP PMLPPSQPEH SSSEEAPSRT IPTACVRPTH PLRSFANPLL
(1351) PPPMSAIEPK VPYTPLLSQP GPTLPKTHVK TASLGLAGKA RSPLLPVSVP
(1401) TAPEVSEESH KPTEDSANVY EQDDLSEQMA SLEGLMKQLN AITGSAF
                                                     (1447)
```

Sequence of exon 1A:
GKDAKEKKQLTYNSSSYISWTVGMYHLFHLHCTCLRHRISLR
Sequence of exon 28A:
VEMEWVKQYAISYYPLLLPASSF

**Domain structure**

| | |
|---|---|
| 1–25 | Potential signal sequence |
| 1098–1122 | Potential transmembrane domain |
| 54–124, 154–219, 254–317, 345–407 | Four Ig-like C2-like domains |
| 426–522, 525–618, 619–716, 722–816, 840–940, 941–1042 | Six fibronectin type III-like domains |
| 61–117, 161–212, 261–310, 352–400 | Potential disulfide bonds |
| 94, 299, 318, 478, 628, 702 | Potential carbohydrate attachment sites |

Conflict: 138, 233, 329 and 421 are missing in Nigro et al. [14].

**Mutations detected in *DCC***

In colorectal carcinomas: Loss of DCC, point mutation in exon 28 (Pro → His), point mutation in intron 13 [17].

In oesophageal carcinomas: $Met^{168} \rightarrow Thr$; $Arg^{201} \rightarrow Gly$ [18]. Mutation detected in a colorectal carcinoma: $Phe^{1375} \rightarrow His$. Increased frequency of mutation appears to correlate with distance of lymph node metastasis from the primary tumour.

**Database accession numbers**

| | *PIR* | *SWISSPROT* | *EMBL/GENBANK* | *REFERENCES* |
|---|---|---|---|---|
| Human *DCC* | A54100, A40098, A38442 | P43146, DCC_HUMAN | Hsdccg X76132, M32286, M32288, M32290, M32292, M63696, M63700, M63702, M63718, M63698 | 17 |
| Mouse *Dcc* | | X85788 | | 6 |

***References***


1 Fearon, E.R. et al. (1990) Science 247, 49–56.
2 Hedrick, L. et al. (1994) Genes Dev. 8, 1174–1183.
3 Turley, H. et al. (1995) Cancer Res. 55, 5628–5631.
4 Ekstrand, B.C. et al. (1995) Oncogene 11, 2393–2402.
5 Reale, M.A. et al. (1994) Cancer Res. 54, 4493–4501.
6 Cooper, H.M. et al. (1995) Oncogene 11, 2243–2254.
7 **Cho, K.R, and Fearon, E.R. (1995) Eur. J. Cancer Part A: Gen. Top. 31, 1055–1060.**

[8] Chenevix-Trench, G. et al. (1992) Oncogene 7, 1059–1065.
[9] Scheck, A.C. and Coons, S.W. (1993) Cancer Res. 53, 5605–5609.
[10] Murty, V.V.V.S. et al. (1994) Oncogene 9, 3227–3231.
[11] Narayanan, R. et al. (1992) Oncogene 7, 553–561.
[12] Gotley, D.C. et al. (1996) Oncogene 13, 787–795.
[13] Devilee, P. et al. (1991) Oncogene 6, 311–315.
[14] Nigro, J.M. et al. (1991) Cell 64, 607–613.
[15] Klingelhutz, A.J. et al. (1993) Oncogene 8, 95–99.
[16] Klingelhutz, A.J. et al. (1995) Oncogene 10, 1581–1586.
[17] Cho K.R. et al. (1994) Genomics 19, 525–531.
[18] Miyake, S. et al. (1994) Cancer Res. 54, 3007–3010.

## Identification

*DPC4* (deleted in pancreatic carcinoma, locus 4)/*SMAD4* was identified by analysing pancreatic carcinomas for convergent sites of homozygous deletion[1].

## Related genes

The *Drosophila Mad* (*mothers against decapentaplegic*) gene and the related *Caenorhabditis elegans Sma* genes are implicated in TGF$\beta$-activated signal transduction pathways. The five homologues of this family identified in humans, mice and *Xenopus* are denoted as *SMAD*, a merger of *Sma* and *Mad*[2]. This family comprises: *SMAD1* (formerly *hMAD1/MADR1, Bsp1, Dwarfin-A, Xmad, Xmad1* or *JV4-1*), *SMAD2* (formerly *hMAD2/MADR2/ Xmad2* or *JV18-1*, *SMAD3* (formerly *hMAD3*), *SMAD4* (formerly *DPC4, hMAD4* or *Xmad4*), *SMAD5* (formerly *Dwarfin-C*)[3]. *SMAD4* has high similarity (~85%) in exons 1, 2 and 11 to *Drosophila melanogaster Mad* and *C. elegans sma-2*; lower similarity (~75%) in exons 8, 9 and 10.

| | ***DPC4/SMAD4*** |
|---|---|
| **Chromosome** | 18q21.1 |
| **Cellular location** | Ubiquitous |
| **Tissue location** | Cytoplasm |

## Protein function

DPC4/SMAD4 and SMAD3 are components of the TGF$\beta$ signalling pathway. In *Drosophila* MAD is necessary for signalling by the *decapentaplegic* (*dpp*) gene product that encodes a growth factor that belongs to the TGF$\beta$ superfamily and is involved in multiple developmentally related cell–cell signalling events. The *mothers against dpp* (*mad*) gene mediates Dpp signalling and homozygous *mad* mutants exhibit defects reminiscent of *dpp* mutant phenotypes. DPC4/SMAD4 and SMAD3 synergize to induce expression of the TGF$\beta$-activated plasminogen activator inhibitor 1 gene and C-terminally truncated versions of the proteins act as dominant negative inhibitors of this response. SMAD3 is phosphorylated in response to TGF$\beta$ and associates with the ligand-receptor complex[3].

### Cancer

*DPC4/SMAD4* is lost or mutated in ~50% of pancreatic carcinomas examined and may be involved in some colon, bladder and biliary tumours. Mutations also occur with low frequency in lung cancers[4].

## Gene structure

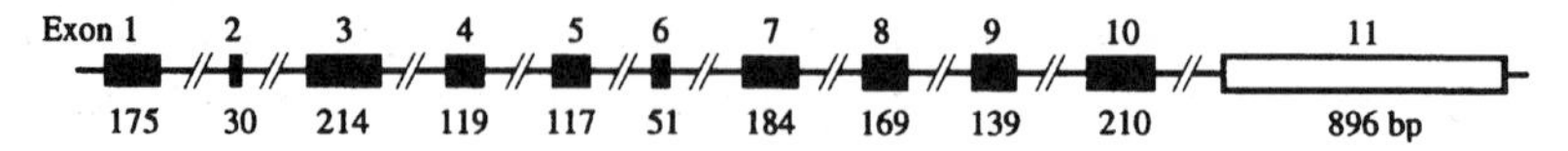


## Amino acid sequence of human SMAD4

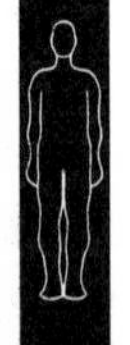

```
  1 MDNMSITNTP TSNDACLSIV HSLMCHRQGG ESETFAKRAI ESLVKKLKEK
 51 KDELDSLITA ITTNGAHPSK CVTIQRTLDG RLQVAGRKGF PHVIYARLWR
101 WPDLHKNELK HVKYCQYAFD LKCDSVCVNP YHYERVVSPG IDLSGLTLQS
151 NAPSSMMVKD EYVHDFEGQP SLSTEGHSIQ TIQHPPSNRA STETYSTPAL
201 LAPSESNATS TANFPNIPVA STSQPASILG GSHSEGLLQI ASGPQPGQQQ
251 NGFTGQPATY HHNSTTTWTG SRTAPYTPNL PHHQNGHLQH HPPMPPHPGH
301 YWPVHNELAF QPPISNHPAP EYWCSIAYFE MDVQVGETFK VPSSCPIVTV
351 DGYVDPSGGD RFCLGQLSNV HRTEAIERAR LHIGKGVQLE CKGEGDVWVR
401 CLSDHAVFVQ SYYLDREAGR APGDAVHKIY PSAYIKVFDL RQCHRQMQQQ
451 AATAQAAAAA QAAAVAGNIP GPGSVGGIAP AISLSAAAGI GVDDLRRLCI
501 LRMSFVKGWG PDYPRQSIKE TPCWIEIHLH RALQLLDEVL HTMPIADPQP
551 LD (552)
```

### Mutations detected in *DPC4*

Mutations in DCC have been detected in pancreatic carcinomas[1], lung cancers[4], head and neck squamous cell carcinoma[6], ovarian carcinomas and one breast carcinoma[5].

Polymorphism: $Ile^{525} \rightarrow Val$[6].

### Database accession numbers

| | *PIR* | *SWISSPROT* | *EMBL/GENBANK* | *REFERENCE* |
|---|---|---|---|---|
| Human *DPC4/SMAD4* | | U44378 | HS443781 | 1 |

### *References*

[1] Hahn, S.A. et al. (1996) Science 271, 350–353.
[2] Derynck, R. et al. (1996) Cell 87, 173.
[3] Zhang, Y. et al. (1996) Nature 383, 168–172.
[4] Nagatake, M. et al. (1996) Cancer Res. 56, 2718–2720.
[5] Schutte, M. et al. (1996) Cancer Res. 56, 2527–2530.
[6] Kim, S.K. et al. (1996) Cancer Res. 56, 2519–2521.

## Identification

E2F was identified through its role in transcription activation of the adenovirus E2 promoter [1–3] and E2F1 was isolated *via* its association with pRb [4–6].

## Related genes

E2F1 is related to E2F2, E2F3, E2F4 and E2F5.

| | ***E2F1*** |
|---|---|
| **Nucleotides (kb)** | 16.6 |
| **Chromosome** | 20q11 |
| **Mass (kDa): predicted** | 46.9 |
| **expressed** | 60 |
| **Cellular location** | Nucleus |

**Tissue distribution**
Ubiquitous. High expression in brain, placenta and lung. E2F1 is amplified and over-expressed in HEL erythroleukaemic cells [7].

## Protein function

The E2F family (E2F1, E2F2, E2F3, E2F4 and E2F5) are a closely related group of transcription factors, each of which heterodimerizes with members of the DP transcription factor family. E2F proteins mediate transcriptional activation of the adenovirus E2 promoter in an E1A-dependent manner. E2F1, E2F2 and E2F3 associate with pRb [8]. E2F1, E2F2, E2F3 and E2F4 dimerize with DP1.

E2F1 can function either as an oncogene or a tumour suppressor gene because it can both *trans*-activate and repress genes required for proliferation but it is not essential for viability or proliferation. When coexpressed with other oncogenes E2F1 transforms cells but its deletion leads to tissue-specific tumour development.

E2F recognizes the adenovirus E2A promoter motif TTTTCGCGCAATT and complexes containing pRb bind to related motifs with high affinity [9]. Similar sequence motifs in the promoters of the *E2F1, MYC, MYCN, MYB, CDC2* [10], dihydrofolate reductase, thymidine kinase and *EGFR* genes appear to be important for regulation of their transcription. E2F/pRb complexes are dissociated by cyclin-dependent kinase-mediated phosphorylation or by E1A, SV40 T antigen or HPV E7 [11–15]. The pRb/E2F complex occurs in $G_1$ in human primary cells and tumour cell lines and functions as a transcriptional repressor in the presence of an additional factor, RBP60 [16,17]. As cells enter S phase, a second E2F1 complex forms containing the pRb-related protein p107 and cyclin A [18]. p107 inhibits E2F-dependent transcription in co-transfection assays [19]. p130 binds stably to E2F4 in arrested cells to suppress E2F4-mediated *trans*-activation: as cells pass the $G_1$–S transition, the levels of p107 and pRb increase and E2F4 switches its association to these regulators[20]. Coexpression of E2F4 effectively overcomes p130-mediated $G_1$ arrest whereas E2F1 is more effective at overcoming pRb-mediated $G_1$ block[21].

E2F1 over-expression leads to inhibition of cyclin D1-dependent kinase activity that is mediated by a p16$^{INK4A}$-related CDKI[22]. Thus E2F1 activity, released by CDK phosphorylation of pRb, activates a negative feedback loop to switch off cyclin D1/CDK.

Phosphorylation of pRb may cause release of E2F at the $G_1$/S transition. However, Rb/E2F complexes are detectable in S phase and it is possible that in Rb/E2F complexes Rb may function as a *trans*-repressor and that phosphorylation of Rb may interfere with this function, rather than with its ability to bind E2F[23].

### Transgenic animals

*E2f-1*$^{-/-}$ mice are viable and fertile. However, with advancing age they show testicular atrophy and exocrine gland dysplasia and develop an unusual range of tumours including sarcomas of the reproductive tract, lung tumours and lymphomas. They also have a defect in T cell development arising from decreased susceptibility of CD4$^+$/CD8$^+$ cells to apoptosis[24,25].

### *In vitro*

E2F1, E2F2 or E2F3 transform immortal but otherwise untransformed rodent cells[26]. The E2F partners DP1 or DP2 can cooperate with *Hras* to transform primary fibroblasts[27]. E2F4, with which p107 associates, transforms rat embryo fibroblasts when over-expressed with oncogenic *Ras*[28].

## Gene structure

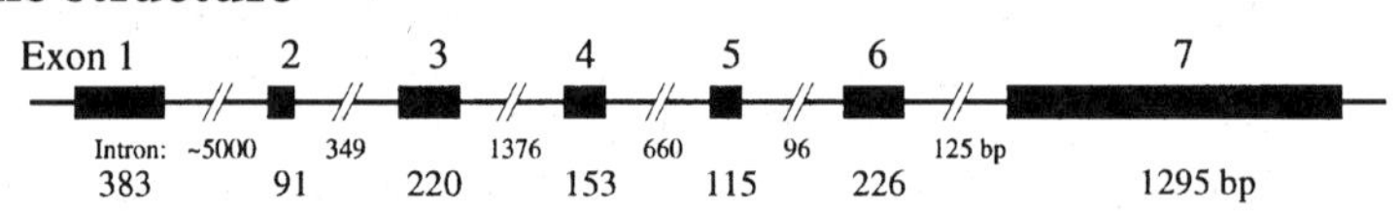


Many of the functional domains of E2F1 are encoded by single exons[29].

## Protein structure

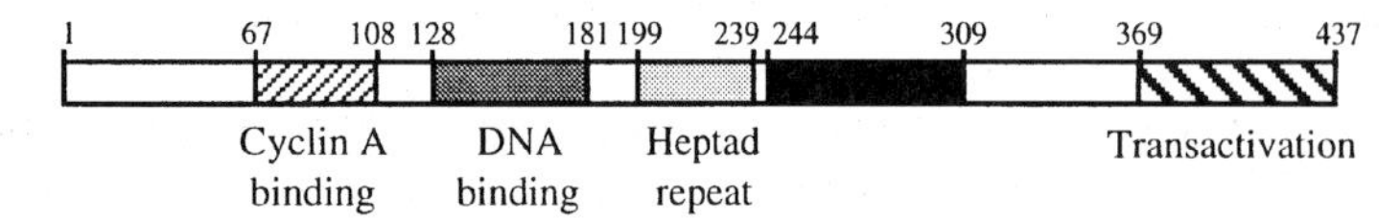


## Amino acid sequence of human E2F1

```
  1 MALAGAPAGG PCAPALEALL GAGALRLLDS SQIVIISAAQ DASAPPAPTG
 51 PAAPAAGPCD PDLLLFATPQ APRPTPSAPR PALGRPPVKR RLDLETDHQY
101 LAESSGPARG RGRHPGKGVK SPGEKSRYET SLNLTTKRFL ELLSHSADGV
151 VDLNWAAEVL KVQKRRIYDI TNVLEGIQLI AKKSKNHIQW LGSHTTVGVG
201 GRLEGLTQDL RQLQESEQQL DHLMNICTTQ LRLLSEDTDS QRLAYVTCQD
251 LRSIADPAEQ MVMVIKAPPE TQLQAVDSSE NFQISLKSKQ GPIDVFLCPE
301 ETVGGISPGK TPSQEVTSEE ENRATDSATI VSPPPSSPPS SLTTDPSQSL
351 LSLEQEPLLS RMGSLRAPVD EDRLSPLVAA DSLLEHVRED FSGLLPEEFI
401 SLSPPHEALD YHFGLEEGEG IRDLFDCDFG DLTPLDF (437)
```

**Domain structure**

| | |
|---|---|
| 89–191 | DNA binding domain |
| 181–185 | Potential nuclear localization signal (underlined) |
| 409–426 | pRb binding domain (underlined). E1A and E7 compete with pRb for binding to this region although it lacks the LXCXE motif |
| 427–431 | CBP binding domain. Rb binding represses both activation domains and CBP/p300 binding stimulates [30] |

**Database accession numbers**

| | *PIR* | *SWISSPROT* | *EMBL/GENBANK* | *REFERENCES* |
|---|---|---|---|---|
| Human *E2F1/RBAP1* | A42997 | RBB3_HUMAN Q01094 | M96577 | 4 |
| | A42998 | | | 5 |
| | | | U47675/6/7 | 29 |

***References***


1. Kovesdi, I. et al. (1986) Cell 45, 219–228.
2. Yee et al. (1987) EMBO J. 6, 2061–2068.
3. Yee et al. (1989) Mol. Cell. Biol. 9, 578–585.
4. Helin, K. et al. (1992) Cell 70, 337–350.
5. Kaelin, W.G. et al. (1992) Cell 70, 351–364.
6. Shan, B. et al. (1992) Mol. Cell. Biol. 12, 5620–5631.
7. Saito, M. et al. (1995) Genomics 25, 130–138.
8. Lees, J.A. et al. (1993) Mol. Cell. Biol. 13, 7813–7825.
9. Ouellette, M.M. et al. (1992) Oncogene 7, 1075–1081.
10. Dalton, S. (1992) EMBO J. 11, 1797–1804.
11. Chellappan, S.P. et al. (1991) Cell 65, 1053–1061.
12. Chellappan, S.P. et al. (1992) Proc. Natl Acad. Sci. USA 89, 4549–4553.
13. Hamel, P.A. et al. (1992) Mol. Cell. Biol. 12, 3431–3438.
14. Pagano, M. et al. (1992) Oncogene 7, 1681–1686.
15. Flemington, E.K. et al. (1993) Proc. Natl Acad. Sci. USA 90, 6914–6918.
16. Weintraub, S.J. et al. (1992) Nature 358, 259–261.
17. Ray, S.K. et al. (1992) Mol. Cell. Biol. 12, 4327–4333.
18. Shirodkar, S. et al. (1992) Cell 68, 157–166.
19. Schwarz, J.K. et al. (1993) EMBO J. 12, 1013–1020.
20. Moberg, K et al. (1996) Mol. Cell. Biol. 16, 1436–1449.
21. Vairo, G. et al. (1995) Genes Dev. 9, 869–881.
22. Khleif, S.N. et al. (1996) Proc. Natl Acad. Sci. USA 93, 4350–4354.
23. Sellers, W.R. et al. (1995) Proc. Natl Acad. Sci. USA 92, 11544–11548.
24. Field, S.J. et al. (1996) Cell 85, 549–561.
25. Yamasaki, L. et al. (1996) Cell 85, 537–548.
26. Xu, G. et al. (1995) Proc. Natl Acad. Sci. USA 92, 1357–1361.
27. Jooss, K. et al. (1995) Oncogene 10, 1529–1536.
28. Beijersbergen, R.L. et al. (1994) Genes Dev. 8, 2680–2690.
29. Neuman, E. et al. (1996) Gene 173, 163–169.
30. Trouche, D. and Kouzarides, T. (1996) Proc. Natl Acad. Sci. USA 93, 1439–1442.

# E-cadherin/CDH1

## Identification

E-cadherin/*CDH1* (also called uvomorulin (*UVO*), *Arc-1* or cell-CAM 120/80) was first isolated from embryonal carcinoma cell membranes[1].

## Related genes

B-cadherin, EP-cadherin, K-cadherin, M-cadherin, N-cadherin (ACAM), P-cadherin, R-cadherin, T-cadherin, U-cadherin, cadherins 4 to 11 and LCAM.

| | **E-cadherin** |
|---|---|
| **Nucleotides (kb)** | 100 |
| **Chromosome** | 16q22.1 |
| **Mass (kDa): predicted** | 97.5 |
| **expressed** | 120 |
| **Cellular location** | Type I plasma membrane protein |
| **Tissue location** | Non-neural epithelial tissues |

## Protein function

Cadherins are a multigene family of transmembrane glycoproteins that mediate $Ca^{2+}$-dependent intercellular adhesion, cytoskeletal anchoring and signalling, and are thought to be essential for the control of morphogenetic processes, including myogenesis[2]. Cadherins preferentially interact with themselves in a homophilic manner in connecting cells.

E-cadherin is involved in the formation of intercellular junctional complexes and in the establishment of cell polarization. E-cadherin associates with three cytoplasmic proteins, $\alpha$-, $\beta$- and $\gamma$-catenin, and its cytoplasmic tail is linked *via* $\alpha$- and $\beta$-catenin, with which the tumour suppressor gene product APC also interacts, to the actin cytoskeleton. $\alpha$-Catenin (or CAP102) is a vinculin-like protein and $\beta$-catenin is a homologue of plakoglobulin.

E-cadherin is the receptor for internalin, a bacterial surface protein essential for the entry of the gram-positive bacterium *Listeria monocytogenes* into epithelial cells[3]. E-cadherin can therefore be involved both in cell–cell adhesion and in bacterial invasion of non-phagocytic cells.

### Cancer

Impaired E-cadherin expression has been detected in 53% of a sample of primary breast cancers[4] although E-cadherin mutations appear to be confined to infiltrative lobular carcinomas, being undetectable in infiltrative ductal or medullary carcinomas[5]. Mutations have also been detected in endometrial and ovarian carcinomas[6]. E-cadherin protein expression is also lost in primary hepatocellular carcinomas, metastatic squamous cell carcinomas and prostate cancers and mRNA levels are lower in squamous cell carcinoma lines than in normal keratinocytes, whereas P-cadherin levels are similar. In gastric carcinomas, however, P-cadherin expression is either downregulated or unstable whereas that of E-cadherin is normal. In bladder carcinoma there is a frequent correlation between decreased E-cadherin expression and increased AMF (gp79) expression that may define individuals at high risk [7]. Hypermethylation of the promoter may be one mechanism of E-cadherin inactivation in human carcinomas [8].

Thus E-cadherin and possibly other members of the family act as invasion suppressors, although some cadherins, for example K-cadherin, are over-expressed in some human cancers [9].

### *In vitro*

Catenins that regulate cadherin function undergo tyrosine phosphorylation in cells transformed by v-*src* that correlates with metastatic potential and may reflect the modulation of cell–cell adhesion [10]. In human gastric cancer cells, $\beta$-catenin associates with and is phosphorylated by HER2. Suppression of tyrosine phosphorylation of $\beta$-catenin reduces metastasis of tumours derived from these cells in nude mice [11].

Loss of E-cadherin-dependent cell–cell adhesion has been detected in a human cancer cell line (HSC-39) that expresses normal E-cadherin but in which mutation of the $\beta$-catenin gene gives rise to a truncated protein [12]. In breast cancer cell lines complete loss or markedly reduced expression of E-cadherin and $\alpha$- and $\beta$-catenin occurs with high frequency [13]. In E-cadherin-negative breast and prostate carcinoma cell lines and in primary breast carcinoma tissue the 5′ CpG island of *CDH1* is densely methylated, although it is unmethylated in normal breast tissue [14]. The loss of cell–cell adhesion caused by the selective downregulation of E-cadherin expression can cause de-differentiation and invasiveness of human carcinoma cells.

Human cell lines derived from bladder, breast, lung and pancreatic carcinomas that have an epithelioid phenotype are non-invasive and express E-cadherin but those with a fibroblastoid phenotype are invasive and have lost E-cadherin expression. In MCF-7/6 breast cancer cells tamoxifen restores the function of E-cadherin on the cell surface and causes an increase in cell aggregation and an inhibition of invasion *in vitro* [15]. The polyunsaturated fatty acid $\gamma$-linolenic acid (GLA) also upregulates expression of CDH1 on cancer cells [16]. Invasiveness is blocked by transfection with E-cadherin cDNA and reinduced by treatment of the transfected cells with anti-E-cadherin monoclonal antibodies. The expression of E-cadherin-specific anti-sense RNA in non-invasive *RAS*-transformed cells with high endogenous E-cadherin expression renders the cells invasive. Furthermore, expression of E-cadherin inhibits the migration of cells into three-dimensional collagen gels and some invasive cells that retain E-cadherin on their surface appear to block its function by expressing enlarged proteoglycans [17].

## Gene structure

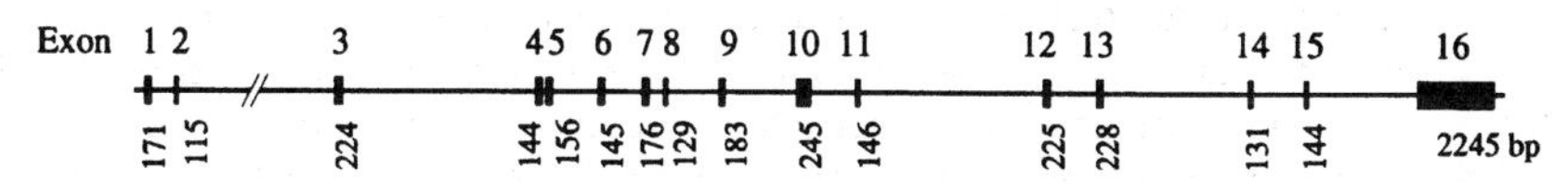


Intron 1 contains a high-density CpG island. Intron 2 is 65 kb in length [18].

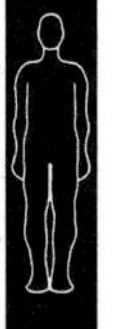

## Amino acid sequence of human E-cadherin

1 MGPWSRSLSA LLLLLQVSSW LCQEPEPCHP GFDAESYTFT VPRRHLERGR
51 VLGRVNFEDC TGRQRTAYFS LDTRFKVGTD GVITVKRPLR FHNPQIHFLV
101 YAWDSTYRKF STKVTLNTVG HHHRPPPHQA SVSGIQAELL TFPNSSPGLR
151 RQKRDWVIPP ISCPENEKGP FPKNLVQIKS NKDKEGKVFY SITGQGADTP
201 PVGVFIIERE TGWLKVTEPL DRERIATYTL FSHAVSSNGN AVEDPMEILI
251 TVTDQNDNKP EFTQEVFKGS VMEGALPGTS VMEVTATDAD DDVNTYNAAI
301 AYTILSQDPE LPDKNMFTIN RNTGVISVVT TGLDRESFPT YTLVVQAADL
351 QGEGLSTTAT AVITVTDTND NPPIFNPTTY KGQVPENEAN VVITTLKVTD
401 ADAPNTPAWE AVYTILNDDG GQFVVTTNPV NNDGILKTAK GLDFEAKQQY
451 ILHVAVTNVV PFEVSLTTST ATVTVDVLDV NEAPIFVPPE KRVEVSEDFG
501 VGQEITSYTA QEPDTFMEQK ITYRIWRDTA NWLEINPDTG AI$^{S}/_{F}$TRAELDR
551 EDFEHVKNST YTALIIATDN GSPVATGTGT LLLILSDVND NAPIPEPRTI
601 FFCERNPKPQ VINIIDADLP PNTSPFTAEL THGASANWTI QYNDPTQESI
651 ILKPKMALEV GDYKINLKLM DNQNKDQVTT LEVSVCDCEG AAGVCRKAQP
701 VEAGLQIPAI LGILGGILAL LILILLLLLF LRRRAVVKEP LLPPEDDTRD
751 NVYYYDEEGG GEEDQDFDLS QLHRGLDARP EVTRNDVAPT LMSVPRYLPR
801 PANPDEIGNF IDENLKAADT DPTAPPYDSL LVFDYEGSGS EAASLSSLNS
851 SESDKDQDYD YLNEWGNRFK KLADMYGGGE DD (882)

### Domain structure

| | |
|---|---|
| 1–27 | Signal sequence |
| 28–154 | Precursor sequence |
| 155–707 | Extracellular domains including calcium binding motifs |
| 708–731 | Transmembrane domain |
| 732–882 | Cytoplasmic domain |
| 838–851 | Serine-rich domain |
| 558, 637, 849 | Potential glycosylation sites |
| 155–262, 263–375, 376–486, 487–593, 594–697 | Repeats (cadherin 1, 2, 3, 4, 5) |

Mutations in breast carcinomas (infiltrative lobular carcinomas): Glu$^{261}$ → stop, Glu$^{386}$ → stop, 4 bp del at cdn 399 → stop$^{416}$, Glu$^{504}$ → stop [5].

Polymorphisms detected at Thr$^{115}$, Thr$^{560}$, Ala$^{692}$, Asn$^{751}$, intron 4 G → C [18].

### Database accession numbers

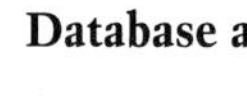

| | *PIR* | *SWISSPROT* | *EMBL/GENBANK* | *REFERENCES* |
|---|---|---|---|---|
| Human *E-cad* | S25141, S37654 | CADE_HUMAN P12830 | Z13009, Z18923, X12790, Z35402–15, Z35489 | 18 |
| | | | HSCDH01, Z35480, L34784–97, L34936/7 | 19 |

### *References*


1 Hyafil, F. et al. (1980) Cell 21, 927–934.
2 **Takeichi, M. (1991) Science 251, 1451–1455.**
3 Mengaud, J. et al. (1996) Cell 84, 923–932.
4 Oka, H. et al. (1993) Cancer Res. 53, 1696–1701.
5 Berx, G. et al. (1995) EMBO J. 14, 6107–6115.
6 Risinger, J.I. et al. (1994) Nature Genet. 7, 98–102.

[7] Otto, T. et al. (1994) Cancer Res. 54, 3120–3123.
[8] Yoshiura, K. et al. (1995) Proc. Natl Acad. Sci. USA 92, 7416–7419.
[9] Xiang, Y.-Y. et al. (1994) Cancer Res. 54, 3034–3041.
[10] Hamaguchi, M. et al. (1993) EMBO J. 12, 307–314.
[11] Shibata, T. et al. (1996) Oncogene 13, 883–889.
[12] Kawanishi, J. et al. (1995) Mol. Cell. Biol. 15, 1175–1181.
[13] Pierceall, W.E. et al. (1995) Oncogene 11, 1319–1326.
[14] Graff, J.R. et al. (1995) Cancer Res. 55, 5195–5199.
[15] Bracke, M.E. et al. (1994) Cancer Res. 54, 4607–4609.
[16] Jiang, W.G. et al. (1995) Cancer Res. 55, 5043–5048.
[17] Vleminckx, K.L. et al. (1994) Cancer Res. 54, 873–877.
[18] Berx, G. et al. (1995) Genomics 26, 281–289.
[19] Bussemakers, M.J.G. et al. (1993) Mol. Biol. Rep. 17,123–128.

## Identification

*MSH2* was cloned by PCR amplification of degenerate oligonucleotide primers that hybridize to highly conserved regions of bacterial *MutS/HexA* and *S. cerevisiae MSH2*[1]. *MSH3* was detected as a transcript initiated from the opposite strand of the 5′ region of the dihydrofolate reductase gene[2]. MSH6/GTBP (G/T binding protein) was isolated from HeLa cells by G/T mismatch affinity chromatography[3,4]. *MLH1* was identified by amplifying human *MutL*-related sequences using degenerate probes[5]. *PMS1* and *PMS2* are homologues of *MutL*[6].

## Related genes

*MSH2, MSH3* and *MSH6* encode members of the *MutS* DNA mismatch repair protein superfamily. *MLH1, PMS1* and *PMS2* are members of the *MutL/HexB* DNA mismatch repair family. There are at least 11 members of the human *PMS* family and *PMS2*-related genes have been detected at 7p12–p13, 7q11 and 7q22 loci[7,8].

| | ***MSH2*** | ***MSH3*** | ***MSH6/GTBP*** | ***MLH1*** |
|---|---|---|---|---|
| **Nucleotides (kb)** | 73 | 160 | | ~58 |
| **Chromosome** | 2p22–21 | 5q11–13 | 2p16 | 3p21.3–p23 |
| **Mass (kDa): predicted** | 105 | | 142 | 84.6 |
| **expressed** | 100 | | 160 | |
| **Cellular location** | Nucleus | Nucleus | Nucleus | Nucleus |

**Tissue distribution**

*MSH2* mRNA is highly expressed in the thymus and testis. MSH2 protein is highly expressed in proliferative cells of the oesophageal and intestinal epithelia, ileum and colon[11,12].

## Protein function

In *E. coli* proteins that participate in DNA mismatch repair include the products of the *MutH, MutL* and *MutS* genes: mutation in any of these genes generates a mutator phenotype in which spontaneous mutagenesis is enhanced at many loci. Human MSH2 is closely related to *S. cerevisiae* MSH2 and it binds to mismatched nucleotides in DNA to provide a target for the excision repair processes during postreplication mismatch repair[13,14]. MSH2 acts as a dimer with G/T binding protein (GTBP)[3,4] and GTBP/MSH6 inactivation has been detected in three hypermutatable cell lines correlating with alterations primarily in mononucleotide tracts[15]. MSH2 in combination with either MSH3 or GTBP/MSH6 repairs one or two base insertions and deletions.

*MLH1* encodes a product homologous to the bacterial mismatch repair protein MutL and similar to the yeast mismatch repair protein MLH1 [5].

**Cancer**

Hereditary non-polyposis colon cancer (HNPCC) is the most common hereditary colon cancer susceptibility for which a candidate gene has been detected, the defective gene being carried by ~1/200 individuals and being responsible for ~5% of all colon cancers. There is an increased risk of extracolonic cancers, principally uterine, ovarian and urothelial, and HNPCC is classified into two subtypes, Lynch syndrome 1 and 2, in which extracolonic cancers are absent or present, respectively. The Muir-Torre syndrome in which there is associated sebaceous cyst tumours and internal malignancy is a further HNPCC variant. Two HNPCC loci have been detected, *MSH2* and *MLH1*. Mismatch repair systems play a major role in preventing mutations during DNA replication and HNPCC patients have germline mutations in *MSH2* at an intronic splice acceptor site presumed to cause aberrant mRNA processing [1]. A variety of other mutations have been detected, suggesting that at least 40% of HNPCC kindreds are associated with germline mutations in *MSH2*, most of which produce drastic alterations in the protein product [16]. Multiple somatic mutations have been detected in *APC* (up to six) and *P53* (up to four in the same patient) following loss of *MSH2* in HNPCC [17].

Microsatellite instability is frequently associated with HNPCC and germline mutations in *MSH2* account for ~50% of these cases. A significant proportion of sporadic colorectal cancers with microsatellite instability have somatic mutations in *MSH2* [18]. Microsatellite instability has been detected in gynaecological sarcomas and a 2 bp deletion in exon 14 of *MSH2*, causing premature termination at codon 796, occurs in mutant uterine sarcoma cell lines [19].

**Transgenic animals**

Homozygous $MSH2^{-/-}$ mice develop normally but after two months of age are susceptible with high frequency to lymphoid tumours that contain microsatellite instabilities [20].

## Gene structure of *MSH2*

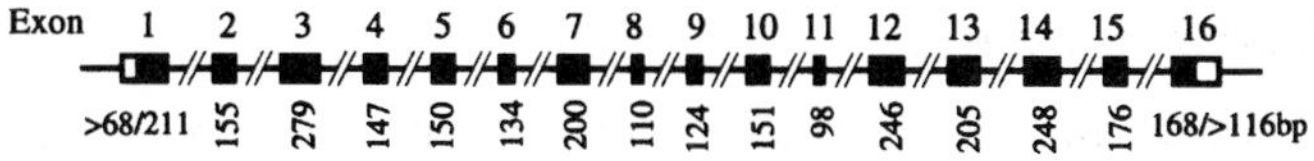

Exon 1 2 3 4 5 6 7 8 9 10 11 12 13 14 15 16

>68/211 155 279 147 150 134 200 110 124 151 98 246 205 248 176 168/>116bp

The human *MSH2* promoter lacks TATA or CAAT boxes but includes consensus binding sites for AP2 (−998 to −1007), SP1-IE (−839 to −844), AP1 (−606 to −612 and −291 to −297), HSP70.6/NF-Y-MHCII (−166 to −170 and −115 to −119) and SP1-HSP70 (−63 to −68) [21].

## Gene structure of *MSH3* [22]

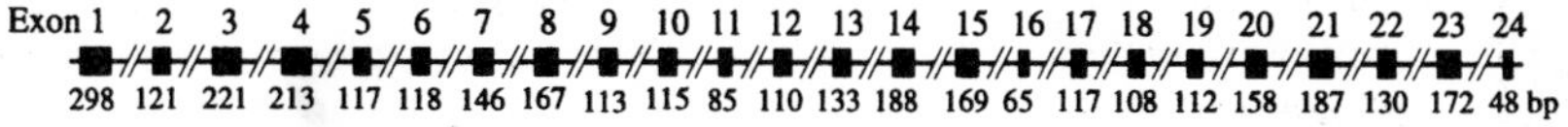

Exon 1 2 3 4 5 6 7 8 9 10 11 12 13 14 15 16 17 18 19 20 21 22 23 24

298 121 221 213 117 118 146 167 113 115 85 110 133 188 169 65 117 108 112 158 187 130 172 48 bp

## Gene structure of *MLH1*

Exon 1 2 3 4 5 6 7 8 9 10 11 12 13 14 15 16 17 18 19

Intron: 2.9 4.0 3.3 2.5 1.6 2.9 0.15 2.3 2.9 2.7 5.5 2.3 12.6 1.9 5.1 0.8 0.3 1.5 kb

137 90 98 73 72 91 42 88 112 93 153 370 148 108 63 164 92 113 167bp

Deletion of exon 13 occurs in over 90% of individuals. The product is expressed as an apparently untranslated mRNA (predicted 68 kDa protein) in lymphocytes [9,23].

## Gene structure of *PMS2*

Exon 1 2 3 4 5 6 7 8 9 10 11 12 13 14 15

Intron: 0.9 1.6 0.2 1.0 0.8 0.8 0.8 0.8 1.6 0.6 0.3 0.7 0.8 1.9 kb

723 140 87 103 184 169 97 100 85 157 861 168 100 171 >144bp

The 15 exons of *PMS2* encode a 16 kb transcript [8].

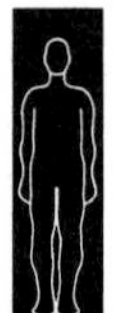

## Amino acid sequence of human MSH2

```
  1 MAVQPKETLQ LESAAEVGFV RFFQGMPEKP TTTVRLFDRG DFYTAHGEDA
 51 LLAAREVFKT QGVIKYMGPA GAKNLQSVVL SKMNFESFVK DLLLVRQYRV
101 EVYKNRAGNK ASKENDWYLA YKASPGNLSQ FEDILFGNND MSASIGVVGV
151 KMSAVDGQRQ VGVGYVDSIQ RKLGLCEFPD NDQFSNLEAL LIQIGPKECV
201 LPGGETAGDM GKLRQIIQRG GILITERKKA DFSTKDIYQD LNRLLKGKKG
251 EQMNSAVLPE MENQVAVSSL SAVIKFLELL SDDSNFGQFE LTTFDFSQYM
301 KLDIAAVRAL NLFQGSVEDT TGSQSLAALL NKCKTPQGQR LVNQWIKQPL
351 MDKNRIEERL NLVEAFVEDA ELRQTLQEDL LRRFPDLNRL AKKFQRQAAN
401 LQDCYRLYQG INQLPNVIQA LEKHEGKHQK LLLAVFVTPL TDLRSDFSKF
451 QEMIETTLDM DQVENHEFLV KPSFDPNLSE LREIMNDLEK KMQSTLISAA
501 RDLGLDPGKQ IKLDSSAQFG YYFRVTCKEE KVLRNNKNFS TVDIQKNGVK
551 FTNSKLTSLN EEYTKNKTEY EEAQDAIVKE IVNISSGYVE PMQTLNDVLA
601 QLDAVVSFAH VSNGAPVPYV RPAILEKGQG RIILKASRHA CVEVQDEIAF
651 IPNDVYFEKD KQMFHIITGP NMGGKSTYIR QTGVIVLMAQ IGCFVPCESA
701 EVSIVDCILA RVGAGDSQLK GVSTFMAEML ETASILRSAT KDSLIIIDEL
751 GRGTSTYDGF GLAWAISEYI ATKIGAFCMF ATHFHELTAL ANQIPTVNNL
801 HVTALTTEET LTMLYQVKKG VCDQSFGIHV AELANFPKHV IECAKQKALE
851 LEEFQYIGES QGYDIMEPAA KKCYLEREQG EKIIQEFLSK VKQMPFTEMS
901 EENITIKLKQ LKAEVIAKNN SFVNEIISRI KVTT (934)
```

### Domain structure

669–676 Potential ATP binding site

### Mutations detected in *MSH2*

A variety of germline mutations have been identified in HNPCC, including skipping of exon 3, 5 or 12, a high proportion of which give rise to truncation or large deletions of the protein [24,25]. Missense mutation of a residue highly conserved in the mismatch repair family has also been detected [16,26] and mutation of MSH2 has also been detected in uterine sarcoma cell lines [19]. A polymorphism has been detected in intron 10 of *MSH2* [27].

### Mutations detected in *MSH3*

A mutation (1148delA) has been detected in *MSH3* in an endometrial cancer that gives rise to loss of the C-terminal 723 amino acids of MSH3 [28].

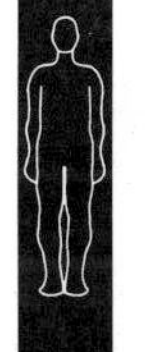

## Amino acid sequence of human MLH1

```
  1 MSFVAGVIRR LDETVVNRIA AGEVIQRPAN AIKEMIENCL DAKSTSIQVI
 51 VKEGGLKLIQ IQDNGTGIRK EDLDIVCERF TTSKLQSFED LASISTYGFR
101 GEALASISHV AHVTITTKTA DGKCAYRASY SDGKLKAPPK PCAGNQGTQI
151 TVEDLFYNIA TRRKALKNPS EEYGKILEVV GRYSVHNAGI SFSVKKQGET
201 VADVRTLPNA STVDNIRSIF GNAVSRELIE IGCEDKTLAF KMNGYISNAN
251 YSVKKCIFLL FINHRLVEST SLRKAIETVY AAYLPKNTHP FLYLSLEISP
301 QNVDVNVHPT KHEVHFLHEE SILERVQQHI ESKLLGSNSS RMYFTQTLLP
351 GLAGPSGEMV KSTTSLTSSS TSGSSDKVYA HQMVRTDSRE QKLDAFLQPL
401 SKPLSSQPQA IVTEDKTDIS SGRARQQDEE MLELPAPAEV AAKNQSLEGD
451 TTKGTSEMSE KRGPTSSNPR KRHREDSDVE MVEDDSRKEM TAACTPRRRI
501 INLTSVLSLQ EEINEQGHEV LREMLHNHSF VGCVNPQWAL AQHQTKLYLL
551 NTTKLSEELF YQILIYDFAN FGVLRLSEPA PLFDLAMLAL DSPESGWTEE
601 DGPKEGLAEY IVEFLKKKAE MLADYFSLEI DEEGNLIGLP LLIDNYVPPL
651 EGLPIFILRL ATEVNWDEEK ECFESLSKEC AMFYSIRKQY ISEESTLSGQ
701 QSEVPGSIPN SWKWTVEHIV YKALRSHILP PKHFTEDGNI LQLANLPDLY
751 KVFERC (756)
```

Alternative splicing of *MLH1* occurs in normal lymphocytes, colon, stomach, breast, bladder and skin in which either exons 9 and 10 are deleted with loss of codons 227–295, or exons 10 and 11 or 9, 10 and 11 are lost giving rise to truncated proteins comprising the N-terminal 264 and 226 amino acids [29].

**Mutations detected in *MLH1***

Missense mutations and loss of the wild-type *MLH1* gene occurs in individuals with chromosome 3-linked HNPCC, indicating that the DNA mismatch repair genes resemble tumour suppressor genes in that two hits are required for the phenotypic effect [30].

Germline mutations identified in *MLH1* in cancer patients who belong to HNPCC pedigrees include missense mutations, intronic mutations affecting splicing, and frameshift mutations resulting in truncation of MLH1 [10,24,31]. Polymorphisms have also been identified.

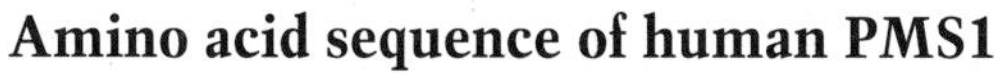

## Amino acid sequence of human PMS1

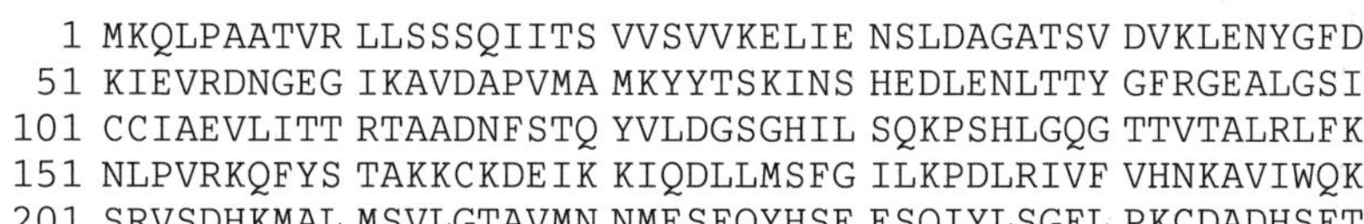

```
  1 MKQLPAATVR LLSSSQIITS VVSVVKELIE NSLDAGATSV DVKLENYGFD
 51 KIEVRDNGEG IKAVDAPVMA MKYYTSKINS HEDLENLTTY GFRGEALGSI
101 CCIAEVLITT RTAADNFSTQ YVLDGSGHIL SQKPSHLGQG TTVTALRLFK
151 NLPVRKQFYS TAKKCKDEIK KIQDLLMSFG ILKPDLRIVF VHNKAVIWQK
201 SRVSDHKMAL MSVLGTAVMN NMESFQYHSE ESQIYLSGFL PKCDADHSFT
251 SLSTPERSFI FINSRPVHQK DILKLIRHHY NLKCLKESTR LYPVFFLKID
301 VPTADVDVNL TPDKSQVLLQ NKESVLIALE NLMTTCYGPL PSTNSYENNK
351 TDVSAADIVL SKTAETDVLF NKVESSGKNY SNVDTSVIPF QNDMHNDESG
401 KNTDDCLNHQ ISIGDFGYGH CSSEISNIDK NTKNAFQDIS MSNVSWENSQ
451 TEYSKTCFIS SVKHTQSENG NKDHIDESGE NEEEAGLENS SEISADEWSR
501 GNILKNSVGE NIEPVKILVP EKSLPCKVSN NNYPIPEQMN LNEDSCNKKS
551 NVIDNKSGKV TAYDLLSNRV IKKPMSASAL FVQDHRPQFL IENPKTSLED
601 ATLQIEELWK TLSEEEKLKY EEKATKDLER YNSQMKRAIE QESQMSLKDG
651 RKKIKPTSAW NLAQKHKLKT SLSNQPKLDE LLQSQIEKRR SQNIKMVQIP
701 FSMKNLKINF KKQNKVDLEE KDEPCLIHNL RFPDAWLMTS KTEVMLLNPY
751 RVEEALLFKR LLENHKLPAE PLEKPIMLTE SLFNGSHYLD VLYKMTADDQ
801 RYSGSTYLSD PRLTANGFKI KLIPGVSITE NYLEIEGMAN CLPFYGVADL
851 KEILNAILNR NAKEVYECRP RKVISYLEGE AVRLSRQLPM YLSKEDIQDI
901 IYRMKHQFGN EIKECVHGRP FFHHLTYLPE TT (932)
```

## Amino acid sequence of human PMS2

```
  1 MERAESSSTE PAKAIKPIDR KSVHQICSGQ VVLSLSTAVK ELVENSLDAG
 51 ATNIDLKLKD YGVDLIEVSD NGCGVEEENF EGLTLKHHTS KIQEFADLTQ
101 VETFGFRGEA LSSLCALSDV TISTCHASAK VGTRLMFDHN GKIIQKTPYP
151 RPRGTTVSVQ QLFSTLPVRH KEFQRNIKKE YAKMVQVLHA YCIISAGIRV
201 SCTNQLGQGK RQPVVCTGGS PSIKENIGSV FGQKQLQSLI PFVQLPPSDS
251 VCEEYGLSCS DALHNLFYIS GFISQCTHGV GRSSTDRQFF FINRRPCDPA
301 KVCRLVNEVY HMYNRHQYPF VVLNISVDSE CVDINVTPDK RQILLQEEKL
351 LLAVLKTSLI GMFDSDVNKL NVSQQPLLDV EGNLIKMHAA DLEKPMVEKQ
401 DQSPSLRTGE EKKDVSISRL REAFSLRHTT ENKPHSPKTP EPRRSPLGQK
451 RGMLSSSTSG AISDKGVLRP QKEAVSSSHG PSDPTDRAEV EKDSGHGSTS
501 VDSEGFSIPD TGSHCSSEYA ASSPGDRGSQ EHVDSQEKAP ETDDSFSDVD
551 CHSNQEDTGC KFRVLPQPTN LATPNTKRFK KEEILSSSDI CQKLVNTQDM
601 SASQVDVAVK INKKVVPLDF SMSSLAKRIK QLHHEAQQSE GEQNYRKFRA
651 KICPGENQAA EDELRKEISK TMFAEMEIIG QFNLGFIITK LNEDIFIVDQ
701 HATDEKYNFE MLQQHTVLQG QRLIAPQTLN LTAVNEAVLI ENLEIFRKNG
751 FDFVIDENAP VTERAKLISL PTSKNWTFGP QDVDELIFML SDSPGVMCRP
801 SRVKQMFASR ACRKSVMIGT ALNTSEMKKL ITHMGEMDHP WNCPHGRPTM
851 RHIANLGVIS QN (862)
```

**Mutations detected in *PMS1* and *PMS2***

*PMS1* and *PMS2* undergo mutations in the germline of HNPCC patients [6].

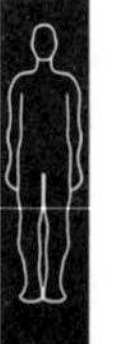

## Amino acid sequence of human MSH6/GTBP

```
   1 MSRQSTLYSF FPKSPALSDA NKASARASRE GGRAAAAPGA SPSPGGDAAW
  51 SEAGPGPRPL ARSASPPKAK NLNGGLRRSV APAAPTSCDF SPGDLVWAKM
 101 EGYPWWPCLV YNHPFDGTFI REKGKSVRVH VQFFDDSPTR GWVSKRLLKP
 151 YTGSKSKEAQ KGGHFYSAKP EILRAMQRAD EALNKDKIKR LELAVCDEPS
 201 EPEEEEEMEV GTTYVTDKSE EDNEIESEEE VQPKTQGSRR SSRQIKKRRV
 251 ISDSESDIGG SDVEFKPDTK EEGSSDEISS GVGDSESEGL NSPVKVARKR
 301 KRMVTGNGSL KRKSSRKETP SATKQATSIS SETKNTLRAF SAPQNSESQA
 351 HVSGGGDDSS RPTVWYHETL EWLKEEKRRD EHRRRPDHPD FDASTLYVPE
 401 DFLNSCTPGM RKWWQIKSQN FDLVICYKVG KFYELYHMDA LIGVSELGLV
 451 FMKGNWAHSG FPEIAFGRYS DSLVQKGYKV ARVEQTETPE MMEARCRKMA
 501 HISKYDRVVR REICRIITKG TQTYSVLEGD PSENYSKYLL SLKEKEEDSS
 551 GHTRAYGVCF VDTSLGKFFI GQFSDDRHCS RFRTLVAHYP PVQVLFEKGN
 601 LSKETKTILK SSLSCSLQEG LIPGSQFWDA SKTLRTLLEE EYFREKLSDG
 651 IGVMLPQVLK GMTSESDSIG LTPGEKSELA LSALGGCVFY LKKCLIDQEL
 701 LSMANFEEYI PLDSDTVSTT RSGAIFTKAY QRMVLDAVTL NNLEIFLNGT
 751 NGSTEGTLLE RVDTCHTPFG KRLLKQWLCA PLCNHYAIND RLDAIEDLMV
 801 VPDKISEVVE LLKKLPDLER LLSKIHNVGS PLKSQNHPDS RAIMYEETTY
 851 SKKKIIDFLS ALEGFKVMCK IIGIMEEVAD GFKSKILKQV ISLQTKNPEG
 901 RFPDLTVELN RWDTAFDHEK ARKTGLITPK AGFDSDYDQA LADIRENEQS
 951 LLEYLEKQRN RIGCRTIVYW GIGRNRYQLE IPENFTTRNL PEEYELKSTK
1001 KGCKRYWTKT IEKKLANLIN AEERRDVSLK DCMRRLFYNF DKNYKDWQSA
1051 VECIAVLDVL LCLANYSRGG DGPMCRPVIL LPEDTPPFLE LKGSRHPCIT
1101 KTFFGDDFIP NDILIGCEEE EQENGKAYCV LVTGPNMGGK STLMRQAGLL
1151 AVMAQMGCYV PAEVCRLTPI DRVFTRLGAS DRIMSGESTF FVELSETASI
1201 LMHATAHSLV LVDELGRGTA TFDGTAIANA VVKELAETIK CRTLFSTHYH
1251 SLVEDYSQNV AVRLGHMACM VENECEDPSQ ETITFLYKFI KGACPKSYGF
1301 NAARLANLPE EVIQKGHRKA REFEKMNQSL RLFREVCLAS ERSTVDAEAV
1351 HKLLTLIKEL (1360)
```

**Mutations detected in *MSH6***

A missense mutation in GTBP/MSH6 (Thr$^{1219}$ → Ile) occurs in the HHUA endometrial carcinoma cell line[28].

**Database accession numbers**

| | PIR | SWISSPROT | EMBL/GENBANK | REFERENCES |
|---|---|---|---|---|
| Human *MLH1* | S43085 | P40692, MLH1_HUMAN | HS07343, U07343, U40978, U40960–U40977 | 5 |
| Human *MSH2* | | P43246, MSH2_HUMAN | HS04045, U04045, | 1 |
| | | | L47583, U04045, U03911 | 16 |
| Human *MSH3* | | | HSMSH3A-T, D61397–D61416 | 22 |
| Human *PMS1* | | | HS13695, U13695 | 6 |
| Human *PMS2* | | | HS13696, U13696 | 6 |
| Human *GTBP* | | | HS289461, U28946 | 4,32 |

***References***


1 Fishel, R. et al. (1993) Cell 75, 1027–1038.
2 Fujii, H. and Shimada, T. (1989) J. Biol. Chem. 264, 10057–10064.
3 Drummond, J.T. et al. (1995) Science 268, 1909–1912.
4 Palombo, F. et al. (1995) Science 268, 1912–1914.
5 Bronner, C.E. et al. (1994) Nature 368, 258–261.
6 Nicolaides, N.C. et al. (1994) Nature 371, 75–80.
7 Horii, A. et al. (1994) Biophys. Res. Commun. 204, 1257–1264.
8 Nicolaides, N.C. et al. (1995) Genomics 30, 95–206.
9 Kolodner, R.D. et al. (1995) Cancer Res. 55, 242–248.
10 Han, H.J. et al. (1995) Hum. Mol. Genet. 4, 237–242.
11 Leach, F.S. et al. (1996) Cancer Res. 56, 235–240.
12 Wilson, T.M. et al. (1995) Cancer Res. 55, 5146–5150.
13 Fishel, R. et al. (1994) Cancer Res. 54, 5539–5542.
14 **MacPhee, D.G. (1995) Cancer Res. 55, 5489–5492.**
15 Papadopoulos, N. et al. (1995) Science 268, 1915–1917.
16 Liu, B. et al. (1994) Cancer Res. 54, 4590–4594.
17 Lazar, V. et al. (1994) Hum. Mol. Genet. 3, 2257–2260.
18 Bubb, V.J. et al. (1996) Oncogene 12, 2641–2649.
19 Risinger, J.I. et al. (1995) Cancer Res. 55, 5664–5669.
20 Reitmair, A.H. et al. (1995) Nature Genet. 11, 64–70.
21 Scherer, S.J. et al. (1996) Hum. Genet. 97, 114–116.
22 Watanabe, A. et al. (1996) Genomics 31, 311–318.
23 Xia, L. et al. (1996) Cancer Res. 56, 2289–2292.
24 Froggatt, N.J. et al. (1996) J. Med. Genet. 33, 726–730.
25 Wijnen, J. et al. (1995) Am. J. Hum. Genet. 56, 1060–1066.
26 Mary, J.L. et al. (1994) Hum. Mol. Genet. 3, 2067–2069.
27 Wijnen, J. et al. (1994) Hum. Mol. Genet. 3, 2268.
28 Risinger, J.I. et al. (1996) Nature Genet. 14, 102–105.
29 Charbonnier, F. et al. (1995) Cancer Res. 55, 1839–1841.
30 Hemminki, A. et al. (1994) Nature Genet. 8, 405–410.
31 Tannergård, P. et al. (1995) Cancer Res. 55, 6092–6096.
32 Nicolaides N.C. et al. (1996) Genomics 31, 395–397.

## Identification

Neurofibromatosis type 1 (NF1) is caused by defects in the *NF1* gene that was identified by sequencing the region in which translocations in neurofibromatosis had been identified [1,2]. The gene responsible for neurofibromatosis type 2 (NF2) was shown to be distinct from *NF1* by linkage studies and *NF2* was isolated from clones derived from a bidirectional cosmid walk in the region of chromosome 22 deleted in NF2 [3,4].

## Related genes

NF1 is highly conserved and has regions of homology with *Saccharomyces cerevisiae* IRA1 and IRA2, mammalian GTPase-activating protein (GAP) and with the microtubule-associated proteins MAP2 and TAU. Seven *NF1*-related loci have been detected (2q21, 2q33–q34, 14q11.2, 15q11.2, 18p11.2, 21q11.2–q21 and 22q11.2 [5].

The *NF2* gene product merlin (moesin-ezrin-radixin-like protein) shares 45–47% amino acid identity with the cytoskeleton-associated proteins from which its name is derived and is a member of the band 4.1 superfamily [4].

| | ***NF1* (neurofibromin)** | ***NF2* (merlin)** |
|---|---|---|
| **Nucleotides (kb)** | 335 | 110 |
| **Chromosome** | 17q11.2 | 22q12 |
| **Mass (kDa): predicted** | 327 | 66 |
| **expressed** | 250 | 72 |
| **Cellular location** | Particulate cellular fraction | Colocalizes with F-actin in the motile regions of fibroblasts and meningioma cells [3,6] |

**Tissue distribution**

Neurofibromin: Highly expressed in brain [7]. An alternatively spliced *NF1* gene product (type II) is widely expressed in vertebrates [8] and has been reported to be differentially expressed in neuronal cells stimulated to differentiate by retinoic acid [9] and in brain tumours [10]. Another alternatively spliced isoform utilizing exon 9br is expressed at high levels in the central nervous system but shows reduced expression in medulloblastomas and oligodendrogliomas [11].

Merlin: Present in muscle and Schwann cells [12]. Detected in *NF2* lymphoblast cell lines. A variety of isoforms occur (see **Gene structure of *NF2***).

## Protein function

Neurofibromin: Contains a GAP-related domain (NF1 GRD) that stimulates the GTPase activity of normal but not oncogenic $p21^{ras}$ [13,14]. In tumour cells from NF1 patients $p21^{RAS}$ is activated even though $p120^{GAP}$ is present[15]. *RAS* appears to promote *NF1*-linked malignancy and *NF1* itself may function as a recessive oncogene, its normal gene product converting RAS to the inactive form. Neurofibromin interacts with the effector domain of RAS, however, and may thus be the target of RAS rather than its regulator (see ***RAS***). Neurofibromin GAP activity is inhibited by arachidonate, phosphatidate or PtdIns(4,5)$P_2$, to which $p120^{GAP}$ is insensitive, and by tubulin[16,17]. Neurofibromin undergoes serine/threonine phosphorylation in cells stimulated by growth factors.

Merlin: The homology of merlin with erythrocyte band 4.1 proteins (N-terminal 200–300 residues, long $\alpha$ helix and a highly charged C-terminus) suggests it may function as a membrane organizing protein. Merlin binds to a number of cellular proteins (p165, p145, p125, p85 and p70) and is constitutively phosphorylated on serine and threonine residues. However, its phosphorylation state does not appear to be modulated during cell stimulation by growth factors and merlin may be involved in signalling growth inhibitory pathways[18].

## Cancer

The most frequent forms of neurofibromatosis are peripheral neurofibromatosis (NF1) and central neurofibromatosis (NF2). NF1 (von Recklinghausen neurofibromatosis) affects ~1 in 3500 individuals and arises in cells derived from the embryonic neural crest, causing benign growths including neurofibromas and café-au-lait spots on the skin, pheochromocytomas and malignant Schwannomas and neurofibrosarcomas. The *NF1* gene is always inherited as a mutant allele, unlike *RB1*. Both *NF1* mRNA and protein expression appear to be increased in all grades of human astrocytomas as a result of positive feedback by RAS which is also constitutively activated in these tumours[19].

*NF2* functions as a recessive tumour suppressor gene and loss of NF2 protein function is a necessary step in Schwannoma pathogenesis. Mutations (incidence 1 in 40 000) give rise to acoustic neuromas, Schwann cell-derived tumours and meningiomas[20] and also occur in some breast carcinomas[21] and malignant mesotheliomas[22].

## Transgenic animals

Mice carrying a null mutation at *Nf-1* die *in utero* with severe malformation of the heart[23]. Heterozygous mutant mice show no obvious abnormalities, despite the fact that, in humans, neurofibromatosis type 1 is an autosomal dominant disorder. However, heterozygous mice are highly susceptible to various tumours, notably pheochromocytoma and myeloid leukaemia[24]. The loss of *Nf-1* renders myeloid cells hypersensitive to GM-CSF manifested by an enhanced and prolonged rise in the level of RAS-GTP[25]. Schwann cells from *Nf-1*-deficient mice have elevated levels of RAS-GTP and inhibited growth in response to glial growth factor[26]. Deletion of $p120^{GAP}$ causes extensive neuronal cell death and synergizes with the deletion of *Nf-1*, indicating that RAS-GAP and neurofibromin act together to regulate RAS activity during embryonic development[27].

***In vitro***

Primary leukaemic cells from children with neurofibromatosis type 1 are hypersensitive to GM-CSF stimulation, showing elevated levels of RAS-GTP, even though cellular levels of GAP remain normal[28]. Thus neurofibromin appears to regulate negatively GM-CSF signalling *via* RAS in haematopoietic cells and the loss of this regulation may be a primary event in the onset of juvenile chronic myelogenous leukaemia. The minimal RAS-binding fragment of NF1 (1441–1496) can suppress v-*Hras*-induced transformation[29].

Over-expression of *NF2* can reverse v-*Hras*-induced transformation *in vitro*[30]. Anti-sense oligodeoxynucleotides against *NF2* DNA suppress synthesis of merlin and cause adherent cells to round-up and detach from their substrate[31].

## Gene structure of *NF1*

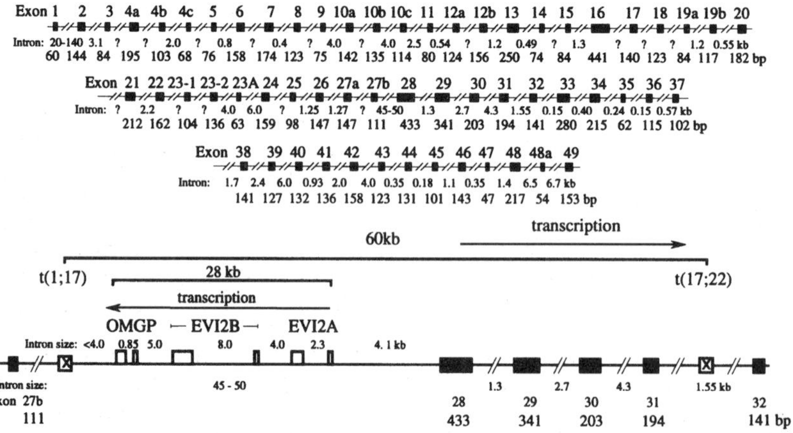

Exon 4 is divided into 4a, 4b and 4c, exon 10 into 10a, 10b and 10c, exon 12 into 12a and 12b, exon 19 into 19a and 19b and exon 27 into 27a and 27b. Exon 23a is an alternatively spliced insertion and two exons derived from exon 23 have therefore been designated 23-1 and 23-2[5].

The lower figure represents two translocations (t(1;17) and t(17;22)) in *NF1* patients (crosses). The breakpoints flank a 60 kb segment of DNA that contains the *EV12A*, *EVI2B* and oligodendrocyte-myelin glycoprotein (*OMGP*) loci, shown as open boxes[32,33]. *EV12A*, *EVI2B* and *OMGP* have the same transcriptional orientation, similar genomic organization and are contained within intron 27b of *NF1*, which is transcribed from the opposite strand. Some but not all of the deletions detected in *NF1* affect the *EV12A*, *EV12B* or *OMPG* genes.

Sequence-specific DNA methylation is common in the *NF1* gene, particularly in CpG dinucleotides, and may contribute to spontaneous germline mutations[34].

## Gene structure of *NF2*

Exon 1 2 3 4 5 6 7 8 9 10 11 12 13 14 15 16
114 126 123 84 69 83 76 135 75 114 123 218 106 128 163 47 bp

A variety of alternatively spliced forms are expressed in various tissues[35,36]. These include lack of exons 2 and 3, lack of exon 3, lack of exon 15, inclusion of a second exon 1 (1′: 39 amino acids) and inclusion of a second exon 15 (15′: 11 amino acids).

## Protein structure

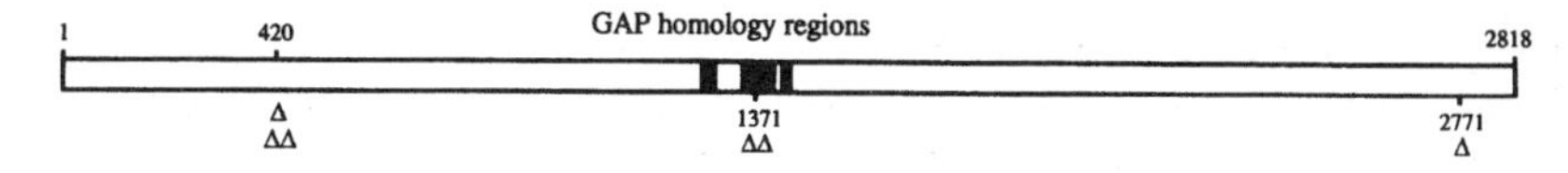


The black boxes indicate regions of GAP family homology. The three arrows mark the insertion point for exon 9br: the double and single arrows indicate other alternative sequence insertion points[37].

## Amino acid sequence of human neurofibromin

1 MAAHRPVEWV QAVVSRFDEQ LPIKTGQQNT HTKVSTEHNK ECLINISKYK
51 FSLVISGLTT ILKNVNNMRI FGEAAEKNLY LSQLIILDTL EKCLAGQPKD
101 TMRLDETMLV KQLLPEICHF LHTCREGNQH AAELRNSASG VLFSLSCNNF
151 NAVFSRISTR LQELTVCSED NVDVHDIELL QYINVDCAKL KRLLKETAFK
201 FKALKKVAQL AVINSLEKAF WNWVENYPDE FTKLYQIPQT DMAECAEKLF
251 DLVDGFAEST KRKAAVWPLQ IILLILCPEI IQDISKDVVD ENNMNKKLFL
301 DSLRKALAGH GGSRQLTESA AIACVKLCKA STYINWEDNS VIFLLVQSMV
351 VDLKNLLFNP SKPFSRGSQP ADVDLMIDCL VSCFRISPHN NQHFKICLAQ
401 NSPSTFHYVL VNSLHRIITN SALDWWPKID AVYCHSVELR NMFGETLHKA
451 VQGCGAHPAI RMAPSLTFKE KVTSLKFKEK PTDLETRSYK YLLLSMVKLI
501 HADPKLLLCN PRKQGPETQG STAELITGLV QLVPQSHMPE IAQEAMEALL
551 VLHQLDSIDL WNPDAPVETF WEISSQMLFY ICKKLTSHQM LSSTEILKWL
601 REILICRNKF LLKNKQADRS SCHFLLFYGV GCDIPSSGNT SQMSMDHEEL
651 LRTPGASLRK GKGNSSMDSA AGCSGTPPIC RQAQTKLEVA LYMFLWNPDT
701 EAVLVAMSCF RHLCEEADIR CGVDEVSVHN LLPNYNTFME FASVSNMMST
751 GRAALQKRVM ALLRRIEHPT AGNTEAWEDT HAKWEQATKL ILNYPKAKME
801 DGQAAESLHK TIVKRRMSHV SGGGSIDLSD TDSLQEWINM TGFLCALGGV
851 CLQQRSNSGL ATYSPPMGPV SERKGSMISV MSSEGNADTP VSKFMDRLLS
901 LMVCNHEKVG LQIRTNVKDL VGLELSPALY PMLFNKLKNT ISKFFDSQGQ
951 VLLTDTNTQF VEQTIAIMKN LLDNHTEGSS EHLGQASIET MMLNLVRYVR
1001 VLGNMVHAIQ IKTKLCQLVE VMMARRDDLS FCQEMKFRNK MVEYLTDWVM
1051 GTSNQAADDD VKCLTRDLDQ ASMEAVVSLL AGLPLQPEEG DGVELMEAKS
1101 QLFLKYFTLF MNLLNDCSEV EDESAQTGGR KRGMSRRLAS LRHCTVLAMS
1151 NLLNANVDSG LMHSIGLGYH KDLQTRATFM EVLTKILQQG TEFDTLAETV
1201 LADRFERLVE LVTMMGDQGE LPIAMALANV VPCSQWDELA RVLVTLFDSR
1251 HLLYQLLWNM FSK*EVELADS MQTLFRGNSL ASKIMTFCFK* VYGATYLQKL
1301 LDPLLRIVIT SSDWQHVSFE VDPTRLEPSE SLEENQRNLL QMTE*KFFHAI*
1351 *ISSSSEFPPQ LRSVCHCLYQ VVSQRFPQNS IGAVGSAMFL RFINPAIVSP*
ΔΔ
1401 *YEAGILD*KKP PPRIE*RGLKL MSKILQSIAN* HVLFTKEEHM RPFNDFVKSN
1451 FDAARRFFLD IASDCPTSDA VNHSLSFISD GNVLALHRLL WNNQEKIGQY

```
1501 LSSNRDHKAV GRRPFDKMAT LLAYLGPPEH KPVADTHWSS LNLTSSKFEE
1551 FMTRHQVHEK EEFKALKTLS IFYQAGTSKA GNPIFYYVAR RFKTGQINGD
1601 LLIYHVLLTL KPYYAKPYEI VVDLTHTGPS NRFKTDFLSK WFVVFPGFAY
1651 DNVSAVYIYN CNSWVREYTK YHERLLTGLK GSKRLVFIDC PGKLAEHIEH
1701 EQQKLPAATL ALEEDLKVFH NALKLAHKDT KVSIKVGSTA VQVTSAERTK
1751 VLGQSVFLND IYYASEIEEI CLVDENQFTL TIANQGTPLT FMHQECEAIV
1801 QSIIHIRTRW ELSQPDSIPQ HTKIRPKDVP GTLLNIALLN LGSSDPSLRS
1851 AAYNLLCALT CTFNLKIEGQ LLETSGLCIP ANNTLFIVSI SKTLAANEPH
1901 LTLEFLEECI SGFSKSSIEL KHLCLEYMTP WLSNLVRFCK HNDDAKRQRV
1951 TAILDKLITM TINEKQMYPS IQAKIWGSLG QITDLLDVVL DSFIKTSATG
2001 GLGSIKAEVM ADTAVALASG NVKLVSSKVI GRMCKIIDKT CLSPTPTLEQ
2051 HLMWDDIAIL ARYMLMLSFN NSLDVAAHLP YLFHVVTFLV ATGPLSLRAS
2101 THGLVINIIH SLCTCSQLHF SEETKQVLRL SLTEFSLPKF YLLFGISKVK
2151 SAAVIAFRSS YRDRSFSPGS YERETFALTS LETVTEALLE IMEACMRDIP
2201 TCKWLDQWTE LAQRFAFQYN PSLQPRALVV FGCISKRVSH GQIKQIIRIL
2251 SKALESCLKG PDTYNSQVLI EATVIALTKL QPLLNKDSPL HKALFWVAVA
2301 VLQLDEVNLY SAGTALLEQN LHTLDSLRIF NDKSPEEVFM AIRNPLEWHC
2351 KQMDHFVGLN FNSNFNFALV GHLLKGYRHP SPAIVARTVR ILHTLLTLVN
2401 KHRNCDKFEV NTQSVAYLAA LLTVSEEVRS RCSLKHRKSL LLTDISMENV
2451 PMDTYPIHHG DPSYRTLKET QPWSSPKGSE GYLAATYPTV GQTSPRARKS
2501 MSLDMGQPSQ ANTKKLLGTR KSFDHLISDT KAPKRQEMES GITTPPKMRR
2551 VAETDYEMET QRISSSQQHP HLRKVSVSES NVLLDEEVLT DPKIQALLLT
2601 VLATLVKYTT DEFDQRILYE YLAEASVVFP KVFPVVHNLL DSKINTLLSL
2651 CQDPNLLNPI HGIVQSVVYH EESPPQYQTS YLQSFGFNGL WRFAGPFSKQ
2701 TQIPDYAELI VKFLDALIDT YLPGIDEETS EESLLTPTSP YPPALQSQLS
2751 ITANLNLSNS MTSLATSQHS PGIDKENVEL SPTTGHCNSG RTRHGSASQV
                                 Δ
2801 QKQRSAGSFK RNSIKKIV (2818)
```

**Domain structure**

| | |
|---|---|
| 1264–1290, 1345–1407 and 1415–1430 | Regions of similarity to the GAP family (italics) |
| 583–586, 815–818, 873–876, 2236–2239, 2573–2576 and 2810–2813 | Six potential cAMP-dependent protein kinase recognition sites (underlined) |
| 2549–2556 | Potential tyrosine kinase recognition site (underlined italics) |

Alternative processing causes insertion of an additional 18 (ASLPCSNSAVFMQ LFPHQ (Δ)) or 21 (ATCHSLLNKATVKEKKENKKS (ΔΔ)) amino acids, respectively. The 21 amino acid insertion decreases GTPase activity [8].

The alternatively spliced isoform encoded by exon 9br has 10 amino acids (ATGLGHKFTS) between residues 420 and 421 [11].

**Mutations detected in *NF1***

A considerable number of mutations in *NF1* have been detected in both familial and sporadic peripheral neurofibromatosis, mainly predicted to cause premature termination [38–43].

Somatic deletions of *NF1* have been detected in benign neurofibromas [44] and deletion of one allele together with a frameshift somatic mutation (4 bp deletion in exon 4b) in the other allele has also been detected [45].

Substitution of Lys[1423], detected in colon adenocarcinoma, myelodysplastic syndrome, anaplastic astrocytoma and NF1, decreases GAPase activity without affecting the binding affinity for RAS-GTP [46,47]. Mutation of Phe[1434] or Lys[1436] in NF1 confers the capacity to suppress the action of oncogenic RAS in v-*ras*-transformed NIH 3T3 cells [48].

*NF1* undergoes mRNA editing: cytidine (nt 2914) is modified to uridine in an Arg (CGA codon at the 5′ end of the GAP-related domain), giving an in-frame stop codon [49]. Editing occurs in normal tissue but is several-fold increased in tumours.

## Protein structure

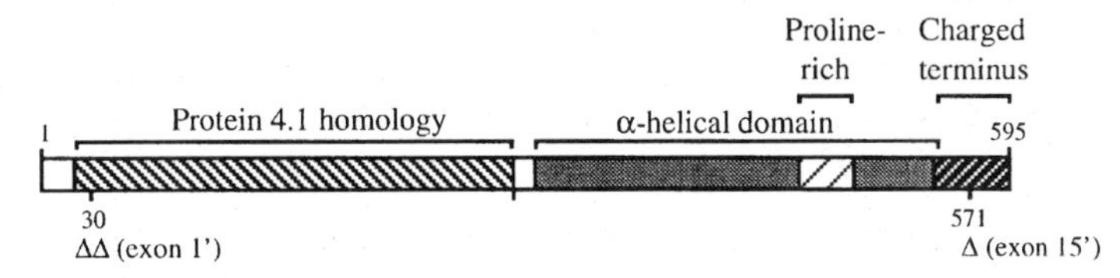


The double and single arrows mark the insertion points for two alternatively spliced exons.

## Amino acid sequence of human merlin

```
  1 MAGAIASRMS FSSLKRKQPK TFTVRIVTMD AEMEFNCEMK WKGKDLFDLV
 51 CRTLGLRETW FFGLQYTIKD TVAWLKMDKK VLDHDVSKEE PVTFHFLAKF
101 YPENAEEELV QEITQHLFFL QVKKQILDEK IYCPPEASVL LASYAVQAKY
151 GDYDPSVHKR GFLAQEELLP KRVINLYQMT PEMWEERITA WYAEHRGRAR
201 DEAEMEYLKI AQDLEMYGVN YFAIRNKKGT ELLLGVDALG LHIYDPENRL
251 TPKISFPWNE IRNISYSDKE FTIKPLDKKI DVFKFNSSKL RVNKLILQLC
301 IGNHDLFMRR RKADSLEVQQ MKAQAREEKA RKQMERQRLA REKQMREEAE
351 RTRDELERRL LQMKEEATMA NEALMRSEET ADLLAEKAQI TEEEAKLLAQ
401 KAAEAEQEMQ RIKATAIRTE EEKRLMEQKV LEAEVLALKM AEESERRAKE
451 ADQLKQDLQE AREAERRAKQ KLLEIATKPT YPPMNPIPAP LPPDIPSFNL
501 IGDSLSFDFK DTDMKRLSME IEKEKVEYME KSKHLQEQLN ELKTEIEALK
551 LKERETALDI LHNENSDRGG SSKHNTIKKL TLQSAKSRVA FFEEL (595)
                                P QAQGRRPICI (590)
```

### Domain structure

| | |
|---|---|
| 1–350 | Moesin-ezrin-radixin homology domain |
| 38, 39 | Exon 1/exon 2 boundary (underlined) |
| 80, 81 | Exon 2/exon 3 boundary (underlined) |
| 121, 122 | Exon 3/exon 4 boundary (underlined) |
| 524, 525 | Exon 14/exon 15 boundary (underlined) |
| 579, 580 | Exon 15/exon 16 boundary (underlined) |
| 580–590 | Alternative C-terminal 11 amino acids following Lys[571] encoded by exon 15′ (italics) [35] |

Sequence encoded by the alternative exon between exons 1 and 2 (exon 1′) [35]: SCSVTLAGVQWRDLGLLQPLPPKFKRFSCLSFPSSWDYR.
Alternative splicing of exon 14 to exon 16 results in 19 C-terminal amino acids following Glu[524]: NSPCRAPSPEWPSLKSSSR.

### Mutations detected in *NF2*

A large number of mutations in *NF2* have been detected that are predicted mainly to cause premature termination. Mutations, caused by C to T transitions, occur with high frequency in five CGA codons (57, 198, 262, 341, 466). In-frame deletions and missense mutations also occur [50,51].

In addition to mutations in vestibular Schwannoma [52,53], mutations in *NF2* occur in melanoma, breast carcinoma, in meningioma precursor cell lines [35], in colorectal carcinomas [36] and in malignant mesotheliomas [22].

A translocation breakpoint on chromosome 22q12.2 detected in NF2 in intron 14 gives rise to a C-terminally truncated protein [54].

### Database accession numbers

| | *PIR* | *SWISSPROT* | *EMBL/GENBANK* | *REFERENCES* |
|---|---|---|---|---|
| Human *NF1* | A35222, A35605 | P21359 | M82814, M89914 | 55 |
| Human *NF2* | | | L11353 | 4 |
| | | | Schwannom Z22664 | 3 |

### *References*


1 Cawthon, R.M. et al. (1990) Cell 62, 193–201.
2 Viskochil, D. et al. (1990) Cell 62, 187–192.
3 Rouleau, G.A. et al. (1993) Nature 363, 515–521.
4 Trofatter, J.A. et al. (1993) Cell 72, 791–800.
5 Li, Y. et al. (1995) Genomics 25, 9–18.
6 Gonzalez-Agosti, C. et al. (1996) Oncogene 13, 1239–1247.
7 Hattori, S. et al. (1992) Oncogene 7, 481–485.
8 Andersen, L.B. et al. (1993) Mol. Cell. Biol. 13, 487–495.
9 Nishi, T. et al. (1991) Oncogene 6, 1555–1559.
10 Suzuki, Y. et al. (1991) Biochem. Biophys. Res. Commun. 181, 955–961.
11 Danglot, G. et al. (1995) Hum. Mol. Genet. 4, 915–920.
12 den Bakker, M.A. et al. (1995) Oncogene 10, 757–763.
13 Bollag, G. and McCormick, F. (1991) Nature 351, 576–579.
14 Dibattiste, D. et al. (1993) Oncogene 8, 637–643.
15 Basu, T.N. et al. (1992) Nature 356, 713–715.
16 Golubic, M. et al. (1991) EMBO J. 10, 2897–2903.
17 Bollag, G. et al. (1993) EMBO J. 12, 1923–1927.
18 Takeshima, H. et al. (1994) Oncogene 9, 2135–2144.
19 Gutmann, D.H. et al. (1996) Oncogene 12, 2121–2127.
20 Ruttledge, M.H. et al. (1994) Nature Genet. 6, 180–184.
21 Bianchi, A.B. et al. (1994) Nature Genet. 6, 185–192.
22 Bianchi, A.B. et al. (1995) Proc. Natl Acad. Sci. USA 92, 10854–10858.
23 Brannan, C.I. et al. (1994) Genes Dev. 8, 1019–1029.
24 Jacks, T. et al. (1994) Nature Genet. 7, 353–361.
25 Largaespada, D.A. et al. (1996) Nature Genet. 12, 137–143.
26 Kim, H.A. et al. (1995) Oncogene 11, 325–335.
27 Henkemeyer, M. et al. (1995) Nature 377, 695–701.
28 Bollag, G. et al. (1996) Nature Genet. 12, 137–143.
29 Fridman, M. et al. (1994) J. Biol. Chem. 269, 30105–30108.
30 Tikoo, A. et al. (1994) Proc. Natl Acad. Sci. USA 91, 23387–23390.
31 Huynh, D.P. and Pulst, S.M. (1996) Oncogene 13, 73–84.
32 Cawthon, R.M. et al. (1991) Genomics 9, 446–460.

[33] Viskochil, D. et al. (1991) Mol. Cell. Biol. 11, 906–912.
[34] Andrews, J.D. et al. (1996) Hum. Mol. Genet. 5, 503–507.
[35] Pykett, M.J. et al. (1994) Hum. Mol. Genet. 3, 559–564.
[36] Arakawa, H. et al. (1994) Hum. Mol. Genet. 3, 565–568.
[37] **Gusella, J.F. et al. (1996) Curr. Opin. Genet. Dev. 6, 87–92.**
[38] Valero, M.C. (1994) Hum. Mol. Genet. 3, 639–641.
[39] Hutter, P. et al. (1994) Hum. Mol. Genet. 3, 663–665.
[40] Legius, E. et al. (1994) Hum. Mol. Genet. 3, 829–830.
[41] Purandare, S.M. et al. (1994) Hum. Mol. Genet. 3, 1109–1115.
[42] Heim, R.A. (1995) Hum. Mol. Genet. 4, 975–981.
[43] Ainsworth, P. et al. (1994) Hum. Mol. Genet. 3, 1179–1181.
[44] Colman, S.D. et al. (1995) Nature Genet. 11, 90–92.
[45] Sawada, S. et al. (1996) Nature Genet. 14, 110–112.
[46] Li, Y. et al. (1992) Cell 69, 275–281.
[47] Gutmann, D.H. et al. (1993) Oncogene 8, 761–769.
[48] Nakafuku, M. et al. (1993) Proc. Natl Acad. Sci. USA 90, 6706–6710.
[49] Skuse, G.R. et al. (1996) Nucleic Acids Res. 24, 478–486.
[50] Sainz, J. et al. (1995) Hum. Mol. Genet. 4, 137–139.
[51] Bourn, D. et al. (1994) Hum. Mol. Genet. 3, 813–816.
[52] Sainz, J. et al. (1994) Hum. Mol. Genet. 3, 885–891.
[53] Sainz, J. et al. (1996) Hum. Genet. 97, 121–123.
[54] Arai, E. et al. (1994) Hum. Mol. Genet. 3, 937–939.
[55] Marchuk, D.A. et al. (1991) Genomics 11, 931–940.

## Identification

p53, encoded by the *TP53* (tumour protein 53)/*P53* gene, was first isolated in immunoprecipitates of large tumour antigen (large T) from SV40-transformed rodent cells [1].

| | *P53* |
|---|---|
| **Nucleotides (kb)** | 12.5 |
| **Chromosome** | 17p13.1 |
| **Mass (kDa): predicted** | 43.5 |
| **expressed** | p53 |
| **Cellular location** | Nucleus |

p53 is detectable at the plasma membrane during mitosis in normal and transformed cells [2,3]. The conformational phenotype (see below) may determine its location [4] and in glioblastomas different mutations at the same codon modulate cytoplasmic/nuclear distribution [5].

**Tissue distribution**
Ubiquitous. In most transformed and tumour cells the concentration of p53 is increased 5- to 100-fold [6] over the minute concentration in normal cells (~1000 molecules/cell), principally due to the half-life of the mutant forms (4 h) compared with that of the wild-type (20 min). High concentrations of p53 protein are transiently expressed in human epidermis and superficial dermal fibroblasts following mild UV irradiation [7]. p53 may play a specific role in murine lens morphogenesis [8]. High levels of p53 occur in senescing human diploid fibroblasts [9] and enhanced *trans*-activation by p53 may directly regulate senescence [10].

## Protein function

p53 is a transcription factor that regulates the normal cell growth cycle by activating transcription of genes that control progression through the cycle and of other genes that cause arrest in $G_1$ when the genome is damaged [11–14]. Mutations in p53 have been detected in many types of tumours: these occur in the central region of the protein required for binding to DNA (the "core domain"). p53 can also promote apoptosis in growth arrested cells. Ten genes have been identified as being differentially expressed during p53-induced apoptosis, including phospholipase C$\beta_4$ and ZFM1 [15]. Hypoxia, occurring in areas of tumours with poor blood supply, promotes p53-dependent apoptosis: cells with mutated p53 are resistant to killing by hypoxia [16]. Purified wild-type, but not mutant p53, also possesses a $Mg^{2+}$-dependent 3′ to 5′ DNA exonuclease capacity. This activity resides in the core domain and raises the possibility that p53 may play a direct role in mismatch repair and/or act as an external proofreading enzyme for cellular DNA polymerases [17]. Proteolysis of p53 occurs when it is bound to damaged DNA and this may activate some functions of p53 [18].

**Transcriptional regulation by p53**

Wild-type p53 has weak sequence-specific DNA binding activity that is strongly enhanced by factors acting on its C-terminal regulatory domain that promote binding with high affinity to a DNA consensus sequence, the p53 response element, comprising of two copies of 5′-PuPuPuC($^{A}/_{T}$)($^{T}/_{A}$)GPyPy-3′ separated by 0–13 bp[19]. Oligomerization of p53 is necessary for sequence specific DNA binding[20] and stacked tetramers of p53 linking separated DNA binding sites through DNA loops have been detected by electron microscopy[21]. However, monomeric variants of wild-type p53 express *trans*-activation activity although they do not activate expression of chromosomal *WAF1*. Nevertheless, monomeric p53 retains the capacity to block transformation by HPV E7 and RAS[22].

| *Factors enhancing* p53 *binding to DNA* | *Sequences to which* p53 *binds* |
|---|---|
| Phosphorylation by PKC/casein kinase II | *WAF1*[23,24] |
| MAb binding | *MDM2*[25] |
| *E. coli* dnaK | *GADD45*[24] |
| Deletion of the C-terminus[26] | *EGFR*[27] |
| C-terminal (369–383) peptide of p53[28] | Human ribosomal gene cluster (RGC)[29] |
| | Thrombospondin-1[30] |
| | TGFα[31] |
| | Basic fibroblast growth factor |
| | *Bax* |
| | Cyclin G[25,32] |
| | *ErbA-1*[33] |
| | *GML* (GPI-anchored molecule-like protein)[34] |
| | Insulin-like growth factor binding protein 3 (IGF-BP3)[35] |
| | Mouse/rat muscle creatine kinase gene (*MCK*)[36] |
| | Mouse endogenous retrovirus-like element (GLN LTR)[37] |
| | Plasminogen activator inhibitor type 1 (PAI-1)[38] |
| | Proliferating cell nuclear antigen (PCNA) |
| | SV40[39] |
| | Human polyomavirus BK[40] |
| | *Sgk* serum/glucocorticoid-inducible serine/threonine protein kinase gene[41] |
| | A26[42] |
| | Fas/APO-1[43] |
| | Human immunodeficiency virus type 1 LTR[44] |

Wild-type p53 inhibits TAT-mediated *trans*-activation of HIV-1 LTR[45]. However, both TAT and the HTLV-I TAX protein inhibit p53 transcription, the latter *via* a bHLH element in the *P53* promoter[46]. These repressive effects of retrovirally encoded regulators of transcription contrast with the mechanism of p53 inactivation by DNA tumour viruses which is *via* oncoproteins that interact with the protein. However, deleted N-terminal (1–223) p53 has been shown to activate CMV-CAT promoter activity[47].

The first 42 amino acids of p53 comprise the minimum region sufficient for the powerful *trans*-activating function in p53/GAL4 fusion proteins[48–50]. Transcriptional activation by these proteins is prevented by mutant p53 proteins or by adenovirus E1B and inhibition of *trans*-activation by p53 correlates with transformation of primary cells by E1B in cooperation with E1A[51]. SV40 T antigen and HPV-16 E6 protein also inhibit *trans*-activation by wild-type p53[52,53]. *Trans*-activation by p53 is enhanced by the presence of a $GC_3$ element (GCCCGGGC) adjacent to the consensus sequence to which a distinct factor may bind[33].

p53 binds *via* its acidic activation domain to the N-terminal region of the 90 kDa product of the murine double minute 2 (*Mdm2*) gene in intact cells and *in vitro* the formation of this complex inhibits p53-mediated *trans*-activation[54–57]. The expression of *Mdm2* is itself induced by p53, e.g. following UV irradiation[58], which may serve to autoregulate p53 activity in normal cells[59]. *MDM2* is amplified in 36% of human sarcomas analysed and the over-expression of this gene may enable p53-regulated growth control to be over-ridden in these tumours. MDM2 also directly contacts E2F1 and stimulates the transcriptional activity of E2F1/DP1: thus MDM2 not only releases a proliferative block by silencing p53 but augments proliferation by stimulating an S phase inducing transcription factor[60].

p53 also regulates the expression of genes encoding products that regulate proteolytic degradation of the extracellular matrix. Thus p53 represses transcription from the human urokinase-type (u-PA) gene and the tissue-type plasminogen activator (t-PA) gene *via* non-DNA binding mechanisms[38]. p53 also activates transcription directly from the plasminogen activator inhibitor type 1 (PAI-1).

**Genes repressed by p53**

p53 suppresses transcription of the following genes that lack a p53 response element.

$\beta$-actin
Basic FGF: transcription activated by mutant p53[67]
*Bcl-2*[68,69]
DNA polymerase $\alpha$
*Fos, Jun, Mybb*[70,71]
Human interleukin 6[72]
Porcine MHC class I gene[72,73]
Proliferating cell nuclear antigen (PCNA) (high levels of wild-type p53 repress transcription of PCNA whereas mutant p53 activates, but low levels of p53 *trans*-activate the human PCNA promoter[74]
Retinoblastoma (*RB1*)[75,76]
P-glycoprotein promoter (regulates the expression of the gene responsible for acquisition of the multidrug resistance (MDR) phenotype)[77]
HPV type 16 and 18 promoters by wild-type, not mutant p53[78]
Human insulin-like growth factor I receptor (activated by mutant p53)[79]
Human $O^6$-methylguanine-DNA methyltransferase gene (MGMT)[80]
Human topoisomerase II$\alpha$[81]
Human TR2 orphan receptor[82]

Repression of some of these genes may be mediated by the binding of wild-type (but not mutant p53) to human TATA-binding protein (TBP) and *in vitro* this inhibits transcription from minimal promoters[61–63]. p53 also interacts with CCAAT binding factor (CBF) to repress transcription from the heat shock protein 70 promoter[64].

It seems probable that there are distinct classes of p53-responsive promoters that respond differentially. Thus, for example, transient transfection of wild-type p53 induces apoptosis, whereas the mutant form p53Ala143 does not, even though it activates transcription from many p53-responsive promoters (e.g. *WAF1*). However, p53Ala143 is a poor activator of *BAX*, which possesses an imperfect p53 consensus binding sequence. The resultant low levels of BAX relative to that of BCL2, may account for the prolongation of cell survival (see Chapter 5)[65,66].

## Proteins associating with p53

| *Cellular proteins* | *Viral proteins* |
|---|---|
| ABL[83] | Adenovirus type 5 E1B[84] |
| CCAAT binding factor (CBF)[64] | Adenovirus E4orf6 protein[85] |
| Heat shock protein 70 (HSP70)[86,87] | HPV E6/p53 interaction mediator (E6-AP)[88] |
| Cellular oncoprotein MDM2[54–57] | Epstein–Barr nuclear antigen 5 (EBNA-5)[90] |
| MDMX (MDM2-related)[89] | Epstein–Barr BZLF1 (Z) protein[92] |
| Replication protein A (RPA)[91] | Human papillomavirus type 16/18 E6 (HPV 16/18 E6)[52] |
| Transcription factor ERCC3[93] | Simian virus 40 large T antigen (SV40 T Ag)[53] |
| E2F1 and DP1[97,98] | Hepatitis B virus X protein[94,95] |
| RAD51/RecA[99] | Human cytomegalovirus IE84[96] |
| Spot-1[100] | |
| Transcription factor SP1[101] | |
| TATA binding protein (TBP)[61–63] | |
| TFIIB and TFIID: $TAF_{II}40$ and $TAFI_{II}60$ (components of the TFIID complex)[102] | |
| TFIIH-associated factors XPD (Rad3), XPB[93,94] | |
| UREB1 rat DNA binding protein[103] | |
| CSB[104] | |
| Wilms' tumour protein (WT1)[105] | |
| p45, p56, p70 and other proteins in non-small cell lung carcinoma cells[106] | |
| Human proteins 53BP1 and 53BP2[107] | |
| p38, p42 (with mutant p53)[108] | |
| Nuclear matrix attachment region/scaffold attachment region (MAR/SAR) DNA elements[109] | |

## Kinases phosphorylating p53

- Casein kinase I[110]
- Casein kinase II[111–113]
- Cyclin-dependent kinases[24]
- MAP kinase
- JNK1-like kinase[114]
- Protein kinase C
- DNA-activated kinase[115]
- p42 large T antigen-activated kinase[116]
- RAF1 *in vitro*[117].

**P53 and the cell growth cycle**

When 3T3 cells are stimulated by TPA or serum there is a 10- to 20-fold rise in *P53* mRNA within 6 h. p53 protein is phosphorylated *in vitro* by cyclin D1/CDK4, cyclin B/CDC2, cyclin A/CDK2 or cyclin E/CDK2. However, cyclin D1/CDK4 and cyclin E/CDK2 appear to cause very weak phosphorylation whereas S and $G_2$/M CDKs exert a strong phosphorylating action, promoting a conformational change that markedly stimulates sequence-specific binding. Phosphorylation by either cyclin B/CDC2 or cyclin A/CDK2 strongly stimulates p53 binding to the *WAF1* and *GADD45* promoters, both of which contain p53 response elements. WAF1 is a key regulator of normal proliferation that binds to and inhibits a wide variety of cyclin/CDKs. *WAF1* and *GADD45* expression is not induced by mutant p53[24].

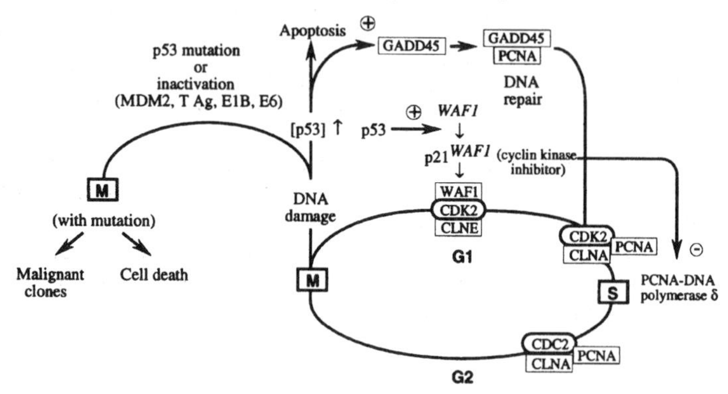


**Apoptosis**

The accumulation of p53 may merely inhibit the cell cycle until damaged DNA has been repaired or, if sustained, cause the activation of programmed cell death. The latter mechanism is consistent with the finding that expression of wild-type p53 (or pRb) can mediate apoptosis in the absence of appropriate differentiation or proliferation signals and that extended survival without growth factors is conferred by the expression of mutant forms of p53 that act as dominant negative inhibitors of the wild-type protein[118]. Increased levels of p53, relative to those that cause growth arrest, are necessary to induce apoptosis, although DNA damage may promote cell death at lower levels of p53. Induction of *WAF1* by p53 occurs in growth arrest but apoptosis may be induced by transcriptionally inactive p53. However, the full apoptotic response requires both the sequence-specific *trans*-activation domains and the C-terminal regulatory domains of p53[119]. The inactivation of p53 may thus permit replication of damaged DNA and promote the development of malignant cell clones, as occurs with high frequency in *P53* null mice and in patients with Li–Fraumeni syndrome.

Coexpression of *Myc* and *Bcl-2* can totally overcome p53-induced apoptosis and cell cycle arrest by altering the subcellular trafficking of p53, causing it to remain in the cytosol during $G_1$[120]. In pituitary cells immortalized with SV40 T Ag, p53 drives apoptosis following irradiation but appears to do so by a mechanism independent of transcription or translation[121].

In 70Z/3 pre-B cells $\gamma$-irradiation causes accumulation of p53 and, in addition to apoptosis, cell differentiation manifested by the expression of $\kappa$ light chain immunoglobulin as the cells accumulate in $G_2$ phase [122]. This and data from Burkitt's lymphoma cells suggest that, in the absence of functional p53, the primary point of entry of cells into apoptosis following DNA damage is at $G_2/M$ [123]. Fibroblasts from $P53^{-/-}$ embryos exposed to spindle inhibitors undergo multiple rounds of DNA replication, forming tetraploid and octaploid cells [124]. This suggests that p53 may be involved in a mitotic checkpoint ensuring maintenance of diploidy.

**GADD45**

The *GADD* (growth arrest on DNA damage) genes are induced in a wide variety of mammalian cells by DNA damaging agents or other causes of growth arrest [125,126] and wild-type but not mutant p53 induces *GADD45* transcription *via* a p53 response element. GADD45 binds to PCNA, a protein that is involved in both DNA replication (see above) and repair [127]. GADD45 stimulates DNA excision repair *in vitro* and inhibits entry into S phase and regulation of its expression by p53 provides another mechanism by which p53 protects the cell from replication of damaged DNA.

p53 interacts directly with human RAD51, a key factor in homologous recombination and recombinatorial DNA repair, and also with the *E. coli* RecA protein which is involved in recombinatorial repair of double-strand breaks and controls the SOS response after UV radiation damage [99]. Thus p53 may be involved in selecting DNA repair pathways and in regulating genetic variation introduced by homologous recombination.

**Cancer**

Chromosome 17p is frequently lost in human cancers and in most tumours that have been examined point mutations have occurred in one allele of the *P53* gene [11,12], over 6000 of which have been detected [128]. Mutations arise with an average frequency of 70% but the incidence varies from zero in carcinoid lung tumours [129] through 30–86% in breast cancers [130–132] to 97% in primary melanomas [133]. Mutations in *P53* correlate strongly with poor prognosis in breast cancer and appear to constitute the best prognostic indicator for recurrence and death [134]. Cervical carcinomas expressing HPV DNA sequences normally coexpress wild-type *P53* mRNA, mutant *P53* being present in the absence of HPV DNA [135], although *P53* mutations do occur in some HPV-associated cancers [136] and mutant *P53* can convert HPV-immortalized cells to a more transformed state [137]. p53 binds to the HPV E6 oncoprotein [88] and this interaction targets the suppressor form of p53 for ubiquitin-mediated degradation [136,138].

In Li–Fraumeni syndrome the mutated gene is transmitted in the germline and this autosomal dominant syndrome is characterized by the occurrence of a variety of mesenchymal and epithelial neoplasms at multiple sites. p53 can positively regulate the expression of thrombospondin-1, a potent inhibitor of angiogenesis, and in fibroblasts from Li–Fraumeni patients the loss of p53 function appears to mediate the switch to an angiogenic phenotype [29]. The malignant progression of astrocytoma towards end-stage glioblastoma is characterized by the early loss of wild-type *P53* activity and by dramatic neovascularization of the neoplasms. Glioblastoma cells that do not express *P53* show strong angiogenic activity, whereas on induction of wild-type, but not mutant *P53*, they secrete glioma-derived angiogenesis inhibitory factor (GD-AIF) which neutralizes the angiogenic factors produced by the parental cells[139].

**In animals**

Disruption of *P53* by proviral insertion of murine leukaemia viruses has been detected in erythroid and lymphoid tumours[140]. The expression of wild-type *P53 via* an adenoviral vector prevents the establishment of tumours derived from implantations of human squamous cell carcinoma of the head and neck in nude mice[141].

***In vitro***

Mutant (but not wild-type) *P53* together with *Ras* transforms primary rat embryo fibroblasts[142] and mutant *P53*, like E1A, can immortalize cells. In transfected primary fibroblasts expression of wild-type *P53* inhibits the ability of mutant *P53* with *Ras* (or E1A plus *Ras*) to cause transformation and the suppression of E1A-mediated transformation occurs by stimulating apoptosis[143]. In cells transformed by *Ras* with *P53*, mutant p53 protein occurs in a trimeric complex with the heat shock protein and wild-type p53.

The transfection of some breast carcinoma, osteosarcoma, colorectal carcinoma and glioblastoma cell lines that carry mutations in *P53* with a wild-type *P53* gene suppresses growth[144–145]. The tumorigenicity of breast carcinoma cell lines that harbour mutations in both *P53* and *RB1* genes is reduced by the expression of wild-type forms of either *P53* or *RB1*[146]. These *in vitro* findings are consistent with the occurrence of *trans*-dominant mutations in *P53*, as are those from analysis of a number of animal and human tumours.

In cells transformed by SV40 or adenovirus and in tumours induced by these viruses, p53 is found in an oligomeric complex with SV40 large T antigen or with adenovirus type 5 E1B 55kDa protein.

In many epithelial cell lines *P53* mutation correlates with loss of TGF$\beta$ responsiveness. TGF$\beta$ may inhibit growth by a p53-independent pathway (e.g. inhibiting CDK4 expression and/or activating INK4B and KIP1 expression), although the overall balance of CDK activity may be the result of combined p53- and TGF$\beta$-mediated pathways[147].

Exposure of human fibroblasts to the putative mutagen nitric oxide (NO) promotes accumulation of wild-type p53 and subsequent downregulation of NO synthesis *via* the action of p53 on the nitric oxide synthase promoter[148].

**Transgenic animals**

Mice homozygous for the null allele develop normally, indicating that p53 is not essential for cell growth control during early development, but are predisposed to spontaneous tumour formation at an early age[149], although a small proportion of animals display defects in neural tube closure[150] and *P53* acts as a suppressor of teratogenesis caused by DNA damaging agents *in utero*[151]. However, tumour development in *P53*-deficient mice is sporadic, indicating that additional genetic or epigenetic events are required. The loss of p53-dependent apoptosis in brain tissue accelerates tumorigenesis and the expression of a dominant negative C-terminal (302–390 amino acids) p53 fragment elicits a *P53*-null phenotype[152].

Fibroblasts from *P53*-deficient mice have accelerated growth properties *in vitro* and are genetically unstable [153]. Crosses of transgenic mice carrying the $Glu^{135} \rightarrow Val$ mutation with $P53^{-/-}$ mice reveal that p53 mutations exert a dominant negative effect on cell growth. Thus, the mutant protein accelerates tumour development in a cellular background of one or both wild-type *P53* genes but has no effect (i.e. shows no gain of function) in p53-deficient cells [154]. The mutant transgene also affects the tumour spectrum, promoting the incidence of lung adenocarcinoma by 10-fold.

Benign skin tumours induced in $P53^{-/-}$ mice have a malignant conversion frequency of ~50% and there is an inverse correlation between $TGF\beta_1$ expression levels in these tumours and the probability of malignant progression [155]. *Ras* and *Myc* induce progression to non-metastatic cancer in the mouse prostate reconstitution model: the complete loss of p53 function leads to metastasis in this system [156].

Mice mutant for both *P53* and *Rb1* genes have reduced viability and develop pinealoblastomas and islet cell tumours [157]. Thus, p53 and pRb can cooperate in the transformation of certain cell types and mutation of p53 may permit cell survival of $pRb^{-/-}$ cells.

Mice expressing the HBV *trans*-activator protein HBx develop liver tumours with 80–90% penetrance [158]. Tumour development correlates with HBV X protein binding to wild-type p53 which prevents nuclear translocation and inhibits its sequence-specific DNA binding *in vitro*. This interaction also prevents association of p53 with the general transcription factor ERCC3 that is involved in nucleotide excision repair [159]. Interferon $\alpha$ suppresses HBV gene expression and treatment of the transgenic mice with IFN-$\alpha$ results in p53 accumulation within the nucleus. Hepatitis B virus oncoprotein HBx blocks p53 binding to XPB and XPD and this interaction with TFIIH-associated factors appears to be part of the mechanism by which HBx inhibits p53-mediated apoptosis [160]. The interaction of p53 with HBx also inhibits HBV replication [161].

## Gene structure

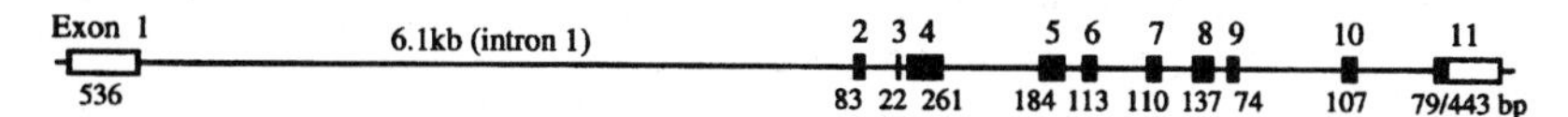


The *P53* promoter sequence is highly conserved between humans, mice and rats, although the murine major transcription start maps ~100 bp upstream from the human start site [162]. *P53* lacks TATA or CAAT sequences but contains a helix-loop-helix consensus binding sequence [163] either downstream of the transcription initiation site (+70 to +75, mouse) or upstream (–29 to –34, human). The untranslated first exon of human *P53* contains a PAX binding site (–21) and in diffuse astrocytomas PAX5 expression inversely correlates with that of P53 [164]. Within the first 100 bp upstream from the start site there are a number of elements that cause strong transcriptional activation: these include a UV response element [165], an NF-$\kappa$B binding site and an element (–167) involved in genotoxic stress-inducible p53 gene expression [165], a region involved in *trans*-activation of the *P53* promoter by p53 itself [166], a b-HLH element (–149) through which *P53* is *trans*-activated by high levels of MYC expression and repressed by high MAX expression[167] and which also binds USF[168], a composite element (–216) that has mutual and exclusive binding capacity for YY1 and NF1[169], a p53 factor 1 (PF1: –291) binding site[73] and a C/EBP site (–345).

Alternative splicing of human *P53* can occur in the 5′ end of the mRNA or may utilize 133 bp from intron 9 that gives rise to a truncated protein having 10 new amino acids[170]. This C-terminally truncated protein does not bind DNA *in vitro* and has greatly reduced transcriptional activity.

A strong promoter present within intron 1[171] and intron 4, essential for *P53* expression in transgenic mice[172], binds a protein that is necessary for transformation by p53[173]. Differentiating mouse erythroleukaemia cells accumulate anti-sense RNA to the first intron that may be involved in the downregulation of *P53* mRNA[174].

Alternatively spliced forms of murine *P53* occur in which amino acids 364–390 are replaced with 17 new amino acids[175] or which contain an additional 96 bases derived from intron 10, the latter being present in normal and epidermal carcinoma cells at ~30% of the level of the normal transcript[176,177].

## Protein structure

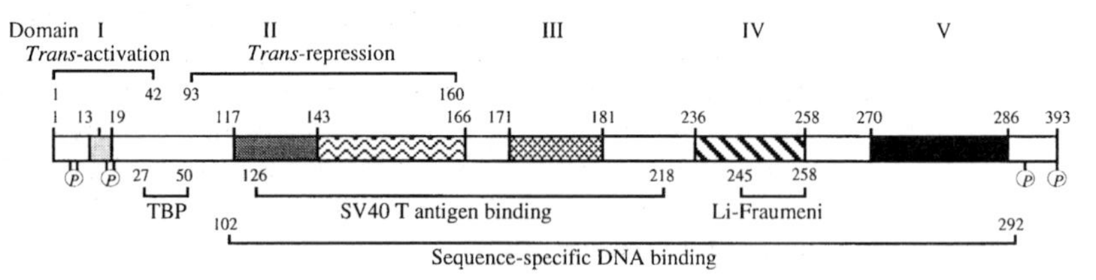


X-ray diffraction reveals that the core domain binds to DNA *via* a large $\beta$ sandwich that acts as a scaffold for three loop-based elements[13,178]. Loop 1 (LSH: loop-sheet-helix) binds to the major groove *via* $Arg^{283}$, $Lys^{120}$, $Cys^{277}$, $Arg^{280}$, $Ala^{276}$ and $Arg^{273}$; Loop 2 (L) binds within the minor groove *via* $Ser^{241}$ and $Arg^{248}$; Loop 3 (LH: loop-helix) packs against L and stabilizes it *via* $Arg^{249}$, $Arg^{175}$ and a tetrahedrally coordinated $Zn^{2+}$ ion.

The crystal structure of the region 325–356 reveals that monomers associate across an anti-parallel $\beta$ sheet and an anti-parallel $\alpha$ helix to form a dimer: two dimers associate across a second parallel helix–helix interface to form the tetramer[179].

Interaction between p53 and 53BP2 involves the loop 3 binding to the SH3 domain of 53BP2 and loop 2 binding to an ankyrin repeat within 53BP2. The six most commonly observed p53 mutations ($Arg^{175} \rightarrow$ His, $Gly^{245} \rightarrow$ Ser, $Arg^{249} \rightarrow$ Ser, $Arg^{175} \rightarrow$ His, $Arg^{282} \rightarrow$ Trp, $Arg^{248} \rightarrow$ Trp, $Arg^{273} \rightarrow$ His) disrupt binding to 53BP2[180].

A large number of *P53* mutations involve a single substitution of a nucleic acid base pair[181] but ~10% of human cancers are characterized by deletions or insertions in this gene[182]. Somatic mutations have been detected in most exons but the majority are clustered in four domains, highly conserved among vertebrates, involving exons 5–10 (amino acids 120–290). Exons 7 and 8 are the most frequent site of germline mutations. Mutations in p53 may (i) be of the dominant negative type when the protein overrides the action of the suppressor wild-type p53 or (ii) result in the loss of suppressor function or (iii) result in a protein that functions as a tumour promoter [183,184]. Mutations producing an abnormal protein that inhibits the function of its normal allelic gene product by formation of mutant/wild-type protein complexes are called *trans*-dominant mutations. The residues most frequently mutated in cancers are all at or near the protein–DNA interface and may be classified as (i) those directly contacting DNA (e.g. $Arg^{248}$ and $Arg^{273}$, the most frequently mutated residues in p53) and (ii) those having a stabilizing role (e.g. $Arg^{175}$ and $Arg^{249}$). The resultant mutant proteins generally fail to bind MAb PAb246 (which recognizes an epitope between amino acids 88 and 109 of wild-type p53), bind to DNA or to SV40 T antigen with much reduced affinity but bind with high affinity to the HSP70 protein *via* the C-terminal 28 amino acids of p53 [86]. However, many mutations in codons 175, 248 and 273 do not affect MAb or HSP70 binding. There is a strict correlation between *trans*-activation capacity and inhibition of cell proliferation although some mutations do not affect *trans*-activation capacity or MAb or HSP70 binding [185].

Mutations at codons 141 or 175 abolish *trans*-activation potential whereas mutations at 248 or 273 result in substantial enhancement [186]. In hepatocellular carcinomas (HCCs) there is a mutational hotspot at codon 249 [187,188] that may be caused by exposure to hepatitis B virus (HBV) and aflatoxin $B_1$. Other mutations occur in HCCs from patients who have not been exposed to HBV or $AFB_1$ that frequently cause overexpression of the mutant protein [189]. Germline mutations in intron 5 and exon 7 that create truncated proteins have been detected in families with early onset breast-ovarian cancer [190,191].

**Functional classes of p53 mutations [153]**

| | *Wild-type* | *Pseudo wild-type* | *Loss of function* | *Dominant negative* | *Dominant oncogene* |
|---|---|---|---|---|---|
| | wt p53 | $Leu^{175}$ | Null p53 | $Val^{135}$ | $His^{175}$ |
| Loss of tumour suppressor function: | No | No | Yes | Yes | Yes |
| Dominant negative: | No | No | No | Yes | Yes |
| Gain of function: | No | No | No | No | Yes |

The $Glu^{135} \rightarrow$ Val mutation exerts a dominant negative effect on cell growth and in myeloid leukaemia cells suppresses apoptosis induced by the overexpression of *Myc* [192]; $Arg^{175} \rightarrow$ His shows dominant negative activities *in vitro* and gain of function in $P53^{-/-}$ cells by augmenting tumorigenicity.

## Amino acid sequence of human P53

```
  (1) MEEPQSDPSV EPPLSQETFS DLWKLLPENN VLSPLPSQAM DDLMLSPDDI
 (51) EQWFTEDPGP DEAPRMPEAA PPVAPAPAAP TPAAPAPAPS WPLSSSVPSQ
(101) KTYQGSYGFR LGFLHSGTAK SVTCTYSPAL NKMFCQLAKT CPVQLWVDST
(151) PPPGTRVRAM AIYKQSQHMT EVVRRCPHHE RCSDSDGLAP PQHLIRVEGN
(201) LRVEYLDDRN TFRHSVVVPY EPPEVGSDCT TIHYNYMCNS SCMGGMNRRP
(251) ILTIITLEDS SGNLLGRNSF EVRVCACPGR DRRTEEENLR KKGEPHHELP
(301) PGSTKRALPN NTSSSPQPKK KPLDGEYFTL QIRGRERFEM FRELNEALEL
(351) KDAQAGKEPG GSRAHSSHLK SKKGQSTSRH KKLMFKTEGP DSD (393)
```

Sequence arising from alternative splicing of human *P53* after exon 9: `(331) DQTSFQKENC (341)` [170].

**Domain structures**

| | |
|---|---|
| 1–75 | Acidic: predicted α helix |
| 22, 23 | Hydrophobic residues ($Leu^{22}$, $Trp^{23}$) critical for transcriptional activity [193] |
| 20–57 and 318–393 | TBP binding regions: *trans*-repression *via* the C-terminal interaction is relieved by adenovirus 13S E1A [194] |
| 6, 9, 15, 37, 315, 392 | Major phosphorylation sites [195]. $Ser^{315}$ phosphorylated by CDC2 and CDK2 [196] |
| 34 | Ser JNK1-like kinase phosphorylation site [114] |
| 75–150 | Proline-rich, hydrophobic |
| 102–292 | Sequence-specific DNA binding, essential for tumour suppressor function [159,197,198] |
| 135–179 and 238–277 | Putative $Zn^{2+}$-binding domains (underlined) [199] |
| 173, 235, 239 | The corresponding cysteines in murine p53 are critically involved in DNA binding and transcription activation [200] |
| 126–218 | T antigen binding region mapped by mutagenesis [201] |
| 287–340 | Casein kinase II binding region [202] |
| 323–393 | C-terminus: highly basic containing helix-coil-helix motifs: forms stable tetramers and is necessary and sufficient for DNA binding: the affinity of the interaction is increased if the DNA is damaged by ionizing radiation [93,198,203] |
| 313–322 | Nuclear localization signal |
| 389 | 5.8S rRNA attachment site [204,205] |
| 392 | Casein kinase II phosphorylation [26] |
| 15, 37 | Phosphorylated *in vitro* by the protein kinase DNA-PK [115] |
| 22, 23 | N-terminal hydrophobic amino acids ($Leu^{22}$ and $Trp^{23}$) required for *trans*-activation |
| 14, 19, 22, 23 | N-terminal hydrophobic amino acids ($Leu^{14}$, $Phe^{19}$, $Leu^{22}$ and $Trp^{23}$) required for human MDM2 binding |
| 22, 23, 24, 25, 26 and 27 | N-terminal hydrophobic amino acids ($Leu^{22}$ and $Trp^{23}$ and also $Lys^{24}$, $Pro^{27}$, $Leu^{25}$ and $Leu^{26}$) required for Ad5 E1B 55 kDa protein binding [206, 207] |
| 323–355 | Sufficient to constitute a strong tetramerization domain |
| 83–323 | Weaker oligomerization domain [208] |
| 322–355 | Tetramerization domain [209] |
| 363–393 | *Trans*-repression domain (I1) |
| 145–170 | *Trans*-repression domain (I2) [210] |
| 378 | PKC phosphorylation site |
| 293–393 | HBx binding region |

372–382 PAb421 epitope within or near which is an *O*-glycosylation site, modification of which confers strong DNA binding[211]

80–290 The core domain together with the C-terminal 60 amino acids confer high-affinity binding by mutant p53 to nuclear matrix attachment region/scaffold attachment region (MAR/SAR) DNA elements

**Mutations detected in *P53***

Of the large number of mutations that occur in human cancers[128], the most common include: Lys$^{132}$ → Asn, Met$^{133}$ → Leu, Phe$^{134}$ → Leu, Cys$^{135}$ → Tyr, Cys$^{141}$ → Tyr, Pro$^{151}$ → Ser, Val$^{157}$ → Ser, Ala$^{159}$ → Pro, Arg$^{175}$ → His, His$^{179}$ → Tyr, Asn$^{239}$ → Ser, Asn$^{247}$ → Ile, Arg$^{248}$ → Thr/Trp, Arg$^{249}$ → Pro/Leu, Val$^{272}$ → Met, Arg$^{273}$ → His/Leu/Pro, Asp$^{281}$ → Gly.

Mutations in breast cancer (30–86% of cases) occur mainly within exons 5–8 (~80%) and none have been detected in exon 1 or in the putative promoter region[130–132,134]. They include missense, frameshift and splice site mutations[212–215]. Two germline mutations detected in breast cancers (Leu$^{257}$ → Gln and a frameshift) give rise to 87 abnormal C-terminal amino acids)[191].

Mutations in ovarian cancer (44–61% of cases) include missense, frameshift, null and splice site mutations[216,217].

Approximately 60% of lung cancers are associated with mutations in P53 and over 500 have been defined[128,129,218–220]. There are several mutational hotspots in lung cancer, notably at codons 157, 248 and 273. Guanine residues in these codons undergo specific adduct formation with benzo[*a*]pyrene, a polycyclic aromatic hydrocarbon component of cigarette smoke that is a potent mutagen and carcinogen[221].

Mutations in colorectal carcinomas[222,223] reveal that point mutations within the conserved domains of p53 occur in more aggressive tumours than those outside these regions: codon 175 mutations are particularly aggressive[224].

Li–Fraumeni germline mutations: The hotspot is in exon 7 (245–258). Some Li–Fraumeni families have also been identified with *P53* mutations outside exon 7[225,226], including a base deletion in codon 215 which causes premature termination giving a protein of 214 *bona fide* N-terminal residues followed by 31 illegitimate amino acids[227], a point mutation (Leu$^{344}$ → Pro) in exon 10 in the tetramerization domain[228] and deletion of exon 10[229]. Germline transmission of replication damaged trinucleotide repeats, elongating the protein by one amino acid (Val$^{216}$ → TrpLeu) has been detected in one family[230].

Multiple somatic mutations have been detected in *P53* (together with *APC* mutations) following loss of *MSH2* in hereditary non-polyposis colon cancer[231].

Mutations have also been detected in acute myelogenous leukaemia (6%)[232], brain tumours (~10%) and malignant astrocytomas (~30%)[233,234], Burkitt's lymphoma cell lines (60%)[235], epithelial carcinomas (48%)[236], oesophageal adenocarcinomas (80%)[237], gastric carcinomas (57%)[238], HBV-positive hepatomas (18%)[239], head and neck cancers (~69%)[240], melanomas (primary) (97%)[133], multiple myeloma (20%)[241], neuroblastoma cell lines

(80%)[242], osteosarcomas (41%)[243], osteosarcoma cell lines (90%)[145], pancreatic carcinoma (40%)[244,245], papillary thyroid carcinomas (50%)[246,247], renal cell carcinoma (79%)[248], rhabdomyosarcoma (45%)[243], squamous cell carcinoma of the larynx (60%)[249], and in Wilms' tumour (73%)[250].

**Database accession numbers**

| | *PIR* | *SWISSPROT* | *EMBL/GENBANK* | *REFERENCES* |
|---|---|---|---|---|
| Human *TP53* | A25224 | P04637 | K03199 | 251 |
| | A25397 | | X01405 | 252 |
| | B25397 | | X02469 | 253 |
| | JT0436 | | | 254 |
| | | | M13114–M13121 | 255 |

**References**


1 Zakut-Houri, R. et al. (1983) EMBO J. 4, 1251–1255.
2 Milner, J. and Cook, A. (1986) Virology 150, 265–269.
3 Shaulsky, G. et al. (1990) Oncogene 5, 1707–1711.
4 Zerrahn, J. et al. (1992) Oncogene 7, 1371–1381.
5 Ali, I.U. et al. (1994) Cancer Res. 54, 1–5.
6 Hassapoglidou, S. et al. (1993) Oncogene 8, 1501–1509.
7 Hall, P.A. et al. (1993) Oncogene 8, 203–207.
8 Pan, H. and Griep, A. (1994) Genes Dev. 8, 1285–1299.
9 Atadja, P. et al. (1995) Proc. Natl Acad. Sci. USA 92, 8348–8352.
10 Bond, J. et al. (1996) Oncogene 13, 2097–2104.
11 **Donehower, L.A. and Bradley, A. (1993) Biochim. Biophys. Acta 1155, 181–205.**
12 **Vogelstein, B. and Kinzler, K.W. (1992) Cell 70, 523–526.**
13 **Milner, J. (1995) Trends Biochem. Sci. 20, 49–51.**
14 **Lane, D.P. (1992) Nature 358, 15–16.**
15 Amson, R.B. et al. (1996) Proc. Natl Acad. Sci. USA 93, 3953–3957.
16 Graeber, T.G. et al. (1995) Nature 379, 88–91.
17 Mummenbrauer, T. et al. (1996) Cell 85, 1089–1099.
18 Molinari, M. et al. (1996) Oncogene 13, 2077–2086.
19 El-Deiry, W.S. et al. (1992) Nature Genet. 1, 45–49.
20 Hainaut, P. et al. (1994) Oncogene 9, 299–303.
21 Stenger, J.E. et al. (1994) EMBO J. 13, 6011–6020.
22 Tarunina, M. et al. (1996) Oncogene 13, 589–598.
23 El-Deiry, W.S. et al. (1993) Cell 75, 817–825.
24 Wang, Y. and Prives, C. (1995) Nature 376, 88–91.
25 Zauberman, A. et al. (1995) Oncogene 10, 2361–2366.
26 Hupp, T.R. et al. (1992) Cell 71, 875–886.
27 Ludes-Meyers, J.H. et al. (1996) Mol. Cell. Biol. 16, 6009–6019.
28 Hecker, D. et al. (1996) Oncogene 12, 947–951.
29 Hupp, T.R. et al. (1995) Cell, 83, 237–245.
30 Dameron, K.M. et al. (1994) Science 265, 1582–1584.
31 Shin, T.H. et al. (1995) Mol. Cell. Biol. 15, 4694–4701.
32 Okamoto, K. et al. (1996) Mol. Cell. Biol. 16, 6593–6602.
33 Shiio, Y. et al. (1993) Oncogene 8, 2059–2065.
34 Furuhata, T. et al. (1996) Oncogene 13, 1965–1970.
35 Buckbinder, L. et al. (1995) Nature 377, 646–649.
36 Zhao, J. et al. (1996) Oncogene 13, 293–302.

[37] Zauberman, A. et al. (1993) EMBO J. 12, 2799–2808.
[38] Kunz, C. et al. (1995) Nucleic Acids Res. 23, 3710–3717.
[39] Bargonetti, J. et al. (1992) Genes Dev. 6, 1886–1898.
[40] Shivakumar, C.V. and Das, G.C. (1996) Oncogene 13, 323–332.
[41] Maiyar, A.C. et al. (1996) J. Biol. Chem. 271, 12414–12422.
[42] Buckbinder, L. (1994) Proc. Natl Acad. Sci. USA 91, 10350–10354.
[43] Owen-Schaub, L.B. et al. (1995) Mol. Cell. Biol. 15, 3032–3040.
[44] Gualberto, A. et al. 1995) Mol. Cell. Biol. 15, 3450–3459.
[45] Li, C.J. et al. (1995) Proc. Natl Acad. Sci. USA 92, 5461–5464.
[46] Uittenbogaard, M.N. et al. (1995) J. Biol. Chem. 270, 28503–28506.
[47] Subler, M.A. et al. (1994) Oncogene 9, 1351–1359.
[48] Fields, S. and Jang, S.K. (1990) Science 249, 1046–1049.
[49] Raycroft, L. et al. (1990) Science 249, 1049–1051.
[50] Unger, T. et al. (1992) EMBO J. 11, 1383–1390.
[51] Yew, P.R. and Berk, A.J. (1992) Nature 357, 82–85.
[52] Mietz, J.A. et al. (1992) EMBO J. 11, 5013–5020.
[53] Band, V. et al. (1993) EMBO J. 12, 1847–1852.
[54] Momand, J. et al. (1992) Cell 69, 1237–1245.
[55] Oliner, J.D. et al. (1993) Nature 362, 857–860.
[56] Chen, J. et al. (1993) Mol. Cell. Biol. 13, 4107–4114.
[57] Haines, D.S. et al. (1994) Mol. Cell. Biol. 14, 1171–1178.
[58] Perry, M.E. et al. (1994) Proc. Natl Acad. Sci. USA 90, 11623–11627.
[59] Juven, T. et al. (1993) Oncogene 8, 3411–3416.
[60] Martin, K. et al. 1995) Nature 375, 691–694.
[61] Seto, E. et al. (1992) Proc. Natl Acad. Sci. USA 89, 12028–12032.
[62] Mack, D.H. et al. (1993) Nature 363, 281–283.
[63] Liu, X. et al. (1993) Mol. Cell. Biol. 13, 3291–3300.
[64] Agoff, S.N. et al. (1993) Science 259, 84–87.
[65] Ludwig, R.L. et al. (1996) Mol. Cell. Biol. 16, 4952–4960.
[66] Friedlander, P. et al. (1996) Mol. Cell. Biol. 16, 4961–4971.
[67] Ueba, T. et al. (1994) Proc., Natl Acad. Sci. USA 91, 9009–9013.
[68] Selvakumaran, M. et al. (1994) Oncogene 9, 1791–1798.
[69] Miyashita, T. and Reed, J.C. (1995) Cell 80, 293–299.
[70] Mercer, W.E. et al. (1991) Proc. Natl Acad. Sci. USA 88, 1958–1962.
[71] Lin, D. et al. (1992) Proc. Natl Acad. Sci. USA 89, 9210–9214.
[72] Santhanam, U. et al. (1991) Proc. Natl Acad. Sci. USA 88, 7605–7609.
[73] Ginsberg, D. et al. (1991) Proc. Natl Acad. Sci. USA 88, 9979–9983.
[74] Shivakumar, C.V. et al. (1995) Mol. Cell. Biol. 15, 6785–6793.
[75] Shiio, Y. et al. (1992) Natl Acad. Sci. USA 89, 5206–5210.
[76] Jackson, P. et al. (1993) Oncogene 8, 589–597.
[77] Zastawny, R.L. et al. (1993) Oncogene 8, 1529–1535.
[78] Desaintes, C. et al. (1995) Oncogene 10, 2155–2161.
[79] Werner, H. et al. (1996) Proc. Natl Acad. Sci. USA 93, 8318–8323.
[80] Harris, L.C. et al. (1996) Cancer Res. 56, 2029–2032.
[81] Ines Sandri, M. et al. (1996) Nucleic Acids Res. 24, 4464–4470.
[82] Lin, D.L. and Chang, C. (1996) J. Biol. Chem. 271, 14649–14652.
[83] Goga, A. et al. (1995) Oncogene 11, 791–799.
[84] Lin, J. et al. (1994) Genes Dev. 8, 1235–1246.
[85] Dobner, T. et al. (1996) Science 272, 1470–1473.
[86] Hainaut, P. and Milner, J. (1992) EMBO J. 11, 3513–3520

[87] Ory, K. et al. (1994) EMBO J. 13, 3496–3504.
[88] Huibregtse, J.M. et al. (1993) Mol. Cell. Biol. 13, 775–784.
[89] Shvarts, A. et al. (1996) EMBO J. 15, 5349–5357.
[90] Szekely, L. et al. (1993) Proc. Natl Acad. Sci. USA 90, 5455–5459
[91] Dutta, A. et al. (1993) Nature 365, 79–82.
[92] Zhang, Q. et al. (1994) Mol. Cell. Biol. 14, 1929–1938.
[93] Wang, X.W. et al. (1994) Proc. Natl Acad. Sci. USA 91, 2230–2234.
[94] Wang, X.W. et al. (1995) Cancer Res. 55, 6012–6016.
[95] Lee, H. et al. (1995) J. Biol. Chem. 270, 31405–31412.
[96] Speir, E. et al. (1994) Science 265, 391–394.
[97] O'Connor, D.J. et al. (1995) EMBO J. 14, 6184–6192.
[98] Sørensen, T.S. et al. (1996) Mol. Cell. Biol. 16, 5888–5895.
[99] Sturzbecher, H.-W. et al. (1996) EMBO J. 15, 1992–2002.
[100] Elkind, N.B. et al. (1995) Oncogene 11, 841–851.
[101] Borellini, F. and Glazer, R.I. (1993) J. Biol. Chem. 268, 7923–7928.
[102] Liu, X. and Berk, A.J. (1995) Mol. Cell. Biol. 15, 6474–6478.
[103] Gu, J. et al. (1995) Oncogene 11, 2175–2178.
[104] Wang, X.W. et al. (1995) Nature Genet. 10, 188–195.
[105] Maheswaran, S. et al. (1995) Genes Dev. 9, 2143–2156.
[106] Srinivasan, R. and Maxwell, S.A. (1996) Oncogene 12, 193–200.
[107] Iwabuchi, K. et al. (1994) Proc. Natl Acad. Sci. USA 91, 6098–6102.
[108] Chen, Y. et al. (1994) Mol. Cell. Biol. 14, 6764–6772.
[109] Müller, B.F. et al. (1996) Oncogene 12, 1941–1952.
[110] Milne, D.M. et al. (1992) Oncogene 7, 1361–1369
[111] Meek, D.W. et al. (1990) EMBO J. 9, 3253–3260.
[112] Herrmann, C.P. et al. (1991) Oncogene 6, 877–884.
[113] Filhol, O. et al. (1992) J. Biol. Chem. 267, 20577–20583.
[114] Milne, D.M. et al. (1995) J. Biol. Chem. 270, 5511–5518.
[115] Lees-Miller, S.P. et al. (1992) Mol. Cell. Biol. 12, 5041–5049.
[116] Müller, E. and Scheidtmann, K.H. (1995) Oncogene 10, 1175–1185.
[117] Jamal, S. and Ziff, E.B. (1995) Oncogene 10, 2095–2101.
[118] Gottlieb, E. et al. (1994) EMBO J. 13, 1368–1374.
[119] Chen, X. et al. (1996) Genes Dev. 10, 2438–2451.
[120] Ryan, J.J. et al. (1994) Proc. Natl Acad. Sci. USA 91, 5878–5882.
[121] Caelles, C. et al. (1994) Nature 370, 220–223.
[122] Aloni-Grinstein, R. et al. (1995) EMBO J. 14, 1392–1401.
[123] Allday, M.J. et al. (1995) EMBO J. 14, 4994–5005.
[124] Cross, S.M. et al. (1995) Science 267, 1353–1356.
[125] Kastan, M.B. et al. (1992) Cell 71, 587–597.
[126] Zhan, Q. et al. (1994) Mol. Cell. Biol. 14, 2361–2371.
[127] Smith, M.L. et al. (1994) Science 266, 1376–1380.
[128] Hainaut, P. et al. (1997) Nucleic Acids Res. 25, 151–157.
[129] Iggo, R. et al. (1990) Lancet 335, 675–679.
[130] Horak, E. et al. (1991) Oncogene 6, 2277–2284.
[131] Varley, J.M. et al. (1991) Oncogene 6, 413–421.
[132] Saitoh, S. et al. (1994) Oncogene 9, 2869–2875.
[133] Akslen, L.A. and Morkve, O. (1992) Int. J. Cancer 52, 13–16.
[134] Kovach, J.S. et al. (1996) Proc. Natl Acad. Sci. USA 93, 1093–1096.
[135] Crook, T. et al. (1991) Oncogene 6, 873–875
[136] Crook, T. and Vousden, K.H. (1992) EMBO J. 11, 3935–3940.

[137] Chen, T.-M. et al. (1993) Oncogene 8, 1511–1518.
[138] Scheffner, M. et al. (1990) Cell 63, 1129–1136.
[139] Meir, E.G. et al. (1994) Nature Genet. 8, 171–176.
[140] Munroe, D.G. et al. (1990) Mol. Cell. Biol. 10, 3307–3313.
[141] Clayman, G.L. et al. (1995) Cancer Res. 55, 1–6.
[142] Finlay, C. et al. (1989) Cell 57, 1083–1093.
[143] Lowe, S.W. et al. (1994) Proc. Natl Acad. Sci. USA 91, 2026–2030.
[144] Baker, S.J. et al. (1990) Science 249, 912–915.
[145] Diller, L. et al. (1990) Mol. Cell. Biol. 10, 5772–5781.
[145] Mercer, W.E. et al. (1990) Proc. Natl Acad. Sci. USA 87, 6166–6170.
[146] Wang, N.P. et al. (1993) Oncogene 8, 279–288.
[147] Blaydes, J.P. et al. (1995) Oncogene 10, 307–317.
[148] Forrester, K. et al. (1996) Proc. Natl Acad. Sci. USA 93, 2442–2447.
[149] Donehower, L.A. et al. (1992) Nature 356, 215–221.
[150] Sah, V.P. et al. (1995) Nature Genet. 10, 175–179.
[151] Nicol, C.J. et al. (1995) Nature Genet. 10, 181–187.
[152] Bowman, T. et al. (1996) Genes Dev. 10, 826–835.
[153] Harvey, M. et al. (1993) Oncogene 8, 2457–2467.
[154] Harvey, M. et al. (1995) Nature Genet. 9, 305–311.
[155] Cui, W. et al. (1994) Cancer Res. 54, 5831–5836.
[156] Thompson, T.C. et al. (1995) Oncogene 10, 869–879.
[157] Williams, B.O. et al. (1994) Nature Genet. 7, 480–484.
[158] Ueda, H. et al. (1995) Nature Genet. 9, 41–47
[159] Wang, Y. et al. (1994) Genes Dev. 7, 2575–2586.
[160] Wang, X.W. et al. (1995) Cancer Res. 55, 6012–6016.
[161] Lee, H. et al. (1995) J. Biol. Chem. 270, 31405–31412.
[162] Tuck, S.P. and Crawford, L. (1989) Mol. Cell. Biol. 9, 2163–2172.
[163] Ronen, D. et al. (1991) Proc. Natl Acad. Sci. USA 88, 4128–4132.
[164] Stuart, E.T. et al. (1995) EMBO J. 14, 5638–5645.
[165] Sun, X. et al. (1995) Mol. Cell. Biol. 15, 4489–4496.
[166] Deffie, A. et al. (1993) Mol. Cell. Biol. 13, 3415–3423.
[167] Roy, B. et al. (1994) Mol. Cell. Biol. 14, 7805–7815.
[168] Reisman, D. and Rotter, V. (1993) Nucleic Acids Res. 21, 345–350.
[169] Furlong, E.E.M. et al. (1996) Mol. Cell. Biol. 16, 5933–5945.
[170] Flaman, J.-M. et al. (1996) Oncogene 12, 813–818.
[171] Reisman, D. (1988) Proc. Natl Acad. Sci. USA 85, 5146–5150.
[172] Lozano, G. and Levine, A.J. (1991) Mol. Carcinog. 4, 3–9.
[173] Beenken, S.W. et al. (1991) Nucleic Acids Res. 19, 4747–4752.
[174] Khochbin, S. and Lawrence, J.-J. (1989) EMBO J. 8, 4107–4114.
[175] Wu, L. et al. (1995) Mol. Cell. Biol. 15, 497–504.
[176] Arai, N. et al. (1986) Mol. Cell. Biol. 6, 3232–3239.
[177] Han, K.-A. and Kulesz-Martin, M.F. (1992) Nucleic Acids Res. 20, 1979–1981.
[178] Cho, Y. et al. (1994) Science 265, 346–355.
[179] Jeffrey, P.D. et al. (1995) Science 267, 1498–1502.
[180] Gorina, S. and Pavletich, N.P. (1996) Science 274, 1001–1005.
[181] **Hollstein, M. et al. (1991) Science 253, 49–53.**
[182] Jego, N. et al. (1993) Oncogene 8, 209–213.
[183] Milner, J. et al. (1991) Cell. Biol. 11, 12–19.
[184] Halazonetis, T. et al. (1993) EMBO J. 12, 1021–1028.

[185] Ory, K. et al. (1994) EMBO J. 13, 3496–3504.
[186] Miller, C.W. et al. (1993) Oncogene 8, 1815–1824.
[187] Hsu, I.C. et al. (1991) Nature 350, 427–428.
[188] Bressac, B. et al. (1991) Nature 350, 429–431.
[189] Volkmann, M. et al. (1994) Oncogene 9, 195–204.
[190] Jolly, K.W. et al. (1994) Oncogene 9, 97–102.
[191] Mazoyer et al. (1994) Oncogene 9, 1237–1239.
[192] Lotem, J. and Sachs, L. (1995) Proc. Natl Acad. Sci. USA 92, 9672–9676.
[193] Lin, J. et al. (1995) Oncogene 10, 2387–2390.
[194] Horikoshi, N. et al. (1995) Mol. Cell. Biol. 15, 227–234.
[195] Wang, Y. and Eckhart, W. (1992) Proc. Natl Acad. Sci. USA 89, 4231–4235.
[196] Price, B.D. et al. (1995) Oncogene 11, 73–80.
[197] Pietenpol, J.A. et al. (1994) Proc. Natl Acad. Sci. USA 91, 1998–2002.
[198] Pavletich, N.P. et al. (1994) Genes Dev. 7, 2556–2564.
[199] Hainaut, P. and Milner, J. (1993) Cancer Res. 53, 1739–1742.
[200] Rainwater, R. et al. (1995) Mol. Cell. Biol. 15, 3892–3903.
[201] Ruppert, J.M. and Stillman, B. (1993) Mol. Cell. Biol. 13, 3811–3820.
[202] Appel, K. et al. (1995) Oncogene 11, 1971–1978.
[203] Reed, M. et al. (1995) Proc. Natl Acad. Sci. USA 92, 9455–9459.
[204] Samad, A. and Carroll, R.B. (1991) Mol. Cell. Biol. 11, 1598–1606.
[205] Fontoura, B.M.A. et al. (1992) Mol. Cell. Biol. 12, 5145–5151.
[206] Lin, J. et al. (1994) Genes Dev. 8, 1235–1246.
[207] Marston, N.J. et al. (1994) Oncogene 9, 2707–2716.
[208] Wang, P. et al. (1994) Mol. Cell. Biol. 14, 5182–5191.
[209] Waterman, J.L.F. et al. (1995) EMBO J. 14, 512–519.
[210] Hsu, Y.-S. et al. (1995) J. Biol. Chem. 270, 6966–6974.
[211] Shaw, P. et al. (1996) Oncogene 12, 921–930.
[212] Carrere, N. et al. (1993) Mol. Genet. 2, 1075.
[213] Hartmann, A. et al. (1995) Oncogene 10, 681–688.
[214] Sun, X.-F. et al. (1996) Oncogene 13, 407–411.
[215] Blasyk, H. et al. (1996) Oncogene 13, 2159–2166.
[216] Milner, B.J. et al. (1993) Cancer Res. 53, 2128–2132.
[217] Skilling, J.S. et al. (1996) Oncogene 13, 117–123.
[218] Takahashi, T. et al. (1991) Oncogene 6, 1775–1778.
[219] Harpole, D.H. et al. (1995) Clin. Cancer Res. 1, 659–664.
[220] Gusterson, B.A. et al. (1991) Oncogene 6, 1785–1789.
[221] Denissenko, M.F. et al. (1996) Science 274, 430–432.
[222] Rodrigues, N.R. et al. (1990) Proc. Natl Acad. Sci. USA 87, 7555–7559.
[223] Tominaga, O. et al. (1993) Oncogene 8, 2653–2658.
[224] Goh, H.-S. et al. (1995) Cancer Res. 55, 5217–5221.
[225] Prosser, J. et al. (1992) Br. J. Cancer 65, 527–528.
[226] Srivastava, S. et al. (1990) Nature 348, 747–749.
[227] Stolzenberg, M.-C. et al. (1994) Oncogene 9, 2799–2804.
[228] Varley, J.M. et al. (1996) Oncogene 12, 2437–2442.
[229] Plummer, S.J. et al. (1994) Oncogene 9, 3273–3280.
[230] Strauss, E.A. et al. (1995) Cancer Res. 55, 3237–3241.
[231] Lazar, V. et al. (1994) Hum. Mol. Genet. 3, 2257–2260.
[232] Fu, L. et al. (1996) EMBO J. 5, 4392–4401.
[233] Mashiyama, S. et al. (1991) Oncogene 6, 1313–1318.
[234] Lang, F.F. et al. (1994) Oncogene 9, 949–954.

[235] Wiman, K.G. et al. (1991) Oncogene 6, 1633–1639.
[236] Moles, J.-P. et al. (1993) Oncogene 8, 583–588.
[237] Gleeson, C.M. et al. (1995) Cancer Res. 55, 3406–3411.
[238] Martin, H.M. et al. (1992) Int. J. Cancer 50, 859–862.
[239] Hosono, S. et al. (1993) Oncogene 8, 491–496.
[240] Ahomadegbe, J.C. et al. (1995) Oncogene 10, 1217–1227.
[241] Portier, M. et al. (1992) Oncogene 7, 2539–2543.
[242] Davidoff, A.M. et al. (1992) Oncogene 7, 127–133.
[243] Mulligan, L.M. et al. (1990) Proc. Natl Acad. Sci. USA 87, 5863–5867.
[244] Ruggeri, B. et al. (1992) Oncogene 7, 1503–1511.
[245] Redston, M.S. et al. (1994) Cancer Res. 54, 3025–3033.
[246] Dongi, R. et al. (1992) In Mutant Oncogenes, Eds N. Lemoine and A. Epenetos. Chapman and Hall, London, pp.187–192.
[247] Nikiforov, Y.E. et al. (1996) Oncogene 13, 687–693.
[248] Oda, H. et al. (1995) Cancer Res. 55, 658–662.
[249] Maestro, R. et al. (1992) Oncogene 7, 1159–1166.
[250] Bardeesy, N. et al. (1994) Nature Genet. 7, 91–97.
[251] Harlow, E. et al. (1985) Mol. Cell. Biol. 5, 1601–1610.
[252] Zakut-Houri, R. et al. (1985) Nature 306, 594–597.
[253] Harris, N. et al. (1986) Mol. Cell. Biol. 6, 4650–4656.
[254] Buchman, V.L. et al. (1988) Gene 70, 245–252.
[255] Matlashewski, G. et al. (1984) EMBO J. 3, 3257–3262.

## Identification

*PTC* was isolated by sequence-sampling cosmids from the nevoid basal cell carcinoma syndrome region[1].

## Related genes

The *PTC* homologue *Patched* (60% amino acid similarity) was originally identified as a segment polarity gene in *Drosophila*. Mouse and chicken homologues have been identified.

| | *PTC* |
|---|---|
| **Nucleotides (kb)** | 34 |
| **Chromosome** | 9q22.3 |
| **Mass (kDa): predicted** | 143 |
| **expressed** | |
| **Cellular location** | Transplasma membrane |
| **Tissue distribution** | Ubiquitous |

## Protein function

PTC controls development by signalling the cell type-specific repression of transcription of genes encoding members of the TGF$\beta$ and WNT families of signalling proteins. *Drosophila* Ptc is the putative receptor for Sonic hedgehog (Shh)[2]. The *Drosophila mothers against dpp* (*mad*) gene is homologous to the tumour suppressor gene *DPC4/SMAD4* and its product mediates signalling by Dpp, a TGF$\beta$ homologue specifically repressed by *ptc*.

### Cancer

Mutations in *PTC* are associated with nevoid basal cell carcinoma syndrome (NBCC), also called basal cell nevus carcinoma or Gorlin's syndrome. NBCC is an autosomal dominant disorder of prevalence 1 in 56 000 characterized by a variety of tumours including fibromas of the ovaries and heart, CNS cancers including medulloblastomas and meningiomas, and most frequently basal cell carcinomas (BCCs).

## Gene structure

Exon 1 1A 1B 2 3 4 5 6 7 8 9 10 11 12 13 14 15 16 17 18 19 20 21 22 23

189 239 193 200 60 92 199 122 148 132 157 97 245 403 310 143 185 281 138 143 101 255 537 bp

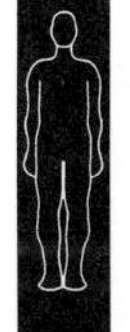

## Amino acid sequence of human PTC

```
   1 MFNPQLMIQT PKEEGANVLT TEALLQHLDS ALQASRVHVY MYNRQWKLEH
  51 LCYKSGELIT ETGYMDQIIE YLYPCLIITP LDCFWEGAKL QSGTAYLLGK
 101 PPLRWTNFDP LEFLEELKKI NYQVDSWEEM LNKAEVGHGY MDRPCLNPAD
 151 PDCPATAPNK NSTKPLDMAL VLNGGCHGLS RKYMHWQEEL IVGGTVKNST
 201 GKLVSAHALQ TMFQLMTPKQ MYEHFKGYEY VSHINWNEDK AAAILEAWQR
 251 TYVEVVHQSV AQNSTQKVLS FTTTTLDDIL KSFSDVSVIR VASGYLLMLA
 301 YACLTMLRWD CSKSQGAVGL AGVLLVALSV AAGLGLCSLI GISFNAATTQ
 351 VLPFLALGVG VDDVFLLAHA FSETGQNKRI PFEDRTGECL KRTGASVALT
 401 SISNVTAFFM AALIPIPALR AFSLQAAVVV VFNFAMVLLI FPAILSMDLY
 451 RREDRRLDIF CCFTSPCVSR VIQVEPQAYT DTHDNTRYSP PPPYSSHSFA
 501 HETQITMQST VQLRTEYDPH THVYYTTAEP RSEISVQPVT VTQDTLSCQS
 551 PESTSSTRDL LSQFSDSSLH CLEPPCTKWT LSSFAEKHYA PFLLKPKAKV
 601 VVIFLFLGLL GVSLYGTTRV RDGLDLTDIV PRETREYDFI AAQFKYFSFY
 651 NMYIVTQKAD YPNIQHLLYD LHRSFSNVKY VMLEENKQLP KMWLHYFRDW
 701 LQGLQDAFDS DWETGKIMPN NYKNGSDDGV LAYKLLVQTG SRDKPIDISQ
 751 LTKQRLVDAD GIINPSAFYI YLTAWVSNDP VAYAASQANI RPHRPEWVHD
 801 KADYMPETRL RIPAAEPIEY AQFPFYLNGL RDTSDFVEAI EKVRTICSNY
 851 TSLGLSSYPN GYPFLFWEQY IGLRHWLLLF ISVVLACTFL VCAVFLLNPW
 901 TAGIIVMVLA LMTVELFGMM GLIGIKLSAV PVVILIASVG IGVEFTVHVA
 951 LAFLTAISDK NRRAVLALEH MFAPVLDGAV STLLGVLMLA GSDFDFIVRY
1001 FFAVLAILTI LGVLNGLVLL PVLWSFFGPY PEVSPANGLN RLPTPSPEPP
1051 PSVVRFAMPP GHTHSGSDSS DSEYSSQTTV SGLSEELRHY EAQQGAGGPA
1101 HQVIVEATEN PVFAHSTVVH PESRHHPPSN PKQQPHLDSG SLPPGRQGQQ
1151 PRRDPPRKGL WPPLYRPRRD AFEISTEGHS GPSNRARWGP RGARSHNPRN
1201 PTSTAMGSSV PGYCQPITTV TASASVTVAV HPPPVPGPGR NPRGGLCPGY
1251 PETDHGLFED PHVPFHVRCE RRDSKVEVIE LQDVECEERP RGSSSN(1296)
```

### Domain structure

Twelve putative transmembrane domains.

### Mutations detected in *PTC*

Germline (NBCC) mutations in *PTC*: 9 bp insertion (exon 15) causing insertion of Pro-Asn-Ile after codon 815; 11 bp deletion (exon 15) creating a stop codon after amino acid 813; C1081T (exon 8) → stop; del 804–840 (exon 6); G1148A → stop (exon 8); 2047ins CT (exon 13).

Sporadic NBCC mutations: 2000insC (exon 13); 2583delC (exon 15).

Mutations in *PTC* in sporadic BCC not associated with BCNS: $Leu^{175}$ → Phe (exon 3); CC1081TT (exon 8) → stop; del 2704–2717 (exon 16) [3].

### Database accession numbers

| | *PIR* | *SWISSPROT* | *EMBL/GENBANK* | *REFERENCES* |
|---|---|---|---|---|
| Human *PTC* | | U43148 | HSU43148 | 4 |
| | | | U59464 | 3 |

### *References*


1 Hahn, H. et al. (1996) Cell 85, 841–851.
2 Stone, D.M. et al. (1996) Nature 384, 129–134.
3 Johnson, R.L. et al. (1996) Science 272, 1668–1671.
4 Hahn, H. et al. (1996) J. Biol. Chem. 271, 12125–12128.

## Identification

The retinoblastoma gene (*RB1*) provides the classical model for a recessive tumour suppressor gene in that both paternal and maternal copies of the gene must be inactivated for the tumour to develop. For *P53* and some other tumour suppressor genes, mutation at one allele may be sufficient to give rise to the altered cell phenotype. Genetic analysis confirmed Knudson's hypothesis that two successive genomic lesions gave rise to retinoblastoma and that the two events corresponded to inactivation of both copies of a gene located on chromosome 13q. *RB1* itself was subsequently identified following the isolation of a cDNA segment hybridizing to the gene [1].

## Related genes

*RB1* and *RB2* are members of the retinoblastoma gene family. *RB1* has regions of homology with p107 and the transcription factor TFIIB [2,3]. *RB2* encodes p130, the E1A-associated protein that also binds cyclins A and E [4]. *Drosophila RBF* combines several of the structural features of *RB1*, p107 and p130 [5].

| | ***RB1*** |
|---|---|
| **Nucleotides (kb)** | 180 |
| **Chromosome** | 13q14.2 |
| **Mass (kDa): predicted** | 106 |
| **expressed** | 105 |
| **Cellular location** | Nuclear matrix: some mutant forms cytoplasmic |
| **Tissue distribution** | Ubiquitous |

pRb may play a specific role in murine lens morphogenesis [6].

## Protein function

pRb (p105$^{RB1}$) functions as a signal transducer, connecting the cell cycle clock with transcriptional control mechanisms mediating progression through the first two-thirds of $G_1$ phase. pRb binds to double-stranded DNA in a non-sequence-specific manner. The broad transcriptional effects of pRb are mediated by its inhibition of factors such as E2F that are required for the expression of genes involved in DNA replication and by repressing transcriptional activation by RNA polymerases I and III.

The prevention of normal pRb function by E1A, T antigen and E7 appears to be a crucial common step in cell transformation by the corresponding DNA tumour viruses [7]. Over 20 cellular proteins (see table and summary below) have been detected that bind to pRb in a manner similar to T antigen, E1A and E7, many of which are transcription factors. However, only E2F sites have been identified as promoter elements normally targeted by pRb and pRb functions by being selectively recruited to promoters *via* E2F and then inactivating surrounding transcription factors (e.g. MYC, ELF1, PU.1) by blocking their interaction with the basal transcription complex[8].

## Proteins associating with pRb

| *Cellular proteins* | *Viral proteins* |
|---|---|
| Cyclins D2, D3, B1 and C[9–11] | Adenovirus type 5 E1A[7] |
| ATF-2 (transcription factor)[12] | Epstein–Barr nuclear antigens 2 and 5 (EBNA-2, EBNA-5)[13] |
| PU.1 (transcription factors)[14] | Human papillomavirus type 16 E7 (HPV16 E7)[15] |
| $TAF_{II}250$[16] | Simian virus 40 large T antigen (SV40 T Ag)[15] |
| E2F1, E2F2, E2F3 (see ***E2F1***) | |
| UBF[17] | |
| BRM (glucocorticoid receptor-mediated transcriptional activator)[18] | |
| ELF-1[8] | |
| Heat shock cognate protein (hsc73)[19] | |
| MDM2[20] | |
| NF-IL-6[21] | |
| RAK nuclear tyrosine kinase[22] | |
| RBAP-2[23,24] | |
| RBAP46[25], RBAP48[26] | |
| RBP1 and RBP2[27,28] | |
| RIZ (Zn finger protein)[29] | |
| RBQ-1/RBBP6[30] | |
| Lamins A and C[31] | |
| ABL, MYC and MYCN[32] | |
| $p34^{CDC2}$[33] | |
| p48[34] | |
| Protein phosphatase PP-1α2 (type 1 catalytic subunit)[35] | |
| BRG1[36] | |
| Mouse *microphthalmia* (*mi*) gene product[37] | |
| Insulin[38] | |

pRb acts through a motif (the retinoblastoma control element, RCE) to repress *MYC* transcription in human keratinocytes (see ***MYC***), to repress transcription of *FOS* and of the human p107 promoter[39], to enhance transcription of insulin-like growth factor II and to regulate either positively or negatively the expression of *TGFB1* depending on cell type (see ***FOS***). pRb also promotes insulin receptor gene expression[40]. pRb autoregulates its own transcription (see below) and stimulates transcription of cyclin D1 (see below), endothelin 1 and the proteoglycans versican and PG40. pRb also negatively regulates expression of $p34^{CDC2}$[41] and studies of $Rb\text{-}1^{-/-}$ mice have revealed at least 15 genes that are repressed by pRb including thrombospondin[42].

Three "retinoblastoma control proteins" (p80, SP1 (p95) and SP3 (p115)) specifically bind to the RCE and pRb specifically interacts with SP1 or SP3 to "super-activate" their *trans*-activating capacity, for example in activating *JUN* transcription[43,44].

## pRb and cell proliferation

pRb contains at least 12 distinct serine or threonine phosphorylation sites and phosphorylation of pRb appears to be essential for progression through the cell cycle, inhibition of pRb phosphorylation causing cell cycle arrest[45–48]. Hypo-phosphorylated pRb associates with the ubiquitous transcription factor E2F and this complex is dissociated by E1A, SV40 T antigen or HPV E7[15,49,50] or by phosphorylation of pRb by cyclin D1/CDK4, cyclin E/CDK2 or cyclin A/CDK2[51–54]. The pRb/E2F complex occurs in $G_1$ in human primary cells and tumour cell lines and functions as a transcriptional repressor in the presence of an additional factor, RBP60[55,56]. pRb binds with high affinity to E2F1 to control its intracellular location. The complex associates with DNA as a repressor unit comprised of a DNA binding component (E2F1/DP1) and an active repressor domain that is part of pRb[57]. One target of the pRb/E2F complex is the *CDC2* promoter[41]. As the cells enter S phase, a second E2F complex forms containing the pRb-related protein p107 and cyclin A[58,59].

A CDC2-related kinase (PITALRE) also phosphorylates pRb although, in contrast to CDC2, it does not phosphorylate histone H1[60]. In a variety of human tumour cell lines each of the D-cyclins (D1, D2 and D3) form complexes with CDK4 and its close relative CDK6 but these complexes are not detectable in cells having a mutated *RB1* gene or in which the function of pRb is compromised by the presence of a DNA tumour virus oncoprotein. Thus, as well as being a potential substrate for D-cyclin kinases, pRb contributes to the formation or stability of such complexes[61] and directly regulates transcription of cyclin D1[62]. Complementation studies in yeast indicate that D cyclins collaborate with cyclin E in the hyperphosphorylation of pRb[63].

Three physiological signals (TGF$\beta$, cAMP and contact inhibition) are known to block pRb phosphorylation by activating a CDK inhibitor (CDKI) that associates either with a CDK or with a cyclin/CDK complex.

The C-terminus (792–928) of pRb binds to MDM2 and complexation of pRb with MDM2 relieves pRb suppression of E2F transactivating function[64]. MDM2 also directly contacts E2F1 and stimulates the transcriptional activity of E2F1/DP1: thus MDM2 not only releases a proliferative block by silencing p53 but augments proliferation by stimulating an S phase-inducing transcription factor[65].

pRb also appears to mediate growth suppression by specifically blocking the activation of transcription by RNA polymerase I by binding directly to

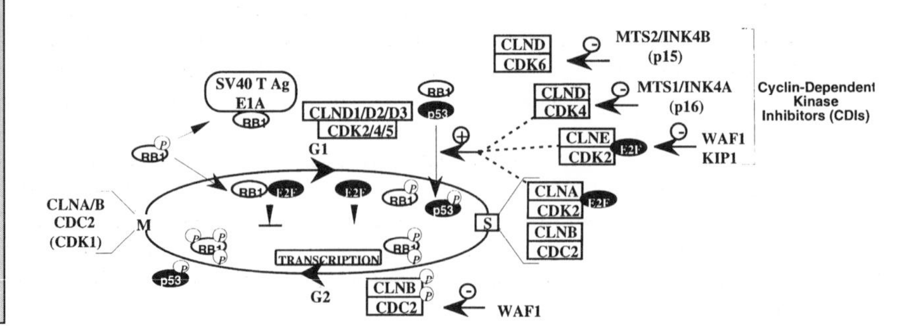

UBF which binds to the pol I promoter and interacts directly with the polymerase[17]. pRb also represses most if not all genes transcribed by RNA polymerase III. Naturally occurring mutations and deletions in the Rb pocket disrupt pol III repression[66]. The capacity of pRb to function as a negative regulator of cell growth is consistent with the findings from transgenic studies[67].

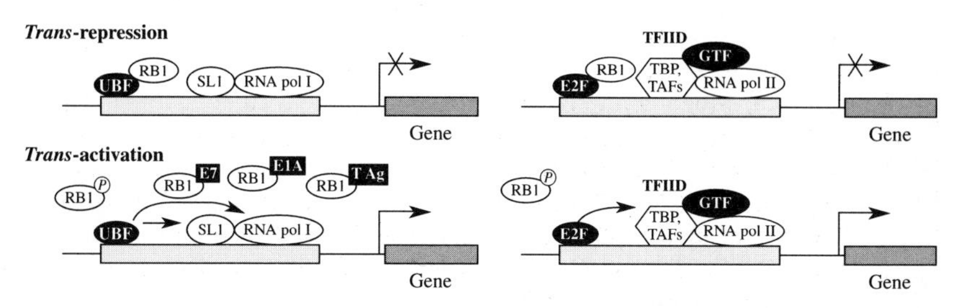


Phosphorylation of pRb may cause release of E2F at the $G_1$/S transition. However, Rb/E2F complexes are detectable in S phase and it is possible that in Rb/E2F complexes Rb may function as a *trans*-repressor and that phosphorylation of Rb may interfere with this function without necessarily affecting E2F binding[68].

## Cancer

Retinoblastoma is a rare hereditary disease, occurring in 1 child in 20 000, that affects retina cell precursors. In 60% of the cases the condition is termed sporadic, when there is no family history of the disease and a single tumour occurs in one eye; in the remaining 40% of cases (familial or germinal retinoblastoma) tumours are bilateral and more than one independently derived tumour is frequently present. The disease is caused by the loss of both copies of the gene by inactivating mutations resulting in null alleles. Familial retinoblastoma occurs because one of the alleles was mutated at conception with the other undergoing somatic mutation. In sporadic retinoblastoma both mutations occur somatically within one cell to give rise to a tumour clone[69,70]. Independent sporadic mutations have been detected within one family[71].

The *RB1* gene is defective in all retinoblastomas and in a number of other cancers. Inactive *RB1* alleles are very common in small cell lung carcinoma, and they occur in 20–30% of non-small cell lung cancers[72], bladder and pancreatic carcinomas and in human breast carcinomas and cell lines derived therefrom[73–76]. Individuals with inherited retinoblastoma are also susceptible to malignant tumours in mesenchymal tissues, often osteosarcomas or soft tissue sarcomas.

The neoplastic phenotype of retinoblastoma, osteosarcoma, small cell lung carcinoma or human prostate carcinoma cells carrying inactivated *RB1* genes can be suppressed by transfection of a cloned *RB1* gene[77] and transfection of the *RB1* gene also arrests the growth of normal cells[78]. pRb suppresses transformation and metastasis induced in fibroblasts by *Neu*.

### Senescence

Depletion of pRb by the use of an anti-sense oligomer in fibroblasts that are about to senesce extends the lifespan of the cells [79], an effect potentiated by co-treatment with anti-sense *P53* oligomer. SV40-immortalized fibroblasts become senescent when expression of T antigen is inhibited [80]. Senescent hamster embryo cells express only unphosphorylated pRb [81] which inhibits proliferation.

### Transgenic animals

Mice carrying a homozygous mutation in the retinoblastoma gene die before the 16th day of gestation, indicating that *Rb-1* is essential for normal mouse development but not for cell proliferation and differentiation in the early stages of organogenesis [48,49,82,83]. Heterozygous mice appear normal but almost invariably develop spontaneous pituitary tumours within the first 11 months [84,85]. Fibroblasts from *Rb-1*$^{-/-}$ mice have a decreased $G_1$ phase and markedly increased levels of cyclin E and CDK2 activity [86]. The expression in transgenic mice of T antigen, which binds to pRb, in the retina causes hereditable ocular tumours apparently identical to those of retinoblastoma [87]. When pRb is inactivated by the expression of HPV E7, photoreceptor cells undergo apoptosis, retinoblastoma developing only when *P53* expression is suppressed [88]. pRb also decreases apoptosis in SAOS-2 cells exposed to radiation and pRb may have a general role in protecting cells from apoptosis, induced for example by interferon $\gamma$ or TGF$\beta_1$, that is inhibited by proteins such as E1A, E2F or MYC [89–91].

### *In vitro*

p107 associates with E2F4 which, when over-expressed with oncogenic *Ras*, transforms rat embryo fibroblasts [92]. E2F1, E2F2 or E2F3 transform immortal but otherwise untransformed rodent cells [93].

## Gene structure

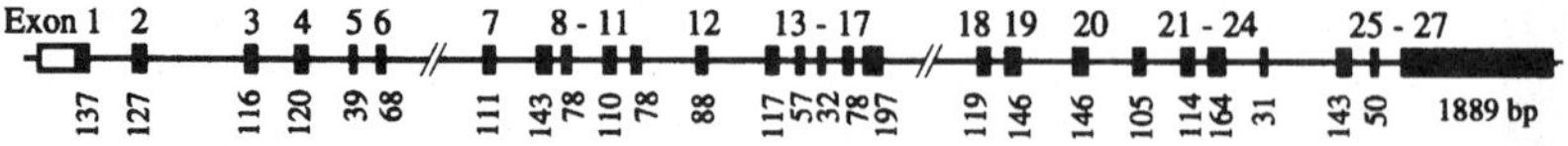

Exon 1 2 3 4 5 6 // 7 8 - 11 12 13 - 17 // 18 19 20 21 - 24 25 - 27

137 127 116 120 39 68 111 143 78 110 78 88 117 57 32 78 197 119 146 146 105 114 164 31 143 50 1889 bp

## Transcriptional regulation

The region extending 600 bases upstream from the initiation codon includes potential AP1, E2F/DRTF1 and ATF (CREB family) binding sites and three SP1 sites to which RBF1 (distinct from SP1) binds. RBF-1 recognizes a sequence (5′-GGCGGAAGT-3′) specific for ETS family transcription factors [94]. BCL3 promotes *RB1* transcription by interaction with E4TF1 that binds to the (+72 to +85) CGGAAGT site [95]. E2F1 *trans*-activates the *RB1* promoter *via* an E2F recognition sequence. *RB1* expression has been shown to be negatively regulated by its own gene product acting *via* E2F [96] and to be positively autoregulated *via* ATF-2 [97]. Point mutations in the E2F site that inhibit E2F binding stimulate the *RB1* promoter activity, indicating that a silencer factor(s) may bind at this site [98].

During myogenic differentiation of murine cells MyoD enhances *Rb-1* gene transcription although it does not bind directly to an E-box DNA sequence motif present in the *Rb-1* promoter[99].

## Structure of human pRb

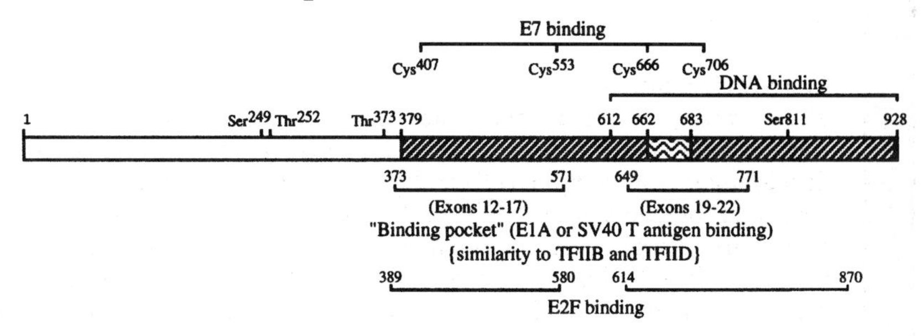


The E1A and SV40 T antigen binding regions overlap sites of naturally occurring mutations, notably in exon 21 (bladder carcinoma and small cell lung carcinoma), and these regions also mediate E2F binding[100,101].

## Amino acid sequences of human pRb

```
  (1) MPPKTPRKTA ATAAAAAAEP PAPPPPPPPE EDPEQDSGPE DLPLVRLEFE
 (51) ETEEPDFTAL CQKLKIPDHV RERAWLTWEK VSSVDGVLGG YIQKKKELWG
(101) ICIFIAAVDL DEMSFTFTEL QKNIEISVHK FFNLLKEIDT STKVDNAMSR
(151) LLKKYDVLFA LFSKLERTCE LIYLTQPSSS ISTEINSALV LKVSWITFLL
(201) AKGEVLQMED DLVISFQLML CVLDYFIKLS PPMLLKEPYK TAVIPINGSP
(251) RTPRRGQNRS ARIAKQLEND TRIIEVLCKE HECNIDEVKN VYFKNFIPFM
(301) NSLGLVTSNG LPEVENLSKR YEEIYLKNKD LDARLFLDHD KTLQTDSIDS
(351) FETQRTPRKS NLDEEVNVIP PHTPVRTVMN TIQQLMMILN SASDQPSENL
(401) ISYFNNCTVN PKESILKRVK DIGYIFKEKF AKAVGQGCVE IGSQRYKLGV
(451) RLYYRVMESM LKSEEERLSI QNFSKLLNDN IFHMSLLACA LEVVMATYSR
(501) STSQNLDSGT DLSFPWILNV LNLKAFDFYK VIESFIKAEG NLTREMIKHL
(551) ERCEHRIMES LAWLSDSPLF DLIKQSKDRE GPTDHLESAC PLNLPLQNNH
(601) TAADMYLSPV RSPKKKGSTT RVNSTANAET QATSAFQTQK PLKSTSLSLF
(651) YKKVYRLAYL RLNTLCERLL SEHPELEHII WTLFQHTLQN EYELMRDRHL
(701) DQIMMCSMYG ICKVKNIDLK FKIIVTAYKD LPHAVQETFK RVLIKEEEYD
(751) SIIVFYNSVF MQRLKTNILQ YASTRPPTLS PIPHIPRSPY KFPSSPLRIP
(801) GGNIYISPLK SPYKISEGLP TPTKMTPRSR ILVSIGESFG TSEKFQKINQ
(851) MVCNSDRVLK RSAEGSNPPK PLKKLRFDIE GSDEADGSKH LPGESKFQQK
(901) LAEMTSTRTR MQKQKMNDSM DTSNKEEK (928)
```

### Domain structures

| | |
|---|---|
| 10–18 | Poly-Ala |
| 20–29 | Poly-Pro |
| 37–140 | Promote insulin receptor gene expression *via* interaction with GA and GC boxes[110] |
| 301–372 | hsc73 (heat shock cognate protein) binding site |
| 373–771 | Binding pocket (binds T antigen and E1A) |
| 373–579 | Domain A |

| | |
|---|---|
| 379–928 | The minimal region necessary for growth suppression by pRb[111] |
| 407, 553, 666 and 706 | Four of eight Cys residues in the "binding pocket": they are involved in E7 binding |
| 580–639 | Spacer |
| 612–928 | DNA binding activity[112] |
| 640–771 | Domain B. The A and B pockets can interact directly to form the repressor motif that binds to E2F proteins to inhibit their *trans*-activating function[113] |
| 662–683 | Leucine zipper motif (bold, underlined) |
| 249–255, 354–359, 372–376, 567–568, 607–615, 787–791, 794–798, 806–814, 820–824, 826–827 | Potential serine/threonine phosphorylation sites conserved between human and mouse (underlined) |
| 792–928 | MDM2 binding region |

**Mutations detected in *RB1***

Major deletions in the gene occur in 15–40% of retinoblastomas, point mutations being scattered throughout the gene with exons 3, 8, 18 and 19 preferentially altered. More subtle effects are presumed to occur in all other cases, which may include (i) hypermethylation in the 5′ region of the gene which inhibits binding of ATF-like and RBF1 transcription factors[102], (ii) point mutations in the ATF and SP1 sites that are known to cause hereditary retinoblastoma[103,104], (iii) a point mutation that results in the loss of exon 21 and inactivation of the protein, (iv) $Cys^{706}$ to $Phe^{706}$ mutation in a small cell lung carcinoma line resulting in an under-phosphorylated protein that does not bind to SV40 T Ag or E1A[105] but retains the capacity to bind MYC and MYCL proteins[56,106] and (v) loss of 103 nucleotides in the promoter preventing expression of *RB1* in human prostate tumours[107]. The germline point mutation $Arg^{661} \rightarrow$ Trp has been detected in a family with incomplete penetrance of familial retinoblastoma: the protein retains the wild-type properties of nuclear localization, capacity to undergo hyperphosphorylation and suppress growth of $Rb^-$ cells but has defective binding pocket activity[108]. Three mutations have been detected in exons 8 and 18 ($Arg^{255} \rightarrow$ Stop; $Arg^{578} \rightarrow$ Stop) and in intron 16[109].

**Database accession numbers**

| | *PIR* | *SWISSPROT* | *EMBL/GENBANK* | *REFERENCES* |
|---|---|---|---|---|
| Human *RB1* | A39947 | RB_HUMAN, P06400 | M15400, M28419 | 114,115 |
| | JS0276 | | | 74 |
| | | | M33647 | 116 |
| | | | HSRB1G, X16439 | |

***References***


1 **Knudson, A.G. (1993) Proc. Natl Acad. Sci. USA 90, 10914–10921.**

2 Ewen, M.E. et al. (1991) Cell 66, 1155–1164.

3 Zhu, L. et al. (1993) Genes Dev. 7, 1111–1125.

4 Li, Y. et al. (1993) Genes Dev. 7, 2366–2377.

[5] Du, W. et al. (1996) Genes Dev. 10, 1206–1218.
[6] Pan, H. and Griep, A. (1994) Genes Dev. 8, 1285–1299.
[7] Whyte, P. et al. (1988) Nature 334, 124–129.
[8] Weintraub, S.J. et al. (1995) Nature 375, 812–815.
[9] Dowdy, S.F. et al. (1993) Cell 73, 499–511.
[10] Ewen, M.E. et al. (1993) Cell 73, 487–497.
[11] Kato, J. et al. (1993) Genes Dev. 7, 331–342.
[12] Kim, S.-J. et al. (1992) Nature 358, 331–334.
[13] Szekely, L. et al. (1993) Proc. Natl Acad. Sci. USA 90, 5455–5459.
[14] Ouellette, M.M. et al. (1992) Oncogene 7, 1075–1081.
[15] Chellappan, S.P. et al. (1992) Proc. Natl Acad. Sci. USA 89, 4549–4553.
[16] Shao, Z. et al. (1995) Proc. Natl Acad. Sci. USA 92, 3115–3119.
[17] Cavanaugh, A.H. et al. (1995) Nature 374, 177–180.
[18] Singh, P. et al. (1995) Nature 374, 562–565.
[19] Inoue, A. (1995) J. Biol. Chem. 270, 22571–22576.
[20] Xiao, Z.-X. et al. (1995) Nature 375, 694–698.
[21] Chen, P.-L. et al. (1996) Proc. Natl Acad. Sci. USA 93, 465–469.
[22] Craven, R.J. et al. (1995) Cancer Res. 55, 3969–3972.
[23] Helin, K. et al. (1992) Cell 70, 337–350.
[24] Kaelin, W.G. et al. (1992) Cell 70, 351–364.
[25] Huang, S. et al. (1991) Nature 350, 160–162.
[26] Qian, Y.-W. and Lee, E.Y.-H.P. (1995) J. Biol. Chem. 270, 25507–25513.
[27] Otterson, G.A. et al. (1993) Oncogene 8, 949–957.
[28] Fattaey, A.R. et al. (1993) Oncogene 8, 3149–3156.
[29] Buyse, I.M. et al. (1995) Proc. Natl Acad. Sci. USA 92, 4467–4471.
[30] Sakai, Y. et al. (1995) Genomics 30, 98–101.
[31] Ozaki, T. et al. (1994) Oncogene 9, 2649–2653.
[32] Hateboer, G. et al. (1993) Proc. Natl Acad. Sci. USA 90, 8489–8493.
[33] Hu, Q. et al. (1992) Mol. Cell. Biol. 12, 971–980.
[34] Qian, Y.-W. et al. (1993) Nature 364, 648–652.
[35] Durfee, T. et al. (1993) Genes Dev. 7, 555–569.
[36] Dunaief, J.L. et al. (1994) Cell 79, 119–130.
[37] Yavuzer, U. et al. (1995) Oncogene 10, 123–134.
[38] Radulescu, R.T. et al. (1995) Biochem. Biophys. Res. Commun. 206, 97–102.
[39] Zhu, L. et al. (1995) Mol. Cell. Biol. 15, 3552–3562.
[40] Shen et al. (1995) J. Biol. Chem. 270, 20525–20529.
[41] Dalton, S. et al. (1992) EMBO J. 11, 1797–1804.
[42] Rohde, M. et al. (1996) Oncogene 12, 2393–2401.
[43] Chen, L.I. et al. (1994) Mol. Cell. Biol. 14, 4380–4389;
[44] Udvadia, A.J. et al. (1995) Proc. Natl Acad. Sci. USA 92, 3953–3957.
[45] Schonthal, A. and Feramisco, J.R. (1993) Oncogene 8, 433–441.
[46] Alberts, A.S. et al. (1993) Proc. Natl Acad. Sci. USA 90, 388–392.
[47] Ludlow, J.W. et al. (1993) Oncogene 8, 331–339.
[48] Zhang, W. et al. (1992) Biochem. Biophys. Res. Commun. 184, 212–216.
[49] Hamel, P.A. et al. (1992) Mol. Cell. Biol. 12, 3431–3438.
[50] Pagano, M. et al. (1992) Oncogene 7, 1681–1686.
[51] Suzuki-Takahashi, I. et al. (1995) Oncogene 10, 1691–1698.
[52] Lin, B.T.-Y. et al. (1991) EMBO J. 10, 857–864.
[53] Williams, R.T. et al. (1992) Oncogene 7, 423–432.

[54] Hu, Q. et al. (1992) Mol. Cell. Biol. 12, 971–980.
[55] Weintraub, S.J. et al. (1992) Nature 358, 259–261.
[56] Ray, S.K. et al. (1992) Mol. Cell. Biol. 12, 4327–4333.
[57] Zacksenhaus, E. et al. (1996) EMBO J. 15, 5917–5927.
[58] Shirodkar, S. et al. (1992) Cell 68, 157–166.
[59] Schwarz, J.K. et al. (1993) EMBO J. 12, 1013–1020.
[60] Grana, X. et al. (1994) Proc. Natl Acad. Sci. USA 91, 3834–3838.
[61] Bates, S. et al. (1994) Oncogene 9, 1633–1640.
[62] Müller, H. et al. (1994) Proc. Natl Acad. Sci. USA 91, 2945–2949.
[63] Hatakeyama, M. et al. (1994) Genes Dev. 8, 1759–1771.
[64] Xiao, Z.-X. et al. (1995) Nature 375, 694–698.
[65] Martin, K. et al. (1995) Nature 375, 691–694.
[66] White, R.J. et al. (1996) Nature 382, 88–90.
[67] Williams, B.O. et al. (1994) Nature Genet. 7, 480–484.
[68] Sellers, W.R. et al. (1995) Proc. Natl Acad. Sci. USA 92, 11544–11548.
[69] **Goodrich, D.W. and Lee, W.H. (1993) Biochim. Biophys. Acta 1155, 43–61.**
[70] **Zacksenhaus, E. et al. (1993) Adv. Cancer Res. 61, 115–141.**
[71] Bia, B. and Cowell, J.K. (1995) Oncogene 11, 977–979.
[72] Reissmann, P.T. et al. (1993) Oncogene 8, 1913–1919.
[73] Horowitz, J.M. et al. (1990) Proc. Natl Acad. Sci. USA 87, 2775–2779.
[74] Friend, S.H. et al. (1987) Proc. Natl Acad. Sci. USA 84, 9059–9063.
[75] Weichselbaum, R.R. et al. (1988) Proc. Natl Acad. Sci. USA 85, 2106–2109.
[76] Ruggeri, B. et al. (1992) Oncogene 7, 1503–1511.
[77] Ookawa, K. et al. (1993) Oncogene 8, 2175–2181.
[78] Fung, Y.-K.T. et al. (1993) Oncogene 8, 2659–2672.
[89] Hara, E. et al. (1991) Biochem. Biophys. Res. Commun. 179, 528–534.
[80] Shay, J.W. et al. (1991) Exp. Cell Res. 196, 33–39.
[81] Futreal, P.A. and Barrett, J.C. (1991) Oncogene 6, 1109–1113.
[82] Lee, E.Y.-H.P. et al. (1992) Nature 359, 288–294.
[83] Jacks, T. et al. (1992) Nature 359, 295–300.
[84] Hu, N. et al. (1994) Oncogene 9, 1021–1027.
[85] Nikitin, A.Y. and Lee, W.-H. (1996) Genes Dev. 10, 1870–1879.
[86] Herrera, R.E. et al. (1996) Mol. Cell. Biol. 16, 2402–2407.
[87] Windle, J.J. et al. (1990) Nature 343, 665–669.
[88] Howes, K.A. et al. (1994) Genes Dev. 8, 1300–1310.
[89] Haas-Kogan, D.A. et al. (1995) EMBO J. 14, 461–472.
[90] Berry, D.E. et al. (1996) Oncogene 12, 1809–1819.
[91] Fan, G. et al. (1996) Oncogene 12, 1909–1919.
[92] Beijersbergen, R.L. et al. (1995) Genes Dev. 9, 1340–1353.
[93] Xu, G. et al. (1995) Proc. Natl Acad. Sci. USA 92, 1357–1361.
[94] Savoysky, E. et al. (1994) Oncogene 9, 1839–1846.
[95] Shiio, Y. et al. (1996) Oncogene 12, 1837–1845.
[96] Shan, B. et al. (1994) Mol. Cell. Biol. 14, 299–309.
[97] Park, K. et al. (1994) J. Biol. Chem. 269, 6083–6088.
[98] Ohtani-Fujita, N. et al. (1994) Oncogene 9, 1703–1711.
[99] Martelli, F. et al. (1994) Oncogene 9, 3579–3590.
[100] Qian, Y. et al. (1992) Mol. Cell. Biol. 12, 5363–5372.
[101] Hiebert, S.W. et al. (1993) Mol. Cell. Biol. 13, 3384–3391.
[102] Ohtani-Fujita, N. et al. (1993) Oncogene 8, 1063–1067.
[103] Sakai, T. et al. (1991) Nature 353, 83–86.

[104] Cowell, J.K. et al. (1996) Oncogene 12, 431–436.
[105] Kaye, F.J. et al. (1990) Proc. Natl Acad. Sci. USA 87, 6922–6926.
[106] Kratzke, R.A. et al. (1992) J. Biol. Chem. 267, 25998–26003.
[107] Bookstein, R. et al. (1990) Proc. Natl Acad. Sci. USA 87, 7762–7766.
[108] Kratzke, R.A. et al. (1994) Oncogene 9, 1321–1326.
[109] Blanquet, V. et al. (1994) Hum. Mol. Genet. 3, 1185–1186.
[110] Shen, W. et al. (1995) J. Biol. Chem. 270, 20525–20529.
[111] Qin, X.-Q. et al. (1992) Genes Dev. 6, 953–964.
[112] Wang, N.P. et al. (1990) Cell Growth Differ. 1, 233–239.
[113] Chow, K.N.B. and Dean, D.C. (1996) Mol. Cell. Biol. 16, 4862–4868.
[114] Lee W.-H. et al. (1987) Science 235, 1394–1399.
[115] Lee W.-H. et al. (1987) Nature 329, 642–645.
[116] T'Ang, A. et al. (1989) Oncogene 4, 401–407.

# TGFBR1, TGFBR2

## Identification

The transforming growth factor $\beta$ (TGF$\beta$) family of peptide growth factors includes five highly conserved genes (*TGFB1* to *TGFB5*), the products of which form homodimers of ~25 kDa [1–3]. Three cell surface proteins have been identified by their ability to bind and be cross-linked by radioiodinated TGF$\beta$: type I and type II receptors (*TGFBR1* and *TGFBR2*) and type III TGF$\beta$ receptor (also called $\beta$-glycan, detected in rats), bind TGF$\beta_1$ and TGF$\beta_2$ with high affinity and may concentrate ligand for presentation to the type II receptor [4].

## Related genes

The cysteine-rich extracellular domain has a region of relatively high identity with DAF-1 and the activin receptor (ActRII) but no homology with EGFR or PDGFR. Porcine and human TGFBR2 are 88% identical.

| | ***TGFBR1*** | ***TGFBR2*** |
|---|---|---|
| **Chromosome** | 9q33–34.1 | 3p21.3–22 |
| **Mass (kDa): predicted** | 44 | 60 |
| **expressed** | 53 | 80 |
| **Cellular location** | Plasma membrane | Plasma membrane |
| **Tissue distribution** | Ubiquitous | Ubiquitous |

## Protein function

The two TGF$\beta$ receptors that mediate the actions of members of the TGF$\beta$ family are transmembrane serine/threonine kinases. TGF$\beta$ binds directly to TGFBR2, a constitutively active kinase, and is then recognized by TGFBR1 which is directly phosphorylated and activated by TGFBR2 [5,6]. TGFBR2 is multiply phosphorylated both by itself and by other kinases, irrespective of ligand binding, whereas the phosphorylation state of TGFBR1 is dramatically increased within 2 min of TGF$\beta$ addition by the action of TGFBR2. The receptor complex required for signalling by TGF$\beta$ comprises TGFBR1 homodimers that form heterodimers with TGFBR2 and cells defective in either receptor are refractory to any TGF$\beta$ effects [7].

It is evident from this mechanism that the signalling pathways activated by TGF$\beta$ emanate from TGFBR1, although as yet the nature of these pathways is unresolved. However, there is evidence that RAS is involved and that TGF$\beta$ may elevate GTP-RAS levels [8]. Thus, activated RAS proteins are equivalent to TGF$\beta$ during myogenic differentiation [9], to the TGF$\beta$-related factor activin during mesoderm formation [10] and to the effect of genes activated by TGF$\beta$ in cardiac muscle [11,12]. Co-transfection of TGF$\beta$-responsive reporter genes and *Ras* expression vectors into cardiac myocytes and mink lung epithelial cells indicates that dominant inhibitory RAS mutants (Asn$^{17}$) inhibit TGF$\beta$-dependent expression whereas constitutively active RAS (Arg$^{12}$) upregulates expression[13]. RAS also links TGF$\beta_1$ or TGF$\beta_2$ signalling to the activation of ERK1[14].

A member of the mitogen-activated protein kinase kinase kinase (MAPKKK) family, TAK1 (TGF$\beta$-activated kinase 1), is involved in the signalling pathway activated by members of the TGF$\beta$ superfamily, and TAB1 and TAB2 interact with TAK1[15]. The MAP kinase ERK1 and the small GTP binding protein RAC may be involved in the signalling pathway[16]. SMAD1/hMAD1/MADR1, a human homologue of *Drosophila* Mad, is a transcription factor that is activated by ligand binding to TGF$\beta$ receptors and translocates to the nucleus upon phosphorylation[17]. Other MAD-related human genes have been identified and somatic mutations have been detected in one (*MADR2*) in colorectal carcinomas[18–20]. DPC4/SMAD4 and SMAD3 are also components of the TGF$\beta$ signalling pathway (see ***DPC4/SMAD4***).

At least two TGFBR1 subunits can be present in the same TGF$\beta$-induced receptor complex. Mutations have been detected in the kinase domain of TGFBR1 that either inactivate the kinase or prevent ligand-induced TGFBR1 phosphorylation and thus activation by TGFBR2. A kinase-defective TGFBR1 can functionally complement an activation-defective TGFBR1 mutant by rescuing its TGFBR2-dependent phosphorylation[21]. The immunophilin FKBP12 interacts specifically with TGFBR1 to inhibit signalling and is released following ligand-induced TGFBR1 phosphorylation of TGFBR1[22].

In endothelial cells endoglin is the most abundant TGF$\beta$ binding protein, associating with TGFBR1 or TGFBR2 in the presence of TGF$\beta$. Mutations in endoglin are responsible for the autosomal dominant disorder hereditary haemorrhagic telangiectasia (HHT) which results in multisystemic vascular dysplasia and recurrent haemorrhage[23]. Farnesyl transferase-$\alpha$ interacts with TGFBR1 but it is not evident that this is functionally significant[24].

## General properties of TGF$\beta$

TGF$\beta$ is a ubiquitously expressed paracrine polypeptide, three highly homologous forms (TGF$\beta_1$ (*TGFB1*), TGF$\beta_2$ (*TGFB2*), and TGF$\beta_3$ (*TGFB3*)) having been detected in humans (normal plasma concentrations: 4.1 ng/ml (TGF$\beta_1$), <0.2 ng/ml (TGF$\beta_2$), <0.1 ng/ml (TGF$\beta_3$)) and other mammals. Dinucleotide repeat polymorphism giving rise to five alleles has been detected in *TGFB2*[25]. Each isoform inhibits growth of a wide range of normal and transformed cells (epithelial, endothelial, fibroblast, neuronal, lymphoid and haematopoietic), lung epithelial cells and keratinocytes being most susceptible. In some cells (fibroblasts, osteoclasts) TGF$\beta$ can act as a mitogen, probably by stimulating the release of autocrine factors (e.g. PDGF, see below). However, in general, TGF$\beta$ inhibits cell cycle progression by lengthening or arresting the $G_1$ phase[26], although in vascular smooth muscle cells it greatly lengthens $G_2$ with no significant effect on $G_1$. TGF$\beta$ can also positively or negatively affect differentiation, for example, initiating growth arrest which precedes differentiation of epithelial cells[27] or under certain conditions repressing MyoD and myogenin transcription and the differentiation of myoblasts[28,29]. TGF$\beta_1$ is a potent inhibitor of several differentiated functions of adrenal cells and this effect has been reversed by treatment with anti-sense oligonucleotide complementary to TGF$\beta_1$ mRNA[30].

Consistent with these observations and supplementary to its effects on the expression of extracellular matrix proteins, TGF$\beta$ represses a variety of growth-related genes in various cell types (*KC, JE, JUN, JUNB, MYC, MYCN,* TGF$\beta$), although in some primary cells TGF$\beta$ prevents serum-stimulated DNA synthesis without blocking the activation of *JE* and *MYC*[31]. However, TGF$\beta$ induces sustained expression of *Fos* mRNA in EL2 rat fibroblasts and NIH 3T3 fibroblasts although in the latter cells FOS protein remains undetectable[32]. In most cell types (epithelial cells, endothelial cells, keratinocytes, colon/breast carcinoma cells) it seems probable that transcriptional repression of *MYC* may be a crucial component of the growth-inhibiting effects of TGF$\beta$. This may be generally mediated by pRb, which acts through the retinoblastoma control element (RCE) to repress *MYC* transcription in human keratinocytes, to repress *FOS* transcription, to enhance transcription of *JUN* and insulin-like growth factor II (IGF-II) and to autoregulate either positively or negatively the expression of *TGFB1* itself depending on cell type. Thus, TGF$\beta$ does not suppress *MYC* in DNA tumour virus-transformed cells when pRb is inactivated by tumour virus oncoproteins but, when added to lung epithelial cells in late $G_1$, TGF$\beta$ inhibits pRb phosphorylation (permissive for progression through the cell cycle) and arrests growth. pRb is phosphorylated *in vitro* by cyclin D1/CDK4, cyclin E/CDK2 and cyclin A/CDK2 and phosphorylation by any of these complexes disrupts E2F/pRb complexes and reduces transcriptional repression. In keratinocytes and epithelial cells TGF$\beta_1$ regulates the activity of cyclin-CDK/CDC complexes by suppressing the synthesis of CDK4 and CDK2 and by inhibiting cyclin A synthesis[33,34]. Constitutive synthesis of CDK4, but not CDC2, overcomes TGF$\beta_1$ growth inhibition.

TGF$\beta$ induces expression of *PDGFA* and *PDGFB* that are responsible for the autocrine stimulation of DNA synthesis in smooth muscle cells and immortalized cell lines caused by TGF$\beta$. A similar autocrine loop may occur in melanoma cells which secrete TGF$\beta$, PDGFA and PDGFB: however, these cells have also been shown to secrete TGF$\alpha$, bFGF GM-CSF, melanoma growth stimulatory activity (MGSA), IL-1, IL-6 and IL-8 and the roles of these cytokines as autocrine factors and in modulating the behaviour of normal host cells remain unclear[35]. High concentrations of TGF$\beta$ inhibit expression of *PDGFRA* and thus mitogenesis. In fibroblasts TGF$\beta$ induces increased glycolysis, amino acid uptake, Ins$P_3$ accumulation, prostaglandin $E_1$ synthesis and increase in $Ca^{2+}$ influx[36], presumably as consequences of the effect of PDGF. In vascular endothelial cells elevation of the concentration of cAMP inhibits TGF$\beta$-induced *PDGFB* expression, as does pertussis or cholera toxin in fibroblasts[37]. However, pertussin toxin (PT) does not block TGF$\beta_1$-stimulated expression of fibronectin and collagen mRNA in these cells, indicating that the TGF$\beta$ receptor (see below) may activate more than one signalling pathway.

TGF$\beta$ repression of the rat stromelysin gene is mediated by a TGF$\beta$ inhibitory element (TIE) in the stromelysin promoter, a TIE binding protein complex being induced by treatment of rat fibroblasts with TGF$\beta$[38]. This protein complex includes FOS and induction of *Fos* by TGF$\beta$ is required for the repressive effects of TGF$\beta$ on stromelysin gene expression. *Fos* induction

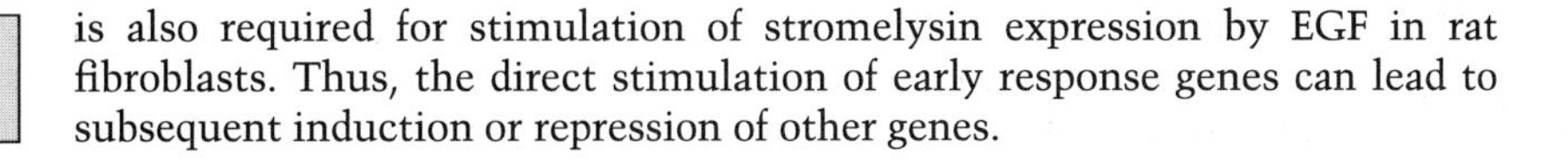


is also required for stimulation of stromelysin expression by EGF in rat fibroblasts. Thus, the direct stimulation of early response genes can lead to subsequent induction or repression of other genes.

## TGF$\beta$ and metastasis

As summarized above, the TGF$\beta$ isoforms are potent inhibitors of proliferation for many types of cell both *in vitro* and *in vivo*. However, TGF$\beta$ is also secreted by a variety of tumour cells and its over-production frequently correlates with abnormal cell adhesion properties and increased metastasis *in vivo*[1–3]. A series of studies using model systems indicates that as tumour cells attain metastatic competence their responsiveness to growth factors may alter. In particular, they may acquire susceptibility to aberrant growth stimulation by TGF$\beta$[39–42] and IL-6 may become a less powerful proliferation inhibitor for more aggressive cell lines[43].

Active and latent forms of TGF$\beta$ have been detected in conditioned medium from rat colon carcinoma cells and murine melanoma cells[44] and in head and neck cancers lymph node lymphocytes adjacent to the tumour express increased TGF$\beta$ mRNA[45]. TGF$\beta$ stimulates the proliferation of a regressive adenocarcinoma cell line[46] and promotes invasiveness of lung tumour cells both *in vitro* and *in vivo*[47,48]. It also modulates detachment *in vitro* of rat prostate cancer cells, which show increased expression of TGF$\beta$ compared with normal cells that may reflect an autocrine mechanism promoting growth[49,50]. Furthermore, a strong correlation between a rapid rate of disease progression and high expression of TGF$\beta_1$ has been detected in human breast cancer[51] although other studies show no significant correlation between elevated levels of TGF$\beta_1$ and advanced metastatic breast cancer[52].

In the majority of human breast cancer cell lines, however, TGF$\beta$ inhibits proliferation, possibly through the stimulation of insulin-like growth factor binding protein (IGFBP)-3 expression[53]. The evidence that growth inhibitory anti-oestrogens (tamoxifen, droloxifen, 3-hydroxytamoxifen) can increase TGF$\beta_1$ synthesis in hormonally responsive cells[54] and that oestradiol and norethindrone promote cell growth but inhibit TGF$\beta_2$ and TGF$\beta_3$ synthesis suggests that TGF$\beta$ may function as a negative autocrine growth factor. Furthermore, in other epithelially derived tumours, loss of TGF$\beta_1$ expression appears to constitute a risk factor for malignant progression[55–57]. Tamoxifen increases levels of TGF$\beta_2$ in the plasma of patients who respond to the anti-oestrogen, consistent with the evidence that TGF$\beta_1$ induces transcription of TGF$\beta_2$[58]. Benign skin tumours induced in *P53*$^{-/-}$ mice have a malignant conversion frequency of ~50% and there is an inverse correlation between TGF$\beta_1$ expression levels in these tumours and the probability of malignant progression[59]. Transgenic mice that over-express TGF$\beta$ do not develop spontaneous tumours and MMTV-TGF$\alpha$/MMTV-TGF$\beta_1$ double transgenics show inhibition of the carcinoma development that is normally caused by expression of the TGF$\alpha$ transgene.

The principal effects of TGF$\beta$ on extracellular matrix proteins are mediated by the stimulated transcription of tissue inhibitors of matrix metalloproteinases (TIMPs), cell adhesion and urokinase-type plasminogen activator (uPA) receptor genes and, in general, the repression of transcription of matrix metalloproteinases (MMPs) and uPA. Thus in the human A549 lung carcinoma cell line, TGF$\beta_1$ causes a large increase in uPAR with a subsequent increase in cell surface plasmin [60]. However, in these cells TGF$\beta_1$ also increases both uPA and plasminogen activator inhibitor-1 (PAI-1) levels and a variety of breast cancer cell lines secrete all three isoforms of TGF$\beta$ together with uPA [61]. Furthermore, in some cells, for example cultured human cervical epithelial cells, TGF$\beta$ increases the transcription and/or enzymatic activity of *MMP-2* and *MMP-9* [62,63]. In human endometrium, the stromelysins are expressed during proliferation-associated remodelling and menstruation-associated breakdown, but they are not expressed during the secretory phase of the cycle: suppression of the epithelial-specific *MMP-7* (matrilysin) is mediated by TGF$\beta$ induced by progesterone [64].

TGF$\beta$ inhibits human T cell and neutrophil binding to endothelium (in contrast to TNF, IL-1 or IFN-$\gamma$) and also inhibits the binding of mouse mastocytoma cells (and spleen, T cells, etc.) but not of melanoma cells. Inhibition of binding is reversed by cycloheximide and blocked by increase in $[Ca^{2+}]_i$ (by ionomycin or thapsigargin) or by the phosphatase inhibitor okadaic acid. Stimulation of PKC or PKA activators has the opposite effect to TGF$\beta$ [65]. Expression of TGF$\beta$ correlates with fibronectin (FN) and FN-R expression in mammary carcinoma cells. TGF$\beta$ protein expression may induce FN/FN-R interaction and it has been reported to correlate with lymph node metastasis [66,67] and to be preferentially expressed at the advancing tumour edges in such metastases [68].

These general observations indicate the inhibitory effect of TGF$\beta$ on cell growth which may be consistent with its frequent association with metastasis if TGF$\beta$ plays a dual role by retarding tumour development but enhancing subsequent progression to malignancy once neoplastic transformation has occurred. This would be consistent with the finding that growth stimulation by TGF$\beta_1$ occurs in advanced colon cancers, including poorly differentiated primary tumours or metastases, whereas TGF$\beta_1$ either inhibits or has no effect on the growth of more differentiated cancers, all types showing the same set of mutations in *APC, DCC, P53* and *RAS* [69]. However, the general conclusion from the available data is that metastasis is probably driven by an appropriate imbalance between facilitating and inhibiting proteins rather than as the result of the activation of a single gene or group of genes.

**Cancer**

*TGFBR2* maps close to or within one of the interstitial deletions that occur in 30–50% of head and neck, breast and small cell lung cancers. Truncated and completely deleted *TGFBR2* mRNAs have been detected in retinoblastoma, hepatoma and gastric cancer cell lines [70–72]. Three mutations in TGFBR2, causing the absence of receptor expression at the cell surface, have been detected in a subset of colon cancer cell lines exhibiting genetic instability caused by mismatch repair [73–75]. Mutations occur in the serine/threonine kinase domain (exons 4–7) in hereditary non-polyposis colorectal carcinoma [76]. Point mutations in TGFBR2 of highly conserved sites in the kinase domain (causing defective autophosphorylation or constitutive kinase activation) have been detected in squamous cell carcinoma lines from head and neck tumours [77]. A point mutation in TGFBR2 has been detected in the clinically aggressive stage of cutaneous T cell lymphoma. Cells isolated from this tumour also express wild-type TGFBR2, indicating that the mutation is of a dominant inhibitory nature, and it appears to block transport of the receptor to the cell surface [78].

In skin biopsies from patients with AIDS-related Kaposi's sarcoma TGFBR1 expression is lost although TGFBR2 is detectable in most samples [79].

Five forms of the TGFBR1-related activin type I receptor have been detected (ALK1-5, activin receptor-like kinase 1-5) and three truncated versions of ALK4 generated by alternative splicing of the kinase domain have been detected in pituitary adenomas. ActRIIB is an activin-specific receptor having a three- to four-fold higher affinity for activin than ActRII and is highly prevalent in pituitary adenomas but expressed at very low levels in normal pituitary tissue [80].

**Transgenic animals**

TGF$\beta_1$ expression in keratinocytes suppresses the development of chemically induced benign tumours. However, TGF$\beta_1$ confers a high frequency of conversion to malignancy in those tumours that do develop that is associated with an epithelial to mesenchymal transition and the formation of spindle cell carcinomas [81].

***In vitro***

Transfection of *TGFBR2* into human breast cancer MCF-7 cells causes reversion of malignancy in nude mice [82] and reduction of tumorigenicity of colon carcinoma cells [83]. Loss of TGFBR1 function occurs in the LNCaP human prostate and colon carcinoma cell lines: transfection with *TGFBR1* cDNA restores TGF$\beta_1$ inhibition of proliferation [84,85].

## Gene structure of *TGFBR2*

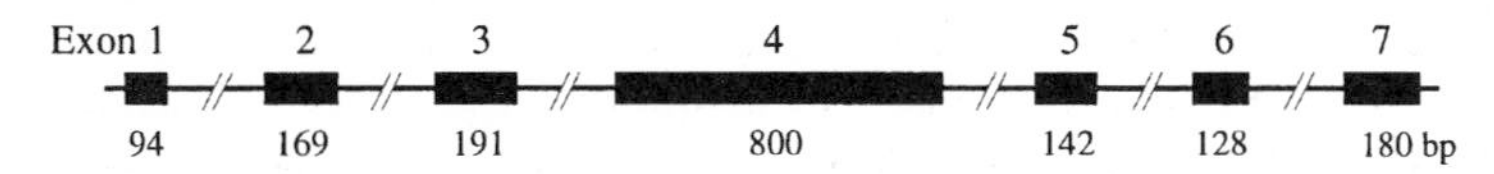


Exon 7 contains a 51 bp untranslated 3′ region [75].

## Amino acid sequence of human TGFBR1

```
  1 MPPSGLRLLL LLLPLLWLLV LTPGRPAAGL STCKTIDMEL VKRKRIEAIR
 51 GQILSKLRLA SPPSQGEVPP GPLPEAVLAL YNSTRDRVAG ESAEPEPEPE
101 ADYYAKEVTR VLMVETHNEI YDKFKQSTHS IYMFFNTSEL REAVPEPVLL
151 SRAELRLLRL KLKVEQHVEL YQKYSNNSWR YLSNRLLAPS DSPEWLSFDV
201 TGVVRQWLSR GGEIEGFRLS AHCSCDSRDN TLQVDINGFT TGRRGDLATI
251 HGMNRPFLLL MATPLERAQH LQSSRHRRAL DTNYCFSSTE KNCCVRQLYI
301 DFRKDLGWKW IHEPKGYHAN FCLGPCPYIW SLDTQYSKVL ALYNQHNPGA
351 SAAPCCVPQA LEPLPIVYYV GRKPKVEQLS NMIVRSCKCS (390)
```

**Domain structure**

| | |
|---|---|
| 1–23 | Putative signal sequence |
| 24–278 | Propeptide |
| 279–390 | TGF$\beta_1$ |

| | |
|---|---|
| 285–294, 293–356, 22–387, 326–389 | Disulfide bonds |
| 355 | Interchain disulfide bond |
| 82, 136, 176 | Consensus *N*-linked glycosylation sites |
| Variants | 10 (L → P), 159 (R → RR)[86] |

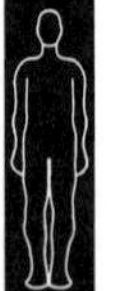

## Amino acid sequence of human TGFBR2

```
  1 MGRGLLRGLW PLHIVLWTRI ASTIPPHVQK SVNNDMIVTD NNGAVKFPQL
 51 CKFCDVRFST CDNQKSCMSN CSITSICEKP QEVCVAVWRK NDENITLETV
101 CHDPKLPYHD FILEDAASPK CIMKEKKKPG ETFFMCSCSS DECNDNIIFS
151 EEYNTSNPDL LLVIFQVTGI SLLPPLGVAI SVIIIFYCYR VNRQQKLSST
201 WETGKTRKLM EFSEHCAIIL EDDRSDISST CANNINHNTE LLPIELDTLV
251 GKGRFAEVYK AKLKQNTSEQ FETVAVKIFP YEEYASWKTE KDIFSDINLK
301 HENILQFLTA EERKTELGKQ YWLITAFHAK GNLQEYLTRH VISWEDLRKL
351 GSSLARGIAH LHSDHTPCGR PKMPIVHRDL KSSNILVKND LTCCLCDFGL
401 SLRLDPTLSV DDLANSGQVG TARYMAPEVL ESRMNLENAE SFKQTDVYSM
451 ALVLWEMTSR CNAVGEVKDY EPPFGSKVRE HPCVESMKDN VLRDRGRPEI
501 PSFWLNHQGI QMVCETLTEC WDHDPEARLT AQCVAERFSE LEHLDRLSGR
551 SCSEEKIPED GSLNTTK (567)
```

### Domain structure

| | |
|---|---|
| 1–23 | Putative signal sequence (cleavage after Thr[23]) |
| 70, 94, 154 | Consensus *N*-linked glycosylation sites |
| 84–101 | Homology with the TGF$\beta$ antagonist fetuin (TGF$\beta$ receptor II homology 1 domain (TRH1): underlined)[87] |
| 167–187 | Transmembrane domain |
| 244–544 | Kinase domain |
| 277 | ATP binding site |
| 379 | Active site |
| 246–542 | Serine/threonine kinase domain |
| 203, 206, 286, 401, 421, 441, 458, 485, 546, 563 | Consensus protein kinase C phosphorylation sites |

### Mutations detected in *TGFBR2*

Deletions within an (A)10 repeat occur in primary colorectal adenocarcinomas, colon and gastric cancer cell lines and endometrial cancers[72–74]. Deletion in both alleles occurs with high frequency. Heterozygous deletions appear to be frequently associated with missense mutations at other locations within *TGFBR2*. Point mutations detected in hereditary non-polyposis colorectal carcinoma include Ser$^{441}$ → Pro, Arg$^{528}$ → Cys and Arg$^{537}$ → Cys/His[75] and in cutaneous T cell lymphoma Asp$^{404}$ → Gly[77]. Mutations have also been detected in colon carcinoma cell lines (a 2 bp insertion at codon 533 causing a frameshift and point mutations (Leu$^{452}$ → Pro and Leu$^{454}$ → Pro))[72–74] and in head and neck squamous cell carcinoma lines[75].

Mutations in *SMAD2* in colorectal carcinoma: 42 bp deletion (codons 345–358)[18]; Arg$^{133}$ → Cys; Leu$^{440}$ → Arg; Pro$^{445}$ → His; Asp$^{450}$ → Glu[19].

## Database accession numbers

| | PIR | SWISSPROT | EMBL/GENBANK | REFERENCES |
|---|---|---|---|---|
| Human *TGFBR2* | A42100 | P37173, TGR2_HUMAN | HSTGFBIIR M85079 | 88 |
| Human *SMAD1* | | | U59423 | 89 |
| Human *SMAD2* | | | U59911, U65019 | 19,20 |
| Human *TGFBR1* | A01395, WFHU2, A22290, A22290, A27513, A27513 | P01137, TGF1_HUMAN | HSTGFBG1 X05839, X05840, X05844, X05849, X05850, | 86,90 91,92 |

## *References*


1 **Kingsley, D.M. (1994) Genes Dev. 8, 133–146.**
2 **Massagué, J. (1990) Annu. Rev. Cell Biol. 6, 597–641.**
3 **Moses, H.L. et al. (1990) Cell 63, 245–247.**
4 Moustakas, A. et al. (1993) J. Biol. Chem. 268, 22215–22218.
5 Wrana, J.L. et al. (1994) Nature 370, 341–347.
6 **Massagué, J. (1996) Cell 85, 947–950**
7 Luo, K. and Lodish, H.F. (1996) EMBO J. 15, 4485–4496.
8 Sadoshima, J.I. and Izumo, S. (1993) EMBO J. 12, 1681–1692.
9 Payne, P.A. et al. (1987) Proc. Natl Acad. Sci. USA 84, 8956–8960.
10 Whitman, M. and Melton, D.A. (1992) Nature 357, 252–254.
11 Parker, T.G. et al. (1990) J. Clin. Invest. 85, 507–514.
12 Brand, T. et al. (1993) J. Biol. Chem. 268, 11500–11503.
13 Abdellatif, M. et al. (1994) J. Biol. Chem. 269, 15423–15426.
14 Hartsough, M.T. et al. (1996) J. Biol. Chem. 271, 22368–22375.
15 Shibuya, H. et al. (1996) Science 272, 1179–1182.
16 Mucsi, I. et al. (1996) J. Biol. Chem. 271, 16567–16572.
17 Liu, F. et al. (1996) Nature 381, 620–623.
18 Zhang, Y. et al. (1996) Nature 383, 168–172.
19 Riggins, G.J. et al. (1996) Nature Genet. 13, 347–349
20 Eppert, K. et al. (1996) Cell 86, 543–552.
21 Weis-Garcia, F. and Massagué, J. (1996) EMBO J. 15, 276–289.
22 Wang, T. et al. (1996) Cell 86, 435–444.
23 McAllister, K.A. et al. (1994) Nature Genet. 8, 345–351.
24 Ventura, F. et al. (1996) J. Biol. Chem. 271, 13931–13934.
25 Weston, M.D. et al. (1994) Hum. Mol. Genet. 3, 1211.
26 Laiho, M. et al. (1990) Cell 62, 175–185.
27 Niles, R.M. et al. (1994) *In Vitro* Cell. Dev. Biol. 30A, 256–262.
28 Cusella De Angelis, M.G. et al. (1994) Development 120, 925–933.
29 Filvaroff, E.H. et al. (1994) Development 120, 1085–1095.
30 Le Roy, C. et al. (1996) J. Biol. Chem. 271, 11027–11033.
31 Sorrentino, V. and Bandyopadhyay, S. (1989) Oncogene 4, 569–574.
32 Liboi, E. et al. (1988) Biochem. Biophys. Res. Commun. 151, 298–305.
33 Slingerland, J.M. et al. (1994) Mol. Cell. Biol. 14, 3683–3694.
34 Ewen, M.E. et al. (1993) Cell 74, 1009–1020.
35 Singh, R.K. et al. (1994) Cancer Res. 54, 3242–3247.
36 Muldoon, L.L. et al. (1988) J. Biol. Chem. 263, 18834–18841.

[37] Howe, P.H. et al. (1990) J. Cell. Physiol. 142, 39–45.
[38] Matrisian, L.M. et al. (1992) Mol. Reprod. Dev. 32, 111–120.
[39] Theodorescu, D. et al. (1991) J. Cell. Physiol. 148, 380–390.
[40] Schwarz, L.C. et al. (1988) Cancer Res. 48, 6999–7003.
[41] Hafez, M.M. et al. (1990) Cell Growth Differ. (AYH) 1, 617–626.
[42] Clark, W. (1991) Br. J. Cancer 64, 631–644.
[43] Chiu, J.J.S. and Cowan, K. (1993) Proc. Am. Assoc. Cancer Res. 34, 55.
[44] Blanckaert, V.D. et al. (1993) Cancer Res. 53, 4075–4081.
[45] Vitolo, D. et al. (1992) Monogr. Natl Cancer Inst. 13, 203–208.
[46] Gregoire, M. et al. (1992) Invasion Metastasis 12, 185–196.
[47] Mooradian, D.L. et al. (1992) J. Natl Cancer Inst. 84, 523–527.
[48] Ueki, N. et al. (1993) Jpn J. Cancer Res. 84, 589–593.
[49] Matuo, Y. et al. (1992) Adv. Exp. Med. Biol. 324, 107–114.
[50] Steiner, M.S. and Barrack, E.R. (1992) Mol. Endocrinol. 6, 15–25.
[51] Gorsch, S.M. et al. (1994) Cancer Res. 52, 6949–6952.
[52] Wakefield, L.M. et al. (1995) Clin. Cancer Res. 1, 29–136.
[53] Oh, Y. et al. (1995) J. Biol. Chem. 270, 13589–13592.
[54] Knabbe, C. et al. (1991) Am. J. Clin. Oncol.: Cancer Clin. Trials 14, Suppl.2, S15–S20.
[55] Glick, A.B. et al. (1993) Proc. Natl Acad. Sci. USA 90, 6076–6080.
[56] Jirtle, R.L. et al. (1993) Cancer Res. 53, 3849–3852.
[57] Eklov, S. et al. (1993) Cancer Res. 53, 3193–3197.
[58] Kopp, A. et al. (1995) Cancer Res. 55, 4512–4515.
[59] Cui, W. et al. (1994) Cancer Res. 54, 5831–5836.
[60] Lund, L.R. et al. (1991) EMBO J. 10, 3399–3407.
[61] Luparello, C. et al. (1993) Differentiation 55, 73–80.
[62] Marti, H.P. et al. (1994) Am. J. Pathol. 144, 82–94.
[63] Agarwal, C. et al. (1994) Cancer Res. 54, 943–949.
[64] Bruner, K.L. et al. (1995) Proc. Natl Acad. Sci. USA 92, 7362–7366.
[65] Bereta, J. et al. (1992) J. Immunol. 148, 2932–2940.
[66] Oda, K. et al. (1992) Acta Pathol. Jpn 42, 645–650.
[67] Walker, R.A. and Dearing, S.J. (1992) Eur. J. Cancer 28, 641–644.
[68] Dalal, B.I. et al. (1993) Am. J. Pathol. 143, 381–389.
[69] Huang, F. et al. (1994) Oncogene 9, 3701–3706.
[70] Kimchi, A. et al. (1988) Science 239, 196–199.
[71] Inagaki, M. et al. (1993) Proc. Natl Acad. Sci. USA 90, 5359–5363.
[72] Park, K. et al. (1994) Proc. Natl Acad. Sci. USA 91, 8772–8776.
[73] Markowitz, S. et al. (1995) Science 268, 1336–1338
[74] Myeroff, L.L. et al. (1995) Cancer Res. 55, 5545–5547.
[75] Parsons, R. et al. (1995) Cancer Res. 55, 5548–5550.
[76] Lu, S.-L. et al. (1996) Cancer Res. 56, 4595–4598.
[77] Garrigue-Antar, L. et al. (1995) Cancer Res. 55, 3982–3987.
[78] Knaus, P.I. et al. (1996) Mol. Cell. Biol. 16, 3480–3489.
[79] Ciernik, I.F. et al. (1995) Clin. Cancer Res. 1, 1119–1124.
[80] Alexander, J.M. et al. (1996) J. Clin. Endocrinol. Metab. 81, 783–790.
[81] Cui, W. et al. (1996) Cell 86, 531–542.
[82] Sun, L. et al. (1994) J. Biol. Chem. 269, 26449–26455.
[83] Wang, J. et al. (1995) J. Biol. Chem. 270, 22044–22049.
[84] Kim, I.Y. et al. (1996) Cancer Res. 56, 44–48.
[85] Wang, J. et al. (1996) J. Biol. Chem. 271, 17366–17371.

[86] Derynck, R. et al. (1985) Nature 316, 701–705.
[87] Demetriou, M. et al. (1996) J. Biol. Chem. 271, 12755–12761.
[88] Lin, H.Y. et al. (1992) Cell 68, 775–785 [erratum: Cell 70, following 1068].
[89] Liu, F. et al. (1996) Nature 381, 620–623.
[90] Derynck, R. et al. (1987) Nucleic Acids Res. 15, 3188–3189.

## Identification

The *VHL* gene was identified by positional cloning of the region of chromosome 3p that is deleted in Von Hippel–Lindau disease [1].

## Related genes

VHL is highly conserved and has a short region of similarity with elongin A but no significant homology to other known proteins, although it includes eight copies of a tandemly repeated pentamer similar to that present in the surface membrane protein of *Trypanosoma brucei.*

| | ***VHL*** |
|---|---|
| **Nucleotides (kb)** | <20 |
| **Chromosome** | 3p25–p26 |
| **Mass (kDa): predicted** | 24.2 |
| **expressed** | 17/31/28/26/24 |
| **Cellular location** | Nuclear in sparse cultures: cytosolic in dense cultures [2] |

**Tissue distribution**

Widely expressed. Only one of the 6.0 or 6.5 kb mRNA species is expressed in fetal brain and fetal kidney: both are expressed in adult tissues.

## Protein function

VHL binds specifically to the elongin B and C subunits of elongin (SIII), a heterotrimer (transcriptionally active subunit A and regulatory subunits B and C) that activates transcription by RNA polymerase II [3,4]. Elongin C and also VBP-1 bind to the C-terminal region: the actin-binding protein filamin and the HIV TAT binding protein 1 also associate with VHL [5].

The repeated acidic domain resembling that of the *Trypanosoma brucei* surface membrane protein suggests that VHL may be involved in signal transduction or cell–cell contacts. VHL immunoprecipitates with 9 and 16 kDa proteins [6].

**Cancer**

Von Hippel–Lindau disease is a dominantly inherited familial cancer syndrome with variable expression and with age-dependent penetrance that predisposes individuals most frequently to haemangioblastomas of the central nervous system and retina, renal cell carcinoma and pheochromocytoma. The incidence at birth is at least 1/36 000. *VHL* is one of the major tumour suppressor genes in human renal cell carcinomas, particularly in the clear cell subtype renal cell carcinoma. Usually one allele is mutated (or methylated) and the other deleted.

In Von Hippel–Lindau disease-associated and sporadic haemangioblastomas VEGF is secreted by tumour stromal cells and interacts with its high-affinity receptors, FLT1 and KDR [7].

### *In vitro*

Expression of wild-type VHL suppresses growth of A498 and UMRC6 renal carcinoma cells *in vitro*[8].

## Gene structure



The promoter does not contain TATA or CCAAT boxes: transcription is initiated around a putative SP1 binding site ~60 bp upstream from the first AUG codon in the mRNA. Other putative binding sites (for nuclear respiratory factor 1 and PAX) occur upstream[9].

## Amino acid sequence of human VHL

```
  1 (PRLRYNSLRC WRILLRTRTA SGRLFPRARS ILYRARAKTT EVDSGARTQL
 51 RPASDPRIPR RPARVVWIAE G)MPRRAENWD EAEVGAEEAG VEEYGPEEDG
                                            *       ***
101 GEESGAEESG PEESGPEELG AEEEMEAGRP RPVLRSVNSR EPSQVIFCNR
151 SPRVVLPVWL NFDGEPQPYP TLPPGTGRRI HSYRGHLWLF RDAGTHDGLL
201 VNQTELFVPS LNVDGQPIFA NITLPVYTLK ERCLQVVRSL VKPENYRRLD
251 IVRSLYEDLE DHPNVQKDLE RLTQERIAHQ RMGD (213/284)
```

### Domain structure

| | |
|---|---|
| 14–53 | Tandemly repeated acidic domain Gly-X-Glu-Glu-X (underlined) (of 213 sequence) |
| 2–5 and 57–61 | Putative bipartite nuclear localization signal[2] |
| 157–172 | Region binding to elongin B and C: frequently altered by germline mutations in VHL kindreds. A peptide corresponding to this region inhibits binding whereas a mutant form ($Cys^{162} \rightarrow Phe$) does not[10] |
| 157–189 | p10 and p14 binding region[4] |

### Mutations detected in *VHL*

Major deletions or insertions in the gene occur with high frequency (>12%) in VHL patients and germline deletions within each of the three exons occur[11,12]. Forty different germline mutations in 55 unrelated kindreds encompassing missense, nonsense, frameshift insertions/deletions, in-frame deletions and a splice donor site mutation have been described[13]. The two most common are missense mutations at codon 238 (167 in 213 amino acid form) which confer a high risk of pheochromocytoma.

One intragenic 8 bp frameshift insertion and two intragenic deletions (removing amino acids 153–154 (***) or $Ile^{146}$ (*)) have also been detected in VHL families. In sporadic renal cell carcinomas mutations are frequent (57%) and varied[14], and four small intragenic frameshift deletions and one nonsense mutation have been detected. The effects of these mutations are the replacement of residues 246–284 with 28 new amino acids, the replacement of residues 238–284 with 32 new residues, the replacement of residues 212–284 with 62 new residues or the generation of a truncated protein caused by mutation in codon 254. Deletion at position –2 in the 5′ splice donor site of exon 2 gives rise to two mRNAs, each encoding predicted premature termination has been detected [15,16]. A variety of germline mutations have been detected in Japanese VHL patients with and without pheochromocytoma [17].

Somatic mutations have been detected in 56% of clear cell renal carcinomas including 15 deletions, three insertions, three missense mutations, and one nonsense mutation. Nineteen of these mutations are predicted to produce truncation of the VHL protein. These mutations mainly occurred in the last one-third region of exons 1, 2 and 3. In addition, loss of heterozygosity of *VHL* was observed in 84% of 19 informative clear cell renal carcinomas. No somatic mutations were detected in eight non-clear cell carcinomas [18,19].

Mutations in VHL have also been detected in lung cancer ($Gly^{177} \rightarrow Asp$), mesothelioma ($Leu^{160} \rightarrow His$) [20] and in sporadic central nervous system haemangioblastomas ($Trp^{159} \rightarrow Ser$, $Leu^{206} \rightarrow Phe$, 12 nt deletion at nt 663) [21].

Hypermethylation of a normally unmethylated CpG island in the 5′ region of *VHL*, detected in 19% of one sample of renal cell carcinomas, silences the gene [22].

**Database accession numbers**

| | *PIR* | *SWISSPROT* | *EMBL/GENBANK* | *REFERENCES* |
|---|---|---|---|---|
| Human *VHL* | | | L15409 | 1 |
| | | U19763 | | 9 |

***References***


1 Latif, F. et al. (1993) Science 260, 1317–1320.
2 Lee, S. et al. (1996) Proc. Natl Acad. Sci. USA 93, 1770–1775.
3 Duan, D.R. et al. (1995) Science 269, 1402–1406.
4 Kishida, T. et al. (1995) Cancer Res. 55, 4544–4548.
5 Tsuchiya, H. et al. (1996) Cancer Res. 56, 2881–2885.
6 Duan, D.R. et al. (1995) Proc. Natl Acad. Sci. USA 92, 6459–6463.
7 Wizigmann-Voos, S. et al. (1995) Cancer Res. 55, 1358–1364.
8 Chen, F. et al. (1995) Cancer Res. 55, 4804–4807.
9 Kuzmin, I. et al. (1995) Oncogene 10, 2185–2194.
10 Kibel, A. et al. (1995) Science 269, 1444–1446.
11 Richards, F.M. et al. (1994) Hum. Mol. Genet. 3, 595–598.
12 Richards, F.M. et al. (1995) Hum. Mol. Genet. 4, 2139–2143.
13 Crossey, P.A. et al. (1994) Hum. Mol. Genet. 3, 1303–1308.
14 Gnarra, J. et al. (1994) Nature Genet. 7, 85–90.
15 Kishida, T. et al. (1994) Hum. Mol. Genet. 3, 1191–1192.
16 Loeb, D.B. et al. (1994) Hum. Mol. Genet. 3, 1423–1424.
17 Kondo, K. et al. (1995) Hum. Mol. Genet. 4, 2233–2237.
18 Shuin, T. et al. (1994) Cancer Res. 54, 2852–2855.
19 Foster, K. et al. (1994) Hum. Mol. Genet. 3, 2169–2173.
20 Sekido, Y. et al. (1994) Oncogene 9, 1599–1604.
21 Kanno, H. et al. (1994) Cancer Res. 54, 4845–4847.
22 Herman, J.G. et al. (1994) Proc. Natl Acad. Sci. USA 91, 9700–9704.

# WT1

## Identification

*WT1* was identified by genetic analysis of the region of chromosome band 11p13 in which constitutional heterozygous deletions correlate with hereditary predisposition to Wilms' tumour [1–3].

## Related genes

*WT1* is a member of the early growth response family (*EGR1, EGR2, EGR3, EGR4*).

| | ***WT1*** |
|---|---|
| **Nucleotides (kb)** | 50 |
| **Chromosome** | 11p13 |
| **Mass (kDa): predicted** | 49 |
| **expressed** | 52/54 |
| **Cellular location** | Nucleus |

**Tissue distribution**

*WT1* is expressed in the developing kidney, gonads, spleen and mesothelium and brain [3–5]. +KTS forms of WT1 (see below) associate preferentially with interchromatin granules in which snRNPs characteristically localize. –KTS forms have a distribution similar to that of classical transcription factors (SP1 or TFIIB), suggesting that the isoforms have distinct functions that may include post-transcriptional RNA processing [6].

## Protein function

WT1 proteins are members of the zinc finger protein transcription factor family that may play a crucial role in normal genitourinary development [7].

Two alternative splicing events generate four alternative products: (i) exon 5 encodes a 17 amino acid serine-rich insert between the proline-rich regulatory domain and the zinc finger domain [WT1 + 17AA], (ii) three amino acids (KTS) may be inserted between the third and fourth zinc fingers [WT1 + KTS]. The predominant variant in all cells that express WT1 is [WT1 + KTS + 17AA] [8]. [WT1 + KTS] binds to GNGNGGGNGNS (S = G or C) and [WT1 – KTS] to GNGNGGGNGNG [9]. [WT1 + KTS + 17AA] is a strong transcriptional repressor binding to the core sequence 5′-GCGGGGGCG-3′ [2,10,11], to which EGR1 binds as a transcriptional activator [12]. [WT1 ± KTS] also binds to the sequence 5′-GGAGAGGGAGGATC-3′ to which EGR1 does not bind [13]. A second consensus binding sequence (5′-GCGTGGGAGT-3′ (WTE)) identified in mice has a 20- to 30-fold higher affinity for WT1 [14].

WT1 isoforms can interact *in vitro* and *in vivo via* their N-terminal domains. In particular, the germline missense mutations that occur in Denys–Drash syndromes and give rise to severe developmental disorders antagonize transcriptional repression by wild-type WT1 and thus act in a dominant negative manner [15].

Promoters containing WT1 consensus binding sites: *BCL2*, IGF-II (paternally expressed); IGF-I receptor, *CSF1, PDGFA, PAX2, EGR1*, Midkine[16], *Myc*, TGF$\beta$ and retinoic acid receptor $\alpha$ (RAR-$\alpha$1). WT1 represses transcription from the human *BCL2* promoter and from murine *Myc*[17]. Human *PDGFA* transcription is strongly repressed by WT1 interaction with two binding sites that lie 5′ and 3′ relative to the transcription start site[18,19]. When WT1 binds to only one of these sites it functions as a *trans*-activator. *IGF2*, which encodes an autocrine growth factor expressed at high levels in Wilms' tumour, is also repressed by WT1[+KTS][20]. *IGF2* is expressed from the paternal allele in normal human fetal tissue but expression can occur biallelically in Wilms' tumour[21]. Many of the actions of *IGF2* are mediated *via* the IGF1 receptor and WT1 can repress the activity of the *IGF1R* promoter, a key mechanism by which WT1 suppresses cell growth[22]. The *H19* gene, which does not encode a protein, is maternally expressed in normal tissue but can show biallelic expression in Wilms' tumour[23] and transfection of *H19* can suppress tumorigenicity[24]. Both *IGF2* and *H19* are located at 11p15 and the relaxation of genomic imprinting may therefore be involved in the development of Wilms' tumour. In most Wilms' tumours *H19* mRNA is greatly reduced from the levels detected in fetal kidney[25]. *H19* may be involved in muscle cell differentiation and its expression may be correlated with that of *MOS*[26].

WT1 also represses *EGFR* by binding to two TC-rich repeat sequences[27], *MYB* by binding to a GCGGGGGCG sequence also recognized by *EGR1*, *EGR2* and *EGR3* and *NOV* by binding to two TC-rich regions[28]. The G protein G$\alpha_{i2}$ also contains an *EGR1* motif and growth of LLC-PK$_1$ renal cells is inhibited by WT1 repression of this gene[29]. In a variety of cell lines WT1 represses transcription of ornithine decarboxylase (*ODC1*) by binding to G-rich sequences[30]. The murine *syndecan-1* promoter contains several potential WT1 binding sites and is transcriptionally activated by WT1[±KTS][31].

**Cancer**

Wilms' tumour is an embryonal renal neoplasm that occurs in sporadic and familial forms and affects 1 in 10 000 children[32,33], although only three families with the familial form are known. It is the most common solid tumour of childhood. Approximately 2% of Wilms' tumours occur in association with aniridia (cf. 1/50 000 in general population), genitourinary anomalies and mental retardation (WAGR syndrome) in which deletions of 11p13 were first detected. WAGR, Beckwith–Wiedemann and Denys–Drash syndromes confer hereditary susceptibility to Wilms' tumour.

Wilms' tumour, like retinoblastoma, occurs in unilateral or bilateral early onset forms but may involve three loci rather than one. Deletions at 11p13 are associated with the bilateral form but in the sporadic forms allelic loss occurs at 11p15 (*WT2*), exclusively in the maternal allele, and a third locus (16q) may also be involved. Insertion of a normal human chromosome 11 into Wilms' tumour cells causes reversion to normal, non-tumorigenic cells that correlates with the expression of the *QM* gene[34].

Mutations in *WT1, WT2* or at 16q are not involved in the familial form but a genome linkage search in one family indicates that *FWT1* on chromosome 17q is a familial Wilms' tumour predisposition gene [35].

In the desmoplastic small round cell tumour (DSRCT), the translocation t(11;22)(p13;q12) fuses the *EWS* and *WT1* genes so that the N-terminal domain of EWS1 is joined to the DNA binding domain of WT1, including zinc fingers 2–4 [36]. In contrast to the effect of WT1, EWS/WT1 is a strong activator of transcription from the IGF-I-receptor promoter and the expression of this receptor may play an important role in the development of DSRCT [37].

*NOV* RNA has been detected in Wilms' tumours, the level of expression being generally inversely related to that of *WT1* [38].

**Transgenic animals**

Homozygous mutation of the murine *Wt-1* gene causes embryonic lethality arising from the failure of kidney and gonad development [39].

***In vitro***

In transfected cells p53 interacts with WT1 and converts the *trans*-activating capacity of the latter to *trans*-repression. WT1 also exerts a cooperative effect on p53 by enhancing its *trans*-activation of promoters containing an RGC sequence (TGCCT repeats), including the muscle creatine kinase (*MCK*) promoter (see *P53*), but reducing transcriptional repression of TATA-containing promoters. In transfected SAOS-2 cells these WT1 effects inhibit p53-induced apoptosis [40].

WT1 inhibits *ras*-mediated transformation of NIH 3T3 fibroblasts [41]. When expressed in adenovirus-transformed baby rat kidney cells, WT1 lacking the 17 amino acid insert suppresses the tumorigenic phenotype whereas the form lacking both inserts increases tumour growth rate [42].

Microinjection of WT1 blocks serum-induced cell cycle progression into S phase. This activity is abrogated by over-expression of cyclin E/CDK2 or cyclin D1/CDK4, and both CDK2 and CDK4 are downregulated in cells over-expressing WT1 [43].

The functional consequences of expressing WT1 can be affected by the nature of the expression vector used in transfection experiments. Thus, expression driven by the RSV promoter causes activation whereas the CMV promoter-containing WT1 expression vector causes repression of the *Egr-1* promoter [44]. This suggests that cofactors may be required for WT1-mediated *trans*-activation.

## Gene structure

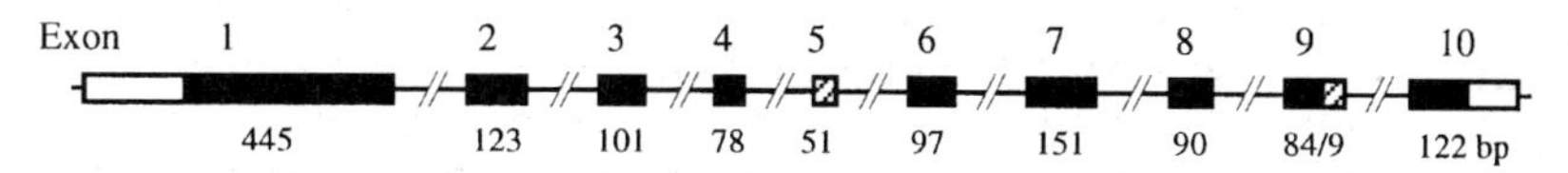


The alternatively spliced exons are cross-hatched. All four possible variants are expressed in normal developing kidney, that including both alternative sequences being most common[8], and also in Wilms' tumours[45]. Variants having a three amino acid insertion at exon 9 cannot bind the *EGR* recognition element[12]. Mutations within intron 6 (that lead to exon skipping, resulting in transcripts either missing exon 6 or exons 5 and 6[46]), or within intron 9 (that prevent alternative splicing) or in the zinc finger domains occur in Denys–Drash syndrome[47] and in Wilms' tumour DNA[48]. Mutations in exons 7 and 8 that generate truncated proteins lacking part of the zinc finger domain have also been detected in unilateral Wilms' tumours[49]. A subset of Wilms' tumours lacks exon 2 and this mutant form of WT1 is a powerful activator of transcription of the *IGF2* gene, as is the $Gly^{201}$ WAGR mutant protein[50].

## Transcriptional regulation

The *WT1* promoter lacks TATA or CCAAT boxes, has a high GC content and contains four transcriptional start sites within a 32 bp region. There are 11 putative SP1 binding sites and potential recognition sites for AP1, AP2, EGR, PAX2, PAX8 and GAGA-like factors[51]. PAX2 and PAX8 *trans*-activate *WT1 via* the element –33 to –71[52,53]. WT1 binds to multiple sites in its own promoter to autoregulate transcription negatively[54]. Tissue-specific expression of *WT1* is modulated by an enhancer located >50 kb 3′ of the promoter[55]. GATA-1, a haematopoietic transcription factor, activates *WT1*[56]. Intron 3 of *WT1* contains a 460 bp transcriptional silencer (~12 kb from the promoter) that contains a full-length *Alu* repeat and represses *WT1* transcription in cells of non-renal origin. A 5′ 148 bp region identical to the first intron of *WIT1* functions as a transcriptional enhancer and contains multiple potential transcription factor binding sites[57]. Intron 1 contains an anti-sense *WT1* promoter that functions in the opposite direction to 5′ promoter. Expression of a *WT1* exon 1 anti-sense mRNA downregulates cellular WT1 protein levels[58] and anti-sense to *WT1* mRNA induces apoptosis in myeloid leukaemia cell lines[59].

The *WT1* transcript can undergo RNA editing in which $U^{839}$ is converted to C, causing $Leu^{281}$ to be replaced by proline[60]. The WT1-$Leu^{281}$ protein represses the EGR1 promoter ~30% more efficiently than WT1-Pro. However, this mechanism appears infrequent in tumorigenesis[61].

## Protein structure

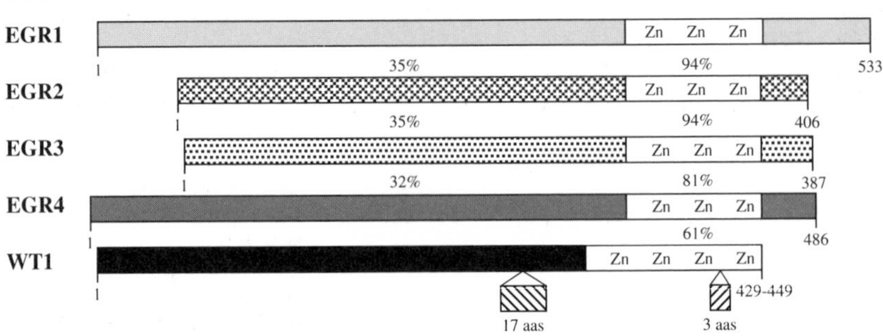

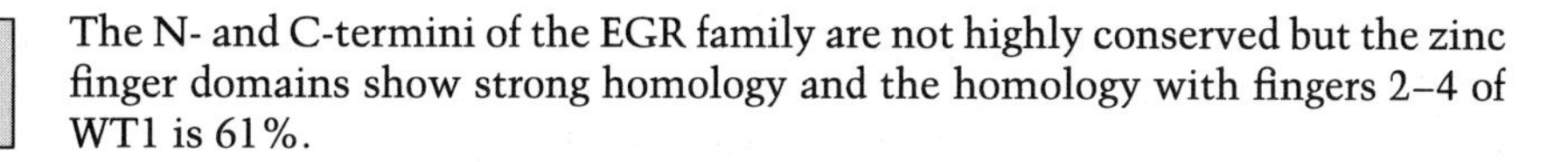

The N- and C-termini of the EGR family are not highly conserved but the zinc finger domains show strong homology and the homology with fingers 2–4 of WT1 is 61%.

## Amino acid sequence of human WT1

```
  1 MGSDVRDLNA LLPAVPSLGG GGGCALPVSG AAQWAPVLDF APPGASAYGS
 51 LGGPAPPPAP PPPPPPPPHS FIKQEPSWGG AEPHEEQCLS AFTVHFSGQF
101 TGTAGACRYG PFGPPPPSQA SSGQARMFPN APYLPSCLES QPAIRNQGYS
151 TVTFDGTPSY GHTPSHHAAQ FPNHSFKHED PMGQQGSLGE QQYSVPPPVY
201 GCHTPTDSCT GSQALLLRTP YSSDNLYQMT SQLECMTWNQ MNLGATLKGV
251 AAGSSSSVKW TEGQSNHSTG YESDNHTTPI LCGAQYRIHT HGVFRGIQDV
301 RRVPGVAPTL VRSASETSEK RPFMCAYPGC NKRYFKLSHL QMHSRKHTGE
351 KPYQCDFKDC ERRFSRSDQL KRHQRRHTGV KPFQCKTCQR KFSRSDHLKT
401 HTRTHTGKTS EKPFSCRWPS CQKKFARSDE LVRHHNMHQR NMTKLQLAL
                                                   (449)
```

**Domain structure**

| | |
|---|---|
| 20–50 | RNA recognition motif [62] |
| 27–83 | Proline-rich domain |
| 85–124 and 181–250 | Function independently with a DNA binding domain to repress or activate transcription, respectively [63] |
| 323–347, 353–377, 383–405, 414–438 | Zinc finger ($C_2H_2$ type) regions (underlined) |
| 250–266 and 408–410 | Alternative splice regions (underlined italics). These introduce 17 amino acids after 249 or three amino acids between the third and fourth zinc fingers. The variant containing the KTS sequence is the predominant form in all cells that express WT1 [8] |
| 84–179 | Mediate *trans*-repression by the 429 amino acid form |
| 180–294 | Mediate *trans*-activation by the 429 amino acid form |
| 1–182 | Major domain required for association with dominant negative forms of WT1 [64] |
| 365, 393 | Phosphorylation (by protein kinase A or protein kinase C) inhibits DNA binding [65] |

**Mutations detected in *WT1***

Mutations detected in Denys–Drash syndrome (missense, mutations in zinc-complexing amino acids and removal of zinc fingers) block DNA binding [66]. They appear to be dominant negative in their mode of action and lead to severe developmental abnormalities of the gonads and kidney as well as to Wilms' tumour.

Mutations in WAGR include $Gly^{201} \rightarrow$ Asp that converts WT1 from a transcriptional repressor to an activator [67] and $Cys^{385} \rightarrow$ Tyr within a zinc finger [68].

**Database accession numbers**

| | *PIR* | *SWISSPROT* | *EMBL/GENBANK* | *REFERENCES* |
|---|---|---|---|---|
| Human *WT1* | A34673, S08273 | P19544 | X51630 | 69 |
| | | | M30393 | 2 |
| | | | M80217–M80221 | 8 |
| | | | M80228 | 70 |
| | | | M80231/32, M74917 | 48 |

*References*


[1] Rose, E.A. et al. (1990) Cell 60, 495–508.
[2] Call, K.M. et al. (1990) Cell 60, 509–520.
[3] Huang, A. et al. (1990) Science 250, 991–994.
[4] Pritchard-Jones, K. et al. (1990) Nature 346, 194–197.
[5] **Rauscher, F.J. (1993) FASEB J. 7, 896–903.**
[6] Larsson, S.H. et al. (1995) Cell 81, 391–401.
[7] Pelletier, J. et al. (1991) Cell 67, 437–447.
[8] Haber, D.A. et al. (1991) Proc. Natl Acad. Sci. USA 88, 9618–9622.
[9] Hewitt, S.M. et al. (1996) J. Biol. Chem. 271, 8588–8592.
[10] Morris, J.F. et al. (1991) Oncogene 6, 2339–2348.
[11] Wang, Z.-Y. et al. (1995) Oncogene 10, 415–422.
[12] Madden, S.L. et al. (1993) Oncogene 8, 1713–1720.
[13] Little, M.H. et al. (1996) Oncogene 13, 1379–1385.
[14] Nakagama, H. et al. (1995) Mol. Cell. Biol. 15, 1489–1498.
[15] Moffett, P. et al. (1995) Proc. Natl Acad. Sci. USA 92, 11105–11109.
[16] Adachi, Y. et al. (1996) Oncogene 13, 2197–2203.
[17] Hewitt, S.M. et al. (1995) Cancer Res. 55, 5386–5389.
[18] Gashler, A.L. et al. (1992) Proc. Natl Acad. Sci. USA 89, 10984–10988.
[19] Wang, Z.-Y. et al. (1993) J. Biol. Chem. 268, 9172–9175.
[20] Drummond, I.A. et al. (1994) Mol. Cell. Biol. 14, 3800–3809.
[21] Ogawa, O. et al. (1993) Nature 362, 749–751.
[22] Werner, H. et al. (1995) Mol. Cell. Biol. 15, 3516–3522.
[23] Rainer, S. et al. (1993) Nature 362, 747–749.
[24] Hao, Y. et al. (1993) Nature 365, 764–767.
[25] Moulton, T. et al. (1994) Nature Genet. 7, 440–447.
[26] Leibovitch, M.P. et al. (1995) Oncogene 10, 251–260.
[27] Englert, C. et al. (1995) EMBO J. 14, 4662–4675.
[28] Martinerie, C. et al. (1996) Oncogene 12, 1479–1492.
[29] Kinane, T.B. et al. (1995) J. Biol. Chem. 270, 30760–30764.
[30] Moshier, J.A. et al. (1996) Nucleic Acids Res. 24, 1149–1157.
[31] Cook, D.M. et al. (1996) Oncogene 13, 1789–1799.
[32] **van Heyningen, V. and Hastie, N.D. (1992) Trends Genet. 8, 16–21.**
[33] **Haber, D.A. and Buckler, A.J. (1992) New Biol. 4, 97–106.**
[34] Dowdy, S.F. et al. (1991) Nucleic Acids Res. 19, 5763–5769.
[35] Rahman, N. et al. (1996) Nature Genet. 13, 461–463.
[36] Ladanyi, M. and Gerald, W. (1994) Cancer Res. 54, 2837–2840.
[37] Karnieli, E. et al. (1996) J. Biol. Chem. 271, 19304–19309.
[38] Martinerie, C. et al. (1994) Oncogene 9, 2729–2732.
[39] Kreidberg, J.A. et al. (1993) Cell 74, 679–691.
[40] Maheswaran, S. et al. (1995) Genes Dev. 9, 2143–2156.
[41] Luo, X.-N. et al. (1995) Oncogene 11, 743–750.
[42] Menke, A.L. et al. (1996) Oncogene 12, 537–546.
[43] Kudoh, T. et al. (1995) Proc. Natl Acad. Sci. USA 92, 4517–4521.
[44] Reddy, J.C. et al. (1995) J. Biol. Chem. 270, 29976–29982.
[45] Brenner, B. et al. (1992) Oncogene 7, 1431–1433.
[46] Schneider, S. et al. (1993) Hum. Genet. 91, 599–604.
[47] Bruening, W. et al. (1992) Nature Genomics 1, 144–148.
[48] Little, M.H. et al. (1992) Proc. Natl Acad. Sci. USA 89, 4791–4795.
[49] Baird, P.N. et al. (1992) Oncogene 7, 2141–2149.

50 Nichols, K.E. et al. (1995) Cancer Res. 55, 4540–4543.
51 Hofmann, W. et al. (1993) Oncogene 8, 3123–3132.
52 Dehbi, M. et al. (1996) Oncogene 13, 447–453.
53 Dehbi, M. and Pelletier, J. (1996) EMBO J. 15, 4297–4306.
54 Rupprecht, H.D. et al. (1994) J. Biol. Chem. 269, 6198–6206.
55 Fraizer, G.C. et al. (1994) J. Biol. Chem. 269, 8892–8900.
56 Wu, Y. et al. (1995) J. Biol. Chem. 270, 5944–5949.
57 Hewitt, S.M. et al. (1995) J. Biol. Chem. 270, 17908–17912.
58 Malik, K.T.A. et al. (1995) Oncogene 11, 1589–1595.
59 Algar, E.M. et al. (1996) Oncogene 12, 1005–1014.
60 Sharma, P.M. et al. (1994) Genes Dev. 8, 720–731.
61 Gunning, K.B. et al. (1996) Oncogene 13, 1179–1185.
62 Kennedy, D. et al. (1996) Nature Genet. 12, 329–332.
63 Wang, Z.-Y. et al. (1995) Oncogene 10, 1243–1247.
64 Reddy, J.C. et al. (1995) J. Biol. Chem. 270, 10878–10884.
65 Ye, Y. et al. (1996) EMBO J. 15, 5606–5615.
66 Little, M. et al. (1995) Hum. Mol. Genet. 4, 351–358.
67 Park, S. et al. (1993) Cancer Res. 53, 4757–4760.
68 Pritchard-Jones, K. et al. (1994) Hum. Mol. Genet. 3, 723–728.
69 Gessler, M. et al. (1990) Nature 343, 774–778.
70 Buckler, A.J. et al. (1991) Mol. Cell. Biol. 11, 1707–1712.

# Section V

# CYCLIN-DEPENDENT KINASE INHIBITORS

# INK4A/MTS1/CDK41/CDKN2, INK4B/MTS2, INK4C, INK4D

## Identification

*INK4A*, *INK4B* and *INK4C* cDNAs were isolated by two-hybrid screening for proteins associating with CDK4 [1]. *INK4D* was similarly cloned using the orphan steroid receptor Nur77/NGFI-B [2].

## Related genes

The four INK4 genes that have been isolated encode proteins containing four tandemly repeated ankyrin motifs. *INK4B* is a gene adjacent to *INK4A* containing a region of 93% identity to *INK4A*. INK4A and INK4B are 44% indentical in the first 50 amino acids: 97% in the next 81. INK4D has 48% identity with INK4A and INK4C and INK4D are ~40% identical.

| | *INK4A* | *INK4B* | *INK4C* | *INK4D* |
|---|---|---|---|---|
| **Nucleotides (kb)** | ~30 | ~30 | Not fully mapped | Not fully mapped |
| **Chromosome** | 9p21 | 9p21 | 1p32 | 19p13 |
| **Mass (kDa): predicted** | 15.8 | 14.7 | 18.2 | 17.7 |
| **expressed** | 16 | 15 | 18 | 19 |
| **Cellular location** | Nucleus | Nucleus | Nucleus | Nucleus |

**Tissue distribution**

Two mRNA forms of *INK4A* are differentially expressed in both a tissue and cell cycle-specific manner [3]. The $\beta$ transcript is ubiquitously expressed: the $\alpha$ form is expressed in a limited range of tissues.

*INK4B* is widely expressed (spleen, lung, brain, heart, liver, colon and kidney [4]. *INK4C* and *INK4D* are expressed ubiquitously in proliferating cells. In macrophages *INK4D* mRNA and protein levels oscillate during the cell cycle, being mimimal during $G_1$ and increasing as cells enter S phase [5]. *INK4D* expression is high in the thymus, spleen, peripheral blood lymphocytes, fetal liver, brain and testes [6].

## Protein function

INK4 (inhibitor of CDK4) proteins are cyclin-dependent kinase inhibitors (CDIs) that block the action of cyclin D-dependent kinase to induce cell cycle arrest. *INK4A* also called multiple tumour suppressor 1 (*MTS1*), *CDK41/INK4A* or *CDKN2*, encodes p16$^{INK4A}$ which binds with 1:1 stoichiometry to CDK4 and CDK6 in competition with cyclin D to block CDK activity [7–10]. Enforced expression of p16$^{INK4A}$ in various cell lines induces $G_1$ arrest [11] by a mechanism that requires functional pRb [12]. However, INK4A levels are increased in cells lacking functional pRb, and over-expression of *INK4A* or inhibition of cyclin D-dependent kinase does not affect $G_1$/S phase progression in pRb-defective cells. pRb represses transcription of *INK4A* [13], the reciprocal of pRb-stimulated cyclin D expression, and there is an inverse correlation between pRb status and the expression of INK4A [14]. The expression of the S phase-regulated gene thymidine kinase is mediated by E2F and in *INK4A*$^{-/-}$ cells the activity of this enzyme is greatly increased: transient over-expression of INK4A establishes normal expression of thymidine kinase paralleled by an increase in the hypophosphorylated form of pRb [15]. The normal function of INK4A may be to downregulate CDK4/CDK6 *via* activation of E2F after inactivation of pRb by phosphorylation. Thus loss of *INK4A*, loss of *RB1* and over-expression of cyclin D1 may be functionally analogous in tumorigenesis.

The human T cell leukaemia virus type I (HTLV-I) TAX oncoprotein binds to INK4A, promoting activation of CDK4 kinase [16] and in a number of cell lines TAX induces expression of *INK4A* whilst suppressing that of *INK4C* [17].

p15$^{INK4B}$ (also called multiple tumour suppressor 2 (*MTS2*)), binds to CDK4 and CDK6, the main catalytic partners for cyclins D1, D2 and D3 [18]. *INK4B* is activated by TGF$\beta$ *via* an SP1 consensus site [19]: the level of mRNA increases 30-fold in cells treated with TGF$\beta$ and this is reflected by a corresponding decrease in CDK6-associated kinase activity.

p18$^{INK4C}$ interacts strongly with CDK6, weakly with CDK4 and inhibits the kinase activity of cyclin-CDKs [20]. Ectopic expression of INK4C (or INK4B) suppresses cell growth in a pRb-dependent manner.

p19$^{INK4D}$ inhibits CDK4 and CDK6 and its expression causes cell cycle arrest in $G_1$ [2,5].

## Cancer

Homozygous deletion of a minimal region including *INK4A* and excluding *INK4B* is the most common genetic event in primary tumours [21] and the *INK4A* locus is rearranged, deleted or mutated in ~75% of tumour cell lines [22], although there are reports that mutations in *INK4A* occur less frequently in primary tumours [23]. Nevertheless, allelic deletions have been reported to occur in 83% of T cell lineage acute lymphoblastic leukaemias (ALL) [24,25], in 50% of precursor B cell ALL [26], between 27 and 85% of pancreatic adenocarcinomas [27,28] and of glioblastomas [29], in ~50% of anaplastic astrocytomas [30], in oesophageal squamous cancers [31], in the early stages of Barrett's oesophagus [32], in non-small cell lung cancers [33,34] and with high frequency in primary bladder cancer [35] and pituitary tumours [36]. There is a low incidence of mutations in *INK4A* and *INK4B* in primary gliomas [37] and in addition to deletion of *INK4A*, CDK4 amplification and increased expression occurs in some glioma cell lines [38].

*INK4A* may be the gene for one type of familial melanoma (MLM): potentially inactivating germline mutations being detected in 75% of MLMs in one study and in <20% in another [39]. However, metastatic development can occur in melanoma cells without the involvement of *INK4A* mutations [40]. *INK4A* deletions also occur in some benign nevus specimens and normal human melanocytes [41] and *INK4B* may be lost without mutation of *INK4A* [42]. Mutant forms are generally defective in interaction with CDK4 and CDK6 [43]. Missense mutations have also been detected in sporadic carcinomas [44,45].

Co-deletion of *INK4A* and *INK4B* occurs with high frequency in NSCLC and tumours having a normal copy number of both genes have been detected with a nonsense mutation in exon 2 of *INK4A* [46]. Similar co-deletion or loss of INK4A expression occurs in advanced mouse skin tumours[47]. *INK4A* is deleted in ~80% of astrocytoma cell lines but only 1 in 30 primary tumours show mutation with loss of the second allele[48]. A high frequency of co-deletion of *INK4A* and *INK4B* has been described in recurrent glial tumours[49]. Hypermethylation and downregulation of *INK4A* may be involved in the development of nasopharyngeal carcinoma[50]. *INK4A* alterations have not been detected in ovarian tumours[51].

Two-thirds of cancer cell lines that carry mutations in cyclin D1 (amplification or translocation) have also lost expression of INK4A[52], indicating that aberrations in steps upstream of pRb can cooperate in tumorigenesis.

Hypermethylation of a G:C-rich region in exon 1 appears to be involved in loss of *INK4A* expression[53,54] and *INK4A* loss may be closely associated with inactivation of 5′-deoxy-5′-methyladenosine phosphorylase in a variety of cancers[55] and co-deletion of the interferon genes *IFNA1* and *IFNB1* occurs in some pancreatic carcinoma cell lines[56]. Hypermethylation associated with inactivation of *INK4B* is common in gliomas and leukaemias[57].

*INK4B* expression correlates inversely with that of *RB1* in a subset of lung cancer cell lines. In ~90% of SCLC samples the *RB1* gene was deleted or defective whilst *INK4B* abnormalities predominated in non-SCLC lines[58]. Both *INK4A* and *INK4B* are homozygously deleted in two-thirds of glioblastoma multiform tumours[59] and co-deletion of these genes occurs frequently in primary malignant mesothelioma[60].

**Transgenic animals**

Mice in which *INK4A* and *ARF* are deleted are viable but develop tumours spontaneously at an early age and are highly susceptible to carcinogens[61].

***In vitro***

The spontaneous immortalization of Li–Fraumeni fibroblasts appears to be associated with the loss of both p53 and INK4A expression and there is an increase in the amount of telomeric DNA[62]. Adenovirus-mediated gene transfer of *INK4A* causes growth arrest of human glioma cell lines[63].

Loss of *INK4A* is necessary but not sufficient for immortalization of human keratinocytes[64].

## Gene structure of *INK4A* and *INK4B*

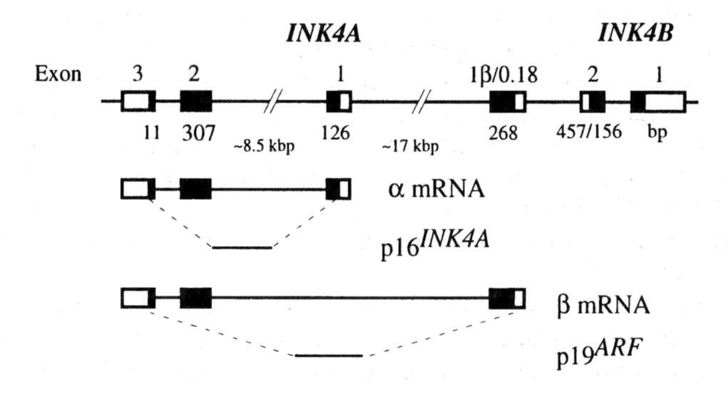

Alternatively spliced forms of *INK4A* (ORF 2) utilize different first exons (E1$\alpha$, E1$\beta$)[65–67]. The $\beta$ transcript incorporates exon 2 coding sequences in an alternative reading frame and has thus been designated p19$^{ARF}$[68]. Ectopic expression of p19$^{ARF}$ induces $G_1$ and $G_2$ phase arrest of rodent fibroblasts. Basal promoter activity is contained within 869 bp of the initiation codon and includes a region activatable by non-phosphorylated Rb[69]. *INK4B* exon 1 is ~2.5 kb upstream of exon 2[3].

## Amino acid sequence of human INK4A/MTS1

```
  (1) MDPAAGSSME PSADWLATAA ARGRVEEVRA LLEAVALPNA PNSYGRRPIQ
 (51) VMMMGSARVA ELLLLHGAEP NCADPATLTR PVHDAAREGF LDTLVVLHRA
(101) GARLDVRDAW GRLPVDLAEE LGHRDVARYL RAAAGGTRGS NHARIDAAEG
(151) PSDIPD (156)
```

**Domain structure**

10–43, 44–75, 77–109, 110–144 Four tandemly repeated ankyrin motifs (first underlined) each of ~32 amino acids in length

**Mutations detected in *INK4A***

A translocation (t(9;14)(p21–p22;q11)) in a B cell type acute lymphoblastic leukaemia juxtaposes part of the constant region of TCR$\alpha$ to exons 2 and 3 of *INK4A*[70]. The putative encoded transcript (132 amino acids, $M_r$ 13.9 kDa, pI 13.2) is entirely divergent from that of INK4A.

Mutations of Asp$^{74}$, Phe$^{81}$, Arg$^{87}$, His$^{98}$, Gly$^{101}$, Arg$^{103}$ or Val$^{126}$ result in proteins defective in inhibiting the catalytic activity of CDK4/D1 kinase[66]. The mutations Glu$^{120}$ $\rightarrow$ Lys, Arg$^{144}$ $\rightarrow$ Cys or deletion from Gly$^{136}$ do not significantly affect capacity of INK4A to inhibit CDK4/cyclin: Gly$^{101}$ $\rightarrow$ Trp reduces inhibitory capacity and mutations His$^{83}$ $\rightarrow$ Tyr and deletion of the fourth ankyrin repeat prevent inhibition[45].

The three variant N-terminal forms (giving rise to 148, 156 or 164 amino acid proteins) are indistinguishable in binding to CDK4 and CDK6. Seven mutations resulting in premature termination have been detected in tumours (Gln$^{50}$, Arg$^{58}$, Glu$^{61}$, Glu$^{69}$, Arg$^{80}$, Trp$^{110}$, Glu$^{120}$ $\rightarrow$ stop) and all of these cause loss of CDK binding. Termination in intron 2 (Asp$^{153}$) retains CDK binding. Mutants Arg$^{87}$ $\rightarrow$ Pro, His$^{98}$ $\rightarrow$ Pro and Ala$^{100}$ $\rightarrow$ Pro have negligible or greatly reduced CDK binding. The mutations Gly$^{101}$ $\rightarrow$ Trp and Val$^{126}$ $\rightarrow$ Asp have been shown by *in vitro* analysis to be temperature sensitive for binding to CDK4 and CDK6 and also for increasing the proportion of $G_1$ cells after transfection[71].

Germline mutations have been detected in melanomas[72–75] and in melanoma cell lines[76].

Mutations also occur in primary astrocytomas[48], primary gliomas[37,77], primary biliary tract cancers[78], primary oesophageal squamous cell carcinomas[79], Barrett's oesophagus[32], pancreatic adenocarcinomas[28], primary non-small cell lung cancer[34,80] and other tumours[81] and in cell lines from chondrosarcomas, fibrosarcoma, leiomyosarcomas, liposarcomas and bladder, prostate and lung carcinomas[34].

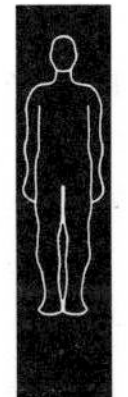

## Amino acid sequence of human INK4B/MTS2

```
  1 MREENKGMPS GGGSDEGLAS AAARGLVEKV RQLLEAGADP NGVNRFGRRA
 51 IQVMMMGSAR VAELLLLHGA EPNCADPATL TRPVHDAARE GFLDTLVVLH
101 RAGARLDVRD AWGRLPVDLA EERGHRDVAG YLRTATGD (138)
```

**Domain structure**

13–39 Incomplete ankyrin motif
73–103 Ankyrin motif

Conflicts: 20–21 SA → TP, 23 missing, 32–34 QLL → HSW (in ref. 18).

**Mutations detected in *INK4B***

Germline mutations (Asn$^{41}$ → Ser) in primary glioma [37]. One somatic nonsense mutation and one silent mutation in primary oesophageal carcinomas [82].

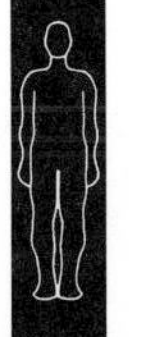

## Amino acid sequence of human INK4C

```
  1 MAEPWGNELA SAAARGDLEQ LTSLLQNNVN VNAQNGFGRT ALQVMKLGNP
 51 EIARRLLLRG ANPDLKDRTG FAVIHDAARA GFLDTLQTLL EFQADVNIED
101 NEGNLPLHLA AKEGHLRVVE FLVKHTASNV GHRNHKGDTA CDLARLYGRN
151 EVVSLMQANG AGGATNLQ (168)
```

**Domain structure**

64–95 Ankyrin motif

**Mutations detected in *INK4C***

Four amino acid substitutions (Arg → Pro, Cys → Phe, Leu → Phe, Val → Ile) have been detected in a breast cancer cell line, together with a polymorphism (Gly$^{114}$ → Gly) [83]. Mutations in *INK4C* appear to be rare in human tumours.

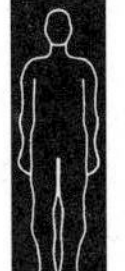

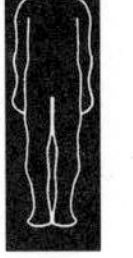

## Amino acid sequence of human INK4D

```
  1 MLLEEVRAGD RLSGAAARGD VQEVRRLLHR ELVHPDALNR FGKTALQVMM
 51 FGSTAIALEL LKQGASPNVQ DTSGTSPVHD AARTGFLDTL KVLVEHGADV
101 NVPDGTGALP IHLAVQEGHT AVVSFLAAES DLHRRDARGL TPLELALQRG
151 AQDLVDILQG HMVAPL (166)
```

**Database accession numbers**

| | *PIR* | *SWISSPROT* | *EMBL/GENBANK* | *REFERENCES* |
|---|---|---|---|---|
| Human *INK4A* | | P42771 | HSCDK4X, L27211, S69804, U12818–20 | 1,7,84–86 |
| Human *INK4B* | | P42772 | U17075, L36844, S69805 | 7,18 |
| Human *INK4C* | | P42773, CDN6_HUMAN | HS17074, U17074 | 87 |
| Human *INK4D* | | | HS403431, U40343 | 20 |
| Mouse *Ink-4C* | | | MM19596, U19596 | 5 |
| Human *INK4D* | | | U20498, U19597 | 2 |
| Mouse *Ink-4D* | | | U20497 | 2 |

***References***


1 Serrano, M. et al. (1993) Nature 366, 704–707.
2 Chan, F.K.M. et al. (1995) Mol. Cell. Biol. 15, 2682–2688.

[3] Stone, S. et al. (1995) Cancer Res. 55, 2988–2994.
[4] Quelle, D.E. et al. (1995) Oncogene 11, 635–645.
[5] Hirai, H. et al. (1995) Mol. Cell. Biol. 15, 2672–2681.
[6] Okuda, T. et al. (1995) Genomics 29, 623–630.
[7] Kamb, A. et al. (1994) Science 264, 436–440.
[8] Nobori, T. et al. (1994) Nature 368, 753–756.
[9] Della Ragione, F. et al. (1996) J. Biol. Chem. 271, 15942–15949.
[10] **Larsen, C.-J. (1996) Oncogene 12, 2041–2044.**
[11] Serrano, M. et al. (1995) Science 267, 249–252.
[12] Medema, R.H. et al. (1995) Proc. Natl Acad. Sci. USA 92, 6289–6293.
[13] Li, Y. et al. (1994) Cancer Res. 54, 6078–6082.
[14] Parry, D. et al. (1995) EMBO J. 14, 503–511.
[15] Hengstschläger, M. et al. (1996) Oncogene 12, 1635–1643.
[16] Suzuki, T. et al. (1996) EMBO J. 15, 1607–1614.
[17] Akagi, T. et al. (1996) Oncogene 12, 1645–1652.
[18] Hannon, G.J. and Beach, D. (1994) Nature 371, 257–261.
[19] Li, J.-M. et al. (1995) J. Biol. Chem. 270, 26750–26753.
[20] Guan K.L. et al. (1996) Mol. Biol. Cell 7, 57–70.
[21] Cairns, P. et al. (1995) Nature Genet. 11, 210–212.
[22] Liu, Q. et al. (1995) Oncogene 10, 1061–1067.
[23] Spruck, C.H. et al. (1994) Nature 370, 183–184.
[24] Hebert, J. et al. (1994) Blood 84, 4038–4044.
[25] Haidar, M.A. et al. (1995) Blood 86, 311–315.
[26] Kees, U.R. et al. (1996) Oncogene 12, 2235–2239.
[27] Huang, L. et al. (1996) Cancer Res. 56, 1137–1141.
[28] Caldas, C. et al. (1994) Nature Genet. 8, 27–32.
[29] Schmidt, E.E. et al. (1994) Cancer Res. 54, 6321–6324.
[30] Walker, D.G. et al. (1995) Cancer Res. 55, 20–23.
[31] Liu, Q. et al. (1995) Oncogene 10, 619–622.
[32] Barrett, M.T. et al. (1996) Oncogene 13, 1867–1873.
[33] Washimi, O. et al. (1995) Cancer Res. 55, 514–517.
[34] Nakagawa, K. et al. (1995) Oncogene 11, 1843–1851.
[35] Williamson, M.P. et al. (1995) Hum. Mol. Genet. 4, 1569–1577.
[36] Woloschak, M. et al. (1996) Cancer Res. 56, 2493–2496.
[37] Li, Y.-J. et al. (1995) Oncogene 11, 597–600.
[38] He, J. et al. (1995) Cancer Res. 55, 4833–4836.
[39] Lukas, J. et al. (1995) Nature 375, 503–506.
[40] Luca, M. et al. (1995) Oncogene 11, 1399–1402.
[41] Wang, Y. and Becker, D. (1996) Oncogene 12, 1069–1075.
[42] Glendening, J.M. et al. (1995) Cancer Res. 55, 5531–5535.
[43] Reymond, A. and Brent, R. (1995) Oncogene 11, 1173–1178.
[44] Ranade, K. et al. (1995) Nature Genet. 10, 114–116.
[45] Yang, R. et al. (1995) Cancer Res. 55, 2503–2506.
[46] Xiao, S. et al. (1995) Cancer Res. 55, 2968–2971.
[47] Linardopoulos, S. et al. (1995) Cancer Res. 55, 5168–5172.
[48] Ueki, K. et al. (1994) Hum. Mol. Genet. 3, 1841–1845.
[49] Saxena, A. et al. (1996) Oncogene 13, 661–664.
[50] Lo, K.-W. et al. (1996) Cancer Res. 56, 2721–2725.
[51] Rodabaugh, K.J. et al. (1995) Oncogene 11, 1249–1254.
[52] Lukas, J. et al. (1995) Cancer Res. 55, 4818–4823.

[53] Otterson, G.A. et al. (1995) Oncogene 11, 1211–1216.
[54] Fueyo, J. et al. (1996) Oncogene 13, 1615–1619.
[55] Della Ragione, F. et al. (1995) Oncogene 10, 827–833.
[56] Chen, Z.-H. et al. (1996) Cancer Res. 56, 1083–1090.
[57] Herman, J.G. et al. (1996) Cancer Res. 56, 722–727.
[58] Otterson, G.A. et al. (1994) Oncogene 9, 3375–3378.
[59] Jen, J. et al. (1994) Cancer Res. 54, 6353–6358.
[60] Xiao, S. et al. (1995) Oncogene 11, 511–515.
[61] Serrano, M. et al. (1996) Cell 85, 27–37.
[62] Rogan, E.M. et al. (1995) Mol. Cell. Biol. 15, 4745–4753.
[63] Fueyo, J. et al. (1996) Oncogene 12, 103–110.
[64] Loughran, O. et al. (1996) Oncogene 13, 561–568.
[65] Stone, S. et al. (1995) Oncogene 11, 987–991.
[66] Wick, S.T. et al. (1995) Oncogene 11, 2013–2019.
[67] Mao, L. et al. (1995) Cancer Res. 55, 2995–2997.
[68] Quelle, D.E. et al. (1995) Cell 83, 993–1000.
[69] Hara, E. et al. (1996) Mol. Cell. Biol. 16, 859–867.
[70] Duro, D. et al. (1996) Cancer Res. 56, 848–854.
[71] Parry, D. and Peters, G. (1995) Mol. Cell. Biol. 16, 3844–3852.
[72] Holland, E.A. et al. (1995) Oncogene 11, 2289–2294.
[73] Liu, L. et al. (1995) Oncogene 11, 405–412.
[74] Walker, G.J. et al. (1995) Hum. Mol. Genet. 4, 1845–1852.
[75] Borg, Å. et al. (1996) Cancer Res. 56, 2497–2500.
[76] Pollock, P.M. et al. (1995) Oncogene 11, 663–668.
[77] Kyritsis, A.P. et al. (1996) Oncogene 12, 63–67.
[78] Igaki, H. et al. (1995) Cancer Res. 55, 3421–3423.
[79] Yoshida, S. et al. (1995) Cancer Res. 55, 2756–2760.
[80] Packenham, J.P. et al. (1995) Clin. Cancer Res. 1, 687–690.
[81] Okamoto, A. et al. (1994) Proc. Natl Acad. Sci. USA 91, 11045–11049.
[82] Suzuki, H. et al. (1995) Hum. Mol. Genet. 4, 1883–1887.
[83] Zariwala, M. et al. (1996) Oncogene 12, 451–455.
[84] Hayashi, N. et al. (1994) Biochem. Biophys. Res. Commun. 202, 1426–1430.
[85] Hussussian, C.J. et al. (1994) Nature Genet. 8, 15–21.
[86] Okamoto, A. et al. (1995) Cancer Res. 55, 1448–1451.
[87] Guan, K.-L. et al. (1994) Genes Dev. 8, 2939–2952.

## Identification

p27[KIP1] protein was isolated from Mv1 Lu mink epithelial cells arrested in $G_1$ of the cell cycle by contact inhibition or by TGFβ [1].

## Related genes

KIP1, KIP2 and WAF1/CIP are members of the KIP/CIP family of mammalian cyclin-dependent kinase inhibitors. *KIP1, KIP2* and *WAF1* are highly homologous and have similar genomic structure, suggesting their common ancestry. The N-terminus of KIP1 is 42% identical to that of WAF1.

| | ***KIP1*** |
|---|---|
| **Chromosome** | 12p13 |
| **Mass (kDa): predicted** | 22 |
| **expressed** | 27 |
| **Cellular location** | Nucleus |
| **Tissue distribution** | Ubiquitous |

*KIP1* mRNA levels are constant during the cell cycle whereas levels of protein increase rapidly at exit from the cycle and in quiescent cells [2].

## Protein function

KIP1 (Cdk inhibitory protein 1, also called ICK and PIC2) is a cyclin-dependent kinase inhibitor (CDI) that interacts with CDK2, CDK4 and CDK6 to prevent their activation by CDK-activating kinase and *in vitro* KIP1 inhibits cyclin D/E/A and B kinases. Like WAF1, KIP1 binds more avidly to CLN/CDKs than to CDKs alone. Over-expression blocks cells in $G_1$ and the cAMP-induced block in $G_1$ of mouse macrophages is probably caused by increased levels of KIP1. In T cells the immunosuppressant rapamycin prevents the decrease in KIP1 levels normally caused by IL-2. In $G_0$ KIP1 is predominantly monomeric, with some associated with cyclin A/E-CDK2 and it represses the activity of cyclin A/CDK complexes [3]. During $G_1$ all KIP1 is sequestered by D1/CDK4 which, depending on the stoichiometry of KIP1 binding, may be active or inactive. KIP1 then redistributes to cyclin A/E-CDK2 as cells enter S phase.

KIP1 activity is increased by TGFβ or cell–cell contact and may be involved in TGFβ-mediated inhibition of pRb phosphorylation and arrest. TGFβ-induced increase in INK4B, which competes with D cyclins for binding to CDK4/CDK6, causes displacement of KIP1 from CDK4 to CDK2, leading to inhibition of E/CDK2 and $G_1$/S arrest [4].

KIP1 also binds to a cyclin D/CDK5 complex. However, in neurones CDK5 is associated with a non-cyclin activator, p35, and this complex is not inhibited by KIP1. CDK5 is thought to control cytoskeletal functions in postmitotic neurones [5].

### Cancer

Two polymorphisms and a nonsense mutation have been detected in KIP1. Heterozygous deletions commonly affect 12p12–13 and may therefore influence the concentration of KIP1 protein. The absence of KIP1 mutations has been noted both in leukaemias and in a variety of solid tumours [6–8].

Many transformed cells fail to arrest in response to TGF$\beta$, possibly because of mutations in the TGF$\beta$ receptor, which may reflect alteration in KIP1 and/or INK4B function.

### Transgenic animals

*Kip-1*$^{-/-}$ mice have enhanced growth and in particular the thymus, pituitary and adrenal glands are enlarged. The animals often spontaneously develop pituitary tumours. Ovarian follicle development is impaired and females are infertile. Cell cycle arrest induced by TGF$\beta$, rapamycin or contact inhibition is unimpaired [9–11].

### *In vitro*

The levels of KIP1 protein are markedly reduced following PDGF stimulation of quiescent BALB/c 3T3 cells which promotes a decrease in the amounts of KIP1-associated cyclins E, D1, D2 and D3 [12]. Anti-sense cDNA to *Kip-1* increases cyclin D1, cyclin A and dihydrofolate reductase expression, together with DNA synthesis in fibroblasts and the cells appear unable to leave the cell cycle [13].

## Gene structure

The genomic structure of *KIP1* is similar to that of *WAF1* [6].

## Protein structure

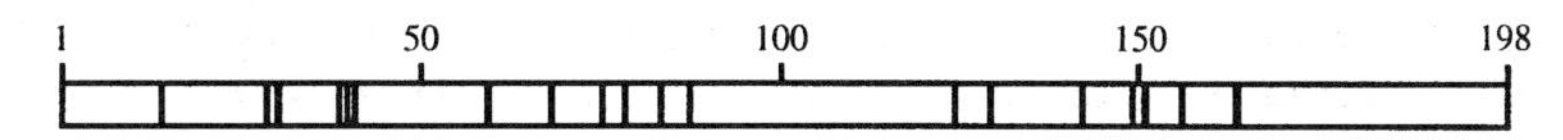


The vertical lines indicate regions of identity with WAF1. The crystal structure of the complex reveals that KIP1 binds to a groove on the conserved cyclin box of cyclin A and to the N-terminal lobe of CDK2 and that it also mimics ATP and inserts into the catalytic cleft [14].

## Amino acid sequence of human KIP1

```
  1 MSNVRVSNGS PSLERMDARQ AEHPKPSACR NLFGPVDHEE LTRDLEKHCR
 51 DMEEASQRKW NFDFQNHKPL EGKYEWQEVE KGSLPEFYYR PPRPPKGACK
101 VPAQESQDVS GSRPAAPLIG APANSEDTHL VDPKTDPSDS QTGLAEQCAG
151 IRKRPATDDS STQNKRANRT EENVSDGSPN AGSVEQTPKK PGLRRRQT
                                                   (198)
```

**Domain structure**

25–50 Cyclin A binding region (underlined)
28–88 Homology with WAF1
32–34 LFG motif conserved in all KIP/CIP/WAF1 family members
53–93 CDK2 binding region
152–166 Putative nuclear localization signal (underlined)
187–190 CDC2 kinase consensus site (underlined)

**Mutations detected in *KIP1***

Mutation in breast cancer: CAG → TAG, $Gln^{104}$ → stop [15].

KIP1 polymorphism: GTC → GGC, $Val^{109}$ → Gly (exon 1b); ACG → ACA, $Thr^{142}$ → Thr [6,15].

**Database accession numbers**

| | PIR | SWISSPROT | EMBL/GENBANK | REFERENCES |
|---|---|---|---|---|
| Human *KIP1* | | P46527 | U10906 | 1 |
| Mouse *Kip-1* | | P46414 | U10440, U09968 | 1.16 |

***References***


1 Polyak, K. et al. (1994) Cell 78, 56–66.
2 Hengst, L. and Reed, S.I. (1996) Science 271, 1861–1864.
3 Resnitzky, D. et al. (1995) Mol. Cell. Biol. 15, 4347–4352.
4 Reynisdóttir, I. et al. (1995) Genes Dev. 9, 1831–1845.
5 Lee, M.-H. et al. (1996) Proc. Natl Acad. Sci. USA 93, 3259–3263.
6 Pietenpol, J.A. et al. (1995) Cancer Res. 55, 1206–1210.
7 Stegmaier, K. et al. (1996) Cancer Res. 56, 1413–1417.
8 Ponce-Castaneda, M.V. et al. (1995) Cancer Res. 55, 1211–1214.
9 Nakayama, K. (1996) Cell 85, 707–720.
10 Kiyokawa, H. (1996) Cell 85, 721–732.
11 Fero, M.L. (1996) Cell 85, 733–744.
12 Agrawal, D. et al. (1996) Mol. Cell. Biol. 16, 4327–4336.
13 Rivard, N. et al. (1996) J. Biol. Chem. 271, 18337–18341.
14 Russo, A.A. et al. (1996) Nature 382, 325–331.
15 Spirin, K.S. et al. (1996) Cancer Res. 56, 2400–2404.
16 Toyoshima, H., and Hunter, T. (1994) Cell 78, 67–74.

## Identification

Mouse KIP2 was isolated using a two-hybrid screen to identify proteins binding to cyclins and cyclin-dependent kinases. Human *KIP2* was isolated by low stringency hybridization of a mouse probe to a human library [1].

## Related genes

KIP2, KIP1 and WAF1/CIP are highly homologous members of the KIP/CIP family of mammalian cyclin-dependent kinase inhibitors.

| | ***KIP2*** |
|---|---|
| **Nucleotides (kb)** | <2.2 |
| **Chromosome** | 11p15.5 |
| **Mass (kDa): predicted** | 38 |
| **expressed** | 57 |
| **Cellular location** | Nucleus |

**Tissue distribution**

During mouse embryogenesis KIP2 is coexpressed with WAF1 in skeletal muscle, brain, heart, lungs, eyes, cartilage and tongue muscle. In adult tissues, it is widely expressed in brain, kidney, heart, lung, liver and skeletal muscle [1].

## Protein function

*KIP2* is a cyclin-dependent kinase inhibitor (CDI) that is a potent inhibitor of cyclin E/CDK2, cyclin A/CDK2, cyclin E/CDK3 and cyclin D2/CDK4/CDK6. Its over-expression arrests cells in $G_1$, as does that of *KIP1*. Unlike *WAF1, KIP2* is not regulated by p53.

In growing cells KIP1 may be sequestered by D/CDK4 which, depending on the stoichiometry of KIP1 binding, may be active or inactive.

**Cancer**

The chromosomal region within which *KIP2* resides (11p15.5) is a common site for loss of heterozygosity in some types of sarcoma, Wilms' tumours and tumours associated with the Beckwith–Wiedemann syndrome. However, there is no evidence for mutations of *KIP2* in most of these tumours [2], although two mutations have been detected in Beckwith–Wiedemann syndrome ($Gln^{47} \rightarrow$ stop; $Phe^{265} \rightarrow stop^{274}$) that probably inactivate KIP2 [3].

In both humans and mice the paternal allele of *Kip-2* is transcriptionally repressed and methylated: genomic imprinting of a CDI gene suggests that inactivation of the maternal allele could contribute to tumorigenesis by acceleration of cell cycle progression [4]. In humans the gene is imprinted with maternal expression and the maternal alleles are selectively lost in a high proportion of lung cancers [5]. Four types of 12 bp in-frame deletions in the

PAPA repeat region have been detected in breast, liver and bladder carcinomas at approximately twice the frequency with which they occur in normal individuals [6]. Polymorphisms have also been detected (C to T at codons 44 and 185) [6,7].

## Gene structure

Alternative splicing of intron 1 generates three transcripts [6]. The GC-rich promoter has a TATA element and two CAT elements at −70, −160 and −210 bp, respectively, 5′ of the putative start site. There are 11 SP1 binding sites (−500 to −160 bp).

## Protein structure

Mouse KIP2 has four structurally distinct domains: N-terminal CDK inhibitory, proline-rich, acidic repeat and a C-terminal domain conserved with KIP1. Human KIP2 has conserved the N- and C-terminal domains but replaced the internal regions by Pro-Ala repeats.

## Amino acid sequence of human KIP2

```
  1 MERLVARGTF PVLVRTSACR SLFGPVDHEE LSRELQARLA ELNAEDQNRW
 51 DYDFQQDMPL RGPGRLQWTE VDSDSVPAFY RETVQVGRCR LLLAPRPVAV
101 AVAVSPPLEP AAESLDGLEE APEQLPSVPV PAPASTPPPV PVLAPAPAPA
151 PAPVAAPVAA PVAVAVLAPA PAPAPAPAPA PAPVAAPAPA PAPAPAPAPA
201 PAPAPDAAPQ ESAEQGANQG QRGQEPLADQ LHSGISGRPA AGTAAASANG
251 AAIKKLSGPL ISDFFAKRKR SAPEKSSGDV PAPCPSPSAA PGVGSVEQTP
301 RKRLR (316)
```

### Domain structure

18–80 N-terminal CDK inhibitory domain with homology to WAF1 and KIP1
131–205 Proline-alanine repeat domain (underlined)
291–302 C-terminal QT domain conserved with KIP1

The major transcript has an ORF with a translation start site 12 amino acids downstream from the full length start site [6].

### Database accession numbers

| | PIR | SWISSPROT | EMBL/GENBANK | REFERENCES |
|---|---|---|---|---|
| Human *KIP2* | | P49918 | HSKIP2, U22398 | 1 |
| | | | D64137 | 6 |
| | | | U48869 | 7 |
| Mouse *Kip-2* | | | U22399 | 1 |

***References***

1. Matsuoka, S. et al. (1995) Genes Dev. 9, 650–662.
2. Orlow, I. et al. (1996) Cancer Res. 56, 1219–1221.
3. Hatada, I. et al. (1996) Nature Genet. 14, 171–173.
4. Hatada, I. and Mukai, T. (1995) Nature Genet. 11, 204–206.
5. Kondo, M. et al. (1996) Oncogene 12, 1365–1368.
6. Tokino, T. et al. (1996) Human Genet. 97, 625–631.
7. Reid, L.H. et al. (1996) Cancer Res. 56, 1214–1218.

## Identification

WAF1 was identified as a cyclin D1-associated protein in anti-cyclin D1 immunoprecipitates [1].

## Related genes

INK4 and KIP/CIP proteins comprise two families of mammalian cyclin-dependent kinase inhibitors. The KIP/CIP family is composed of three structurally related members, WAF1/CIP, KIP1 and KIP2. *WAF1, KIP1* and *KIP2* are highly homologous and have similar genomic structure, suggesting their common ancestry. $p27^{Xic1}$ is a *Xenopus* cyclin-dependent kinase inhibitor with homology to WAF1 and KIP1/KIP2 [2].

| | ***WAF1*** |
|---|---|
| **Nucleotides** | Not fully mapped |
| **Chromosome** | 6p21 |
| **Mass (kDa): predicted** | 18 |
| **expressed** | 21 |
| **Cellular location** | Nuclear |

**Tissue distribution**

*WAF1* mRNA in normal cells is maximal between 3 and 9 h after serum, EGF or TPA stimulation and minimal during S phase [3], being activated by a MAP kinase pathway [4]. It is expressed in cells undergoing either $G_1$ arrest (e.g. caused by $\gamma$-irradiation, when WAF1 expression correlates with inhibition of E/CDK2 and A/CDK2 and failure to accumulate hypophosphorylated pRb [5,6]) or apoptosis by p53-dependent mechanisms [7]. However, WAF1 is fully dispensable for p53-mediated apoptosis [8]. WAF1 is expressed during differentiation of neuroblastoma cells, blocking cyclin E/CDK activity: inhibition of expression promotes apoptosis [9].

## Protein function

*WAF1* (wild-type p53-activated fragment 1, also called SDI1, CIP1, CAP20 or PIC1) is a cyclin-dependent kinase inhibitor (CDI) and is a key regulator of normal proliferation, the expression of which is induced by wild-type but not mutant p53 acting *via* a p53 response element. $p21^{WAF1}$ binds to and inhibits cyclin D/CDK4, E/CDK2, A/CDK2 and CDC2. The interaction is with the cyclins (D1, D2, D3, E, A and, more weakly B): WAF1 (and also KIP1) do not bind significantly to isolated kinase subunits [10]. WAF1 is present in a quaternary complex with a cyclin, a cyclin-dependent kinase (CDK) and proliferating cell nuclear antigen (PCNA, the processivity unit of DNA polymerase $\delta$, the principal replicative DNA polymerase) in normal human cells, but not in many tumour cells. WAF1 thus functions as a CDK inhibitor to block progression through the cell cycle, directly inhibiting PCNA-dependent DNA replication in the absence of a cyclin/CDK, blocking the ability of PCNA to activate DNA polymerase $\delta$[11] and preventing interaction of PCNA with GADD45, with which it also directly interacts[12,13]. WAF1 also causes FEN1, a 5′–3′ exonuclease essential for the degradation of RNA primer–DNA junctions, to dissociate from PCNA[14]. These quaternary complexes are active at a 1:1 stoichiometry of WAF1 binding but are inactivated when >1 molecule of WAF1 is bound and WAF1 may function as an assembly factor for cyclin/CDK complexes. D/CDKs and E/CDKs regulate the $G_1$/S checkpoint: A/CDK2 is required for ongoing DNA synthesis. Most transformed cells contain WAF1/cyclin/CDK complexes. The inhibition of cyclin-dependent kinase activity by WAF1 inhibits phosphorylation of pRb and hence E2F activation in late $G_1$ but WAF1 also suppresses E2F *trans*-activation capacity in a pRb-independent manner[15]. WAF1 also inhibits the stress-activated protein kinases (SAPKs or JNKs), but not extracellullar signal-regulated kinases (ERKs)[16]. WAF1 binds to the regulatory subunit of protein kinase CK2 and downregulates its activity[17].

Terminal differentiation of M1 myeloblastic leukaemia cells is associated with induction of myeloid differentiation primary response (MyD) genes: these include *MyD118* that is related to *GADD45*. MyD118 also interacts with PCNA and WAF1. However, *MyD118* but not *GADD45* is induced by TGF$\beta$ whereas the converse pattern is induced by wild-type p53[18]. WAF1 (or INK4A) expression increases muscle-specific gene expression.

The inhibition by MYC of p53-induced $G_1$ arrest does not involve increased expression of CDKs or cyclins but appears to arise from induction of a heat-labile inhibitor of WAF1[19].

## Cancer

Four WAF1 mutations have been detected in primary prostate cancer[20]. Mutants in WAF1 have not been detected in primary sarcomas but a mutation (Arg$^{94}$ → Trp) has been detected in breast carcinoma that impairs the CDK inhibitory activity of WAF1[21]. Mutations also occur in transitional cell carcinoma, including Leu$^{89}$ → Gln, Gly$^{96}$ → Thr and putative truncating frameshifts[22]. Two common variants occur in both tumours and normal cells: their frequency of appearance is greater in tumours expressing wild-type *P53*[23]. Over-expression of WAF1 has been detected in the majority of a sample of human gliomas[24] and of non-small cell lung carcinomas[25].

*WAF1* expression can be induced by p53-independent mechanisms (e.g. by treatment with TPA, serum, PDGF, FGF, okadaic acid, butyric acid, retinoic acid, vitamin $D_3$, IL-6, IFN-$\gamma$, TGF$\beta$ or G-CSF) in human breast carcinoma cell lines[26] and other tumour-derived cells[27] and its overexpression causes apoptosis in these cells[28]. Constitutive expression of *Mybb* can bypass p53-induced WAF1-mediated $G_1$ arrest by a mechanism dependent on the DNA binding domain of MYBB[29].

## Transgenic animals

Mice lacking p21$^{WAF1}$ develop normally and do not acquire spontaneous malignancies. However, embryonic fibroblasts are defective in ability to arrest in $G_1$ in response to DNA damage, having a phenotype intermediate between that of wild-type cells and p53$^{-/-}$ cells[30].

Over-expression of p21 in murine hepatocytes arrests cell cycle progression, giving rise to aberrant tissue organization and increased mortality[31].

### *In vitro*

Over-expression of WAF1 arrests the growth of chicken embryo fibroblasts transformed by v-*src*, *Ras*, *Mos* or *Myc*[32]. In human colon adenocarcinoma cell lines the p53-mediated $G_1$ arrest induced by DNA damage is completely abrogated in WAF1-deficient cells and the cells arrest in a $G_2$-like state, subsequently undergoing additional S phases without intervening mitoses, followed by apoptosis[33]. *WAF1*-deficient human colon carcinoma cells are defective in DNA repair capacity[34].

In NIH 3T3 cells expressing TRKA, nerve growth factor (NGF) greatly increases levels of WAF1 protein and causes cell cycle arrest[35].

In immortalized human keratinocytes TGF$\beta$ induces WAF1 expression by a p53-independent pathway[36]. As TGF$\beta$ can also induce KIP1 and INK4B it evidently exerts multiple effects on growth controlling pathways. Over-expression of *WAF1* suppresses proliferation *in vitro* and tumorigenicity *in vivo*[37]. In rat keratinocytes ultraviolet B (UVB, 290–320 nm) irradiation induces WAF1 expression causing $G_1$/S phase arrest[38].

In a number of cell lines the human T cell leukaemia virus type I (HTLV-I) TAX oncoprotein strongly induces expression of *WAF1*[39].

## Transcriptional regulation

The promoter includes a TATA box, SP1 consensus binding sequences and an almost perfect p53 consensus binding sequence. A 10 bp promoter (−74 to −83) sequence is required for TGF$\beta$ *trans*-activation to which factors including SP1 and SP3 bind[40]. The CCAAT/enhancer binding protein $\alpha$ (C/EBP$\alpha$) activates the *WAF1* promoter and also causes post-translational stabilization of WAF1 protein[41]. Three potential SIE (sis-inducible element) sites contain the palindromic sequence TTCNNNGAA (at −640, −2540 and −4183 nt relative to the TATA promoter site) and each binds STAT proteins. Inhibition of cell growth (by EGF or IFN-$\gamma$) correlates with the activation of STATs and STAT1 and STAT3 directly induce *WAF1* transcription[42].

## Protein structure

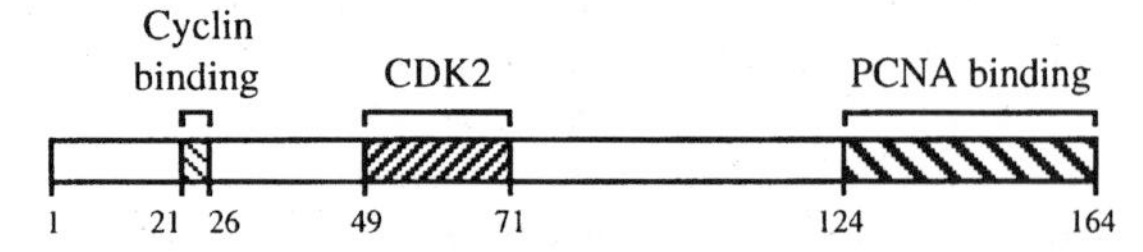


Mutation of WAF1 at amino acids 21 and 24 abrogates the capacity to suppress tumour cell growth. Over-expression of cyclin D or cyclin E can partially overcome growth suppression by wild-type WAF1[43]. Conformational disorder in the N-terminus, revealed by NMR spectroscopy, may promote the diversity of binding to cyclin/CDK complexes[44].

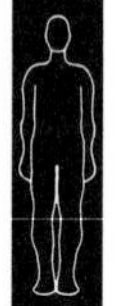

## Amino acid sequence of human WAF1

```
  1 MSEPAGDVRQ NPCGSKACRR LFGPVDSEQL SRDCDALMAG CIQEARERWN
 51 FDFVTETPLE GDFAWERVRG LGLPKLYLPT GPRRGRDELG GGRRPGTSPA
101 LLQGTAEEDH VDLSLSCTLV PRSGEQAEGS PGGPGDSQGR KRRQTSMTDF
151 YHSKRRLIFS KRKP (164)
```

## Domain structure

13–41 Putative zinc finger

18–30 CDK/cyclin inhibitory domain: region of high homology with KIP1 and KIP2 (underlined: LFGPVD is absolutely conserved). This region includes residues 17–24 that have been designated cyclin binding motif (Cy1)[45]

21, 24 Essential for cyclin D and E binding (D1/CDK4, E/CDK2)

141–156 Putative nuclear localization signal

53–58 K site, essential for CDK2 binding

152–159 Second cyclin binding motif (underlined: Cy2). Cyclins A and E associate with Cy1; cyclin D or CDK4 do not bind alone but the complex associates *via* Cy1[45]

## Database accession numbers

| | *PIR* | *SWISSPROT* | *EMBL/GENBANK* | *REFERENCES* |
|---|---|---|---|---|
| Human *WAF1/CDN1* | S39357 | P38936 | U09579, L25610, U03106, S67388 | 46–48 |
| Mouse *WAF1/CDN1* | A49438 | P39689 | | 47 |

## *References*


1. Xiong, Y. et al. (1992) Cell 71, 505–514.
2. Su, J.Y. et al. (1995) Proc. Natl Acad. Sci. USA 92, 10187–10191.
3. Li, Y. et al. (1994) Oncogene 9, 2261–2268.
4. Liu, Y. et al. (1996) Cancer Res. 56, 31–35.
5. Dulic, V. et al. (1994) Cell 76, 1013–1023.
6. Slebos, R.J.C. et al. (1994) Proc. Natl Acad. Sci. USA 91, 5320–5324.
7. El-Deiry, W.S. et al. (1994) Cancer Res. 54, 1169–1174.
8. Attardi, L.D. et al. (1996) EMBO J. 15, 3693–3701.
9. Poluha, W. et al. (1996) Mol. Cell. Biol. 16, 1335–1341.
10. Hall, M. et al. (1995) Oncogene 11, 1581–1588.
11. Waga, S. et al. (1994) Nature 369, 574–578.
12. Chen, I.-T. et al. (1995) Oncogene 11, 1931–1937.
13. Kearsey, J.M. et al. (1995) Oncogene 11, 1675–1683.
14. Chen, J. et al. (1996) Proc. Natl Acad. Sci. USA 93, 11597–11602.
15. Dimri, G.P. et al. (1996) Mol. Cell. Biol. 16, 2987–2997.
16. Shim, J. et al. (1996) Nature 381, 804–807.
17. Götz, C. et al. (1996) Oncogene 13, 391–398.
18. Vairapandi, M. et al. (1996) Oncogene 12, 2579–2594.
19. Hermeking, H. et al. (1995) Oncogene 11, 1409–1415.
20. Gao, X. et al. (1995) Oncogene 11, 1395–1398.
21. Balbín, M. et al. (1996) J. Biol. Chem. 271, 15782–15786.
22. Malkowicz, S.B. et al. (1996) Oncogene 13, 1831–1837.
23. Mousses, S. (1995) Hum. Mol. Genet. 4, 1089–1092.
24. Jung, J.-M. et al. (1995) Oncogene 11, 2021–2028.
25. Marchetti, A. et al. (1996) Oncogene 12, 1319–1324.
26. Sheikh, M.S. et al. (1994) Oncogene 9, 3407–3415.
27. Zeng, Y.-X. and El-Deiry, W.S. (1996) Oncogene 12, 1557–1564.
28. Sheikh, M.S. et al. (1995) Oncogene 11, 1899–1905.
29. Lin, D. et al. (1994) Proc. Natl Acad. Sci. USA 91, 10079–10083.
30. Deng, C. et al. (1995) Cell 82, 675–684.

[31] Wu, H. et al. (1996) Genes Dev. 10, 245–260.
[32] Givol, I. et al. (1995) Oncogene 11, 2609–2618.
[33] Waldman, T. et al. (1996) Nature 381, 713–716.
[34] McDonald, E.R. et al. (1996) Cancer Res. 56, 2250–2255.
[35] Decker, S.J. (1995) J. Biol. Chem. 270, 30841–30844.
[36] Datto, M.B. et al. (1995) Proc. Natl Acad. Sci. USA 92, 5545–5549.
[37] Chen, Y.Q. et al. (1995) Cancer Res. 55, 4536–4539.
[38] Petrocelli, T. et al. (1996) Oncogene 12, 1287–1296.
[39] Akagi, T. et al. (1996) Oncogene 12, 1645–1652.
[40] Datto, M.B. et al. (1995) J. Biol. Chem. 270, 28623–28628.
[41] Timchenko, N.A. et al. (1996) Genes Dev. 10, 804–815.
[42] Chin, Y.E. et al. (1996) Science 272, 719–722.
[43] Lin, J. et al. (1996) Mol. Cell. Biol. 16, 1786–1793.
[44] Kriwacki, R.W. et al. (1996) Proc. Natl Acad. Sci. USA 93, 11504–11509.
[45] Chen, J. et al. (1996) Mol. Cell. Biol. 16, 4673–4682.
[46] Harper, J.W. et al. (1993) Cell 75, 805–816.
[47] El-Deiry, W.S. et al. (1993) Cell 75, 817–825.
[48] Xiong, Y. et al. (1993) Nature 366, 701–704.

# Section VI

# DNA TUMOUR VIRUSES

## INTRODUCTION

Within the genomes of some DNA tumour viruses genes have been identified the products of which will transform cells, that is, the oncogene product has evolved to stimulate cells after viral infection. These include simian vacuolating virus 40 (SV40), the human BK and JC papovaviruses, polyomavirus and adenovirus, each of which can transform a variety of types of cell *in vitro* and can be tumorigenic in appropriate host animals. These viruses are not oncogenic in humans although their oncoproteins function at least in part by binding to and thus inhibiting the normal activity of the products of tumour suppressor genes in target cells (p53 and pRb). A similar strategy is followed by the oncogenic human papillomaviruses (see below). The Epstein–Barr herpesvirus carries one gene that is consistently expressed in Burkitt's lymphoma and another that is required for immortalization of primary B cells by EBV and is tumorigenic *in vitro*. For other DNA viruses (e.g. hepatitis B virus (HBV) and the herpesviruses HSV-1, HSV-2 and human cytomegalovirus) there is circumstantial evidence for their association with the development of a variety of human cancers and HBV is established as a causative agent in the vast majority of primary hepatocellular carcinomas that are responsible for up to 1 million deaths per annum world wide [1,2]. However, the genomes of these viruses contain no known oncogenes, although the hepatitis B virus protein HBx elevates expression of the TATA binding protein and *trans*-activates a variety of promoters commonly involved in transformation [3–5], activates the RAS-RAF pathway [6], accelerates progression through $G_0/G_1$ and $G_2/M$ of the cell cycle by enhancing CDK2 and CDC2 activity [7] and causes tumours in transgenic mice [8]. The T cell transforming herpesvirus saimiri ($\gamma$ herpesvirus group) contain ORFs with transformation capacity and one of these (STP-C) induces epithelial tumours when expressed in transgenic mice [9].

### *References*


[1] **Harris, C.C. (1990) Cancer Cells 2, 146–148.**

[2] **zur Hausen, H. (1991) Science 254, 1167–1173.**

[3] Kekule, A.S. et al. (1993) Nature 361, 742–745.

[4] Schlüter, V. et al. (1994) Oncogene 9, 3335–3344.

[5] Wang, H.-D. et al. (1995) Mol. Cell. Biol. 15, 6720–6728.

[6] Doria, M. et al. (1995) EMBO J. 14, 4747–4757.

[7] Benn, J. and Schneider, R.J. (1995) Proc. Natl Acad. Sci. USA 92, 11215–11219.

[8] Kim, C.-M. et al. (1991) Nature 351, 317–320.

[9] Murphy, C. et al. (1994) Oncogene 9, 221–226.

# Papillomaviruses

There are over 60 human papillomavirus (HPV) genotypes [1] that colonize various stratified epithelia including the skin and oral and genital mucosa and induce the formation of self-limiting, benign tumours known as papillomas (warts) or condylomas. Viruses increase the division rate of infected stem cells in the epithelial basal layer and the viral genome replicates episomally in concert with the cell genome. As the keratinocytes terminally differentiate, synthesis of virally encoded late proteins causes cell death. Very infrequently this mechanism defaults and malignant tumours develop [2–4]. The genomes comprise "early" ORFs (E1–E8) and "late" ORFs (L1 and L2).

The human *PE5L* gene and mouse HC1 are related to human papillomavirus type 18 E5 [5].

| | *E5* | *E6* | *E7* |
|---|---|---|---|
| **Mass (kDa): predicted** (type 16/18) | 9.4/8.3 | 19.2/18.8 | 11.0/11.9 |
| **expressed** | 44 | 18 | 20 |
| **Cellular location** | Plasma membrane (homodimers) | Nucleus | Nucleus |
| **Protein function** | Activates receptors for PDGFB, EGF and CSF1 | Binds p53 | Binds pRb |

**Cancer**

Eleven HPVs have been shown to be commonly associated with human tumours and DNA sequences of HPV16 or HPV18 are expressed in ~90% of all cervical carcinomas and HPV16 DNA is present in 50% of anal cancers. HPV types 6 and 11 are frequently present together with types 16 and 18 in genital infections but are rarely detected in cervical malignancies. Thus infection with certain types of HPV constitutes an increased risk of the onset of some cancers. The HPV16 promoter is *trans*-activated by MYB [6] and the HPV16 enhancer has binding sites for a variety of ubiquitous transcription factors (OCT1, NFA, TEF2, NF1 and AP1). The HPV16 long control region contains three binding sites for the cellular transcriptional repressor YY1. Point mutations in this region have been detected in episomal HPV16 DNAs from cervical cancers that result in enhanced promoter activity [7].

**E5**

Expression of the HPV16 E5 gene transforms 3T3-A31 cells and enhances their tumorigenicity. The transcription of *Fos* in 3T3-A31 cells in response to EGF, PDGF or serum is increased by expression of E5 [8]. This suggests that E5 may contribute to the early stages of tumorigenesis by enhancing proliferation in HPV16-infected cells.

**E6 and E7**

E6 and E7 DNAs are detectable in all HPV-associated tumours. E2 is usually interrupted or deleted and the late genes are often missing. In some classes of tumours (e.g. cervical carcinomas) the E1, E6 and E7 ORFs of HPV DNA are usually integrated into the host chromosome (the ORFs of E2, E3, E4 and E5 are lost) whereas in others (e.g. squamous cell carcinomas) they are present in extra-chromosomal form[9]. The expression of anti-sense E6/E7 renders HPV-containing cervical carcinoma cells non-tumorigenic[10].

The expression of HPV type 16 E6 and E7 oncogenes in the skin of transgenic mice promotes the development of epidermal cancers[11] in a process that may involve the action of autocrine growth factors[12]. Expression of either HPV16 E6 or E7 genes in non-metastatic cell lines renders these cells metastatic in nude mice[13]. Testicular tumours are induced in male transgenic mice expressing HPV16 E6/E7 in which the activation of *KIT* by its ligand is essential for tumorigenesis to develop[14].

HPV16 E6 and E7 proteins cooperate to immortalize human fibroblasts, keratinocytes or mammary epithelial cells[15,16]. Prolonged growth *in vitro* produces malignant clones from such immortalized cells[15]. The expression of E6 in early passage human keratinocytes or mammary epithelial cells activates telomerase by a mechanism that appears to be p53-independent. Keratinocytes expressing E6 have an extended lifespan but are not immortalized as a result of telomerase activation[18].

The coexpression of v-*ras* with E7 transforms primary cells[19] and v-*ras* and E6 immortalize primary cells[20]. E7 functions as an immortalizing oncogene in that there is a continued requirement for its expression, together with *Ras*, to maintain the transformed phenotype[21]. However, this does not imply that E6 functions as a RAS protein. Binding of E7 to the retinoblastoma protein is necessary but not sufficient for co-transformation[22], although mutational analysis of the cottontail rabbit papillomavirus E7 protein indicates that its association with pRb is not necessary for the viral induction of warts[23]. E7 also associates with $p33^{CDK2}$ and cyclin A[24]. E7 is a weak mitogen for Swiss 3T3 fibroblasts in which its expression induces the appearance of the transcriptionally active form of E2F[25]. E7 also forms complexes with JUN, JUNB, JUND and FOS and can *trans*-activate JUN-induced transcription[26].

HPV16 and 18 E6 enhance the degradation rate of p53 *via* specific binding (see Section IV, ***P53***) and a fusion protein of the N-terminal half of HPV16 E7 and full-length HPV16 E6 promotes the degradation of pRb[27]. E6 also associates with other cellular proteins (p33, p75, p81, p100, p182, p212)[28] and with E6BP (identical to ERC-55), a putative calcium-binding protein located in the endoplasmic reticulum[29].

**Bovine papillomavirus 1 (BPV-1)**

BPV-1 replicates episomally in fibroblasts *in vitro* and transforms the cells: replication with high efficiency requires the E1, E6 and E7 gene products. The E5 gene encodes a hydrophobic, 44 amino acid (7 kDa) protein that causes tumorigenic transformation of rodent fibroblasts[30]. In NIH 3T3 cells this involves activation of EGF and CSF1 receptors, independently of the presence of their ligands[31]. It is the smallest known oncoprotein and microinjection into the nucleus of the 13 C-terminal amino acids of E5 activates DNA synthesis. The 30 amino acid, N-terminus contains two cysteine residues essential for homodimerization and biological activity. In two cell lines, E5 has been shown to form an activating complex with the PDGFRB and to stimulate DNA synthesis[32]. Activation of its transforming capacity may require association with a 16 kDa cellular protein[33].

The transmembrane domain may be largely replaced by that of NEU. Transformation by E5 appears to require dimerization capacity and presentation of a negatively charged residue at the extracellular side of the membrane[34].

## Amino acid sequence of HPV16 E5

1 MTNLDTASTT LLACFLLCFC VLLCVCLLIR PLLLSVSTYT SLIILVLLLW
51 ITAASAFRCF IVYIIFVYIP LFLIHTHARF LIT (83)

## Amino acid sequence of HPV18 E5

1 MLSLIFLFCF CVCMYVCCHV PLLPSVCMCA YAWVLVFVYI VVITSPATAF
51 TVYVFCFLLP MLLLHIHAIL SLQ (73)

## Amino acid sequence of HPV16 E6

1 MHQKRTAMFQ DPQERPRKLP QLCTELQTTI HDIILECVYC KQQLLRREVY
51 DFAFRDLCIV YRDGNPYAVC DKCLKFYSKI SEYRHYCYSL YGTTLEQQYN
101 KPLCDLLIRC INCQKPLCPE EKQRHLDKKQ RFHNIRGRWT GRCMSCCRSS
151 RTRRETQL (158)

**Domain structure**

37–73 and 110–146 Zinc finger domains (underlined)

## Amino acid sequence of HPV18 E6

1 MARFEDPTRR PYKLPDLCTE LNTSLQDIEI TCVYCKTVLE LTEVFEFAFK
51 DLFVVYRDSI PHAACHKCID FYSRIRELRH YSDSVYGDTL EKLTNTGLYN
101 LLIRCLRCQK PLNPAEKLRH LNEKRRFHNI AGHYRGQCHS CCNRARQERL
151 QRRRETQV (158)

**Domain structure**

32–68 and 105–141 Zinc finger domains (underlined)

## Amino acid sequence of HPV16 E7

1 MHGDTPTLHE YMLDLQPETT DLYCYEQLND SSEEEDEIDG PAGQAEPDRA
51 HYNIVTFCCK CDSTLRLCVQ STHVDIRTLE DLLMGTLGIV CPICSQKP (98)

**Domain structure**

| | |
|---|---|
| 21–30 | pRb binding region (underlined). An additional domain (amino acids 60–98) appears necessary to inhibit complex formation between pRb and E2F[35]. |
| 58–61 and 91–94 | C-XX-C motifs |
| 31 and 32 | Casein kinase II phosphorylation sites. Mutation in these regions inhibits transforming potential, although negative charge at these sites is not essential for pRb binding[36]. HPV16 E7 shows significant sequence homology with E1A, T Ag, MYC and the yeast mitotic regulator *cdc25*. |

## Amino acid sequence of HPV18 E7

```
  1 MHGPKATLQD IVLHLEPQNE IPVDLLCHEQ LSDSEEENDE IDGVNHQHLP
 51 ARRAEPQRHT MLCMCCKCEA RIKLVVESSA DDLRAFQQLF LNTLSFVCPW
101 CASQQ (105)
```

**Domain structure**

| | |
|---|---|
| 24–33 | pRb binding region (underlined) |
| 63–66 and 98–101 | C-XX-C motifs |

## Amino acid sequence of BPV-1 E5

```
  1 MPNLWFLLFL GLVAAMQLLL LLFLLLFFLV YWDHFECSCT GLPF (44)
```

**Domain structure**

| | |
|---|---|
| 6–30 | Transmembrane domain (underlined) |
| 32–44 | Cellular DNA synthesis induction |

**Database accession numbers**

| | *PIR* | *SWISSPROT* | *EMBL/GENBANK* | *REFERENCES* |
|---|---|---|---|---|
| HPV16 E5 | A30016 | P06927 | K02718 | 37,38 |
| HPV18 E5 | F26251 | P06792 | X05015 | 39 |
| HPV16 E6 | A30682 | P03126 | K02718 | 37 |
| HPV18 E6 | G26251 | P06463 | X04354, X05015, M20325, M26798 | 39–43 |
| HPV16 E7 | | P03129 | K02718 | 37,44 |
| HPV18 E7 | H26251 | P06788 | X05015 | 39 |
| | | | M20324 | 41 |
| | | | M20325, M26798 | 42 |
| BPV-1 E5 | B18151 | P06928 | X02346 | 45 |
| | F18151, S12366 | | M20219 | 33,46 |


*References*

[1] de Villiers, E.-M. (1989) J. Virol. 63, 4898–4903.
[2] **Campo, M.S. (1992) J. Gen. Virol. 73, 217–222.**
[3] **DiMaio, D. (1991) Adv. Cancer Res. 56, 133–159.**
[4] **Galloway, D.A. and McDougall, J.K. (1989) Adv. Virus Res. 37, 125–171.**
[5] Geisen, C. et al. (1995) Hum. Mol. Genet. 4, 1337–1345.
[6] Nürnberg, W. et al. (1995) Cancer Res. 55, 4432–4437.
[7] May, M. et al. (1994) EMBO J. 13, 1460–1466.
[8] Leechanachai, P. et al. (1992) Oncogene 7, 19–25.
[9] Cullen, A.P. et al. (1991) J. Virol. 65, 606–612.
[10] von Knebel Doeberitz, M. et al. (1988) Cancer Res. 48, 3780–3786.
[11] Lambert, P.F. et al. (1993) Proc. Natl Acad. Sci. USA 90, 5583–5587.
[12] Auewarakul, P. et al. (1994) Mol. Cell. Biol. 14, 8250–8258.
[13] Chen, L. et al. (1993) Proc. Natl Acad. Sci. USA 90, 6523–6527.
[14] Kondoh, G. et al. (1995) Oncogene 10, 341–347.
[15] Band, V. et al. (1990) Proc. Natl Acad. Sci. USA 87, 463–467.
[16] Hudson, J.B. et al. (1990) J. Virol. 64, 519–526.
[17] Hurlin, P.J. et al. (1991) Proc. Natl Acad. Sci. USA 88, 570–574.
[18] Klingelhutz, A.J. (1996) Nature 380, 79–82.
[19] Durst, M. et al. (1989) Virology 173, 767–771.

[20] Storey, A. and Banks, L. (1993) Oncogene 8, 919–924.
[21] Crook, T. et al. (1989) EMBO J. 8, 513–519.
[22] Banks, L. et al. (1990) Oncogene 5, 1383–1389.
[23] Defeo-Jones, D. et al. (1993) J. Virol. 67, 716–725.
[24] Tommasino, M. et al. (1993) Oncogene 8, 195–202.
[25] Morris, J.D.H. et al. (1993) Oncogene 8, 893–898.
[26] Antinore, M.J. et al. (1996) EMBO J. 15, 1950–1960.
[27] Scheffner, M. et al. (1992) EMBO J. 2425–2431.
[28] Keen, N. et al. (1994) Oncogene 9, 1493–1499.
[29] Chen, J.J. et al. (1995) Science 269, 529–531.
[30] Settleman, J. et al. (1989) Mol. Cell. Biol. 9, 5563–5572.
[31] Martin, P. et al. (1989) Cell 59, 21–32.
[32] Petti, L., and DiMaio, D. (1992) Proc. Natl. Acad. Sci. USA 6736–6740.
[33] Goldstein, D.J. and Schlegel, R. (1990) EMBO J. 9, 137–145.
[34] Meyer, A.N. et al. (1994) Proc. Natl Acad. Sci. USA 91, 4634–4638.
[35] Huang, P.S. et al. (1993) Mol. Cell. Biol. 13, 953–960.
[36] Firzlaff, J.M. et al. (1991) Proc. Natl Acad. Sci. USA 88, 5187–5191.
[37] Seedorf, K. et al. (1985) Virology 145, 181–185.
[38] Bubb, V. et al. (1988) Virology 163, 243–246.
[39] Cole, S.T. and Danos, O. (1987) J. Mol. Biol. 193, 599–608.
[40] Matlashewski, G. et al. (1986) J. Gen. Virol. 67, 1909–1916.
[41] Inagaki, Y. et al. (1988) J. Virol. 62, 1640–1646.
[42] Schneider-Gaedicke, A. and Schwarz, E. (1986) EMBO J. 5, 2285–2292.
[43] Grossman, S.R. and Laimins, L.A. (1989) Oncogene 4, 1089–1093.
[44] Phelps, W.C. et al. (1988) Cell 53, 539–547.
[45] Schlegel, R. et al. (1986) Science 233, 464–467.
[46] Petti, L. et al. (1991) EMBO J. 10, 845–856.

# Epstein–Barr virus (EBV)

Epstein–Barr virus (EBV) is a herpesvirus of high incidence in humans that binds to the CR2 receptor for complement factor iC3b present only on B lymphocytes, follicular dendritic cells and B- and T-derived cell lines [1]. After infection the viral genome circularizes through its terminal repeats and in immortalized cells is latently maintained as an episomal molecule. Approximately 11 genes are expressed in latent infections [2]. Five regions of the genome encode latent polyadenylated transcripts and these give rise to six different nuclear proteins (Epstein–Barr nuclear antigens (EBNAs) 1, 2, 3A, 3B and 3C and a leader protein (EBNA-LP)) and three viral membrane proteins (latent membrane proteins, LMPs 1, 2A and 2B). The switch to expressing most of the other genes is mediated by the virally encoded transcription factor Z (also called BZLF1, EB1, ZEBRA or Zta). Z transcription can be enhanced by TPA, cross-linked cell surface Ig or by its own protein product. The Z protein has a basic DNA binding domain homologous to that of FOS that binds to AP1 sites and activates transcription of the EBV early genes *BSLF2* and *BMLF1* in some cell types (e.g. HeLa cells) but not others (e.g. Jurkat cells). Z also has a coiled coil-like dimerization domain, shares homology with C/EBP and binds to the same CCAAT sequence as C/EBP [3]. However, in Jurkat or Raji cells (an $EBV^+$ B cell line), Z interacts synergistically with MYB to activate the BMRF1 promoter [4]. Z/MYB binds through the interaction of Z with a 30 bp region containing an AP1 consensus sequence. In epithelial cells Z causes arrest in $G_0/G_1$ as a consequence of inducing expression of P53, WAF1 and KIP1, resulting accumulation of hypophosphorylated pRb [5].

EBV also activates transcription from the HIV-1-LTR. EBV encodes a homologue of interleukin 10 which may act to promote transformation, in part by suppressing the synthesis of interferon $\gamma$ by infected B cells [6].

The EBV early gene *BMLF1* codes for a nuclear phosphoprotein, EB2, that transforms established cell lines and primary rat fibroblasts [7].

| | ***EBNA-1*** (strain B95-8) | ***EBNA-2*** (strain B95-8) | ***LMP1*** (strain B95-8) |
|---|---|---|---|
| **Mass (kDa): predicted** | 56 | 52.5 | 42 |
| **expressed** | p80 | p82 | p62 |
| **Cellular location** | Nucleus: free in nucleoplasm: some association with chromatin but little with the nuclear matrix | Nuclear matrix | Transmembrane; localized in patches |

## Protein function

### Cancer

EBV causes infectious mononucleosis (glandular fever) and is associated with Burkitt's lymphoma (BL), nasopharyngeal carcinoma (NPC) and Hodgkin's lymphoma. The endemic (African) form of BL is usually associated with Epstein–Barr virus whereas the sporadic form is normally EBV$^-$.

All BL tumours express only the EBNA-1 EBV gene. High titres of antibodies directed against viral proteins precede the appearance of tumours by several months. The fully malignant phenotype requires the development of a BL clone and a chromosome translocation involving *MYC* (see ***MYC***). NPCs always express EBNA-1 and some (~65%) express LMP1 [8].

EBV infection of primary B cells *in vitro* causes little or no virus production but gives rise to an immortal lymphoblastoid cell line (LCL) that is not tumorigenic. LCL cells express six EBNAs, LMP1, LMP2A and LMP2B, two small RNAs (EBERs) and two terminal proteins (TP1 and TP2). LCLs are good targets for lysis by T cells, in contrast to BL cells (see below). During LCL cell immortalization EBNA-5, p53 and pRb are expressed at high levels but only cells with low p53 and focal expression of EBNA-5 in nuclear bodies become immortalized [9]. LCLs expressing the full complement of latent viral genes are very sensitive to DNA damaging agents such as cisplatin. In addition to accumulation of p53, the response includes induction of *MDM2* and *WAF1* and within 24 h the majority of the cells undergo apoptosis [10].

### EBNA-1

EBNA-1 does not have a demonstrated transforming capacity *in vitro* but its expression in transgenic mice gives rise to B cell lymphomas [11]. It is a *trans*-activating factor required for replication from the latency origin of replication (*oriP*) which is also an EBNA-1-dependent enhancer. EBNA-1 thus maintains replication of the EBV episome in the latent cycle. EBNA-1 protein is inadequately processed for MHC class I recognition and thus escapes immune surveillance.

### EBNA-2

EBNA-2 is a transcription factor involved in the latent cycle that binds pRb and is phosphorylated by casein kinase II. $Ser^{469}$ in EBNA-2 corresponds to those in E7 and T Ag that are phosphorylated by casein kinase II. EBNA-2 is one of the first genes expressed after primary B cell infection *in vitro* and it *trans*-activates LMP1, LMP2A, LMP2B and a lymphoid-specific enhancer in the *Bam*H1 C promoter of EBV [12]. EBNA-2 also regulates *CD23*, *CD21* [13] and *FGR* [14]. EBNA-2 is essential for EBV immortalization of human B cells and regulates transcription from several viral and cellular promoters. It has a region of partial homology with the pRb-binding domain of Ad5 E1A, SV40 T Ag and HPV16 E7 [15].

Evidence that EBNA-2 is required for transformation: (i) deletion of EBNA-2 in the P3HR-1 or Daudi EBV genome is associated with the unique inability of these viruses to transform B cells, (ii) transformation-competent recombinants between P3HR-1 and another EBV genome have a restored EBNA-2 coding region, (iii) EBNA-2 confers the ability of Rat-1 cells to grow in media with low serum concentration [16] and (iv) expression of EBNA-2 in kidney tubule cells induces adenocarcinomas [17].

In LCLs and B cell lymphoma lines EBNA-2 suppresses the Igμ gene, down-regulating IgM and *MYC* expression. In Burkitt's lymphoma cell lines with t(8;14) translocations, EBNA-2 also represses Igμ and *MYC*, inducing growth

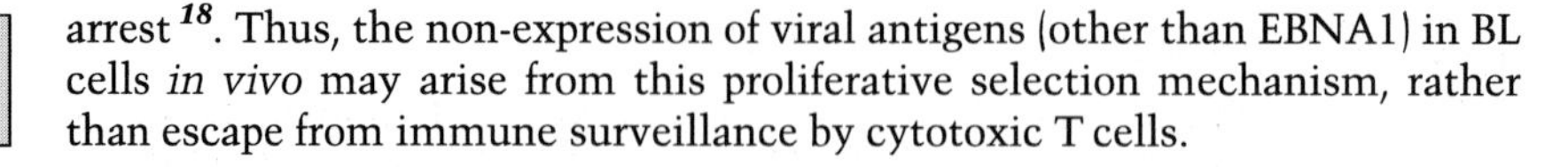

arrest[18]. Thus, the non-expression of viral antigens (other than EBNA1) in BL cells *in vivo* may arise from this proliferative selection mechanism, rather than escape from immune surveillance by cytotoxic T cells.

**LMP1**

LMP1 is a transforming protein encoded by *BNLF1* that reduces serum dependency, contact inhibition and anchorage-dependent growth and increases tumorigenicity in rodent fibroblasts[19,20]. LMP1 expression is *trans*-activated by EBNA-2 *via* a 142 bp *cis*-acting element[21]. However, in nasopharyngeal carcinoma cells and in cells from Hodgkin's lymphoma LMP1 transcription is independent of viral products and may be activated by cAMP[22]. The short half-life of the protein (2–5 h), its localized distribution on the cell surface and association with the cytoskeleton are necessary for its transforming function[23]. EBV transformation causes a major reorganization of intermediate filaments and microtubules and LMP1 appears in secondary lysosomes together with ubiquitin-protein conjugates and HSP70[24]. Transient expression of LMP1 causes upregulation of *CD21, CD23, ICAM1* and *LFA1* and induces DNA synthesis in human B cells[25].

Human B cells and Burkitt lymphoma cells are protected from apoptosis by the expression of LMP1 which activates transcription of *BCL2*[26]. BCL2 protein is 25% identical to a 149 amino acid region of the EBV immediate early *BHRF1* gene[27]. *BHRF1* is abundantly expressed early in the lytic life cycle but is not required for EBV-induced transformation of B lymphocytes *in vitro*. However, BHRF1 may be essential for infected cell survival when *BCL2* is not induced by LMP1. In BL cells the *BCL2* and *BHRF1* gene products are equally effective in rescuing the cells from apoptosis[28]. Mutations introduced into BHRF1 can give rise to a protein with the capacity to promote proliferation[29].

LMP1 and LMP2A associate in the plasma membrane. LMP2A is phosphorylated and, in EBV-transformed B lymphocytes, forms a complex with the LYN tyrosine kinase[30]. LMP1 interacts with LMP1-associated protein (LAP1) which is homologous to the murine tumour necrosis factor receptor-associated factor 2 (TRAF2). LMP1 expression causes LAP1 and EB16 (the human homologue of a second TRAF {TRAF1} which is induced by EBV infection) to localize to LMP1 clusters in lymphoblast plasma membranes[31].

The cumulative effect of the expression of the latent EBV proteins appears to be to activate growth regulating pathways involved in normal B cell stimulation[32], including the expression of a variety of surface antigens (the EBV receptor (C3d or CR2 (*CD21*)), *CD23*, LFA-1$\beta$ (*CD18*) and LFA-3 (*CD58*)).

Fusion of BL cells and non-tumorigenic EBV-immortalized B-lymphoblastoid cells suppresses the malignant phenotype of the BL cell line, despite the fact that the hybrid cells contain EBV and express deregulated *MYC*[33]. These observations may reflect the action of tumour necrosis factor (TNF) or a related protein. The synthesis of TNF by tumour cells correlates with reduced tumorigenicity and invasiveness[34]. A TNF-like gene may be disrupted in BL with complementation occurring in the hybrid cells.

Point mutations and a 30 bp deletion have been detected in LMP1 in EBV$^+$ Hodgkin's disease and in other lymphoproliferative disorders[35,36].

## Amino acid sequence of EBNA-1

```
  1 MSDEGPGTGP GNGLGEKGDT SGPEGSGGSG PQRRGGDNHG RGRGRGRGRG
 51 GGRPGAPGGS GSGPRHRDGV RRPQKRPSCI GCKGTHGGTG AGAGAGGAGA
101 GGAGAGGGAG AGGGAGGAGG AGGAGAGGGA GAGGGAGGAG GAGAGGGAGA
151 GGGAGGAGAG GGAGGAGGAG AGGGAGAGGG AGGAGAGGGA GGAGGAGAGG
201 GAGAGGAGGA GGAGAGGAGA GGGAGGAGGA GAGGAGAGGA GAGGAGAGGA
251 GGAGAGGAGG AGAGGAGGAG AGGGAGGAGA GGGAGGAGAG GAGGAGAGGA
301 GGAGAGGAGG AGAGGGAGAG GAGAGGGGRG RGGSGGRGRG GSGGRGRGGS
351 GGRRGRGRER ARGGSRERAR GRGRGRGEKR PRSPSSQSSS SGSPPRRPPP
401 GRRPFFHPVG EADYFEYHQE GGPDGEPDVP PGAIEQGPAD DPGEGPSTGP
451 RGQGDGGRRK KGGWFGKHRG QGGSNPKFEN IAEGLRALLA RSHVERTTDE
501 GTWVAGVFVY GGSKTSLYNL RRGTALAIPQ CRLTPLSRLP FGMAPGPGPQ
551 PGPLRESIVC YFMVFLQTHI FAEVLKDAIK DLVMTKPAPT CNIRVTVCSF
601 DDGVDLPPWF PPMVEGAAAE GDDGDDGDEG GDGDEGEEGQ E (641)
```

**Domain structure**

87–352 Glycine/alanine-rich region

## Amino acid sequence of EBNA-2

```
  1 MPTFYLALHG GQTYHLIVDT DSLGNPSLSV IPSNPYQEQL SDTPLIPLTI
 51 FVGENTGVPP PLPPPPPPPP PPPPPPPPPP PPPPPPPPSP PPPPPPPPPP
101 QRRDAWTQEP SPLDRDPLGY DVGHGPLASA MRMLWMANYI VRQSRGDRGL
151 ILPQGPQTAP QARLVQPHVP PLRPTAPTIL SPLSQPRLTP PQPLMMPPRP
201 TPPTPLPPAT LTVPPRPTRP TTLPPTPLLT VLQRPTELQP TPSPPRMHLP
251 VLHVPDQSMH PLTHQSTPND PDSPEPRSPT VFYNIPPMPL PPSQLPPPAA
301 PAQPPPGVIN DQQLHHLPSG PPWWPPICDP PQPSKTQGQS RGQSRGRGRG
351 RGRGRGKGKS RDKQRKPGGP WRPEPNTSSP SMPELSPVLG LHQGQGAGDS
401 PTPGPSNAAP VCRNSHTATP NVSPIHEPES HNSPEAPILF PDDWYPPSID
451 PADLDESWDY IFETTESPSS DEDYVEGPSK RPRPSIQ (487)
```

**Domain structure**

59–100 Poly-proline region

458–474 pRb and casein kinase II recognition region, homologous to those in HPV E7, adenovirus E1a and SV40 T antigen (underlined)

## Amino acid sequence of LMP1

```
  1 MEHDLERGPP GPRRPPRGPP LSSSLGLALL LLLLALLFWL YIVMSDWTGG
 51 ALLVLYSFAL MLIIIILIIF IFRRDLLCPL GALCILLLMI TLLLIALWNL
101 HGQALFLGIV LFIFGCLLVL GIWIYLLEML WRLGATIWQL LAFFLAFFLD
151 LILLIIALYL QQNWWTLLVD LLWLLLFLAI LIWMYYHGQR HSDEHHHDDS
201 LPHPQQATDD SGHESDSNSN EGRHHLLVSG AGDGPPLCSQ NLGAPGGGPD
251 NGPQDPDNTD DNGPQDPDNT DDNGPHDPLP QDPDNTDDNG PQDPDNTDDN
301 GPHDPLPHSP SDSAGNDGGP PQLTEEVENK GGDQGPPLMT DGGGGHSHDS
351 GHGGGDPHLP TLLLGSSGSG GDDDDPHGPV QLSYYD (386)
```

**Domain structure**

1–24 and 187–386 Cytoplasmic domains

25–44, 52–72, 77–97, 105–125, 139–159, 166–186 Six potential transmembrane domains (underlined)

242–386 P25 peptide

## Amino acid sequence of LMP2A/LMP2B

```
  1 MGSLEMVPMG AGPPSPGGDP DGYDGGNNSQ YPSASGSSGN TPTPPNDEER
 51 ESNEEPPPPY EDPYWGNGDR HSDYQPLGTQ DQSLYLGLQH DGNDGLPPPP
101 YSPRDDSSQH IYEEAGRGSM NPVCLPVIVA PYLFWLAAIA ASCFTASVST
151 VVTATGLALS LLLLAAVASS YAAAQRKLLT PVTVLTAVVT FFAICLTWRI
201 EDPPFNSLLF ALLAAAGGLQ GIYVLVMLVL LILAYRRRWR RLTVCGGIMF
251 LACVLVLIVD AVLQLSPLLG AVTVVSMTLL LLAFVLWLSS PGGLGTLGAA
301 LLTLAAALAL LASLILGTLN LTTMFLLMLL WTLVVLLICS SCSSCPLSKI
351 LLARLFLYAL ALLLLASALI AGGSILQTNF KSLSSTEFIP NLFCMLLLIV
401 AGILFILAIL TEWGSGNRTY GPVFMCLGGL LTMVAGAVWL TVMSNTLLSA
451 WILTAGFLIF LIGFALFGVI RCCRYCCYYC LTLESEERPP TPYRNTV (497)
```

### Domain structure

| | |
|---|---|
| 1–497 | Membrane protein LMP2A. |
| 120–497 | Membrane protein LMP2B. Separate mRNAs are transcribed to give rise to two proteins the shorter of which (LMP2B) lacks 119 N-terminal amino acids[37] |
| 122–141, 150–168, 178–198, 208–235, 242–259, 267–288, 300–316, 321–339, 355–373, 392–411, 419–443, 450–470 | Eleven potential transmembrane domains (underlined) |

### Database accession numbers

| | *PIR* | *SWISSPROT* | *EMBL/GENBANK* | *REFERENCES* |
|---|---|---|---|---|
| EBNA-1 | A03773 | P03211 | V01555 | 38 |
| | | | M13941 | 37,39 |
| EBNA-2 | | P12978 | V01555 | 38,39 |
| LMP1 (B95-8) | A03794 | P03230 | V01555 | 20.38,40 |
| LMP1 (Raji) | C28918 | P13198 | M20868 | 41 |
| LMP2A/2BA | 30178 | P13285 | V01555 | 42 |
| | | | Y00835, M24212 | 37 |

### *References*


1 **Chee, M. and Barrell, B. (1990) Trends in Genet. 6, 86–91.**
2 **Middleton, T. et al. (1991) Adv. Virus Res. 40, 19–55.**
3 Kouzarides, T. et al. (1991) Oncogene 6, 195–204.
4 Kenney, S.C. et al. (1992) Mol. Cell. Biol. 12, 136–146.
5 Cayrol, C. and Flemington, E.K. (1996) EMBO J. 15, 2748–2759.
6 Stuart, A.D. et al. (1995) Oncogene, 11, 1711–1719.
7 Corbo, L. et al. (1994) Oncogene 9, 3299–3304.
8 Rowe, M. et al. (1987) EMBO J. 6, 2743–2751.
9 Szekely, L. et al. (1995) Oncogene 10, 1869–1874.
10 Allday, M.J. et al. (1995) EMBO J. 14, 1382–1391.
11 Wilson, J.B. et al. (1996) EMBO J. 15, 3117–3126.
12 Abbot, S.D. et al. (1990) J. Virol. 64, 2126–2134.
13 Wang, F. et al. (1990) J. Virol. 64, 2309–2318.
14 Knutson, J.C. (1990) J. Virol. 64, 2530–2536.
15 Inoue, N. et al. (1991) Virology 182, 84–93.
16 Sample, J. et al. (1986) Proc. Natl Acad. Sci. USA 83, 5096–5100.
17 Törnell, J. et al. (1996) Oncogene 12, 1521–1528.

[18] Jochner, N. et al. (1996) EMBO J. 15, 375–382.
[19] Wang, D. et al. (1988) J. Virol. 62, 2337–2346.
[20] Baichwal, V.R. and Sugden, B. (1988) Oncogene 2, 461–467.
[21] Tsang, S.-F. et al. (1991) J. Virol. 65, 6765–6771.
[22] Fåhraeus, R. et al. (1994) EMBO J. 13, 6041–6051.
[23] Martin, J. and Sugden, B. (1991) J. Virol. 65, 3246–3258.
[24] Laszlo, L. et al. (1991) J. Pathol. 164, 203–214.
[25] Peng, M. and Lundgren, E. (1992) Oncogene 7, 1775–1782.
[26] Okan, I. et al. (1995) Oncogene 11, 1027–1031.
[27] Marchini, A. et al. (1991) J. Virol. 65, 5991–6000.
[28] Henderson, S. et al. (1993) Proc. Natl Acad. Sci. USA 90, 8479–8483.
[29] Theodorakis, P. et al. (1996) Oncogene 12, 1707–1713.
[30] Burkhardt, A.L. et al. (1992) J. Virol. 66, 5161–5167.
[31] Mosialos, G. et al. (1995) Cell 80, 389–399.
[32] Calender, A. et al. (1990) Int. J. Cancer 46, 658–663.
[33] Wolf, J. et al. (1990) Cancer Res. 50, 3095–3100.
[34] Vanhaesebroeck, B. et al. (1991) Cancer Res. 51, 2229–2238.
[35] Knecht, H. et al. (1995) Oncogene 10, 523–528.
[36] Itakura, O. et al. (1996) Oncogene 13, 1549–1553.
[37] Sample, J. et al. (1989) J. Virol. 63, 933–937.
[38] Baer, R. et al. (1984) Nature 310, 207–211.
[39] Petti, L. et al. (1990) Virology 176, 563–574.
[40] Moorthy, R. and Thorley-Lawson, D.A. (1990) J. Virol. 64, 829–837.
[41] Hatfull, G. (1988) Virology 164, 334–340.
[42] Laux, G. et al. (1988). EMBO J. 7, 769–774.

# Index

Note: Where an entry includes references to both a gene and its protein product, the entry is given in the gene form (i.e. italics) unless specifically stated. **Bold** entries indicate those described in detail in section III.

**T**

QMW LIBRARY
(MILE END)